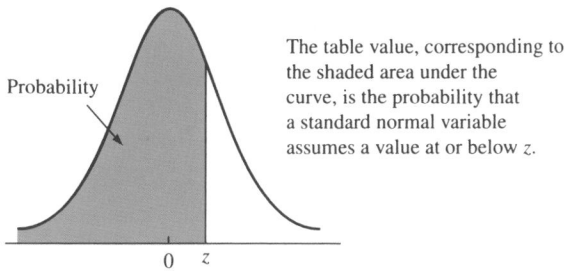

Probability

The table value, corresponding to the shaded area under the curve, is the probability that a standard normal variable assumes a value at or below z.

0 z

Second decimal place in z

z	.00	.01	.02	.03	.04	.05	.06	.07	.08	.09
0.0	.5000	.5040	.5080	.5120	.5160	.5199	.5239	.5279	.5319	.5359
0.1	.5398	.5438	.5478	.5517	.5557	.5596	.5636	.5675	.5714	.5753
0.2	.5793	.5832	.5871	.5910	.5948	.5987	.6026	.6064	.6103	.6141
0.3	.6179	.6217	.6255	.6293	.6331	.6368	.6406	.6443	.6480	.6517
0.4	.6554	.6591	.6628	.6664	.6700	.6736	.6772	.6808	.6844	.6879
0.5	.6915	.6950	.6985	.7019	.7054	.7088	.7123	.7157	.7190	.7224
0.6	.7257	.7291	.7324	.7357	.7389	.7422	.7454	.7486	.7517	.7549
0.7	.7580	.7611	.7642	.7673	.7704	.7734	.7764	.7794	.7823	.7852
0.8	.7881	.7910	.7939	.7967	.7995	.8023	.8051	.8078	.8106	.8133
0.9	.8159	.8186	.8212	.8238	.8264	.8289	.8315	.8340	.8365	.8389
1.0	.8413	.8438	.8461	.8485	.8508	.8531	.8554	.8577	.8599	.8621
1.1	.8643	.8665	.8686	.8708	.8729	.8749	.8770	.8790	.8810	.8830
1.2	.8849	.8869	.8888	.8907	.8925	.8944	.8962	.8980	.8997	.9015
1.3	.9032	.9049	.9066	.9082	.9099	.9115	.9131	.9147	.9162	.9177
1.4	.9192	.9207	.9222	.9236	.9251	.9265	.9279	.9292	.9306	.9319
1.5	.9332	.9345	.9357	.9370	.9382	.9394	.9406	.9418	.9429	.9441
1.6	.9452	.9463	.9474	.9484	.9495	.9505	.9515	.9525	.9535	.9545
1.7	.9554	.9564	.9573	.9582	.9591	.9599	.9608	.9616	.9625	.9633
1.8	.9641	.9649	.9656	.9664	.9671	.9678	.9686	.9693	.9699	.9706
1.9	.9713	.9719	.9726	.9732	.9738	.9744	.9750	.9756	.9761	.9767
2.0	.9772	.9778	.9783	.9788	.9793	.9798	.9803	.9808	.9812	.9817
2.1	.9821	.9826	.9830	.9834	.9838	.9842	.9846	.9850	.9854	.9857
2.2	.9861	.9864	.9868	.9871	.9875	.9878	.9881	.9884	.9887	.9890
2.3	.9893	.9896	.9898	.9901	.9904	.9906	.9909	.9911	.9913	.9916
2.4	.9918	.9920	.9922	.9925	.9927	.9929	.9931	.9932	.9934	.9936
2.5	.9938	.9940	.9941	.9943	.9945	.9946	.9948	.9949	.9951	.9952
2.6	.9953	.9955	.9956	.9957	.9959	.9960	.9961	.9962	.9963	.9964
2.7	.9965	.9966	.9967	.9968	.9969	.9970	.9971	.9972	.9973	.9974
2.8	.9974	.9975	.9976	.9977	.9977	.9978	.9979	.9979	.9980	.9981
2.9	.9981	.9982	.9982	.9983	.9984	.9984	.9985	.9985	.9986	.9986
3.0	.9987	.9987	.9987	.9988	.9988	.9989	.9989	.9989	.9990	.9990
3.1	.9990	.9991	.9991	.9991	.9992	.9992	.9992	.9992	.9993	.9993
3.2	.9993	.9993	.9994	.9994	.9994	.9994	.9994	.9995	.9995	.9995
3.3	.9995	.9995	.9995	.9996	.9996	.9996	.9996	.9996	.9996	.9997
3.4	.9997	.9997	.9997	.9997	.9997	.9997	.9997	.9997	.9997	.9998
3.5	.9998									
4.0	.99997									
4.5	.999997									
5.0	.9999997									

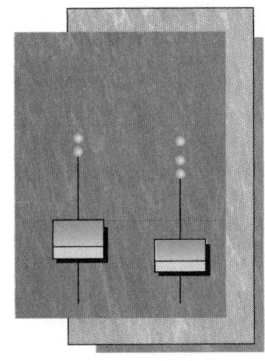

Exploring Statistics

A Modern Introduction to Data Analysis and Inference

SECOND EDITION

LARRY J. KITCHENS

Appalachian State University

An Alexander Kugushev Book

DUXBURY PRESS

An imprint of Brooks/Cole Publishing Company

I(T)P ® An International Thomson Publishing Company

Pacific Grove, CA   Albany, NY   Bonn   Boston   Cincinnati   Detroit   Johannesburg   London
Madrid   Melbourne   Mexico City   New York   Paris   Singapore   Tokyo   Toronto   Washington

Developmental Editor   *Alan Venable*
Assistant Editor   *Cynthia Mazow*
Editorial Assistant   *Martha O'Connor*
Production   *Susan L. Reiland*
Print Buyer   *Karen Hunt*
Permissions Editor   *Peggy Meehan*
Copy Editor   *Carol Reitz*
Cover Designer   *Lucy Lesiak*
Cover Image   *Franz Lazi, FPG International*
Interior Designer   *Stuart D. Paterson, Image House*
Compositor   *Integre Technical Publishing Company, Inc.; G&S Typesetters, Inc.*
Printer   *Phoenix Book Technology Park/Hagerstown*

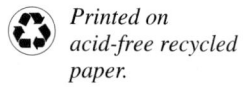

Printed on acid-free recycled paper.

Printed in the United States of America
    6 7 8 9 10

For more information, contact Duxbury Press at Brooks/Cole Publishing Company.

Brooks/Cole Publishing Company
511 Forest Lodge Road
Pacific Grove, CA 93950, USA

International Thomson Publishing Europe
Berkshire House 168-173
High Holborn
London, WC1V 7AA, England

Thomas Nelson Australia
102 Dodds Street
South Melbourne 3205
Victoria, Australia

Nelson Canada
1120 Birchmount Road
Scarborough, Ontario
Canada M1K 5G4

International Thomson Publishing Southern Africa
Building 18, Constantia Park
240 Old Pretoria Road
Halfway House, 1685 South Africa

International Thomson Editores
Campos Eliseos 385, Piso 7
Col. Polanco
11560 México D.F. México

International Thomson Publishing GmbH
Königswinterer Strasse 418
53227 Bonn, Germany

International Thomson Publishing Asia
221 Henderson Road
#05-10 Henderson Building
Singapore 0315

International Thomson Publishing Japan
Hirakawacho Kyowa Building, 3F
2-2-1 Hirakawacho
Chiyoda-ku, Tokyo 102, Japan

Library of Congress Cataloging-in-Publication Data

Kitchens, Larry J.
    Exploring statistics : a modern introduction to data analysis and inference /
Larry J. Kitchens.— 2nd ed.
        p.   cm.
    Includes index.
    ISBN 0-534-36351-2
    1. Statistics.   I. Title.
QA276.12.K586   1996
519.5—dc21   96-48250

To Anita and our children,

Christopher, Stephen, Joseph, and Carolyn

Without their love and support this book would
not have been possible.

Contents

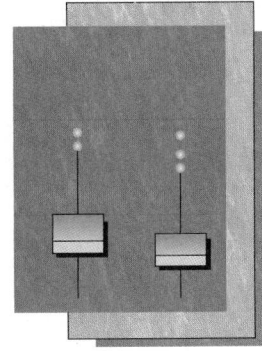

Preface

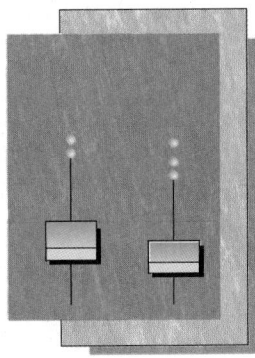

Objectives

I have written *Exploring Statistics* to make statistics interesting and understandable for students who have had no previous exposure to the subject. The language is informal, yet the book retains the proper rigor to ensure correct understanding of the concepts. I have sought to make the book modern by thoroughly integrating the uses and benefits of computer technology and the most current statistical techniques as they apply to the analysis of data. The material should be accessible to anyone who has the maturity of those who have successfully completed a course in elementary algebra.

In developing this book, I have adhered to a series of principles that constitute sound, modern pedagogy for a beginning statistics course:

- Because working with data is of central importance to a modern statistics course, I place foremost emphasis on data analysis, using the most modern tools.
- I believe that the student should be actively involved in learning; thus I stress hands-on experiences, working with real data, and discovery through graphical displays and computer simulations.
- I see the student as a consumer and. to a certain extent, a producer of statistics. As a consumer, the student learns to critically appraise statistical information produced by others, predominantly through analyzing computer printouts. As a producer of statistics, the student learns how to begin analyzing data for the purpose of extracting important information and selecting proper inference procedures.
- I follow a logical progression from the simplest topics to the more difficult ones.
- I introduce modern statistical techniques and procedures.
- I aim to make the concepts, definitions, and methodology work together in a reasonable and interesting manner.

Important Features

Data Analysis Data analysis, a critical study by which we extract information from data, is central to the modern study of statistics. Consequently, it is the central focus of this book. Data analysis is not confined to the descriptive sections, but rather pervades the book as the main theme.

Implementation of Data Analysis Armed with an assortment of tools and techniques, the data analyst blends an approach with a process for detecting patterns and structure in the data. A main element in the approach is flexibility. Because no two data sets are alike, there is no standard procedure that works best all the time. As we examine data for structure, we must be able to respond to different patterns and unexpected features that may prompt new questions. It is in this spirit that much of the material is developed.

Data analysis provides us with the process for determining the structure of the data, a structure that should direct us toward the proper inference procedure. In the descriptive statistics sections, for example, we learn that for highly skewed distributions, the median is a more representative measure of the center of the distribution. This leads the student to reason that we need inference procedures for the population median when we detect skewness. The book approaches the univariate inference problem by asking the student to use the exploratory tools to describe the shape of the underlying population distribution, to decide which parameter best describes the characteristic of interest, and then to choose a statistic that best estimates the parameter. This basic process of data analysis is presented in detail in Chapter 6.

Emphasis on Interpretation Because of the wide availability of statistical software packages for the analysis of data, I place little emphasis on numerical computation. Instead, I spend time on understanding the basic concepts of statistics such as distribution, central tendency, variability, association, randomness, and significance. The student is taught to recognize outliers and influential observations, the value of a thorough residual analysis, and the purpose of studying sampling distributions. Using a hands-on approach, this book teaches students to visualize the data, identify patterns, summarize the results, and then draw conclusions.

The Use of Real Data Data sets incorporated in this book are interesting to the student and relevant in today's society. Even those drawn from professional journals are such that the student understands the intent of the study and what is being measured. The disastrous space shuttle *Challenger* accident, data on AIDS testing, velocities of galaxies, worldwide seismic activity, ratings of NFL quarterbacks, and Charles Darwin's famous study of cross-fertilized and self-fertilized plants are just a few of the situations that are presented in examples and exercises. With over 400 *different* data sets, I have endeavored to show statistics as an exciting discipline, dealing with the real problems of the contemporary world.

Examples and Exercises An abundance of examples relevant to a wide range of disciplines is given so that applications in a variety of settings can be seen. Each section ends with a set of exercises to reinforce the ideas just presented. At the end of each chapter a group of review exercises (including computer exercises) tie together the chapter concepts. Finally, at the end of every third chapter in the Unit Review, a unit test including computer exercises and discussion exercises is presented to address issues that span the three previous chapters. As described above, many examples and exercises have been taken from actual surveys, reports, and experiments.

Latest Statistical Techniques Discovery through graphical displays is implemented throughout the book. Boxplots, histograms, normal probability plots, and scatterplots are used to visualize data in both descriptive and inferential situations. Nonparametric alternatives to the classical t and F procedures are integrated throughout the book. The p-value approach to hypothesis testing is used in the inference portion. A complete analysis of residuals accompanies the regression chapter. Optional material includes the bootstrap procedure, the resistant line, and robust inference procedures.

Computer Use The book includes computer printouts for the student to analyze and interpret. Many exercises with significantly large data sets lend themselves to computer solutions. All chapters and the unit summaries conclude with a section of computer exercises. Throughout the book, "Computer Tips" are given to guide the student through the steps required to execute the computer analysis. Additionally, computer simulations are used to illustrate population distributions and sampling distributions. I have chosen Minitab to illustrate concepts and perform calculations. However, any statistical software package with good graphing capabilities may be used.

Probability Coverage The intent of this book is not to offer a mini-course on probability. However, because random behavior is fundamental to inferential statistics, I believe that students should have a basic understanding of elementary probability topics. For example, independence of events is an extremely important topic when discussing the binomial distribution and the analysis of contingency tables. I have limited the discussion of probability to only those topics relevant to the study of inference.

Procedural Assumptions In the inference sections, particular attention is given to the rationale for using each statistical procedure. The assumptions that are necessary and indications for when the procedure is appropriate are carefully outlined and are a main focus of the book. The students are continually asked to use the exploratory tools to check assumptions and validate models.

Pedagogical Help

- Each chapter begins with a preview that gives the chapter direction and links it to preceding chapters.
- A "Statistical Insight" from a general-interest news article is given to capture the student's curiosity and to illustrate the topics that are covered as the chapter unfolds. The "Insight" is revisited in the exercises at the end of the chapter, bringing the application full circle.
- At the end of each chapter, all key concepts are summarized, followed by a list of "Learning Goals" and a set of "Questions for Review."
- At the end of every third chapter, a "Unit Review" is given to pull together the main ideas previously stressed and point out important cautions and issues that apply to the practice of statistics. It discusses concepts that span more than one chapter and ends with a unit test.
- Chapter 6 is pivotal. It describes the basic philosophy of data analysis and gives many illustrative examples with real data.

- Over 1,400 exercises and over 400 data sets give the student a wide variety of applications of statistics.

ORGANIZATION

The book can be used in a one-quarter, one-semester, or two-quarter course in beginning statistics with data analysis. For the shorter-duration course, the instructor has a degree of flexibility in choosing topics. The tools of the data analyst are developed in the first four chapters and then used collectively in Chapters 5 and 6 to describe sampling distributions and the important features of a population distribution. Chapter 6, usually not found in introductory books, bridges descriptive data analysis and inference. The remaining Chapters 7–12 are devoted to statistical inference. If one chooses to cover Chapter 10 or 11, then, of course, Chapter 3 is required. On the other hand, if regression analysis and the analysis of categorical data are omitted, then Chapter 3 is optional. Additionally, the material in Chapter 4 could be abbreviated by those who wish to give an intuitive approach to probability. For a typical course, however, most would choose to cover Chapters 1–8 and then selectively choose from Chapters 9–12 to complete the course.

A COMPLETE PEDAGOGICAL PACKAGE

MINITAB Lab Manual A lab manual, written by Dr. Alan Arnholt, gives a thorough introduction to Minitab and provides numerous laboratory projects that can be assigned on a chapter-by-chapter basis throughout the course. *The Student Edition of Minitab for Windows* is also offered as a bundled package with the text.

Statistics Tutor Introductory statistics students often experience a degree of anxiety and apprehension about the course. The *Statistics Tutor,* written by Dr. Anita Kitchens, begins with a proven approach to studying that includes practical study skills, a discussion of different ways of thinking, and a discussion of the effects of self-beliefs on behavior. Additionally, for each chapter the *Statistics Tutor* contains a chapter summary, important definitions, a glossary of terms, two practice tests, and complete solutions for all odd-numbered exercises in *Exploring Statistics*.

Data Diskette All 400+ data sets are provided in Minitab, ASCII, SAS, and StataQuest formats on the data diskette. The data sets can also be downloaded from the Internet at location http://www.thomson.com/duxbury.

Instructor's Manual This manual gives complete solutions to every exercise in *Exploring Statistics*.

Test Bank The Test Bank, written by Alice Fugate, contains true–false, matching, short answer, and analysis problems that ask students to analyze data presented in a life-like context.

ACKNOWLEDGMENTS

I would like to express my appreciation to the following reviewers for their valuable assistance in the preparation of the manuscript for this second edition:

Mary Sue Beersman
Northeast Missouri State

Ken Brown
College of San Mateo

Judith Eckstrand
San Francisco State

James Guffey
Truman State University

Gary Kersting
Sacramento City College

Brian Marx
Louisiana State University

James Paige
Wayne State College

Robert Raymond
University of St. Thomas

Susan Schwartz
Private consultant

Thomas Short
Villanova University

Dana Thomas
University of Alaska at Fairbanks

Bruce Trumbo
California State University, Hayward

I am also grateful to all students and instructors, especially those at Appalachian State University, who studied and taught from the first edition of *Exploring Statistics*. Their input was instrumental in the development of this edition. Thanks also go to Jeff Hildebrand for checking the solutions, Alice Fugate for producing the test bank, Alan Arnholt for producing the lab manual, Susan Reiland for her diligent work during the production stage, and to my wife, Anita, for her work on the *Statistics Tutor*. A special thank you goes to Alex Kugushev for believing in the basic concept of this book and for his unending support and encouragement.

1

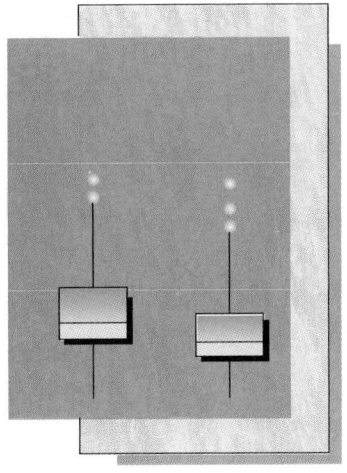

COLLECTING AND UNDERSTANDING DATA

Statistical training is necessary and important for many reasons. In almost any area of work—business, medicine, education, the social and physical sciences—you must be able to read, interpret, and apply the results of a statistical analysis of research data. Obviously, one introductory course will not make you a research statistician, but it will enable you to understand the statistical terminology and techniques that are used. This chapter introduces basic definitions and concepts so that you, the student, will better understand how statistical information is produced, used, and evaluated.

CONTENTS

STATISTICAL INSIGHT

TOP 20 DISCOVERIES IN SCIENCE

In 1984, *Science* magazine polled the leading U.S. scientists and asked them what they believed were the most important scientific discoveries made since 1900. The magazine concluded that these discoveries have most changed our lives:

Antibiotics	The laser
Atomic fission	Networks
The big-bang theory	Pesticides
Birth control pills	Plant breeding
Blood types	Plastics
The computer	Statistics
DNA	The Taung skull
Drugs for mental illness	Television
Einstein's theory of relativity	The transistor
The IQ test	The vacuum tube

Statistics, though listed as a separate discovery, is a common component in many of the other listed discoveries. The IQ test is analyzed with statistical tools. Data from plant breeding experiments are analyzed with statistics. The effects of penicillin and other antibiotics on humans are studied with statistics. The birth control pill and drugs for mental illness are studied by comparing the effects on a control group and an experimental group. Professionals in almost all fields of study use statistics to analyze their data. H. G. Wells understood the importance of the proper analysis of data as he pointed out a century ago: "Statistical thinking will one day be as necessary for efficient citizenship as the ability to read and write."

1.1 ESSENTIAL ELEMENTS OF STATISTICS

Modern-day *statistics* involves the collection, organization, interpretation, and presentation of numerical information. A statistical problem may be as complicated as a study of the interrelations of several different quantities in the treatment of a disease or as simple as a graph of the batting averages for a Little League team. In fact, when we first think of statistics, we may think of sport statistics like batting averages, free throw averages, or yards rushing. We might also think of the results of a national poll that are published in the local newspaper, or government figures on unemployment that are presented on television. Applications of statistics appear in almost all fields of study; no matter what career you pursue, you probably will be called upon to interpret numerical information. Studying statistics will help you to understand the information and to reach correct conclusions.

A statistical problem involves studying some *characteristic* associated with a group of objects commonly called **experimental units** or **subjects**.

- An article in *Money* magazine (April 1993, p. 132) reports, "Don't Get Cheated by Supermarket Scanners." A 2-month survey by *Money* concluded that scanners may overcharge at 30% of all stores and cost consumers from $1 billion to $2.5 billion annually. In some cases, scanner errors account for more than half of a supermarket's profits. The article goes on to tell you, the consumer, how to beat the system.

What is an experimental unit in this example? Consumer is a reasonable choice, but *Money* magazine surveyed scanners, not consumers. In this example, an experimental unit (or subject) is a *scanner*.

A characteristic being studied that is associated with each experimental unit in a statistical problem is called a **variable**. The collection of numerical values (called **observations**) associated with the variables in the study is called a **data set**.

> ■ In an effort to appeal to the diet-conscious, McDonald's has produced a nutrition chart listing calories, protein, carbohydrates, total fat, saturated fat, cholesterol, and sodium for each item on its menu. The concerned consumer can very easily count calories or avoid high-cholesterol items on a McDonald's menu. Complete information is posted in restaurant lobbies and is available through the company's Nutrition Information Center.

Here we might say that a consumer is an experimental unit, but McDonald's is studying the calories, protein, carbohydrates, and so on for each item on its menu. A McDonald's item is an experimental unit, and calories, protein, and carbohydrates are three of the seven variables in this study. Figure 1.1 (page 4) is McDonald's nutrition chart that lists the seven variables with their associated observations.

An **experimental unit** (or **subject**) is the smallest entity that is of interest in a statistical study.

A **variable** is any characteristic that can be measured on each experimental unit in a statistical study.

An **observation** is a value that the variable assumes for a single unit.

The collection of observations assumed by the variables in the study is called a **data set**.

One objective of *data analysis* is to organize and summarize the information in a data set to make it more comprehensible. Often this area of statistics is referred to as *descriptive statistics*. With today's computer capabilities, data analysis is simplified to the point that all calculations are handled by the machine and informative graphs are drawn that illustrate the data in a fashion that previously was not possible.

Exhibit 1.1 (page 5) gives a Minitab worksheet (**McDonald.mtw**) that includes the observations associated with each sandwich for the seven variables in the McDonald's data set. Figure 1.2 (page 5) shows a graph for the variable Calorie. Without the aid of a computer, computations and graphs like this one would be very time-consuming to produce. Having the computer available to perform the computations, we are able to obtain greater insight into the data and draw meaningful conclusions.

FIGURE 1.1

NUTRITION INFORMATION PER SERVING	Calories	Protein (g)	Carbohydrates (g)	Total Fat (g)	Saturated Fat (g)	Cholesterol (mg)	Sodium (mg)
SANDWICHES/FRENCH FRIES							
Hamburger	255	12	30	9	3	37	490
Cheeseburger	305	15	30	13	5	50	725
Quarter Pounder®	410	23	34	20	8	85	645
Quarter Pounder® with Cheese	510	28	34	28	11	115	1110
McLean Deluxe™	320	22	35	10	4	60	670
McLean Deluxe™ with Cheese	370	24	35	14	5	75	890
Big Mac®	500	25	42	26	9	100	890
Filet-O-Fish®	370	14	38	18	4	50	730
McChicken®	415	19	39	20	4	50	830
Chicken Fajitas (1)	190	11	20	8	2	35	310
Small French Fries	220	3	26	12	2.5	0	110
Medium French Fries	320	4	36	17	3.5	0	150
Large French Fries	400	6	46	22	5	0	200
CHICKEN McNUGGETS/SAUCES							
Chicken McNuggets® (6 pieces)	270	20	17	15	3.5	55	580
Hot Mustard Sauce	70	0	8	3.6	0.5	5	250
Barbeque Sauce	50	0	12	0.5	0.1	0	340
Sweet 'N Sour Sauce	60	0	14	0.2	0	0	190
Honey	45	0	12	0	0	0	0
SALADS/SALAD DRESSINGS							
Chef Salad	170	17	8	9	4	111	400
Chunky Chicken Salad	150	25	7	4	1	78	230
Garden Salad	50	4	6	2	0.6	65	70
Side Salad	30	2	4	1	0.3	33	35
Croutons	50	1	7	2	0.5	0	140
Bacon Bits	15	1	0	1	0.5	1	95
Bleu Cheese (1/2 oz) 5 servings/packet†	50	0	1	4	1	7	150
Lite Vinaigrette (1/2 oz) 4 servings/packet†	12	0	2	0.5	0.1	0	60
Ranch (1/2 oz) 4 servings/packet†	55	0	1	5	1	5	130
Red French Reduced Calorie (1/2 oz) 4 servings/packet†	40	0	5	2	0.3	0	115
1000 Island (1/2 oz) 5 servings/packet†	45	0	4	3	1	8	100
BREAKFAST							
Egg McMuffin®	280	18	28	11	4	235	710
Sausage McMuffin®	345	15	27	20	7	57	770
Sausage McMuffin® with Egg	430	21	27	25	8	270	920
English Muffin with Spread	170	5	26	4	1	0	285
Sausage Biscuit	420	12	32	28	8	44	1040
Sausage Biscuit with Egg	505	19	33	33	10	260	1210
Bacon, Egg & Cheese Biscuit	440	15	33	26	8	240	1215
Biscuit with Biscuit Spread	260	5	32	13	3	1	730
Sausage	160	7	0	15	5	43	310
Scrambled Eggs (2)	140	12	1	10	3	425	290
Hash Brown Potatoes	130	1	15	7	1	0	330
Hotcakes with Margarine & Syrup (2 pats)	440	8	74	12	2	8	685
Breakfast Burrito	280	12	21	17	4	135	580
Cheerios®	80	3	14	1	0.2	0	210
Wheaties®	90	2	19	1	0.1	0	220
Fat-Free Apple Bran Muffin	180	5	40	0	0	0	200
Apple Danish	390	6	51	17	4	25	370
Iced Cheese Danish	390	7	42	21	6	47	420
Cinnamon Raisin Danish	440	6	58	21	5	34	430
Raspberry Danish	410	6	62	16	3	26	310
DESSERTS/MILK SHAKES							
Vanilla Lowfat Frozen Yogurt Cone (3 oz)	105	4	22	1	0.5	3	80
Strawberry Lowfat Frozen Yogurt Sundae	210	6	49	1	0.5	5	95
Hot Fudge Lowfat Frozen Yogurt Sundae	240	7	50	3	2	6	170
Hot Caramel Lowfat Frozen Yogurt Sundae	270	7	59	3	1.5	13	180
Apple Pie	260	2	30	15	4	6	240
McDonaldland® Cookies	290	4	47	9	1	0	300
Chocolaty Chip Cookies	330	4	42	15	4	4	280
Chocolate Lowfat Milk Shake	320	11	66	1.7	0.7	10	240
Strawberry Lowfat Milk Shake	320	11	67	1.3	0.6	10	170
Vanilla Lowfat Milk Shake	290	11	60	1.3	0.6	10	170
BEVERAGES							
1% Lowfat Milk (8 fl oz)	110	9	12	2	1.6	10	130
Orange Juice (6 fl oz)	80	1	19	0	0	0	0
Grapefruit Juice (6 fl oz)	80	1	19	0	0	0	0
Apple Juice (6 fl oz)	90	0	23	0	0	0	5
Coca-Cola Classic® (16 oz cup)	190	0	50	0	0	0	20
diet Coke® (16 oz cup)	1	0	0.4	0	0	0	40
Sprite® (16 oz cup)	190	0	48	0	0	0	20
Orange Drink (16 oz cup)	180	0	44	0	0	0	20

†Salad dressings utilize the standard serving size of 1/2 oz (1 tbsp.) for comparative purposes. Soft drinks were analyzed with ice, sizes 12, 16, 21, and 32 oz. Analysis performed by Hazleton Laboratories America, Inc.

> An *inference* can be described as a generalization drawn from incomplete information.

Drawing generalizations or conclusions that extend beyond the collected data is the branch of statistics called *inferential statistics*. With inferential statistics we make generalizations about a larger body, called the *population*, based on the collected data, which we call the *sample*.

The **population** is the collection of all objects or items that are of interest in a statistical study. The individual objects in the population are the experimental units or subjects.

A **sample** is a finite portion (subset) of the population that is used to study the characteristics of concern in the population.

EXHIBIT 1.1

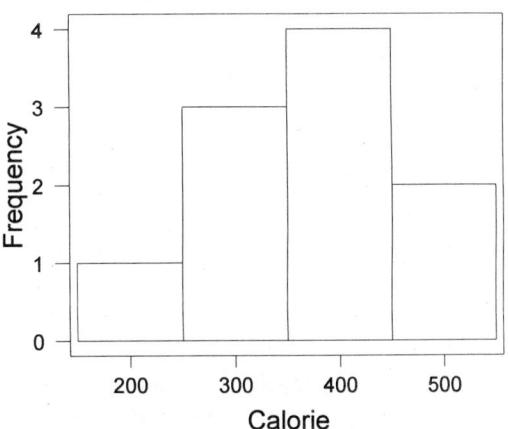

	C1-A	C2	C3	C4	C5	C6	C7	C8	C9	C10	C11	C12
	Sandwich	Calorie	Protein	Carbohyd	TotalFat	Saturatd	Choleste	Sodium				
1	hamburger	255	12	30	9	3	37	490				
2	cheeseburger	305	15	30	13	5	50	725				
3	qt-pounder	410	23	34	20	8	85	645				
4	qt-pounder/ch	510	28	34	28	11	115	1110				
5	mclean	320	22	35	10	4	60	670				
6	mclean/cheese	370	24	35	14	5	75	890				
7	big mac	500	25	42	26	9	100	890				
8	filet fish	370	14	38	18	4	50	730				
9	Mcchicken	415	19	39	20	4	50	830				
10	chick-fajitas	190	11	20	8	2	35	310				

FIGURE 1.2

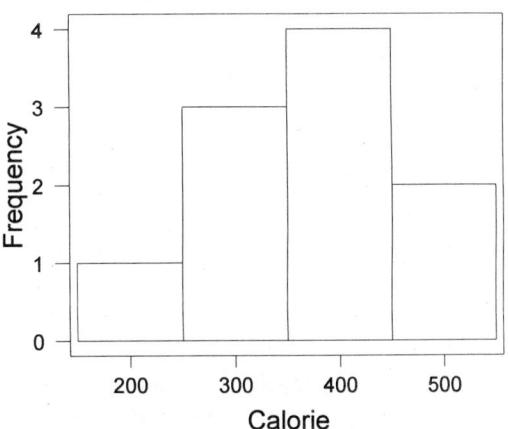

Before taking a sample, one must clearly define the population. For example, if we are interested in families that are poverty stricken, will we consider only those on welfare or should we attempt to identify those whose net worth is below a certain level? If we are to make generalizations from the sample, then the population must be explicitly described in order to obtain a sample that provides accurate information about the population.

In some situations, the population of concern may not be the same as the sampled population. If we have determined that the population of interest is all poverty-stricken families whose net worth is below a certain level, it may be that we are able to sample only those who are receiving welfare benefits. The population of all poverty-stricken families is called the *target population*, and the population of welfare recipients is the *sampled population*.

The **target population** is the population of interest in the study. The **sampled population** is the population from which the sample is obtained.

If we are interested in all poverty-stricken families but sample only those who receive welfare, then the target population is larger than the sampled population because not all poverty-stricken families receive welfare. But the target population can also be smaller than the sampled population. For example, we might be interested in the opinions of those who will vote in a coming election. Not knowing who will vote, we sample the much larger population of all registered voters. Where the target and sampled populations differ, we must be concerned about how the characteristics of the two differ.

A listing of all the subjects in the sampled population is sometimes referred to as a *sampling frame*, or just a *frame*. If there is a frame and it agrees closely with the target population, then it is possible to take a *census*.

A **census** is a sample consisting of the entire population.

It might be said that a census is an *attempt* to sample the entire population.

Because we are concerned with information about the population, why not always take a census? The census is appropriate in certain circumstances, but usually a sample is preferred for the following reasons:

Cost: Measuring all subjects in a population costs more, so a sample is more economical because it contains fewer subjects.

Time: Because the population is usually very large, just the time involved in measuring all subjects often prohibits the study. For example, a survey of voter preference that takes 6 months to collect is of little value. Opinions are likely to change over that time.

Destructive testing: In some studies, just the act of taking a measurement destroys the unit on which the observation is taken. For example, measuring the useful lifetime of a battery involves operating the battery until it fails. In this case, taking a census depletes the inventory.

Inaccessible population units: Suppose one wishes to study the effectiveness of a new miracle drug to prevent the common cold. All people who may someday use this miracle drug cannot be determined and thus are not available for observation. A sample must be used for the study.

Inaccuracy of a census: Because a census is such a difficult undertaking, systematic biases (a tendency to underestimate or overestimate) sometimes appear. Accounting for millions of people is going to lead to some inaccuracies simply because of the magnitude of the job. It is much easier, and maybe more accurate, to analyze a sample rather than the entire population.

Thus, a census is rarely practical, and we often resort to using a sample. Clearly, any conclusions that are made about a population are only as good as the sample. Anyone can interview a portion of the public and get responses, but do the responses truly represent the attitudes of the entire body from which the data come? Later we will look at different methodologies for obtaining a sample that represents the population, but first we look at some examples that identify the population, describe the sample, and list some variables of interest.

EXAMPLE 1.1

A television news commentator recently reported that a Gallup poll of 1,850 adults revealed that 38% of the nation approved of the President's foreign policy. The *population* is all adults in the nation. The *sample* is the 1,850 who were surveyed. A *variable* of interest is the opinion of the subject on foreign policy. The 38% is an *inference* about all adults based on the sample.

EXAMPLE 1.2

An admissions office at a university is interested in using this year's freshman class to develop an admissions formula for all new applicants. The *population* consists of all students who would apply for admission. The *sample* is this year's freshman class. *Variables* of interest might be an achievement test score and high school rank.

EXAMPLE 1.3

A scientist is investigating the effectiveness of a new drug to relieve the symptoms of the common cold. She administers the drug to 100 adults. The *population* is all adults who would try the drug in the future. The *sample* is the 100 chosen adults. *Variables* of interest might be the age of the subject and the time required for the drug to relieve a particular cold symptom. The scientist would *infer* that the average time of relief for the sample of 100 adults would approximate the average time of relief for a subject in the population.

EXAMPLE 1.4

To determine the percent of men and women who smoke, the Surgeon General mailed questionnaires to 5,000 people across the United States. The *population* is all U.S. residents. The *sample* consists of those of the 5,000 who returned their questionnaire. A *variable* of interest is the smoking habits of a subject. The Surgeon General would *infer* that the percentage of male and female smokers in the United States is approximated by the percentages found in the sample.

Exploratory Data Analysis

We have considered examples where we have a specific population in mind and we intend to use the collected data from a sample to infer certain facts about the population. There may be situations, however, where the population is not clearly identified or we have not formulated our questions in a sample–population context. We wish to study a *batch* of observations with no particular inference in mind. (We use the term *batch* to mean that the data may not be a random sample from a specified population.) This is where *exploratory data analysis* (EDA) is most useful. It gives us the tools and strategies to examine a collection of data for structure, patterns, or unusual behavior whether or not we have a particular analysis in mind. As we learn more about the data, we are able to formulate better questions, which may ultimately lead to a formal inference.

EXAMPLE 1.5

The general manager of the Cleveland Indians baseball team wishes to investigate the possibility of signing a new rookie pitcher. Aware of salary caps and other restrictions imposed by league rules, he wishes to study the current salaries of all team members. His accountant enters the 1995 salaries in a Minitab worksheet (see Exhibit 1.2) on her computer, names it **Cleve.mtw**, and creates a simple graph of the data called a *dotplot*. The graph, shown in Exhibit 1.3, consists of a scale for the salaries (in units of $1,000)

EXHIBIT 1.2

	C1-A	C2	C3	C4	C5	C6	C7
↓	position	salary					
1	p	4500.0					
2	of	4300.0					
3	2b	3625.0					
4	1b	3000.0					
5	ss	2850.0					
6	c	2550.0					
7	of	1875.0					
8	p	1800.0					
9	p	1450.0					
10	1b	1075.0					
11	p	900.0					
12	3b	825.0					
13	p	787.0					
14	p	750.0					
15	p	725.0					
16	of	600.0					
17	p	500.0					
18	ss	450.0					
19	c	450.0					
20	of	362.5					
21	p	350.0					
22	p	350.0					

MINITAB - CLEVE.MTW
File Edit Manip Calc Stat Graph Editor Window Help
Data

EXHIBIT 1.3   `MTB > DotPlot 'salary'.`

`Character Dotplot`

with a dot for the salary of each player. Now the accountant is able to study the salaries for structure, patterns, or unusual behavior.

Notice that many of the salaries are in the 0–1,000 area, but then the salaries spread out on the right end of the graph.

We defined a variable as a characteristic that can be measured on each subject in a statistical study. It is the *variation* exhibited by the variable that is of interest to the statistician. If, in the preceding example, all ballplayers received the same salary, then there would be no need for the study. The owner is interested in studying the current salaries of all team members because of the variation in their salaries. The dotplot in Exhibit 1.3 is a picture of the variation. It is an illustration of the *distribution* of the variable.

The **distribution** of a variable specifies the distinct values that the variable assumes and how often these values occur. The distribution illustrates the pattern of the variation in the data.

As you progress through the textbook, keep in mind that measurements will vary from one individual to the next, and one of our main goals is to picture the distribution of the variable with a dotplot or some other graphical display.

C O M P U T E R   T I P

Exhibit 1.4 (page 10) outlines the steps to create the dotplot. First select **Graph,** click on **Character Graphs,** and then select **Dotplot.** When the Dotplot window appears, highlight C2 salary and click on **Select** and then click **OK.** A quicker way is to enter

MTB > **Dotplot 'salary'**

at the MTB prompt as seen in the Session window in Exhibit 1.4.

In this section, we have introduced some basic definitions as well as the concept of exploratory data analysis. Whether we have a formal inference in mind or not, exploratory data analysis should be the first step in processing data. Using numerical quantities and graphs, we inspect the data for patterns, trends, unexpected features, and even errors that may alter our plan of attack. In the next few chapters, you will learn about the methods and techniques that are necessary for the exploration of data.

EXHIBIT 1.4

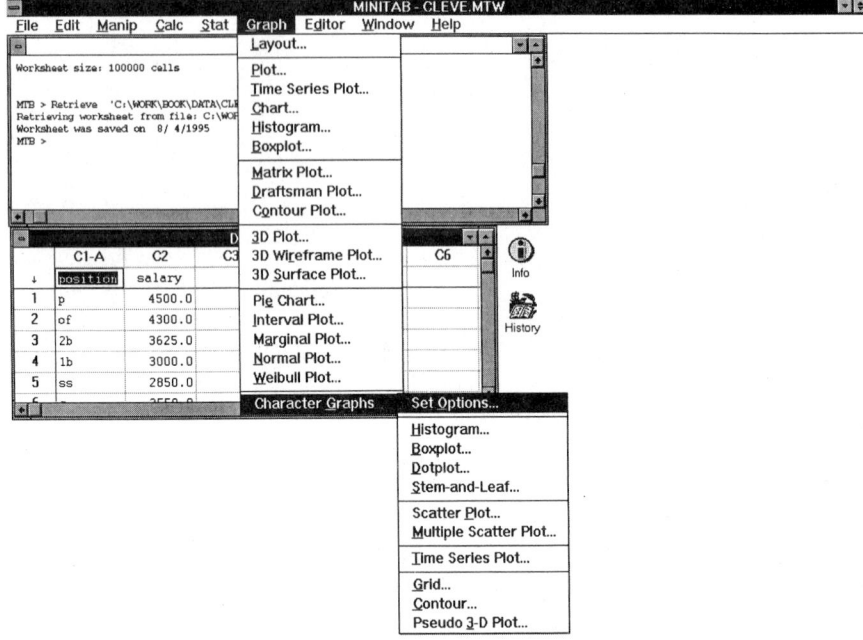

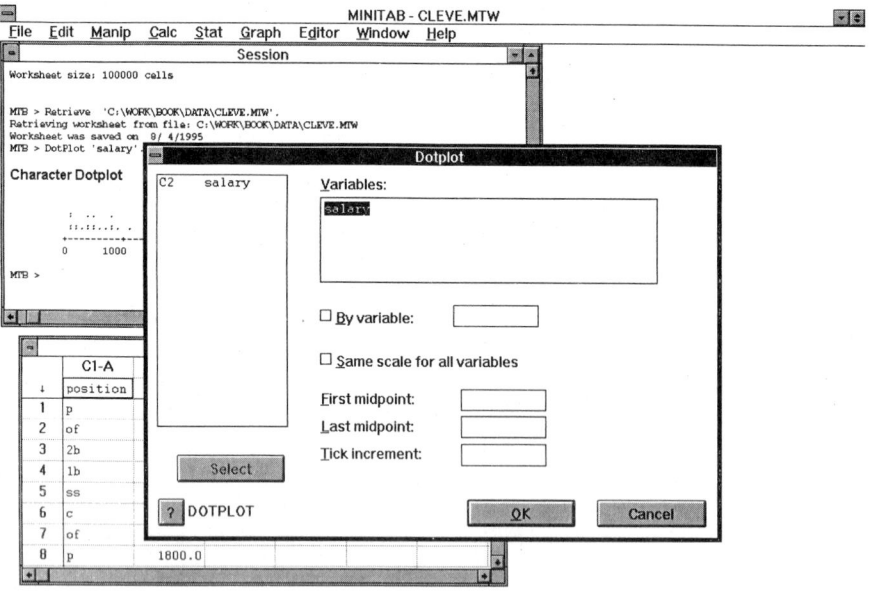

EXERCISES 1.1

1.1   To investigate the attitudes of college students about abortion, a statistics class selected 500 students from the college and obtained their opinions on abortion. What is an experimental unit? What population would this study generalize to? How would you select the 500 students?

1.2   From a poll of 1,000 residents of a neighborhood, it was found that 460 agreed with the neighborhood zoning ordinance. What is an experimental unit? What is the population of interest? What are some of the variables that should be measured?

1.3   Out of 1,000 accidents investigated by the highway patrol, 526 were found to be alcohol-related. What is an experimental unit? Is the population clearly identified in this study? What are some of the variables that should be measured? Is the sample size 1,000 or 526?

1.4   To estimate the percent of students who experience an emotional tragedy during the school year, a psychology professor polled all of her classes. Identify the population of interest. Is this a reasonable sample if she wishes to generalize the results?

1.5   The admissions office at a university sampled 200 freshmen to determine whether the university was their first choice. Identify the population of interest. What are some of the variables that should be measured? How would you select the 200 freshmen?

1.6   Advice columnist Ann Landers asked her readers to clip a questionnaire about sexual attitudes out of the newspaper, complete it, and return it to her. What population would this study generalize to? Is this a reasonable sample if she wishes to generalize the results?

1.7   Explain why a sample is preferred to a census in the following situations:
   a. Testing the effectiveness of an experimental drug that is supposed to reduce pain
   b. Determining the potential market for a new product by distributing free samples
   c. Estimating the amount of timber in a forest
   d. Estimating the lifetime of a flashlight battery

1.8   According to the 1995 *World Almanac and Book of Facts*, the number of deaths due to major aircraft disasters in the 10-year period from 1974 through 1983 was 5,811. This compares with 3,520 deaths in the previous 10-year span from 1964 through 1973.
   a. Does this mean that it is becoming more dangerous to travel by air?
   b. Is the population of air travelers during the first 10-year period comparable to the population of air travelers during the second 10-year period?
   c. Is the total number of fatalities per year (or per 10-year period) a reasonable variable with which to measure the danger of air travel? Can you think of a more valid measure?

1.9   After surprise inspections at nine North Carolina Kmart stores, the Standards Division of the North Carolina Department of Agriculture imposed $25,000 in penalties against five Kmart stores for scanner error overcharges. Their findings are listed here:

Kmart store	Overcharges
4777	41 out of 300 items randomly selected
3886	23 out of 300 items randomly selected
4956	20 out of 300 items randomly selected
3808	23 out of 450 items randomly selected
7342	15 out of 300 items randomly selected

SOURCE: The North Carolina Department of Agriculture, *Agricultural Review*, Vol. LXX, No. 12, December 1995.

   a. Calculate the error rate for each store.
   b. Construct a dotplot of the error rates.
   c. The average error rate for the five stores is 7.6%. Does this mean that every time you shop at a Kmart, you should expect a 7.6% error rate for the scanner?
   d. Does this information imply that all Kmart stores have scanner error overcharges?

1.10   To estimate the murder rates in major U.S. cities, a law enforcement agency determined the rates in the 12 largest cities in 1994. What is the target population? What is the sampled population?

1.11   To assess the possible success of a new shopping center, a questionnaire was mailed to a sample of people obtained from the mailing list of the local chamber of commerce. The questionnaire asked whether or not the respondent thought the availability of merchandise was adequate in the city. Distinguish between the target population and the sampled population. Do you think that taking every tenth name from the chamber of commerce list is a reasonable way to get a sample? Explain your answer. Other than the chamber of commerce list, can you think of other lists to use for the sample?

1.12   The Conservation Reserve Program (CRP) is a federal program where farmers are paid to take their most fragile lands out of production and perform certain conservation practices over a 10-year period. Annual per-acre rental payments (established by bids) range from $4 to $200, the average being $55 per acre per year. Total cost of the 10-year program is expected to be $20 billion. Two-thirds of the acreage enrolled in CRP is concentrated in the ten states on the following list. Enter the data into a Minitab (or any statistics software package) worksheet.

Worksheet: CRP.mtw

State	Total CRP Enrolled (thousand acres)	Average Rental Value ($/acre)	Total 10-Year Contract Value ($millions)
Texas	4,150	45	1,859
N. Dakota	3,181	42	1,321
Kansas	2,938	57	1,675
Montana	2,854	39	1,124
Iowa	2,225	86	1,920
S. Dakota	2,120	44	940
Colorado	1,978	46	960
Minnesota	1,929	39	1,137
Missouri	1,727	67	1,163
Nebraska	1,423	59	842
U.S. total	36,422	55	19,510

SOURCE: "Confronting the Issues," *NC Farm Bureau News*, July/August 1994, p. 10.

a. Create a dotplot of the average rental values.
b. Create a dotplot of the total CRP acres enrolled.
c. Can you generate the total 10-year contract value from the total CRP enrolled and the average rental value? If so, how would you do it on your worksheet?

1.13   A new urine and blood test can help predict a woman's risk of osteoporosis, the bone-thinning illness that makes older adults vulnerable to fractures (*USA Today*, June 16, 1994). Dr. Michael Kleerekoper of Wayne State University Medical School in Detroit stated that the tests combined with X-rays that measure current bone density "will tell where you are now and where you'll be 5 or 10 years later." Describe the target population in this situation. Can you identify any variables that would be of interest to Dr. Kleerekoper?

1.14   A 1994 study, based on census data, by the Federal Highway Administration shows that commuting times in most urban areas are becoming longer. The list compares the average travel times to work (in minutes) in several major cities.

Worksheet: Commute.mtw

City	1980	1990	City	1980	1990
New York	33.7	31.1	San Diego	19.5	22.2
Washington	27.2	29.5	Cincinnati	21.8	22.1
Houston	25.9	29.5	Cleveland	21.6	22.0
Chicago	26.3	28.1	San Antonio	20.2	21.9
Los Angeles	23.6	26.4	Indianapolis	20.8	21.9
Atlanta	24.9	26.0	Sacramento	19.5	21.8
Baltimore	25.3	26.0	Tampa	20.2	21.8
San Francisco	23.9	25.6	Portland, OR	21.4	21.7
New Orleans	24.5	24.4	Charlotte, NC	19.9	21.6
Seattle	22.8	24.3	Norfolk, VA	21.0	21.6
Boston	23.4	24.2	Kansas City, MO	20.7	21.4
Dallas	22.4	24.1	Minneapolis	20.1	21.1
Miami	22.6	24.1	Columbus, OH	20.1	21.2
Philadelphia	24.0	24.1	Hartford, CT	20.1	20.6
Detroit	22.5	23.4	Milwaukee	18.8	20.0
St. Louis	22.6	23.1	Salt Lake City	20.2	19.8
Phoenix	21.6	23.0	Rochester, NY	19.3	19.7
Orlando	20.3	22.9	Providence, RI	18.3	19.6
Pittsburgh	22.8	22.6	Buffalo	19.3	19.4
Denver	22.0	22.4			

SOURCES: Federal Highway Administration and *USA Today*, June 16, 1994.

a. What is an experimental unit?
b. What variables are measured? What are the values of the variables for Charlotte, NC?
c. How do you suppose the Federal Highway Administration came up with the values?
d. What city had the greatest change in travel time from 1980 to 1990?
e. Create a dotplot of the 1980 data.
f. Create a dotplot of the 1990 data.
g. What general statements can you make about how the commuting times have changed from 1980 to 1990?

1.15   Nearly the whole of the states have now returned their census. I send you the result, which as far as founded on actual returns is written in black ink, and the numbers not actually returned, yet pretty well known, are written in red ink. Making a very small allowance for omissions, we are upwards of four millions; and we know that the omissions have been very great.

Thomas Jefferson (1791)

SOURCE: Thomas Jefferson (1791), Letter to David Humphreys, published in *The Papers of Thomas Jefferson*, ed. Charles T. Cullen, Vol. 22 (Princeton, NJ: Princeton University Press, 1986), p. 62.

Even today the U.S. Census Bureau is still plagued with a large undercount in the national census. In fact, several cities and states sued the U.S. Census Bureau because of undercounts in the 1980 census. (Much of the federal funding to cities and states is based on census counts.) The Census Bureau spends much time and money studying the pattern of the undercount and developing methods to adjust the count. Researchers have attempted to relate the undercount to the percent minority, crime rate, and percent living in poverty. For their study, they divided the United States into 66 areas: 16 central cities, the 12 remainders of states in which the cities are located, and the 38 remaining whole states. In Chapter 11, we will study the relationship in greater detail, but for now create dotplots of the data associated with the four variables listed in the accompanying table. Are there any similarities in the four dotplots?

Worksheet: Census.mtw
minority = proportion black or Hispanic
crimrate = number of reported crimes per 1,000 population
poverty = percent of population living in poverty
undercnt = estimate of 1980 census undercount (in percent)

State	minority	crimrate	poverty	undercnt
Alabama	26.1	49	18.9	−.04
Alaska	5.7	62	10.7	3.35
Arizona	18.9	81	13.2	2.48
Arkansas	16.9	38	19.0	−.74
California	24.3	73	10.4	3.60
Colorado	15.2	73	10.1	1.34
Connecticut	10.8	58	8.0	−.26
Delaware	17.5	68	11.8	−.16
Florida	22.3	81	13.4	2.20
Georgia	27.6	55	16.6	.37
Hawaii	9.1	75	9.9	1.46
Idaho	4.2	48	12.6	1.53
Illinois	8.1	48	7.7	1.69
Indiana	7.1	48	9.4	−.68
Iowa	2.3	47	10.1	−.59
Kansas	7.9	54	10.1	.94
Kentucky	7.7	34	17.6	−1.41
Louisiana	31.4	54	18.6	2.46
Maine	.7	44	13.0	2.06
Maryland	16.7	58	6.8	2.03
Massachusetts	3.8	53	8.5	−.57
Michigan	7.0	61	8.7	.89
Minnesota	2.1	48	9.5	1.57
Mississippi	35.8	34	23.9	1.52
Missouri	7.8	45	11.2	.81
Montana	1.5	50	12.3	1.81
Nebraska	4.8	43	10.7	.36
Nevada	13.0	88	8.7	5.08
New Hampshire	1.0	47	8.5	−1.49
New Jersey	19.0	64	9.5	1.44
New Mexico	38.4	59	17.6	2.69
New York	8.0	48	8.9	−1.48
North Carolina	23.1	46	14.8	1.36
North Dakota	1.0	30	12.6	.35
Ohio	8.9	52	9.6	.97
Oklahoma	8.6	50	13.4	−.12
Oregon	3.9	60	10.7	.93
Pennsylvania	4.8	33	8.8	−.78

(*continued*)

State	minority	crimrate	poverty	undercnt
Rhode Island	4.9	59	10.3	.74
South Carolina	31.0	53	16.6	6.19
South Dakota	.9	32	16.9	.42
Tennessee	16.4	44	16.4	−2.31
Texas	30.6	55	15.0	.27
Utah	4.7	58	10.3	1.14
Vermont	.9	50	12.1	−1.12
Virginia	20.0	46	11.8	1.11
Washington	5.4	69	9.8	1.48
West Virginia	3.9	25	15.0	−.69
Wisconsin	1.7	45	7.9	1.45
Wyoming	5.9	49	7.9	4.01
Baltimore	55.5	100	22.9	6.15
Boston	28.4	135	20.2	2.27
Chicago	53.7	66	20.3	5.42
Cleveland	46.7	101	22.1	5.01
Dallas	41.6	118	14.2	8.18
Detroit	65.4	106	21.9	4.33
Houston	45.1	80	12.7	5.79
Indianapolis	22.5	53	11.5	.31
Los Angeles	44.4	100	16.4	7.52
Milwaukee	27.2	65	13.8	3.17
New York City	44.0	101	20.0	7.39
Philadelphia	41.3	60	20.6	6.41
St. Louis	46.7	143	21.8	3.60
San Diego	23.6	81	12.4	.47
San Francisco	24.8	107	13.7	5.18
Washington, D.C.	72.6	102	18.6	5.93

SOURCE: E. P. Ericksen, J. B. Kadane, and J. W. Tukey, "Adjusting the 1980 Census of Population and Housing," *Journal of the American Statistical Association* 84 (1989): 927–944.

1.16   An issue of the school newspaper reported the results of a survey conducted by the office of student development on the percent of students who use their student government representative to convey their feelings on university issues.
   a. What is the population of interest?
   b. Other than recording whether the respondent uses the representative, what other variables do you think should be measured?
   c. Is this a situation where the target population and the sampled population are the same? Explain.

1.2 SOURCES OF DATA

Every day we read or hear about certain numerical facts that describe our world. Following are examples of data that have been collected and summarized:

- The Commerce Department reported that in May 1994 housing starts rose 2.6% to an annual figure of 1,510,000 units. A 16.7% increase in multifamily starts to an annual figure of 308,000 units accounted for the rise. Starts in single-family units dropped .5% to a 1,202,000 annual figure. (*Wall Street Journal*, June 17, 1994)
- In 1991 the average family paid $4,296, 11.7% of average family income, on health care. (*U.S. News & World Report*, December 23, 1991)
- Nicotine patches can double a smoker's chances of quitting successfully, according to a report in the *Journal of the American Medical Association*. In the study, 27% of the patients ceased smoking while using the patch, whereas 13% of those on placebos quit. (*Journal of the American Medical Association*, June 22, 1994)
- Women who take birth control pills for longer than 4 years are almost twice as likely as nonusers to get breast cancer before they are 50 years old. (*USA Today*, October 2, 1990)
- The average age of public school teachers is 42. The average high school principal earns more than $50,000 per year. (*Career Opportunities News*, Vol. 9, No. 5, 1992)
- It was reported that 70% of women married 5 years or longer have extramarital affairs. [Shere Hite, *The New Hite Report: Women and Love: A Cultural Revolution in Progress* (New York: Knopf, 1987)]
- An MCI/Gallup poll of 417 men and 146 women found that 62% of the men versus 78% of the women found business travel stimulating. Additionally, 30% of the men versus 20% of the women found it boring. (*USA Today*, April 13, 1994)

The sources in these examples are varied: national newspapers and magazines, government agencies or private polling organizations, and specific reports from individual researchers. We might add that some give more reliable results than others.

Preexisting Data

In this book you will learn how to evaluate statistical statements made by others as well as how to use their *preexisting data* to make your own calculations and conjectures. In the third example listed above, the *Journal of the American Medical Association* reported from a 1994 study on the effectiveness of nicotine patches that 27% of the patients ceased smoking while using patches. Where did the reporters get the information to arrive at the 27% figure? How can we be assured of the accuracy of the figure? Generally, statements like this are obtained by analyzing data that are collected from surveys or experiments, and the accuracy depends on how well the data were collected.

The U.S. government is a major collector of data. Within the government, the Census Bureau, the Bureau of Labor Statistics, the Bureau of Justice Statistics, and the Department of Agriculture are the most active collectors of data. The Census Bureau is responsible for compiling information on the basic characteristics of all citizens in the country. Information gathered on individuals and households is used by the federal government, for example, to allocate funds to states and cities for various health and welfare programs. In addition to the decennial census, the Census Bureau conducts the monthly *Current Population Survey* (CPS). Each month some 60,000 households are surveyed to obtain information about the employment status of the U.S. workforce.

The Bureau of Labor Statistics periodically conducts surveys that are used to establish certain quantitative measures of the general growth of the economy. One such item, the *Consumer Price Index* (CPI), is computed monthly and measures changes in the prices of goods and services that are purchased by all consumers.

The Bureau of Justice Statistics annually compiles the *Sourcebook of Criminal Justice Statistics*. The sourcebook is a comprehensive reference volume of existing nationwide statistical data on crime and the criminal justice system. The data are obtained from various operating agencies, academic institutions, research organizations, and public opinion polling firms. A large part of the data are provided by the FBI in its annual *Uniform Crime Reports*. These data are compiled by the FBI from information provided by local law enforcement agencies throughout the United States.

The Department of Agriculture (USDA) produces a multitude of reports that analyze current agricultural conditions, forecast market conditions, and provide economic analysis in the areas of trade, production, rural development, and other topics. Reports from the National Agricultural Statistics Service estimate production, stocks, inventories, dispositions, utilization, and prices of agricultural commodities. The World Agricultural Outlook Board issues regular forecasts of U.S. and world supply and demand prospects for all major agricultural commodities. All of these reports are made available to the public through the many USDA economic agencies.

Much of this statistical information can be found in the *Statistical Abstracts* in your university library. In addition to accessing materials from books, journals, magazines, and newspapers, with a personal computer you can electronically access data through information retrieval systems and on-line databases. The Internet also contains a wealth of statistical information that is easily accessible to anyone with an Internet account.

Other sources that we see in the news everyday are survey results from independent agencies such as the Gallup, Harris, and Roper polling organizations. George Gallup, who received his doctorate in journalism from the University of Iowa in the early 1930s, devised new sampling procedures for estimating newspaper readership. Using those techniques, he predicted Franklin Roosevelt's victory in the 1936 presidential election when others were predicting defeat (see Exercise 1.31 at the end of this section). Gallup has been very successful in subsequent presidential polls with the exception of the 1948 presidential election when Harry Truman defeated Thomas Dewey (see the historical note on surveys at the end of Unit 1).

The A. C. Nielsen Company is another polling organization that is very influential. It provides the Nielsen Rating Service for major television networks by conducting a viewer survey of all network television shows. The "ratings" help the networks decide which shows will be renewed and which will be canceled. Top-rated shows such as the Super Bowl command as much as $1,000,000 for a half-minute commercial advertisement, whereas lower-rated shows sell commercial time for as little as $25,000 a minute. The ratings indicate the percent of viewers watching the show and are obtained from a random sample of about 1,500 homes with televisions.

In addition to government agencies and national polling organizations, we often see data produced by independent testing agencies such as *Consumer Reports*. We see test data on automobiles, televisions, stereos, and the like in magazines such as *Motor Trend* and *Stereo Review*. The medical industry produces data every day from experiments with new drugs and treatments. And, of course, much of the statistics information that we read, see, and hear about is summarized in daily newspapers and reported on television.

As you can see, there is an almost unending supply of preexisting data. Remember, however, that quality data are difficult to obtain, and even with the best of intentions, errors (measurement errors, sampling errors, recording errors) can creep into data. We now look at ways of avoiding those errors when you collect your own data.

Collecting Data   Although much can be gained by summarizing and analyzing data collected by others, we often face the more complex problem of having to collect our own data. To obtain reliable information that will help answer your research questions, follow these steps:

1. Determine the objectives of the study you are undertaking.
2. Choose the variables that you will measure in the study.
3. Decide on an appropriate *design* for producing the data.
4. Collect the data.

In any study, statistical or otherwise, we must understand the problem and know what results we are seeking. Specifying the objectives, then, is of the utmost importance if we are to draw meaningful conclusions from the study.

To accomplish the stated objectives of a statistical study, we must determine what will be measured and how it will be measured. For example, suppose we are interested in studying the issue of substance abuse on campus. The main objective is to determine whether there is a substance abuse problem. We are interested in what substances are used and what percent of the student body uses each, and whether the percentages differ for men and women. A poll of the student body would involve determining the gender of the respondents and their substance use habits. We might also be interested in whether or not the percentages change for the different classes, whether family history of substance use is a factor, and so on. This means that several variables must be measured on each respondent. In order not to overlook an important issue, we should attempt to identify all relevant variables prior to collecting any data. And, if there is some doubt whether to measure a certain quantity, it is best to go ahead and obtain measurements rather than later wishing that it had been measured.

Statistical Designs   Having stated the objectives and determined the variables that will be measured, we turn our attention to the design of the statistical study. Statistical designs generally fall into two categories: surveys and experiments.

Surveys   The poll of the student body described earlier is an example of a survey. In fact, all opinion polls are surveys. Polling organizations such as Gallup, Harris, Roper, ABC/*Washington Post*, and Time/CNN conduct polls almost daily. Most of their polls are "random digit dialing" telephone surveys in which telephone numbers are randomly selected by a computer. In a survey, the already existing opinions or facts about a group of people are solicited. Unlike an experiment, there is no control over respondents' behavior; they are simply asked to respond in a way that reflects their opinion at that point in time. Their responses are then tabulated for analysis.

Experiments   An experiment, on the other hand, attempts to determine a cause-and-effect relationship between two or more variables. The experimenter plays an active role in the study by controlling the environment and by administering a treatment to the subjects. In an experiment, for example, a scientist administers a drug to a group of subjects in an attempt to study its effect on the common cold. After the treatment is applied to the subjects, measurements are taken and then analyzed.

Representative Samples   The design of a statistical study, whether it be a survey or an experiment, also specifies how the data will be collected. In surveys, the primary goal is to obtain a sample whose characteristics match those of the target population under study. If the sample is *representative* of the population, then the results will be accurate and useful. But representative in what sense—IQ, height, weight, age, race? We hope the sample is representative of the target population in those factors that are relevant to the study. If we are conducting a political survey, then the weight of the subject probably is irrelevant; however, the race of the subject is relevant. If the population consists of 30% blacks, then the sample should consist of 30% blacks so that the sample is representative with respect to race.

EXAMPLE **1.6**

Suppose we wish to study the television viewing habits of the general population. Our sample should be representative of the population with respect to variables such as education level, age, income, IQ, and occupation of the subject.

Bias   Statistical studies can lead to incorrect conclusions when the collected data are *not representative* of the population. Such samples are said to be *biased*, and often the results are useless and misleading.

Bias is a systematic tendency of the sample to misrepresent the population.

Suppose we wish to study the attitudes of the adult population toward sexual harassment. If a sample consists of 70% women and 30% men and their attitudes toward sexual harassment differ, then we would say the sample is biased toward women. The male segment of the population is not properly represented in the sample.

Simple Random Sample   One way to obtain a representative sample is to select the sample in such a way that every sample of that size has the same chance of being chosen. The resulting sample is called a *simple random sample*.

A **simple random sample** of size n consists of n elements chosen from the population in such a way that all samples of that size have the same chance of being selected.

The aim of a simple random sample is to obtain a sample that is representative of the population and not biased in any way. There are several ways to obtain a simple random sample. A popular method is to use *physical mixing*, also referred to as *lottery sampling*. Each element in the sampling frame is given an identification tag. The tags are thoroughly mixed in a barrel, and the sample is selected from the barrel, one observation at a time. If the mixing is thorough, then each element that remains in the population has the same chance of being selected at each draw. If n tags are drawn, then that group of n is just as likely as any other group of n, resulting in a simple random sample of size n.

EXAMPLE 1.7   **1970 Draft Lottery**

Although physical mixing seems simple, it is difficult to get a complete mixing. A historical example of a problem with physical mixing was the 1970 draft lottery. The draft lottery was designed to randomly select birth dates. Men born on the first date chosen would be the first drafted, those born on the next date chosen would be drafted next, and so on. The 1970 lottery was for eligible men born in 1952 and earlier. Cylindrical capsules, each with a slip of paper inside corresponding to a day of the year, were placed in a large box. The 31 capsules corresponding to January were placed in the box first, stirred, and then pushed to one side. Next, the 29 February capsules were placed in the box and mixed in with the January capsules. The capsules for succeeding months were placed in the box and mixed with those already in the box. The box was shaken several times and the capsules were drawn one at a time, resulting in an assigned draft number for each of the 366 birth dates (1952 was a leap year). The draft numbers are listed in worksheet **Draft70.mtw** on the data diskette. September 14 had draft number 1, meaning that men with birthdays on September 14 were the very first drafted in 1970. As it turned out, the dates later in the year were the first drawn because their capsules were put in the box last and not mixed as thoroughly as the capsules corresponding to the early months. For example, the January capsules were mixed 11 times with the other capsules, whereas the December capsules were mixed only once. Thus, the likelihood of being drafted was greater for young men born later in the year. In 1971, the draft lottery was turned over to the statisticians at the National Bureau of Standards (see worksheet **Draft71.mtw** for the 1971 results), who used a *table of random digits*.

Prior to the widespread use of computers, tables of random digits were commonly used to randomly select elements from a sampling frame. Today most people use a computer, or in some cases a calculator, to generate random numbers like those found in a table of random digits.

C O M P U T E R   T I P

A computer or calculator can be used to generate random numbers. Many scientific calculators have a random number key, labeled [RND#], that produces a three-digit random number when pressed. If six-digit numbers are desired, then press the key twice and put the two numbers together.

To generate random numbers like those in Exhibit 1.6, go to the Session window and type

MTB > **Random 15 c1;**
SUBC > **Integer 1 10000.**

Or, from **Calc** on the menu bar select **Random Data** followed by **Integer**. Then complete the entries in the Integer Distribution window as illustrated in Exhibit 1.5 and click **OK.**

EXHIBIT 1.5

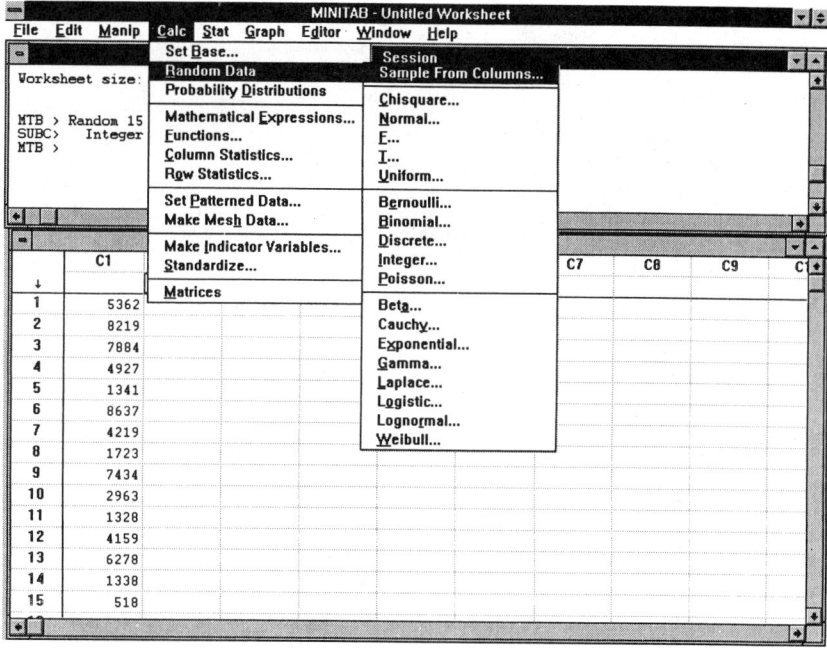

Exhibit 1.6 shows a Minitab worksheet where 15 random numbers, chosen from the integers from 1 to 10,000, are stored in column 1.

EXHIBIT 1.6

```
┌──────────────────────────────────────────────────────────────────────┐
│ ═══              MINITAB - Untitled Worksheet              ▼│◆         │
│ File  Edit  Manip  Calc  Stat  Graph  Editor  Window  Help            │
│ ┌────────────────────────────────────────────────────────────────┐   │
│ │ ═                        Session                       ▼│◆      │   │
│ │ Worksheet size: 4000000 cells                                   │   │
│ │                                                                 │   │
│ │ MTB > Random 15 C1;                                             │   │
│ │ SUBC>   Integer 1 10000.                                        │   │
│ │ MTB > |                                                         │   │
│ └────────────────────────────────────────────────────────────────┘   │
│ ┌────────────────────────────────────────────────────────────────┐   │
│ │ ═                          Data                         ▼│◆     │   │
│ │        C1     C2     C3    C4    C5    C6    C7    C8    C9   C1 │   │
│ │   ↓                                                             │   │
│ │   1    5362                                                     │   │
│ │   2    8219                                                     │   │
│ │   3    7884                                                     │   │
│ │   4    4927                                                     │   │
│ │   5    1341                                                     │   │
│ │   6    8637                                                     │   │
│ │   7    4219                                                     │   │
│ │   8    1723                                                     │   │
│ │   9    7434                                                     │   │
│ │  10    2963                                                     │   │
│ │  11    1328                                                     │   │
│ │  12    4159                                                     │   │
│ │  13    6278                                                     │   │
│ │  14    1338                                                     │   │
│ │  15     518                                                     │   │
│ └────────────────────────────────────────────────────────────────┘   │
└──────────────────────────────────────────────────────────────────────┘
```

It is important to realize that most of the statistical methods presented later in this text are developed with the assumption that the sample is a simple random sample selected by some random procedure like generating random digits with a computer.

EXERCISES 1.2

1.17   A sample of size 10,000 was obtained from the subscription list of *Time* magazine. Identify the most general population for which this sample is representative.

1.18   A sample of size 200 was obtained from a list of students living in dorms at a small midwestern university. Is this sample representative of all students at small midwestern universities? Explain.

1.19   A sample of size 1,000 was obtained from those who attended a rock concert in the Superdome in New Orleans. Do you think that this sample generalizes to the population of all fans of rock music? Explain your answer.

1.20   A sample of size 1,000 was obtained from those who attended a football game in the Superdome. Does this sample generalize to the population of all football fans?

1.21   Explain how random digits generated by a computer can be used to select a random sample of size 15 from your class.

1.22   The net worth of Americans approaching retirement is $238,544. Thirty-eight percent have had high blood pressure at some time in their lives, and 42% say it is at least a little tough to stoop, kneel, or crouch. This information is from a government-funded overview of almost 13,000 Americans aged 51–61 obtained by University of Michigan researchers (*USA Today*, LIFELINE 2/20/96, www.usatoday.com). To what population do you think these results generalize? Are these results from a survey or an experiment?

1.23   With your calculator obtain 15 random numbers between 0 and 1,000.

1.24 With your computer obtain 200 random numbers between 0 and 10,000.

1.25 Locate *Statistical Abstracts* in your library and bring two data sets that are of interest to you to class for discussion.

1.26 The Justice Department estimates that 572,000 women are victims of domestic violence annually. How do you suppose they came up with that number? Did they conduct an experiment? Did they conduct a survey? Did they collect their own data or analyze existing data? If they analyzed existing data, what data would they look for and where would they find it?

1.27 Find the latest issue of the *Uniform Crime Reports* in your library. Determine the crime rates for each of the counties in your state. Enter the data in your computer and construct a dotplot of the crime rates. Are there any counties with unusually high crime rates? Unusually low? Save your data for later analysis.

1.28 From a national newspaper such as *USA Today*, find the Nielsen ratings for last week. Record the number of viewers for each of the top ten shows. Construct a dotplot of the data. Are the values close together or widely scattered? Are there any unusual scores? Save your data for later analysis.

1.29 Explain in a paragraph how you think Nielsen came up with the ratings for the various television shows. Concentrate on what you think are the target population, the sampled population, and important variables, and how the data were obtained. What population do you think Nielsen is trying to generalize to? Do you think the sample is representative?

1.30 Explain how bias might enter into the following statistical study: To evaluate energy consumption for heating residential buildings, a random sample of 50 buildings was monitored for one heating season by a regional electrical utility company.

1.31 *Discussion question—1936 presidential election*: A national survey that yielded unreliable results occurred in the 1936 presidential election. A national magazine of the time, called the *Literary Digest*, had successfully predicted every presidential election since 1916. For the 1936 election, the magazine mailed out 10 million sample ballots to people whose names were taken from subscription lists of magazines, telephone directories, and automobile registrations. More than 2.3 million ballots were returned, the largest number of people ever to respond to a survey. From the responses on the 2.3 million cards, the *Digest* predicted that Alf Landon would defeat Franklin Roosevelt by a margin of 57% to 43%. It predicted that Roosevelt would get only 161 of the 531 electoral votes. As it turned out, Roosevelt received 523 of the 531 electoral votes for a landslide victory with 62% of the popular vote. This marked the end of a 20-year period in which the *Literary Digest* had successfully predicted the presidents. The *Digest* went bankrupt shortly thereafter. Discuss reasons the *Digest*'s prediction was erroneous.

1.3 CRITICALLY APPRAISING DATA

Is it fact, fiction, or speculation? This is a question we should consider when we read about statistical information in the news. For example, the *Hite Report* from a survey of women in 1987 reported that 84% of the respondents were not emotionally satisfied with their marital relationship and that 95% experienced "emotional and psychological harassment" from men. To critically appraise information such as this, we must determine the reliability of the data, which depends on how well the data were collected. Hite's sample consisted of approximately 4,500 returned questionnaires from the 100,000 that she sent out. "So few people responded, it's not representative of any group, except the odd group who agreed to respond," says Donald Rubin, statistics professor and chairman of the Department of Statistics at Harvard University (*Chance*, Summer 1988, pp. 26–31). In fact, most of the statistics community has discounted Hite's figures and concluded that they certainly do not represent the general population of women as originally intended. Her error in sampling occurred by allowing the subjects to *self-select* themselves for the study. In all likelihood, her sample was biased toward those who have strong opinions on sexual issues.

Some issues to consider when assessing the value of statistical data and the reliability of the source are the following:

- What is the purpose of the study? Is there reason for the source to be biased in reporting the results?
- What procedure is used to collect the data?
- Are the data reasonable; would you normally expect such results?
- Are the data useful, relevant, and reported properly?
- Are the graphs misleading?

Is the Source Biased?

In a controversial article ("Does Exposure to Second-Hand Smoke Increase Lung Cancer Risk?" *Chance*, Fall 1993) supported in part by the Tobacco Institute, author Alan J. Gross concludes that "the data relating to whether ETS (environmental tobacco smoke) is a causal agent for lung cancer are at best very weak." In "Letters to the Editor" (*Chance*, Winter 1994), Dorothy Rice and Stanton Glantz state, "The article fails to present an accurate and balanced review of the literature and presents results heavily biased in behalf of the sponsors of the work." In "Letters" (*Chance*, Spring 1994), Gross defends his review of the literature and points out that Glantz should acknowledge that he is the founder of Californians (now Americans) for Nonsmokers Rights. As we read articles and reports, we should ask ourselves if the author has a special interest in the outcome of the study or, on the other hand, is the author giving an unbiased report of the findings.

We should be skeptical of data produced for the extremely competitive field of advertising. Manufacturers are constantly dreaming up schemes to market their products. A very effective method of advertising is to quote statistical data similar to "Four out of five dentists recommend brand A toothpaste." Was this a scientific poll of dentists across the United States by an independent agency, or did the manufacturer of brand A toothpaste poll five preselected dentists? When the toothpaste manufacturer declares its product superior, we must question the motive. If a competitor did the study, would the results be the same?

EXAMPLE 1.8

In 1994 Wal-Mart, the nation's largest retailer, agreed to sign a deal with Michigan's attorney general to stop running misleading advertisements. Wal-Mart's ad for Hills Bros. Coffee said its price was $1.45 lower than at competitor Meijer. As it turned out, Wal-Mart's price was for a $34\frac{1}{2}$-ounce can versus 39 ounces at Meijer. Also, Wal-Mart offered the Dirt Devil hand-held vacuum $15 cheaper than Meijer, but its model came without attachments.

Were the Data Collected Properly?

Were the data obtained by an independent party? Was a scientific procedure used to collect the data? Did the data represent the intended population? Was the sample size adequate? Were the data derived from other information? Each of these questions must be answered when we examine the origin of the data.

Occasionally we receive in the mail a request from our congressional representative soliciting our opinion on a current issue. Presumably the returned information will be compiled and used by the representative. How useful is this information? Is this a proper way to collect information from the constituency? Usually only those who have a strong opinion on the issue will bother to respond, and thus the results will not generalize to the intended population. Another example of improperly collected data is from dial-in phone surveys. Usually the caller is charged a fee of around a dollar to participate and only those with strong opinions will bother to call. Because of the self-selection of the participants, the results of these *convenience samples* will not generalize to any reasonable population and thus are of little value.

EXAMPLE 1.9

A local television station conducts a weekly telephone survey in conjunction with the evening news. On one occasion, the question posed was: Should the salaries of local firefighters be comparable to those of the local police? Callers were able to phone in their responses (yes or no) throughout the evening, and the results were announced on the 11:00 news. How much credibility should be attached to the results of this survey?

SOLUTION

Little or none! First, it was a voluntary survey and only those who felt strongly (firefighters and their families) about the issue would bother to call. Also, there was no way to monitor the number of times a person called. Therefore, those who called did not constitute a representative sample of the residents of the city. It should be added that such polls tend to mislead the general population. The naive citizen believes that the results are reliable when, in fact, they misrepresent the true attitude of the population. The results of this survey most likely would have been biased toward the firefighters.

Many organizations conduct their own polls and try to generalize the results. We would hope that the data are collected by a party that will not gain monetarily from the results. Magazines such as *People*, *Glamour*, and *Ladies' Home Journal* conduct polls of their readers. Do their results reflect the views of all Americans or just their readers? A scientific procedure should always be used in designing an experiment or conducting a survey.

Are the Data Reasonable?

It has been reported that 80% of Miami's economy depends on drugs. How can this extremely high figure be verified? Dealing in drugs is illegal and much of it goes undetected, so there is no way to know what percent of the economy is dependent on drugs.

We should also be aware of the *variability* of measurements. Two people using the same device to measure a patient's blood pressure will probably get different readings.

However, the reading obtained by each would be a valid measurement of the blood pressure if the procedure were carried out properly. In fact, if the measurements were exactly the same, then we would begin to worry, as is illustrated in Example 1.10.

EXAMPLE 1.10

Cyril Burt was an English psychologist who studied the IQ scores of identical twins who were raised apart. A high correlation between the IQs of the separated twins would indicate that heredity is a determining factor in IQ. In 1955, Burt reported a correlation of .771 for 21 pairs of twins. In 1958, he reported a correlation of .771 for "over 30" pairs of twins. In 1966, he reported a correlation of .771 for a sample of 53 pairs of twins. Although correlation will not be studied until Chapter 3, the point here is that it is highly unlikely that the same correlation would be obtained in three different studies. History has discounted Burt's research as useless and perhaps fraudulent (Wade, 1976).

Often we are given only summary information; the *raw data* (original data) are not supplied. It is possible that only favorable results are reported and unfavorable results are ignored. Also, if only summary information is supplied, we should be concerned about misinterpretation.

EXAMPLE 1.11

In discussing energy sources, a national columnist reported that New England supplied only 9.1% of its own energy needs. This figure seems extremely low. In fact, the columnist later corrected himself by stating that New England supplied 9.1% of its energy needs with natural gas.

Are the Data Useful, Relevant, and Reported Properly?

One could argue that all data are relevant. At least they should be relevant to someone; otherwise, why were they collected? But as we view data, we will see that some reported figures are more relevant than others. To the residents of California, the intensity of recent earthquakes is extremely relevant, but the percent of people who favor Coke over Pepsi is probably not that newsworthy. Of course, that percentage is very important to Coca-Cola and Pepsi.

Properly reporting and interpreting data can be difficult. The data in Table 1.1 suggest that in the five southwestern states listed, California and Texas are getting most of the scholarships given in that region in the 30th annual National Merit Scholarship Program.

The truth is that those two states should get most of the scholarships. If we also consider the populations of the five states, we find that California and Texas have approximately 89% of the total population. Thus, they should have approximately 89% of the scholarships. If we look at the rate of scholarships per 100,000 residents, as presented in Table 1.2, we see that California has the lowest rate and New Mexico and Nevada the highest.

TABLE 1.1 Number and Percent of Semifinalists for Five Selected Southwestern States

State	Number of Scholarships	Percent of Scholarships
Arizona	148	6
California	1,294	52
Nevada	49	2
New Mexico	95	4
Texas	893	36
Total	2,479	100

SOURCE: National Merit Scholarship Corp.

TABLE 1.2 Number and Rate of Semifinalists for Five Selected Southwestern States

State	Number of Scholarships	Rate Per 100,000 Residents
Arizona	148	62.87
California	1,294	58.04
Nevada	49	77.77
New Mexico	95	78.38
Texas	893	68.62
Total	2,479	

Are the Graphs Misleading?

The computer age has greatly enhanced the graphic presentation of data. Complicated graphs of data can be produced by almost anyone with access to a personal computer.

> A **bar graph** is a picture consisting of horizontal and vertical axes with rectangles (or rectangular objects) that represent the frequency (or amount) of the categories of a variable. The categories of the variable are listed along one axis and the frequencies along the other.

Figure 1.3 (page 28) is a bar graph that gives the world population (in billions) from 1950 to 2000. Notice that the vertical axis scale starts at 2.5, rather than at 0. This *suppressed zero axis* condition shows a more dramatic increase in the world population

FIGURE 1.3

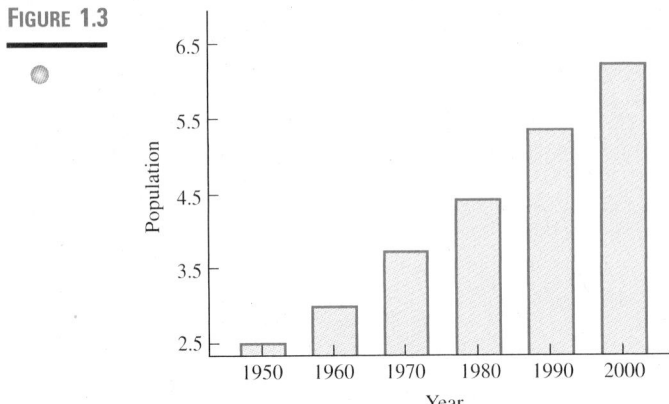

than if we had started the axis at 0. In fact, we can show only a modest growth in world population by stretching the vertical axis as shown in Figure 1.4. You can see that the population growth is then barely apparent.

Figure 1.5 is the proper way to illustrate these data with a bar graph. The vertical scale starts at 0 and is not stretched out so much that it distorts the graph.

Figure 1.6 is an example of a bar graph with the bars replaced by pictures. This is acceptable, but in this case, the data are misleading because both the heights and widths of the "bars" change. Changing both dimensions gives a distorted view of the actual frequencies because we see a difference in the area and not just the height of the bars. For example, if one category has twice the amount of another category, we in effect quadruple the appearance by changing both dimensions. To accurately depict the situation, the frequency or amount of the various categories should be reflected only in the height of the bars.

FIGURE 1.4

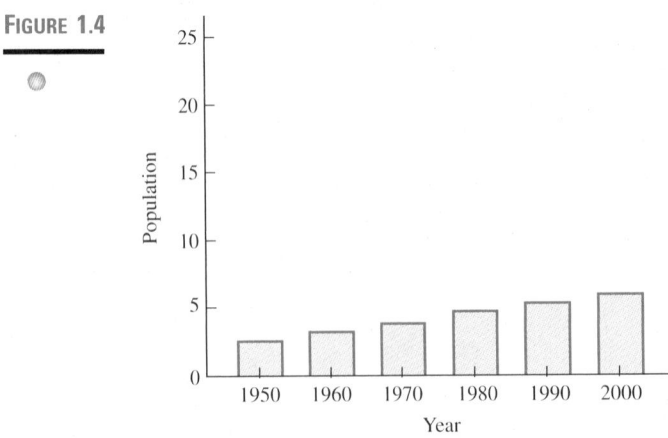

FIGURE 1.5

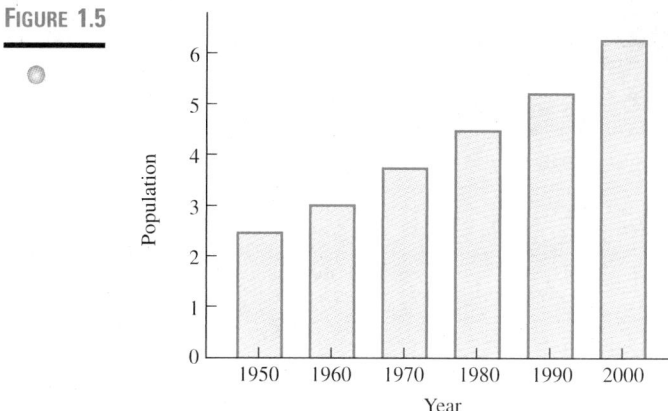

FIGURE 1.6   Per capita income

$14,000
Mississippi

$28,000
Connecticut

C O M P U T E R   T I P

To create a bar graph in Minitab, first choose **Chart** on the **Graph** pull-down menu and then choose the Y and X variables for the display. To adjust the scales of the axes, click on the **Frame** button in the Chart window as shown in Exhibit 1.7 (page 30) and then choose **Min** and **Max**. Set the X and Y Scale Extremes and click on **OK**.

Figure 1.7 (page 31) gives a deceptive view of the U.S. trade deficit with Japan. To dramatize the decline, the graph has been rotated. A rotation in the opposite direction would make the trade deficit appear to *increase* rather than decrease. To complicate matters more, the left-hand scale has been mislabeled. If we are speaking of the trade *deficit*, the scale should not be negative.

EXHIBIT 1.7

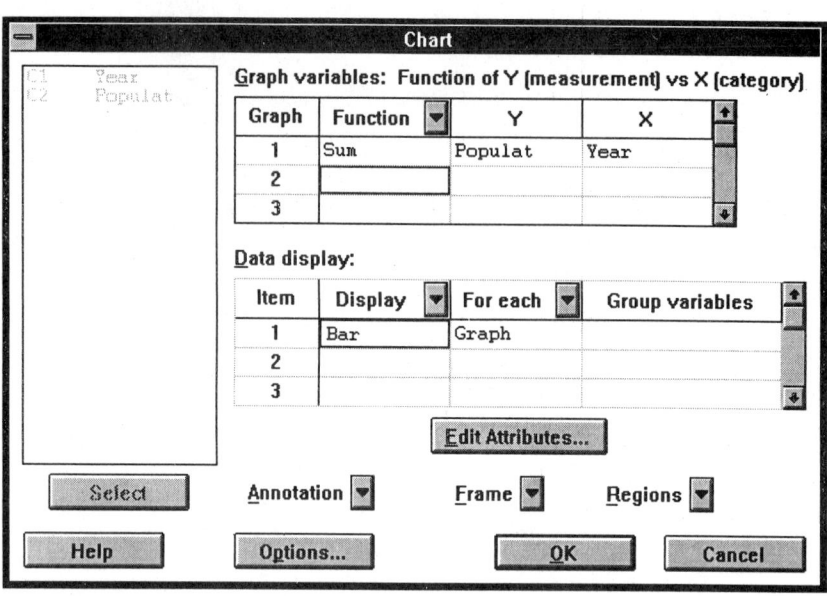

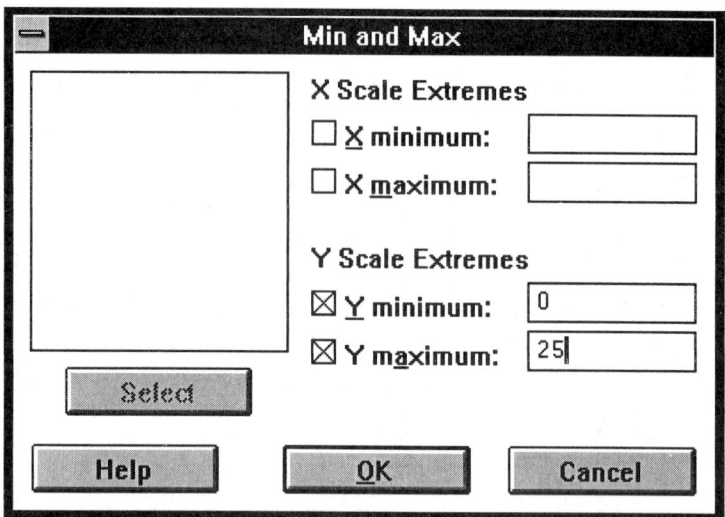

Graphs are the most widely used tools for exhibiting statistical data, and they can be very informative; however, we have seen that they can also be misleading. To get the most out of a graph, we must take a close look at the scales of measurement on both axes and ask whether the data have been reported properly.

FIGURE 1.7

Trade imbalance widens

In the first seven months of 1984, the USA's trade deficit
with Japan nearly reached the record set in all of 1983.

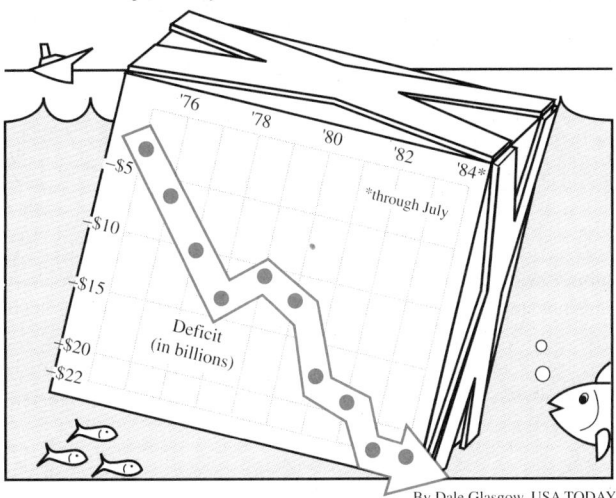

By Dale Glasgow, USA TODAY.

SOURCE: Adapted from *USA Today*, October 9, 1984. Copyright, 1984 *USA Today*.
Reprinted with permission. Data from Data Resources Inc.

EXERCISES 1.3

1.32   The American Cancer Society and the tobacco industry disagree on the results of research concerning the relationship between tobacco and lung cancer. Clearly, the tobacco industry has something at stake. What interest does the American Cancer Society have in this controversy?

1.33   The mayor of a major city reported that 95% of the police force have never taken a bribe. Is this reasonable? Can it be verified?

1.34   The accompanying data, supplied by the Federal Highway Administration, are the numbers of tickets issued for speeding. Is it reasonable that Hawaii has the fewest speeding tickets? Why do California and Texas have the most speeding tickets? What other information would help in interpreting these data? Can they be reported in a better way?

Worksheet: Tickets.mtw

Ala	103,875	Del	24,752	Ind	97,438
Alaska	11,200	D.C.	—	Iowa	135,180
Ariz	197,638	Fla	325,054	Kan	205,639
Ark	67,170	Ga	194,781	Ky	73,758
Calif	1,022,180	Hawaii	4,759	La	171,358
Colo	101,019	Idaho	30,781	Maine	19,212
Conn	67,130	Ill	222,491	Md	129,051

(*continued*)

Mass	208,077	N.M.	150,602	S.D.	33,084
Mich	202,535	N.Y.	247,444	Tenn	147,845
Minn	110,745	N.C.	226,517	Texas	881,673
Miss	180,160	N.D.	45,106	Utah	108,606
Mo	184,284	Ohio	399,636	Vt	32,900
Mont	96,802	Okla	141,495	Va	212,583
Neb	77,755	Ore	99,287	Wash	143,179
Nev	59,885	Pa	218,087	W.Va	54,553
N.H.	31,472	R.I.	27,975	Wis	95,524
N.J.	164,426	S.C.	194,284	Wyo	56,020

SOURCE: Federal Highway Administration.

1.35   The graphic illustrates the number of reports of sexual abuse of children in the years 1977–1982. The figures are staggering. Can you suggest a reason why there was such a dramatic increase over this period of time? Does the number of cases reported give any indication of how many actual cases there were? The sizes of the children inside the rectangles are misleading to a certain extent. Can you explain why?

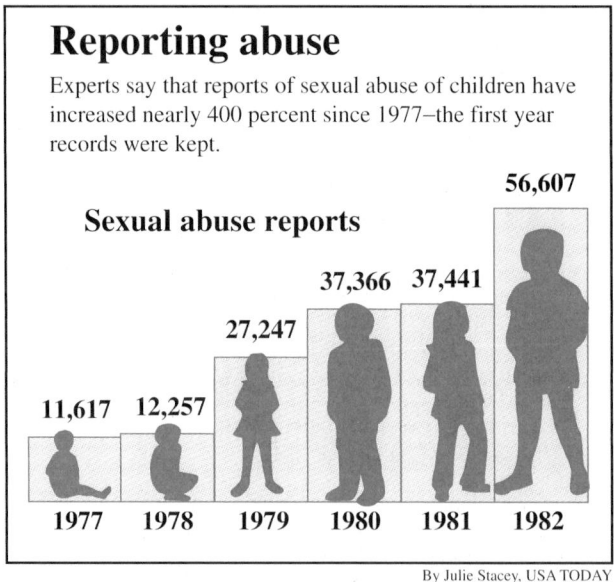

By Julie Stacey, USA TODAY

SOURCE: *USA Today*, September 17, 1984, Copyright 1984 USA Today. Reprinted with permission. Data from the American Humane Association, Denver.

1.36   In her syndicated newspaper column, Ann Landers asked her estimated 70 million readers: Would you be content to be held close and treated tenderly and forget about "the act"? Of the more than 90,000 who responded, 72% said they were content to be hugged. Is this a scientific poll? How reliable is this figure? Can it be generalized to the general population?

1.37   In a move to monitor public opinion, the Democratic Party invited nearly 100 disaffected Clinton supporters and Ross Perot backers in Dayton, Ohio, to watch President Clinton's speech on health care reform. Those in attendance used hand-held dials to register their approval or disapproval on key issues. The president's personal approval rating among the group jumped nearly 50%. Discuss the shortcomings of this method of collecting data to measure public opinion.

1.38   Listed here are the graduation rates for student athletes at schools in the Southeastern Conference:

Worksheet: Graduate.mtw

School	Percent
Alabama	35
Arkansas	42
Auburn	51
Florida	44
Georgia	42
Kentucky	53
Louisiana State	30
Mississippi	46
Mississippi State	53
South Carolina	54
Tennessee	47
Vanderbilt	67

SOURCE: NCAA Graduation Rates Report, 1992.

Construct two bar graphs of the data: one with a suppressed zero axis and one without it. Which one is the proper way to illustrate the data?

1.39   "Numbers can be misleading, expert warns" reads the headline of the article accompanying the graduation rates of the athletes referenced in Exercise 1.38. Clifford Adelman, Director of the U.S. Department of Education's Division of Higher Education, states that students who transfer to another school are counted as not graduating at the school from which they transfer and are not counted at all at the school to which they transfer, whether they graduate or not. How does this cause the numbers to be misleading? Are the actual graduation rates higher or lower than those given? How could the NCAA calculate graduation rates to avoid the problem?

1.40   A *New England Journal of Medicine* paper reported that in a Tennessee experiment, 88 out of 150 motorists stopped for reckless driving tested positive for cocaine, marijuana, or both. Does this mean that more than half of all reckless drivers are high on drugs?

1.41   Diplomatic immunity covers parking tickets, so foreign diplomats do not bother to pay them. The major offenders in New York are listed in the table. Are the Russians the worst offenders? Explain.

Worksheet: Diplomat.mtw

Country	Number of Tickets	Tickets/vehicle/month
Russia	8,138	8.9
Nigeria	2,556	3.3
Israel	2,363	1.6
Indonesia	1,582	1.6
Egypt	1,421	2.7
Bulgaria	1,263	6.6
Brazil	1,260	2.1
S. Korea	1,239	0.8
Ukraine	956	10.6
Venezuela	919	2.2

SOURCE: *Time*, November 8, 1993. Figures are from January to June 1993.

1.42 From a newspaper, magazine, or television show, find an example of where a convenience sample was used. What population do you think the results generalize to? Explain how the sample might be biased. Could a simple random sample be taken in the situation under study?

1.4 SURVEYS, EXPERIMENTS, OBSERVATIONAL STUDIES (OPTIONAL)

Surveys

In many cases the statistical information we see or hear about is derived from a survey. Polling organizations provide us with numbers and charts describing everything from the state of the economy to our personal preferences in soft drinks.

Knowing the source of a sample survey is extremely important. The Gallup, Harris, and Nielsen organizations are considered reliable sources of information, but some organizations are less reliable. For example, in a Senate primary race, a polling firm offered its client two surveys: one was for use with the press for publicity and fund-raising and the other was a confidential survey reporting how the client actually stood in the race (Roll & Cantril, 1972).

Instead of trying to classify a survey organization as a reliable source of information, it is wiser to judge the survey itself. A properly designed survey reports the following information:

- A description of the sampled population
- A description of the method of contact for interviews
- The response rate
- The exact wording of the questions
- The timing of the interview
- The size of the sample (or the margin of error)
- The sampling technique

The following discussion of the preceding topics is not intended to train people to become professional pollsters, but rather to point out possible shortcomings of a survey and to help detect poorly designed surveys. It is also intended to boost confidence in the reliability of the properly designed survey. For example, many people criticize the Nielsen ratings because the sample size appears to be small. But if the sampling is scientific, we will see that the sample size used by Nielsen is certainly adequate.

Population

It is important that the target population be properly identified so that we can determine whether the collected sample is representative of the population. For example, if we wish to estimate the present cost of housing in a major city, is a sample of home loans obtained from savings and loan associations representative of the population? Possibly not, because the population of concern consists of all homes in the city that are for sale. The sample is representative of the population of homes that were mortgaged through a savings and loan. Clearly, we are talking about two different populations: all homes that are for sale and all homes mortgaged through a savings and loan. On the other hand, is a sample of homes listed by the realtors of the city representative? This might be more representative of the population of concern, but not all sales are made through realtors. We could say, however, that the sample is representative of the population of homes sold by realtors in the city. Properly identifying the population might also suggest the method of contact.

Method of Contact

Possible methods of contact are mail, telephone, and personal interview. A mail survey is time-consuming and often has a low response rate. The sample will possibly be biased because a larger percentage of those who feel strongly about the issue will respond. Also, negative or socially unacceptable opinions are more likely to be voiced through mail surveys than through any other type.

Telephone surveys are easily conducted and relatively inexpensive, and they can yield timely results. A word of caution though: 11% of American households do not have telephones, and it is estimated that 16% of all residential phones are unlisted (see Figure 1.8).

FIGURE 1.8   Unlisted phone numbers

The highest and lowest percentage of phone numbers that are unlisted:

Highest–California		*Lowest–Florida*	
Sacramento	64.7%	West Palm Beach	13.2%
Los Angeles	64.6%	Daytona Beach	11.1%
Oakland	64.4%	Sarasota	6.5%

Source: *USA Today*, September 8, 1994.

The personal interview has a higher response rate and generally less bias than either the mailed or telephone survey. However, personal interview surveys are expensive and time-consuming. The interviewer must be trained and paid for time and travel. Care must be taken in choosing the interviewer. He or she should be knowledgeable about the issues raised in the survey and should not be overbearing or suggestive in presenting the questions.

Of the three methods of contact, the personal interview is the best, but the telephone survey is the most widely used because it is timely and relatively inexpensive.

Response Rate

Regardless of the method of contact, one should be aware of the response rate—the number responding to the survey. Recall that for the *Hite Report*, only 4,500 responded out of the 100,000 women who were sent questionnaires. A carefully designed mailed questionnaire can increase the response rate. Properly trained interviewers can increase the response rate in both the telephone and the personal interview.

One must be prepared to re-call subjects who were initially unavailable for an interview. Generally, those people who are not at home in the evening tend to have a different lifestyle from those who are and possibly would respond differently to a survey. Every effort must be made to contact those in the original sample and obtain as high a response rate as possible.

Wording of Questions

The response rate can be increased with carefully designed questions. A survey should not ask too many questions, and ambiguous and confusing questions should be eliminated. A trial run is a good way to identify troublesome questions. Questions should be posed in a neutral sense rather than a negative or positive sense so as to avoid leading the subject. Just the way the question is worded can affect the outcome. For example, when a New York Times/CBS News poll asked people if they favored an amendment "prohibiting abortions," more than 50% said no. But when asked whether they favored one "protecting the life of the unborn child," more than 20% changed their opinion.

EXAMPLE 1.12

During the Vietnam War in 1967, President Johnson ordered the bombing of Hanoi and Haiphong. Two members of Congress conducted independent surveys on the issue. Member A posed the question as follows: Do you approve of the recent decision to extend bombing raids in North Vietnam aimed at the strategic supply depots around Hanoi and Haiphong? Sixty-five percent favored the decision. Member B posed the question as follows: Do you believe the United States should bomb Hanoi and Haiphong? Only 14% favored the decision (Wheeler, 1976).

It is hard to compare the two figures in Example 1.12 because they were not obtained from random samples but rather from specific voting districts. The difference is so dramatic, however, that one is led to believe that the wording of the question had a definite effect on the outcome of the survey.

Timing

The timing of a survey often affects the results. A 1994 Harris Poll of 1,249 adults showed that 30% of Americans supported the idea of having a professional soccer league with games shown on television. It should be pointed out, however, that the poll was conducted shortly after the successful World Cup soccer matches were held in the United States. Three weeks before the World Cup matches, only 25% of those surveyed even knew the event was to be held in the United States and only 15% said they were very interested in watching any of the games on television (*USA Today*, August 25, 1994).

EXAMPLE 1.13

In 1975, a Gallup poll showed that President Ford was ahead of Jimmy Carter in the 1976 presidential race, 51% to 39%. A Harris poll showed that Carter was leading Ford 52% to 41%. The difference in the two predictions can be attributed to the timing of the two surveys. The Harris survey was conducted after a cabinet shakeup, after Ronald Reagan announced he would oppose Ford for the Republican nomination, and before Ford traveled to China. The Gallup poll was conducted after Ford left for Peking. Presidential popularity usually increases after foreign travel (Haack, 1979).

Sample Size

The size of the sample used in a survey is determined by the accuracy desired and the available resources. The larger the sample size, the more accurate the results, but usually only limited amounts of time and money are available. The method of contact is also a factor in the choice of the sample size. Clearly a telephone survey is less expensive than personal interviews. We will see later that the sample size is computed from the desired *margin of error*. If we listen closely when the results of a poll are given, we will hear the newscaster say, "The margin of error was plus or minus 3 percentage points." If it were desirable to decrease the margin of error from 3 to 2 percentage points, the sample size would have to be increased substantially. At some point the gain in accuracy is not enough to justify the larger sample size, which is why most national polls have a sample size of fewer than 2,000 subjects. Gaining one more percentage point of accuracy is not worth the time and cost involved in obtaining a significantly larger sample.

Sampling Techniques

The bias introduced in a sampling procedure is called *sampling bias*. To reduce the effects of sampling bias, the respondents to a survey must be randomly selected. We must avoid situations that allow the respondents to be *self-selected*, as is the case in most convenience samples. The *simple random sample*, described in Section 1.2, eliminates bias by giving each element in the population an equal chance of being chosen. For some surveys, however, the simple random sample may not be desirable or economically feasible. We now describe three sampling procedures that are commonly used in survey sampling.

Systematic Sample   Suppose we have a sampling frame of 10,000 students listed by their Social Security numbers. If we start at a random place within the first 100 names on the list and then select every one-hundredth name, the result is a *systematic random sample* of size 100.

It is possible to use systematic sampling without having a tangible sampling frame. In conjunction with a census, to obtain a 10% sample of homes, the U.S. Bureau of the Census might start at a random place and select every tenth home for further study. In this manner, it is possible to sample the households in a city without knowing how many households there are in the city.

Systematic sampling should be avoided when the data in the population are of a periodic nature. For example, suppose we want to sample the daily receipts of a large department store. A one-in-seven systematic sample would result in a biased sample because the sample would consist of the daily receipts for only one particular day of the week. This problem could be avoided by taking a one-in-ten systematic sample. However, it is best to use some method other than systematic sampling when the population is periodic.

Stratified Sample   A simple random sample of size 1,000 could have 600 women and 400 men. If the population is evenly divided among women and men, this sample is not representative of that population. To obtain a representative sample, we need to take two random samples, one of 500 women and the other of 500 men. We divide the population into the two strata (women and men) and select random samples from each stratum. The resulting sample is a *stratified random sample*. The strata can be geographical regions, religious preference, race, income bracket, political party, or gender as mentioned above.

The stratification can be based on any variable relevant to the survey. Stratification will guarantee that the sample is similar to the population in those characteristics on which one chooses to stratify.

EXAMPLE 1.14

Suppose bank officials wish to sample a bank's savings accounts. Suppose, furthermore, that they know that 5% of the accounts are over $50,000, 20% are between $10,000 and $50,000, 25% are between $5,000 and $10,000, and 50% are below $5,000. Using stratification, they can assure themselves that a sample of 100 accounts will have accounts of all possible categories.

Often, stratified random sampling is accompanied by *proportional* sample size. This means, for example, that if stratum 1 is twice as large as stratum 2, then the random sample from stratum 1 will be twice as large as the sample from stratum 2. That is, the sample size for a stratum is proportional to the population size in that stratum.

EXAMPLE 1.15

Using proportional sampling in Example 1.14, the bank officials would sample so as to have 5 of the desired 100 accounts from the over-$50,000 category. Similarly, 20 accounts would be from the $10,000-to-$50,000 category, 25 from the $5,000-to-$10,000 category, and the remaining 50 accounts from the below-$5,000 category. Not only does the sample consist of accounts from all possible categories, but also the sample sizes are proportional to the stratum sizes.

To use proportional sampling it is necessary that the stratum sizes be known. This information can be obtained only from a census. Thus, it is important that up-to-date census information be available.

Cluster Sample Often the expense of sampling is a major factor in a statistical survey. *Cluster sampling* is an economical way of selecting a sample by choosing groups of subjects, called *clusters*, and then surveying every individual unit in the cluster. Suppose we wish to sample the residents of a large city. With cluster sampling, the entire city can be divided by a city map into blocks (clusters) of households. Then a simple random sample of clusters can be obtained, and each resident of each selected cluster will be interviewed. Clearly, this is a more cost-effective method than traveling around the city looking for individuals in a simple random sample. On the other hand, cluster sampling may be susceptible to sampling bias. A certain ethnic group may not be represented in the sample simply because its cluster is not selected. Also, the population units in a cluster usually are very homogeneous (similar age, income, educational background, recreational interests, etc.), and thus a cluster sample is less informative than a simple random sample of the same size. With cluster sampling we may be unable to get a true cross section of the population.

Experiments

Generally the sample survey is used to describe the characteristics of an existing population. Experimentation is concerned with investigating a population that has been altered for study. Unlike in survey sampling, the experimenter actively intervenes by controlling the environment and administering *treatments* to the subjects in order to study their effects on a *response variable*. The simplest form of an experiment is one in which a single treatment is applied to the subjects in a study. A configuration of the design is

Pool of subjects ⇒ Treatment ⇒ Observations

Confounding Variables

The aim of such an experiment is to evaluate the effects that the treatment has on the experimental units. For this design to be useful, however, we must make sure that no outside or *extraneous* variable is *confounded* with the treatment.

> A **confounding variable** is one whose effect on the response variable is mixed up with the effect of the treatment.

EXAMPLE 1.16

To study the effect that a particular drug might have on blood pressure, a physician selects a random sample of her patients, administers the drug, and then measures their blood pressures. Any number of extraneous variables such as age, gender, and weight could be confounded with the treatment on the blood pressures of the patients. She cannot determine to what extent a given patient's blood pressure is affected by the experimental drug (treatment) and to what extent it is affected by their age, gender, and weight.

Comparative Experiment

The *comparative experiment* is used to avoid the confounding of an experimental factor with extraneous factors. In the comparative experiment, we randomly divide a group of subjects into two equivalent groups, with one group receiving the treatment and the other usually serving as a *control group*. A configuration of the design is

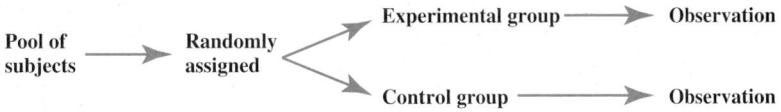

The control group is treated the same as the experimental group except that it does not receive the treatment; then, any observed difference between the two groups should be due to the treatment. Extraneous factors should affect both groups the same, and thus their effects are nullified.

It is possible to have a comparative experiment without a control. Suppose we were comparing two drugs. We could assign drug A to one group and drug B to the other group. The two experimental groups serve as controls for each other. Any observed difference in the two groups should be due to the effects of the different drugs.

Even well-designed comparative experiments can have problems, as the next example illustrates.

EXAMPLE 1.17

A group of senior citizens was randomly divided into two groups. One group was given daily doses of vitamin C, and the other group was given no treatment. After the winter, the vitamin C group reported fewer colds than did the group with no treatment. The experimenter concluded that vitamin C helps prevent colds. In a similar study, however, patients were given a placebo (a dummy tablet) and were told it was vitamin C. They had fewer colds than those who were given vitamin C but thought it was a placebo. ("Is Vitamin C Really Good for Colds?" *Consumer Reports*, February 1976, pp. 68–70)

It is quite possible that the vitamin C group in the first study experienced the *placebo effect*—a psychological response to a neutral treatment in which the subject has confidence. It is likely that the subjects who received vitamin C *knew* that they should have fewer colds.

Double-blind Experiment   To combat the placebo effect, experimenters conduct *double-blind studies*. In a double-blind experiment, neither the subject nor the one administering the treatment knows who is receiving the actual treatment and who is receiving the placebo. The double-blind procedure is used in most studies involving medical trials. Subconsciously, the diagnosing physician may be influenced by knowledge of what treatment the subject receives. For this reason, he or she is kept ignorant of who is receiving the treatment and who is receiving the placebo. For reasons discussed previously, the subjects do not know whether they are receiving the treatment or the placebo. Only the director of the study knows who is receiving which treatment.

Randomization   To reduce possible bias in the comparative experiment, subjects are randomly assigned to the groups. It is hoped that *randomization* will create equivalent groups prior to the experiment. For example, to study the effect of alcohol on reaction time, we certainly would not want our subjects to choose which group (control or experimental) they would enter. Even if the subjects are homogeneous (alike in all characteristics relevant to the experiment), they do have some differences. Random allocation tends to distribute these differences evenly between the two groups.

Matched Pairs Design   Randomization tends to average out the differences and produce nearly equivalent groups for comparison; however, there is no guarantee that it will. For example, suppose 20 subjects are to be randomly assigned to two experimental groups of ten subjects each. It is possible that a randomized procedure would put ten subjects of high intelligence in one group and ten subjects of low intelligence in the other group. If it is important that the two groups be equivalent as far as intelligence is concerned, then the randomization procedure has failed. The problem can be avoided by assigning subjects according to the *matched pairs design*.

In the matched pairs design, the 20 subjects are divided into ten intelligence groups (pairs) of two subjects each. The two subjects in each matched pair should be as much alike as possible prior to the experiment and are then randomly allocated to the two experimental groups. The end result is that each of the groups has subjects from each of the ten intelligence groups, and thus the groups are equivalent as far as intelligence of the subjects is concerned. A configuration of the design is shown here:

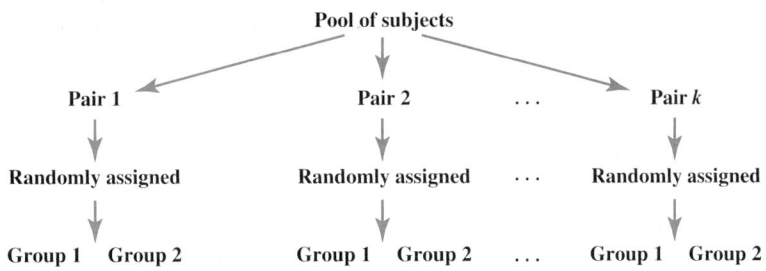

Note that each pair consists of two matched subjects who are randomly assigned to the two treatment groups. Because each pair is similar prior to the experiment, any observed difference is then due to the treatment.

EXAMPLE 1.18

A clothing manufacturer would like to compare the durability of a newly designed line of children's clothing with that of its existing line of clothing. Ten sets of identical twin children are chosen for the experiment. One of each set of twins is randomly selected to wear the new clothing, and the other will wear the old type. After a period of time, the durability of the two types of clothing will be evaluated. In this example, the sets of twins make up the matched pairs and the new and old lines of clothing are the two groups. A configuration of the design follows:

Twin set	1	2	3	4	5	6	7	8	9	10
Clothing type A	x	x	x	x	x	x	x	x	x	x
Clothing type B	y	y	y	y	y	y	y	y	y	y

A value of x denotes a measure of durability for type A, and a value of y denotes the measure of durability for type B. Identical twins are used as matched pairs because it is assumed that they will give approximately the same treatment to the two sets of clothing.

In a modification of this design, the *before–after design*, a group of subjects serve as their own match and observations are taken both before and after the treatment:

Pool of subjects $\Rightarrow$ Observation $\Rightarrow$ Treatment $\Rightarrow$ Observation

The *randomized block design* (covered in more advanced textbooks) is a generalization of the matched pairs design where there are b blocks and g groups and the blocks can represent any variable that has a confounding effect on the response variable.

Observational Studies

In many situations it is not possible to allocate subjects randomly to groups to study the effects of a treatment. For example, if we wish to study the effects of income on one's spending habits, it is not possible to assign subjects randomly to different income groups. However, we can observe the spending habits of people who already fall into different income groups and then make statistical generalizations about how income affects one's spending habits.

> An **observational study** (also called a quasi-experiment) is an experiment in which one observes how a treatment has already affected the subjects.

The most obvious observational study that comes to mind relates to the smoking/nonsmoking issue. Clearly, we *cannot* randomly allocate subjects to smoking and nonsmoking groups, yet we can compare the incidence of lung cancer for those who smoke and those who do not smoke. Subjects are not assigned to the treatment (smoking) by the experimenter, but rather they have assigned themselves to the smoking and nonsmoking groups. In other words, the smoking and nonsmoking groups already exist and we are able to compare their rates of lung cancer with an observational study.

A major distinction between an observational study and a designed experiment is that an observational study can only reveal an *association* between the treatment and response variable, whereas designed experiments can establish *causation* (cause and effect). The association that we see in an observational study may be due to confounding variables whose effects cannot be reduced by randomization as they can in the designed experiment.

EXAMPLE 1.19

A university would like to compare the salaries of male and female faculty members. It is not possible to assign the subjects randomly to a gender. Therefore, the two groups (male and female) have to be compared, as they are, in an observational study.

Suppose that a difference is observed between the salaries of men and women. Does this mean that the university should be accused of sex discrimination? Possibly not. It is possible, and very likely, that the years of experience differ for the two groups, which could account for the discrepancy in the salaries. We would say that the variable years of experience is confounded with gender on the salaries of the professors.

In the preceding example, it is possible to control for the confounding variable by comparing the salaries of men and women who have similar years of experience. In other observational studies, extraneous variables are so confounded with the treatment that it is impossible to compare groups.

Ethical Issues

Human Experimentation

Serious ethical questions should be raised when experiments are conducted on human subjects.

If a specific treatment has the potential of curing a disease, should it be withheld from some subjects for the sake of having a control group?

Should an experimenter deliberately expose humans to a substance that is suspected of being detrimental to their health?

The human subject should be made aware of the possible consequences of the experiment. In fact, the federal government requires that the *informed consent* of the subjects be obtained in order to use them in an experiment. Subjects are told of the possible risks and benefits and asked whether they will consent to random assignment to the experimental and control groups.

Animal Experimentation

Experiments conducted on animals are less restricted than those done on humans. We may purposely expose rats to a substance that is suspected of causing cancer to learn whether it indeed causes cancer, and then we can infer the same for the human population. But can we generalize to the human population? Critics of animal experimentation say no. They contend that the results are not valid because a laboratory animal reacts differently to a drug than a human does. Moreover, when a substance is tested, the laboratory animal is frequently given a massive overdose to reduce the duration of the experiment. The drawback is that the animal may develop cancer as a result of the massive overdose. Even so, many feel that if a substance causes cancer in a laboratory animal (regardless of the dose), it should not be given to humans. In fact, the Delaney Amendment of 1958 requires that the Food and Drug Administration (FDA) outlaw additives that are found to produce cancer in humans or *animals*.

EXAMPLE **1.20**

To evaluate the sugar substitute saccharin, rats were fed a diet consisting of 5% saccharin. A second group of like rats was fed and treated the same except that they received no saccharin. The results of the experiment showed that the treatment group had significantly more bladder tumors than did the control group. Saccharin was then classified as a carcinogen—a substance that produces cancer in laboratory animals. Some respected scientists criticize the saccharin experiment because of the overdose given. They contend that the large amounts of saccharin irritated the bladders of the rats, causing tumors.

EXERCISES 1.4

1.43   To obtain opinions on a local bond issue, an interviewer randomly selected households throughout the city. At one of his chosen households, no one is at home. Should he interview the neighbor who is at home?

1.44   In a house-to-house survey of a small city, the interviewer selected every corner house and asked the head of the household how he or she intended to vote on a municipal bond for upgrading the sewer system. Will there be sampling bias in the results? Why?

1.45   To get the public's opinion of nuclear reactors, a utility company mailed a questionnaire to all of its customers. Along with the questionnaire was a pamphlet describing nuclear reactors as a potential source of energy. List a strong point and a weak point of such a questionnaire and comment on whether the results of the questionnaire will represent the true feelings of the population.

1.46   What is a double-blind experiment?

1.47   What is meant by confounding a variable with the treatment?

1.48   While evaluating a treatment with an experiment, how can confounding be avoided or at least reduced?

1.49   Describe the placebo effect.

1.50   Suppose we wish to compare two different methods of teaching first-graders to read. At the beginning of the school year, we will have two first-grade teachers choose the students for their respective reading instructions. At the end of the school year, all students will be given a reading test, and then the two methods will be evaluated on the basis of the test scores.
     a. What is an experimental unit?
     b. What is the treatment?
     c. What is the response variable?
     d. Are any extraneous variables confounded with the treatment?
     e. Is there possible bias in this experiment?
     f. Has randomization been used to select the samples?
     g. Is there a control group? Is one needed?

1.51   "Since over half of all fatal traffic accidents involve alcohol, don't you think the penalties for drunk driving should be increased?" Why is this question not appropriate for a questionnaire on the drunk driving issue? Should the "over half" figure even be mentioned at all? How might the question be changed so that it is appropriate for the questionnaire?

1.52   A public and a private university conducted independent surveys of their respective student bodies on the issue of drug abuse. Will the results of the two surveys be similar? Why or why not? Can the results be combined to represent all college students?

1.53   In a questionnaire to constituents, a member of Congress posed the following question:
     To deal with the problems in Central America, the United States should
     A. increase both economic and military aid to those countries that protect the rights of their citizens.
     B. continue or increase economic assistance but stop all military assistance.
     C. stop all U.S. aid.
Does this question lead the respondent in any way? Should it be reworded not to lead the respondent? If so, how would you reword it?

1.54   From a study of 31,604 pregnancies, the *Journal of the American Medical Association* reported that babies whose mothers had one or two alcoholic drinks daily weighed an average of about 3 ounces less than those of nondrinking mothers.
     a. Is this an observational study or a randomized experiment?
     b. Are there any potential confounding variables?

1.55   To study the effects of television violence on children, an experimenter randomly assigned 3-year-old children to two groups. One group viewed a half-hour program filled with violent acts. The other group viewed a half-hour segment of *Mister Rogers' Neighborhood*. The numbers of violent acts engaged in by the children in the hour following the viewing were recorded for the two groups.

a. Is this an observational study or a randomized experiment?

b. What is the treatment? What is the response variable?

c. Are any extraneous variables potentially confounded with the treatment?

1.56   Explain why a control group is advisable in the following studies:

a. Seventy percent of those who took a pain reliever experienced relief from pain within 3 hours after taking the pill.

b. Farmers who used a new crop fertilizer experienced a 10% increase in crop yield over the yield of the preceding year.

1.57   Classify each of the following as either a randomized experiment or an observational study:

a. A medical lab compared the incidence of lung cancer in people who smoke with that of those who do not smoke.

b. A building contractor used insulation type A in half the houses he built and insulation type B in the other half. Afterward the energy efficiency of each house was measured.

c. The birth weights of newborn babies of mothers who drank alcohol were compared with the birth weights of babies of nondrinking mothers.

d. Students in a tenth-grade class were divided randomly into two groups and taught two different methods of typing. A typing test was then given to compare the two methods.

1.58   Of the examples listed in Exercise 1.57, which do you think might have a biased sample?

1.59   Classify each of the following as either a sample survey or an experiment:

a. A national television network asked voters to indicate for whom they voted as they exited the polling booth.

b. Terminally ill patients were divided into two groups, with one group receiving medication A and the other group receiving medication B. After a period of time, each subject's improvement was assessed.

c. Two thousand employees of a large corporation were asked whether they preferred the new health insurance plan proposed by the president or the one provided by their employer.

1.60   How would you select a sample of 100 students from your student body so that you would get representatives of each class? What sampling technique would you use?

1.61   The Weather Channel asked viewers to call 1-900-WEATHER to give their opinion on how the channel was doing in predicting the weather. How reliable will the results of the survey be? Will the sample be a random sample of Weather Channel viewers?

1.62   Refer back to Exercise 1.16 to the survey conducted by the office of student development on the percent of students who use their student government representative to convey their feelings on university issues.

a. Explain in some detail how you would take a sample for this survey.

b. Identify the sampling procedure if a personal interview was conducted with randomly selected students as they left the cafeteria.

c. What is the sampling procedure if a survey was sent to every student on the second floor of four randomly selected dorms?

d. What is the sampling procedure if a survey was sent to the first name on each page of the student directory?

1.63   A marketing analyst is asked to study the buying habits of shoppers at a national chain store. Suppose there are 200 stores around the country.

a. Describe the population of interest.

b. Describe a sampling procedure for obtaining a representative sample.

c. Are the target population and sampled population the same? Explain.

d. Give a variable of interest and describe a typical observation.

1.64   Who are the safer drivers, men or women? Would a comparison of the accident rates for men and women suffice for determining who is safer? Are the two groups exposed to the same type of hazards? Describe an experiment that would compare the two groups. Describe an observational study that would compare the two groups. Which is more practical?

1.65   To study the effects of unemployment on the divorce rate, a sociologist selected a random sample of unemployed people from the local unemployment agency and a sample of employed people from his graduation class. He then compared the divorce rates for the two groups.
    a. Is this an observational study or a randomized experiment?
    b. Identify the treatment and the response variable.
    c. Are there any extraneous variables that might have an effect on the divorce rate that could not be separated from the effect of employment status?

1.66   Sixty subjects were randomly divided into two groups to study the effects of alcohol on reaction time. The members of one group consumed a specified amount of alcohol, and members of the other group had a nonalcoholic beverage. The reaction times of both groups were measured before and after the beverage.
    a. What is the treatment?
    b. What is the response variable?
    c. What is an experimental unit?
    d. What is the purpose of one group drinking a nonalcoholic beverage?
    e. Are any extraneous variables possibly confounded with the treatment in this study?
    f. Is this study an observational study or a randomized experiment?

1.67   A popular sampling procedure conducted by the television networks on the day of an election is *exit polling*. Randomly selected people are asked who they voted for as they exit preselected polling booths around the country. Is this systematic, convenience, stratified, or cluster sampling? Explain.

1.68   A survey of 11,631 students in grades 9–12 in schools across the United States were asked by the Centers for Disease Control whether they had carried a weapon (gun, knife, club, etc.) into school in the last 30 days.
    a. What is the population of interest?
    b. How do you think the 11,631 students were selected?
    c. What method of contact do you think was used?
    d. Explain how you would select 11,631 students across the United States and how you would collect the data.

1.69   Which method of contact for a survey sample would you recommend in the following situations?
    a. The local radio station wants to know what percent of the county's residents listen to the farm report.
    b. The state medical examiner wants to know what percent of the public is aware of the Poison Control Center.
    c. A U.S. senator wants to know the opinions of constituents on several national issues.
    d. The city planner wants to know the opinions of local restaurant owners on a new sign ordinance.

1.70   Identify the sampling technique in the following cases:
    a. To obtain a sample of books in the library, the librarian randomly selects a book from the first rack. He then selects a book from that same position from each rack in the library.
    b. To obtain a sample of voters from the county, ten voters are randomly selected from each precinct.
    c. A local television station asks people to send in their name on a postcard to win a free trip to Disney World. The postcards are placed in a large cage that is rotated several times, and then the winning card is selected.
    d. To investigate the attitudes of the general population on the subject of nuclear weapons, *Time* magazine takes a random sample of its readers.

1.71   A large department store has its charge accounts listed alphabetically by the customer's last name. Accountants want to estimate the total amount of unpaid balances. Should systematic, stratified, or simple random sampling be used? Discuss the advantages of your choice.

1.5   SUMMARY AND REVIEW

KEY CONCEPTS

✓   This book describes methods of collecting, organizing, interpreting, and presenting numerical information. *Descriptive statistics* is the means by which the collected data are organized and summarized. *Inferential statistics* draws conclusions about a population from the data in a sample.

✓   A *variable* is a characteristic associated with each unit in a population. An *observation* is a single value assumed by a variable.

✓   A *population* is the collection of all objects of interest to the statistician. A *sample* is a finite subset of the population. A *census* is a sample consisting of the entire population.

✓   The *target population* and the *sampled population* may differ, but it is hoped that their characteristics of interest are similar.

✓   Often the statistician is interested in analyzing *preexisting* data that have been collected by others. Statisticians may also collect their own data. In so doing, they should follow these steps:

- Determine the objectives of the study.
- Choose the variables to be measured.
- Identify an appropriate design for producing the data.
- Collect the data.

✓   Two important areas of statistics are sample surveys and experiments. A *survey* studies the opinions of the subjects in an existing population. An *experiment* is concerned with a population that has been altered for study by the investigator.

✓   When critically appraising data, we should be concerned with these factors:

- The source
- The collection procedures
- The usefulness
- The presentation

✓   These factors should be considered when survey results are studied:

- Population
- Method of contact
- Response rate
- Wording of questions
- Timing of the interview
- Sample design

✓   The goal of sampling is to obtain a sample that is representative of the population.

✓ The different types of sampling techniques are:

- Convenience
- Simple random
- Systematic
- Stratified
- Cluster

✓ An *experiment* attempts to establish a cause-and-effect relationship between two variables. The *treatment* is controlled or manipulated by the experimenter. The *response variable* is the variable that changes, if at all, because of the treatment. A *control group* is the group that does not get the treatment. An *extraneous variable* is a variable, outside of the experiment, with an effect that might be *confounded* with the treatment.

✓ The *comparative experiment* with *random allocation* to groups is used to avoid confounding. An *observational study* is a study of the effect that a treatment has on subjects when random assignment is not possible.

LEARNING GOALS   Having completed this chapter, you should be able to:

1. Understand the concept of a population. *Section 1.1*
2. Understand the concept of a sample. *Section 1.1*
3. Distinguish between the target population and the sampled population. *Section 1.1*
4. Know what a census is and know why a sample is usually more practical. *Section 1.1*
5. Identify a variable and understand how it is used in a statistical study. *Section 1.1*
6. Know several sources of preexisting data. *Section 1.2*
7. Know the basic ideas of collecting data so that the sample is representative of the population. *Section 1.2*
8. Be able to recognize misleading graphs in the literature. *Section 1.3*
9. Understand the concept of bias as it relates to a survey or an experiment. *Section 1.4*
10. Understand the details of a sample survey and how it is used to describe a population. *Section 1.4*
11. Identify the various sampling techniques. *Section 1.4*
12. Understand the details of an experiment and how it is used to study the effects of a treatment on subjects. *Section 1.4*

QUESTIONS FOR REVIEW   Use the following problems to test your skills:

1.72   Earlybird Airline is interested in the possibility of opening a new route between Charlotte and Dallas. A survey concerning the issue was sent to 2,000 past customers.
a. Describe the population of interest.
b. What is the sample?
c. Give a variable of interest.

1.73   A stock market investor is interested in oil stocks. She collects last year's price/earnings ratios on ten selected oil stocks.
a. Describe the population of interest.
b. What is the sample?
c. Give a variable of interest.

1.74   A newspaper article reported the number of farmers in each state who went bankrupt in 1984. For example, Texas had 428 farmers go out of business and North Carolina had 230. Is this a meaningful measure to compare the numbers of bankrupt farmers in the various states? Can you think of a more valid measure for comparison?

1.75   A study determined that education has a significant effect on one's health. One year more in schooling lowers the chance of death by .4 percentage point. Give a variable you think might be confounded with the effect of education on health.

1.76   Give two reasons a sample is preferred to a census.

1.77   Give a reason the Nielsen ratings might be biased.

1.78   A study was designed to see whether SAT scores can be used to predict success in college.
    a. What is the treatment?
    b. What is the response variable?
    c. What is an experimental unit?
    d. Give a variable you think might be confounded with the treatment.

Multiple Choice (1.79–1.84)

1.79   In 1936 the *Literary Digest* (a popular magazine of the 1920s and 1930s) mailed out 10 million sample ballots for the upcoming presidential election. From the 2.3 million ballots returned, the *Digest* predicted that Alf Landon would defeat Franklin D. Roosevelt by a 3-to-2 margin. Of course, FDR won the election. Mailed or voluntary questionnaires are
    A. a very reliable source of data in a survey.
    B. easy to administer and almost always represent the population.
    C. the most widely used method of collecting survey data.
    D. usually biased because interested people have a high response rate.

1.80   In an effort to survey the attitudes of former graduates, the registrar randomly selects 100 names from each of the five past graduating classes. The sampling technique used is
    A. convenience sampling.
    B. simple random sampling.
    C. systematic random sampling.
    D. stratified random sampling.

1.81   Among the following, the most reliable method of obtaining a simple random sample is with
    A. random digits.
    B. physical mixing in a bowl.
    C. a telephone book.
    D. a convenience sample.

1.82   A Gallup poll of 1,500 people stated that 83% of Americans opposed preferential treatment for women and minorities in college admission and job placement. This figure is
    A. inaccurate because only 1,500 people were asked.
    B. biased because there is no guarantee of women and minorities being in the sample.
    C. meaningless because college admission and job placement are unrelated.
    D. None of the above

1.83   A census is
    A. an attempt to sample the whole population.
    B. rarely practical.
    C. sometimes less reliable than a scientific sample.
    D. All of the above
    E. None of the above

1.84   A psychology teacher uses his class as a sample of university students for a study he is conducting. The sampling technique used is

A. stratified sampling.
B. systematic sampling.
C. convenience sampling.
D. lottery sampling.

Fill in the blank (1.85–1.88)

1.85   A _____ is controlled or manipulated by the experimenter.

1.86   A _____ experiment is one in which neither the subject nor the one measuring the response knows who is receiving the treatment.

1.87   Two variables are _____ when the effects of the two cannot be separated.

1.88   In a well-designed experiment, the control group is given a _____ so that the response is to the treatment rather than to the idea of the treatment.

1.89   True or false?
   a. An experiment can establish causation.
   b. A survey can establish causation.
   c. An association between two variables means that one causes the other to happen.
   d. A placebo is always required for a comparative experiment.
   e. The double-blind experiment should increase bias.

1.90   Of the methods of contact in a sample survey, give the one that fits each of the following descriptions:
   a. Lowest response rate
   b. Highest response rate
   c. Most reliable
   d. Least reliable
   e. Most expensive to conduct
   f. Least expensive to conduct
   g. Most often used

1.91   To study the effects of exercise on the risk of heart disease, an investigator wishes to compare the incidences of heart disease in bus drivers and pedestrian police officers in New York City. He selects subjects so that the ages of the two groups are very similar and each subject has been on the job for at least 10 years.
   a. What is the treatment?
   b. What is the response variable?
   c. What is an experimental unit?
   d. Name two possible confounding variables.
   e. Is this an observational study or a randomized experiment?

1.92   To obtain the opinions of the residents of an agricultural state on a decision by the president to phase out all farm supports, a survey was sent to a random sample of 1,000 registered voters in the state by the Department of Agriculture.
   a. What is the target population?
   b. What is the sampled population?
   c. Is this a survey or an experiment?
   d. Do you think the results will be reliable?

1.93   The following question was on the survey referred to in the preceding problem:

"Don't you agree that a farmer should be allowed to raise as much of any crop as he chooses without any restrictions or supports from the federal government?"

Should this question be revised in order not to lead the respondent? If so, how would you revise it?

1.94 A university is interested in the success its placement office has in placing graduates in their chosen fields. A survey concerning the issue was sent to 500 past graduates.
   a. Describe the population of interest.
   b. What is the sample?
   c. Give a variable of interest.

1.95 Identify the correct sampling procedure used in each of the following cases:
   a. Twenty classes are randomly selected across campus, and each student in the classes is given a survey to complete.
   b. Samples of sizes 100, 80, 70, and 60 are randomly selected from the freshman, sophomore, junior, and senior classes, respectively.
   c. Every 50th person is selected from a list of registered voters.
   d. The winning number is selected from a revolving barrel.
   e. Five taxpayers are randomly selected from each county in the state.

1.96 A Latin teacher claims that studying Latin increases one's verbal skills because, in that school, the average SAT verbal score for Latin students is 532 and for those not taking Latin it is only 489.
   a. Is this an observational study or a randomized experiment?
   b. Are there any confounding variables?
   c. How reliable is this comparison?

1.97 A survey on pets and people was conducted by *Psychology Today*. In comparing the demographics of the pet owners and nonowners, the study found that (i) 34% of the pet owners had incomes of more than $40,000, compared with 25% for the nonowners; and (ii) 49% of the owners were married, compared with 33% of the nonowners. *Psychology Today* pointed out that the pet owners who responded were more satisfied with their lives than the nonowners. Does this mean that owning a pet enhances one's quality of life? Explain.

1.98 Does the following question lead the respondent?

"The existence of the textile industry in the United States is threatened by foreign imports. Don't you agree that the government should limit foreign imports of textile materials?"

If so, how should it be reworded not to lead the subject?

1.99 A sample of 500 men and 500 women were asked how often they experience depression. Is this a sample survey or an experiment?

COMPUTER EXERCISES

1.100 Retrieve the worksheet **Commute.mtw** that gives 1980 and 1990 commuting times in 39 major cities (see Exercise 1.14 in Section 1.1). Construct *side-by-side dotplots* of the 1980 and 1990 commuting times. Select both variables and click on the **Same scale for all variables** box so that the two dotplots have a common scale for comparison purposes. Are there any unusual observations in either dotplot? Briefly describe how commuting times have changed from 1980 to 1990.

1.101 Exhibit 1.5 demonstrates how to randomly generate data into a column of the worksheet. Determine the number of students at your college or university and assume that you have a complete numerical listing of all the students. With a computer, generate 200 random numbers corresponding to 200 students on the list. You now have a random sample of 200 students on your campus. How would you go about locating the 200 students in order to get their opinions for a survey on a controversial issue?

1.102 Retrieve the graduation rates, given in worksheet **Graduate.mtw**, for student athletes at schools in the Southeastern Conference (see Exercise 1.38 in Section 1.3). Use the code variable for the names of the schools and the Chart command to create a bar graph of the percent who graduate. Use Exhibit 1.7 for a guideline on how to construct the bar graph.

1.103 The worksheet **Baseball.mtw** contains 1992 salaries for four major league teams: Baltimore Orioles, Boston Red Sox, California Angels, and Cleveland Indians. Construct *side-by-side dotplots* of the salaries for the four teams. Select all of the variables and click on the **Same scale for all variables** box so that the dotplots have a common scale for comparison purposes. Are there unusual observations in any dotplot? How do the salaries of the Cleveland Indians compare with the salaries of the other teams? Which team pays its players more?

1.104 Given here are the graduation rates for student athletes and nonathletes who entered schools in the Big Ten Conference in 1984–85:

Worksheet: BigTen.mtw

School	Student	Athlete
Illinois	78	67
Indiana	53	58
Iowa	62	64
Michigan	81	66
Michigan State	66	64
Minnesota	34	44
Northwestern	87	81
Ohio State	51	55
Penn State	73	63
Purdue	68	60
Wisconsin	66	62

SOURCE: NCAA Graduation Rates Report, 1992.

Construct *side-by-side dotplots* of the graduation rates for the athletes and nonathletes by selecting both variables and clicking on the **Same scale for all variables** box so that the dotplots have a common scale for comparison purposes. Based on the dotplots, what can you say about the differences between the two distributions? From the dotplots can the graduation rates for athletes and nonathletes be compared for a single university?

1.105 Retrieve the data on parking tickets given diplomats in New York that is given in worksheet **Diplomat.mtw** (see Exercise 1.41 in Section 1.3). Use the code variable for the countries and the Chart command to create a bar graph of the rate of tickets per vehicle per month. Does the chart tell you who are the worst offenders? Who are the least offenders?

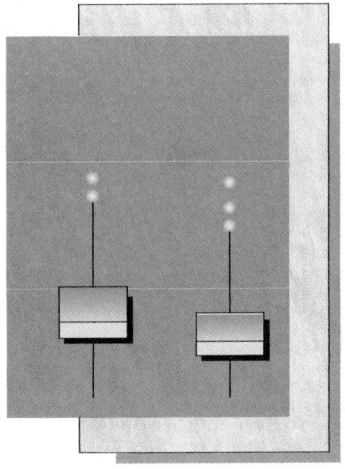

2

ORGANIZING AND SUMMARIZING UNIVARIATE DATA

The survey sample and the experiment, which we considered in Chapter 1, are two general areas of statistics. In a survey, sample data are used to draw conclusions about a population. In an experiment, data are used to investigate a cause-and-effect relationship between two or more variables. In each, the data are important; they must be collected, organized, and summarized.

Having completed Chapter 1, you should have a good understanding of how data are collected. Now we study the different types of data and see how they are organized and numerically summarized.

CONTENTS

STATISTICAL INSIGHT

JOB MARKET FOR COLLEGE GRADS

What type of job do you expect to get when you graduate? How long do you think it will take you to find a job? What salary do you expect? Many factors will affect your answers to these questions. Certainly your grade point average, related work experience in internships and co-ops, verbal and written communication skills, and interviewing skills will affect how long it takes you to get a job and the salary you will receive. Another determining factor, of course, is your academic major. Listed here are the starting salaries in 28 different majors for 1995 graduates with a bachelor's degree.

Worksheet: Job.mtw

Agriculture	$24,134	Natural Resources	$22,554
Accounting	27,787	Financial Admin.	26,630
General Business Admin.	23,760	Hotel/Restaurant Mgt.	23,713
Marketing/Sales	24,607	Personnel Admin.	22,923
Retailing	22,002	Advertising	21,627
Communications	21,640	Journalism	20,587
Telecommunications	20,680	Education	22,685
Chemical Engineering	40,341	Civil Engineering	29,547
Computer Science	32,446	Electrical Engineering	34,979
Industrial Engineering	33,348	Mechanical Engineering	35,369
Human Ecology	21,053	Liberal Arts	20,860
Chemistry	28,386	Geology	28,414
Mathematics	26,415	Physics	27,087
Nursing	29,868	Social Science	22,333

SOURCE: Michigan State University, East Lansing.

By examining the data, one can see that the largest reported salary is $40,341 for chemical engineering and the smallest is $20,587 for journalism, but other than that, not much can be said about the general distribution of the salaries across the different majors. What is a typical salary that a college graduate should expect to receive? Will it be closer to $40,000 or $20,000, or somewhere in between? How many majors have a starting salary greater than $30,000?

These are just a few of the many questions that can be asked about a data set. (See Exercise 2.156 for a comprehensive analysis of these data.) In this chapter, we investigate ways to summarize the data so that questions like these can be answered. Also, several methods of graphically displaying data are presented. The objective is to present the data in a manner that will convey as much information as possible. First we look at the different types of data.

2.1 TYPES OF DATA

Data Sets   A *data set* is a set of measurements or observations taken on a group of objects or subjects. Several *variables* may be listed, as in the data set in Table 2.1, which gives the gender, major, class, grade point average (GPA), and Scholastic Assessment Test (SAT–math and SAT–verbal) scores for a group of undergraduate students.

TABLE 2.1

Data set for seven undergraduate students

Gender	Major	Class	GPA	SATM	SATV
Male	Physics	Junior	3.2	620	460
Male	Political Science	Soph.	2.8	410	550
Female	Psychology	Junior	3.4	460	620
Male	Psychology	Soph.	2.2	400	530
Female	Sociology	Senior	3.5	560	670
Female	Mathematics	Junior	2.6	540	470
Male	Biology	Soph.	3.1	580	520

The column headings—Gender, Major, Class, GPA, SATM, and SATV—represent different variables measured on each subject. The various descriptive words or numbers that they assume for the subjects are the *observations* or the *values* of the variables. The data set is a *multivariate* data set because several variables have been measured on the same unit.

A **univariate data set** is a data set in which one measurement (variable) has been made on each experimental unit.

A **bivariate data set** is a data set in which two measurements (variables) have been made on each experimental unit.

A **multivariate data set** is a data set in which several measurements (variables) have been made on each experimental unit.

Types of Variables

Notice that there is a basic difference between the values of the first three variables (Gender, Major, Class) and the values of the last three variables (GPA, SATM, SATV). The last three are *numerical* and the first three are not.

A **numerical variable** (also called a *quantitative* or *measurement* variable) is a variable whose values are numbers obtained by a count or measurement.

The following are examples of numerical variables:

Grade point average

Weight

Age

Family income

Birth rate

Number of children in a family

Number of suicides in a given year

Number of voters who favor a certain issue

Number of successfully treated patients

Number of A's in a class

Numerical variables are further classified as either *discrete* or *continuous*.

A **discrete variable** is a numerical variable that can assume a finite number or at most a countable infinite number of values. (*Countable* means you can associate the values with the counting numbers 1, 2, 3, . . .; that is, the values can be counted.)

Normally, discrete variables are the result of some type of *counting process*. The last five numerical variables listed above are all discrete. The number of children in a family, for example, is represented by a whole number (0, 1, 2, 3, . . .), as opposed to all possible numbers in an interval of the real number line.

A **continuous variable** is a numerical variable that can assume an infinite number of values associated with the numbers on an interval of the real number line.

Normally, continuous variables are the result of some *measurement process*. The first five numerical variables listed above are all continuous. Weight, for example, is a continuous variable because it can assume any value greater than zero.

Not all variables necessarily represent quantities, however.

A **categorical variable** (also called a *qualitative* variable) is a variable whose values are classifications or categories.

The following are examples of categorical variables:

Gender

Major

Classification

Political party affiliation

Occupation

Religious preference

Marital status

Employment status

Although categorical variables are nonnumerical, we can assign a number code to the values that the variable can assume. For example, the variable Gender can be coded 0 for male and 1 for female. The variable Class can be coded:

Freshman–1   Sophomore–2   Junior–3   Senior–4

The coding of categorical variables is very useful when we wish to store the data in a computer, but remember that it is only a code. Assigning a numerical code does not make a categorical variable numerical. Arithmetic performed on such codes, such as averaging, is meaningless.

Figure 2.1 illustrates the different types of data and how they are measured.

FIGURE 2.1

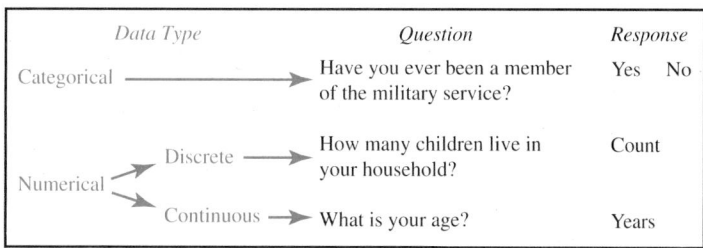

EXAMPLE 2.1

Classify the following as either categorical or numerical. If numerical, further classify as discrete or continuous.

    a. Divorce rate

    b. Opinion on a political issue

    c. Number of hospitals that have a trauma center

SOLUTION

    a. The divorce rate is measured in different ways. It can be given as the ratio of divorces to marriages in a given year. For example, in 1990, there were 2,448,000 marriages and 1,175,000 divorces, so the divorce rate was 48% of the 1990 marriages. Another way to measure the divorce rate is to give the number of divorces per 1,000 people. In 1990, there were 4.7 divorces per 1,000. At the same time, the marriage rate was 9.8 per 1,000 people, which is consistent with the ratio being approximately 48%. (SOURCE: National Center for Health Statistics, *Monthly Vital Statistics Report*, Vol. 39, No. 13, August 28, 1991.) Regardless of the way it is measured, the divorce rate is in fact measured and hence is a numerical variable. Generally all rates (death rate, abortion rate, etc.) are considered to be continuous.

    b. Opinion is not a measurement but rather a classification, such as for or against; hence, it is categorical. This is not to be confused with the variable that gives the *number* of those who favor an issue. The number of voters who favor an issue is a discrete numerical variable.

c. The number of hospitals that have a trauma center is a *count* variable and thus a discrete numerical variable.

The amount of information contained in a data set depends on how the variables are measured. For example, recording income in dollars is more informative than recording it simply as low, medium, or high. Income recorded in dollars is numerical, and income recorded as low, medium, and high is categorical. Knowing the amount of information contained in the data may indicate possible procedures to use for analyzing the data.

EXAMPLE 2.2

Identify the following variables as either categorical or numerical. Comment on the amount of information conveyed in each.

a. The length (in hours) of a baseball game
b. Colors of paint in the inventory of a paint company
c. Ranks of personnel in the military

SOLUTION

a. Time is a quantity that is measured and thus is a numerical variable. A 4-hour game is twice as long as a 2-hour game; thus, ratios and differences of time are relevant. Any sort of arithmetic operation may be performed on this variable.

b. Color is simply an attribute and thus a categorical variable. Although a numerical code can be assigned to different colors, the values are not numerical and thus numerical calculations are not meaningful.

c. Rank in the military is not numerical, although it does exhibit a definite ordering. It is a categorical variable.

Time Series Data

Business and economic data, such as the daily Dow-Jones Average or the quarterly earnings of a company, are often observed at regular time intervals. Any set of data recorded at given time intervals is called *time series data*. The data can be presented numerically or graphically. The bar graph of the world population from 1950 through 2000 in Figure 1.5 in Chapter 1 is a typical graphical presentation of time series data.

Time series data are frequently presented in the form of a curve, as in Figure 2.2. A curve is preferable to a bar graph when the time increments are close together. Note that the 0 on the vertical axis is suppressed; that is, the graph does not begin at 0 on the vertical axis. In Section 1.3, we warned against the suppressed zero axis on graphs. This is acceptable here, however, because we are graphing only one variable; giving the rest of the graph (from 0 to 3400) would not be informative and would simply take up space. Avoid the suppressed zero axis on graphs that involve comparisons, like Figure 2.3, where multiple time series are graphed on the same coordinate system.

FIGURE 2.2   Dow-Jones Industrial Averages

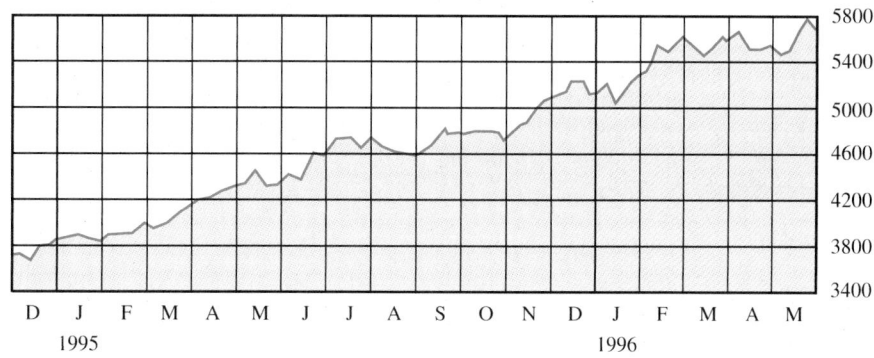

SOURCE: Wall Street Journal, June 7, 1996.

FIGURE 2.3

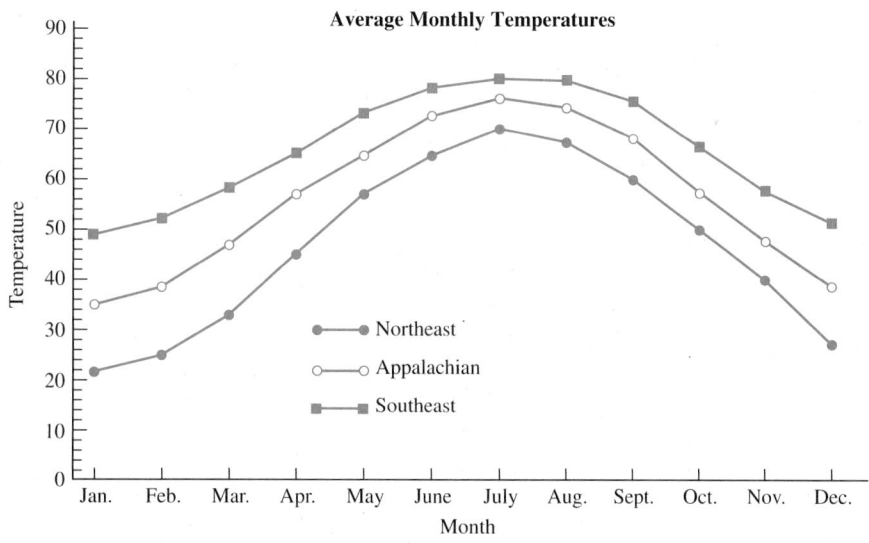

SOURCE: *Weather in U.S. Agriculture*, Statistical Bulletin 834, U.S. Dept. of Agriculture, January 1992, pp. 12-17.

Statistically analyzing time series data is generally more complicated than analyzing ordinary data that are not time dependent because factors such as *trends* and *cycles* in the data must be investigated. But, as is illustrated in the following example, even simple graphs may uncover important facts about the data.

EXAMPLE 2.3

It can be traced back as far as 28 B.C. that humans have observed dark spots on the sun. In 1848 Rudolf Wolf of Zurich developed the "Wolfer sunspot number" based on the number of groups of sunspots, the total number of sunspots, and a constant associated with the observatory that made the original observation. The number, which represents the average number of sunspots during each year, has been reconstructed as far back as 1700. The Wolfer sunspot numbers from 1700 through 1979 are stored in worksheet: **Sunspot.mtw** [data obtained from Tong (1983, p. 280)]. This time series data set has become one of the most widely studied time series of all time.

Figure 2.4 is a plot of the sunspot data. At first, we see only a bunch of dots. But if we look closer, we see a dip around 1800 and one around 1900. Furthermore, it appears that there is a dip around 1700. Will there be one around 2000?

FIGURE 2.4   Sunspot data

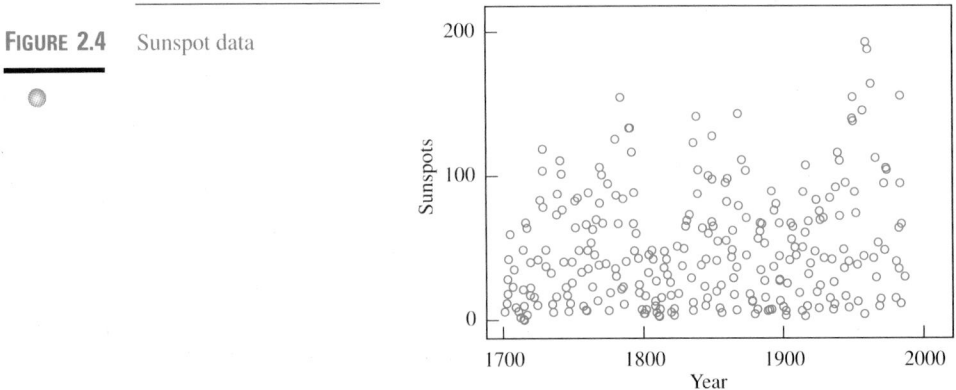

Figure 2.5 shows the same data with the dots connected in time order. Now we see more detail. Not only do we see the dips at the beginning of each century, but we also count nine peaks each 100 years. Moreover, the pattern of the peaks in the first 100 years seems to repeat in the second 100 years. This is consistent with the conjecture made by time series analysts, who have suggested that the sunspot data are cyclical (tend to repeat) and have a period of 11 years.

FIGURE 2.5   Sunspot data

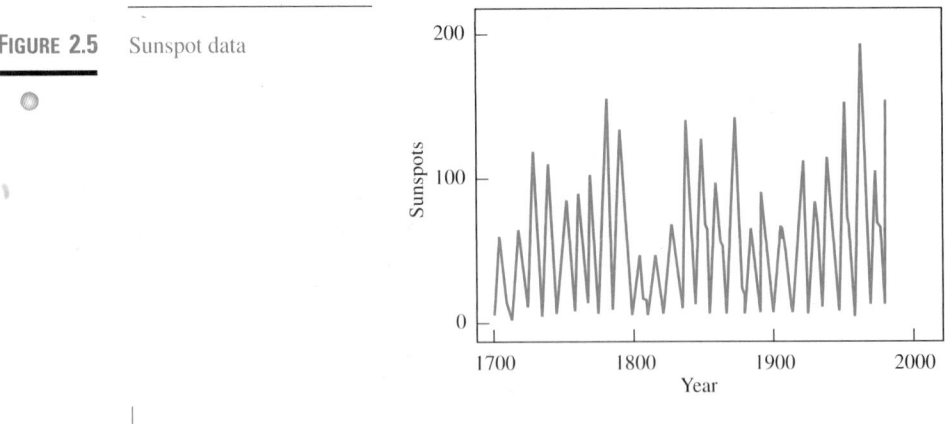

It is well beyond the scope of this book to investigate time series data further. Therefore, for the remainder of the book we will analyze *cross-sectional* data where the observations are taken at one point in time and are assumed to be independent of one another.

EXERCISES 2.1

For Exercises 1–4:
  a. Describe the population of interest.
  b. Describe how you might obtain a random sample.
  c. Give three categorical variables that would be of interest.
  d. Give three numerical variables that would be of interest and classify them as either continuous or discrete.

2.1   You are interested in studying the welfare recipients of a state.

2.2   You are interested in evaluating a new treatment for heart disease.

2.3   You are interested in studying the effectiveness of computers as a learning tool for third-graders.

2.4   You are interested in studying the effect of alcohol on teenage crime.

2.5   Classify the following as categorical or numerical. If a variable is numerical, further classify it as discrete or continuous.
  a. Age of freshmen U.S. senators
  b. Faculty rank
  c. Weight of newborn babies
  d. Per capita income of residents of a state
  e. Murder rate in a major city
  f. Number of students in a classroom
  g. Brand of television set

2.6   A doctor is keeping a record of patients who are in need of surgery. For each patient, she records gender, age, marital status, occupation, type of surgery needed, and urgency of the surgery. Classify each of these variables as categorical or numerical.

2.7   A university placement office keeps track of past graduates of the university. For each former student, they record gender, major, grade point average at graduation, number of uses of the placement service, and type of employment at graduation. Classify each of these variables as categorical or numerical.

2.8   Consider the following 1993 robbery rates (per 100,000 people) for selected cities with populations exceeding 250,000 people:

City	Rate
Arlington, Texas	252.4
Boston, Mass.	736.8
Cleveland, Ohio	849.7
Detroit, Mich.	1,332.4
Fresno, Calif.	757.7
Louisville, Ky.	509.2
San Jose, Calif.	146.5

SOURCE: U.S. Department of Justice, *Sourcebook of Criminal Justice Statistics*, 1994.

a. Is robbery rate categorical or numerical?

b. Is the variable city categorical or numerical?

c. Is robbery rate a continuous or discrete variable?

d. Reclassify the robbery rates as low (less than 300), medium (between 300 and 1,000), or high (greater than 1,000). Does this affect the amount of information in the data set?

2.9 A psychologist wishes to investigate the threshold reaction time (in seconds) for persons subjected to emotional stress.

a. Is threshold reaction time categorical or numerical?

b. Is the variable continuous or discrete?

2.10 A sociologist wishes to study the occupations and salaries of Vietnam veterans discharged prior to 1974.

a. Is the variable occupation categorical or numerical?

b. Is the variable salary categorical or numerical?

2.11 Discuss the difference between categorical and numerical data.

2.12 Identify each of the following as either categorical or numerical:

a. Phone number

b. Social Security number

c. Age

d. Street address

e. Gross annual income

f. Number of years of education

2.13 Identify each of the following as either continuous or discrete variables:

a. Number of hours children watch television

b. Suicide rate

c. Number of violent crimes

d. Housing cost

e. Number of adults over age 65

f. Unemployment rate

2.14 Give a continuous and a discrete variable that might be of interest in a survey of city residents on the issue of gun control.

2.15 A recent study on toxic waste sites showed that the percentage of minorities living in communities with commercial hazardous waste sites rose from 25% in 1980 to almost 31% in 1993. The following list gives hazardous waste sites near minority communities:

Worksheets: Toxic.xls (Excel) and Toxic.mtw (Minitab)

State	Region	Hazardous waste sites	Sites with above average minority	Percent minorities
Alabama	south	12	10	26.8
Alaska	west	0	0	0
Arizona	west	8	8	28.8
Arkansas	south	5	4	17.7
California	west	54	53	42.9
Colorado	west	9	5	19.8
Connecticut	northeast	6	4	16.4
Delaware	northeast	0	0	0
District of Columbia	northeast	1	1	71.4
Florida	south	13	12	27.3
Georgia	south	16	13	29.8
Hawaii	west	1	1	68.4

(continued)

State	Region	Hazardous waste sites	Sites with above average minority	Percent minorities
Idaho	west	2	1	8.3
Illinois	midwest	30	14	25.7
Indiana	midwest	16	7	10.6
Iowa	midwest	1	0	4.3
Kansas	midwest	8	6	12
Kentucky	south	12	3	8.4
Louisiana	south	12	10	34.6
Maine	northeast	2	0	2.1
Maryland	south	3	2	31.1
Massachusetts	northeast	12	5	12.7
Michigan	midwest	23	12	18.1
Minnesota	midwest	5	0	6.7
Mississippi	south	4	3	37
Missouri	midwest	13	5	13.2
Montana	west	1	0	8.5
Nebraska	midwest	3	1	7.9
Nevada	west	4	2	21.9
New Hampshire	northeast	0	0	0
New Jersey	northeast	18	10	26.4
New Mexico	west	3	3	50.4
New York	northeast	19	7	31
North Carolina	south	17	14	25.2
North Dakota	midwest	0	0	0
Ohio	midwest	43	15	13.2
Oklahoma	south	6	5	20
Oregon	west	8	1	9.8
Pennsylvania	northeast	19	2	12.4
Rhode Island	northeast	2	1	2
South Carolina	south	16	13	31.5
South Dakota	midwest	0	0	0
Tennessee	south	19	12	17.5
Texas	south	28	24	39.4
Utah	west	4	0	9.3
Vermont	northeast	1	0	2
Virginia	south	10	7	24.3
Washington	west	19	11	13.9
West Virginia	south	2	0	4.2
Wisconsin	midwest	20	3	9.1
Wyoming	west	0	0	0

SOURCE: *USA Today,* August 25, 1994.

a. Identify the experimental units.
b. Is this a univariate, bivariate, or multivariate data set?
c. How many variables are listed?
d. State whether each of the variables is numerical or categorical.
e. Classify those variables that are numerical as either discrete or continuous.

2.16   The Alan Guttmacher Institute is a not-for-profit research group based in New York City that studies reproductive health issues including teen sexuality, contraceptive use, prenatal care, AIDS, and abortion. On the issue of abortion, they are interested in the following variables:

a. Where the abortion is performed
b. Race of subject
c. Age of subject
d. Socioeconomic background of subject
e. Marital status of subject
f. Number of previous children

What is an experimental unit? Classify each variable as numerical or categorical. Classify the numerical variables as either discrete or continuous.

2.17 The worksheet gives the daily highs, lows, and closing prices of the Dow-Jones Industrial Average for a 2-week period ending January 19, 1996. Construct time series graphs for all three prices (high, low, close) on the same coordinate system similar to Figure 2.3.

Worksheet: Stocks.mtw

Date	Mon	Tue	Wed	Thu	Fri	Mon	Tue	Wed	Thu	Fri
High	5207	5214	5128	5066	5090	5079	5089	5098	5125	5194
Close	5198	5130	5033	5065	5061	5044	5088	5067	5124	5185
Low	5180	5116	5015	5020	5023	5037	5019	5052	5066	5112

SOURCE: *Wall Street Journal*, January 15 and January 22, 1996.

2.18 Divorce rate in this exercise is recorded as the number of divorces for every 1,000 people. In 1960 the rate was 2.2, in 1970 it was 3.5, in 1980 it was 5.2, and in 1990 it was 4.7. (SOURCE: U.S. Department of Health and Human Services, *Monthly Vital Statistics Report*, Vol. 39, No. 13, August 28, 1991.) Illustrate these time series data in a bar graph.

2.19 The following table gives the life expectancies of men and women in the United States. Compare these time series by graphing both on the same coordinate system.

Worksheet: Life.mtw

Year	Men	Women
1920	53.6	54.6
1930	58.1	61.6
1940	60.8	65.2
1950	65.6	71.1
1960	66.6	73.1
1970	67.1	74.7
1980	70.0	77.5
1990	71.8	78.8

SOURCE: National Center for Health Statistics.

2.20 The Consumer Price Index (CPI) is a measure of the cost of goods in the United States. The table gives the changes in the CPI from the previous year from 1979 to 1994. Illustrate these time series data with a bar graph.

Worksheet: CPI.mtw

Year	1979	1980	1981	1982	1983	1984	1985	1986
CPI	13.3	12.5	8.9	3.8	3.8	3.9	3.8	1.1

Year	1987	1988	1989	1990	1991	1992	1993	1994
CPI	4.4	4.4	4.6	6.1	3.1	2.9	2.7	2.3

SOURCE: Bureau of Labor Statistics.

2.21   The worksheet lists the winning times (in seconds) for the men's 1,500-meter run in the Olympics from 1896 to 1992. (No Olympics were held in 1916, 1940, and 1944 because of world wars; in 1980, the United States and some other countries boycotted the Olympics.)

Worksheet: Track15.mtw

Year	Time	Year	Time	Year	Time
1896	273.2	1932	231.2	1972	216.3
1900	246.2	1936	227.8	1976	219.2
1904	245.4	1948	229.8	1980	218.4
1908	243.4	1952	225.2	1984	212.5
1912	236.8	1956	221.2	1988	216.0
1920	241.8	1960	215.6	1992	220.1
1924	233.6	1964	218.1		
1928	233.2	1968	214.9		

SOURCE: *The World Almanac and Book of Facts*, 1995.

a. Draw a time series plot of the data. How did you deal with the 3 years that had no data?
b. Do you find any unusual behavior in the data?
c. Are the data cyclical? Is there a trend?

2.2   DISPLAYING CATEGORICAL DATA

Frequency Table

Data are often organized and displayed in the form of tables and graphs for the purpose of illustrating their distribution. For categorical data, the form often used is a frequency count of each of the categories.

> A table that lists the different categories of categorical data and the corresponding frequencies with which they occur is called a **frequency table**.

The following question was presented to 500 adult Americans in a telephone poll conducted by Time/CNN on August 12, 1993:

Has the amount of crime in your community increased in the past 5 years?

Increased _____   Decreased _____   Remained the same _____

The results of the survey are presented in the frequency table in Table 2.2 (page 66).
Note that the *frequency* is simply a count of the number of subjects who fall into the different categories.

TABLE 2.2   Time/CNN telephone poll of 500 adult Americans

Has the amount of crime in your community increased in the past 5 years?

Response	Frequency
Increased	305
Decreased	25
Remained the same	150
No response	20
Total	500

The frequencies may be divided by the total count to obtain *relative frequencies*. From Table 2.2, we get the following relative frequencies:

305/500 = .61 or 61% answered increased

25/500 = .05 or 5% answered decreased

150/500 = .30 or 30% answered remained the same

20/500 = .04 or 4% did not respond

Table 2.3 is the completed relative frequency table.

TABLE 2.3   Time/CNN telephone poll of 500 adult Americans

Has the amount of crime in your community increased in the past 5 years?

Response	Frequency	Relative Frequency
Increased	305	.61
Decreased	25	.05
Remained the same	150	.30
No response	20	.04
Total	500	1.00

Relative frequency is important because its value is independent of the size of the data set. This is necessary when we compare two (or more) data sets of different size.

EXAMPLE 2.4   A sample of 1,000 university students who live in dorms were surveyed in regard to their housing on campus. They were asked to rate their current housing on the following scale:

1—very desirable

2—desirable

3—sufficient

4—livable

5—undesirable

The results showed that 120 students chose category 1, 180 chose category 2, 360 chose category 3, 240 chose category 4, and 100 chose category 5. Another sample of 400 students who live off campus were given the same survey. Of that group, 32 chose category 1, 60 chose category 2, 160 chose category 3, 120 chose category 4, and 28 chose category 5. Summarize these data in a table listing the frequencies and relative frequencies for the two groups.

SOLUTION

TABLE 2.4   Student opinion of living conditions

> The relative frequency may be stated as a decimal or as a percent. It is a simple matter of moving the decimal point to get from one to the other.

| | Dorm | | Off Campus | |
Opinion	Frequency	Relative Frequency	Frequency	Relative Frequency
very desirable	120	12%	32	8%
desirable	180	18	60	15
sufficient	360	36	160	40
livable	240	24	120	30
undesirable	100	10	28	7
Total	1,000	100%	400	100%

Note in Table 2.4 that twice as many students (240 versus 120) who live in dorms classified their living conditions as "livable," yet a greater percentage (30% versus 24%) of off-campus students classified their conditions as "livable."

Although rates, such as the crime rate per 1,000 persons, are not frequency counts, they can be treated as frequencies and displayed in a frequency table as shown in Table 2.5 on page 68.

TABLE 2.5

Estimated rate (per 1,000 persons 12 or older) of personal victimization in the United States, 1992

Type of Victimization	Rate per 1,000
Rape and attempted rape	.7
Robbery	5.9
Aggravated assault	9.0
Simple assault	16.5
Personal larceny with contact	2.3
Personal larceny without contact	56.8

Source: Bureau of Justice Statistics, U.S. Department of Justice, *Criminal Victimization in the United States*, 1992, p. 6.

Bar Graph

Tables of numbers are sometimes difficult to interpret. In those cases, a picture might better illustrate the distribution of the data. The bar graph, introduced in Section 1.3, may be used to give a graphical illustration of the data from a frequency table.

EXAMPLE 2.5

According to an FBI report, the number of hate crimes rose to 7,684 in 1993. Of those reported, 4,168 were motivated by race, 1,189 were motivated by religion, 806 were sexually oriented, 583 were motivated by ethnicity, and 938 had an unknown motivation. Summarize these data in a bar graph.

SOLUTION

The vertical axis must be scaled to above 4,168, the highest frequency. Bars are constructed at the five motivation categories. The completed graph in Figure 2.6 can be

FIGURE 2.6

Excel bar graph of hate crimes reported to the FBI in 1993

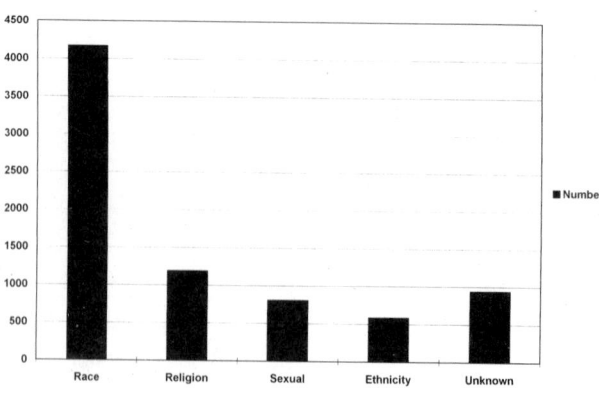

constructed by hand, but in this case the software package Microsoft Excel was used to construct the graph. Notice how much more race is a motivating factor in hate crimes than any of the other categories.

The height of the bars in a bar graph can be measured in either frequency or relative frequency. The bars of the bar graph can be either vertical as in Figure 2.6 or horizontal as in Figure 2.7.

FIGURE 2.7   Horizontal bar graph of the data in Example 2.5

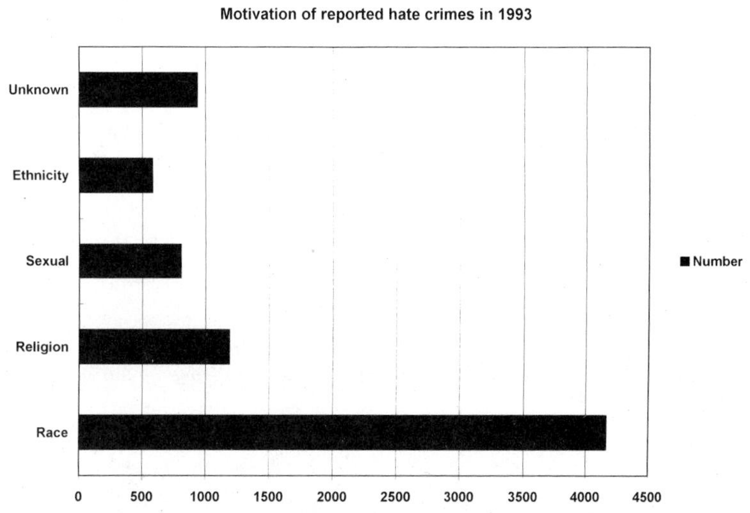

> The bars in a bar graph should always have the same width so that the difference in frequencies is reflected only in the height of the bars.

Motivation of reported hate crimes in 1993

SOURCE: Federal Bureau of Investigation.

Pie Chart

The pie chart is another useful way to exhibit categorical data. A circle, or pie, is divided into pieces corresponding to the categories of the variable so that the size (angle) of the slice is proportional to the relative frequency of the category. Normally the relative frequency (as a percent) is shown in each slice. Figure 2.8 (page 70) is an Excel pie chart that exhibits the motivations of hate crimes in 1993 reported in Example 2.5. The pie chart shows that more than half (55%) of all hate crimes are motivated by race.

The pie chart and bar graph can also be used to compare groups. Just remember that the comparison should be based on relative frequencies instead of frequencies.

FIGURE 2.8   Excel pie chart of hate crimes reported to the
FBI in 1993

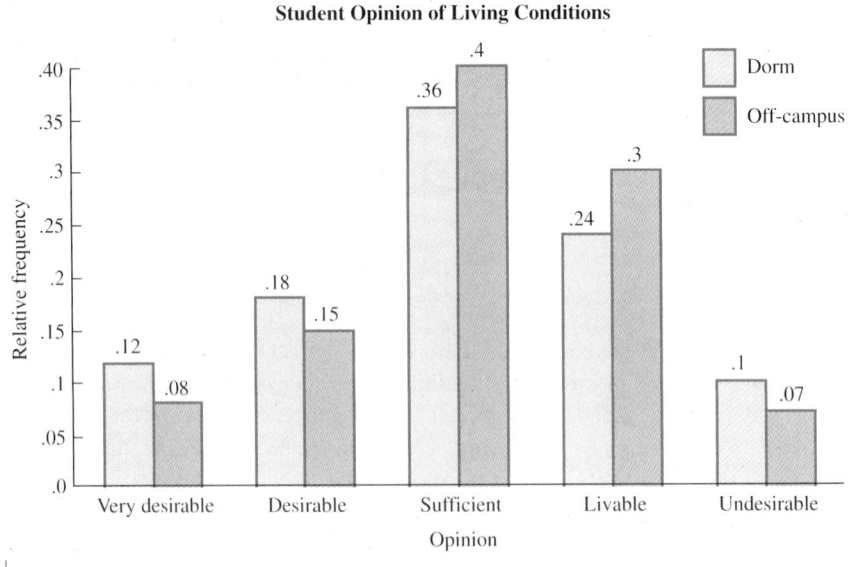

Hate Crimes Reported in 1993 by Motivation

SOURCE: Federal Bureau of Investigation.

EXAMPLE 2.6   Using the relative frequencies for the two groups of students in Example 2.4, we can draw the comparison bar graph in Figure 2.9. Notice the title at the top that identifies the data, the legend on the right side that identifies the two groups, and the vertical axis that is labeled Relative Frequency. The relative frequencies printed at the top of the bars are optional but often help to clarify the data.

FIGURE 2.9

C O M P U T E R   T I P

Microsoft Excel is a spreadsheet software package that operates on personal computers. When bar graphs and pie charts are constructed, the data are entered into the worksheet and highlighted. Click on the **Chart Wizard**, and the program will walk you through the steps to construct a bar graph or pie chart. Once a graph is constructed, it is a simple matter to change the format of the graph. For instance, to convert from vertical bars to horizontal bars, click on **Chart Type** and select the horizontal bars.

EXERCISES 2.2

2.22 In a random sample of college students, it was found that 124 had blue eyes, 150 had brown eyes, 15 had green eyes, and 103 had hazel eyes. Display these data in a frequency table with relative frequencies.

2.23 Draw a bar graph of the data in Exercise 2.22.

2.24 Construct a pie chart for the data in Exercise 2.22.

2.25 The accompanying pie charts were given to compare the educational backgrounds of parents of entering freshmen at a medium-sized state university. Notice that it is difficult to compare the mothers' and fathers' educational backgrounds. Convert the data (worksheet: ParentEd.xls) to a comparison bar graph that better illustrates the comparison of their educational backgrounds.

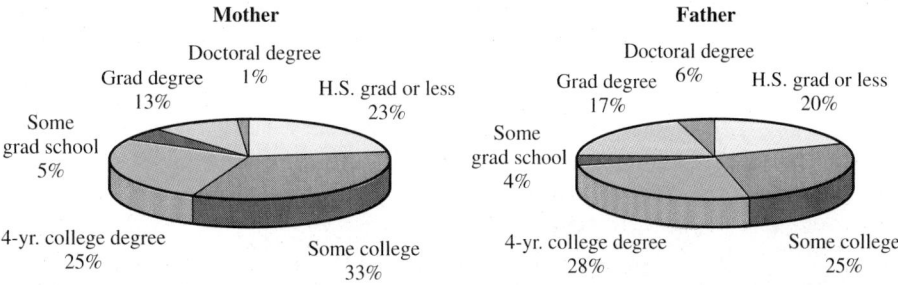

SOURCE: *Student Life and Learning*, Vol. 4, No. 2, Student Development, Appalachian State University, Boone, NC.

2.26 In the second quarter of 1994, personal income rose 1.9% from the first quarter of 1994. Following is the percent change from the first to second quarter for each state and the District of Columbia:

Worksheet: Income.mtw

Ala	1.3	Calif	3.9
Alaska	.9	Colo	1.5
Ariz	1.8	Conn	1.5
Ark	.7	Del	2.3

(*continued*)

Fla	2.2	N.J.	1.9
Ga	1.7	N.M	2.0
Hawaii	1.2	N.Y	1.5
Idaho	1.5	N.C.	1.9
Ill	1.6	N.D.	1.6
Ind	1.6	Ohio	1.0
Iowa	.1	Okla	1.5
Kan	1.7	Ore	1.9
Ky	2.1	Pa	1.7
La	2.0	R.I.	2.0
Maine	1.5	S.C.	1.2
Md	1.7	S.D.	-.5
Mass	1.8	Tenn	1.5
Mich	1.5	Texas	1.5
Minn	.2	Utah	2.1
Miss	1.4	Vt	1.2
Mo	1.9	Va	2.0
Mont	1.1	Wash	1.7
Neb	2.2	W.Va	1.2
Nev	2.4	Wis	1.5
N.H.	1.1	Wyo	1.7
D.C.	1.5		

SOURCE: U.S. Department of Commerce.

Categorize the percent changes into eight classes as follows:

Class 1	% change ≤ .5
Class 2	.5 < % change ≤ 1.0
Class 3	1.0 < % change ≤ 1.5
⋮	
Class 8	3.5 < % change ≤ 4.0

Make a frequency count of the states that fall into the classes and give the relative frequencies. From the frequency table construct a bar graph.

2.27 The Justice Department says that one of every 185 women, a total of 572,032, are victims of domestic abuse every year. The rates of domestic violence per 1,000 women by age groups are given here:

Worksheet: Domestic.mtw

Age group	Rate
12–19	5.8
20–24	15.5
25–34	8.8
35–49	4.0
50–64	.9

SOURCE: U.S. Department of Justice.

a. Illustrate these data in a bar graph.
b. Is it possible to determine from these rates the actual number of women who are victims of domestic abuse in each of the age groups? What additional information is needed?

2.28   Illustrate the data on domestic violence from Exercise 2.27 in a pie chart.

2.29   According to the Energy Information Administration, the United States produces 30% of the world's nuclear power. France produces 17%, Japan produces 11%, Germany produces 7%, and Russia produces 6%. Illustrate these data in a bar graph. Do these percents sum to 100%? How will you deal with this in the bar graph?

2.30   How has the cost of owning a car changed since 1984? The worksheet gives the percentages of total costs for 1984 and 1993. Illustrate this comparison in a graph. What category has seen the greatest change?

Worksheet: Owncar.mtw

	1984	1993
Depreciation/Interest	45.6%	56.2%
Gasoline	30.3	16.9
Insurance	11.7	14.3
Maintenance	7.8	9.3
Other	4.6	3.3

SOURCE: Runzheimer International and *USA Today*, September 27, 1994.

2.31   In 1992 there were 5,440 crimes reported in Abilene, Texas. Of the 5,440 crimes, 4 were murder, 89 were forcible rape, 136 were robbery, 630 were aggravated assault, 1,439 were burglary, 2,946 were larceny theft, 166 were motor vehicle theft, and 30 were arson. (SOURCE: *Uniform Crime Reports*, U.S. Department of Justice.) Arrange these data (worksheet: Abilene.mtw) in a frequency table that includes relative frequencies.
   a. What crime was most prevalent?
   b. What percent of all crimes involved theft?

2.32   The SAT scores for 900 female and 800 male college freshmen are classified as low, medium, or high:

Range	Men	Women
High	190	250
Medium	430	520
Low	180	130

   a. Draw a bar graph illustrating the frequencies of the SAT scores for men.
   b. Draw a bar graph illustrating the frequencies of the SAT scores for women.
   c. Can these two bar graphs be compared as drawn?
   d. If the bar graphs are not comparable, how should they be drawn so that they can be compared?

2.33   From the data in Exercise 2.15 of Section 2.1, determine the total number of hazardous waste sites and the total number in above average minority areas in the four regions of the country. Draw a bar graph illustrating the number of waste sites and the number near minority areas for the four regions.

2.34   The market shares of major supermarket chains are illustrated in the pie chart and bar graph on page 74. Which format do you think is easier to interpret? Which gives more information? Are horizontal bars acceptable or should they be vertical?

Worksheet: Supermkt.mtw

Market Shared by Major Supermarket Chains

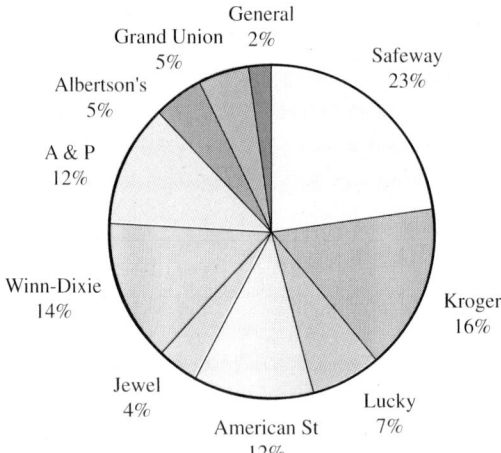

Number of Stores Operated by Major Supermarket Chains

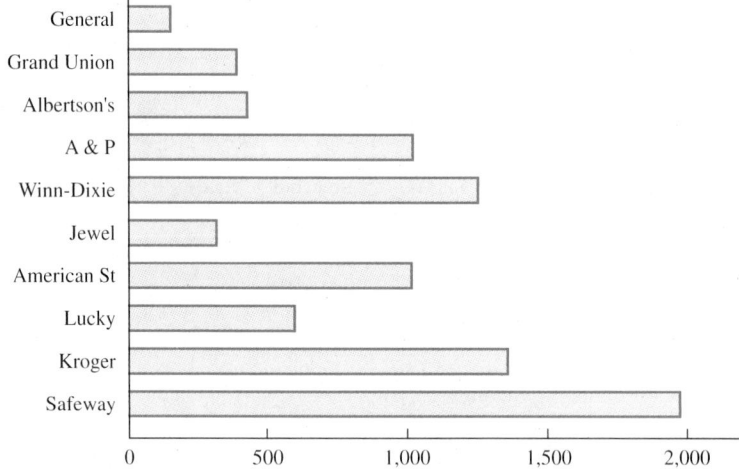

2.35   Women received 53% of all the bachelor's degrees awarded in 1990 compared with only 43% in 1970. Moreover, women are entering male-dominated fields. In 1990, for example, 13.8% of all engineering degrees were awarded to women compared with less than 1% in 1970. The table lists the percentages of bachelor's degrees awarded to women in different fields in 1970 and 1990.

    a. Arrange these data in a comparison bar graph.

    b. What areas show a dramatic change from 1970 to 1990?

    c. Would the differences show up better in comparison pie charts?

    d. Does a graphical display illustrate the differences better than the table?

Worksheet: Degree.mtw

Field	Percent awarded, 1970	Percent awarded, 1990
Health	78.0	84.3
Education	75.0	78.1
Foreign Languages	73.4	73.4
Psychology	43.3	71.5
Fine Arts	57.3	67.5
Life Sciences	27.8	50.7
Business	8.7	46.7
Social Science	37.1	44.2
Physical Sciences	13.6	31.2
Engineering	.7	13.8
All fields	43.1	53.2

SOURCE: U.S. Department of Health and Human Services, National Center for Education Statistics.

2.36   Housing of American Indians on the 314 national reservations is often less than desirable. For instance, 20% of American Indian households on reservations do not have indoor plumbing, 18% do not have complete kitchen facilities, and more than half do not have a telephone. Heating their homes is also a serious issue. The table gives the primary heating source for all U.S. homes, homes on reservations, and homes of American Indians who are not living on reservations.

Worksheet: Heat.mtw

Heating fuel	American Indians on reservations	All U.S. households	American Indians not on reservations
Utility gas	16%	48%	51%
LP bottled gas	22	9	6
Electricity	19	25	26
Fuel oil	6	9	12
Wood	34	8	4
Other fuel or none	4	1	1

SOURCE: Bureau of the Census, *Housing of American Indians on Reservations*, Statistical Brief 95-11, April 1995.

a. Does each of the columns sum to 100%? Should they? If they do not, why not?
b. Construct a comparison bar graph including all three groups.
c. How do American Indian households, both on and off reservations, compare with all U.S. households? Describe what you see in the bar graph. Are the differences easier to see in the bar graph or in the table?

2.3   DISPLAYING NUMERICAL DATA

Stem and Leaf Plot

The *stem and leaf plot* is a quick and easy way to display the distribution of numerical data. It is extremely useful for arranging the observations from smallest to largest so that specific locations within the data set can be found. Example 2.7 describes the construction of a stem and leaf plot.

EXAMPLE 2.7

A psychologist wishes to test a new method to improve rote memorization by college students. A sample of 20 college students were taught the new technique and then asked to memorize a list of 100 word phrases. The following numbers of correct word phrases were recorded for the 20 students:

Worksheet: Rote.mtw

84 59 82 78 74 96 44 76 85 66
77 91 62 54 72 65 84 38 76 70

By observation we see that the scores are two-digit numbers that range from the 30s to the 90s. To construct the display, we divide each observation into a stem and a leaf. In this example, the digits in the tens place of the numbers become the stems, and the digits in the units place become the leaves of the stem and leaf plot. A vertical line is drawn to separate the stems from the leaves.

Figure 2.10(a) shows only the stems and the first observation (84) plotted. Note that only the 4 of 84 is plotted on the 8 stem. Figure 2.10(b) shows the completed stem and leaf plot after all 20 leaves (corresponding to the 20 observations) are properly plotted.

FIGURE 2.10

```
3 |                3 | 8              3 | 8
4 |                4 | 4              4 | 4
5 |                5 | 9 4            5 | 4 9
6 |                6 | 6 2 5          6 | 2 5 6
7 |                7 | 8 4 6 7 2 6 0  7 | 0 2 4 6 6 7 8
8 | 4              8 | 4 2 5 4        8 | 2 4 4 5
9 |                9 | 6 1            9 | 1 6
```

(a) one observation   (b) completed   (c) ordered
plotted   stem and leaf plot   stem and leaf plot

The stem and leaf plot leaves the data intact for future calculations. Some of those calculations involve finding specific locations in the data; this would be relevant only if the data were ordered from smallest to largest. The stem and leaf plot allows us to do this easily. All that is required is to reorder the leaves of each stem from smallest to largest. Figure 2.10(c) is the resulting *ordered* stem and leaf plot. Rotating the stem and leaf plot 90 degrees counterclockwise, as in Figure 2.11, we see that the leaves form a graphical picture that shows how the data are *distributed* across a number line.

FIGURE 2.11

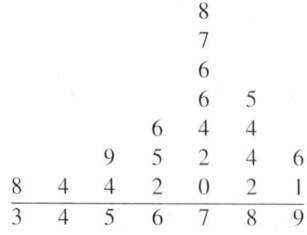

```
                     8
                     7
                     6
                     6     5
               6     4     4
         9     5     2     4     6
   8     4     4     2     0     2     1
   3     4     5     6     7     8     9
```

The next examples illustrate how the stem and leaf plot can be modified to accommodate data that might not immediately yield themselves to a stem and leaf interpretation.

EXAMPLE 2.8

An ecologist wishes to investigate the level of mercury pollution in a major lake. He catches 25 lake trout and measures the concentration of mercury (measured in parts per million) in each fish. From the following data, construct an ordered stem and leaf plot.

Worksheet: Mercury.mtw

2.2	3.4	3.0	2.6	3.8	1.8	2.8	3.2	3.7
1.4	2.7	3.6	1.9	2.2	3.0	3.3	2.3	
1.7	2.6	3.5	3.0	2.9	3.4	3.1	2.4	

SOLUTION

When confronted with a decimal point, we ignore it while constructing the stem and leaf plot, but understand that the numbers do actually have decimals. Thus, in this problem, we treat the numbers as two-digit numbers that range from 14 to 38. Using stem values of 1, 2, and 3, we get the following stem and leaf plot:

```
1│8 4 9 7
2│2 6 8 7 2 3 6 9 4
3│4 0 8 2 7 6 0 3 5 0 4 1
```

Realizing that the plot is somewhat compact, we can spread it out by constructing a *double-stem* stem and leaf plot. This is done by splitting each stem into two stems. The stem values are then 1, 1, 2, 2, 3, and 3. For each split stem, the first value is used to denote leaves 0 through 4 and the second value denotes leaves 5 through 9. We get the following double-stem stem and leaf plot and the ordered double-stem stem and leaf plot:

```
      Leaf unit = .10                 Leaf unit = .10

1│4                          1│4
1│8 9 7                      1│7 8 9
2│2 2 3 4                    2│2 2 3 4
2│6 8 7 6 9                  2│6 6 7 8 9
3│4 0 2 0 3 0 4 1            3│0 0 0 1 2 3 4 4
3│8 7 6 5                    3│5 6 7 8
double-stem                  ordered double-stem
```

Note: Leaf unit = .10 means that each leaf value is multiplied by .10 in order to reinsert the decimal place.

Minitab and most other statistical packages can produce stem and leaf plots.

EXAMPLE 2.9

Homicide rates for major cities are measured as the number of homicides per 100,000 inhabitants. Following are the homicide rates in 1992 for 30 cities east of the Mississippi River with over 250,000 population:

Worksheet: City.mtw

City	Rate	City	Rate
Atlanta, GA	48.2	Baltimore, MD	44.3
Birmingham, AL	48.8	Boston, MA	12.7
Buffalo, NY	23.0	Charlotte, NC	24.2
Chicago, IL	33.1	Cincinnati, OH	13.3
Cleveland, OH	30.6	Columbus, OH	17.6
Detroit, MI	57.0	Indianapolis, IN	17.8
Jacksonville, FL	18.5	Louisville, KY	14.2
Memphis, TN	28.0	Miami, FL	34.2
Milwaukee, WI	22.7	Minneapolis, MN	15.9
Nashville, TN	17.5	Newark, NJ	31.3
New York, NY	27.1	Norfolk, VA	29.3
Philadelphia, PA	26.5	Pittsburgh, PA	11.8
St. Louis, MO	57.4	St. Paul, MN	11.8
Tampa, FL	16.8	Toledo, OH	12.7
Virginia Beach, VA	5.7	Washington, DC	75.2

SOURCE: U.S. Department of Justice, *Sourcebook of Criminal Justice Statistics*, 1993, p. 368.

Use Minitab to construct a stem and leaf plot for these data.

SOLUTION

The data are keyed into the Minitab worksheet in a column named **Rate**. We can then use either the Session command or the pull-down menu to obtain the stem and leaf plot.

C O M P U T E R   T I P

In the Session window at the MTB> prompt, type

   MTB> **Stem 'Rate'**

or in Menu mode, first select **Graph**, click on **Character Graphs**, and then select **Stem-and-Leaf**. When the Stem-and-Leaf window appears, highlight the column Rate and click on **Select** and then click **OK**. (See Exhibit 1.4 for an illustration of the different windows.) The output appears in Exhibit 2.1.

EXHIBIT 2.1

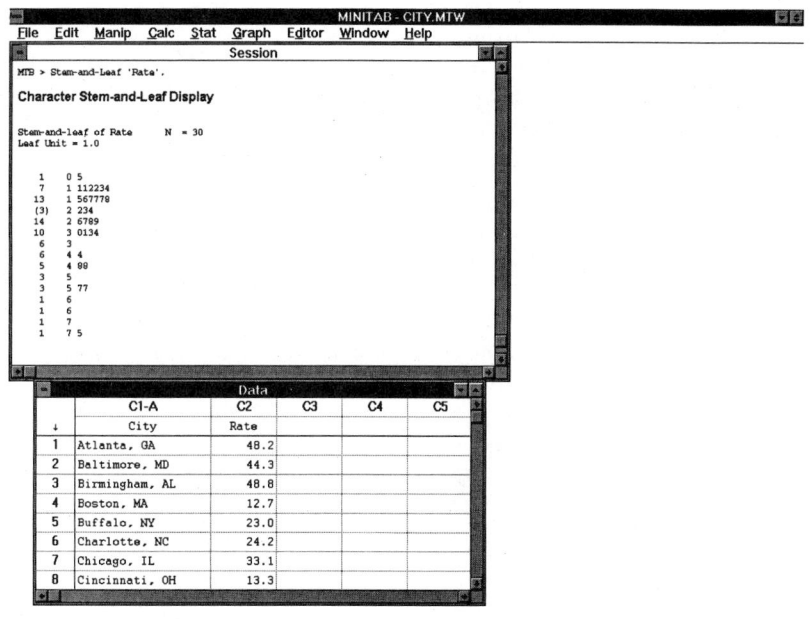

The following example illustrates the treatment of multidigit leaves.

EXAMPLE 2.10

The National Advisory Committee on Criminal Justice Standards and Goals studies the cost and resource implications of correctional standards for halfway houses. Reports are compiled each year so that state and local governments have cost information related to the activities carried out by halfway houses. One variable reported is the yearly per-bed rental costs. That variable is recorded for a random sample of 35 halfway houses:

Worksheet: Halfway.mtw

472	303	280	282	417	400	257	205	384
264	317	76	643	480	136	250	100	732
317	264	384	750	402	422	373	325	313
749	791	196	891	283	52	186	693

Construct a stem and leaf plot of these data.

SOLUTION

The obvious stems are the values corresponding to the hundreds place of the number, and therefore the leaves are the remaining two digits of the number.

The leaves are
separated by
commas so that,
for example, 76
is not
interpreted as a
7 and a 6.

```
0 | 76, 52
1 | 36, 00, 96, 86
2 | 80, 82, 57, 05, 64, 50, 64, 83
3 | 03, 84, 17, 17, 84, 73, 25, 13
4 | 72, 17, 00, 80, 02, 22
5 |
6 | 43, 93
7 | 32, 50, 49, 91
8 | 91
```

Exhibit 2.2 shows the Minitab stem and leaf plot of the same data. Notice that the data have been truncated and only single-digit leaves are given. For example, the last observation, 891, has been rounded to 890 and is exhibited with stem value 8 and leaf value 9. "Leaf Unit = 10" multiplies each entry by 10 so that 89 becomes 890. Although it appears that we have lost significant digits, the values remain unrounded internal to the computer. Minitab simply truncates the observations for display purposes.

EXHIBIT 2.2   Character Stem-and-Leaf Display

```
Stem-and-leaf of cost       N  = 35
Leaf Unit = 10

    2      0 57
    6      1 0389
   14      2 05566888
   (8)     3 01112788
   13      4 001278
    7      5
    7      6 49
    5      7 3459
    1      8 9
```

Occasionally, one or two scores may be far removed from the rest of the data, in which case it is not realistic to continue the stems all the way to those values. These extreme values, called *outliers*, are handled by putting them in a class by themselves with stem value either Hi (for high) or Lo (for low) depending on which side of the data it is on. Consider the 1990 state birth rates given in the next example.

outliers

EXAMPLE 2.11   Following are the 1990 live birth rates per 1,000 population in the United States:

Worksheet: Birth.mtw

Ala	16.2	Colo	16.0
Alaska	21.8	Conn	16.1
Ariz	18.9	Del	17.1
Ark	14.7	Fla	15.3
Calif	20.7	Ga	17.6

(continued)

Hawaii	18.1	N.M.	18.3
Idaho	16.0	N.Y.	16.8
Ill	16.4	N.C.	15.8
Ind	15.1	N.D.	16.0
Iowa	13.9	Ohio	15.1
Kan	15.4	Okla	14.3
Ky	15.2	Ore	15.9
La	16.5	Pa	14.2
Maine	13.1	R.I.	15.6
Md	15.9	S.C.	15.9
Mass	16.0	S.D.	15.2
Mich	16.9	Tenn	15.6
Minn	15.5	Texas	19.2
Miss	16.4	Utah	21.6
Mo	16.0	Vt	14.0
Mont	14.2	Va	15.6
Neb	15.0	Wash	15.8
Nev	18.1	W.Va	12.6
N.H.	15.0	Wis	14.8
N.J.	15.5	Wyo	13.9
D.C.	36.8		

SOURCE: U.S. Deptartment of Health and Human Services, *Monthly Vital Statistics Report*, Vol. 39, No. 13, August 28, 1991.

Exhibit 2.3 shows a Minitab stem and leaf plot of the birth rates.

EXHIBIT 2.3

Stem and leaf plot of 1990 state birth rates

```
Character Stem-and-Leaf Display

Stem-and-leaf of Birth      N  = 51
Leaf Unit = 1.0

    4      1 2333
  (24)     1 444444555555555555555555
   23      1 66666666666677
    9      1 88889
    4      2 011
    1      2
    1      2
    1      2
    1      2
    1      3
    1      3
    1      3
    1      3 6
```

Notice that Minitab extended the stems all the way down to include the birth rate of 36.8 for the District of Columbia. An alternative stem and leaf plot uses the Hi, Lo scheme as follows:

```
Stem and leaf of birth     N = 51
Leaf unit = 1.0

  1 | 2 3 3 3
  1 | 4 4 4 4 4 4 5 5 5 5 5 5 5 5 5 5 5 5 5 5 5 5 5 5
  1 | 6 6 6 6 6 6 6 6 6 6 6 7 7
  1 | 8 8 8 8 9
  2 | 0 1 1
    ————————
 Hi | 36.8
```

Also notice that Minitab created a *five-stem* stem and leaf plot; that is, stems 1, 2, and 3 are divided into five different categories. The first level corresponds to leaves 0 and 1, the next level corresponds to leaves 2 and 3, followed by leaves 4 and 5, 6 and 7, and finally leaves 8 and 9.

Certainly, there are other possibilities for stem and leaf plots that we have not considered. With a little practice, you will be able to come up with a meaningful stem and leaf plot for almost any data set.

Histograms

The *histogram* is a graphical means of illustrating the distribution of numerical data. It can be constructed from the stem and leaf plot; each stem defines an interval of values called a *class*. The *class limits* are the smallest and largest possible values for the leaves of that stem. Returning to the rote memorization problem given in Example 2.7 and the stem and leaf plot in Figure 2.10, we see that the class limits for the stem value 3 are 30 and 39, for stem value 4 they are 40 and 49, and so on. Once the class limits are determined, the data can be formulated in a *grouped frequency table* (grouped in the sense that the numerical data are grouped into various classes) such as Table 2.6.

TABLE 2.6   Number of correct word phrases in a rote memorization study

Class Limits	Class Boundaries	Frequency	Relative Frequency
30–39	29.5–39.5	1	1/20
40–49	39.5–49.5	1	1/20
50–59	49.5–59.5	2	2/20
60–69	59.5–69.5	3	3/20
70–79	69.5–79.5	7	7/20
80–89	79.5–89.5	4	4/20
90–99	89.5–99.5	2	2/20

The *class boundaries* are obtained by lowering the lower class limits by .5 and raising the upper class limits by .5. Thus, class limits 30 and 39 become class boundaries 29.5 and 39.5, respectively. The class boundaries are given so that the classes are continuous;

that is, the first class runs from 29.5 to 39.5, and immediately the second class picks up at 39.5 and runs to 49.5. Then the third class starts at 49.5, and so on.

The histogram can now be constructed from the grouped frequency table. The class boundaries are scaled off on the horizontal axis. Bars are constructed over each class boundary so that the height of each bar is the frequency of the class, which is marked off on the vertical axis. Figure 2.12 is the completed histogram.

FIGURE 2.12

Number of correct word phrases in a rote memorization study

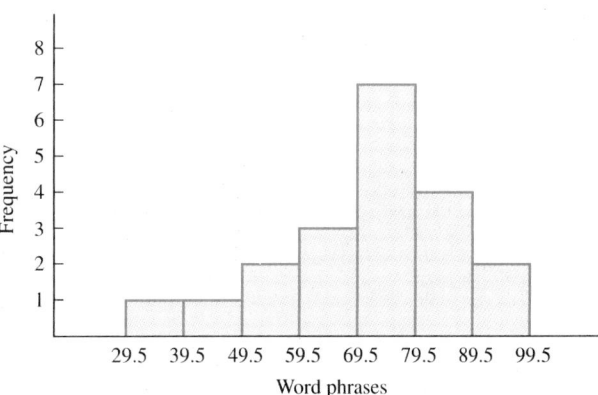

The histogram gives a picture of the distribution of the data. It is very similar to the bar graph presented in Section 2.2 for categorical data. The basic difference is that the histogram is for numerical data, which are continuous, and consequently the bars are joined at the class boundaries. (Recall that the bars are separated in a bar graph.)

COMPUTER TIP

A histogram is created with Minitab in the same manner as the stem and leaf plot described earlier. The data are entered into a column of the worksheet, such as C1, and then the following command is given in the Session window at the MTB prompt:

MTB> **histogram of C1**

If you prefer the Menu mode, first select **Graph**, click on **Histogram**, select the variable to be graphed, and then click on **OK**. The Minitab histogram in Figure 2.13 (page 84) illustrates the yearly per-bed rental costs at the halfway houses given in Example 2.10. Notice that the midpoints of the classes are labeled in place of the class boundaries.

FIGURE 2.13   Yearly per-bed rental cost
(see Example 2.10)

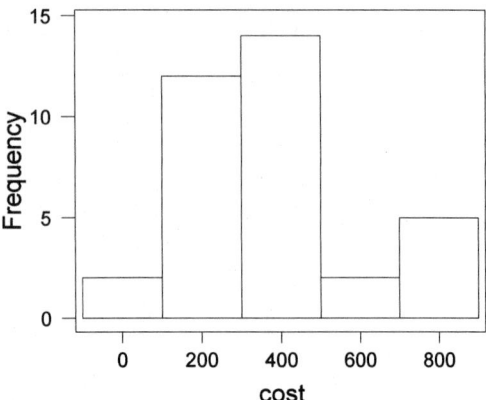

Just as in the case of the bar graph, the vertical axis of the histogram can be scaled in relative frequency, as depicted in Figure 2.14.

FIGURE 2.14   Number of correct word
phrases in a rote
memorization study

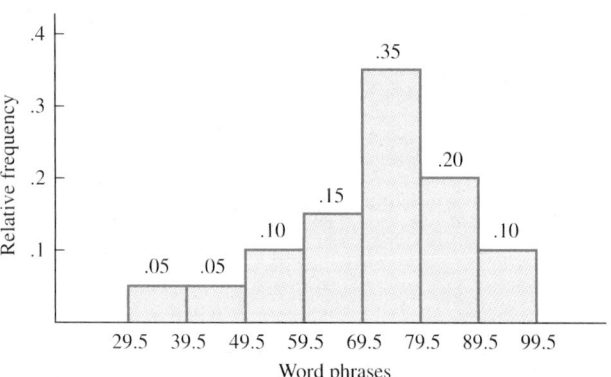

The *relative frequency histogram* is important for two reasons. The first reason is that it removes the dependence of the histogram on the size of the data set. This is important when we are comparing two or more distributions. For example, do women score higher than men on a rote memorization task? If considerably more women are tested than men, then it is not realistic to compare their frequencies. On the other hand, the relative frequencies of the two groups are independent of their sizes and thus a fair comparison can be made.

The second important reason for the relative frequency histogram is that it can be used in a *probability* sense. From Figure 2.14 (or Table 2.6) we see that the relative frequency of the class with boundaries from 79.5 to 89.5 is 4/20. So the percent of subjects who scored somewhere between 79.5 and 89.5 is $4/20 = .2 = 20\%$. The percent that scored somewhere between 69.5 and 89.5 is $7/20 + 4/20 = 11/20 = .55 = 55\%$.

Later you will see that the relative frequency histogram obtained from a sample can be used to approximate the distribution of the entire population of scores. If the relative frequency histogram given in Figure 2.14 accurately describes the population distribution, then we can say that 55% of the population will score somewhere between 69.5 and 89.5. Alternately, we can say that the chance or probability of a randomly selected individual scoring between 69.5 and 89.5 is 55%.

The stem and leaf plot preserves the data and is more quickly constructed than the histogram. This is not to say that you should never use a histogram, however. The histogram is certainly appropriate if you are presenting the data in a completed report or if the data set is very large. But for preliminary analysis and exploratory analysis, you should use the stem and leaf plot. We might say that the histogram is the formal presentation of the data and the stem and leaf plot is the informal presentation of the data.

Polygons

The *frequency polygon* (or *relative frequency polygon*) is another useful tool for describing the distribution of numerical data. It is a graph of connecting line segments that correspond to the frequencies of the various classes. A histogram can be easily converted to a frequency polygon by connecting the midpoints of the tops of the bars with straight line segments. On the end bars, the midpoints are connected down to the horizontal axis to the midpoints of the classes that would have been adjacent to those end classes. When the vertical axis represents relative frequencies, the graph is called a relative frequency polygon. Figure 2.15 is a relative frequency polygon constructed from the relative frequency histogram in Figure 2.14.

FIGURE 2.15 Number of correct word phrases in a rote memorization study

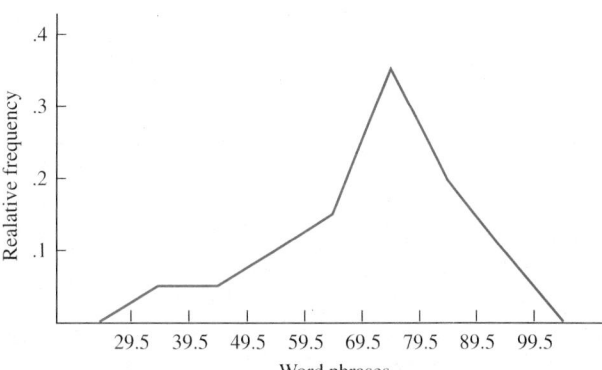

Like the relative frequency histogram, the relative frequency polygon is useful for comparing distributions. For example, to compare rote memorization scores for men and women, relative frequency polygons for both men and women may be graphed on the same coordinate axis system, as illustrated in Figure 2.16 (page 86). Remember, if we are comparing groups, we must compare the relative frequencies if the group sizes are different.

FIGURE 2.16 Number of correct word
phrases for men and
women in a rote
memorization study

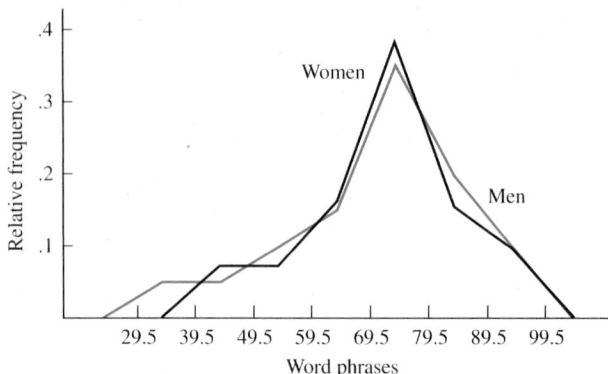

Frequency Curve

We have just shown how a stem and leaf plot, a histogram, and a frequency polygon can be used to display the data associated with a variable. Once the scores are displayed, we notice the pattern of variation exhibited by the distribution. When the number of observations is very large and the class limits are reduced in size, the distribution takes on the appearance of a continuous curve similar to what might be expected if the entire population of values were graphed. Usually the entire population is not available; however, a stem and leaf plot, histogram, or frequency polygon obtained from a representative sample should closely approximate the shape of the *population frequency curve*. The frequency polygon in Figure 2.15 suggests the population frequency curve in Figure 2.17.

FIGURE 2.17 Number of correct word
phrases in a rote
memorization study

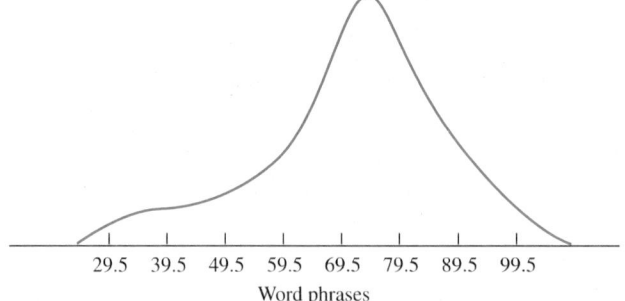

There are several different types of shapes that distributions assume. Later we will investigate and classify those different shapes, but first we study several quantities that summarize the distribution, such as numerical measures of its variation and where the center is located.

EXERCISES 2.3

2.37 Following are 20 scores on the Miller Personality Test:

Worksheet: Miller1.mtw

22 21 16 26 22 23 31 25 20 25
33 17 27 29 25 22 30 18 23 25

Arrange the data in an ordered five-stem stem and leaf plot.

2.38 Fifty families were interviewed and the the following numbers of dependent children recorded:

Worksheet: Depend.mtw

3 2 2 4 1 1 2 3 4 1
2 0 1 2 1 0 4 2 1 0
0 1 3 0 3 2 2 3 0 3
2 5 0 1 2 1 4 3 0 5
2 0 1 1 2 6 1 2 1 5

Arrange the data in an ordered stem and leaf plot.

2.39 The numbers of defective items produced by the 20 employees in a small business are given here:

Worksheet: Defectiv.mtw

7 6 10 9 8 7 7 6 8 8
10 7 6 8 8 9 10 9 9 8

Arrange the data in an ordered stem and leaf plot.

2.40 The numbers of days that the 20 employees in Exercise 2.39 were absent during the year are listed here:

Worksheet: Absent.mtw

1 0 4 3 2 0 0 2 0 2
10 0 0 2 3 4 5 0 2 1

Arrange the data in a stem and leaf plot.

2.41 A group of 23 students participated in a psychology experiment. Their scores are the numbers of correct responses:

Worksheet: Psych.mtw

12 14 18 7 11 15 8 15 10 14 19
14 6 13 14 12 16 10 9 12 15 8 17

Arrange the data in a stem and leaf plot.

2.42 Following are the daily profits of 20 newsstands. Construct an ordered double-stem stem and leaf plot.

Worksheet: Newstand.mtw

$81.32 61.47 64.90 70.88 76.02 75.06 76.73
64.21 74.92 77.56 58.01 68.05 73.37 75.41
59.41 65.43 74.76 76.51 65.10 76.02

2.43   An exercise program had 30 members who did exercises 5 days a week for 1 month. Following are the weight losses of the 30 members (a negative number indicates that the person gained weight):

Worksheet: Exercise.mtw

5	15	3	−4	8	7	5	10	−3	2
−5	9	5	−4	10	6	−2	7	4	3
−12	−6	11	4	8	−10	9	5	5	2

Construct an ordered stem and leaf plot of the weight losses.

2.44   The prices of regular unleaded gasoline were obtained from 25 service stations around a major city. Construct an ordered stem and leaf plot of the following data:

Worksheet: Gasoline.mtw

$1.16	1.22	1.20	1.22	1.18	1.20	1.20	1.19	1.19
1.18	1.20	1.19	1.20	1.19	1.25	1.19	1.20	
1.22	1.19	1.21	1.20	1.19	1.20	1.18	1.21	

2.45   Following are the weights of 25 soccer players:

Worksheet: Soccer.mtw

144	162	197	173	183	129	209	190	117
160	179	177	154	132	151	159	175	
154	148	166	184	157	162	150	136	

Arrange the data in a stem and leaf plot. Construct a frequency histogram from the stem and leaf plot with interval widths of 10 and the first lower class limit at 110.

2.46   Construct a stem and leaf plot of the following data, which are the maximum numbers of sit-ups completed by the participants in an exercise class after 1 month in the program:

Worksheet: Situp.mtw

24	31	54	62	36	28	37	55	18	27
58	32	37	41	55	39	56	42	29	35

2.47   Create a frequency histogram from the stem and leaf plot constructed in Exercise 2.46.

2.48   Construct a frequency polygon of the following data, which are the test scores on the first exam in a beginning biology class:

Worksheet: Biology.mtw

87	79	94	60	75	94	77	83	68	74
82	73	63	75	77	83	92	57	64	53
53	82	73	90	55	68	72	88	65	78

2.49   A random sample of 24 high school seniors were given a college entrance exam that has a maximum score of 100. Construct a stem and leaf plot of the following scores:

Worksheet: Entrance.mtw

64	75	81	43	69	75	86	58	63	66	82	62
79	91	83	55	68	74	48	66	84	77	73	59

2.50   Construct a frequency histogram of the data in Exercise 2.49.

2.51   Construct a frequency polygon of the data in Exercise 2.49.

2.52   To estimate the number of trees on a tree farm, a farmer divided the farm into 1,000 small grids. He then randomly selected 20 grids and counted the numbers of trees, with the following results:

Worksheet: Trees.mtw

```
 81  96  87  83  99  64  77   63  93  84
102  68  94  81  70  84  92  109  74  86
```

Construct a stem and leaf plot and a frequency histogram of the data.

2.53   Construct a relative frequency histogram from the following stem and leaf plot:

```
2 | 3 4 8
3 | 2 4 3 8 4 7
4 | 1 8 9 2 6 7 3 0
5 | 4 5 2 7 5
6 | 7 3 0 8
7 | 5 2 7
8 | 3
```

2.54   A study of toxic waste sites in Exercise 2.15 of Section 2.1 gave the percentage of minorities living in communities with commercial hazardous waste sites (worksheet: **Toxic.mtw**). To better understand the data, construct a double-stem stem and leaf plot of all sites and then a double-stem stem and leaf plot of the number of sites near minority communities.

2.55   Following are cholesterol levels from a sample of 62 subjects from the Framingham Heart Study:

Worksheet: Framingh.mtw

```
393  353  334  336  327  300  300  308  283  285  270  270  272
278  278  263  264  267  267  267  268  254  254  254  256  256
258  240  243  246  247  248  230  230  230  230  231  232  232
232  234  234  236  236  238  220  225  225  226  210  211  212
215  216  217  218  200  202  192  198  184  167
```

SOURCE: R. D'Agostino, et al., "A Suggestion for Using Powerful and Informative Tests of Normality," *The American Statistician* 44 (1990): 316–321.

a. Construct a stem and leaf plot of the data .
b. From the stem and leaf plot, define class limits and class boundaries and formulate the data in a grouped frequency table.
c. From the grouped frequency table, construct a histogram.
d. Construct a frequency polygon.

2.56   The following data are the survival times in weeks for 20 male rats that were exposed to a high level of radiation. Arrange these data in a stem and leaf plot.

Worksheet: rat.mtw

```
152  152  115  109  137  88  94   77  160  165
125   40  128  123  136  101  62  153   83   69
```

SOURCE: J. Lawless, *Statistical Models and Methods for Lifetime Data* (New York: Wiley, 1982).

2.57   A high-volume drug screen was designed to find compounds that reduce low-density lipoproteins (LDL) cholesterol in quail. The treatment group of quail were fed a special diet mixed with a drug compound over a specified period of time. The placebo group of quail were fed the same special diet for the same period of time but without the drug compound. Following are the plasma LDL levels for the two groups:

Worksheet: Quail.mtw

Placebo

64	49	54	64		97	66	76	44	71	89
70	72	71	55		60	62	46	77	86	71

Treatment

40	31	50	48	152	44	74	38	81	64

SOURCE: J. McKean and T. Vidmar, "A Comparison of Two Rank-Based Methods for the Analysis of Linear Models," *The American Statistician* 48 (1994): 220–229.

Organize the data in a *back-to-back stem and leaf plot*. This is two stem and leaf plots (for the two groups) drawn back to back with a common stem. Notice that there is one very large value in the treatment group. The researchers pointed out that this was typical with most of the data in the study. How do you deal with it in the stem and leaf plot?

2.58   Stamp collectors will tell you that the price of any stamp is based on its relative scarcity, and its scarcity depends on a number of factors. One unusual factor is the thickness of the paper that the stamp is printed on. Paper is sold by weight per ream, where a ream contains about 500 pages. Over 100 years ago, when paper was very expensive, manufacturers would keep on hand some extra thick sheets to replace thin sheets in an underweight ream. The goal was to keep the weight up but, at the same time, supply as few sheets as possible. Consequently, a particular stamp could have been printed on a variety of thicknesses of paper (several different reams, each having its own padded sheets).

Mexico began issuing stamps in 1856. Through 1879 almost all stamps carried the image of national hero Miguel Hidalgo y Costilla. One such stamp is the *1872 Hidalgo issue*, which was in circulation until 1874. Today it is estimated that there are fewer than 100,000 of the original 4,772,000 printed stamps. If you buy one, how can you determine whether it is printed on thick, medium, or thin paper? The *1872 Hidalgo* carries a rich history. Between 1979 and 1981, five major collections of the 1872 issue were purchased by the late Walton van Winkler, who measured the thickness (in millimeters) of each of his 485 stamps. If his collection is a representative sample of all the 1872 Hidalgos, then by using his measurements, we can investigate the distribution of the thicknesses of the stamps. The worksheet is a frequency table of van Winkler's measurements.

Worksheet: Stamp.mtw

Thickness	Freq.	Thickness	Freq.	Thickness	Freq.
.060	1	.074	10	.088	2
.061	0	.075	20	.089	10
.062	0	.076	18	.090	9
.063	0	.077	11	.091	3
.064	2	.078	23	.092	5
.065	1	.079	42	.093	6
.066	1	.080	37	.094	3
.067	0	.081	15	.095	2
.068	1	.082	18	.096	3
.069	7	.083	7	.097	7
.070	26	.084	3	.098	5
.071	20	.085	2	.099	5
.072	32	.086	2	.100	15
.073	11	.087	1	.101	9

(continued)

Thickness	Freq.	Thickness	Freq.	Thickness	Freq.
.102	8	.112	5	.122	2
.103	7	.113	0	.123	2
.104	2	.114	3	.124	0
.105	5	.115	3	.125	2
.106	4	.116	0	.126	0
.107	3	.117	1	.127	0
.108	7	.118	0	.128	1
.109	7	.119	4	.129	3
.110	11	.120	3	.130	1
.111	4	.121	1	.131	1

SOURCE: A. Izenman and C. Sommer, "Philatelic Mixtures and Multimodal Densities," *Journal of the American Statistical Association* 83 (1988): 941–953.

a. Regroup the frequency table into fewer groups, where the classes go from .060 to .064, .065 to .069, .070 to .074, and so on.
b. Using a modified frequency table, draw a histogram of the distribution from part **a**.
c. Describe in detail what you have learned about the distribution of the thicknesses of the 1872 issue.
d. Suppose your stamp has a thickness of .116. What percent of the Hidalgos weigh more than yours?

2.4 SUMMARY MEASURES OF LOCATION

One of the main objectives of statistics is to draw generalizations about a population based on the data collected from a sample. In most cases it is difficult to work with the complete distribution of values; thus, summary measures are introduced to help answer our statistical questions. For example, the comparison of the effects of a certain drug on the reaction time of women to that of men may simplify to a comparison of two numerical values that correspond to the centers of the two distributions of reaction times. These summary measures can apply either to a sample or to the entire population.

A **parameter** is a numerical summary measure of a population distribution.

Typically, the values of the parameters associated with a population are unknown. We must select a random sample from the population and use summary measures of the sample to estimate the unknown population parameter values.

A **statistic** is a numerical quantity calculated from the observations in a sample.

Corresponding to each parameter of a population there is a statistic that is its sample counterpart. If the summary measure refers to the entire population, it is called a *parameter*; if it is obtained from information in the sample, it is called a *statistic*.

Among the summary measures are *measures of location* and *measures of variability*. We now look at measures of location, and then we investigate measures of variability in the next section.

Center

A measure of location is a quantity that locates a particular position in the distribution. We first consider measures of the center of the distribution.

Mean

The most common measure of center is the *mean*, which locates the balance point of the distribution.

> The **population mean**, denoted by the Greek letter μ, is the numerical value that locates the *balance point*, also called the *center of mass*, of the population distribution.
>
> If a sample consists of observations $y_1, y_2, y_3, \ldots, y_n$, then the **sample mean** is
> $$\bar{y} = (y_1 + y_2 + \cdots + y_n)/n = \Sigma\, y_i/n.$$

Figure 2.18 illustrates that μ is the center of gravity of the frequency curve, the point about which the distribution would balance.

FIGURE 2.18   Location of the population mean

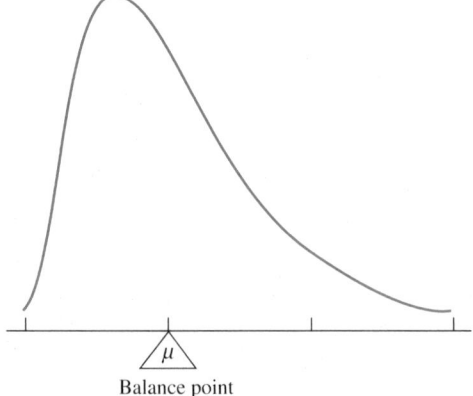

Balance point

If the entire population were available and finite, the population mean μ would be calculated the same way as the sample mean. However, the entire population is rarely available, and thus the value of μ remains unknown. Only $\bar{y}$ is calculated (since the sample is available) and its value serves as an approximation of μ.

EXAMPLE 2.12

On January 28, 1986, the U.S. space shuttle *Challenger* exploded shortly after liftoff. The Presidential Commission that investigated the accident concluded that the disaster was caused by the failure of an O-ring, which resulted in a combustion gas leak through a field joint on the rocket booster. It was determined that the O-rings did not seal properly at low temperatures. Following are the recorded temperatures on the 24 launches of the shuttle *previous* to the accident:

Worksheet: Challeng.mtw

66 70 69 80 68 67 72 73 70 57 63 70
78 67 53 67 75 70 81 76 79 75 76 58

SOURCE: S.R. Dalal, E.B. Fowlkes, and B. Hoadley, "Risk Analysis of the Space Shuttle: Pre-Challenger Prediction of Failure," *Journal of the American Statistical Association* 84, No. 408 (1989): 945–957.

Calculate the sample mean temperature.

SOLUTION

The sample mean is calculated by first finding the total of all observations:

$$66 + 70 + 69 + \cdots + 76 + 58 = 1,680$$

and then dividing the sum by n, the number of observations. We have

$$\bar{y} = \frac{1,680}{24} = 70$$

which is the average temperature of all shuttle launches prior to the accident.

Just as the population mean balances the population distribution, the sample mean balances the observations in a sample. For example, for illustration purposes, consider the five observations: 31, 55, 75, 78, and 81. By placing a fulcrum at the average, which is $\bar{y} = (31 + 55 + 75 + 78 + 81)/5 = 64$, we see in Figure 2.19 that the five observations would balance. If the observation 81 were replaced with 91, for example, then $\bar{y}$ would shift to the right to 66 in order to continue to balance the data.

FIGURE 2.19

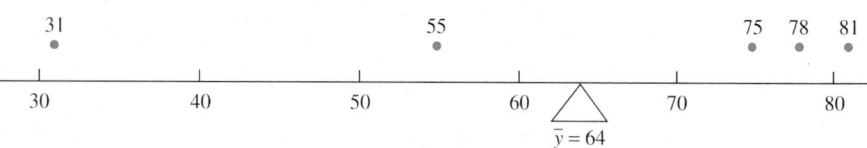

$\bar{y} = 64$

EXAMPLE **2.13**   Researchers have investigated lead absorption in children of parents who worked in a factory where lead is used to make batteries. Following are the levels of lead in the children's blood (in μg/dl of whole blood):

Worksheet: Lead.mtw column 2

```
38  23  41  18  37  36  23  62  31  34  24
14  21  17  16  20  15  10  45  39  22  35
49  48  44  35  43  39  34  13  73  25  27
```

SOURCE: D. Morton, et al., "Lead Absorption in Children of Employees in a Lead-Related Industry," *American Journal of Epidemiology* 155 (1982): 549–555.

Organize the data in a stem and leaf plot and calculate the sample mean.

SOLUTION

The stem and leaf plot is a double-stem stem and leaf:

```
1 | 0 3 4
1 | 5 6 7 8
2 | 0 1 2 3 3 4
2 | 5 7
3 | 1 4 4
3 | 5 5 6 7 8 9 9
4 | 1 3 4
4 | 5 8 9
5 |
5 |
6 | 2
6 |
7 | 3
```

Averaging the 33 observations, we find

$$\bar{y} = 31.85$$

However, the stem and leaf plot shows that the observation 73 and possibly the observation 62 are rather distant from the rest of the data. Because $\bar{y}$ must balance the data, these extreme scores cause $\bar{y}$ to assume an unusually large value.

Given the two unusually large observations in Example 2.13, an obvious question is: Should we always use the mean to represent the center of a distribution? Are there alternatives to $\bar{y}$ that are better measures of the center of a distribution that has extreme observations? Consider the distribution in Figure 2.20, which has unusually thick tails. It is commonly referred to as a *long-tailed distribution*.

FIGURE 2.20   A long-tailed distribution

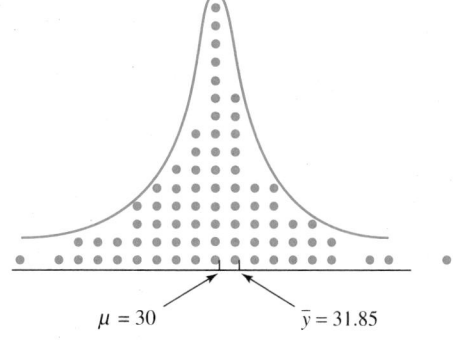

$\mu = 30$ $\bar{y} = 31.85$

Trimmed Mean

A distribution with long tails tends to occasionally produce extreme observations, as is indicated by the far right observation in the graph. Such outliers (observations that are remote or stand apart from the remainder of the data) tend to pull the value of $\bar{y}$ toward them and thus give a distorted measure of the center of the distribution. We see in Figure 2.20 that $\mu = 30$, but because of the extreme score in the right tail, $\bar{y} = 31.85$, which is not a good approximation of μ. A better measure of center is a *trimmed mean*.

> A **$p\%$ trimmed population mean**, denoted by μ_T, is obtained by trimming $p\%$ of the distribution off both ends and finding the mean of the remaining distribution.
>
> A **$p\%$ trimmed sample mean**, denoted by $\bar{y}_T$, is found by calculating $p\%$ of n, trimming that many observations off both ends of the ordered data, and then calculating the mean of the remaining observations. If $p\%$ of n is not a whole number, *round up* to the next whole number.

EXAMPLE 2.14

Find the 10% trimmed mean for the lead absorption data in Example 2.13.

SOLUTION

Remember to order the data before trimming.

Ten percent of 33 observations is 3.3; thus, we trim four observations off each end of the ordered stem and leaf plot. We trim 10, 13, 14, and 15 off the low end and 73, 62, 49, and 48 off the high end of the ordered stem and leaf plot to obtain the trimmed stem and leaf plot:

```
1 | 6 7 8
2 | 0 1 2 3 3 4
2 | 5 7
3 | 1 4 4
3 | 5 5 6 7 8 9 9
4 | 1 3 4
4 | 5
```

The average of the remaining 25 observations is

$$\bar{y}_T = 30.68$$

which is more than 1 unit less than $\bar{y}$ (31.85) and might be a more representative measure of the center of the distribution.

Clearly, trimming minimizes the effect of outliers. In Example 2.13, the 73 is definitely an outlier and has a measurable impact on the ordinary sample mean. Once it is trimmed off (Example 2.14), it has no effect on the trimmed mean, $\bar{y}_T$. In fact, the 10% trimmed mean is not affected by the lower or upper 10% of the observations.

Median

Another measure of location that is not affected by outliers is the *median*.

> The **population median**, denoted by θ, is the numerical value that divides the population distribution in half.
>
> If the sample observations $y_1, y_2, \ldots, y_n$ are arranged in order from smallest to largest, the **sample median**, denoted by M, is the middle observation if n is odd, or the average of the two middle observations if n is even. In either case, the median is located at the position $(n + 1)/2$ in the ordered data set.

Figure 2.21 illustrates that the population median, θ, is the value that separates the lower 50% of the distribution from the upper 50% of the distribution.

FIGURE 2.21

Location of the population median

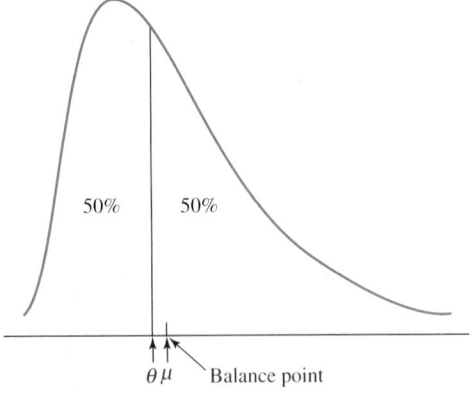

> If the right tail of the distribution is longer than the left tail, then the mean will be greater than the median. The opposite is true if the left tail is longer than the right tail; that is, the mean will be less than the median.

The median will not balance the distribution unless the distribution is symmetrical. For symmetrical distributions, Figure 2.22 shows that the mean and median are the same, in which case either will balance the distribution.

FIGURE 2.22 The mean and median coincide in symmetric distributions

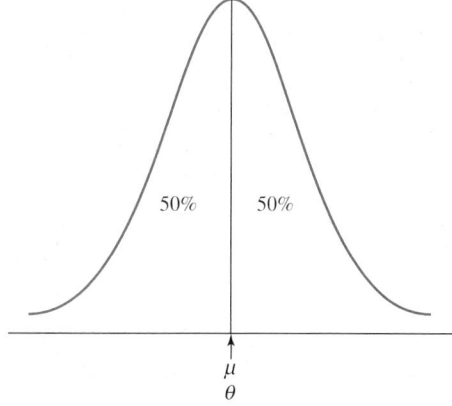

50%   50%

μ
θ

The sample median is located at the center of the sample after the values have been ordered from smallest to largest. It is found by counting from the smallest observation to the position $(n + 1)/2$ in the ordered sample. If n is odd, then $(n + 1)/2$ is a whole number. Simply count to that position in the ordered sample and you have found the median.

EXAMPLE 2.15

The space shuttle launch temperatures from Example 2.12 are presented in the accompanying double-stem stem and leaf plot. The temperature at liftoff on the day of the *Challenger* accident was 31 degrees and is also included in the data set.

```
3 | 1
3 |
4 |
4 |
5 | 3
5 | 7 8
6 | 3
6 | 6 7 7 7 8 9
7 | 0 0 0 0 2 3
7 | 5 5 6 6 8 9
8 | 0 1
```

Find the sample median temperature.

SOLUTION

By counting the leaves in the stem and leaf plot, we find that $n = 25$. Therefore, the median is located at position

$$\frac{n + 1}{2} = \frac{25 + 1}{2} = 13$$

Counting leaves until we get to the 13th observation, we find that the sample median is $M = 70$.

It is interesting that the median is the same as the value of $\bar{y}$ we found in Example 2.12 when the accident temperature (31 degrees) was not included. When 31 is included in the data set, the average becomes $\bar{y} = 68.44$, which is 1.56 degrees lower than the median temperature. You can see how one score can have a measurable impact on the value of the sample mean.

If n is even, then $(n + 1)/2$ will not be a whole number but will have a .5 decimal part. For example, if $n = 24$, then

$$\frac{n + 1}{2} = \frac{24 + 1}{2} = 12.5$$

The sample median is then the average of the two observations adjacent to the median location. In this case, the median is the average of the 12th and 13th observations after the data have been ordered from smallest to largest.

EXAMPLE 2.16

Find the sample median of the space shuttle launch temperatures excluding the 31 degrees recorded on the day of the accident.

SOLUTION

Excluding the 31 degrees, we have the following stem and leaf plot:

```
5 | 3
5 | 7 8
6 | 3
6 | 6 7 7 7 8 9
7 | 0 0 0 0 2 3
7 | 5 5 6 6 8 9
8 | 0 1
```

Because $n = 24$, the median is located at position 12.5. Counting down the leaves of the ordered stem and leaf plot, we find that the 12th observation is 70 and the 13th observation is also 70, so, the median is

$$M = \frac{70 + 70}{2} = 70$$

which is the same as $\bar{y} = 70$. Notice that the median is the same with or without the outlier observation, 31 degrees. The value of $\bar{y}$, however, changed from 70 to 68.44 when the 31 degrees is included.

The preceding examples illustrate that the sample mean may give a distorted measure of the center of a distribution that has outliers. Its value is pulled toward the extreme observations, whereas the sample median is not affected by the outliers. Because the median, like the trimmed mean, is not affected by the extreme observations, we say it is *resistant* to the influence of outliers.

Quartiles

Just as the population median divides the population distribution in half, the population *quartiles* divide it into quarters.

> The **first quartile**, denoted by θ_1, is the numerical value that divides the lower half of the population distribution in half. The **third quartile**, denoted by θ_3, is the value that divides the upper half of the population distribution in half.
>
> The **first** and **third sample quartiles**, Q_1 and Q_3, are similarly defined for samples. The median is the **second quartile**, Q_2.

Figure 2.23 illustrates that the first quartile, the median, and the third quartile divide the population distribution into four quarters. It is easy to see that θ_1 is the median of the lower half of the distribution and θ_3 is the median of the upper half of the distribution.

FIGURE 2.23   Location of the median and both quartiles

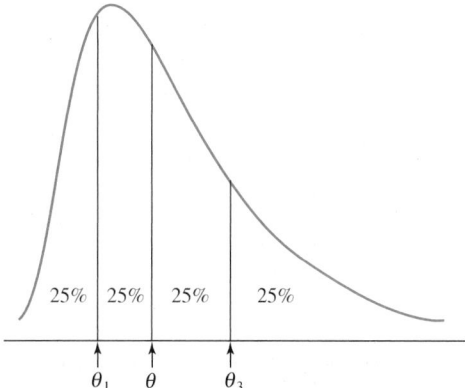

After the data are ordered, the sample quartiles can be found by finding the sample median of the lower and upper halves of the data. Because the median is located at position $(n + 1)/2$, we can take its *integer part*, add 1, and divide by 2 to find the location of the quartiles. "Taking the integer part" means that if $(n + 1)/2$ has a .5 decimal, we just drop it off before adding 1 and dividing by 2.

The first sample quartile, Q_1, is found by counting the observations from the *lower* end of the ordered data until we get to the observation in the quartile position. The third sample quartile, Q_3, is found by counting the observations from the *higher* end of the ordered data until we get to the observation in the quartile position. As in finding the median, if the quartile position has a .5 decimal part, we average the two observations on either side.

EXAMPLE 2.17

Find Q_1 and Q_3 for the space shuttle data in Example 2.16. The stem and leaf plot is repeated here for convenience:

```
5 | 3
5 | 7 8
6 | 3
6 | 6 7 7 7 8 9
7 | 0 0 0 0 2 3
7 | 5 5 6 6 8 9
8 | 0 1
```

SOLUTION

Recall that $(n + 1)/2 = 12.5$. The integer part is 12, and thus the quartile position is $(12 + 1)/2 = 6.5$. Thus Q_1 is the average of the sixth and seventh observations from the lower end, and Q_3 is the average of the sixth and seventh observations from the higher end of the ordered stem and leaf plot. We find

$$Q_1 = \frac{67 + 67}{2} = 67 \quad \text{and} \quad Q_3 = \frac{76 + 75}{2} = 75.5$$

Five-Number Summary Diagram

The quartiles (including the median) along with the highest and lowest scores are often listed in a *five-number summary diagram*. The following diagram is for the space shuttle data given in Example 2.16:

Low	Q_1	M	Q_3	High
53	67	70	75.5	81

EXAMPLE 2.18

Summarize the lead absorption data from Example 2.13 in a five-number summary diagram. The stem and leaf plot is repeated for convenience:

```
1 | 0 3 4
1 | 5 6 7 8
2 | 0 1 2 3 3 4
2 | 5 7
3 | 1 4 4
3 | 5 5 6 7 8 9 9
4 | 1 3 4
4 | 5 8 9
5 |
5 |
6 | 2
6 |
7 | 3
```

SOLUTION

There are $n = 33$ scores, so the position of the median is

$$\frac{n+1}{2} = \frac{33+1}{2} = 17$$

Counting down to the 17th score in the stem and leaf plot, we find

$$M = 34$$

Because the integer part of 17 is 17, the quartiles are located in position $(17 + 1)/2 = 9$ of the ordered stem and leaf plot. Thus,

$$Q_1 = 21 \quad \text{and} \quad Q_3 = 39$$

Summarizing in a five-number summary diagram, we have:

Low	Q_1	M	Q_3	High
10	21	34	39	73

C O M P U T E R   T I P

The measures of location covered in this section can be easily obtained with the aid of a computer. To illustrate, the lead absorption data in Example 2.13 have been entered into the Minitab worksheet **Lead.mtw**. At the MTB> prompt in the Session window, type

> MTB> **Describe 'exposed'**

(Note that column 2 is named 'exposed'; otherwise, type Describe C2.)

In Menu mode under **Stat**, select **Basic Statistics** followed by **Descriptive Statistics**. When the Descriptive Statistics window appears, **Select** the variable and click on **OK**.

The output in Exhibit 2.4 illustrates the Describe command.

EXHIBIT 2.4

```
MTB > Describe 'exposed'.

Descriptive Statistics

Variable      N    Mean   Median   TrMean   StDev   SEMean
exposed      33   31.85    34.00    30.79   14.41     2.51

Variable    Min     Max       Q1       Q3
exposed   10.00   73.00    20.50    40.00
```

The first three entries computed by Minitab (see Exhibit 2.4) are the sample size N, the Mean, and the Median. TrMean is a 5% trimmed mean. The two descriptive statistics, StDev and SEMean, will be discussed later. Finally, the Describe command produces the minimum and maximum scores and the first and third quartiles.

Note: The Describe command in Minitab uses a formula for Q_1 and Q_3 that is slightly different from the formula that we use. The Minitab command LValues, covered next, computes the quartiles the same way that we have presented. Minitab distinguishes between the two by referring to the latter as *hinges*. For a more thorough discussion of the difference between hinges and quartiles, consult the lab manual.

Other Measures of Location

In addition to the quartiles, Q_1 and Q_3, which identify the lower and upper fourths of the data, Minitab can find summary measures that locate the lower and upper eighths, the lower and upper sixteenths, and even the lower and upper thirty-seconds, depending on the size of the data set. These quantities, called *letter values*, are helpful in describing certain characteristics of a distribution.

COMPUTER TIP

To find the letter values for the variable 'exposed' in the lead absorption data, at the MTB> prompt in the Session window, type

MTB> **LVals 'exposed'**

In Menu mode under **Stat**, select **EDA** and click on **Letter values**. When the Letter Values window appears, **Select** the variable and click on **OK**.
The output in Exhibit 2.5 follows the LVals command.

EXHIBIT 2.5   MTB > LVals 'exposed'.

Letter Value Display

	DEPTH	LOWER	UPPER	MID	SPREAD
N=	33				
M	17.0	34.000		34.000	
H	9.0	21.000	39.000	30.000	18.000
E	5.0	16.000	45.000	30.500	29.000
D	3.0	14.000	49.000	31.500	35.000
C	2.0	13.000	62.000	37.500	49.000
	1	10.000	73.000	41.500	63.000

The resulting display in Exhibit 2.5 is called a *quantile summary diagram*. The first column produced by the LVals command is DEPTH, which gives the positions of the letter values in the ordered data set. Next are the letter values. They include the median M, the lower and upper hinges labeled H (we call them quartiles Q_1 and Q_3), the lower and upper eighths labeled E, the lower and upper sixteenths labeled D, the lower and upper thirty-seconds labeled C, and finally the minimum and maximum values in the data set.

By averaging the lower and upper letter values, we get the *midsummary statistics*. These values, under the column labeled MID, are additional measures of the center of a data set and can be used to assess the shape of the distribution. The average of Q_1 and Q_3 is called the *midquarter* or midQ. The average of the lower and upper eighths is called the midE, and so on. The average of the two extremes, low and high, is called the *midrange* or midR. Because the median is already a midsummary, it is included in the list.

Finally, LVals computes five measures of spread (also called measures of variability), which are covered in the next section.

EXERCISES 2.4

2.59   Consider the following data:

   19   22   34   28   18   24   16   25   27   31

   a. Find the mean and the 10% trimmed mean.
   b. Change the 34 to 54 and recompute the mean and trimmed mean. Explain what happened.
   c. Suppose the 16 is changed to 10. Will the mean change? Will the trimmed mean change?

2.60   You are given the following data:

   215   247   169   210   242   198   184   275   170   211

   a. Find the mean, median, and the 20% trimmed mean.
   b. If the 169 and 170 are both changed to 180, what will happen to the values of the mean, median, and trimmed mean?
   c. If the four lowest scores are changed to 180 and the four highest scores are changed to 220, what will happen to the values of the mean, median, and trimmed mean?

2.61   A test to measure aggressive tendencies was given to a group of teenage boys who were members of a street gang. The test is scored from 10 to 50, with a high score indicating more aggression.

   Worksheet: Aggress.mtw

   38   27   44   39   41   26   35   45   39   28
   16   37   11   36   33   46   42   37   40
   19   24   29   32   34   31   30   32   43

   a. Find the mean and the 10% trimmed mean.
   b. Subtract 10 from each score and recompute the mean and trimmed mean. Explain what happened.
   c. Describe, in general, the effect on the mean and trimmed mean if a constant is subtracted from or added to each score.

2.62   Construct a five-number summary diagram for the data given in Exercise 2.61. Include the midsummary statistics. What effect does subtracting a constant from each score have on the midsummaries?

2.63   The Tennessee Self-Concept Scale (a test to measure one's self-confidence) was given to the group of teenage boys in Exercise 2.61. The following scores resulted:

Worksheet: Concept.mtw

```
26  19  23  27  24  33  25  29  14  30
20  25   5  18   7  28  31  37  28
 3  20  25  45  29  22  41  34  22
```

a. Find the mean and the 20% trimmed mean.
b. Divide each score by 10 (multiply by .1) and recompute the mean and the 20% trimmed mean. What was the effect?
c. Describe, in general, the effect on the mean and trimmed mean if each score is multiplied or divided by a constant.

2.64 Construct a five-number summary diagram for the data given in Exercise 2.63. Include the midsummary statistics. What effect does multiplying each score by .1 have on the midsummaries?

2.65 A class of 30 fifth-graders completed a standardized reading test and got the following scores:

Worksheet: Reading.mtw

```
86  103   92  115   94  102  123   81  108   93
97  105   73   94  117   83   99  101   98   94
48  106  100  134   98  149   95   67  102  107
```

Summarize the data in a five-number summary diagram.

2.66 The average temperature of the 24 space shuttle launches previous to the *Challenger* accident was found to be 70 degrees in Example 2.12. The temperature at liftoff on the day of the *Challenger* accident was 31 degrees. What effect does including 31 degrees in the data set in Example 2.12 have on $\bar{y}$?

2.67 Following are the miles per gallon in city driving for the ten most fuel-efficient cars in 1995:

Worksheet: Mpgcity.mtw

Model	MPG
Honda Civic HB VX[1]	47
Honda Civic HB VX	44
Geo Metro[1]	44
Suzuki Swift[1]	44
Honda Civic[1]	42
Honda Civic	40
Geo Metro	39
Suzuki Swift	39
Ford Aspire[1]	36
Honda Civic Del Sol	35

[1] With shift indicator lights.
SOURCE: EPA.

Find the mean and median miles per gallon. Is there much difference between the two? Would a trimmed mean be much different from the mean or median?

2.68 Using 1990 census data, American Demographics, Inc., identified 20 of the fastest growing, wealthiest, and most educated counties in the country. Following is the median household income for each of the counties:

Worksheet: County.mtw

Douglas, CO	$51,864
Fayette, GA	50,187
Fort Bend, TX	42,808
Howard, MD	54,407
Loudoun, VA	52,210
Shelby, AL	36,851
Prince William, VA	49,370
Chesterfield, VA	43,603
Dakota, MN	42,218
Williamson, TN	43,612
Hamilton, IN	45,747
Rockingham, NH	41,880
Washington, MN	44,120
Delaware, OH	37,895
Hunterdon, NJ	54,661
Chester, PA	45,642
Somerset, NJ	55,566
Jefferson, CO	39,084
Saratoga, NY	36,635
Olmsted, MN	35,788

SOURCE: *Wall Street Journal*, March 8, 1994.

a. Find the mean and the 10% trimmed mean of the median incomes of the 20 counties.
b. Construct a five-number summary diagram for the data.
c. Calculate the midsummary statistics for the data. How do they compare to the mean and the trimmed mean found in part **a**?

2.69   Following are the 1990 populations of the 20 counties given in Exercise 2.68:

Worksheet: County.mtw

60,391	62,415	225,421	187,328	86,129
99,358	215,686	209,274	275,227	81,021
108,936	245,845	145,896	66,929	107,776
376,396	240,279	438,430	181,276	106,470

Find the mean population size and the median population size. Is there much difference between the two? Why would there be a difference between the two values? (*Hint*: Draw a dotplot of the data.)

2.70   In Exercise 2.52 in Section 2.3, a tree farmer divided his farm into 1,000 small grids, randomly selected 20 grids, and found the following numbers of trees in the 20 grids:

Worksheet: Trees.mtw

81	96	87	83	99	64	77	63	93	84
102	68	94	81	70	84	92	109	74	86

Calculate the average number of trees per grid. Can you use this figure to estimate how many trees the farmer has on his farm? Explain.

2.71   You were asked in Exercise 1.27 of Section 1.2 to collect data on the crime rates of counties in your state and store them in your computer.
    a. Recall those crime rates and use the Describe command to calculate summary measures.
    b. Use the LValues command to calculate a quantile summary diagram.

2.72   You were asked in Exercise 1.28 of Section 1.2 to collect data on the number of viewers of the top ten television shows from the Nielsen ratings and store them in your computer.
    a. Recall the data and use the Describe command to calculate summary measures.
    b. Use the LValues command to calculate a quantile summary diagram.

2.73   The number of hazardous waste sites discussed in Exercise 2.15 in Section 2.1 is stored in worksheet: **Toxic.mtw.**
    a. Recall the data and use the Describe command to calculate summary measures on the number of sites and on the number of sites near minority communities.
    b. Use the LValues command to calculate quantile summary diagrams for both columns.

2.74   The cholesterol levels of 62 subjects in the Framingham Heart Study are stored in worksheet: **Framingh.mtw**.
    a. Recall the worksheet and use the Describe command to compute summary statistics.
    b. Use the LValues command to calculate a quantile summary diagram.

2.75   The survival times in weeks for 20 rats exposed to high levels of radiation are given here:

Worksheet: Rat.mtw

152	152	115	109	137	88	94	77	160	165
125	40	128	123	136	101	62	153	83	69

SOURCE: J. Lawless, *Statistical Models and Methods for Lifetime Data*, (New York: Wiley, 1982).

Find the mean, median, and 10% trimmed mean survival time.

2.76   A treatment group and a control group of quail were fed a special diet; the treatment group received a drug compound that should reduce levels of low-density lipoproteins. From the following plasma LDL levels, compute summary statistics for each group. How much difference do you find between the means of the two groups? Do you think the drug compound reduced the levels of low-density lipoproteins?

Worksheet: Quail.mtw

Placebo

64	49	54	64	97	66	76	44	71	89
70	72	71	55	60	62	46	77	86	71

Treatment

40	31	50	48	152	44	74	38	81	64

SOURCE: J. McKean, and T. Vidmar, "A Comparison of Two Rank-Based Methods for the Analysis of Linear Models," *The American Statistician* 48 (1994): 220–229.

2.77   Expand your data set on the Nielsen ratings by including all shows that were rated last week in a statistical worksheet (such as Minitab). Record the network (ABC, NBC, CBS, or Fox) in one column and the Nielsen rating in a second column. Obtain the overall average rating for each of the networks. Which network had the highest average rating?

2.5   SUMMARY MEASURES OF VARIABILITY

Measures of central location are important because they describe a "typical" observation in the data set. Not all observations are typical; in fact, most deviate from the center in some way. The amount of deviation from the center is an important consideration when we are investigating the properties of a data set.

For example, suppose that two patients in a hospital have their heart rates measured every 4 hours. These are the rates over a 24-hour period:

Patient A: 68, 70, 69, 70, 71, 72

Patient B: 65, 85, 90, 65, 55, 60

Using a calculator, we can verify that for both patients the average heart rate is 70, but there is a difference between the heart rates of the two patients. The Minitab dotplots, graphed on the same axis in Exhibit 2.6, illustrate that patient A's rate is stable but patient B's rate probably needs to be monitored more closely. Clearly, the difference is that the observations in the data set for patient B are more *spread out* from the mean than those in the data set for patient A.

EXHIBIT 2.6   Minitab dotplots of heart rates

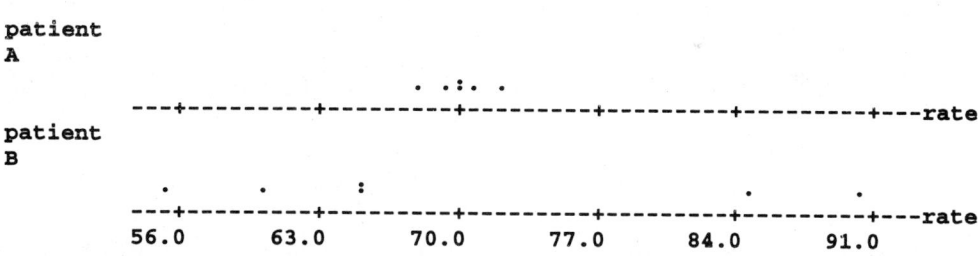

```
Character Dotplot


patient
A
                                  . .:. .
     ---+---------+---------+---------+---------+---------+---rate
patient
B
           .         .        :                   .        .
     ---+---------+---------+---------+---------+---------+---rate
       56.0      63.0      70.0      77.0      84.0      91.0
```

Range

This concept of spread (or variability) is an extremely important characteristic of a data set. The simplest measure of variability is the *range*.

> The **range** is the highest measurement minus the lowest measurement, $H - L$.

The range of heart rates for patient A is $72 - 68 = 4$ and for patient B, $90 - 55 = 35$. This demonstrates that patient B's rate is *more variable* than that of patient A.

The population range applies to the entire population, and the sample range applies to the sample. On a 10-point quiz, the maximum possible is 10 and the minimum possible is 0; thus, the population range is $10 - 0 = 10$. On the other hand, if the highest score in a sample is 8 and the lowest score is 3, then the sample range is only $8 - 3 = 5$.

Standard Deviation and Variance

The range is a useful measure of variability for small data sets. For large data sets, however, a more sensitive measure of variability is needed. The most widely accepted measure of dispersion is the *standard deviation*. To obtain the standard deviation, we must first find the *variance*.

> The **population variance**, σ^2, is the average squared distance of all measurements from the population mean.
>
> The **sample variance**, s^2, is the average squared distance of the sample values from the sample mean. It is calculated with the formula
>
> $$s^2 = \frac{\Sigma(y_i - \bar{y})^2}{n - 1}$$
>
> The expression in the numerator is referred to as a **sum of squares**, which measures the *total deviation* of all the data. It is denoted as
>
> $$SS = \sum(y_i - \bar{y})^2$$

Sum of Squares   The difference $(y_i - \bar{y})$ in the expression for the sum of squares SS is called a *deviation* from the mean. It represents how far the observation y_i is from $\bar{y}$. Figure 2.24 shows the deviations from the mean for the data considered in Figure 2.19. Because $\bar{y}$ is the average of all the measurements and is somewhere in the middle of the data, there are both positive and negative deviations from the mean. The two deviations below the mean are -33 and -9, and the three deviations above the mean are $+11$, $+14$, and $+17$. Because $\bar{y}$ is the balance point of the data, the negative deviations cancel the positive deviations, and consequently the sum of the deviations is zero $(-33 - 9 + 11 + 14 + 17 = 0)$. The *squared* deviations (used in finding the sum of squares), however, do not sum to zero and thus give a meaningful measure of how variable the data are about the mean. In this case, the sum of squares is

$$SS = (-33)^2 + (-9)^2 + (+11)^2 + (+14)^2 + (+17)^2 = 1,776$$

FIGURE 2.24   Deviations from the mean

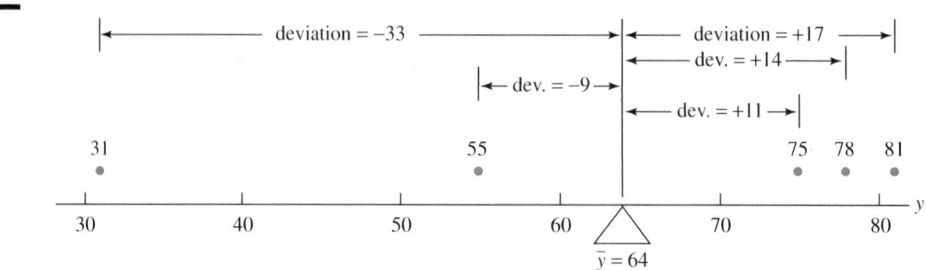

Degrees of Freedom   We find the *average* of the sum of squares, the sample variance, by dividing by $n - 1$ rather than by n for two reasons. First, the deviations always sum to zero, so any one of the deviations can be found from the other $n - 1$ deviations. Thus, the value of the sum of squares depends on only the $n - 1$ deviations that are *free* to vary from one sample to the next. We say that the sum of squares has $n - 1$ *degrees of freedom*. Second, we divide by $n - 1$ because a denominator of n would give a sample variance that tends to underestimate the population variance. Dividing by $n - 1$ makes the sample variance a little larger and a better estimate of the population variance. More will be said about this in Chapter 6.

Because the sample variance comes from a sum of squares, it is in units that are the square of the original units of measurement. That is, if we originally measured our data in miles, then s^2 is in miles squared. To restore the original unit of measure, we define the *standard deviation* as the square root of the variance.

In either case, population or sample, the positive square root of the variance is called the **standard deviation**. The population standard deviation is denoted by σ. The sample standard deviation is given by

$$s = \sqrt{\frac{\Sigma(y_i - \bar{y})^2}{n - 1}} = \sqrt{\frac{SS}{n - 1}}$$

EXAMPLE 2.19

Calculate the sample standard deviation of the heart rate data for the two patients given at the beginning of this section:

Patient A: 68, 70, 69, 70, 71, 72

Patient B: 65, 85, 90, 65, 55, 60

SOLUTION

It was determined that the average heart rate is $\bar{y} = 70$ for both patients. The deviation from the mean for a particular observation is found by subtracting the mean from the observation. The deviations from the mean for the six observations for patient A are

Heart rate	68	70	69	70	71	72
Deviations	-2	0	-1	0	$+1$	$+2$

A negative deviation indicates that the score is below the mean, and a positive deviation indicates that the score is above the mean. Notice that the deviations sum to zero.

The sum of the squared deviations is:

$$SS_A = (-2)^2 + (0)^2 + (-1)^2 + (0)^2 + (+1)^2 + (+2)^2 = 10$$

(Because we have two groups, we use subscripts A and B to distinguish between them.) Thus, the standard deviation is

$$s_A = \sqrt{\frac{10}{5}} = \sqrt{2} = 1.414$$

For patient B, the deviations from the mean are

Heart rate	65	85	90	65	55	60
Deviations	-5	$+15$	$+20$	-5	-15	-10

Notice that these deviations from the mean are much larger than those for patient A, indicating that the data are much more spread out. Again, notice that the deviations sum to zero.

The sum of the squared deviations is

$$SS_B = (-5)^2 + (15)^2 + (20)^2 + (-5)^2 + (-15)^2 + (-10)^2 = 1,000$$

The standard deviation is

$$s_B = \sqrt{\frac{1,000}{5}} = \sqrt{200} = 14.142$$

which is much larger (ten times greater) than the standard deviation of the heart rate for patient A.

In Example 2.19, there were six measurements for each of the two patients and consequently six deviations from the mean. Suppose, for example, that the -10 deviation in the sample for patient B is unknown. Because the first five deviations sum to $+10$, we know that the unknown deviation must be -10. That is, we need to know only five of the deviations to calculate the standard deviation. In other words, each of the above standard deviations has 5 degrees of freedom.

Interpreting the Standard Deviation

Interpreting the standard deviation s is not as easy as, say, interpreting the mean as the balance point of the distribution. As we saw in Example 2.19 when we compared two data sets, a larger value of s in one sample reflects greater variation of the observations from the mean than in the other sample. But how do we interpret variability for a single set of data?

One way is to indicate the percent of the data that is within a specified number of standard deviations of the mean. For example, what percent of the distribution is within one standard deviation of the mean? Within two standard deviations of the mean? The answer, of course, depends on the shape of the distribution. If the frequency distribution is bell-shaped, then the *empirical rule* gives the approximate percentages of the distribution that lie within one, two, and three standard deviations of the mean. Theoretically, this rule applies to population distributions, but it gives a reasonable approximation for any bell-shaped data. We state it here as it applies to sample data.

Empirical Rule Applied to Sample Data

If a stem and leaf plot, histogram, or similar descriptive tool has a bell-shaped appearance, then:

1. Approximately 68% of the measurements will fall within one standard deviation of the mean. The boundaries are $\bar{y} \pm s$.
2. Approximately 95% of the measurements will fall within two standard deviations of the mean. The boundaries are $\bar{y} \pm 2s$.
3. Essentially all the measurements will fall within three standard deviations of the mean. The boundaries are $\bar{y} \pm 3s$.

For distributions that are not necessarily bell-shaped—in fact, for any shape of distribution—we have Chebyshev's rule.

Chebyshev's Rule

Regardless of the shape of the distribution, we have:

1. At least $\frac{3}{4}$ of the measurements will fall within two standard deviations of the mean. The boundaries are $\bar{y} \pm 2s$.
2. At least $\frac{8}{9}$ of the measurements will fall within three standard deviations of the mean. The boundaries are $\bar{y} \pm 3s$.

EXAMPLE 2.20

The recorded temperatures on the 24 launches previous to the *Challenger* accident are given here in a stem and leaf plot. Calculate the mean and the standard deviation and use them to give an interpretation of the amount of variability in the data using either the empirical rule or Chebyshev's rule.

```
5 | 3
5 | 7 8
6 | 3
6 | 6 7 7 7 8 9
7 | 0 0 0 0 2 3
7 | 5 5 6 6 8 9
8 | 0 1
```

SOLUTION

Previously the mean was found to be $\bar{y} = 70$. With a hand-held calculator, we find $s = 7.2$. From the stem and leaf plot, it appears that the data are somewhat bell-shaped, so we apply the empirical rule, which says that approximately 68% of the measurements should be between

$$\bar{y} - s \quad \text{and} \quad \bar{y} + s$$

which is between

$$70 - 7.2 = 62.8 \quad \text{and} \quad 70 + 7.2 = 77.2$$

We find that 17 of the 24 observations are between 62.8 and 77.2. Thus, the actual percent, $\frac{17}{24} = 70.8\%$, is very close to the specified 68%.

The empirical rule further states that approximately 95% of the measurements should be between

$$\bar{y} - 2s \quad \text{and} \quad \bar{y} + 2s$$

which is between

$$70 - 2(7.2) = 55.6 \quad \text{and} \quad 70 + 2(7.2) = 84.4.$$

All but one of the scores fall between 55.6 and 84.4. The actual percent, $\frac{23}{24} = 95.8\%$, is only slightly greater than the 95% stated by the empirical rule.

Finally, virtually all measurements are between

$$\bar{y} - 3s \quad \text{and} \quad \bar{y} + 3s$$

or between

$$70 - 3(7.2) = 48.4 \quad \text{and} \quad 70 + 3(7.2) = 91.6$$

Thus, the data suggest that almost always the launch temperature is somewhere between 48.4 degrees and 91.6 degrees. What does this say about the 31 degrees on the day of the launch of the *Challenger*?

Relative Standing

In Example 2.20, we observed that 31 degrees on the day of the launch of the *Challenger* was an unusually low temperature. Just how low is it? When 31 degrees is included in the data set, the average becomes $\bar{y} = 68.44$ and the standard deviation becomes $s = 10.53$. (Notice how much larger s is when the outlier is included.) To evaluate a single score, such as the 31 degrees, we calculate its *z-score* (sometimes called a *standard score*), which gives its *relative standing* with respect to the other data.

The **z-score** corresponding to a particular observation is given by

$$z = \frac{\text{observation} - \text{mean}}{\text{standard deviation}}$$

It gives the number of standard deviations the observation is from the mean.

We find the z-score for 31 degrees to be

$$z = \frac{31 - 68.44}{10.53} = -3.56$$

A negative z-score indicates that the observation is below the mean. Here we see that 31 is 3.56 standard deviations below the mean. It is generally agreed that any observation with a z-score greater than 3 in absolute value is an outlier. Thus, 31 degrees is an outlier representing an unusually low temperature.

z-scores can also be used to compare the relative standings of observations from different data sets that have different means and standard deviations. For example, if you wish to compare your test score on a statistics exam with a friend's test score in another class, you would compare the two corresponding z-scores.

We saw in Section 2.4 that outliers on *one* tail of a distribution tend to have an adverse effect on the value of $\bar{y}$ (its value is pulled toward the long tail). The standard deviation is also extremely sensitive to outliers. Example 2.21 shows that if the distribution has *symmetrical* long tails, then the sample mean is not distorted (the two long tails tend to cancel each other's effect) but the standard deviation, s, is severely affected by the long tails.

EXAMPLE 2.21

The following data are standardized reading scores for a group of 30 fifth-graders. Calculate $\bar{y}$ and s. Then recalculate $\bar{y}$ and s after three observations are trimmed from each end of the sample.

Worksheet: Reading.mtw

Ordered stem and leaf plot

```
 4 | 8
 5 |
 6 | 7
 7 | 3
 8 | 1 3 6
 9 | 2 3 4 4 4 5 7 8 8 9
10 | 0 1 2 2 3 5 6 7 8
11 | 5 7
12 | 3
13 | 4
14 | 9
```

Trimmed stem and leaf plot

```
 8 | 1 3 6
 9 | 2 3 4 4 4 5 7 8 8 9
10 | 0 1 2 2 3 5 6 7 8
11 | 5 7
```

SOLUTION

Using the complete data set and a calculator, we find

$$\bar{y} = 98.8 \quad \text{and} \quad s = 18.92$$

For the trimmed data, we find

$$\bar{y}_T = 98.75 \quad \text{and} \quad s_T = 8.79$$

The long tails do not affect the value of $\bar{y}$, but s is more than halved when the data are trimmed.

The calculation of $\bar{y}_T$ above shows the correct way to find a 10% trimmed mean; however, we will see later that if we wish to calculate a "trimmed" standard deviation, it is not enough simply to calculate the standard deviation of the trimmed sample as has been done here.

Other Measures of Variability

Q-Spread

Another measure of variability that is *resistant* to the influence of outliers can be obtained from the quartiles in the five-number summary diagram.

> The **Q-spread** is the distance between the first and third sample quartiles, $Q_3 - Q_1$.
>
> The corresponding population **q-spread** is similarly defined using the population quartiles in place of the sample quartiles.

EXAMPLE 2.22

The following five-number summary diagram is created from the reading scores in Example 2.21. Verify that it is correct and then find the Q-spread and the range.

Low	Q_1	M	Q_3	High
48	93	98.5	106	149

SOLUTION

The Q-spread is the difference between Q_3 and Q_1, so

$$Q\text{-spread} = Q_3 - Q_1 = 106 - 93 = 13$$

The range is the difference between the high and low scores, so

$$\text{Range} = H - L = 149 - 48 = 101$$

Notice how much larger the range is than the Q-spread. This is because of the extremely long tails of the distribution. In other words, the Q-spread is not affected by (is resistant to) the long tails, but the range is severely affected.

C O M P U T E R   T I P

The two Minitab commands, Describe and LVals, covered in the Computer Tip in the previous section will produce the measures of variability described in this section.

StDev, produced by the Describe command and illustrated in Exhibit 2.4, is the standard deviation, s. The first entry under the column SPREAD in the LVals command, illustrated in Exhibit 2.5, is the Q-spread, and the last entry is the range.

Interquartile Range

Recall that the Describe command in Minitab computes the quartiles with a formula that is slightly different from the formula that we use. The hinge values computed with the LVals command are the same as our quartiles, Q_1 and Q_3. To avoid confusion, we will refer to the difference between the quartiles produced by the Describe command as the *interquartile range, IQR*. Because of the way it is computed, the IQR will always be slightly larger than the Q-spread.

In addition to the Q-spread and range, the LVals command calculates spreads based on all of the letter values. For example, the E-spread is the difference between the upper and lower eighths, the D-spread is the difference between the upper and lower sixteenths, and so on. Exhibit 2.7 illustrates the spread values produced by LVals for the reading scores in Example 2.21. Notice that Q-spread $= 13$ and range $= 101$ are the values found in Example 2.22. Additionally we get E-spread $= 34$, D-spread $= 58.5$, and C-spread $= 84$.

EXHIBIT 2.7   Letter Value Display

	DEPTH	LOWER	UPPER	MID	SPREAD
N=	30				
M	15.5		98.500	98.500	
H	8.0	93.000	106.000	99.500	13.000
E	4.5	82.000	116.000	99.000	34.000
D	2.5	70.000	128.500	99.250	58.500
C	1.5	57.500	141.500	99.500	84.000
	1	48.000	149.000	98.500	101.000

E X E R C I S E S   2.5

2.78   Consider the following data:

7   12   10   9   22

a. Calculate the deviations from the mean and check that they sum to zero.
b. Calculate the sum of squares.
c. Calculate the standard deviation.

2.79   You are given the following data:

6.8   5.7   9.2   8.4   7.4

a. Calculate the deviations from the mean and check that they sum to zero.
b. Calculate the sum of squares.
c. Calculate the standard deviation.

2.80   The average for the following data is 27. Calculate the total variability of the data by finding the sum of squares. Divide the sum of squares by its degrees of freedom and find s^2. Verify the answer with your calculator. (Remember that most calculators calculate the standard deviation, which you will need to square to get the variance, s^2.)

30   18   35   22   36   18   25   31   28

2.81   A standardized math test is given to 30 students in the tenth grade. Their scores are listed here:

Worksheet: Math.mtw

44   49   62   45   51   59   57   55   70   64
54   58   65   75   43   42   67   63   71   54
60   53   40   49   52   50   54   61   42   38

Construct a stem and leaf plot of the data. Do you think that the curve of data is bell-shaped? If so, interpret the amount of variability in the data using the empirical rule; otherwise, use Chebyshev's rule.

2.82   Following are test scores for two beginning statistics classes:

Worksheet: Statisti.mtw

Class 1
81   73   86   90   75   80   75   81   85   87   83
75   70   65   80   76   64   74   86   80   83
67   82   78   76   83   71   90   77   81   82
Class 2
87   77   66   75   78   82   82   71   79
73   91   97   89   92   75   89   75   95
84   75   82   74   77   87   69   96   65

Construct back-to-back stem and leaf plots (see Exercise 2.57 in Section 2.3) and compare the standard deviations of the two classes. Is the variability exhibited in the stem and leaf plots depicted in the values of the two standard deviations?

2.83   Each of the two statistics classes given in Exercise 2.82 had students who scored 82. Determine which student (the one in class 1 or the one in class 2) scored higher relative to the class.

2.84   How much does the estimated worth of a National Football League team vary from one team to another? Following are the estimated worths (in $ millions), computed by *Financial World* magazine:

Worksheet: Football.mtw

Team	Worth	Team	Worth
Arizona	146	Atlanta	148
Buffalo	164	Chicago	160
Cincinnati	142	Cleveland	165
Dallas	190	Denver	147
Detroit	138	Green Bay	141
Houston	157	Indianapolis	141

(continued)

Team	Worth	Team	Worth
Kansas City	153	L.A. Raiders	146
L.A. Rams	148	Miami	161
Minnesota	147	N.Y. Giants	176
N.Y. Jets	142	New England	142
New Orleans	154	Philadelphia	172
Pittsburgh	143	San Diego	142
San Francisco	167	Seattle	148
Tampa Bay	142	Washington	158

SOURCE: *Financial World*, May 10, 1994.

Construct a stem and leaf plot of the data. Based on the shape of the stem and leaf plot, decide whether an interpretation of the standard deviation should be based on the empirical rule or Chebyshev's rule. Calculate the average and the standard deviation, and give your interpretation.

2.85   Construct a five-number summary diagram for the worth of the NFL teams in Exercise 2.84. Find the Q-spread and the range. What, in the data, would cause the large difference between the Q-spread and the range?

2.86   The Insurance Institute of Highway Safety claims that car bumpers are getting worse. They crashed 22 midsize four-door sedans into barriers four times each at 5 miles per hour and totaled the damage. Construct a stem and leaf plot of the data. Calculate the average repair cost and the standard deviation and give an interpretation based on the empirical rule. What percent of the actual data fall within one standard deviation of the mean? Is this to be expected according to the empirical rule? Which do you think is more appropriate for interpreting the standard deviation, the empirical rule or Chebyshev's rule?

Worksheet: Bumpers.mtw

Car	Repair cost	Car	Repair cost
Honda Accord	$ 618	Chevrolet Cavalier	$ 795
Toyota Camry	1,304	Saturn SL2	1,308
Mitsubishi Galant	1,340	Dodge Monaco	1,456
Plymouth Acclaim	1,500	Chevrolet Corsica	1,600
Pontiac Sunbird	1,969	Oldsmobile Calais	1,999
Dodge Dynasty	2,008	Chevrolet Lumina	2,129
Ford Tempo	2,247	Nissan Stanza	2,284
Pontiac Grand Am	2,357	Buick Century	2,381
Buick Skylark	2,546	Ford Taurus	3,002
Mazda 626	3,096	Oldsmobile Ciera	3,113
Pontiac 6000	3,201	Subaru Legacy	3,266
Hyundai Sonata	3,298		

SOURCE: Insurance Institute of Highway Safety, 1991.

2.87   Arrange the results from Exercise 2.86 in a five-number summary diagram. Find the midQ and the midRange for the repair costs. There is very little difference between the median and the midQ. Can you explain why?

2.88   A social scientist examines the records at several hospitals and finds the following sample of ages of women at the birth of their first child:

Worksheet: Firstchi.mtw

```
30  18  35  22  23  22  36  24  23  28  19
23  25  24  33  21  24  19  33  23  19  32
21  18  36  21  25  17  21  24  39  22  23
18  22  28  18  15  25  21  23  26  38  24
20  36  27  21  28  26  22  28  33  18  17
21  15  20  16  21  23  15  20  38  16  24
42  22  24  24  20  17  26  39  22  21  28
20  29  14  25  20  19  17  21  24  26
```

Calculate the average age and the standard deviation and give an interpretation based on the empirical rule. What percent of the actual data fall within one standard deviation of the mean? Within two standard deviations of the mean? Are these percentages consistent with the empirical rule?

2.89   For the data in Exercise 2.88, construct a five-number quantile summary diagram. Examine the data. What characteristic in the data would cause the median and the midQ to be the same? Why is the midRange larger than the median and the midQ?

2.90   You were asked in Exercise 1.27 of Section 1.2 to collect data on the crime rates of counties in your state and store them in your computer. Recall those crime rates into your computer worksheet.
   a. Construct a stem and leaf plot. Is the distribution of the data reasonably bell-shaped so that the empirical rule applies?
   b. Calculate the mean and standard deviation (see Exercise 2.71 in Section 2.4) and give an interpretation based on either the empirical rule or Chebyshev's rule.

2.91   You were asked in Exercise 1.28 of Section 1.2 to collect data on the number of viewers of the top ten television shows from the Nielsen ratings and store them in your computer. Recall the data into your computer worksheet.
   a. Construct a stem and leaf plot. Is the distribution of the data reasonably bell-shaped so that the empirical rule applies?
   b. Calculate the mean and standard deviation (see Exercise 2.72 in Section 2.4) and give an interpretation based on either the empirical rule or Chebyshev's rule.

2.92   The number of hazardous waste sites discussed in Exercise 2.15 in Section 2.1 is stored in worksheet: **Toxic.mtw**. Recall the data into your computer worksheet.
   a. Construct a stem and leaf plot of the number of sites.
   b. Calculate the mean and standard deviation (see Exercise 2.73 in Section 2.4).
   c. What percent of the actual data fall within one standard deviation of the mean? Within two standard deviations of the mean? Is this to be expected according to the empirical rule? Which do you think is more appropriate for interpreting the standard deviation, the empirical rule or Chebyshev's rule? Why?

2.93   Review the letter value display for the hazardous waste sites data found in Exercise 2.73 in Section 2.4. Notice that the midRange is much greater than the median and the range is much greater than either the Q-spread or the E-spread. What characteristic in the data would cause this to happen?

2.94   The cholesterol levels of 62 subjects in the Framingham Heart Study are stored in **Framingh.mtw**. Recall the data into your worksheet.
   a. Construct a stem and leaf plot of the cholesterol levels.
   b. Calculate the mean and standard deviation (see Exercise 2.74 in Section 2.4).
   c. What percent of the actual data fall within one standard deviation of the mean? Within two standard deviations of the mean? Within three standard deviations of the mean? Is this to be expected according to the empirical rule? Which do you think is more appropriate for interpreting the standard deviation, the empirical rule or Chebyshev's rule? Why?

2.95   The survival times in weeks for 20 rats exposed to high levels of radiation are listed here:

Worksheet: Rat.mtw

152 152 115 109 137   88 94   77 160 165
125   40 128 123 136 101 62 153   83   69

SOURCE: J. Lawless, *Statistical Models and Methods for Lifetime Data* (New York: Wiley, 1982).

a. Find the mean and standard deviation of the survival times.
b. What percent of the actual data are within one standard deviation of the mean? Within two standard deviations of the mean? Is this to be expected according to the empirical rule? Which do you think is more appropriate for interpreting the standard deviation, the empirical rule or Chebyshev's rule? Why?
c. The longest survival time was 165 weeks. Just how large is this score relative to the rest of the data?

2.6   DESCRIBING THE SHAPE OF A DISTRIBUTION

We have seen that the *shape* of a frequency distribution affects the values of the summary measures. Later, when we discuss statistical inference, we will see that the shape of the distribution is important in determining what statistical procedures are best suited for the available data. For these reasons, it is important that we be able to classify the shape of a distribution. Two important characteristics of the shape of a distribution are *symmetric* and *skewed*.

Symmetric Distributions

A frequency distribution is said to be **symmetric** if the scores below the center of the distribution are a mirror image of the scores above the center.

The best known symmetric distribution is the *normal distribution*, illustrated by the frequency curve in Figure 2.25. It is commonly referred to as the *bell curve* and provides

FIGURE 2.25   Normal distribution frequency curve

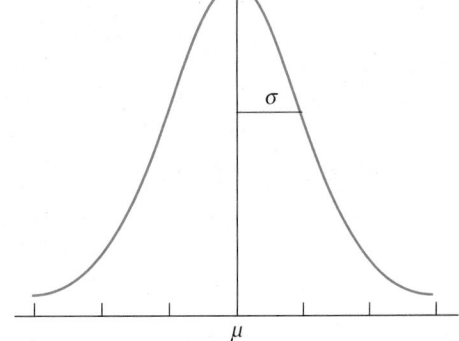

a good model for many distributions of real data. Properties of the normal distribution will be studied in Chapter 4.

When we model real data, we do not insist on a perfect model but one that is closely approximated by the distribution.

EXAMPLE **2.23**

Radiocarbon dating is a method of determining the age of archaeological sites. Shown here is a stem and leaf plot of the "ages" recorded as years before 1983 (the data were obtained in 1983)—that is, B.C. + 1983—for a sample of 23 observations taken from one site (Ceramic Phase 4) from the archaeological excavation of the Danebury iron-age hill fort. [Sources: Cunliffe (1984) and Naylor and Smith (1988)]

Worksheet: Archaeo.mtw column 4

```
Stem-and-leaf of phase 4    N = 23
Leaf Unit = 10

      1     19   0
      3     19   89
      6     20   034
      9     20   669
     (6)    21   012234
      8     21   5677
      4     22   0
      3     22   6
      2     23   0
      1     23   7
```

The stem and leaf plot is not perfectly symmetric, but it is close enough that we should investigate further before ruling out symmetry. A few more observations falling in the right places may help us conclude that the distribution of radiocarbon dates is indeed symmetric and possibly even normally distributed.

Skewed Distributions

The stem and leaf plot in Figure 2.26 of the homicide rates in major cities from Example 2.9 is markedly nonsymmetric. A distribution like this where the frequency curve has one tail longer than the other is called *skewed*.

A frequency distribution whose *left* tail is longer than the right is called **skewed left**. If the *right* tail is longer than the left, then it is called **skewed right**.

FIGURE 2.26

Stem and leaf plot of the homicide rates in 30 major cities

Worksheet: City.mtw

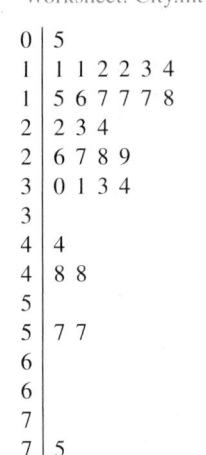

```
0 | 5
1 | 1 1 2 2 3 4
1 | 5 6 7 7 7 8
2 | 2 3 4
2 | 6 7 8 9
3 | 0 1 3 4
3 |
4 | 4
4 | 8 8
5 |
5 | 7 7
6 |
6 |
7 |
7 | 5
```

The stem and leaf plot of the homicide rates tails off to the right (toward the larger values); hence, we classify the distribution as *skewed right*. The frequency curve for the population from which this sample came might look like Figure 2.27.

FIGURE 2.27

A skewed right frequency distribution

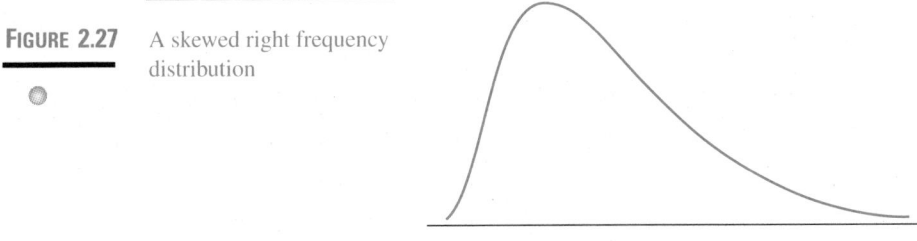

Figure 2.28 shows a *skewed left* distribution.

FIGURE 2.28

A skewed left frequency distribution

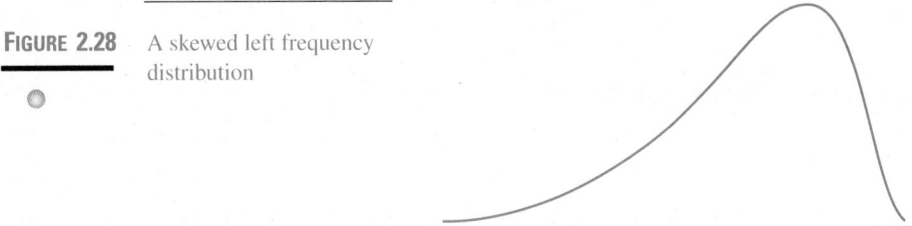

Midsummary Analysis

The midsummary statistics computed from the letter values (see Section 2.4) can help us detect symmetry and skewness. If the middle 50% of a distribution is skewed right, then the upper quartile, Q_3, will be farther from the median than the lower quartile, Q_1. As a result, the value of the midQ (the average of Q_1 and Q_3) will be greater than the median, M. Furthermore, if the middle 75% of a distribution is skewed right then the upper eighth will be farther from the median than the lower eighth, resulting in the midE being larger than the midQ, which in turn is larger than the median. Continuing with the remaining letter values, we have the rules given in the following box.

Midsummary Analysis

1. If the distribution is nearly symmetric, then the midsummaries will be nearly equal.
2. If the distribution is skewed right, then the midsummaries will become progressively larger as we scan down from the median to the range.
3. If the distribution is skewed left, then the midsummaries will become progressively smaller as we scan down from the median to the range.

The midsummaries for the homicide rates in Figure 2.26 are given in the letter value display in Exhibit 2.8. The midsummaries, under the column MID, get progressively larger as we scan down the list. This is supporting evidence that the distribution is heavily skewed right.

EXHIBIT 2.8 Letter Value Display

	DEPTH	LOWER	UPPER	MID	SPREAD
N=	30				
M	15.5	23.600		23.600	
H	8.0	15.900	33.100	24.500	17.200
E	4.5	12.700	48.500	30.600	35.800
D	2.5	11.800	57.200	34.500	45.400
C	1.5	8.750	66.300	37.525	57.550
	1	5.700	75.200	40.450	69.500

It is clear from the stem and leaf plot in Figure 2.26 that the distribution of homicide rates is skewed right, but in some situations it is not so obvious and it always helps to have supporting evidence, such as the midsummary analysis, for our conjectures about the shape of a distribution.

EXAMPLE 2.24

In Example 2.23 we were uncertain about the symmetry of the distribution of radiocarbon ages. Check the symmetry by conducting a midsummary analysis.

SOLUTION

Exhibit 2.9 is a letter value display of the data. The closeness of these midsummaries support the conjecture that the radiocarbon dates are symmetric.

EXHIBIT 2.9   Letter Value Display

	DEPTH	LOWER	UPPER	MID	SPREAD
N=	23				
M	12.0		2120.000	2120.000	
H	6.5	2050.000	2165.000	2107.500	115.000
E	3.5	1995.000	2230.000	2112.500	235.000
D	2.0	1980.000	2300.000	2140.000	320.000
	1	1900.000	2370.000	2135.000	470.000

As mentioned, the normal distribution is the most familiar distributional shape one will encounter. It is the standard with which other population distributions are compared. In fact, if a distribution is not skewed but symmetric, then the length of its tails can be compared with the length of the tails of the normal distribution.

> A symmetric distribution is called **short-tail** if the tails of its frequency curve drop off more rapidly than the tails of a normal curve. It is called **long-tailed** if the tails of its frequency curve are somewhat longer than the tails of a normal curve.

Figure 2.29 depicts a short-tailed and a long-tailed distribution, both compared with a normal frequency curve.

FIGURE 2.29   Short-tailed, long-tailed, and normal distributions

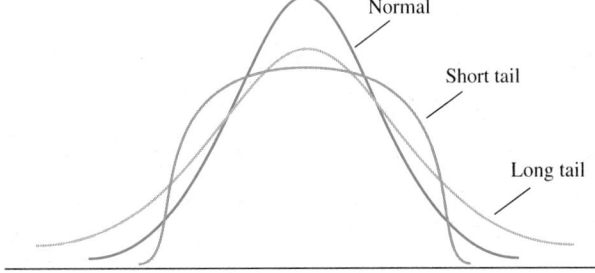

Normal

Short tail

Long tail

Boxplots

The stem and leaf plot, histogram, and dotplot all provide information about the general shape of a frequency distribution. Additionally, the midsummary analysis can be used to check for the symmetry and skewness of the distribution. We now introduce the *boxplot*, which provides an alternative view of this information as well as information about the tails of the distribution by identifying outliers in the data.

Theory has shown that $\bar{y}$ and s perform favorably as measures of the center and the amount of variability when the frequency distribution is normal in shape. As suggested in Section 2.5, however, the presence of outliers tends to inflate s and possibly distort $\bar{y}$. Therefore, it is necessary that we have a method for detecting outliers so that, if necessary, summary measures other than $\bar{y}$ and s can be investigated.

Construction of the Boxplot

The boxplot is constructed based on information in the five-number summary diagram. First, a box is constructed above a number line that extends from Q_1 to Q_3, with a line drawn inside the box at the location of the median. Whiskers (line segments) are extended out from each side of the box to the largest and smallest observations in the data set *when there are no outliers*. In the case of outliers, the whiskers extend to the largest and smallest *non*-outliers in the data set.

Procedure Steps for Construction of Boxplot

1. Draw a number line that includes the range of observations.
2. Above the line, draw a box extending from Q_1 to Q_3.
3. Inside the box, draw a line at the median.
4. To identify outliers, compute lower and upper fences, which are located at $1.5(Q\text{-spread})$ below Q_1 and above Q_3. That is, the lower fence is

$$f_L = Q_1 - 1.5(Q\text{-spread})$$

and the upper fence is

$$f_U = Q_3 + 1.5(Q\text{-spread})$$

5. Observations located beyond the fences are classified as outliers and are identified with an asterisk (*).
6. If there are no outliers, extend horizontal line segments (whiskers) from the ends of the box to the smallest and largest observations. If there are outliers, extend the whiskers to the smallest and largest non-outliers.

Following is the stem and leaf plot and the five-number summary for the space shuttle data from Example 2.16:

```
5 | 3
5 | 7 8
6 | 3
6 | 6 7 7 7 8 9
7 | 0 0 0 0 2 3
7 | 5 5 6 6 8 9
8 | 0 1
```

Low	Q_1	M	Q_3	High
53	67	70	75.5	81

Figure 2.30 shows the construction of the box that extends from Q_1 to Q_3 with the line drawn at the median.

FIGURE 2.30   Partially constructed boxplot of the space shuttle data

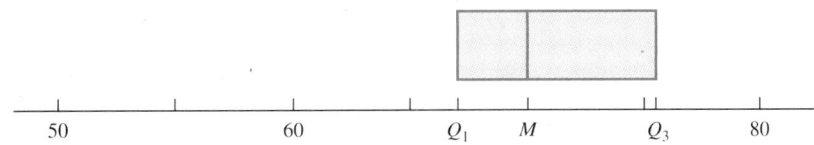

To identify possible outliers, we compute the lower and upper fences using the Q-spread, which was defined as $Q_3 - Q_1$, or 8.5:

$$f_L = Q_1 - 1.5(Q\text{-spread}) = 67 - 12.75 = 54.25$$
$$f_U = Q_3 + 1.5(Q\text{-spread}) = 75.5 + 12.75 = 88.25$$

No observations fall beyond the upper fence, 88.25; hence, the upper whisker extends to 81, the largest observation. Because the observation 53 is below the lower fence, 54.25, it is classified as an outlier, and the lower whisker extends to 57, the smallest non-outlier. Figure 2.31 is the completed boxplot.

FIGURE 2.31   Boxplot of the space shuttle data

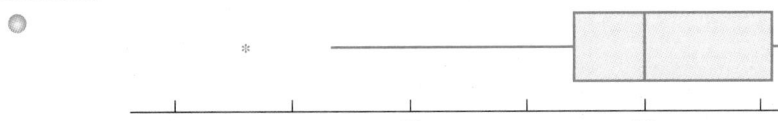

Interpretation of the Boxplot

To examine the length of the tails and classify the shape of a frequency distribution, we must fully understand the information given in a boxplot. The box portion of the boxplot contains the middle 50% of the data, with the median dividing it in half. The position of the median line gives some indication about the shape of the middle of the distribution. If the median line is close to the center of the box, then the middle 50% of the distribution is symmetric. If the median line is toward one end of the box, then the middle 50% of the distribution is skewed in the opposite direction.

The lengths of the whiskers (tails) of the boxplot give some indication about the symmetry or skewness of the rest of the data. If both whiskers are about the same length, the distribution without outliers is symmetric. If one whisker is longer than the other, the distribution is skewed in that direction. Outliers beyond a long whisker are even stronger evidence that the distribution is skewed in that direction.

An *extreme outlier* is an outlier that is more than 3(Q-spread) beyond the quartiles, Q_1 and Q_3, and is identified on the boxplot with a small circle. For a sample drawn from a normal population, one would expect to see about seven outliers out of every 1,000 observations and no more than two extreme outliers out of every 1 million observations. If more outliers than this are observed in a sample, then there is evidence that the distribution is not normal. Certainly, no extreme outliers are expected from a sample of 30 observations drawn from a normal population.

EXAMPLE 2.25

The reading scores for the group of fifth-graders from Example 2.21 are given in the ordered stem and leaf plot in Figure 2.32. The five-number summary diagram from Example 2.22 is also given. Construct a boxplot and comment on the shape of the distribution.

FIGURE 2.32

```
 4| 8
 5|
 6| 7                                5-number summary diagram
 7| 3
 8| 1 3 6                  Low    Q1     M      Q3     High
 9| 2 3 4 4 4 5 7 8 8 9     48     93    98.5   106    149
10| 0 1 2 2 3 5 6 7 8
11| 5 7                            Q-spread = 106 - 93 = 13
12| 3      fL = Q1 - 1.5(Q-spread) = 93 - 1.5(13) = 73.5
13| 4      fU = Q3 + 1.5(Q-spread) = 106 + 1.5(13) = 125.5
14| 9
```

SOLUTION

The boxplot has been drawn in Figure 2.33.

FIGURE 2.33

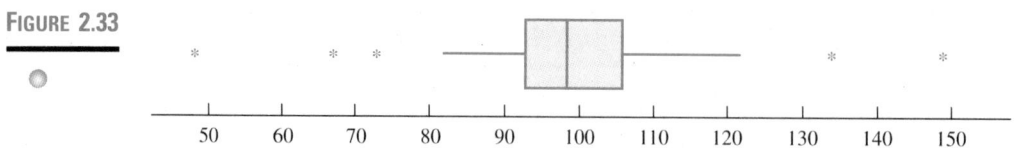

The lower fence is 73.5, so 48, 67, and 73 are outliers. The lower whisker extends down to 81, the smallest non-outlier. The upper fence is 125.5, so 134 and 149 are outliers. The upper whisker extends up to 123, the largest non-outlier. The three outliers on the lower end, the two outliers on the upper end, and the rather symmetric distribution are indicative of a **long-tailed** distribution, which, as you have seen, may distort some of the more commonly used statistical measures.

FIGURE 2.34   Boxplot with extreme outliers

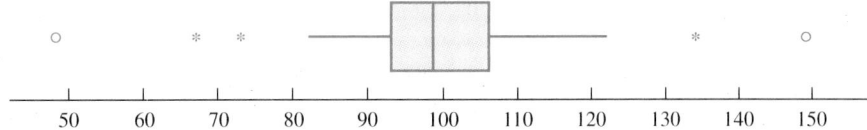

In Example 2.25, 3(Q-spread) = 39; 39 points below Q_1 is $93 - 39 = 54$ and 39 points above Q_3 is $106 + 39 = 145$. Therefore, 48 (which is below 54) and 149 (which is above 145) are extreme outliers. If we wish to identify extreme outliers, the boxplot should be revised as shown in Figure 2.34. In most applications, however, it is enough just to recognize outliers and not necessarily distinguish between mild and extreme outliers.

Character Graph Boxplots and High-resolution Boxplots

Minitab can construct two different types of boxplots. One is a character graph boxplot that is based on the Q-spread and appears in the Session window, and the other is a high-resolution boxplot based on the IQR.

The size of the box, the lengths of the tails, and the identification of outliers are all directly related to the values of the quartiles Q_1 and Q_3. Because the Q-spread and the IQR are based on different definitions of quartiles, a character graph boxplot and a high-resolution boxplot may look different. Recall that the IQR is based on quartiles produced by the Describe command and the Q-spread is based on quartiles produced by the LVals command. Also, the IQR is, in most cases, larger than the Q-spread. Consequently, the box portion of the boxplot is larger and the fences for identifying outliers are farther out in the tails of a high-resolution boxplot. As a result, it is possible that the high-resolution boxplot will identify fewer observations as outliers than will the character graph boxplot.

C O M P U T E R   T I P

To produce a character graph boxplot, select **Graph** followed by **Character Graphs** and then click on **Boxplot**. When the Boxplot window appears, **Select** the variable to be graphed and then click **OK**. The resulting boxplot, which appears in the Session window, is constructed as described above with the extreme outliers identified.

To produce a high-resolution boxplot, first select **Graph** and then click on **Boxplot**. When the Boxplot window appears, **Select** the variable to be graphed and then click **OK**. The resulting boxplot is based on the IQR instead of the Q-spread (see below) and no distinction is made between mild and extreme outliers.

If you prefer the Session mode, after entering the data into, say, column C1 of the worksheet, at the MTB> prompt give the command

MTB> **boxplot C1**

The resulting boxplot will be a high-resolution boxplot.

Figure 2.35 is a high-resolution boxplot of the reading scores from Example 2.25. Notice that only two observations on the lower tail are identified as outliers, and although observations 48 and 149 are extreme outliers, they are not identified as such by the high-resolution graph.

FIGURE 2.35   High-resolution boxplot of reading scores

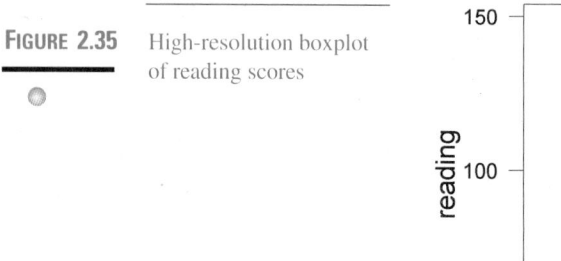

The boxplot is designed to give a general view of the shape of a distribution and for comparing two or more distributions. In most cases, the finer details of the boxplot will not be an issue, so whether you prefer a character graph boxplot or a high-resolution boxplot, it will not matter.

Side-by-Side Boxplots

Example 2.26 shows how *side-by-side boxplots* can be used to compare distributions.

EXAMPLE 2.26

A consumer group wishes to compare four brands of bathroom scales. A standard weight of 100 pounds is placed on each of the scales five different times. The resulting readings are found in worksheet **Scales.mtw**. Compare the results with side-by-side boxplots.

SOLUTION

To generate side-by-side boxplots, it is best to have the data stored in stacked form; that is, one column of the worksheet contains the data and a second column contains a code indicating the brand of bathroom scales. In the Boxplot window of Minitab, select the column containing the data in the Y (measurement) blank and select the column containing the brand in the X (category) blank. Click **OK** and the display shown in Figure 2.36 appears.

Notice that brands C and D have a median value very close to the 100-pound target, but brands A and B are off the mark. The measurements from brand D are much less variable than the measurements from brand C. We should conclude that brand D is the best because the measurements are accurate and precise. But what about brands B and C? Which would you choose if you were purchasing a bathroom scale? Although brand C has a median very close to the target of 100 pounds, it has very low precision; that is,

FIGURE 2.36

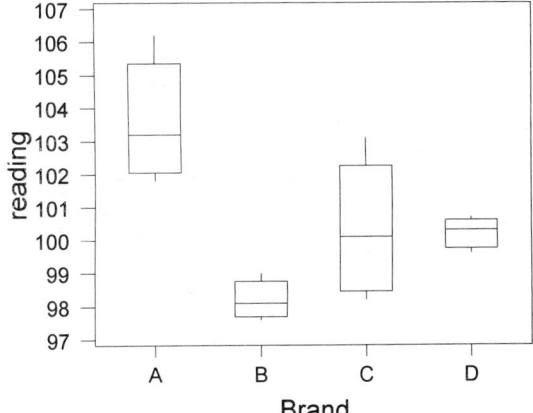

there is no consistency in the measurements. On the other hand, brand B is very precise but tends to give low measurements. If we know the bias, which in this case appears to be about -2 pounds, we can easily adjust the scales up 2 pounds and get very accurate readings. Thus, with a minor adjustment, brand B should be as good as brand D. More will be said about bias and precision in Chapter 6.

Multimodal Distributions

When we are considering the shape of a distribution, another possibility is a *multimodal distribution*.

> A distribution is called **multimodal** if its frequency curve has two or more peaks. A **mode** is the value associated with a peak.

Figure 2.37 shows a bimodal distribution. Drawing conclusions about multimodal populations can be dangerous because the modes are often the result of some extraneous variable being confounded with the data. If the extraneous variable can be identified, then perhaps the population can be separated into two or more populations and analyzed separately.

FIGURE 2.37 A bimodal frequency distribution

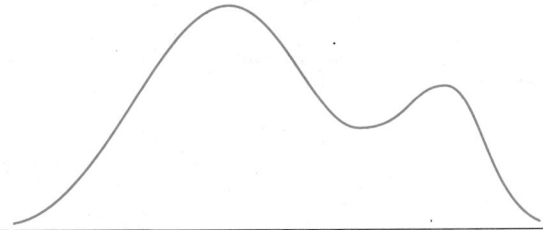

EXAMPLE 2.27   Let us reexamine the lead absorption data first presented in Example 2.13. The stem and leaf plot is reproduced here:

```
1 | 0 3 4
1 | 5 6 7 8
2 | 0 1 2 3 3 4
2 | 5 7
3 | 1 4 4
3 | 5 5 6 7 8 9 9
4 | 1 3 4
4 | 5 8 9
5 |
5 |
6 | 2
6 |
7 | 3
```

The stem and leaf plot appears to have two prominent peaks: one centered at about 20 and the other centered at about 35. The two extreme values, 62 and 73, might also suggest another mode at 67 or 68. If we are not convinced that this is a multimodal distribution, we should classify it as being skewed right. If, on the other hand, we feel rather confident that there are indeed two or even three modes in these data, then our next job is to determine what caused the different modes. Can we find any characteristic in common for the children with "low" levels of lead and for those children with "high" levels of lead? Insufficient information is provided here to answer this question, but the experimenter conducting the study should explore further by asking questions:

Did the parents of the children with low levels (or high levels) work in the same area of the factory?

What were the specific jobs of the parents? Were some parents exposed for greater lengths of time?

What kind of living conditions did the children have outside the factory?

By further analyzing data we hope to identify the extraneous factors that contribute to the different modes.

You have just seen that the boxplot and other descriptive tools provide some guidelines in determining the general shape of a distribution. Once the shape is identified, you can locate the center of the distribution with the mean or median and get a numerical measure of the spread of the distribution with the standard deviation or the Q-spread. These numerical quantities, or summary measures as they are frequently called, characterize the distribution by distinguishing it from other distributions of the same type. We will discuss more about population distribution shapes and their associated parameters and what summary measures to compute in Chapter 6.

EXERCISES 2.6

2.96 The aggressive tendency scores for the group of teenage boys in Exercise 2.61 in Section 2.4 are repeated here. Construct a double-stem stem and leaf plot and a boxplot of the scores and comment on the distributional shape. You may want to use the five-number quantile summary diagram from Exercise 2.62 of Section 2.4.

Worksheet: Aggress.mtw

```
38  27  44  39  41  26  35  45  39  28
16  37  11  36  33  46  42  37  40
19  24  29  32  34  31  30  32  43
```

2.97 The midsummary statistics for the data in Exercise 2.96 are given in the following letter value display:

Letter Value Display

	DEPTH	LOWER	UPPER	MID	SPREAD
N=	28				
M	14.5	34.500		34.500	
H	7.5	28.500	39.500	34.000	11.000
E	4.0	24.000	43.000	33.500	19.000
D	2.5	17.500	44.500	31.000	27.000
C	1.5	13.500	45.500	29.500	32.000
	1	11.000	46.000	28.500	35.000

Based on the midsummaries, how do you classify the distributional shape? Compare your answer with the answer obtained in Exercise 2.96.

2.98 The Tennessee Self-Concept Scale scores for the group of teenagers in Exercise 2.96 that were previously analyzed in Exercises 2.63 and 2.64 of Section 2.4 are repeated here. Construct a stem and leaf plot and a boxplot of the self-criticism scores and comment on the distributional shape. You may want to use the five-number summary diagram from Exercise 2.64 of Section 2.4.

Worksheet: Concept.mtw

```
26  19  23  27  24  33  25  29  14  30
20  25   5  18   7  28  31  37  28
 3  20  25  45  29  22  41  34  22
```

2.99 The midsummary statistics for the data in Exercise 2.98 are given in the following letter value display:

Letter Value Display

	DEPTH	LOWER	UPPER	MID	SPREAD
N=	28				
M	14.5	25.000		25.000	
H	7.5	20.000	29.500	24.750	9.500
E	4.0	14.000	34.000	24.000	20.000
D	2.5	6.000	39.000	22.500	33.000
C	1.5	4.000	43.000	23.500	39.000
	1	3.000	45.000	24.000	42.000

Based on these values, how do you classify the shape of the distribution? Compare your answer with the answer obtained in Exercise 2.98.

2.100   Following are the starting salaries for 25 new Ph.D. psychologists selected randomly from universities in the East and South. Construct a stem and leaf plot and a boxplot, and comment on the distributional shape.

Worksheet: Salary.mtw

27,900	28,300	14,400	25,200	28,900
23,700	31,000	20,100	23,100	26,600
21,200	29,400	29,100	20,900	28,700
29,200	24,600	22,000	29,000	24,800
23,300	34,000	24,200	23,300	30,400

2.101   The midsummary statistics for the data in Exercise 2.100 are given in the following letter value display:

Letter Value Display

	DEPTH	LOWER	UPPER	MID	SPREAD
N=	25				
M	13.0	25200.000		25200.000	
H	7.0	23300.000	29000.000	26150.000	5700.000
E	4.0	21200.000	29400.000	25300.000	8200.000
D	2.5	20500.000	30700.000	25600.000	10200.000
C	1.5	17250.000	32500.000	24875.000	15250.000
	1	14400.000	34000.000	24200.000	19600.000

Based on these values, how do you classify the shape of the distribution? Compare your answer with the answer obtained in Exercise 2.100.

2.102   The reaction times of 30 senior citizens (over 65 years of age) applying for driver's license renewals were measured, with the following results:

Worksheet: Senior.mtw

93	105	66	94	64	98	109	71	86	31
101	128	97	85	96	60	94	42	64	99
84	107	77	55	98	80	90	79	90	96

Construct a stem and leaf plot and a boxplot, and comment on the distributional shape.

2.103   The midsummary statistics for the data in Exercise 2.102 are given in the following letter value display:

Letter Value Display

	DEPTH	LOWER	UPPER	MID	SPREAD
N=	30				
M	15.5	90.000		90.000	
H	8.0	71.000	98.000	84.500	27.000
E	4.5	62.000	103.000	82.500	41.000
D	2.5	48.500	108.000	78.250	59.500
C	1.5	36.500	118.500	77.500	82.000
	1	31.000	128.000	79.500	97.000

Based on these values, how do you classify the shape of the distribution? Compare your answer with the answer obtained in Exercise 2.102.

2.104   In Exercise 2.86 in Section 2.5, we investigated the repair costs for midsize four-door sedans that were crashed into barriers at 5 miles per hour. Following are a letter value display and a boxplot of those repair costs. Based on this information, classify the shape of the distribution.

Letter Value Display

	DEPTH	LOWER	UPPER	MID	SPREAD
N=	23				
M	12.0	2129.000		2129.000	
H	6.5	1478.000	2774.000	2126.000	1296.000
E	3.5	1306.000	3157.000	2231.500	1851.000
D	2.0	795.000	3266.000	2030.500	2471.000
	1	618.000	3298.000	1958.000	2680.000

Character Boxplot

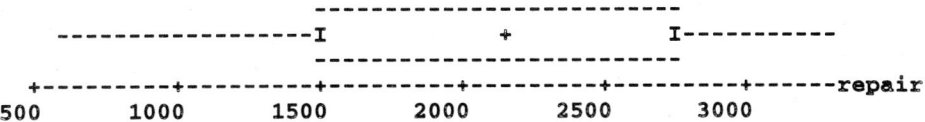

2.105   Shown here are side-by-side boxplots of the test scores for students in two separate statistics classes given in Exercise 2.82 in Section 2.5 (worksheet: **Statisti.mtw**). Compare the two distributions. Is one more variable than the other? Which class tends to have the higher scores? Which class has the higher median?

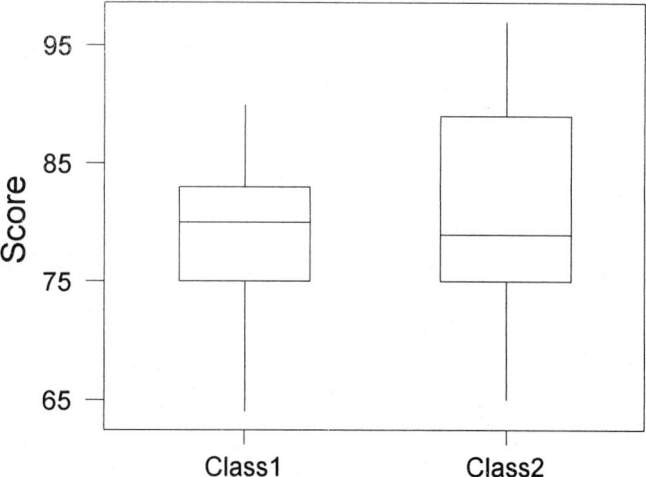

2.106   Shown here are side-by-side boxplots of the plasma LDL levels for a treatment group and a control group of quail that were fed a special diet. The data were given in Exercise 2.76 in Section 2.4 and are in worksheet: **Quail.mtw**. The treatment group received a drug compound that should reduce low-density lipoproteins. Based on the boxplots, do you think there is evidence that the drug compound reduced the LDL levels? What is the effect of the outlier in the treatment group on the average? In these data, do you think we should compare the averages or the medians?

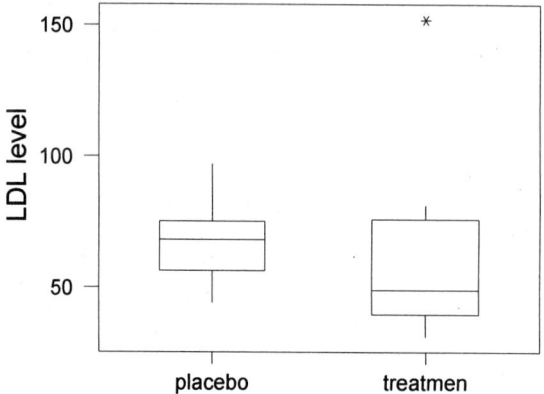

2.107   The accompanying boxplot gives the worth (in $ millions) of football teams in the National Football League as computed by *Financial World* magazine. The data are given in Exercise 2.84 in Section 2.5 and worksheet: **Football.mtw**. How do you classify the shape of the distribution? What do the outlier and the long tail on the high side of the boxplot do to the value of the mean worth? What is the effect on the median? Which is more representative of a typical score, the median or mean?

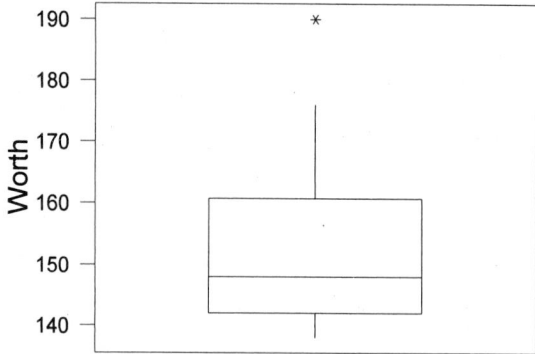

2.108   Following is a list of murder rates per 100,000 for 30 cities selected from the South:

Worksheet: South.mtw

12	10	10	13	12	12	14	7	16	18
8	29	12	14	33	10	6	18	11	25
8	16	14	11	10	20	14	11	12	13

a. Construct a stem and leaf plot.

b. Complete a midsummary analysis.

c. Construct a boxplot.

d. Based on the results obtained above, comment on the shape of the distribution.

2.109   Violent crimes are the offenses of murder, forcible rape, robbery, and aggravated assault. Following are the rates per 100,000 inhabitants for each of the 50 states and the District of Columbia for 1993:

Worksheet: Crime.mtw

Ala	871.7	Ky	535.5	N.D.	83.3
Alaska	660.5	La	984.6	Ohio	525.9
Ariz	670.8	Maine	130.9	Okla	622.8
Ark	576.5	Md	1,000.1	Ore	510.2
Calif	1,119.7	Mass	779.0	Pa	427.0
Colo	578.8	Mich	770.1	R.I.	394.5
Conn	495.3	Minn	338.0	S.C.	944.5
Del	621.2	Miss	411.7	S.D.	194.5
D.C.	2,832.8	Mo	740.4	Tenn	746.2
Fla	1,207.2	Mont	169.9	Texas	806.3
Ga	733.2	Neb	348.6	Utah	290.5
Hawaii	258.4	Nev	696.8	Vt	109.5
Idaho	281.4	N.H.	125.7	Va	374.9
Ill	977.3	N.J.	625.8	Wash	534.5
Ind	508.3	N.M.	934.9	W.Va	211.5
Iowa	278.0	N.Y.	1,122.1	Wis	275.7
Kan	510.8	N.C.	681.0	Wyo	319.5

SOURCE: U.S. Department of Justice, Bureau of Justice Statistics, *Sourcebook of Criminal Justice Statistics*, 1993, p. 366.

a. Construct a stem and leaf plot. Is there anything unusual about the stem and leaf plot?

b. Conduct a midsummary analysis. The range is much larger than the Q-spread. What specifically caused this to happen?

c. Construct a boxplot. Are there any outliers?

d. Based on the above, summarize the shape of the distribution.

2.110   AFDC (Aid to Families with Dependent Children) is a program funded by federal and state governments to provide assistance to needy families. Following are the average monthly payments per person for families in each of the 50 states and the District of Columbia:

Worksheet: Aid.mtw

Ala	$ 57.16	Hawaii	213.96	Mich	144.37
Alaska	253.54	Idaho	111.68	Minn	167.91
Ariz	114.23	Ill	108.82	Miss	42.29
Ark	68.22	Ind	86.83	Mo	91.20
Calif	199.57	Iowa	134.72	Mont	113.24
Colo	110.86	Kan	119.10	Neb	113.75
Conn	199.30	Ky	78.04	Nev	104.13
Del	119.63	La	56.29	N.H.	158.41
D.C.	138.33	Maine	144.74	N.J.	127.27
Fla	99.63	Md	119.26	N.M.	103.54
Ga	90.78	Mass	192.13	N.Y.	197.61

(*continued*)

N.C.	88.86	R.I.	182.19	Vt	192.31
N.D.	126.34	S.C.	67.05	Va	99.37
Ohio	113.72	S.D.	103.79	Wash	174.26
Okla	104.17	Tenn	59.12	W.Va	85.56
Ore	143.45	Texas	56.91	Wis	155.37
Pa	125.61	Utah	123.48	Wyo	120.93

SOURCE: U.S. Department of Health and Human Services, 1993.

a. Construct a stem and leaf plot. Is anything unusual in the stem and leaf plot?
b. Complete a midsummary analysis. Is there a definite pattern to the midsummaries?
c. Construct a boxplot. Are there any outliers?
d. Based on the results obtained above, comment on the shape of the distribution.

2.111 Following is a double-stem stem and leaf plot of the incomes (rounded to the nearest $100) of assistant professors at a small university with an engineering school:

```
27 | 0 3
27 | 5 5 6 8
28 | 0 0 1 2 2 3 4
28 | 5 5 5 7 8 8 9 9
29 | 0 1 2 2 4
29 | 5 5 6 8
30 | 0 0 1 4
30 | 5 6 7 9 9
31 | 0 0 1 2 3 4 4
31 | 5 5 6 6 7 8
32 | 0 0 3
32 | 5 5
33 | 0 1
33 | 5
```

a. There are two peaks in the stem and leaf plot. How do you classify the shape?
b. Can you think of any characteristics of assistant professors that would cause their salaries to be lumped into two different groups?
c. Is it reasonable to average these scores to get a measure of the typical salary of an assistant professor at this university? Why or why not?

2.7 SUMMARY AND REVIEW

KEY CONCEPTS

✓ A data set is a listing of the observed values of one or more variables. A data set with one variable listed is called *univariate*. When two variables are measured on each subject, the data set is called *bivariate*. When several variables are measured on each subject, the data set is *multivariate*.

✓ Variables are classified as either *numerical* or *categorical*. Numerical variables are classified as *discrete* or *continuous*.

✓ Data recorded over time are called *time series* data.

✓   Categorical data can be listed along with the frequency at which they occur in a table called a *frequency table*. The *relative frequency* for a category is obtained by dividing the frequency of the category by the total frequency.

✓   The frequency table for categorical data can be graphed as a *bar graph* or *pie chart*.

✓   Numerical data can be displayed in a *stem and leaf plot*. The data in a stem and leaf plot can be easily transformed into a *histogram* or a *frequency polygon*. An advantage of the stem and leaf plot is that the data are left intact.

✓   Populations and samples are characterized by summary measures. A *parameter* is a summary measure associated with a population. A *statistic* is a summary measure associated with a sample. The stem and leaf plot is used to organize sample data so that the statistics can be calculated more easily.

✓   Summary measures are often classified as *measures of location* and *measures of variability*.

✓   Among the measures of location are those that measure the center of the distribution, including the *mean*, the *trimmed mean*, and the *median*. Other measures of location are *quartiles*, *letter values*, and the *high* and *low*. From these, other measures of center can be found. They are called midsummaries and include the *midQ* and the *midrange*.

✓   Measures of variability include the *range*, the *standard deviation*, and the *Q-spread*. Standard deviations can be interpreted with the *empirical rule* and *Chebyshev's rule*.

✓   The *boxplot* is a graphical tool to picture the data and detect possible *outliers* in the data.

✓   A *skewed* distribution is a distribution with one tail significantly longer than the other. It is said to be skewed in the direction of the long tail. A *symmetric short-tailed* distribution has tails that drop off more rapidly than the tails of a normal curve. A *symmetric long-tailed* distribution has tails longer than the tails of a normal curve. A *multimodal* distribution has more than one peak. The *bimodal* distribution with its two peaks is commonly observed in real data. The occurrence of more than one mode in a data set may be an indication of a nonhomogeneous factor in the data. If at all possible, the factor should be identified.

LEARNING GOALS   Having completed this chapter, you should be able to:

1. Distinguish between a numerical and a categorical variable. *Section 2.1*
2. Distinguish between a discrete and a continuous variable. *Section 2.1*
3. Construct a time series graph. *Section 2.1*
4. Formulate data in a frequency table. *Section 2.2*
5. Construct bar graphs. *Section 2.2*
6. Construct stem and leaf plots. *Section 2.3*
7. Draw a histogram from a stem and leaf plot. *Section 2.3*
8. Construct a frequency polygon from a histogram. *Section 2.3*
9. Decide whether a numerical quantity is a parameter or a statistic. *Section 2.4*
10. Compute the measures of location described in this chapter. *Section 2.4*
11. Construct a five-number summary diagram. *Section 2.4*
12. Compute the measures of variability described in this chapter. *Section 2.5*

13. Give an interpretation of standard deviation using the empirical rule or Chebyshev's rule. *Section 2.5*

14. Construct a boxplot. *Section 2.6*

15. Classify the shape of a distribution. *Section 2.6*

Questions for Review

Use the following problems to test your skills:

2.112 Identify the following variables as either categorical or numerical. If numerical, then further identify them as continuous or discrete.

   a. Amount of tar in a cigarette

   b. SAT scores for entering freshmen

   c. Brands of breakfast cereal

   d. Number of gold medals won by U.S. athletes

   e. One's ratings of his or her favorite television shows

   f. Number of defective computer parts

2.113 A survey by the American Council of Life Insurance of 516 adults on the issue of birth control gave the following results:

Birth Control	Percent
Completely for it	50
Somewhat for it	28
Somewhat against it	10
Completely against it	11
Undecided	1

Illustrate these data in a bar graph.

2.114 The money that Americans spend on alcoholic beverages is distributed as follows:

Beverage	Percent
Beer	52
Wine	23
Liquor	25

Illustrate these data in a pie chart.

2.115 Classify the following variables as categorical or numerical. If numerical, further classify them as discrete or continuous.

   a. Number of vehicles owned by a family

   b. Homicide rate in a major city

   c. Socioeconomic status

   d. Concentration of PCB in a chemical spill

   e. Duration of a kidney transplant operation

2.116 Discuss the difference between continuous and discrete variables and give two examples of each.

2.117 Classify each of the following variables as numerical or categorical:

   a. Race

   b. IQ

   c. Divorce rate

   d. Hair color

   e. Social Security number

   f. Lawyer's fee

2.118 Calculate the mean for the following sample of nine measurements: 8, 12, 18, 16, 9, 10, 2, 8, 7. Verify that the deviations from the mean sum to zero. Calculate the standard deviation.

2.119   For the following data, construct a stem and leaf plot that corresponds to the ages of 25 executives:

Worksheet: Executiv.mtw

35	45	63	42	59	45	50	62	36
64	50	26	51	54	45	59	57
64	28	38	48	42	61	54	60

2.120   Construct a frequency table and draw a histogram of the data in Exercise 2.119.

2.121   Of 15,856 violent crimes reported in Atlanta in 1992, there were 198 murders, 627 forcible rapes, 5,824 robberies, and 9,207 aggravated assaults. (SOURCE: U.S. Department of Justice, *Sourcebook of Criminal Justice Statistics*, 1993, p. 369.) Organize these data in a relative frequency table and then construct a bar graph.

2.122   A survey of 1,000 adults (conducted by R. H. Bruskin Associates) found that 720 thought the legal drinking age should be 21, 50 said 20 years old, 60 said 19 years old, 110 said 18 years old, and 60 were undecided.

    a. Organize these data in a relative frequency table.
    b. Display the data in a bar graph.
    c. Display the data in a pie chart.

2.123   On December 2, 1982, Barney Clark became the first human to receive an artificial heart. He was in surgery for 7 hours. William Schroeder, the second recipient on November 25, 1984, was in surgery for 6.5 hours. On February 17, 1985, Murray Haydon, the third artificial heart recipient, was in surgery for 3.5 hours. Suppose you were keeping records on the artificial heart recipients.

    a. List five relevant variables.
    b. Classify the variables as categorical or numerical.

2.124   In reference to Exercise 2.123, 15 artificial heart transplants were performed, and the durations (in hours) were recorded as follows:

Worksheet: Artifici.mtw

| 7.0 | 6.5 | 3.5 | 3.8 | 3.1 | 2.8 | 2.5 | 2.6 |
| 2.4 | 2.1 | 1.8 | 2.3 | 3.1 | 3.0 | 2.5 |

    a. Organize these data in an ordered stem and leaf plot. Excluding the durations associated with the first two operations (7.0 and 6.5), what do the remaining data look like?
    b. Calculate the mean, median, and standard deviation of all 15 observations. Next calculate the mean, median, and standard deviation without the two extreme observations. What impact does removing the two outliers have on the three statistics?
    c. For a new patient who will undergo the surgery, how long should he or she expect to be in surgery? What should he or she be told about the variability in durations for the surgery?

2.125   A government study showed that of all burglaries, 13% took place when someone was home. Is 13% a parameter or a statistic?

2.126   In 1990, 28.7 million women had full-time-year-round jobs. Assuming we are interested in only the 1990 data, is 28.7 million a parameter or a statistic?

2.127   The military would like to know whether black soldiers feel they are treated fairly. After all, 60% of the military is black. A sample of 2,000 black soldiers were interviewed, and 11% of those interviewed by whites said they were treated unfairly and 35% of those interviewed by blacks said they were treated unfairly.

    Answer the following true or false:
    a. The 60% is a statistic.
    b. The 2,000 is a parameter.
    c. The 11% is a statistic.

     d.   The 35% is a statistic.

     e.   The 35% should be discarded because it is not realistic.

     f.   The sample size of 2,000 is way too small to determine anything about the issue.

2.128   a. Name three statistics that measure the center of a data set.

        b. Name three statistics that measure the variability of a data set.

2.129   Following are the education levels (years of formal education) of nine workers at a plant:

        8   8   1   12   14   9   12   14   12

        a.   Find the mean and the median.

        b.   Find the 10% trimmed mean.

        c.   Find the standard deviation.

2.130   Following are the numbers of reported serious reactions per million doses of vaccines in 11 southern states:

        Worksheet: Vaccine.mtw

Alabama	3.9
Arkansas	27.8
Florida	5.0
Georgia	73.5
Louisiana	24.8
Mississippi	12.1
North Carolina	54.9
Oklahoma	35.3
South Carolina	11.2
Tennessee	138.8
Texas	9.2

        SOURCE: Centers for Disease Control, Atlanta, Georgia.

        Organize these data in an ordered stem and leaf plot and construct a five-number summary diagram.

2.131   Suppose you are conducting a survey of registered voters in your voting district.

        a.   Describe the population of interest.

        b.   Describe how you might obtain a random sample.

        c.   Give three categorical variables that are of interest.

        d.   Give three numerical variables that are of interest and classify them as either continuous or discrete.

2.132   Identify each of the following numerical variables as either continuous or discrete:

        a.   Blood pressure

        b.   Suicide rate

        c.   Number of apartments in a city

        d.   Gross national product (GNP)

        e.   Number of trials to complete a task

2.133   A market analyst is studying the percents of the computer market held by the major manufacturers. She records each corporation name, the area of the market it is active in, (e.g., personal computers, software, etc.), the assets of the company, and the percent of the market it controls. Classify corporate name, market area, assets, and market percent as categorical or numerical.

2.134   The 251,012 poisons reported to 16 poison control centers that served 11% of the U.S. population in 1983 were distributed as follows:

Worksheet: Poison.mtw

Type of poison	Number reported
Drugs	150,857
Cleaning agent	22,347
Plants	22,326
Cosmetics	13,192
Insecticides	8,438
Alcohol	9,201

SOURCE: Centers for Disease Control, Atlanta, Georgia.

Organize these data in a relative frequency bar graph.

2.135   Classify the following variables as categorical or numerical:
   a.   Baseball free-agent salary
   b.   Number of "gold" records by recording stars
   c.   Type of disability
   d.   Number of native Americans in health service professions
   e.   Health service occupations

2.136   To evaluate the side effects of a drug, 70 people were asked to do a certain task after receiving a standard dose of the drug. Following are the numbers of trials necessary to complete the task:

Worksheet: DGeffect.mtw

Trials	Frequency
18	2
17	4
16	9
15	7
14	5
13	7
12	8
11	10
10	12
9	6
Total	70

   a.   Draw a histogram of the data.
   b.   Calculate the mean, median, standard deviation, and Q-spread for the numbers of trials to complete the task.

2.137   Classify the following numerical variables as continuous or discrete:
   a.   Amount of milk consumed daily by teenagers
   b.   Weekend box-office receipts for new releases
   c.   Number of cable television companies
   d.   Daily temperature
   e.   Daily tons of garbage

2.138   Modern technology, regulations, malpractice premiums, and physicians fees all contribute to the spiraling cost of health care. Following are the fees for an appendectomy for a random sample of 20 hospitals in North Carolina:

Worksheet: Append.mtw

| $3,821 | 3,981 | 3,931 | 5,498 | 5,582 | 6,046 | 4,257 | 4,591 | 4,775 | 6,163 |
| 3,840 | 4,053 | 4,844 | 6,266 | 4,026 | 2,478 | 4,347 | 5,104 | 4,673 | 6,389 |

SOURCE: North Carolina Medical Database Commission, August 1994.

Calculate the mean, median, 10% trimmed mean, and standard deviation for the physicians fees.

2.139   Following are the test grades in a beginning statistics class:

Worksheet: Grades.mtw

76	73	81	65	83	90	77	60	67	76
84	72	57	64	71	78	87	76	92	75
58	65	79	86	95	81	71	68	74

Construct a histogram of the test scores. What is the general shape of the histogram?

2.140   Following are the weight gains for a herd of steers after 60 days in a feedlot:

Worksheet: Steer.mtw

| 203 | 189 | 193 | 185 | 216 | 229 | 194 | 197 | 162 | 184 |
| 204 | 210 | 179 | 183 | 199 | 172 | 206 | 225 | 197 | 184 |

Perform an analysis of the data to determine the shape of the distribution. Then give a reasonable measure of the center and the spread of the distribution.

2.141   Classify the following as either categorical or numerical:
   a.   Parent's occupation
   b.   Family size
   c.   ISBN catalog number used by libraries for books
   d.   Type of weapon carried to school
   e.   Discount rate on airline tickets

2.142   Classify the following numerical variables as continuous or discrete:
   a.   Level of calcium in one's diet
   b.   Grade point average
   c.   Time to complete a term paper
   d.   Number of rental cars returned on time
   e.   Number of weapons carried to school

2.143   Following are reading scores on the California Achievement Test for a group of third-graders:

Worksheet: CAT.mtw

| 48 | 54 | 51 | 73 | 66 | 45 | 70 | 73 |
| 62 | 58 | 59 | 54 | 54 | 68 | 50 | 49 |

   a.   Organize the data in a useful ordered stem and leaf plot.
   b.   Complete a five-number summary diagram to include the midsummaries and spreads.
   c.   Perform a midsummary analysis. Does the distribution appear to be skewed or symmetric?
   d.   Construct a boxplot. Is your analysis in part **c** consistent with the boxplot?

2.144   In the finishing department of a furniture company, the quality of the finish is classified as excellent, good, fair, needs improvement, and redo. Sixty-five items were classified as follows: 17 excellent, 22 good, 13 fair, 5 needs improvement, and the rest redo. Organize these data in a frequency table and draw a bar graph.

COMPUTER   2.145
EXERCISES

Following are the weekly rentals for 45 apartments in a large metropolitan area:

Worksheet: Rentals.mtw

100	130	130	305	175	155	150	95	295
210	80	270	135	130	335	230	235	75
90	285	65	345	110	135	185	300	70
250	125	180	150	305	170	95	90	145
90	160	130	80	490	235	75	60	425

a. Organize the data in a frequency histogram. What shape does it appear to have?
b. Compute the descriptive statistics. Is there much difference between the mean and the median? Are any scores more than three standard deviations from the mean?
c. Compute a letter value display. Is there a pattern to the midsummaries? What do they tell about the distribution?

2.146   Following are the numbers of inmates per 100,000 population in the jails across the United States in 1983 and in 1993. Some states have integrated jail–prison systems and were omitted from the data.

Worksheet: Jail.mtw

	1983	1993		1983	1993
Ala	113	169	Mont	50	81
Alaska	8	na	Neb	53	105
Ariz	99	184	Nev	105	215
Ark	69	117	N.H.	50	100
Calif	166	222	N.J.	80	192
Colo	88	177	N.M.	96	189
Conn	na	na	N.Y.	91	164
Del	na	na	N.C.	57	129
D.C.	456	292	N.D.	36	57
Fla	137	250	Ohio	66	105
Ga	178	328	Okla	67	127
Hawaii	na	na	Ore	87	125
Idaho	61	135	Pa	85	160
Ill	77	124	R.I.	na	na
Ind	66	145	S.C.	82	157
Iowa	29	57	S.D.	45	87
Kan	55	111	Tenn	128	282
Ky	100	180	Texas	97	307
La	192	377	Utah	56	102
Maine	49	57	Vt	na	na
Md	107	188	Va	103	225
Mass	57	131	Wash	84	141
Mich	84	132	W.Va	52	97
Minn	47	81	Wis	64	156
Miss	97	184	Wyo	66	105
Mo	76	96	U.S. total	98	178

Note: The District of Columbia jail population declined between 1983 and 1993 because the Occoquan complex was reclassified from a jail to a prison.
SOURCE: U.S. Department of Justice, Bureau of Justice Statistics, *Jails and Jail Inmates 1993–94*, April 1995.

a. Construct a stem and leaf plot of the 1983 data. Are there any unusual observations?
b. Construct a stem and leaf plot of the 1993 data. Are there any unusual observations?
c. Construct dotplots of the data from the two years using a common number scale. Do you see any differences in the two distributions?
d. Construct boxplots of the data from the two years. Draw them side by side using a common number scale. How do you describe the two shapes? Do you see any differences between the two boxplots?

2.147   Following are the birth places, ages at inauguration, and ages at death of the first 42 presidents of the United States:

Worksheet: Presiden.xls (Excel)   Presiden.mtw (Minitab)

President	Birth	Inaug. Age	Age at Death
G. Washington	VA	57	67
J. Adams	MA	61	90
T. Jefferson	VA	57	83
J. Madison	VA	57	85
J. Monroe	VA	58	73
J. Q. Adams	MA	57	80
A. Jackson	SC	61	78
M. Van Buren	NY	54	79
W. Harrison	VA	68	68
J. Tyler	VA	51	71
J. Polk	NC	49	53
Z. Taylor	VA	64	65
M. Fillmore	NY	50	74
F. Pierce	NH	48	64
J. Buchanan	PA	65	77
A. Lincoln	KY	52	56
A. Johnson	NC	56	66
U. S. Grant	OH	46	63
R. Hayes	OH	54	70
J. Garfield	OH	49	49
C. Arthur	VT	50	57
G. Cleveland	NJ	47	71
B. Harrison	OH	55	67
G. Cleveland	NJ	55	71
W. McKinley	OH	54	58
T. Roosevelt	NY	42	60
W. Taft	OH	51	72
W. Wilson	VA	56	67
W. Harding	OH	55	57
C. Coolidge	VT	51	60
H. Hoover	IA	54	90
F. D. Roosevelt	NY	51	63
H. Truman	MO	60	88
D. Eisenhower	TX	62	78
J. Kennedy	MA	43	46
L. Johnson	TX	55	64
R. Nixon	CA	56	81
G. Ford	NE	61	

(continued)

President	Birth	Inaug. Age	Age at Death
J. Carter	GA	52	
R. Reagan	IL	69	
G. Bush	MA	64	
W. Clinton	AR	46	

a. Construct a pie chart of the birth states of the presidents. What state has had the most presidents? Of the 50 states, how many have never produced a president?

b. Construct stem and leaf plots of the inaugural ages and the ages at death. Are they similar in shape?

c. Construct dotplots of the inaugural ages and the ages at death with a common number line. Does this provide additional information about how the two distributions are related?

2.148   The *Census of Retail Sales* reported the following per-resident sales for groceries in the fifty states in 1982:

Worksheet: Grocerie.mtw

Ala	$ 889.96	Mont	1,113.82
Alaska	1,503.37	Neb	850.29
Ariz	1,121.08	Nev	1,306.81
Ark	861.05	N.H.	1,246.98
Calif	1,041.24	N.J.	1,000.89
Colo	1,158.22	N.M.	1,030.74
Conn	1,020.07	N.Y.	856.79
Del	1,005.58	N.C.	961.20
Fla	1,103.52	N.D.	788.22
Ga	929.16	Ohio	960.65
Hawaii	1,008.88	Okla	1,125.49
Idaho	1,023.89	Ore	951.77
Ill	844.76	Pa	896.59
Ind	902.64	R.I.	826.27
Iowa	973.15	S.C.	929.84
Kan	942.26	S.D.	826.93
Ky	917.41	Tenn	938.84
La	1,082.77	Texas	1,154.98
Maine	1,057.85	Utah	921.90
Md	989.76	Vt	1,099.43
Mass	914.01	Va	1,014.82
Mich	856.63	Wash	1,074.46
Minn	866.88	W.Va	979.60
Miss	865.05	Wis	887.16
Mo	913.57	Wyo	1,189.96

SOURCE: U.S. Census Bureau and U.S. Department of Agriculture.

a. Construct a stem and leaf plot of the per-resident sales of groceries. Are there any unusually large observations?

b. Calculate summary statistics. Is there much difference between the mean and the median?

c. Construct a dotplot of the data. Without the largest score, how do you classify the shape of the distribution?

d. Is there a reasonable explanation for the largest score? Should it remain in the data set or should we remove it before calculating summary statistics?

2.149   Following are the percents of the population of each state over the age of 65 in 1985:

Worksheet: Elderly.mtw

Ala	16.8	Mont	16.3
Alaska	4.5	Neb	18.6
Ariz	17.1	Nev	12.7
Ark	19.8	N.H.	15.9
Calif	14.2	N.J.	16.8
Colo	11.9	N.M.	13.7
Conn	17.1	N.Y.	16.9
Del	14.8	N.C.	15.2
D.C.	15.3	N.D.	17.8
Fla	22.8	Ohio	16.4
Ga	13.9	Okla	16.4
Hawaii	12.5	Ore	17.6
Idaho	16.0	Pa	18.8
Ill	16.2	R.I.	18.9
Ind	16.1	S.C.	14.1
Iowa	19.4	S.D.	19.5
Kan	18.1	Tenn	16.4
Ky	16.3	Texas	13.2
La	13.8	Utah	12.3
Maine	18.0	Vt	16.2
Md	13.8	Va	13.7
Mass	17.6	Wash	15.4
Mich	15.5	W.Va	18.0
Minn	17.1	Wis	17.6
Miss	17.1	Wyo	11.5
Mo	18.6		

SOURCE: U.S. Department of Commerce.

a. Construct a stem and leaf plot of the percents of residents in the states over 65 years of age. Are there any unusually large or small observations?

b. Construct a dotplot of the data. Without the largest and smallest observations, what does the shape look like? Can you explain the largest and smallest observations?

c. Calculate the descriptive statistics. Is there much difference between the mean, median, and trimmed mean? Which do you think we should use to represent the center of the data?

d. Complete a letter value display. What do the midsummaries tell you about symmetry? What do they tell you about the tails of the distribution?

2.150   A lottery agent sells lottery tickets in states that have legalized lotteries. The amount the agent earns depends on the number of tickets he or she sells. Following are the average annual sales commissions paid to lottery agents in 18 states that have lotteries:

Worksheet: Lottery.mtw

State	Number of lottery agents	Average sales commission
Arizona	2,100	$ 2,228
Colorado	2,500	4,880
Connecticut	3,000	3,200
District of Columbia	850	3,986

(continued)

State	Number of lottery agents	Average sales commission
Delaware	324	5,247
Illinois	7,710	3,320
Maine	1,629	490
Maryland	1,239	20,399
Massachusetts	3,600	5,279
Michigan	7,293	4,758
New Hampshire	1,150	574
New Jersey	4,000	12,000
New York	12,449	4,281
Ohio	5,400	3,704
Pennsylvania	7,700	8,013
Rhode Island	883	7,667
Vermont	750	307
Washington	4,711	1,804

a. Construct a stem and leaf plot. What shape does it have?
b. Calculate the descriptive statistics for the sales commissions. Compare the mean, median, and trimmed mean. Is there much difference between them?
c. If this distribution is skewed right, then the mean should be larger than the median. Is it? By how much?
d. Construct a letter value display. Is there any pattern to the midsummaries? Does this agree with the conjecture that the distribution is skewed right?

2.151 The fertility rate is the number of births a woman can expect in her child-bearing years. An average rate of 2.12 is needed to keep the population constant. Following are the fertility rates in all states including the District of Columbia:

Worksheet: Fertilit.mtw

Ala	1.9	Mass	1.5
Alaska	2.3	Mich	1.8
Ariz	2.1	Minn	1.9
Ark	2.0	Miss	2.2
Calif	1.9	Mo	1.9
Colo	1.8	Mont	2.1
Conn	1.5	Neb	2.0
Del	1.8	Nev	1.8
D.C.	1.5	N.H.	1.7
Fla	1.7	N.J.	1.6
Ga	1.9	N.M.	2.2
Hawaii	2.1	N.Y.	1.6
Idaho	2.5	N.C.	1.6
Ill	1.9	N.D.	2.1
Ind	1.8	Ohio	1.8
Iowa	2.0	Okla	2.0
Kan	2.0	Ore	1.8
Ky	1.9	Pa	1.6
La	2.2	R.I.	1.5
Maine	1.7	S.C.	1.8
Md	1.6	S.D.	2.4

(*continued*)

Tenn	1.7	Wash	1.8
Texas	2.1	W.Va	1.8
Utah	3.2	Wis	1.9
Vt	1.7	Wyo	2.4
Va	1.6		

SOURCE: Population Reference Bureau.

Find the mean, median, 10% trimmed mean, and standard deviation for the fertility rate data.

2.152   Data collected from the states by the Bureau of Justice Statistics indicate that violent offenders released from state prisons in 1992 served 48% of their sentence. Following are the sentences (in months) of a sample of 41 prisoners convicted of a homicide offense:

Worksheet: Sentence.mtw

117	188	172	145	173	159	123	136	115	190	158
147	117	196	169	160	135	163	176	155	126	
133	150	146	157	137	180	134	135	128	209	
144	164	185	168	121	195	149	148	216	139	

SOURCE: U.S. Department of Justice, Bureau of Justice Statistics, *Prison Sentences and Time Served for Violence*, NCJ-153858, April 1995.

a. Construct a stem and leaf plot of the data.
b. Find the mean, median, 10% trimmed mean, and standard deviation for the data.
c. Use the empirical rule to find limits for the length of 95% of the prison sentences. Does it seem that there is a wide disparity in sentence lengths for these prisoners?

2.153   Disposal of toxic waste is a major concern for U.S. manufacturing plants. One measure of the environmental performance of a plant is its "toxic intensity," which is the pounds of toxic waste released (into the environment or captured and transferred to another site) per $1,000 of shipments of its product. Successful plants release fewer pounds of toxins per $1,000 of shipments than the average of other plants that manufacture the same product. Following are the toxic intensities measured for plants whose product line is herbicidal preparations:

Worksheet: Disposal.mtw

1.45	1.38	4.37	2.97	1.06	.44	2.20	1.25	2.23	3.96
9.12	3.49	5.64	1.74	1.69	4.33	.28	2.43	5.46	3.57
2.03	1.39	3.35	4.33	2.68	2.48	.82	3.47	.33	

SOURCE: Bureau of the Census, *Reducing Toxins*, Statistical Brief SB/95-3, February 1995.

a. Display the data in an ordered stem and leaf plot. Do any plants have an unusually high toxic intensity?
b. Find the mean, median, and standard deviation. Is there much difference between the mean and median? Which do you think better represents the toxic intensity for these plants?
c. Use the empirical rule to find limits for the toxic intensity for 95% of these plants. How many plants fall outside these limits?

2.154   You are to prepare a report on the toxic intensity of herbicidal preparation plants. Use the data in Exercise 2.153 to construct a frequency table and histogram for your report.

2.155   The average speed (in miles per hour) for the winners of the Indianapolis 500 for the years 1961–1994 are as follows:

Worksheet: Indiapol.mtw

Year	Speed	Year	Speed
1961	139.130	1978	161.363
1962	140.293	1979	158.899
1963	143.137	1980	142.862
1964	147.350	1981	139.085
1965	151.388	1982	162.026
1966	144.317	1983	162.117
1967	151.207	1984	163.621
1968	152.882	1985	152.982
1969	156.867	1986	170.722
1970	155.749	1987	162.175
1971	157.735	1988	144.809
1972	162.962	1989	167.581
1973	159.036	1990	185.984
1974	158.589	1991	176.457
1975	149.213	1992	134.477
1976	148.725	1993	157.207
1977	161.331	1994	160.872

SOURCE: *The World Almanac and Book of Facts*, 1995, p. 931.

Retrieve the data and construct a time series plot. Compare the trend in speeds for the first 14 years with the trend in the last 20 years. Can any generalizations be made about the speeds over the entire 34-year period?

2.156   STATISTICAL INSIGHT REVISITED   The starting salaries for the different majors listed in the Statistical Insight at the beginning of the chapter are stored in worksheet: **Job.mtw**. To analyze the data, retrieve the data set and first construct a stem and leaf plot. From the stem and leaf plot, how do you classify the shape of the distribution? Are there any unusual observations? If so, what major do they correspond to? Next create a dotplot of the data. Is there any clustering of the data? That is, are the data multimodal? If so, can you tell how they are clustering? Create a letter value display. Do the midsummaries tell you anything different from what you have already observed? Use the Describe command to calculate the different summary statistics. Are the mean and median different? Does the standard deviation seem large? To evaluate the spread of the distribution, should we use the empirical rule or Chebyshev's rule? Choose one and give your summary. Finally, construct a boxplot of the data. Given the previous analysis, is the boxplot what you expected?

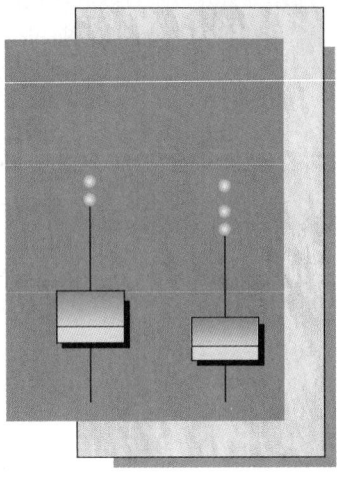

3

BIVARIATE DATA: STUDYING RELATIONSHIPS BETWEEEN VARIABLES

In Chapter 2, we studied characteristics of univariate data. Now we turn to relationships between two variables. For example, is there a relationship between the birth weights of children and their adult heights? Is there a relationship between students' grades and their study habits? Furthermore, if there is a relationship, what form does it take? How strong is the relationship? To answer these questions, our first task is to organize and summarize the bivariate data. Graphical tools like the *scatterplot* are used to examine the data and help establish an association between the variables. The *correlation coefficient*, a summary statistic, is then used to measure numerically the degree of association between the variables.

In Section 3.1, we look at relationships between two categorical variables. In Section 3.2, we illustrate relationships between two numerical variables and between a categorical and a numerical variable. The remainder of the chapter is devoted to the study of relationships between two numerical variables.

CONTENTS

STATISTICAL INSIGHT

DOES WATCHING TELEVISION AFFECT MATH SCORES?

For many years educators have suggested that the social conditions that exist at home contribute heavily to the decline in children's academic performance. A report, *America's Smallest School: The Family* by the Educational Testing Service, provides additional information that suggests that watching television has a serious effect on school performance. In the report, state averages on a national math test were compared with the percent of students in the state who watch 6 or more hours of television per day. The following data were collected for eighth-graders in 37 states, two territories, and Washington, DC:

Worksheet: TV.mtw

State	Percent	Test	State	Percent	Test
Ala	18	253	Neb	9	276
Alaska	na		Nev	na	
Ariz	12	258	N.H.	7	273
Ark	20	256	N.J.	13	270
Calif	11	256	N.M.	11	256
Colo	9	267	N.Y.	17	261
Conn	12	270	N.C.	21	251
Del	18	262	N.D.	6	281
Fla	19	255	Ohio	11	264
Ga	17	258	Okla	14	263
Hawaii	23	252	Ore	9	272
Idaho	7	272	Pa	10	262
Ill	14	261	R.I.	12	261
Ind	11	267	S.C.	na	
Iowa	8	278	S.D.	na	
Kan	na		Tenn	na	
Ky	14	256	Texas	15	258
La	19	246	Utah	na	
Maine	na		Vt	na	
Md	19	261	Va	16	264
Mass	na		Wash	na	
Mich	14	264	W.Va	16	256
Minn	7	276	Wis	8	274
Miss	na		Wyo	7	272
Mo	na		Virg Is	27	218
Mont	6	279	Guam	20	233
D.C.	33	233			

SOURCE: Educational Testing Services, 1992.

In this chapter, we study the association that might exist between two variables like these. We measure the association with the correlation coefficient, and if a relationship exists, we construct a model that relates the two variables. See Exercise 3.98 in Section 3.6 for an analysis of these data.

3.1 CONTINGENCY TABLES

Categorical Versus Categorical Variables

Just as in the univariate case, when we study the relationship between two or more variables, we need to determine whether they are numerical or categorical variables. If both variables are categorical, their values can be displayed in a two-way frequency table called a *contingency table*. The word *contingency* is used because if there is an association between the two variables, then we can say that the values assumed by one variable are *contingent* (dependent) on the values of the other variable.

EXAMPLE 3.1

When customers take their cars back to the dealership for the 15,000-mile servicing, do they get more than they bargained for? A survey by *U.S. News & World Report* found that many dealers routinely sell far more maintenance services than manufacturers recommend. Customers pay for unnecessary inspections, cleaning, lubrication, and adjustments. Researchers randomly selected 122 dealerships in seven metropolitan areas: Philadelphia, Miami, Chicago, Dallas–Fort Worth, Denver, Los Angeles, and Seattle. Their findings are summarized in Table 3.1.

TABLE 3.1 Service rendered by 122 auto dealers

Worksheet: Dealers.xls

Type of dealership	Number that replace parts before they are needed	Number that perform only services recommended by manufacturer	Total
Honda	19	2	21
Saturn	4	15	19
Ford	8	13	21
Dodge	11	10	21
Mazda	12	9	21
Toyota	3	16	19
Total	57	65	122

Analyzing data in a contingency table involves comparing the percents in the various categories. For example, it is easy to see that $57/122 = 46.7\%$ of the dealerships unnecessarily replace parts. Furthermore, $21/122 = 17.2\%$ of the dealerships in the study are Honda dealerships, and $19/122 = 15.6\%$ are Saturn dealerships. This information is obtained from the *marginal totals*, which are the row and column totals.

More interesting, however, is whether the type of service rendered is *contingent* on the type of dealership. If not, we would say that the service and the type of dealership are *independent* and thus the service is the same across all dealerships. If service and dealership are independent, we would expect $57/122$ or 46.7% of all dealers to perform unnecessary replacements. In the case of Honda, that would be 9.8 dealers (46.7% of 21

Honda dealers), and in the case of Saturn, that would be 8.9 dealers (46.7% of 19 Saturn dealers). However, the actual numbers are 19 and 4, respectively. The large discrepancies, 19 versus 9.8 and 4 versus 8.9, suggest that service is contingent (dependent) on the type of dealership.

In Example 3.1, there are six types of dealerships and two types of service rendered; hence, we call the table a 6 × 2 contingency table. In a 6 × 2 contingency table, there are 12 cells for the 12 possible combinations of the two categorical variables.

If the two variables under consideration are independent, we found the *expected* numbers in the cells in Example 3.1 by taking 46.7% of 21 and 46.7% of 19. An alternative approach is to use the following formula:

$$\text{expected} = \frac{\text{row total} \times \text{column total}}{\text{grand total}}$$

For example, the first row total is 21, the first column total is 57, and the grand total is 122; thus, the expected number for that cell is $(21 \times 57)/122 = 9.8$. Table 3.2 shows all the expected cell counts in parentheses.

TABLE 3.2   Service rendered by 122 auto dealers

Type of dealership	Number that replace parts before they are needed		Number that perform only services recommended by manufacturer		Total
Honda	19	(9.8)	2	(11.2)	21
Saturn	4	(8.9)	15	(10.1)	19
Ford	8	(9.8)	13	(11.2)	21
Dodge	11	(9.8)	10	(11.2)	21
Mazda	12	(9.8)	9	(11.2)	21
Toyota	3	(8.9)	16	(10.1)	19
Total	57		65		122

As noted in Example 3.1, the large discrepancies between the actual counts and the expected counts for the Honda and Saturn dealerships suggest an association between the two categorical variables. We do not see that discrepancy for the Ford, Dodge, or Mazda dealerships. An obvious question then is: Just how large should the discrepancies be to declare an association between two categorical variables? This question will be addressed fully in Chapter 10 when we study the chi-square distribution and its associated test of independence.

The association between the service and the type of dealership is also illustrated in the *segmented bar graph* in Figure 3.1 on page 154. The segments show the frequency counts in the two service categories for each of the dealerships. Notice the difference between the Honda and Saturn dealerships.

FIGURE 3.1   Segmented bar graph of the service rendered by 122 auto dealers

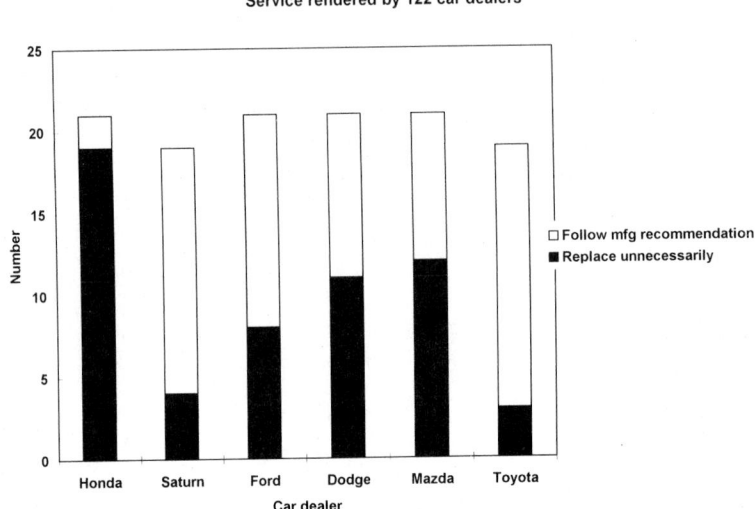

C O M P U T E R   T I P

Figure 3.1 was created with the Chart Wizard in Microsoft Excel. Enter the data in the Excel worksheet as follows:

Type of dealership	Replace unnecessarily	Follow mfg recommendation
Honda	19	2
Saturn	4	15
Ford	8	13
Dodge	11	10
Mazda	12	9
Toyota	3	16

To create the chart, simply follow the five steps of the Chart Wizard in Excel.

Bivariate data can also be displayed in a regular bar graph.

EXAMPLE 3.2

A local election is to be held on the legalization of sales of alcoholic beverages. City residents were asked: Should the sale of alcoholic beverages within the city limits be legalized? The survey yielded the results shown in Table 3.3. From the contingency table, display the data in a bar graph.

TABLE 3.3

	Yes	No	No Opinion	Total
Men	55	16	4	75
Women	53	26	9	88
Total	108	42	13	163

SOLUTION

Figure 3.2 is the bar graph showing the survey results.

FIGURE 3.2 Opinions on the sale of alcoholic beverages

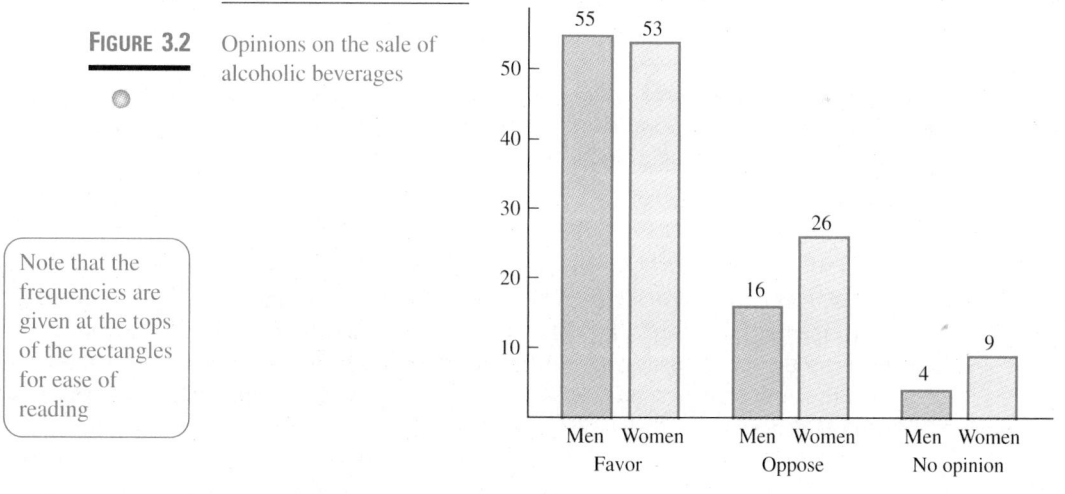

Note that the frequencies are given at the tops of the rectangles for ease of reading

The following example illustrates that when we compare two or more categories in a contingency table, it is important to compare the relative frequencies and not the frequencies.

EXAMPLE 3.3

A sample of 1,200 university students consisting of 700 men and 500 women was surveyed on the issue of the possession of marijuana. They were asked: Do you believe that the possession of a small amount of marijuana is a criminal offense? Of the 516 who responded yes, 294 were men; of the 648 who responded no, 385 were men; and of the 36 with no opinion, 21 were men. Summarize these data in a frequency table including relative frequencies for both men and women.

SOLUTION

Table 3.4 gives the frequencies and relative frequencies. Although fewer women (222) than men (294) responded yes, the relative frequencies show that a greater percent of women (.444) than men (.420) responded yes.

TABLE 3.4

Response	Frequency		Relative Frequency	
	Men	Women	Men	Women
Yes	294	222	.420	.444
No	385	263	.550	.526
No Opinion	21	15	.030	.030
Total	700	500	1.000	1.000

We can analyze three categorical (qualitative) variables in a three-way contingency table by using a two-way table of two of the variables at each level of the third variable. Consider the following example.

EXAMPLE 3.4

On April 22, 1987, the U.S. Supreme Court upheld Georgia's death penalty system by a 5-to-4 margin. The case involved the conviction of a black man for the murder of a white policeman in a 1978 robbery. The issue in question is whether race is a determining factor in the death penalty. David Baldus, a professor of law, and George Woodworth, a professor of statistics, presented the following three-way contingency table as part of their evidence:

		Race of Defendant				
		White		Black		
		White victim	Black victim	White victim	Black victim	Total
Death	Yes	58	2	50	18	128
Sentence	No	687	62	178	1,420	2,347
	Total	745	64	228	1,438	2,475

SOURCE: "Supreme Court Ruling on Death Penalty," *Chance* 1 (1988): 7–8.

Newspaper accounts of the decision noted that the court upheld the death penalty *despite* the statistical evidence of racial disparity in its application. With just a few simple calculations, we can understand the statistical evidence presented by Baldus and Woodworth. First, we collapse the table down to the following two-way table involving the variables race of the victim and death sentence:

		White victim	Black victim	Total
Death Sentence	Yes	108	20	128
	No	865	1,482	2,347
	Total	973	1,502	2,475

We now see that the percent of time the death penalty is given when the victim is white is $108/973 = 11.1\%$, and the percent of time the death penalty is given when the victim is black is only $20/1,502 = 1.3\%$. This suggests that the death penalty is almost ten times more likely to be given when the victim is white than when the victim is black. Based on these data, it appears that Baldus and Woodworth had a strong case. Other interesting percentages are addressed in the exercises.

EXERCISES 3.1

3.1 From the three-way contingency table concerning Georgia's death penalty presented in Example 3.4 determine the following:
   a. What percent of the time is the death penalty given when the defendant is black and the victim is white?
   b. What percent of the time is the death penalty given when the defendant is white and the victim is black?
   c. What do the answers to parts **a** and **b** suggest about racial disparity in the application of the death penalty?

3.2 A sample of 120 employed people are classified according to their gender and occupation. There are 48 women: 9 are blue-collar workers, 23 are white-collar workers, 4 are farm workers, and the rest are classified as other. Of the 72 men, 31 are blue-collar workers, 22 are white-collar workers, 15 are farm workers, and the rest are classified as other. Arrange these data in a bivariate contingency table.
   a. What percent are women?
   b. What percent of the women are blue-collar workers?
   c. What percent of the men are blue-collar workers?
   d. What percent are female blue-collar workers?
   e. Why are the answers to parts **b** and **d** different?

3.3 Draw a bar graph depicting the frequency of men and women in each occupation group for the data in Exercise 3.2.

3.4 The following data are from a poll by George Gallup, on people's attitude toward treatment of the possession of small amounts of marijuana as a criminal offense in the United States in 1980. The question was: Do you think the possession of small amounts of marijuana should or should not be treated as a criminal offense?

	Should be treated as criminal offense	Should not be treated as a criminal offense	No opinion
National	43%	52%	5%
Gender			
Male	42	53	5
Female	44	51	5
Education			
College	30	67	3
High School	45	50	5
Grade School	58	33	9
Age			
18–24	27	67	6
25–29	26	70	4
30–49	45	52	3
50 and older	54	39	7
Religion			
Protestant	49	47	4
Catholic	39	55	6

SOURCE: George H. Gallup, *The Gallup Opinion Index Report No. 179* (Princeton, NJ: The Gallup Poll, July 1980), p. 15.

    a. Using only the levels of the gender variable, draw a relative frequency bar graph illustrating the opinions on the issue of possession of marijuana.

    b. Using only the education level, draw a relative frequency bar graph showing the opinions on the issue of possession of marijuana.

    c. Using only the age level, draw a relative frequency bar graph illustrating the opinions on the issue of possession of marijuana.

3.5   One hundred psychology majors were classified according to gender and class level. Ten were lower division women, 20 were upper division women, 40 were lower division men, and 30 were upper division men.

    a. Arrange the data in a bivariate table.

    b. Calculate the marginal totals.

    c. What percent of the women are lower division?

    d. What percent of the men are lower division?

    e. What percent of the lower division are women?

    f. Are the lower division students more likely to be women or men?

    g. Are the men more likely to be lower or upper division students?

3.6   The SAT scores for 900 female and 800 male college freshmen were classified as low, medium, or high as follows:

Range	Gender	
	Men	Women
High	190	250
Medium	430	520
Low	180	130
Total	800	900

    a. Draw a bar graph illustrating the frequency of the SAT score ranges for men.

    b. Draw a bar graph illustrating the frequency of the SAT score ranges for women.

    c. Can these two bar graphs be compared as drawn?

3.7 In Example 3.1, the service rendered by car dealerships was classified as replacing parts unnecessarily or performing services recommended by manufacturers. From the data presented in Table 3.1, determine the following:

    a. Of the dealerships that replace parts unnecessarily, what percent are foreign dealerships?

    b. Of the dealerships that replace parts unnecessarily, what percent are nonforeign? Which are more likely to replace parts unnecessarily, foreign or nonforeign car dealerships?

    c. Of the foreign dealerships, what percent replace parts unnecessarily?

    d. Should the answers to parts **a** and **c** be the same? Why or why not?

3.8 A survey of 1,000 men and 1,000 women revealed that 100 of the men and 500 of the women had no job. Furthermore, 650 of the men and 400 of the women had one job. The rest had more than one job.

    a. Arrange this information in a bivariate contingency table.

    b. Construct an employment bar graph for the men.

    c. Construct an employment bar graph for the women.

    d. Can these two bar graphs be compared as drawn?

3.9 A survey of 500 students, 100 faculty members, and 30 administrators revealed that 360 students oppose the present parking policy, 60 faculty oppose it, and only 5 administrators oppose it. Arrange the data in a bivariate table by attitude (favor, oppose) and status (student, faculty, administrator).

    a. What percent of those sampled oppose the present parking policy?

    b. What percent of the students oppose the policy?

    c. What percent are faculty and oppose the policy?

    d. Why does the following graph not give a realistic comparison of the opinions of the students, faculty, and administration on the parking policy?

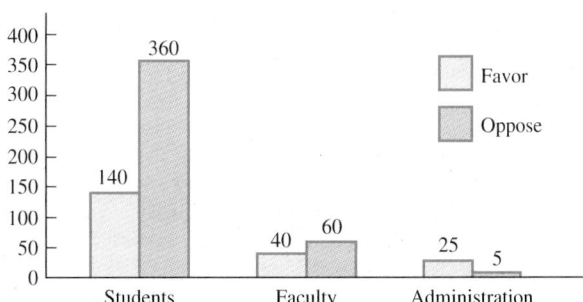

    e. Construct a bar graph that will give a realistic comparison of the opinions of students, faculty, and administrators on the parking policy.

3.10 Using the data in Example 3.1, collapse the dealerships into two categories, foreign cars and American cars. Compute the expected counts as in Table 3.2. Are the expected counts close to the actual counts? If not, which ones differ? Does there appear to be an association between the two variables?

3.11 Hodgkin's disease is a cancer of the lymph nodes. In one study, 538 patients with the disease were classified by histological type and by their response to treatment after 3 months. The histological types are LP = lymphocyte predominance, NS = nodular sclerosis, MC = mixed cellularity, and LD = lymphocyte depletion. The relationship between histological type and response to treatment may be examined from the table.

		Response			
		Positive	Partial	None	Total
	LP	74	18	12	104
Histological	NS	68	16	12	96
type	MC	154	54	58	266
	LD	18	10	44	72
Total		314	98	126	538

SOURCE: I. Dunsmore, and F. Daly, *Statistical Methods, Unit 9, Categorical Data*, Milton Keynes: The Open University, 18.

a. Of patients with histological type LP, what percent had a positive response?
b. Of patients with positive responses, what percent have histological type LP?
c. Why are the percentages in parts **a** and **b** so different?
d. What percent of all patients had no response to treatment?

3.12 The data in the table were collected on 1,398 children to determine whether carriers of the bacterium *Streptococcus pyogenes* have larger tonsils than noncarriers.

		Carrier Status		
		Carrier	Noncarrier	Total
	Normal	19	497	516
Tonsil size	Large	29	560	589
	Very large	24	269	293
Total		72	1,326	1,398

SOURCE: W. J. Krzanowski, *Principles of Multivariate Analysis*, (Oxford: Oxford University Press, 1988), p. 269.

a. Of the children with very large tonsils, what percent are carriers?
b. Of the children with normal tonsils, what percent are carriers?
c. Of all carriers, what percent have very large tonsils and what percent have normal tonsils?
d. Does the evidence presented here show that there is a relationship between tonsil size and the carrier status of these children?

3.2 SCATTERPLOTS

Numerical Versus Numerical Variables

Example 3.1 showed that when two variables are categorical, their relationship can be displayed in a contingency table. When the variables are numerical, such bivariate data are displayed in a *scatterplot*, as illustrated in the following example.

EXAMPLE **3.5**   It has been suggested that there is a relationship between sleep deprivation and the ability to complete simple tasks. To evaluate this hypothesis, 12 people were asked to solve simple tasks after having been without sleep for 15, 18, 21, and 24 hours. The number of tasks completed in 10 minutes after having been deprived of sleep was recorded for the 12 subjects, three at each of the four deprivation levels. The data were recorded as follows:

Subject	1	2	3	4	5	6	7	8	9	10	11	12
Hours without sleep	15	15	15	18	18	18	21	21	21	24	24	24
Tasks completed	13	9	15	8	12	10	5	8	7	3	5	4

To construct a scatterplot, a rectangular coordinate system is drawn, with the number of hours without sleep recorded on the horizontal scale (x-axis) and the number of tasks completed on the vertical scale (y-axis). Then each of the pairs (x, y) of observations is plotted as a point in the xy-plane. Figure 3.3 is a scatterplot of the above data.

FIGURE 3.3   Scatterplot of sleep deprivation versus completion of tasks

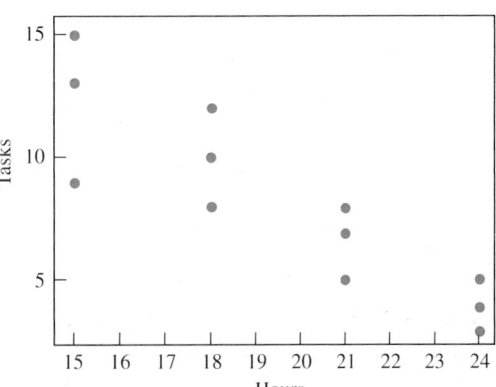

A scatterplot is useful for displaying trends in the data and revealing any association that might exist between the two variables of the bivariate data. If the two variables are related, we then ask:

What is the form of the relationship?

How strong is the relationship?

Can we predict the value of one variable from the other?

It appears from the scatterplot in Figure 3.3 that there is a downward *linear* (straight line) trend between sleep deprivation and the number of tasks performed. This means that the longer one goes without sleep, the fewer tasks one can complete in a 10-minute period. A relationship like this, where one variable tends to decrease as the other increases, is called a *negative* relationship. In a *positive* relationship, higher values of one variable tend to be associated with higher values of the other variable. In constructing scatterplots, if the two variables are such that one variable "depends" on the other, then it is recommended that the dependent values be graphed on the vertical axis.

Bivariate Outliers

Univariate outliers were defined as observations that are far removed from the remainder of the data. In a scatterplot, if an observation pair is far removed from the overall pattern of the remaining data, then it is called a *bivariate outlier*. Care must be exercised when graphing data that have obvious outliers.

EXAMPLE 3.6

If there are outliers in the data, it is a good idea to always graph the data both with and without the outliers.

The number of convictions reported by U.S. attorneys' offices varies widely from district to district. Some claim to be understaffed and unable to prosecute all crimes. Worksheet: **Attorney.mtw** gives a comparison of 88 districts by staff per 1 million population and convictions per 1 million population.

Figure 3.4 is a scatterplot of the staff per million by convictions per million. Because the point $(212, 838)$, Washington, DC, is an outlier (located in the upper right-hand corner), the scale for both the x- and y-axes must be extended to graph the point. Consequently, however, the rest of the data are so compact that they are difficult to interpret. Figure 3.5 is a scatterplot of the same data without the outlier, and you can see that there is a moderate upward linear trend. Note also that there is a large degree of scatter in the graph, indicating that the association between the two variables is weak.

FIGURE 3.4   U.S. attorney data with Washington, DC

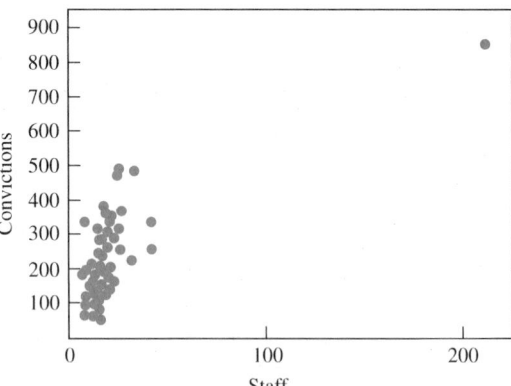

In the next section, we measure the strength of the association between two variables with the *correlation coefficient*. There we will see that outliers like the Washington, DC, observation can have a dramatic effect on the value of the correlation.

FIGURE 3.5 U.S. attorney data without Washington, DC

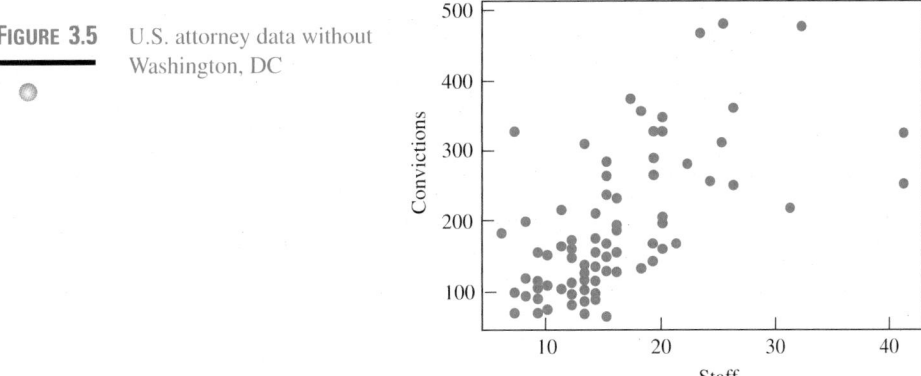

Figure 3.6 shows that an observation can be a *bivariate outlier* but not an outlier in either of the other variables when considered separately. The identified observation is clearly removed from the main pattern of the data, and yet it is still within the range of the *x* data and within the range of the *y* data.

FIGURE 3.6

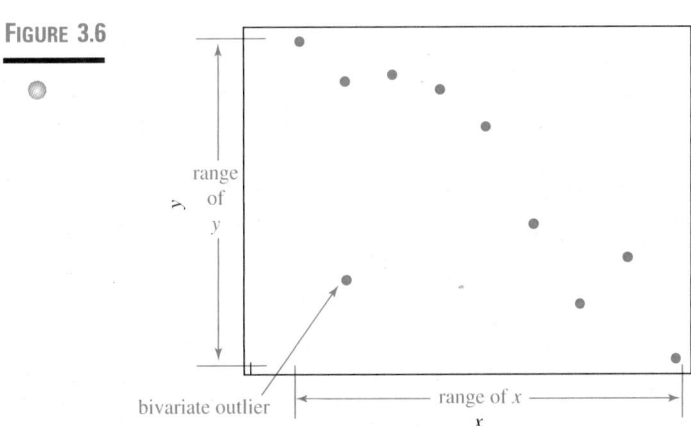

Transforming Data

Occasionally a simple transformation of the data, such as taking a logarithm or square root of each observation, can reveal relationships between variables that might otherwise go undetected.

EXAMPLE 3.7

Are the brain weights of animals related to their body weights? In other words, does it require a larger brain to govern a heavier body? The brain weights (in grams) and the body weights (in kilograms) of 28 animals are given on page 164.

Worksheet: Brain.mtw

Species	Body wt (kg)	Brain wt (g)
Mountain Beaver	1.35	8.1
Cow	465.0	423.0
Gray wolf	36.33	119.5
Goat	27.66	115.0
Guinea pig	1.04	5.5
Diplodocus	11,700.0	50.0
Asian elephant	2,547.0	4,603.0
Donkey	187.1	419.0
Horse	521.0	655.0
Potar monkey	10.0	115.0
Cat	3.3	25.6
Giraffe	529.0	680.0
Gorilla	207.0	406.0
Human	62.0	1,320.0
African elephant	6,654.0	5,712.0
Triceratops	9,400.0	70.0
Rhesus monkey	6.8	179.0
Kangaroo	35.0	56.0
Hamster	.12	1.0
Mouse	.023	.4
Rabbit	2.5	12.1
Sheep	55.5	175.0
Jaguar	100.0	157.0
Chimpanzee	52.16	440.0
Brachiosaurus	87,000.0	154.5
Rat	.28	1.9
Mole	.122	3.0
Pig	192.0	180.0

SOURCE: P. Rousseeuw and A. Leroy, *Robust Regression and Outlier Detection* (New York: Wiley, 1987).

FIGURE 3.7

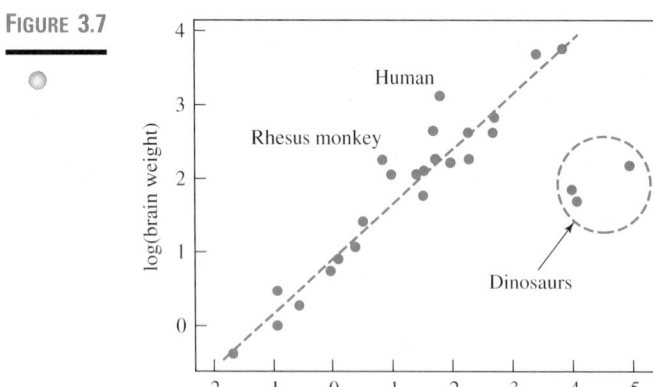

Because of the large variation in body weights (from .023 to 87,000 kg) and brain weights (from .4 to 5,712 g), a standard scatterplot would be so distorted that no reasonable pattern could be detected. After applying logarithms (base 10) to the data, however, we see a definite pattern in Figure 3.7. Notice the main scatter about the dotted line in the graph. Also notice the three outliers associated with the dinosaurs. They are bivariate outliers, but they are not univariate outliers with respect to either of the individual variables.

Categorical Versus Numerical Variables

A variation of the scatterplot can be used to display bivariate data in which one variable is numerical and the other is categorical. This is illustrated in the following example.

EXAMPLE **3.8**

At the end of a semester, a university professor asked her students to grade her teaching on a scale from A to F. She also asked them to record their present grade point averages. Graph the data that were obtained:

Teacher's rating	B	C	C	A	B	D	C	C	C	F	A	C	D	B	B	F
Grade point average	3.2	2.4	3.6	4.0	3.6	1.2	3.2	4.0	2.0	.4	3.6	3.2	2.8	2.4	4.0	1.6

SOLUTION

To graph these data, a coordinate system is drawn with the teacher's grades on the horizontal scale and the student's GPAs on the vertical scale (see Figure 3.8). As in a scatterplot, the data are plotted by placing a dot at the intersection of the grade (rating) and the GPA for each of the 16 students.

FIGURE **3.8** Scatterplot of teacher's rating versus GPA

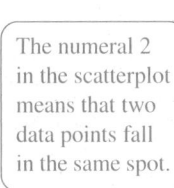

The numeral 2 in the scatterplot means that two data points fall in the same spot.

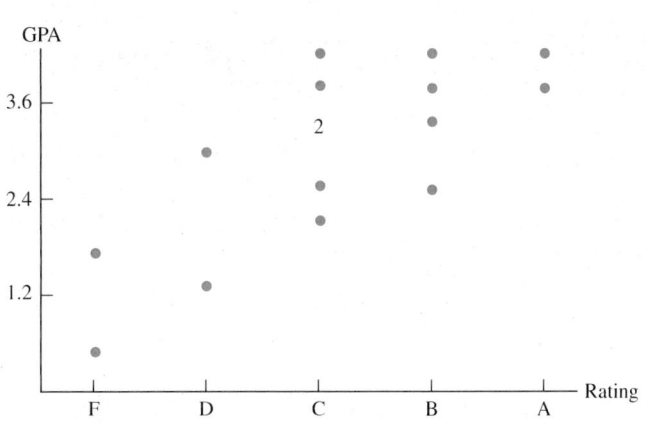

Note that, as might be expected, there is an upward trend; students who rate the teacher higher tend to have higher GPAs. To observe a trend like this, it is necessary that we order the categorical variable (teacher's rating) from smallest to largest; otherwise, we could list the categories in the opposite direction and observe a downward trend.

If there is a large amount of data, boxplots of the numerical data can be constructed at each category of the categorical variable. The resulting graph of side-by-side boxplots provides a comparison of the different distributions of data at each category of the categorical variable. The boxplots may give a more informative view of the data without all the scatter in a scatterplot.

EXAMPLE 3.9

Suppose the university professor in Example 3.8 had a large lecture class of 250 students. A scatterplot of individual grade point averages versus the teacher's ratings, as in Example 3.8, would be unusually cluttered. With the aid of a computer, however, it is a simple matter to construct side-by-side boxplots of the student GPAs for each of the ratings F, D, C, B, and A. Figure 3.9 is such a graph created with Minitab from the worksheet: **Ratings.mtw**.

FIGURE 3.9   GPA distributions versus teacher's ratings

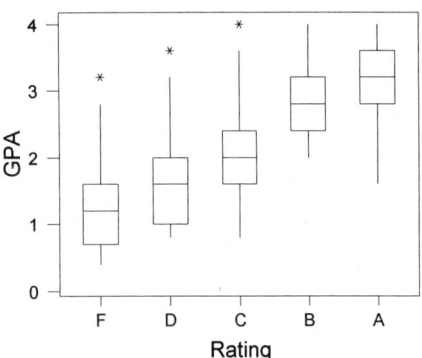

C O M P U T E R   T I P

Scatterplots are easily constructed with Minitab.

Menu mode: After the data are stored in the worksheet, select the **Graph** pull-down menu and click on **Plot**. At the Plot window, highlight the Y variable and click **Select**; then highlight the X variable and click **Select**; then click on **OK**.

Session window: To construct a scatterplot in the Session window, at the MTB prompt, type

MTB> **Plot 'Y'*'X'**

where Y and X are the names of the two columns to be graphed.

EXERCISES 3.2

3.13 For the following data sets construct a scatterplot and comment on the relationship between X and Y:

Worksheet: Ex3-13.mtw

a.
X	2	1	3	2	3	1
Y	4	0	6	6	8	2

b.
X	1	3	4	6	8	9	11	14
Y	1	2	4	4	5	7	8	9

c.
X	1	2	4	5	4	2	6	3
Y	5	11	25	39	27	12	50	17

d.
X	−3	−1	0	1	3
Y	12	7	6	4	1

3.14 The amount of energy consumed in a home is related to the size of the home. From the following data, construct a scatterplot and comment on the relationship between the variables. Does it matter which variable is on the x-axis of the scatterplot?

Worksheet: Energy.mtw

Size of home (sq ft)	Kilowatt-hours per month
2,820	1,975
2,500	1,952
2,350	1,894
2,000	1,841
1,950	1,769
1,875	1,674
1,740	1,590
1,650	1,505
1,490	1,386
1,350	1,220
1,270	1,089
1,200	1,042

3.15 A study was made to relate aptitude test scores to productivity in a factory after employees had 3 months of personnel training. The following results were obtained by testing eight randomly selected applicants and later measuring their productivity. Draw a scatterplot of aptitude scores and productivity and comment on the relationship.

Worksheet: Aptitude.mtw

Applicant	1	2	3	4	5	6	7	8
Aptitude score	9	17	13	19	20	23	12	15
Productivity	23	35	29	33	40	38	25	31

3.16   The robbery rate varies from precinct to precinct. From the following data, construct a scatterplot and comment on the relationship between the robbery rate and the percent of low-income residents in the precinct.

Worksheet: Precinct.mtw

Precinct	1	2	3	4	5	6	7	8
Robbery rate	20	41	165	88	60	120	65	81
Percent low income	4.9	7.1	10.1	11.8	13.5	14.8	16.2	11.2

3.17   A record of maintenance costs is kept on nine cash registers in a major department store chain. Construct a scatterplot and comment on the relationship between the variables.

Worksheet: Register.mtw

Age (years)	6	7	1	3	6	2	5	4	3
Cost (dollars)	92	181	23	40	126	35	86	72	51

3.18   It is suspected that the life span of a particular electronic component used in a spacecraft is dependent on the heat it experiences. Six such components were tested at various levels of heat. Construct a scatterplot and comment on the relationship between the variables.

Worksheet: Lifespan.mtw

Heat (°C)	50	100	150	200	250	300
Life span (hours)	875	884	762	424	365	128

3.19   The table gives PSAT and SAT scores for a group of high school seniors. Construct a scatterplot of SAT versus PSAT scores and comment on the relationship between the variables.

Worksheet: Psat.mtw

Student	1	2	3	4	5	6	7
PSAT	760	1150	820	1060	950	1320	750
SAT	920	1100	1050	1340	1060	1500	920

3.20   A study of the association between reading ability and IQ scores was conducted by a reading coordinator in a large public school system. A random sample of 14 eighth-grade students were given a reading achievement test and an IQ test. Their scores are recorded in the table:

Worksheet: Readiq.mtw

Reading score	42	35	61	28	48	46	59	21	47	29	65	37	35	53
IQ score	105	110	122	92	112	100	120	85	125	96	130	90	107	120

a. Organize the data in a scatterplot.
b. Are there any outliers in the data?
c. Does it appear that a straight line would fit the data reasonably well?
d. Sketch in a straight line that you think fits the data.

3.21   To understand some of the problems facing the American farmer, construct a scatterplot of the following data corresponding to the price of a bushel of wheat and the national weekly earnings of production workers:

Worksheet: Wheat.mtw

Year	1980	1981	1982	1983	1984
Wheat price	$3.91	3.65	3.55	3.54	3.39
Week earnings	$235.10	255.20	267.26	280.70	292.86

Year	1985	1986	1987	1988	1989
Wheat price	$3.08	2.42	2.57	3.72	3.72
Week earnings	$299.09	304.85	312.50	322.02	334.24

Year	1990	1991	1992	1993	1994
Wheat price	$2.61	3.00	3.24	3.26	3.39
Week earnings	$345.35	353.98	363.61	373.64	381.26

SOURCE: *The World Almanac and Book of Facts*, 1995.

a. Prior to 1987 there was a definite downward trend in the relationship between the price of wheat and weekly earnings. Does that trend appear linear?
b. What happened in 1988, 1989, and 1990?
c. What has happened since 1990?
d. Does there seem to be any natural trend in the price of wheat as it relates to weekly earnings?
e. Name two variables that you think would affect the price of wheat.

3.22 Construct a scatterplot of the following data corresponding to pay raises from the previous year for salaried employees and the inflation rates:

Worksheet: Inflatio.mtw

Year	1976	1977	1978	1979	1980	1981	1982	1983	1984
Raise	7.3%	8.0	8.4	8.3	8.1	8.9	5.9	4.4	3.7
Inflation	4.9%	6.7	9.0	13.3	12.5	8.9	3.8	3.8	3.9

Year	1985	1986	1987	1988	1989	1990	1991	1992	1993
Raise	3.0	2.2	2.5	3.3	4.1	3.6	3.1	2.4	2.5
Inflation	3.8	1.1	4.4	4.4	4.6	6.1	3.1	2.9	2.7

SOURCE: Bureau of Labor Statistics.

a. Construct a scatterplot of the pay raises versus the inflation rates.
b. Describe the pattern that you see in the scatterplot.
c. Describe the pattern when the inflation rate is below 6%.
d. What happens when the inflation rate goes above 6%? In particular, when the inflation rate is between 8% and 14%, what can you say about pay raises?

3.23 Using the data in Exercise 3.22, code the inflation rate as low if it is less than 5%, as middle if it is greater than or equal to 5% and less than 8%, and as high if it is greater than or equal to 8%.
a. Create side-by-side boxplots of the pay increase for the three categories of inflation rate.
b. What do you observe about the variability in pay raises for the three categories of inflation rate?

3.24 J. D. Power and Associates ranks cars and trucks every year according to the number of problems reported during the first 3 months of ownership. The following data are obtained from Power's ratings and represent the number of problems reported per 100 cars:

Worksheet: JDPower.mtw

Car	1994	1995	Car	1994	1995
Infiniti	75	55	Lexus	54	60
Acura	101	64	Honda	92	71
Toyota	69	74	Mercedes	91	79
BMW	114	82	Geo	134	82
Nissan	99	83	Volvo	108	86
Subaru	136	88	Buick	91	90
Cadillac	104	91	Saturn	78	91
Lincoln	76	100	Oldsmobile	109	101
Chevrolet	138	105	Mazda	115	106
Pontiac	138	110	Mercury	86	115
Ford	112	116	Jaguar	144	132
Plymouth	122	137	Dodge	144	144
Saab	180	155	Mitsubishi	150	164
Eagle	155	168	Volkswagen	158	183
Hyundai	193	195			

SOURCE: *USA Today*, May 25, 1995.

a. Construct a scatterplot of the 1995 ratings versus the 1994 ratings.
b. Are there any outliers in the data?
c. Does it appear that a straight line would fit the data reasonably well?
d. Sketch in a straight line that you think fits the data.

3.25 The brain weights and body weights of the 28 animals listed in Example 3.7 are in worksheet: **Brain.mtw**.
   a. Construct a scatterplot of brain weight versus body weight. Do you see the problem with graphing the original data?
   b. Construct a scatterplot of log(brain wt) versus log(body wt). By transforming each variable, have you produced a scatterplot that shows a definite relationship between the two variables?

3.26 In World War II, the actual number of German submarines sunk each month by the U.S. Navy did not exactly match the number reported by the Navy. Following are the results for 16 months:

Worksheet: Submarin.mtw

Month	1	2	3	4	5	6	7	8
Reported by Navy	3	2	4	2	5	5	9	12
Actual count	3	2	6	3	4	3	11	9

Month	9	10	11	12	13	14	15	16
Reported by Navy	8	13	14	3	4	13	10	16
Actual count	10	16	13	5	6	19	15	15

SOURCE: F. Mosteller, S. Fienberg, and R. Rourke, *Beginning Statistics with Data Analysis* (Reading, MA: Addison-Wesley, 1983).

a. Construct a scatterplot with the actual count on the vertical (y) axis.
b. Is there a reasonable linear pattern to the data? Is there a positive or negative trend?
c. Are there any outliers in the data?

3.3 CORRELATION

If two variables are related in such a way that the points in a scatterplot tend to fall in a straight line, then we say that there is an association between the variables and that they are linearly correlated. In this section, we discuss how to measure numerically the degree of association (the *correlation*) between the two variables.

Chapter 2 introduced the concept of summarizing a univariate data set with summary statistics. For example, the mean and median are two different quantities that summarize the center of the distribution. Furthermore, the amount of variability in the data is summarized with the sample variance, s^2.

In summarizing bivariate data, we will use the subscripts x and y to distinguish between the two sample variances. That is, we measure the variability of each sample with

$$s_x^2 = \frac{\sum(x_i - \bar{x})^2}{n - 1} \quad \text{and} \quad s_y^2 = \frac{\sum(y_i - \bar{y})^2}{n - 1}$$

These two measures of variability are computed independently of each other and thus tell us nothing about how the two variables are related. To assess a possible relationship, we need a measure of the *covariability* between the two variables.

Covariance

Given n observations $(x_1, y_1), (x_2, y_2), \ldots, (x_n, y_n)$ of the variables x and y, the **covariance between x and y** is

$$s_{xy} = \frac{\sum(x_i - \bar{x})(y_i - \bar{y})}{n - 1}$$

Notice that instead of squaring the x deviations from their mean or the y deviations from their mean, we simply multiply the two deviations together. Also, the divisor of s_{xy} is $n - 1$, the same as the divisor of the sample variances.

As its name suggests, the *covariance* is a measure of how two variables vary together. For instance, as the values of one variable increase, do the corresponding values of the other variable also increase? If they vary in such a way that the scatterplot pattern falls near a straight line, then we say that the variables are *linearly* related. The covariance measures the linear dependence between the two variables.

If there is no linear association between the two variables, then the covariance will be zero. On the other hand, for some data sets, the covariance can get very large and difficult to interpret.

Pearson Correlation Coefficient

To get a more meaningful measure of the linear association between two variables, we can standardize the covariance by dividing it by the standard deviations of the two samples. The result, the *Pearson correlation coefficient*, will always be between -1 and $+1$ and measures the strength of the linear association between the variables x and y.

The **Pearson correlation coefficient**, denoted by r, is given by

$$r = \frac{s_{xy}}{s_x s_y} = \frac{\Sigma(x_i - \bar{x})(y_i - \bar{y})}{\sqrt{\Sigma(x_i - \bar{x})^2 \, \Sigma(y_i - \bar{y})^2}}$$

(The second expression for r is obtained by canceling $n - 1$ in the numerator and denominator.)

The **population correlation coefficient** is denoted by the Greek letter ρ (rho).

EXAMPLE 3.10

A laboratory wishes to study the relationship between the dose of a growth stimulant and weight gain in laboratory animals. Seven animals of the same gender, age, and size are selected and randomly assigned to one of seven doses of the growth stimulant. The following weight gains are recorded:

Dose	.0	1.0	2.0	3.0	4.0	5.0	6.0
Weight gain	1.0	1.2	2.0	2.4	3.4	4.9	5.1

Plot the data in a scatterplot and calculate the correlation between the two variables.

SOLUTION

To plot the data, we place the dose of the growth stimulant on the x-axis and the weight gain on the y-axis. The data, shown in Figure 3.10, appear to fall close to a straight line, suggesting a strong linear correlation. The value of r should bear this out.

FIGURE 3.10

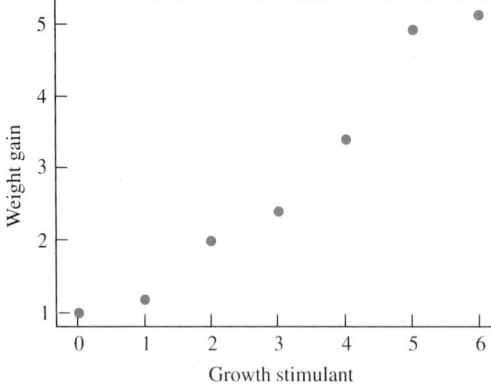

From the x and y data, we find

$$\bar{x} = 3 \qquad s_x = 2.16 \qquad \bar{y} = 2.857 \qquad s_y = 1.665$$

Table 3.5 lists the computations for the covariance. Therefore,

$$s_{xy} = \frac{\Sigma(x - \bar{x})(y - \bar{y})}{n - 1} = \frac{21.1}{6} = 3.5167$$

$$r = \frac{s_{xy}}{s_x s_y} = \frac{3.5167}{(2.16)(1.665)} = .978$$

This is a very high correlation coefficient and consistent with the scatterplot.

	x	y	$x - \bar{x}$	$y - \bar{y}$	$(x - \bar{x})(y - \bar{y})$
TABLE 3.5	0	1.0	-3	-1.857	5.571
	1	1.2	-2	-1.657	3.314
	2	2.0	-1	$-.857$	.857
	3	2.4	0	$-.457$	.000
	4	3.4	1	.543	.543
	5	4.9	2	2.043	4.086
	6	5.1	3	2.243	6.729
				Total	21.1

Interpretation of r

To properly interpret the correlation coefficient, you must understand the basic properties of r:

- The value of r measures the strength of the *linear* relationship between x and y and will always be between -1 and $+1$.
- The closer r is to either -1 or $+1$, the stronger the linear relationship between x and y. In fact, points that fall exactly on a straight line have a correlation of $+1$ if the line has positive slope and -1 if the line has negative slope.
- If r is zero, then x and y are not linearly related. They may be related, but the relationship is not a straight line.
- The value of r does not change when the units of measurement are changed.

Figure 3.11 (page 174) illustrates scatterplots of data for four values of r. In the first scatterplot, where $r = +.05$, there is very little relationship between x and y. The second scatterplot, where $r = +.99$, shows an almost perfect straight-line relationship between x and y. The third and fourth scatterplots show the same degree of association with $r = +.7$ and $r = -.7$. The only difference between the two is that when $r = +.7$ there is a positive linear relationship between x and y (as one variable increases, the other one also increases), and when $r = -.7$ there is a negative linear relationship (as one variable increases, the other decreases).

Caution: There could be a nonlinear relationship between x and y and r could still be near zero. Figure 3.12 on page 174 illustrates a perfect quadratic relationship between x and y and yet the correlation is zero. This is a reminder that Pearson's correlation coefficient measures *linear* relationships.

FIGURE 3.11   Scatterplots for various values of r

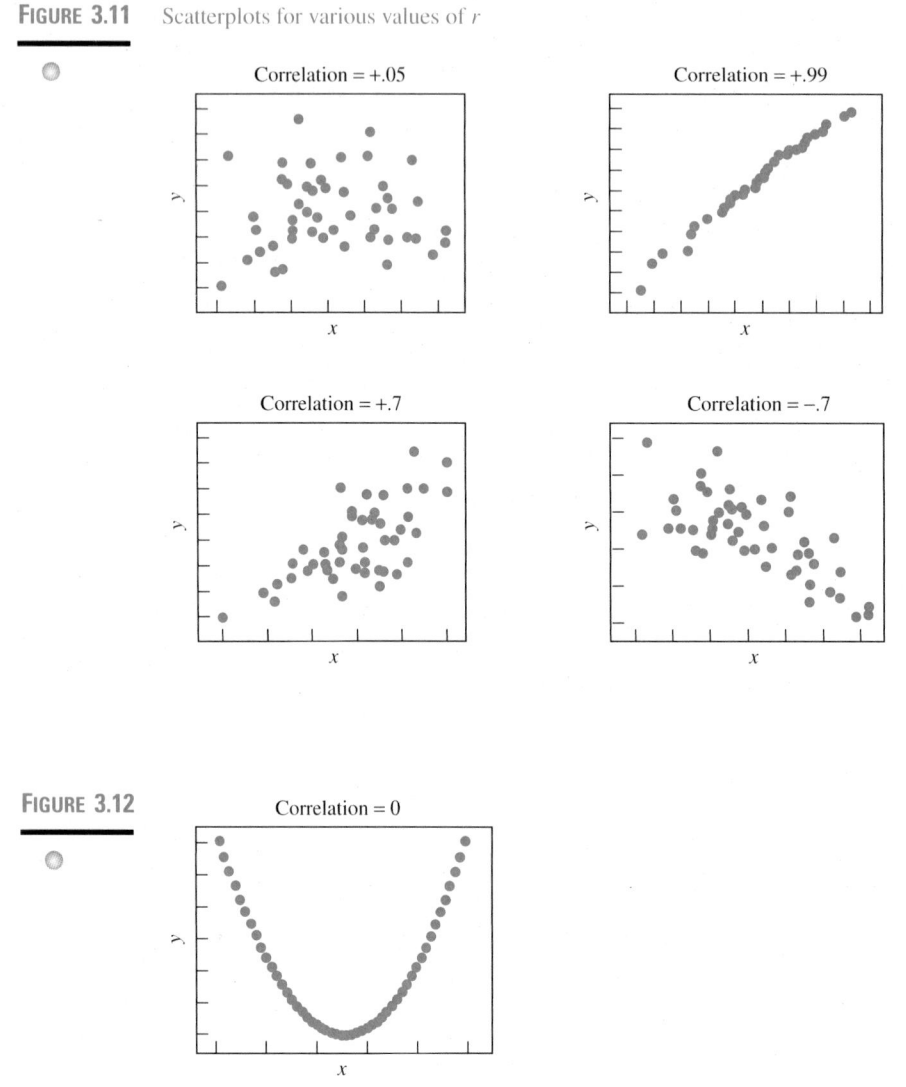

FIGURE 3.12

The following example shows two variables that are highly correlated.

EXAMPLE 3.11   What's in a brand name? *Financial World* magazine, using its own complex formula, estimated how much the following brand names would be worth in cash. How are these estimated values related to the company's revenue? The table gives the brand name, its value in billions of dollars, and the company's revenue in billions:

Worksheet: Name.mtw

Brand name	Value	Revenue	Brand name	Value	Revenue
Marlboro	31.2	15.4	Wrigley's	1.5	1.0
Coca-Cola	24.4	8.4	Schweppes	1.4	1.3
Budweiser	10.1	6.2	Tampax	1.4	.6
Pepsi-Cola	9.6	5.5	Heinz	1.3	.8
Nescafe	8.5	4.3	Quaker	1.2	1.1
Kellogg	8.4	4.7	Colgate	1.2	1.1
Winston	6.1	3.6	Gordon's	1.1	.6
Pampers	6.1	4.0	Hermes	1.0	.5
Camel	4.4	2.3	Kleenex	.8	.7
Campbell	3.9	2.4	Carlsberg	.7	.8
Nestle	3.7	6.0	Haagen-Dazs	.6	.5
Hennessy	3.0	.9	Fisher-Price	.6	.6
Heineken	2.7	3.5	Nivea	.6	.9
Johnnie Walker	2.6	1.5	Sara Lee	.5	.8
Louis Vuitton	2.6	.9	Oil of Olay	.5	.6
Hershey	2.3	2.6	Planters	.5	.7
Guinness	2.3	1.8	Green Giant	.4	1.0
Barbie	2.2	.8	Jell-o	.4	.3
Kraft	2.2	2.8	Band-Aid	.2	.2
Smirnoff	2.2	1.0	Ivory	.2	.4
Del Monte	1.6	2.3	Birds Eye	.2	.3

SOURCE: *Financial World*, August 12, 1992.

Construct a scatterplot and find the correlation between value and revenue.

SOLUTION

Figure 3.13 shows a scatterplot of the data. It appears that the data fall close to a straight line. From Minitab the correlation is found to be +.94, which is a very high positive correlation.

FIGURE 3.13   Brand data

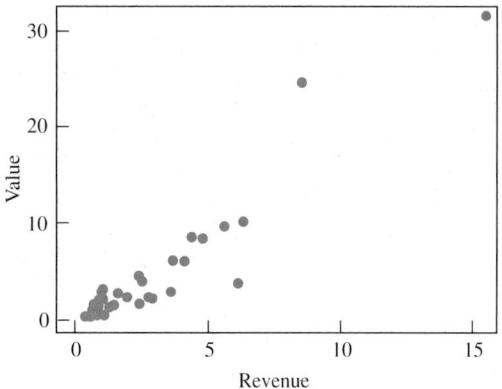

C O M P U T E R   T I P

To correlate data in two columns of the Minitab worksheet, first choose the **Stat** menu, select **Basic Statistics**, and click on **Correlation**. When the Correlation window appears, **Select** the columns and click on **OK**. If more than two columns are selected, a matrix of correlations appears. More will be said about this in Chapter 11.

There are occasions when we think that two variables are correlated but in fact the data do not support our intuition.

EXAMPLE 3.12   Are complaints with airlines correlated with their ability to arrive on time? Listed here are the percentages of on-time arrivals and the numbers of complaints per 1,000 passengers for 11 major airlines:

Worksheet: Airline.mtw

Airline	Percent on time	Complaints (per 1,000)
Alaska	91.1	5.4
American	85.8	3.6
American West	90.8	4.0
Continental	87.2	4.6
Delta	85.7	4.6
Northwest	91.1	4.3
Pan Am	88.3	4.6
Southwest	93.5	3.6
TWA	88.4	5.4
United	87.2	4.9
USAir	87.3	4.4

SOURCE: U.S. Department of Transportation, 1991.

FIGURE 3.14   Airline data

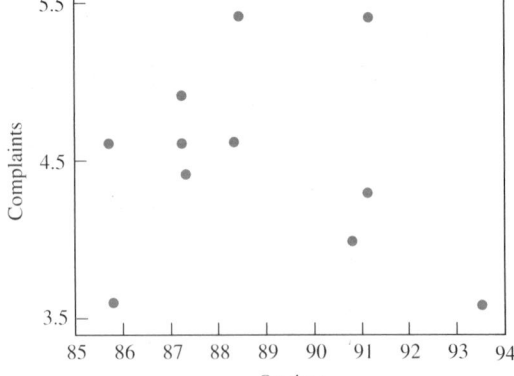

Figure 3.14 is the scatterplot of the data; the correlation is found to be $-.15$. The correlation coefficient is negative, which agrees with our intuition that as the percentage of on-time arrivals increases, the number of complaints should go down. However, the correlation of $-.15$ is not nearly as strong as one would suspect.

Effect of Outliers   It was pointed out in Section 3.2 that outliers make it difficult to construct a scatterplot. Similarly, outliers sometimes have a dramatic effect on the value of the correlation coefficient.

EXAMPLE 3.13   Figure 3.15 is a plot of the airline data given in Example 3.12 with one additional observation. Suppose there is one more airline that is on time 100% of the time and has only 1 complaint per 1,000 passengers. With the addition of that one point, the correlation jumps from the previous value of $-.15$ to $-.74$. Therefore, caution should be exercised when interpreting correlation with the presence of outliers.

FIGURE 3.15   Airline data with the addition of $(100, 1)$

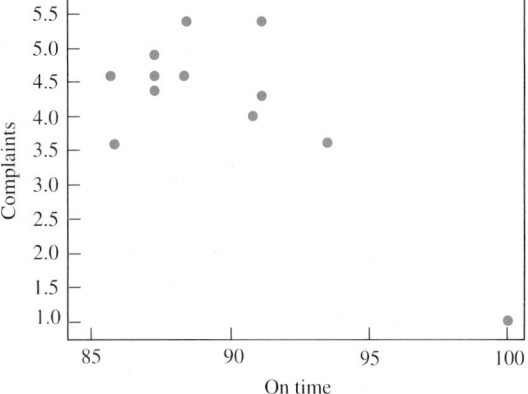

Spearman's Rank Correlation Coefficient   A measure of correlation for categorical data that can be ranked is *Spearman's rank correlation coefficient*, r_S. Given the sample $(x_1, y_1), (x_2, y_2), \ldots, (x_n, y_n)$, ranks of 1 to n are assigned to the x's and to the y's separately. For tied observations, the average of the ranks that would have been assigned had there been no ties is given to each tied observation. The formula for r_S is the same as the formula for Pearson's correlation coefficient except that it is applied to the ranks of the data.

EXAMPLE 3.14   A married couple with children was asked to rank ten factors in raising children from the most important (10) to the least important (1). From the collected data in Table 3.6 on page 178, compute Spearman's rank correlation coefficient.

TABLE 3.6

Factor	Husband's Ranking	Wife's Ranking
1	6	6
2	3	3
3	1	2
4	7	9
5	2	1
6	8	7
7	4	5
8	9	8
9	5	4
10	10	10

SOLUTION

The data are already in the form of ranks; therefore, we apply Pearson's correlation formula to the ranks. After entering the ranks into two columns of the worksheet and executing the Correlation command, we get:

Correlations (Pearson)

Correlation of husband and wife = 0.939

Although the command says Pearson correlation, because we have applied Pearson's formula to the ranks, we call the resulting correlation a Spearman rank correlation coefficient. Based on these data, there is very high agreement between husbands and wives on the main factors in raising children.

Spearman's rank correlation can also be applied to numerical data. All that is required is that the data for each variable be ranked from smallest to largest, and then the formula is applied to the ranks. Again, if there are tied observations (having the same value), average ranks are assigned.

Like the median, Spearman's rank correlation is resistant to the influence of outliers. It treats all observations equally, so that an outlier has no more effect on the outcome than any other observation. Additionally, Spearman's correlation identifies both linear and nonlinear relationships. For example, a weight lifter makes great strides in the amount he can lift early in his training career. Later in his training, however, he may be able to add only ounces to the amount he can lift from week to week. Figure 3.16 illustrates that this is not a linear relationship; it is called a *monotonic* relationship. Because of the upward trend, it can be further classified as monotonic increasing. Unlike Pearson's correlation coefficient, Spearman's correlation coefficient can measure certain nonlinear trends.

FIGURE 3.16

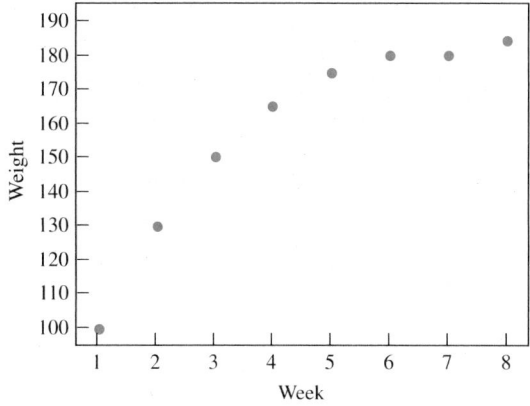

EXAMPLE 3.15

A math teacher suspects that her students' algebra test scores are related to the amount of test anxiety they have prior to the test. With an anxiety scale that ranges from 0 to 100 (the higher the score, the more anxiety exhibited), she scores each student's anxiety.

Worksheet: Algebra.mtw

Anxiety Score	Anxiety Rank	Math Score	Math Rank
61	5	76	5
47	2	84	8
82	9	51	2
65	6	79	6
72	7	71	4
84	10	28	1
55	4	87	9
49	3	82	7
75	8	60	3
12	1	98	10

a. Calculate Pearson's correlation coefficient.

b. Calculate Spearman's correlation coefficient.

c. Comment on the two correlations.

SOLUTION

The scatterplot in Figure 3.17 (page 180) demonstrates a monotonic decreasing pattern.

a. From the anxiety scores and the math scores, we find

$$r = -.8361$$

b. From the anxiety ranks and the math ranks, we find

$$r_s = -.9515$$

FIGURE 3.17

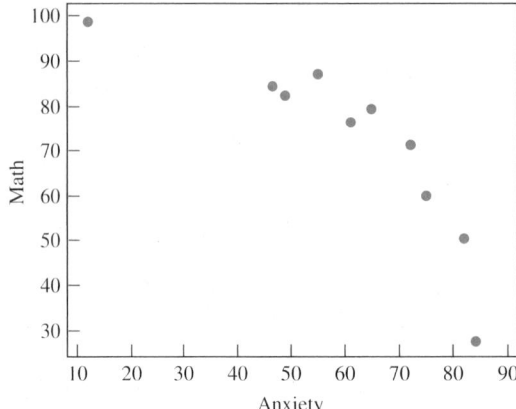

c. The last anxiety score of 12 is unusually low (an outlier), which decreases the Pearson correlation. The exact numerical value of the score is not important; what is important is that it is the lowest anxiety score and we would expect the student to score high on the algebra test. Spearman's correlation ignores the fact that the score is a 12 and simply looks at it as the lowest score, a rank of 1, and is not affected by its magnitude. It is for this reason that Spearman's correlation coefficient of $-.9515$ is the more representative value of the association between the anxiety scores and the algebra scores.

C O M P U T E R   T I P

As previously stated, Spearman's correlation is computed with the Pearson formula applied to the ranks instead of to the actual data. With Minitab it is very easy to convert scores to ranks. Under the **Manip** menu, select **Rank**. When the Rank window appears, **Select** the column you wish to convert to ranks and store it in the **Rank data in** box. Next specify the column that you wish to **Store ranks in** and then click on **OK**. To calculate Spearman's correlation, simply correlate the two columns containing the ranked data.

Correlation and Causation

A large positive correlation between two variables means that large values of one variable tend to be associated with large values of the other variable. This does not necessarily mean that the large values of the first variable *caused* the large values of the other variable. Correlation measures the association between the variables, not the causal effect. If one were to measure the relationship between shoe size and vocabulary for elementary school children, for example, there would probably be a high positive correlation. This certainly does not mean that bigger feet cause an increase in one's vocabulary. The strong association that we observe between the two variables is because both variables are related to a third variable—in this case, age. As children get older, their feet grow and their vocabulary increases. The third variable that is causing the observed correlation between the two variables is called a **lurking variable**.

Lurking variable

Just because two variables are related through a lurking variable does not mean that we should ignore the association between them. Universities all over the country use SAT scores for admission purposes because of the association between SAT scores and college grade point averages. Several lurking variables explain the association between SAT scores and grade point averages—namely, motivation, study habits, home environment, and, of course, intelligence. The highly motivated student with good study habits will achieve a higher SAT score and, most likely, a higher grade point average. It is much easier for the university to measure SAT scores than to measure motivation, study habits, home environment, and intelligence. By measuring the SAT score, they are, in essence, measuring all the other variables at the same time.

About the Notation

When we introduced sample variance in Chapter 2, we identified the numerator of s^2, called a sum of squares, as an alternative way of measuring variability. For example, if we have two data sets of the same size, we can compare the amount of variability between the two by simply comparing

$$SS_x = \sum (x_i - \overline{x})^2 \quad \text{and} \quad SS_y = \sum (y_i - \overline{y})^2$$

The main reason we divide a sum of squares by its degrees of freedom is to get an *average* amount of variability. On some occasions we will compare s_x to s_y, and on other occasions we will compare SS_x to SS_y.

The same applies to the numerator of s_{xy}; that is, we can measure the covariability of x and y with

$$SP_{xy} = \sum (x_i - \overline{x})(y_i - \overline{y})$$

We call SP_{xy} the sum of products of x with y. From this notation it is easy to see that the correlation between x and y is given by

$$r = \frac{SP_{xy}}{\sqrt{SS_x SS_y}}$$

EXERCISES 3.3

3.27   The following data sets appeared in Exercise 3.13 in Section 3.2:

Worksheet: Ex3-13.mtw

a.
X	2	1	3	2	3	1
Y	4	0	6	6	8	2

b.
X	1	3	4	6	8	9	11	14
Y	1	2	4	4	5	7	8	9

c.
X	1	2	4	5	4	2	6	3
Y	5	11	25	39	27	12	50	17

d.
X	−3	−1	0	1	3
Y	12	7	6	4	1

Calculate Pearson's correlation coefficient between X and Y for each of the data sets. Refer to the scatterplots constructed in Exercise 3.13 in Section 3.2. In each case, does the correlation coefficient seem consistent with the scatterplot?

3.28   Construct a scatterplot of the following data:

Worksheet: Correlat.mtw

X	42	61	12	71	52	48	74	65	53	63	55	94	19
Y	75	49	95	64	83	84	38	58	81	47	78	51	93

a. Do you see a linear trend? Is it positive or negative?
b. Sketch a straight line through the data. Is there more variability in the Y data when X is small than when X is large?
c. Are there any data points that you would classify as outliers?
d. Is the correlation positive or negative?
e. Calculate Pearson's correlation coefficient. Is the value what you expected?

3.29   Refer to the scatterplot in Exercise 3.28. Do you think there is a need to calculate Spearman's correlation coefficient instead of Pearson's correlation coefficient? Explain why or why not. Convert the data to ranks and calculate Spearman's correlation. Compare the results with Pearson's correlation from Exercise 3.28. Is there much difference between the two correlations?

3.30   Karl Pearson, who developed the correlation coefficient, gave the following historic data set on the heights (in inches) of brothers and sisters:

Worksheet: Pearson.mtw

Family	Height of brother	Height of sister
1	71	69
2	68	64
3	66	65
4	67	63

(continued)

Family	Height of brother	Height of sister
5	70	65
6	71	62
7	70	65
8	73	64
9	72	66
10	65	59
11	66	62

SOURCE: K. Pearson, and A. Lee, "On the Laws of Inheritance in Man," *Biometrika* 2 (1902–3): 357.

a. Construct a scatterplot of the data. Is there a linear trend? Is there much scatter in the data?
b. Calculate the Pearson correlation coefficient between the two variables.

3.31 The following data compare a child's age with the number of gymnastic activities she is able to complete successfully:

Worksheet: Gym.mtw

Age	2	3	4	4	5	6	7	7
Number of activities	5	5	6	3	10	9	11	13

a. Construct a scatterplot and sketch in the straight line that fits the data.
b. Calculate the correlation coefficient. Is its value consistent with the appearance of the scatterplot?
c. Is the relationship positive or negative? Is that what you would normally expect?

3.32 Recent studies suggest that smoking during pregnancy affects the birth weights of newborn infants. To study this issue further, a sample of 16 women smokers were asked to estimate the average number of cigarettes they smoke per day. Following are their estimates and the birth weights of their children:

Worksheet: Cigarett.mtw

Cigarettes	22	16	4	19	42	8	12	30	14	16	5	20	32	2	15	48
Birth weight	6.4	7.2	8.1	6.9	6.1	8.4	7.6	6.5	8.4	8.1	8.5	6.6	6.0	7.9	7.1	5.5

a. Construct a scatterplot of the data.
b. Does there appear to be a linear trend in the data? Is it positive or negative?
c. Calculate Pearson's correlation coefficient. Is it positive or negative? What does this mean in terms of the linear trend?

3.33 From the appearance of the scatterplot in Exercise 3.32, do you think we should consider Spearman's rank correlation coefficient? Rank the data and calculate Spearman's correlation. Compare the results with Pearson's correlation from Exercise 3.32. Is there much difference between the two correlations?

3.34 A group of 15 football recruits were ranked by two coaches as to their potential value to the team:

Worksheet: Recruit.mtw

Player	1	2	3	4	5	6	7	8	9	10	11	12	13	14	15
Coach A	3	12	9	1	5	14	8	10	2	13	6	4	11	15	7
Coach B	5	10	7	1	8	15	9	12	3	13	4	2	11	14	6

a. To correlate the two coaches' evaluations of the football recruits, should we use Pearson's correlation coefficient or Spearman's rank correlation coefficient?

b. Explain the difference between the ways you would calculate the two correlation coefficients.

c. Perform the calculations and report the correlation between the two coaches' evaluations.

3.35 A professor ranked students in her history class at the beginning of the course based on their performance in a group discussion. Following are her rankings along with the final grades for the students:

Worksheet: History.mtw

Student	1	2	3	4	5	6	7	8	9
Ranking	6	3	2	8	4	9	1	5	7
Grade	72	91	90	68	89	55	95	84	62

a. To correlate the grade with the professor's ranking, should we use Pearson's correlation coefficient or Spearman's rank correlation coefficient?

b. Having answered part **a**, explain why the other correlation is not appropriate.

c. Perform the calculations and report your correlation.

3.36 In Exercise 3.15 in Section 3.2, you constructed a scatterplot relating aptitude test scores and productivity in a factory. To measure the strength of the linear relationship exhibited in the scatterplot, calculate the correlation coefficient (worksheet: **Aptitude.mtw**). Is the value you obtained consistent with the appearance of the scatterplot? Do you think a straight line will fit the data reasonably well?

3.37 There is an extreme outlier corresponding to Marlboro in the data in Example 3.11 (worksheet: **Name.mtw**). Remove the outlier and recalculate the correlation. Is there a significant change in the value of the correlation? Explain what effect this outlier has on the correlation coefficient.

3.38 Exercise 3.21 in Section 3.2 presented the prices of wheat and the national weekly earnings of production workers for the years 1980–1994. Based on the scatterplot from that exercise, do you think there is a high linear correlation between the two variables? Explain you answer. Should we be looking for a linear trend in these data?

3.39 In Exercise 3.22 in Section 3.2, the private pay raise from the previous year was compared with the inflation rate for the years 1976–1993. From the scatterplot in that exercise, does there appear to be a linear trend? Will the correlation coefficient tell you something about the linear trend? From the data, stored in worksheet: **Inflatio.mtw**, calculate the correlation coefficient and evaluate the linear trend.

3.40 Violent crimes are the offenses of murder, forcible rape, robbery, and aggravated assault. Following are the rates per 100,000 inhabitants for each of the 50 states and the District of Columbia for 1983 and 1993:

Worksheet: Crime.mtw

State	1983	1993	State	1983	1993	State	1983	1993
Ala	416.0	871.7	Fla	826.7	1,207.2	La	640.9	984.6
Alaska	613.8	660.5	Ga	456.7	733.2	Maine	159.6	130.9
Ariz	494.2	670.8	Hawaii	252.1	258.4	Md	807.1	1,000.1
Ark	297.7	576.5	Idaho	238.7	281.4	Mass	576.8	779.0
Calif	772.6	1,119.7	Ill	553.0	977.3	Mich	716.7	770.1
Colo	476.4	578.8	Ind	283.8	508.3	Minn	190.9	338.0
Conn	375.0	495.3	Iowa	181.1	278.0	Miss	280.4	411.7
Del	453.1	621.2	Kan	326.6	510.8	Mo	477.2	740.4
D.C.	1,985.4	2,832.8	Ky	322.2	535.5	Mont	212.6	169.9

(continued)

State	1983	1993	State	1983	1993	State	1983	1993
Neb	217.7	348.6	Ohio	397.9	525.9	Texas	512.2	806.3
Nev	655.2	696.8	Okla	423.4	622.8	Utah	256.0	290.5
N.H.	125.1	125.7	Ore	487.8	510.2	Vt	132.6	109.5
N.J.	553.1	625.8	Pa	342.8	427.0	Va	292.5	374.9
N.M.	686.8	934.9	R.I.	355.2	394.5	Wash	371.8	534.5
N.Y.	914.1	1,122.1	S.C.	616.8	944.5	W.Va	171.8	211.5
N.C.	409.6	681.0	S.D.	120.0	194.5	Wis	190.9	275.7
N.D.	53.7	83.3	Tenn	402.0	746.2	Wyo	237.2	319.5

SOURCE: Department of Justice, *Uniform Crime Reports for the U.S. in 1983 and 1993*, pp. 52–63 and 366.

a. Construct a scatterplot of the 1993 crime rate versus the 1983 crime rate.
b. Are there any outliers in the data?
c. Would the removal of outliers change the relationship between the 1983 crime rate and the 1993 crime rate?
d. Would the removal of outliers have an effect on the correlation coefficient?
e. Calculate the correlation both with and without outliers and compare the results.

3.41   In Exercise 3.24 in Section 3.2, a scatterplot was constructed of the 1994 and 1995 ratings of cars by J. D. Power and Associates. The scatterplot showed a modest amount of scatter, and it appeared that a straight line would fit the data reasonably well. From the data, stored in worksheet: **JDPower.mtw**, calculate the correlation between the 1994 and 1995 ratings. Based on the scatterplot and the correlation, do you still think that a straight line fits the data?

3.42   The following four data sets illustrate that data can be very similar in some respects yet very different in others. For each of the data sets, construct a scatterplot and calculate the correlation. Comment on the relationship that you see between X and Y in each data set.

Worksheet: Anscombe.mtw

a.
X	10	8	13	9	11	14	6	4	12	7	5
Y	8.04	6.95	7.58	8.81	8.33	9.96	7.24	4.26	10.84	4.82	5.68

b.
X	10	8	13	9	11	14	6	4	12	7	5
Y	9.14	8.14	8.74	8.77	9.26	8.10	6.13	3.10	9.13	7.26	4.74

c.
X	10	8	13	9	11	14	6	4	12	7	5
Y	7.46	6.77	12.74	7.11	7.81	8.84	6.08	5.39	8.15	6.42	5.73

d.
X	8	8	8	8	8	8	8	8	8	8	19
Y	6.58	5.76	7.71	8.84	8.47	7.04	5.25	5.56	7.91	6.89	12.50

SOURCE: F. J. Anscombe, "Graphs in Statistical Analysis," *American Statistician* 27 (1973): 17–21.

3.4   LEAST SQUARES REGRESSION

Regression Analysis

With regression analysis we continue our study of the relationships among numerical variables. In Section 3.2, we saw that the scatterplot is a graphical procedure for visually detecting relationships in bivariate data. In Section 3.3, we measured the association between variables with the correlation coefficient. Regression analysis gives the tools to numerically describe the relationship between variables so that predictions can be made. Unlike correlation, with regression we must distinguish between the *response (dependent) variable* and the *predictor (independent) variable*.

> A **response variable**, also called a **dependent variable**, is the variable we wish to predict or describe based on the values of another variable. The **predictor variable**, also called the **independent variable**, is the variable that is used to predict the response variable. In the context of regression, the response variable is labeled y and the predictor variable will be labeled x.

EXAMPLE **3.16**   Grades in high school and college are measured with the cumulative grade point average (GPA). College admissions officers attempt to predict the success of applicants based on an assumed relationship between the high school and college GPAs. Here, college GPA is assumed to be the response variable y, and high school GPA is the predictor variable x.

To investigate a possible relationship between variables x and y, we generally set the predictor variable x at values $x_1, x_2, \ldots, x_n$ and then observe the corresponding values of y, denoted by $y_1, y_2, \ldots, y_n$. For example, to study the relationship between the stopping distance of an automobile and the speed at which it is traveling, we might measure the stopping distance when the initial speed is set at 20, 30, 40, and 50 miles per hour. Of course, multiple observations can be taken at the different x values. Note, however, it is not absolutely necessary that x be "set" at values $x_1, x_2, \ldots, x_n$. It is acceptable to observe both x and y. For example, we do not set the high school GPA (x) and then observe the college GPA (y). We simply observe both the high school and college GPAs of each student in the sample. Whether we set the values of x or not, the data consist of the n pairs of observations $(x_1, y_1), (x_2, y_2), \ldots, (x_n, y_n)$ from which the relationship is evaluated.

One of the most common relationships that might exist between two variables is a straight line. To investigate the relationship, the first step is to plot the data in a scatterplot. If a linear relationship seems plausible, the value of the correlation coefficient, r, will shed light on the predictive power of the straight line.

EXAMPLE 3.17

Following are the high school GPAs and the college GPAs at the end of the freshman year for 10 different students:

Worksheet: Gpa.mtw

Student	1	2	3	4	5	6	7	8	9	10
High School GPA	2.7	3.1	2.1	3.2	2.4	3.4	2.6	2.0	3.1	2.5
College GPA	2.2	2.8	2.4	3.8	1.9	3.5	3.1	1.4	3.4	2.5

Graph the data in a scatterplot and then comment on the relationship between the two variables.

SOLUTION

The college GPA is the response variable and is labeled on the vertical axis. The scatterplot is shown in Figure 3.18. It appears that the college GPA increases as the high school GPA increases. In fact, the dots appear to cluster along a straight line. Pearson's correlation coefficient is found to be $+.844$, which indicates that a straight line is a reasonable relationship between the two variables.

FIGURE 3.18   Scatterplot of high school GPA versus college GPA

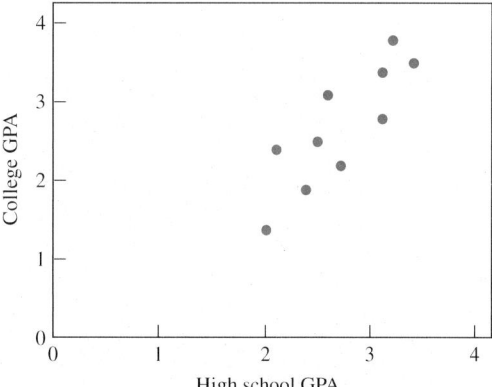

Fitting a Straight Line to Data: The Method of Least Squares

The equation of a straight line (illustrated in Figure 3.19 on page 188) is

$$y = b_0 + b_1 x$$

where b_0 is the y-intercept and b_1 is the slope of the line. The y-intercept is the value of y when $x = 0$, and the slope gives the change in y relative to the change in x.

FIGURE 3.19   Equation of a straight line

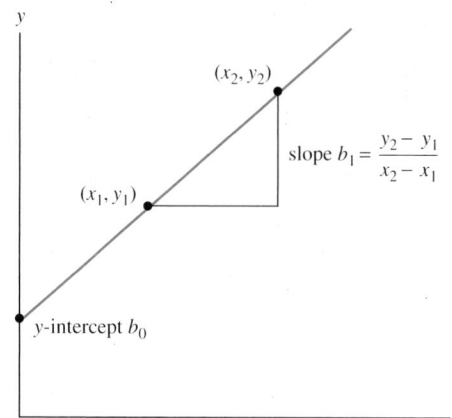

To graph a straight line, it is only necessary to find two points that satisfy the equation and then connect the points with a straight line.

Error

In Example 3.17, we observed that the data corresponding to high school GPA (x) and college GPA (y) appeared to cluster along a straight line. If the points had fallen exactly on a straight line, then there would be an exact straight line equation relating the variables. We could then say that there is no *error* in obtaining y from x. However, there is no reason to expect the points of the scatterplot to fall perfectly on a straight line. There could still be a linear relationship between the variables when the points fall off the straight line because of unknown error. For example, two people with the same high school GPAs most likely will end up with different college GPAs because of other variables such as study habits, motivation, or parents' attitude about college. When these other variables cannot be measured, or are chosen not to be measured, their effects are lumped together into what we call *error*.

From the equation of the line that best fits the data,

$$\hat{y} = b_0 + b_1 x$$

we can compute a *predicted* y for each value of x and then measure the error of the prediction. We use $\hat{y}$ (read "y hat") as the predicted y to distinguish it from the actual y.

From the data point (x_i, y_i) the observed value of y is y_i and the **predicted value of y** is obtained by the equation

$$\hat{y}_i = b_0 + b_1 x_i$$

The error of the prediction (also called the *residual*) is the difference in the actual y_i and the predicted $\hat{y}_i$.

The **residual** associated with the data point (x_i, y_i) is

$$e_i = y_i - \hat{y}_i$$

The residuals associated with the data points are depicted as vertical line segments in Figure 3.20. When we draw a line that comes as close as possible to all the observations, some of the residuals are positive (those above the line) and some are negative (those below the line). In fact, the sum of the residuals is zero because the line is positioned so that the negative residuals cancel the positive residuals. To summarize these residuals into a single value that measures the total error, we square (to avoid the problem of canceling residuals) the residuals and sum them over all cases. The resulting value is called the *sum of squares due to error*.

FIGURE 3.20   Fitted line and residuals

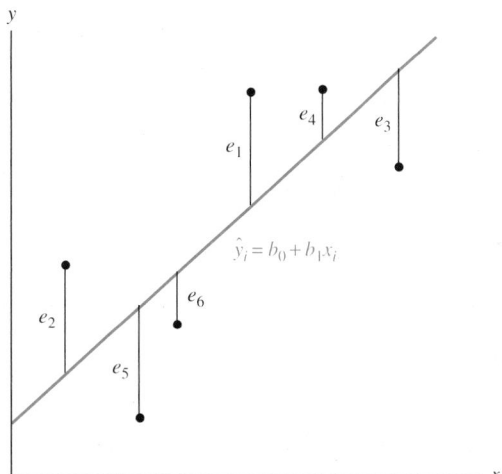

The **sum of squares due to error** (also called the *residual sum of squares*) is given by

$$SSE = \sum (y_i - \hat{y}_i)^2$$

Finding b_0 and b_1   To find the equation of the line that best fits the data, called the *regression line*, we must find values for b_0 and b_1. Substituting for $\hat{y}_i$, we have

$$SSE = \sum (y_i - b_0 - b_1 x_i)^2$$

which clearly depends on b_0 and b_1. The *method of least squares* chooses the values of b_0 and b_1 so that the value of SSE is as small as possible. (This is desirable because SSE is a combined measure of the discrepancy between the actual y and the predicted $\hat{y}$.) The formulas for the values of b_0 and b_1, obtained via the method of least squares, follow.

The **least squares regression line** is

$$\hat{y} = b_0 + b_1 x$$

where

$$b_1 = \frac{s_{xy}}{s_x^2} = \frac{\Sigma(x_i - \bar{x})(y_i - \bar{y})}{\Sigma(x_i - \bar{x})^2}$$

and

$$b_0 = \bar{y} - b_1\bar{x}$$

(The second expression for b_1 is obtained by canceling $n - 1$ in the numerator and denominator.)

EXAMPLE 3.18

Find the least squares regression line $\hat{y} = b_0 + b_1 x$ for the GPA data in Example 3.17. Also find the correlation between the two variables.

SOLUTION

From the data on high school GPA (x), we find

$$\bar{x} = 2.71 \quad \text{and} \quad s_x^2 = .2277$$

From the data on college GPA (y), we find

$$\bar{y} = 2.70 \quad \text{and} \quad s_y^2 = .58$$

The numerator of b_1 is the covariance s_{xy}, which is also used in the numerator of the correlation coefficient r. Following the procedure given in the previous section (or using a computer), we find $s_{xy} = .3067$. Substituting in these values, we find

$$b_1 = \frac{s_{xy}}{s_x^2} = \frac{.3067}{.2277} = 1.347$$

$$b_0 = \bar{y} - b_1\bar{x} = 2.70 - (1.347)(2.71) = -.950$$

The regression line that relates college GPA to high school GPA is therefore

$$\hat{y} = -.950 + 1.347x$$

Figure 3.21 shows the scatterplot with a graph of the straight line. The correlation between x and y is given by

$$r = \frac{s_{xy}}{s_x s_y} = \frac{.3067}{\sqrt{.2277}\sqrt{.58}} = .844$$

which supports the fact that the data points cluster tightly about the regression line.

FIGURE 3.21

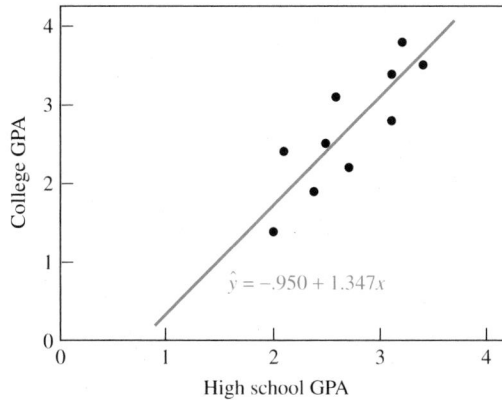

$\hat{y} = -.950 + 1.347x$

High school GPA

EXAMPLE 3.19   Find the residuals for the ten data points in Example 3.17 and then calculate SSE.

SOLUTION

First we substitute the x_i values into the predicting equation found in Example 3.18: $\hat{y} = -.950 + 1.347x$. The result is the predicted $\hat{y}_i$'s. They, along with the residuals, are listed in Table 3.7.

TABLE 3.7

x_i	y_i	$\hat{y}_i$	e_i
2.7	2.2	2.687	$-.487$
3.1	2.8	3.225	$-.425$
2.1	2.4	1.878	.522
3.2	3.8	3.360	.440
2.4	1.9	2.282	$-.382$
3.4	3.5	3.629	$-.129$
2.6	3.1	2.552	.548
2.0	1.4	1.744	$-.344$
3.1	3.4	3.225	.175
2.5	2.5	2.417	.083

Squaring the residuals and summing, we find that

$$\text{SSE} = \sum(y_i - \hat{y}_i)^2$$
$$= (-.487)^2 + (-.425)^2 + \cdots + (.083)^2$$
$$= 1.5023$$

Because of the method of least squares, we can be assured that for the original data, (x_i, y_i), the values of b_0 and b_1 and thus the predicting equation are such that the above value of SSE is as small as it possibly can be.

COMPUTER TIP

To have the computer construct the least squares regression line, store the bivariate data, (x_i, y_i), in two columns of the worksheet. For example, store the x values in C1 and the y values in C2. From the **Stat** menu, select **Regression** and click on **Regression** again. From the Regression window, **Select** the **Response** variable (y) and **Select** the **Predictor** variable (x). If you wish to compute the residuals, click on **Residuals** in the Storage boxes. Finally click on **OK**.

Prediction and Extrapolation

A very important reason for finding a regression line is to predict new values of the response variable from different given values of the predictor variable.

EXAMPLE 3.20

In Example 3.11 the correlation between the value of a brand name and a company's revenue was found to be +.94. Find the least squares regression line to predict the value of a brand name from the company's revenue. Use the equation to predict values of brand names when the revenues are $5 billion, $10 billion, and $20 billion.

SOLUTION

Following is a computer printout of the regression analysis of value using revenue as the predictor variable:

```
Regression Analysis
The regression equation is
value = - 0.889 + 2.02 revenue

Predictor      Coef       Stdev      t-ratio         p
Constant    -0.8889      0.4174       -2.13      0.039
revenue      2.0244      0.1158       17.49      0.000

s = 2.096 R-sq = 88.4% R-sq(adj) = 88.1%
```

Analysis of Variance

```
SOURCE        DF        SS       MS       F       p
Regression     1     1344.1   1344.1   305.82   0.000
Error         40      175.8      4.4
Total         41     1519.9
```

Unusual Observations

Obs.	revenue	value	Fit	Stdev. Fit	Residual	St.Resid
1	15.4	31.200	30.287	1.553	0.913	0.65 X
2	8.4	24.400	16.116	0.779	8.284	4.26R
11	6.0	3.700	11.257	0.539	-7.557	-3.73R

R denotes an obs. with a large st. resid.
X denotes an obs. whose X value gives it large influence.

The printout is much more involved than we need at this time. Later, in Chapter 11, we will learn more about this output. Notice, however, that the regression equation is

$$\text{value} = -0.889 + 2.02 \text{ revenue}$$

Substituting the values 5, 10, and 20 in for revenue and computing value yield the following predicted values of brand names when the revenues are $5, $10, and $20 billion:

Revenue	Predicted Value
$5 billion	$-.889 + 2.02(5) = \$9.211$ billion
$10 billion	$-.889 + 2.02(10) = \$19.311$ billion
$20 billion	$-.889 + 2.02(20) = \$39.511$ billion

Remember that the regression line is obtained from the observed sample. Using the regression line to predict outside the range of the observed data is called *extrapolation*. Caution should be exercised when you attempt to extrapolate the line beyond the range of the data, however. We have no information about how the variables might be related beyond the observed data. In Example 3.20, predicting the value of a brand name for a revenue of $20 billion is probably not wise. The maximum observed revenue is $15.4 billion for Marlboro, which in itself is way beyond the rest of the data. Marlboro is a very special brand name and should be viewed as an outlier. There is no information to suggest that the straight line continues in the same pattern beyond $15 billion.

Assessing the Fit of a Line

Having found the least squares regression line, we next ask: How well does it fit the data? That is, does the line effectively describe the relationship between x and y, or would some other relationship be better?

Because the residuals measure the discrepancy between the line and the observed data, a plot of the residuals versus the predictor variable graphically illustrates the deficiencies of the line as a model of the x-y relationship.

EXAMPLE 3.21

Figure 3.22 is a plot of the residuals found in Example 3.19 against the predictor variable. Notice that the residuals oscillate in a somewhat random pattern about the horizontal line at residual = .0. This pattern is typical of data that do not deviate substantially from the model under study. It appears, in this case, that the straight line has explained most of the trend in these data.

FIGURE 3.22

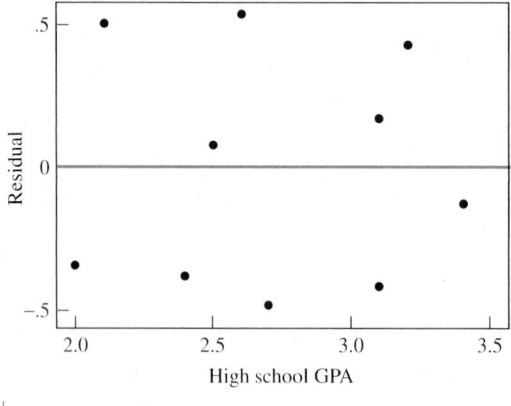

High school GPA

Often the straight-line model adequately describes the relationship between x and y and the residual plot bears this out, as in the previous example. However, there can be relationships that are not linear, as in the following example.

EXAMPLE 3.22

The tensile strength of Kraft paper (in pounds per square inch) was measured for different percentages of hardwood in the batch of pulp that was used to produce the paper.

Worksheet: hardwood.mtw

Tensile	6.3	11.1	20.0	24.0	26.1	30.0	33.8	34.0	38.1	39.9
Hardwood	1.0	1.5	2.0	3.0	4.0	4.5	5.0	5.5	6.0	6.5

Tensile	42.0	46.1	53.1	52.0	52.5	48.0	42.8	27.8	21.9
Hardwood	7.0	8.0	9.0	10.0	11.0	12.0	13.0	14.0	15.0

SOURCE: Joglekar, G. et al., "Lack-of-Fit Testing When Replicates Are Not Available," *The American Statistician* 43:3(1989): 135–143.

Does a straight line fit the data well enough that the tensile strength of the paper can be predicted from the percentage of hardwood?

SOLUTION

Figure 3.23 shows the residual plot after a straight line is fit to the data. It is clear that the pattern of the residuals is not random. It exhibits a nonlinear pattern called *curvilinear*.

Often the scatterplot will reveal such nonlinear relationships, but if one overlooks the scatterplot and continues to fit a straight line to the data, the resulting residual plot will magnify the pattern and, as in this case, suggest that the straight line is an inadequate model.

FIGURE 3.23 Residual Plot

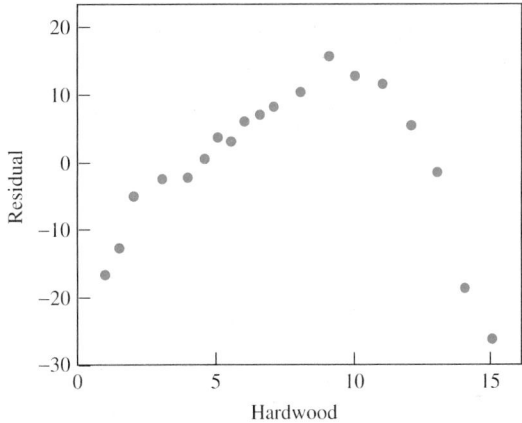

Effects of Outliers

Outliers, if present in the data, can have an undue effect on the regression line.

EXAMPLE 3.23

In Example 3.7, we observed a linear relationship (after taking logarithms) between brain weights and body weights of animals. Figure 3.24 is a scatterplot of the log data (worksheet: **Brain.mtw**) with the regression line drawn in as a solid line. Notice how the line is pulled down toward the three outliers associated with the dinosaur observations. This is a major concern with least squares regression analysis. It is clearly not resistant

FIGURE 3.24

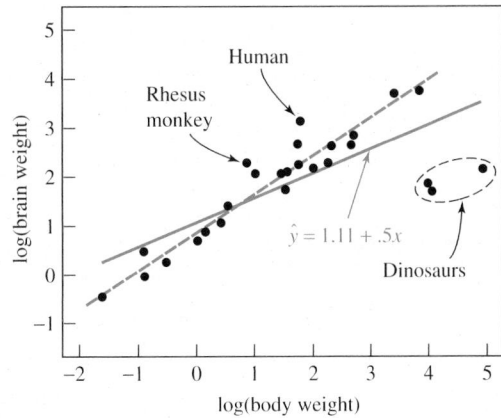

to outliers. In this example, the dotted line is perhaps a more accurate description of the relationship between the two variables.

Curvilinear relationships, effects of outliers, and the analysis of residuals used to check the adequacy of the linear model will be covered in Chapter 11.

Coefficient of Determination

Another useful way to evaluate how well the straight line fits the data is with the correlation coefficient. Recall that it measures the strength of the linear association between x and y. Interpreting its value, however, can be difficult. One interpretation involves its square, r^2. Treating y as *depending* on x, we have that r^2 is the proportion of the variability in y that is explained by x through a linear relationship. The quantity r^2 is referred to as the *coefficient of determination*.

> The **coefficient of determination**, r^2, is the percent of the variability in the dependent variable that is explained by the independent variable.

EXAMPLE 3.24

Calculate the coefficient of determination for the data in Example 3.17 and give an interpretation.

SOLUTION

In Example 3.18, we found that $r = .844$, and thus the coefficient of determination is $r^2 = .712$. Thus, 71.2% of the variability in college GPAs is explained by the linear relationship that they have with high school GPAs.

EXERCISES 3.4

3.43 Identify the response and predictor variables in the following studies:
   a. A study of the relationship between air pollution and high blood pressure
   b. A study of the relationship between mental retardation and lead poisoning
   c. A study of the relationship between the percent of drivers who wear seat belts and the death rate in automobile accidents
   d. A study of the relationship between the suicide rate and alcohol consumption among teenagers
   e. A study of the relationship between per capita income and public education expenditures of the states

3.44 Graph the following straight lines:
   a. $y = 2 + 3x$
   b. $y = -2.6 + 4x$
   c. $y = 7.1 - 8.2x$

3.45 For the following data sets (from Exercise 3.13 in Section 3.2), find the least squares regression equations:

Worksheet: Ex3-13.mtw

a.
X	2	1	3	2	3	1
Y	4	0	6	6	8	2

b.
X	1	3	4	6	8	9	11	14
Y	1	2	4	4	5	7	8	9

c.
X	1	2	4	5	4	2	6	3
Y	5	11	25	39	27	12	50	17

d.
X	−3	−1	0	1	3
Y	12	7	6	4	1

3.46 From the data in Exercise 3.45, find the predicted y for the following values of x:
   a. $x = 0, x = 4$
   b. $x = 2, x = 5, x = 10, x = 15$
   c. $x = 0, x = 5, x = 8$
   d. $x = -2, x = 2, x = 4$

3.47 Find the residuals associated with the least squares equations found in parts **a** and **d** of Exercise 3.45.

3.48 Find the sum of squares due to error associated with the least squares equations found in parts **a** and **d** of Exercise 3.45 (see Exercise 3.47).

3.49 The scatterplot you constructed in Exercise 3.14 in Section 3.2 showed how the amount of energy consumed in a home is related to the size of the home. The scatterplot showed that a straight line is a reasonable fit for the data. Furthermore, the correlation between the size of the home and the kilowatts consumed per month is +.94, which again suggests that the straight line is a reasonable fit. Does the following residual plot support this discussion?

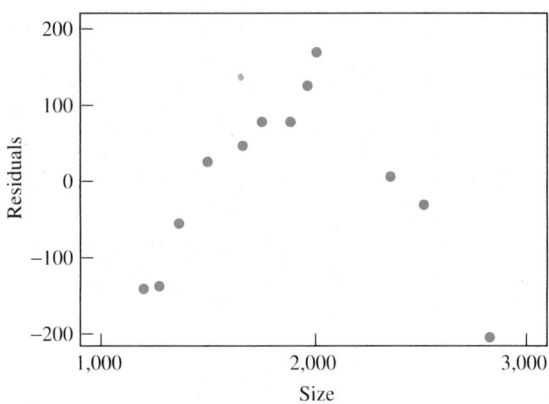

3.50   Data relating aptitude test scores to productivity in a factory are given here (and in Exercise 3.15 of Section 3.2):

Worksheet: Aptitude.mtw

Applicant	1	2	3	4	5	6	7	8
Aptitude score	9	17	13	19	20	23	12	15
Productivity	23	35	29	33	40	38	25	31

a. Use the data to find the least squares regression equation.
b. Find the residuals associated with the data.
c. Calculate SSE.

3.51   The accompanying plot shows the residuals associated with the least squares regression equation found in Exercise 3.50. The correlation coefficient was found to be $+.936$ in Exercise 3.36 in Section 3.3. What is the coefficient of determination? With this information is it reasonable to assume that the straight line found in Exercise 3.50 is the correct equation relating aptitude and productivity?

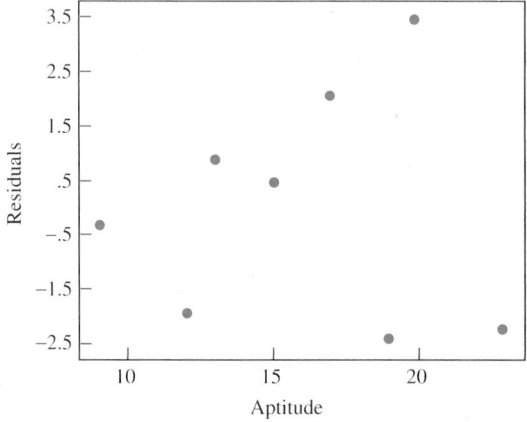

3.52   Data relating robbery rate and the percent of low-income residents in a precinct are given here (and in Exercise 3.16 of Section 3.2):

Worksheet: Precinct.mtw

Precinct	1	2	3	4	5	6	7	8
Robbery rate	20	41	165	88	60	120	65	81
Percent low income	4.9	7.1	10.1	11.8	13.5	14.8	16.2	11.2

a. Find the least squares regression equation.
b. Sketch the line on the scatterplot. Does the line fit the data?
c. The correlation coefficient is only .388. What is the coefficient of determination? What does this say about the fit of the line to the data?

3.53   Data corresponding to maintenance costs on nine cash registers and their ages are recorded here (and in Exercise 3.17 of Section 3.2):

Worksheet: Register.mtw

Age (years)	6	7	1	3	6	2	5	4	3
Cost (dollars)	92	181	23	40	126	35	86	72	51

a. Find the least squares regression equation.
b. Find the residuals associated with the points (1, 23) and (5, 86).

3.54 Shown here is a residual plot associated with the least squares regression equation found in Exercise 3.53. Have we properly described the relationship between the ages and maintenance cost of the cash registers?

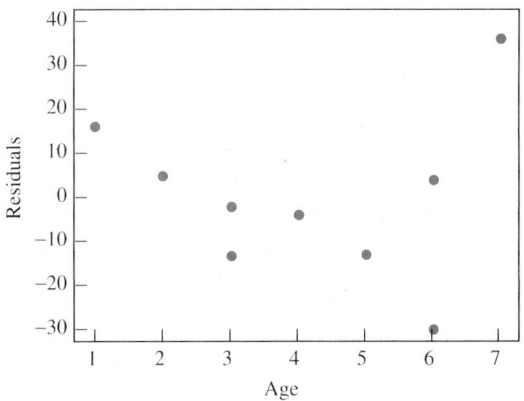

3.55 Data corresponding to the life span of a particular electronic component and the heat it is exposed to are given here (and in Exercise 3.18 of Section 3.2):

Worksheet: Lifespan.mtw

Heat (°C)	50	100	150	200	250	300
Life span (hours)	875	884	762	424	365	128

a. Find the least squares regression equation.
b. Find the residuals.
c. Calculate SSE.

3.56 Plot the residuals found in Exercise 3.55 against the predictor variable, heat. Is there a pattern to the residuals, or do they appear random?

3.57 The accompanying computer printout is associated with the J. D. Power and Associates ratings of 1994 and 1995 cars described in Exercise 3.24 in Section 3.2 and Exercise 3.41 in Section 3.3.

```
Correlation of 1994 and 1995 = 0.822

Regression Analysis

The regression equation is
1995 = 2.2 + 0.910 1994

Predictor      Coef      Stdev     t-ratio        p
Constant       2.22      14.64        0.15     0.880
1994         0.9098     0.1213        7.50     0.000

s = 21.72       R-sq = 67.6%       R-sq(adj) = 66.4%
```

(continued)

Analysis of Variance

SOURCE	DF	SS	MS	F	p
Regression	1	26537	26537	56.26	0.000
Error	27	12735	472		
Total	28	39272			

Unusual Observations

Obs.	1994	1995	Fit	Stdev.Fit	Residual	St.Resid
29	193	195.00	177.82	10.17	17.18	0.90 X

X denotes an obs. whose X value gives it large influence.

a. What is the correlation?
b. What is the least squares regression equation?
c. What is the coefficient of determination?
d. If you square the correlation coefficient found in part **a**, do you get the coefficient of determination?
e. What additional information would help in determining whether the straight line is a reasonable fit to the data?

3.58   Study the accompanying scatterplot that compares the crime rates (violent crimes per 100,000 population) for the different states to the percents of the population in the states without a high school degree.

Worksheet: Educat.mtw

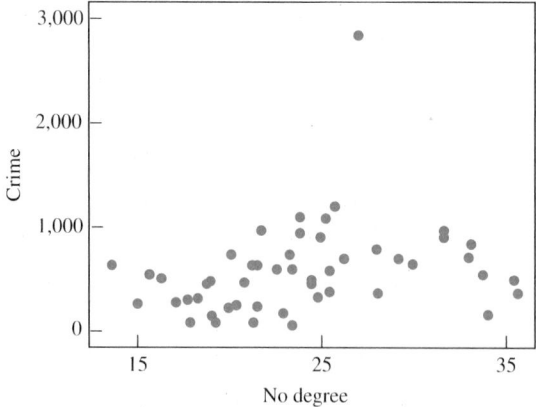

a. Is there a trend in the data?
b. Is the one outlier an outlier with respect to the percent without a degree or with respect to the crime rate or with respect to both variables?
c. Is it appropriate to fit the least squares regression line to these data? Explain.

3.59   The accompanying data give the number of icebergs sighted each month south of Newfoundland and south of the Grand Banks in 1920. Assess the relationship between the two.

Worksheet: Iceberg.mtw

Month	Jan	Feb	Mar	Apr	May	Jun	Jul	Aug	Sep	Oct	Nov	Dec
Newfoundland	3	10	36	83	130	68	25	13	9	4	3	2
Grand Banks	0	1	4	9	18	13	3	2	1	0	0	0

SOURCE: N. Shaw, *Manual of Meteorology*, Vol. 2 (London: Cambridge University Press, 1942), p. 7, and F. Mosteller, and J. W. Tukey, *Data Analysis and Regression* (Reading, MA: Addison-Wesley, 1977).

3.60   Reconsider the four data sets given in worksheet: **Anscombe.mtw**. The scatterplots of the four data sets are very different, yet we learned in Exercise 3.42 in Section 3.3 that the correlation coefficient is the same for all four sets of data. Calculate the least squares regression equation for each data set. What do you find out about the four equations? Do you consider the results unusual?

3.5   A ROBUST ALTERNATIVE TO LEAST SQUARES: THE RESISTANT LINE (OPTIONAL)

As we fit a straight line to a data set, the method of least squares tries to keep the line close to every data point. Thus, outliers, if present, tend to have an undue influence on the fit. In an effort to prevent outliers from distorting the analysis, we can construct the *resistant line*. Because the median is resistant to outliers, the construction of the resistant line is based on medians.

For the construction, the data are divided into three groups of approximately the same size. If the sample size n is not divisible by 3 and has a remainder of 1, the extra point is placed in the middle group. If the remainder is 2, then the two points are placed in the first and third groups. If some points have the same x value, they all must go in the same group. Thus, it may not be possible to divide the sample into three groups of exactly the same size; however, one must attempt to even them out as much as possible.

After we determine the groups, we find the median of the x data and the median of the y data within each group. That is, we find the (x, y) pair of medians in each of the three groups. They summarize the behavior of the data in their respective groups. We label these *summary points* as (x_L, y_L), (x_M, y_M), and (x_R, y_R) for the left, middle, and right groups, respectively. Figure 3.25 on page 202 illustrates the lines that divide the three groups and the locations of the summary points.

The equation of the resistant line can be written as

$$\hat{y}_R = b_0 + b_1 x$$

where b_1 is the slope and b_0 is the y-intercept. Having found the summary points, we can construct the line that has the same slope as a line that passes through the left and right median points by computing the slope with the equation

Often the resistant line can be drawn on a scatterplot by simply eyeballing the location.

$$b_1 = \frac{y_R - y_L}{x_R - x_L}$$

FIGURE 3.25   Locations of summary points

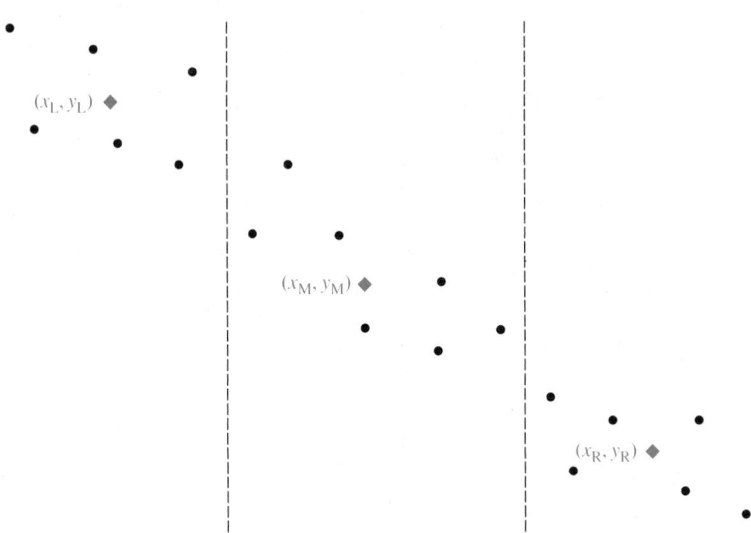

A line can be adjusted up or down by changing the y-intercept. To force the line through the point (x_0, y_0), the y-intercept is computed as $b_0 = y_0 - b_1 x_0$. For the resistant line, however, we compute the y-intercept as an average for all three summary points:

$$b_0 = \tfrac{1}{3}[(y_L + y_M + y_R) - b_1(x_L + x_M + x_R)]$$

EXAMPLE 3.25   Construct the resistant line for the following data set:

x	5.2	7.7	10.8	12.5	13.3	14.1	12.0	16.4
y	2.7	18.2	28.9	27.7	24.6	30.9	35.4	28.8
x	15.3	17.4	17.7	18.4	19.6	2.6	11.6	8.3
y	39.1	33.6	44.3	39.8	36.4	20.4	24.2	24.4

SOLUTION

First we order the data according to the x variable and then find the medians of the three groups for the x data and for the y data:

x	y	
2.6	20.4 ←	
5.2	2.7	
→ 7.7	18.2	$(x_L, y_L) = (7.7, 20.4)$
8.3	24.4	
10.8	28.9	
11.6	24.2	
12.0	35.4	
→ 12.5	27.7 ←	
→ 13.3	24.6	
14.1	30.9 ←	
15.3	39.1	
16.4	28.8	
17.4	33.6	
→ 17.7	44.3	$(x_R, y_R) = (17.7, 36.4)$
18.4	39.8	
19.6	36.4 ←	

12.9 (for the middle group, x); 29.3 (for the middle group, y); $(x_M, y_M) = (12.9, 29.3)$

> There are six points in the middle group; therefore, the median is the average of the third and fourth observation.

From the formulas for b_1 and b_0, we have

$$b_1 = \frac{y_R - y_L}{x_R - x_L} = \frac{36.4 - 20.4}{17.7 - 7.7} = \frac{16.0}{10} = 1.6$$

$$b_0 = \frac{1}{3}[(y_L + y_M + y_R) - b_1(x_L + x_M + x_R)]$$

$$= \frac{1}{3}[(20.4 + 29.3 + 36.4) - 1.6(7.7 + 12.9 + 17.7)]$$

$$= 8.27$$

Therefore, the resistant line is given by

$$\hat{y}_R = 8.27 + 1.6x$$

Figure 3.26 (page 204) gives the scatterplot with the summary points and the resistant line graphed.

In comparison, the least squares line for the data in Example 3.25 is

$$\hat{y} = 7.6 + 1.66x$$

(As a point of interest, the correlation is $r = .8079$.) We see that the regression line has a larger slope and a lower y-intercept, which is because of the outlier at $(5.2, 2.7)$. The least squares equation is pulled toward the point, whereas the outlier has no effect on the resistant line.

FIGURE 3.26   Scatterplot with resistant line

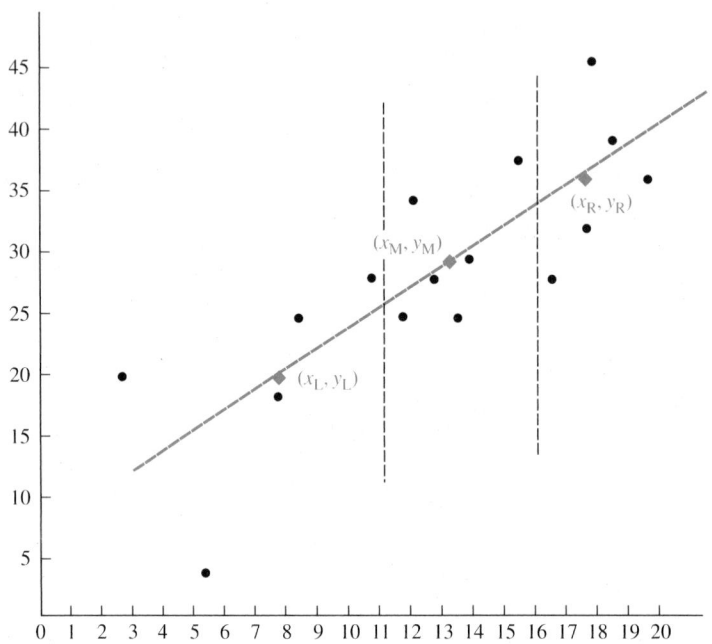

The residuals for the resistant line are computed in the same way as the residuals for the least squares line. That is, the residual for the data point (x_i, y_i) is

$$r_i = y_i - \hat{y}_i$$

where $\hat{y}_i$ is the predicted $\hat{y}$ from the resistant line.

The purpose of investigating residuals is to detect patterns that are not explained by the regression line. In the case of the resistant line, the residuals can be used to "polish" the line so that it better fits the data. For a thorough discussion of polishing the fit, see Velleman and Hoaglin (1981).

EXERCISES 3.5

3.61   Construct the resistant line for the following data set:

x	2.4	3.5	4.7	5.3	6.4	7.2	8.1	9.3	10.6	12.1	13.3	14.5
y	−14.2	−4.3	−2.9	−1.3	2.4	3.5	4.4	5.7	6.5	8.2	9.3	16.3

3.62   Construct the resistant line for the following data set:

x	1.2	1.4	1.6	1.8	2.0	2.2	2.4	2.6	2.8	3.0
y	.44	.65	.63	.71	.64	.83	.81	.88	.85	2.32

3.63   Construct the resistant line for the following data set:

x	120	140	150	160	170	180	190	200	210	220	230
y	32	30	24	21	18	15	33	11	8	6	4

3.64   In Exercise 3.15 of Section 3.2, you were asked to draw a scatterplot of data relating aptitude test scores to productivity in a factory. Review the scatterplot. Are there any outliers in the data? Will the resistant line differ significantly from the least squares regression line? Use the same data (reproduced here) to construct the resistant line. Compare your results with the least squares solution found in Exercise 3.50 of Section 3.4.

Worksheet: Aptitude.mtw

Applicant	1	2	3	4	5	6	7	8
Aptitude score	9	17	13	19	20	23	12	15
Productivity	23	35	29	33	40	38	25	31

3.65   In Exercise 3.16 of Section 3.2, you were to construct a scatterplot of data relating the robbery rate and the percent of low-income residents in a precinct. Review the scatterplot. Are there any outliers in the data? Will the resistant line differ significantly from the least squares regression line? Use the same data (reproduced here) to construct the resistant line. Compare your results with the least squares solution found in Exercise 3.52 of Section 3.4.

Worksheet: Precinct.mtw

Precinct	1	2	3	4	5	6	7	8
Robbery rate	20	41	165	88	60	120	65	81
Percent low income	4.9	7.1	10.1	11.8	13.5	14.8	16.2	11.2

3.66   In Exercise 3.17 of Section 3.2, you were to construct a scatterplot of the maintenance costs on nine cash registers and their ages. Review the scatterplot. Are there any outliers in the data? Will the resistant line differ significantly from the least squares regression line? The rules for constructing the resistant line say that if two observations have the same x value, they should go in the same group. Which group, low or middle, should have the two observations with an age of 3 years? Does it make a difference in the equation of the resistant line? Construct the resistant line and compare it with the least squares solution found in Exercise 3.53 of Section 3.4.

Worksheet: Register.mtw

Age (years)	6	7	1	3	6	2	5	4	3
Cost (dollars)	92	181	23	40	126	35	86	72	51

3.67   In Exercise 3.66, suppose that the first register of age 3 has a maintenance cost of $200 instead of $40. What effect does this have on the equation of the least squares regression line? What effect does it have on the resistant line? Should the observation (3, 200) be classified as an outlier?

3.68   In Exercise 3.18 of Section 3.2, the scatterplot for the data corresponding to the life span of a particular electronic component and the heat it is exposed to shows a very clear linear pattern with no outliers. Furthermore, the correlation is $-.964$. Is there any need to investigate a resistant line? Explain.

3.69   The scatterplots of the Anscombe data sets (worksheet: **Anscombe.mtw**) are shown on page 206. In which case(s) should we use the resistant line to describe the relationship between the response and predictor variables?

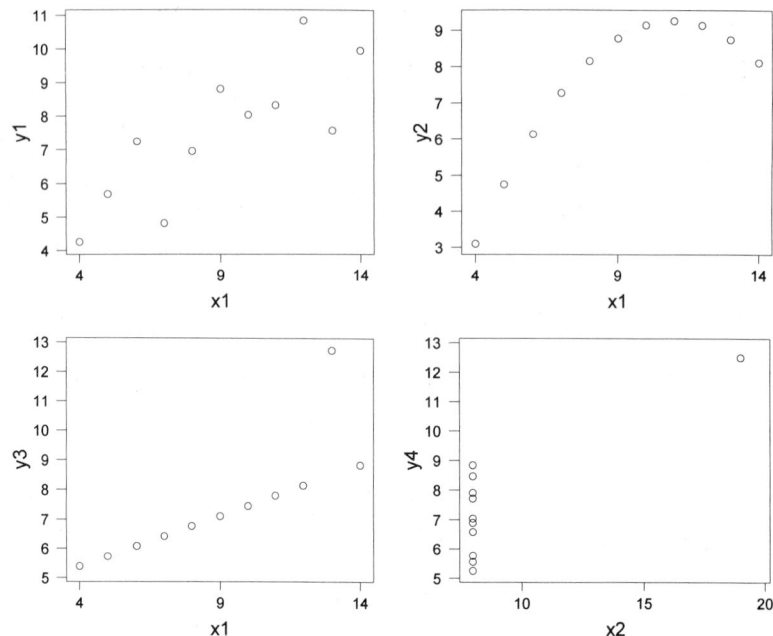

3.70   Worksheet: **Brain.mtw** contains the brain weights and body weights of 28 animals (see Example 3.7). In Example 3.23, we noticed that the least squares regression line is pulled down toward the three outliers that correspond to the dinosaur observations. Using the logarithms of the two variables, calculate the resistant line and compare it with the least squares line found in Example 3.23.

3.6 Sᴜᴍᴍᴀʀʏ ᴀɴᴅ Rᴇᴠɪᴇᴡ

Kᴇʏ Cᴏɴᴄᴇᴘᴛs

✓ Bivariate data in which both variables are categorical are displayed in a *contingency table*. Numerical bivariate data can be displayed in a *scatterplot*.

✓ A *correlation* is a summary measure of the degree of association between two variables. Two correlations considered in this chapter are *Pearson's correlation coefficient* and *Spearman's rank correlation coefficient*. Pearson's correlation is applicable to numerical data. Spearman's correlation is more applicable to categorical data that can be ordered, yet it can be applied to numerical data.

✓ In regression analysis, the independent variable is called the *predictor variable* and the dependent variable is called the *response variable*.

✓ The equation of the least squares regression line is

$$\hat{y} = b_0 + b_1 x$$

where

$$b_1 = \frac{s_{xy}}{s_x^2} = \frac{\Sigma(x_i - \overline{x})(y_i - \overline{y})}{\Sigma(x_i - \overline{x})^2}$$

and

$$b_0 = \overline{y} - b_1\overline{x}$$

✓ The predicted value of y is
$$\hat{y}_i = b_0 + b_1 x_i$$

✓ The residual associated with the data point (x_i, y_i) is

$$e_i = y_i - \hat{y}_i$$

✓ The sum of squares due to error is

$$\text{SSE} = \sum(y_i - \hat{y}_i)^2$$

✓ Examining residuals is a way of assessing the fit of a line.

✓ The coefficient of determination is r^2, the square of the correlation coefficient.

✓ The resistant line is a robust alternative to the least squares regression line. The slope of the resistant line is
$$b_1 = \frac{y_R - y_L}{x_R - x_L}$$

The y-intercept is

$$b_0 = \frac{1}{3}[(y_L + y_M + y_R) - b_1(x_L + x_M + x_R)]$$

LEARNING GOALS Having completed this chapter, you should be able to:

1. Organize data in a contingency table. *Section 3.1*
2. Convert the data in a contingency table to a bar graph. *Section 3.1*
3. Construct a scatterplot. *Section 3.2*
4. Evaluate the relationship between two variables from a scatterplot. *Section 3.2*
5. Compute Pearson's and Spearman's correlation coefficients. *Section 3.3*
6. Obtain the equation of a least squares regression line. *Section 3.4*
7. Calculate residuals and predicted values. *Section 3.4*
8. Assess the fit of a straight line equation. *Section 3.4*
9. Calculate the coefficient of determination. *Section 3.4*
10. Calculate the equation of the resistant line. *Section 3.5*

QUESTIONS FOR REVIEW Use the following problems to test your skills:

3.71 If there is a significant association between two variables, then the variables are
A. confounded
B. correlated

C. biased

D. independent

3.72 _____ measures the strength of the association between two variables.

3.73 Graph the equation $y = 15.3 - 4.7x$. Find the residual for the point $(2, 3.8)$.

3.74 Identify the independent variable and dependent variable in the following studies:
   a. A study involving the yield of a wheat crop and the annual amount of rainfall
   b. A study of health care costs and the number of new AIDS cases
   c. A study of the crime rate and poverty rate

3.75 Two hundred people were interviewed to determine whether they think children of working mothers are adequately cared for. Their opinions are given here:

	Men	Working Mothers	Nonworking Mothers
Yes	28	44	18
No	32	14	64

   a. What percent of the sample are women?
   b. What percent believe the children are adequately cared for?
   c. What percent of the working mothers feel that the children are adequately cared for?
   d. Of those who think children are not adequately cared for, what percent are women?

3.76 Draw a bar graph depicting the relative frequency of those who responded yes and no among the men, working mothers, and nonworking mothers from the data in Exercise 3.75.

3.77 A group of students selected randomly from the student body was asked to identify their majors. There were 58 women, of whom 9 were science majors, 15 were business majors, 22 were education majors, and the rest were liberal arts majors. There were 44 men, of whom 17 were science majors, 15 were business majors, 5 were education majors, and the rest were liberal arts majors.
   a. Arrange the data in a bivariate frequency table.
   b. What percent are science majors?
   c. What percent of the women are education majors?
   d. What percent of the education majors are women?

3.78 Draw a segmented bar graph illustrating the number of men and women in each of the four majors using the bivariate frequency table from Exercise 3.77.

3.79 Consider the following data:

Worksheet: Ex3-79.mtw

X	3	5	4	7	9	2	6	2	5
Y	15	11	10	8	4	16	9	14	10

   a. Draw a scatterplot.
   b. Calculate the correlation between the two variables.
   c. Find the least squares regression line.
   d. Find the resistant line.
   e. What differences do you notice between the regression line and the resistant line?

3.80 Listed here are the 1993 and 1992 rankings of our favorite breeds of dogs:

Worksheet: Dogs.mtw

	Ranking	
Dog	**1993**	**1992**
Labrador	1	1
Rotweilers	2	2
Shepherd	3	4
Spaniel	4	3
Retrievers	5	6
Poodles	6	5
Beagles	7	7
Dachshunds	8	8
Dalmatians	9	15
Shetland	10	9
Pomeranians	11	12
Yorkshire	12	14
ShihTzu	13	11
Schnauzers	14	13
Chows	15	10
Chihuahuas	16	16
Boxers	17	17
Huskies	18	18
Doberman	19	20
Springer	20	19

SOURCE: *The World Almanac and Book of Facts*, 1995.

Find the correlation between the 1993 and 1992 rankings. Is this a Pearson's or Spearman's correlation?

3.81 It is suspected that the more people there are in a family, the less is the cost per person per week for groceries. To evaluate this, a marketing institute randomly sampled 20 families:

Worksheet: Family.mtw

Number in family   2   2   1   3   4   3   2   4   1   3   5   2   2   3   4   1   2   6   3   2
Cost per person   78 85 88 76 72 74 79 69 79 75 68 82 78 72 76 90 84 67 77 79

a. Draw a scatterplot of the number in the family and the cost per person for food.
b. Find the correlation between the number in the family and the cost per person for food.
c. Is there a linear trend to the data? Is it positive or negative?
d. Find the least squares regression line that fits the data.
e. Predict the cost per person for groceries for a family of four.

3.82 A group of college students diagnosed as having dyslexia were asked to attend a reading workshop. They were given a reading test that was scored as the number of words read per minute. The following information was also recorded for each student:

Worksheet: Dyslexia.mtw

Words/min	**Age**	**Gender**	**L/R (handed)**	**Weight (lb)**	**Height (in)**	**Children in family**
165	21	M	L	165	70	2
201	18	F	R	115	66	1
75	19	F	R	138	65	4
124	19	M	R	187	72	3

(*continued*)

Words/min	Age	Gender	L/R (handed)	Weight (lb)	Height (in)	Children in family
105	20	F	L	100	61	2
143	18	M	R	210	71	1
126	19	M	L	178	69	1
92	20	M	R	155	68	3

a. Identify each variable as either numerical or categorical.
b. Identify the numerical variables as either continuous or discrete.
c. Draw a scatterplot of words/minute by handedness.

3.83   The following data compare a child's age with the number of gymnastic activities he or she was able to complete successfully:

Worksheet: Gym.mtw

Age	2	3	4	4	5	6	7	7
Number of activities	5	5	6	3	10	9	11	13

a. Calculate the least squares estimates of b_0 and b_1.
b. Construct a scatterplot and draw in the regression line.
c. Calculate the correlation between age and the number of activities.
d. Can you identify any lurking variables that explain the correlation between the two variables?

3.84   Suppose the following display represents 500 automobile accidents that occurred in a large city:

	No fatalities	At least one fatality
Involved alcohol	68	142
No alcohol	194	96

a. Fill in the marginal totals.
b. What percent of the accidents involved alcohol?
c. What percent of the accidents with at least one fatality involved alcohol?
d. What percent of the non-alcohol-related accidents had at least one fatality?
e. What percent of the accidents involved alcohol and had at least one fatality?

3.85   In Exercise 3.84, compute the expected counts for each of the cells and judge whether there is an association between alcohol and fatalities.

3.86   Shown here is the scatterplot given in Figure 3.13 that describes the relationship between the value of a brand name product and the company's revenue. There are three outliers in the data.

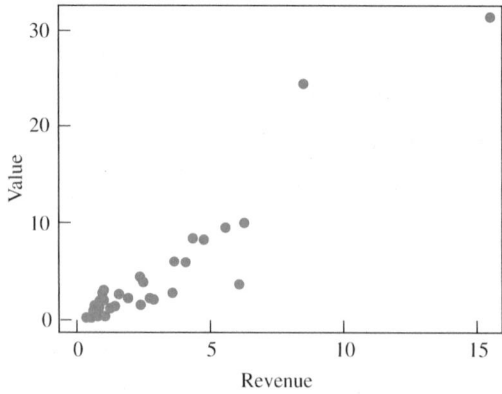

Refer to Example 3.11 and label each outlier by its brand name. Describe in detail what effect the outliers will have on the least squares regression equation. Which one will have the most effect on the equation of the regression line? Should one, two, or all three outliers be removed before you calculate the least squares regression equation?

3.87   In 1990, 53.8% of American Indian adults living on reservations were high school graduates or had even more education. This compares with 75% in the general population. Is the educational attainment of American Indians related to their income or to their poverty rate? Following are the percents of persons 25 years and older with a high school diploma or higher, the per capita income, and the percent in poverty on ten reservations:

Worksheet: Indian.mtw

Reservation	Percent high school	Per capita income	Poverty rate
Blackfeet, MT	66.3	$4,718	50.1
Hopi, AZ	62.6	4,566	49.4
Rosebud, SD	59.3	3,739	60.4
Zuni Pueblo, AZ-NM	55.4	3,904	52.5
Pine Ridge, NE-SD	55.2	3,115	66.6
San Carlos, AZ	49.4	3,173	62.5
Fort Apache, AZ	48.3	3,805	52.7
Papago, AZ	47.3	3,113	65.7
Navajo, AZ-NM-UT	41.1	3,735	57.8
Gila River, AZ	37.3	3,176	64.4

a. Construct a scatterplot of per capita income against the percent with a high school diploma.
b. Construct a scatterplot of poverty rate against percent with a high school diploma.
c. Does the scatterplot in either part **a** or **b** show a distinct linear relationship?
d. Calculate the correlations between all three variables.
e. What does the correlation between the percent with a high school diploma and the per capita income say about the relationship between the two?

3.88   Four hundred employees of a medium-sized company were asked to mark their satisfaction with their job as happy, OK, or unhappy and indicate their marital status. The accompanying contingency table shows the numbers in the various categories:

		Marital Status				
		Single	Married	Divorced	Widowed	Total
	Happy	46	46	28	5	125
Satisfaction	**OK**	30	64	42	18	154
with Job	**Unhappy**	12	42	62	5	121
	Total	88	152	132	28	400

a. What percent of the married employees are happy with their jobs?
b. What percent of those who are happy with their jobs are married?
c. What percent of the employees are married and happy with their jobs?
d. Which of the four groups is most happy with their jobs?

3.89   Graduation rates for all students and for student athletes who entered schools in the Big Ten Conference in 1984–85 are given at the top of page 212.

Worksheet: BigTen.mtw

School	All students	All athletes
Illinois	78%	67%
Indiana	53	58
Iowa	62	64
Michigan	81	66
Michigan State	66	64
Minnesota	34	44
Northwestern	87	81
Ohio State	51	55
Penn State	73	63
Purdue	68	60
Wisconsin	66	62

SOURCE: NCAA Graduation Rates Report, 1992.

a. Create a scatterplot of the graduation rates of students versus student athletes.
b. Calculate the correlation between the two columns.
c. What does a high correlation between these two variables mean?
d. Should we treat one variable as the response and the other as the predictor? Or is this just a correlation problem and not a regression problem?

COMPUTER EXERCISES

3.90 Following are anxiety scores before a major math test and the math test scores for a group of tenth-grade students:

Worksheet: Anxiety.mtw

Anxiety	12	16	6	22	17	14	8	27	18	19
Math test	75	56	91	48	69	73	88	65	72	65

Anxiety	11	9	21	3	15	17	22	19	13	5
Math test	88	94	48	99	78	65	58	52	70	90

a. Compute the correlation between the anxiety scores and the test scores.
b. Find the least squares regression line that relates the anxiety scores and the test scores.
c. Store the residuals in a column and plot them against the anxiety scores. Do the residuals exhibit a random pattern?

3.91 Listed here are the numbers of inmates per 100,000 population in the jails across the United States in 1983 and in 1993. (Some states have integrated jail–prison systems and were omitted from the data.)

Worksheet: Jail.mtw

	1983	1993		1983	1993
Ala	113	169	Fla	137	250
Alaska	8	na	Ga	178	328
Ariz	99	184	Hawaii	na	na
Ark	69	117	Idaho	61	135
Calif	166	222	Ill	77	124
Colo	88	177	Ind	66	145
Conn	na	na	Iowa	29	57
Del	na	na	Kan	55	111
D.C.	456	292	Ky	100	180

(continued)

	1983	1993		1983	1993
La	192	377	Ohio	66	105
Maine	49	57	Okla	67	127
Md	107	188	Ore	87	125
Mass	57	131	Pa	85	160
Mich	84	132	R.I.	na	na
Minn	47	81	S.C.	82	157
Miss	97	184	S.D.	45	87
Mo	76	96	Tenn	128	282
Mont	50	81	Texas	97	307
Neb	53	105	Utah	56	102
Nev	105	215	Vt	na	na
N.H.	50	100	Va	103	225
N.J.	80	192	Wash	84	141
N.M.	96	189	W.Va	52	97
N.Y.	91	164	Wis	64	156
N.C.	57	129	Wyo	66	105
N.D.	36	57	U.S. Total	98	178

Note: The District of Columbia jail population declined between 1983 and 1993 because the Occoquan complex was reclassified from a jail to a prison.

SOURCE: U.S. Department of Justice, Bureau of Justice Statistics Bulletin, *Jails and Jail Inmates 1993–94*, April 1995.

a. Construct a scatterplot of the 1983 and 1993 numbers.
b. Identify any outliers. Is it reasonable to remove the outlier?
c. Without the outlier, do you see a linear trend?
d. Would the outlier have an appreciable effect on the least squares regression line? What about the resistant line?
e. Calculate the regression line using all available data.
f. Remove the outlier and calculate the least squares regression equation.
g. Compare r^2 for the two regression lines found in parts **e** and **f**.

3.92 In 1984, the U.S. Department of Education ranked the states in several categories related to public schools. A scoring method was devised by the Gannett News Service based on the department's rankings of such quantities as pupil–teacher ratios, dropout rates, per capita income spent on education, and teacher salaries. The table gives the composite scores along with the reported SAT scores for each of the 50 states and the District of Columbia:

Worksheet: Gannett.mtw

State	Score	Verbal SAT	Math SAT	State	Score	Verbal SAT	Math SAT
Ala	16.0	467	503	Hawaii	50.7	395	474
Alaska	88.3	443	471	Idaho	21.6	480	512
Ariz	18.6	469	509	Ill	25.1	463	518
Ark	15.8	482	521	Ind	14.2	410	454
Calif	36.3	421	476	Iowa	19.0	519	570
Colo	38.9	465	514	Kan	38.0	502	549
Conn	54.7	436	468	Ky	14.7	479	518
Del	60.0	433	469	La	18.0	472	508
D.C.	59.5	397	426	Maine	20.8	429	463
Fla	19.7	423	467	Md	46.8	429	468
Ga	13.5	392	430	Mass	42.9	429	467

(*continued*)

State	Score	Verbal SAT	Math SAT	State	Score	Verbal SAT	Math SAT
Mich	53.1	461	515	Ore	61.7	435	472
Minn	52.8	481	539	Pa	45.9	425	462
Miss	19.5	480	512	R.I.	66.6	424	461
Mo	16.9	469	512	S.C.	17.0	384	419
Mont	71.1	469	512	S.D.	50.6	520	566
Neb	57.0	493	548	Tenn	14.5	486	523
Nev	17.9	442	489	Texas	15.4	413	453
N.H.	33.4	448	483	Utah	34.6	503	542
N.J.	60.4	418	458	Vt	67.4	437	470
N.M.	29.3	487	527	Va	21.8	428	466
N.Y.	26.6	424	470	Wash	26.7	463	505
N.C.	16.3	395	432	W.Va	32.0	466	510
N.D.	35.6	500	554	Wis	57.0	475	532
Ohio	26.2	460	508	Wyo	93.4	489	545
Okla	18.7	484	525				

SOURCE: *USA Today*, January 2, 1985, and The College Board.

Correlate the composite score with each of the SAT scores. Because there is no uniform evaluation system from state to state, the composite score is not totally reliable in rating the states. For this reason, it is suggested that Spearman's rank correlation coefficient be computed. Evaluate the Gannett News Service rating based on the state SAT scores.

3.93   Worksheet: **Toxic.mtw** lists the numbers of hazardous waste sites in each state. The regions of the country they are in are also given. Construct side-by-side boxplots for the number of hazardous waste sites by the region of the country. Compare the variability in the four boxplots. Which region has the least variability in the number of waste sites? Which region has the most variability? Are there any outliers in any region? If so, identify them. Are the median numbers of sites in the regions about the same?

3.94   The *IAAF/ATFS Track and Field Statistics Handbook* for the 1984 Los Angeles Olympics listed the national records for women in the 100-meter, 200-meter, and 400-meter races. [SOURCE: Dawkins (1989)]

   a. From worksheet: **Track.mtw**, construct a scatterplot of the 100-meter and 200-meter records.
   b. Construct a scatterplot of the 100-meter and 400-meter records.
   c. Which scatterplot has the greater variability? Why?
   d. Which set of variables, the one in part **a** or the one in part **b**, will have the higher correlation?
   e. Calculate the two correlations to verify your answer to part **d**.
   f. Does a straight line describe the relationship between the 100- and 200-meter records? How about between the 100- and 400-meter records?

3.95   The following data are the mean annual levels of Lake Victoria Nyanza for the years 1902–1921 and the numbers of sunspots in each year:

Worksheet: Victoria.mtw

Year	Level	Sunspot
1902	−10	5
1903	13	24
1904	18	42
1905	15	63
1906	29	54

(*continued*)

Year	Level	Sunspot
1907	21	62
1908	10	49
1909	8	44
1910	8	19
1911	1	6
1912	−11	4
1913	−3	1
1914	−2	10
1915	4	47
1916	15	57
1917	35	104
1918	27	81
1919	8	64
1920	3	38
1921	−5	25

SOURCE: N. Shaw, *Manual of Meteorology*, Vol. 1 (London: Cambridge University Press, 1942), p. 284, and F. Mosteller and J. W. Tukey, *Data Analysis and Regression* (Reading, MA: Addison-Wesley, 1977).

a. Identify the response and predictor variables.
b. Construct a scatterplot of the data with the predictor variable on the horizontal axis.
c. Do you detect a linear trend in the data? Is it positive or negative?
d. Are there any outliers in the data?
e. Calculate the correlation coefficient. Is it Pearson's or Spearman's correlation?
f. Find the least squares regression equation that relates the level of the lake to the number of sunspots.
g. Predict the level of the lake when the number of sunspots is 50. Would you use this equation to predict the level when the number of sunspots is 200? Explain.

3.96 The following data are the scores assigned by two judges to 40 competitors in a synchronized swimming event in the 1986 National Olympic Festival in Houston, Texas:

Worksheet: Swim.mtw

Judge 1	33.1	26.2	31.2	27.0	28.4	28.1	27.0	25.1	31.2	30.1	29.0	27.0	31.2	32.3
Judge 2	32.0	29.2	30.1	27.9	25.3	28.1	28.1	27.3	29.2	30.1	28.1	27.0	33.1	31.2
Judge 1	29.5	29.2	32.3	27.3	26.4	27.3	27.3	29.5	28.4	31.2	30.1	31.2	26.2	27.3
Judge 2	28.4	29.2	31.2	30.1	27.3	26.7	28.1	28.1	29.5	29.5	31.2	31.2	28.1	27.3
Judge 1	29.2	29.5	28.1	31.2	28.1	24.0	27.0	27.5	27.3	31.2	27.0	31.2		
Judge 2	26.4	27.3	27.3	31.2	27.3	28.1	29.0	27.5	29.5	30.1	27.5	29.5		

SOURCE: M. A. Fligner and J. S. Verducci, "A Nonparametric Test for Judges' Bias in an Athletic Competition," *Applied Statistics* 37 (1988): 101–110.

a. Are there any outliers in the data?
b. Is there a linear trend between the ratings of the two judges?
c. Calculate the correlation coefficient between the two ratings. Is the correlation higher or lower than you expected?

  d. Given that scores of synchronized swimmers are rather subjective, should you compute Pearson's correlation or Spearman's correlation?

3.97   Example 1.7 in Chapter 1 described the 1970 and 1971 lotteries for young men who were drafted during the Vietnam War era. In each case, every day of the year was randomly assigned a draft number. For example, July 5 was randomly assigned draft number 188 in 1970. All eligible men born on July 5 were then in the 188th group to be drafted. Worksheets: **Draft70.mtw** and **Draft71.mtw** contain the complete listings of draft numbers for the 2 years. For the 1970 data, construct side-by-side boxplots for the 12 months. Next, with a pen or pencil connect the medians for the months on the graph. Do you notice any trend? Go back to Example 1.7 and read how the draft numbers were "randomly" assigned. Based on the *median trace*, do you believe that the draft numbers were randomly distributed across the months? Now do the same for the 1971 data.

3.98   Statistical Insight Revisited   The Statistical Insight at the beginning of the chapter compared national math test scores with the percents of students in the state who watch more than 6 hours of television per day. Use that data (worksheet: **TV.mtw**) to do the following:

  a. Construct a scatterplot of test scores and percents who watch more than 6 hours of television per day.

  b. Do you detect a linear pattern in the data? Is it a positive trend or a negative trend?

  c. Would you classify any data points as outliers?

  d. Calculate the correlation between the two variables. Is it positive or negative? Would you classify it as "high"?

  e. Find the least squares regression equation between the test scores and the percents watching television. Be sure to store the residuals in a column.

  f. Plot the residuals against the percent watching television. Is the residual pattern random?

CAUTIONS IN
GATHERING
DATA

Carefully identify your population. Before you attempt to sample a population, you should know exactly who is in the population and who is not. You should know the difference between the target population and the sampled population, if a difference exists. If you cannot define your population, then it is impossible to determine whether you have a representative sample.

Be aware of the shortcomings of a census. Often they are cost prohibitive and very time-consuming. In most cases, a carefully designed experiment or a properly conducted survey will provide just as reliable results as a census.

If possible, avoid the use of convenience samples and discount studies that use convenience samples. Always check the source of pre-existing data. If you collect your own data, know the design for producing the data.

Understand that satistical studies often lead to incorrect conclusions because the data are *biased*. Bias can enter data inadvertently when the experimenter fails to consider all issues relevant to the study. For example, care should be exercised when data are collected by random digit dialing. Nationally less than 25% of all possible phone numbers are associated with residential housing. Furthermore, 5% of all residents do not have phones. *Nonresponse* is a serious issue for those who rely on voluntary surveys. Having obtained the data, the experimenter calculates percentages of certain demographic characteristics of the respondents and attempts to match them up to those published in censuses. If they agree in certain areas such as gender, age, income, and political views, then the experimenter declares that the sample is representative of the population and treats the sample as having been randomly selected from the population. It is very difficult to obtain a representative sample in this fashion unless the response rate is very high.

Know the important variables in your study. Know which ones are independent and which are extraneous and which one is your response (dependent) variable. Know what the variables measure and be sure you are using them properly. If you are comparing the values of a variable measured on two different groups, make sure they are comparable. For example, is it safer to travel by automobile or plane? What variable would you use to compare the dangers of traveling by the two modes of travel? Is it reasonable to compare the numbers of deaths each year for the two modes of transportation or should you also consider the number of miles traveled? These are just a few of the questions to address in a simple study like this one.

OUTLIERS

Recall that an outlier is an observation that lies beyond the rest of the data. Because it is an extreme value, it may have an adverse effect on certain statistics. We say that a statistic is *sensitive* to outliers if the computed value of the statistic is significantly dependent on the outlier. If not—that is, if the computed value of the statistic is not affected by outliers—then we say that the statistic is *resistant* to outliers.

Among the resistant measures are the median, trimmed mean, Spearman rank correlation coefficient, and the spreads, Q-spread and E-spread. The mean, standard deviation, range, and Pearson correlation coefficient are sensitive to outliers. When an outlier is identified, try to determine its origin. It may be due to a miscalculation or misreading an instrument. It could be that you have data from two different groups, in which case it is wise to separate the data into two data sets and analyze them separately. Whatever the reason, always follow up on outliers because they indicate potential difficulty.

INFLUENTIAL
OBSERVATIONS

In the regression setting, an outlier may or may not have an influence on the equation of the regression line. If it is an outlier but still lies in the same plane as the main scatter of the data, then it will not be an *influential observation*. If, however, it is an outlier in only the *x* variable, then it probably is an influential observation and will have a significant effect on the regression equation. Investigate influential observations for validity as suggested above for outliers. Most computer packages have formal diagnostic procedures for identifying and evaluating outliers and influential observations. More will be said about this in Chapter 11.

CAUSATION

Experimentation and survey sampling both can be used to study the relationship between independent and dependent variables. The difference, however, is that survey sampling can show only a relationship between variables, whereas experimentation (in principle) can show causation. If subjects are treated alike except for manipulation of the independent variable, then any change in the dependent variable is caused by the change in the independent variable. Practically speaking, though, it is difficult to show causation. It may be that both the independent and the dependent variables are related through a third extraneous variable. The change observed in the dependent variable may be due to the change in the extraneous variable.

In 1964, the surgeon general reported that smoking can be dangerous to one's health. This conclusion was *not* a result of the surgeon general's establishing a cause-and-effect relationship between smoking and lung cancer. To show that smoking *causes* lung cancer, an experiment would have to be designed in which a homogeneous group of adults is divided into two groups, with one group required to smoke and the other group not allowed to smoke. Furthermore, the two groups would have to be exposed to the same diets, the same environmental conditions, the same working conditions, and so on. After several years, the incidences of lung cancer for the two groups would be compared. Clearly, such an experiment would be impossible to conduct.

It is possible that a genetic or heredity factor causes cancer-prone people to smoke. Then the association seen between smoking and lung cancer is through this third genetic factor. Its effect is confounded with smoking on the cancer rate. The individuals with the genetic factor would be more susceptible to lung cancer regardless of whether they smoked or not. There are data to support the genetic third-factor theory; however, there is more evidence to the contrary. For example, more women are smoking and the rate of lung cancer in women has increased dramatically, which discounts the genetic theory. Many independent studies have established strong associations between smoking and lung cancer. Other experiments have shown that cigarette smoke causes lesions in the skin of laboratory animals, which suggests that smoke could cause lesions in human lungs. Although it has not been proven, the evidence that smoking is a cause of lung cancer is about as strong as it can be with no comparative experiment on humans as described earlier.

LURKING
VARIABLES

Remember, an association between two variables does not establish causation. For example, it has been shown that there is a positive correlation between class attendance and course grades. Does this mean that if you consistently attend class your grades will improve? Not necessarily. It may be that the association we see between class attendance and grades is through a third variable, called a *lurking variable*. A lurking variable is a variable that has an effect on the response but is not included in the study. Motivation, though difficult to measure, may explain the association between class attendance and course grade. A motivated student attends class and a motivated student gets good grades. Just attending class is not enough to get good grades.

EXTRAPOLATION   Extrapolation is drawing conclusions beyond the available data. A classic example of extrapolation is attempting to predict population growth into the next century with current data. What evidence is there that the current growth trend is going to continue in the same pattern in the future? So many factors affect population growth that it is very difficult to predict even 5 to 10 years in the future.

A noted member of Congress stated in a newsletter that in the next 4 years welfare spending will skyrocket to over $500 billion, roughly doubling defense spending. No supporting data were given to justify his claim. Without supporting data, this is just an opinion and not necessarily fact.

Be careful when interpreting predictions. Normally when predictions are given, they predict the average response and do not necessarily apply to individuals. Each time we buy a new car, we are made aware of this as we read the predicted gas consumption for the vehicle. In the fine print it says that miles per gallon for individuals may vary significantly from the stated figures.

In some instances, the experimental units for a study are averages or sums computed from individual units. If this is the case, we should not try to apply the results to the individuals. For example, it is unrealistic to project snowfall in July (or January) based on average annual snowfall for a region.

HISTORICAL
NOTE ON
EXPERIMENTS

Polio Vaccine Study   One of the largest experiments ever conducted on humans took place in 1954 when the Public Health Service experimented with a polio vaccine developed by Jonas Salk (Tanur et al., 1978). The study involved almost 2 million children in grades 1, 2, and 3. The study was done in two parts, one of which involved 750,000 children. In this experiment, 400,000 consented to treatment and 350,000 refused treatment. The 400,000 were then randomly assigned to two groups: a treatment group to get the Salk vaccine and a control group to get a salt-water injection. The subjects did not know whether they were receiving the vaccine or the salt-water solution. Many forms of polio are difficult to diagnose, and in a borderline case the diagnosis could easily be affected by knowledge of whether or not a subject received the treatment. For this reason, the technician was also not told who received the treatment. The study was a double-blind, randomized, controlled experiment with informed consent.

The second part of the study, involving more than 1 million students, was conducted by the National Foundation for Infantile Paralysis (NFIP). They used all the second-grade students who had parental consent as their treatment group (Salk vaccine) and the first- and third-grade students as their control group (salt solution). Even though the sample size was larger, the results in this experiment were less reliable than in the previously described experiment. First, polio is a contagious disease, so the incidence in the second grade could have been much higher (or lower) than in the control group of first- and third-graders. Thus, the study was possibly biased. Second, parental consent was required in the treatment group but not in the control group. It was known that parents with higher incomes tend to consent more readily than lower-income parents. Moreover, children with higher-income parents were more likely to contract polio than children of lower-income parents. This seems the opposite of what one would believe, but remember that polio is a disease of hygiene. Children who live in less sanitary conditions develop antibodies early in childhood that protect them from polio later in life. Again, the study was biased against the vaccine because the subjects in the treatment and control groups had different family backgrounds.

Both studies showed that the Salk vaccine was effective in preventing polio, but the difference in the randomized double-blind experiment was more pronounced. (See the accompanying table.)

Salk vaccine experiment of 1954

| | Randomized Double-blind Experiment | | |
	Treatment	Control	No consent
Sample Size	200,000	200,000	350,000
Rate/100,000	28	71	46
	NFIP Design		
	Vaccine (Grade 2)	Control (Grades 1 and 3)	No consent (Grade 2)
Sample Size	225,000	725,000	125,000
Rate/100,000	25	54	44

SOURCE: Data from Thomas Francis, Jr., *American Journal of Public Health* 45:5 (1955): p. 1–63.

Considering that the vaccine was effective, we might ask whether the vaccine should have been given to all children. Prior to the study, however, it was not known whether the vaccine would be an effective treatment of polio and, in fact, it might well have had an adverse effect on the subjects. So the vaccine should not have been given to all children. Moreover, to get a valid evaluation of the vaccine, a treatment and a control group were necessary. It would have been unethical if it were known that the vaccine (or any treatment for that matter) would be harmful to one's health. In human experimentation, we should not use a treatment if we know it will cause harm to the subjects. Historically, harmful treatments have been given to subjects, but federal regulations now prohibit such studies. Most research institutions have review boards that screen all research involving human subjects.

HISTORICAL NOTE ON SURVEYS

1948 Presidential Election   In the 1948 Truman–Dewey presidential election, all the major pollsters—Gallup, Crossley, and Roper—predicted that Thomas Dewey would defeat Harry Truman. The next table compares their predictions with the actual results.

1948 presidential election

	Truman	Dewey	Others
Gallup Poll	44.5	49.5	6.0
Crossley Poll	44.8	49.9	5.3
Roper Poll	37.1	52.2	10.7
Actual Results	49.5	45.1	5.4

SOURCE: F. Mosteller, *The Pre-election Polls of 1948, Report to the Committee on Analysis of Pre-election Polls and Forecasts* (Washington, D.C.: Social Science Research Council, Bulletin 60, 1949), p. 17.

The major reason for the incorrect prediction of the 1948 election results was that the pollsters failed to account for late shifts in voter preference. The last survey was made 2 weeks prior to the election, and after that, Truman gained enough support to defeat Dewey. Two weeks prior to the election, approximately 15% of the voters were undecided, and it is reported that three-fourths of the undecided vote went to Truman.

From the 1948 election, pollsters have learned to keep a close watch on the "undecided voter" and to revise their predictions up until the last day preceding the election. An important lesson learned in the Roosevelt–Landon election (see Exercise 1.31 on page 23) is that the size of the sample is not nearly as important as sound sampling techniques. The *Literary Digest* had a sample of more than 2.3 million when 1,500 would have sufficed if the sample were truly representative of the population.

UNIT REVIEW

1. True or false?
   a. A simple random sample gives all samples of that size an equal chance of being chosen.
   b. A variable is any characteristic that can be measured on each unit in the population.
   c. Bias means that in repeated sampling, the values obtained tend to be widely scattered or spread out.
   d. Statistical studies have shown that the wording of the questions in a survey has little to do with the outcome.
   e. The target population is always a subset of the sampled population.

Multiple choice (2–5)

2. The Centers for Disease Control conducted a 1990 survey of 11,631 high school students on the issue of carrying a weapon to school (*USA Today*, November 12, 1991). Data from such samples may have
   A. low precision because the sample is too small.
   B. bias because of the use of a random sample.
   C. bias because many students would not tell the truth.
   D. low precision because the population is very large (70 million high school students).

3. The most practical sampling method for conducting the survey of high school students in Exercise 2 is
   A. convenience sampling.
   B. simple random sampling.
   C. stratified sampling.
   D. cluster sampling.

4. Two variables of interest in the survey in Exercise 2 might be
   A. average age of students and IQ of student.
   B. IQ of student and type of weapon carried to school.
   C. type of weapon carried to school and average age of students.
   D. average age of students and average IQ of students.

5. Convenience sampling is discouraged because it usually results in
   A. low precision because it tends to overrepresent certain elements of the population.
   B. bias because it tends to overrepresent or underrepresent elements of the population.
   C. bias because of low precision.
   D. low precision because of bias.

6. The personnel office of a corporation records each of the following variables on each of its employees. Classify each variable as categorical or numerical.
   a. Name   f. Appearance
   b. Gender   g. Years of education
   c. Age   h. Years of service
   d. Race   i. Current salary
   e. Marital status   j. Job difficulty

7. In a class there are ten psychology majors, three sociology, seven criminal justice, two planning, three computer science, and five classified as other. Organize the data in a frequency table and draw a bar graph.

8. Identify each of the following variables as numerical or categorical:
   a. Number of violent crimes
   b. Social Security number
   c. Number of cities with a population greater than 100,000
   d. Employment status
   e. Reaction time to a stimulus

9. The CLEP Subject Examination in Calculus was given to a group of college students. Their scores are listed here:

   Worksheet: CLEP.mtw

   42 45 39 40 46 46 46 39 52 44 42 41 44 46 60 45
   60 38 43 63 45 46 50 41 40 40 41 44 42 44 50 48
   57 47 46 41 42 42 38 47 50 40 50 34 44 52 44

   Construct a stem and leaf plot of the scores. How do you classify the shape of the distribution?

10. Suppose you are studying the characteristics of automobile accidents in your city.
    a. Describe the population of interest.
    b. Give three relevant categorical variables.
    c. Give three relevant numerical variables and classify them as discrete or continuous.

11. A sample of 1,100 people asked to name their favorite sport gave these responses:

Worksheet: Sport.mtw

Sport	Number
Baseball	324
Basketball	149
Football	347
Golf	29
Ice hockey	68
Soccer	92
Tennis	45
Other	46

a. Organize the data in a relative frequency table.
b. Construct a bar graph of the data.

12. A sample of 10,000 homes revealed that the median family income for married couples in 1994 was $41,260. Is $41,260 a parameter or a statistic?

13. The U.S. fertility rate is 1.8 children per woman. If we limit ourselves to the U.S., is 1.8 a parameter or a statistic?

14. Listed here are 1992 per capita incomes for 20 randomly selected counties in North Carolina:

Worksheet: Ncincome.mtw

```
14,722   13,478   17,916   19,203   15,835   17,065   15,012
11,256   14,235   16,260   18,387   15,647   20,534   20,856
16,028   13,148   14,435   16,892   11,522   17,536
```

SOURCE: North Carolina Office of State Planning, State Planning Newsletter, Vol. 1, No. 3, September 1994.

a. Organize the data in an ordered stem and leaf plot.
b. Graph the data in a frequency polygon.
c. How do you classify the shape of the distribution?

15. The following are the degrees of urbanization of ten localities: 63, 56, 32, 56, 48, 45, 45, 96, 57, 72. Find the mean and standard deviation.

16. Classify the following numerical variables as continuous or discrete:
    a. Breaking strength of a wire
    b. Number of dorms on campus
    c. BTUs generated by a heat source
    d. Number of raisins in a box of cereal
    e. California Achievement Test scores for eighth-graders

17. The Council on Plastics and Packaging in the Environment surveyed 1,000 adults about options for getting rid of garbage. Two hundred said burn it, 140 said compost it, 120 said take it to a landfill, 20 said reuse it, 70 gave some other response, 10 refused to answer, and 440 didn't know what to do with it. Organize these data in an appropriate graph.

18. Following are the low temperatures in major cities in the East on November 12, 1994:

Worksheet: Lowtemp.mtw

```
29 32 30 33 34 30 34 31 35 33 39 41 30
41 34 36 37 41 38 44 50 52 37 41 36
```

SOURCE: *USA Today*, November 12, 1994.

a. Organize the data in a meaningful ordered stem and leaf plot.
b. Find the mean score.
c. Find the 10% trimmed mean.
d. Complete a five-number summary diagram including the midsummaries and spreads.
e. How do you classify the shape of the distribution?

19. Classify the following variables as categorical or numerical:
   a. Wind speed
   b. Make of automobile
   c. Number of courses offered at a university
   d. Thickness of a piece of plate glass
   e. Duration of an opera
   f. Rating of pillows for softness

20. The "fasting blood sugar value" was taken from each of nine apparently normal individuals:

   102   103   117   101   101   93   107   87   89

   Calculate the mean and standard deviation.

21. Two independent samples of 400 executives at the 1,000 largest U.S. companies (200 in 1989 and 200 in 1991) surveyed by the Robert Half International Survey organization yielded the following information: 12 in 1989 and 44 in 1991 were fearful of being fired, 108 in 1989 and 90 in 1991 were fearful of losing their job because of acquisition/merger, and 52 in 1989 and 24 in 1991 were concerned with burnout. The remaining in the two samples had concerns other than those listed (*USA Today*, November 12, 1991). Organize these data in a contingency table and a meaningful graph.

COMPUTER EXERCISES

22. According to the Bureau of Labor Statistics, the average single householder spent $3,456 on food in 1985. Following are the annual food expenditures for a random sample of 40 single households in Ohio:

   Worksheet: Food.mtw

$2,845	3,170	2,352	4,978	3,820	2,475	3,160	5,780	2,175	2,648
2,872	4,250	3,970	2,534	6,870	2,734	2,847	4,670	5,176	3,640
2,765	1,180	3,679	3,320	7,580	2,416	3,743	2,830	3,127	3,249
2,648	1,976	2,784	3,869	2,086	5,587	3,420	2,645	8,147	4,367

   Calculate the mean, median, and 10% trimmed mean for the annual food expenditures.

23. In a hospital, a semiprivate room usually accommodates two patients but may contain as many as four. Listed here are the average daily rates for semiprivate rooms across the United States in 1983:

   Worksheet: Roomrate.mtw

Ala	$160.24	Ga	149.83	Md	183.98
Alaska	265.69	Hawaii	227.53	Mass	214.66
Ariz	191.93	Idaho	189.34	Mich	255.00
Ark	143.48	Ill	237.29	Minn	183.33
Calif	275.77	Ind	182.11	Miss	109.66
Colo	209.05	Iowa	175.59	Mo	183.38
Conn	198.48	Kan	181.22	Mont	189.32
Del	214.34	Ky	168.15	Neb	153.31
D.C.	280.41	La	149.30	Nev	238.76
Fla	177.24	Maine	211.26	N.H.	196.42

(*continued*)

N.J.	184.37	Ore	219.60	Utah	177.18
N.M.	191.44	Pa	254.24	Vt	208.07
N.Y.	223.48	R.I.	200.64	Va	163.29
N.C.	141.05	S.C.	133.21	Wash	223.21
N.D.	170.56	S.D.	159.49	W.Va	165.30
Ohio	226.76	Tenn	138.96	Wis	165.05
Okla	165.23	Texas	155.56	Wyo	151.83

SOURCE: *USA Today*, October 29, 1984, Health Insurance Association of America.

a. Construct a stem and leaf plot.
b. Complete a midsummary analysis.
c. Construct a boxplot.
d. Based on the results obtained above, comment on the shape of the distribution.

24. Here is a list of the number of millionaires in each state:

Worksheet: Million.mtw

Ala	4,100	Mont	900
Alaska	1,200	Neb	3,300
Ariz	5,500	Nev	1,500
Ark	2,600	N.H.	2,100
Calif	64,500	N.J.	8,300
Colo	6,900	N.M.	1,100
Conn	9,400	N.Y.	30,900
Del	700	N.C.	3,600
Fla	40,600	N.D.	3,800
Ga	5,800	Ohio	12,600
Hawaii	800	Okla	4,500
Idaho	1,000	Oregon	2,100
Ill	14,500	Pa	25,700
Ind	4,500	R.I.	800
Iowa	3,300	S.C.	2,200
Kan	3,200	S.D.	1,200
Ky	3,500	Tenn	5,300
La	6,200	Texas	39,500
Maine	800	Utah	4,300
Md	8,900	Vt	700
Mass	7,600	Va	4,900
Mich	7,300	Wash	10,400
Minn	17,600	W.Va	600
Miss	2,000	Wis	4,300
Mo	7,700	Wyo	1,300

a. Instead of the number of millionaires, what variable would better illustrate these data?
b. Draw a stem and leaf plot for the data.

25. Given here are the anxiety scores before a major math test for a group of tenth-grade students:

Worksheet: Tenthgrd.mtw

```
12   16   6   22   17   14    8   27   18   19   11
 9   21   3   15   17   22   19   13    5   16
12   14   8    9   15   22   15    6   10   21
```

a. Compute the mean and standard deviation of the data using the Describe command.
b. What other summary measures are computed by the Describe command?
c. Construct a stem and leaf plot and a histogram of the data.
d. Are the data symmetrical? Are the data long-tailed or short-tailed?

26. Listed in the table are the grade point averages, SAT math scores, and final exam grades in college algebra for a group of 20 sophomores. Determine the correlations between all pairs of variables.

Worksheet: Sophomor.mtw

Student	GPA	SAT	Final Exam
1	2.6	510	84
2	2.1	460	77
3	3.5	680	94
4	1.7	390	45
5	2.2	420	71
6	2.9	510	89
7	2.3	370	65
8	3.2	550	90
9	2.3	420	82
10	2.8	470	87
11	3.9	700	99
12	2.2	450	83
13	1.4	380	63
14	2.7	440	75
15	2.8	460	79
16	3.0	520	85
17	1.8	430	70
18	2.0	410	72
19	3.6	650	98
20	2.5	500	75

27. Find Spearman's rank correlation coefficient between the grade point averages and final exam scores from the data in Exercise 26.

28. Construct a scatterplot of the following data:

Worksheet: ExU1-28.mtw

X	81	82	86	84	79	85	91	13
Y	7.4	6.3	4.9	5.3	8.1	5.1	4.2	9.8

a. Is there an outlier in the data?
b. What effect will it have on Pearson's correlation between X and Y?
c. Calculate Pearson's correlation coefficient. Is the correlation higher or lower than you expected?
d. If the outlier is removed, what effect, if any, will it have on Pearson's correlation?

29. Rank the data in Exercise 28 and calculate Spearman's rank correlation coefficient. What effect did the outlier have on Spearman's correlation?

DISCUSSION EXERCISES

30. A new method of taking the temperature of a patient is currently used in some doctors' offices. The nurse takes an instantaneous electronic reading in the ear of the patient. How

does this method compare with the conventional method under the tongue? Describe in detail how you would design a study to compare the two methods.

31. Of concern to everyone in survey research is the problem of nonresponse to surveys. If you are surveying households in a large city, explain what to do if no one is at home when you attempt to survey the residents of a particular home that is located in a public housing complex in the inner-city area? Is it reasonable to substitute the next-door neighbor who is at home?

32. How would you compare the crop production of wheat versus cotton? That is, how would you determine which commodity is the more productive?

33. Suppose you have two job offers when you graduate. One is in Washington, DC, for $40,000 per year and the other is in Charlotte, NC, for $35,000. How would you compare the two salaries? (Assume you know the costs of living in the two areas.)

34. A marketing analyst is asked to study the buying habits of the shoppers at a national chain store. Suppose there are 200 stores around the country. How would you describe the population of interest? Would you attempt a census or a sample of the stores in the chain? If you decide to sample, how would you select the sample? Name two categorical and two numerical variables of interest.

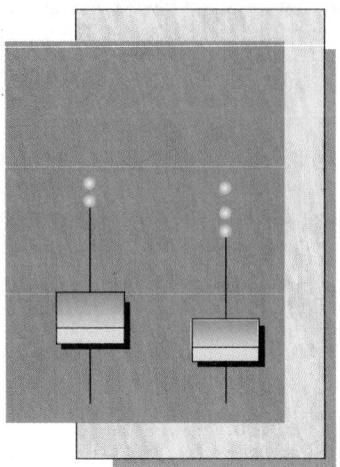

_4_

PROBABILITY AND PROBABILITY DISTRIBUTIONS

Recall that it is usually impractical to investigate an entire population. Therefore, we resort to investigating a sample and infer from the sample to the population. Because the decisions or predictions made about a population are based on sample information, a degree of uncertainty is involved. That uncertainty is measured with probability.

In this chapter, we introduce the vocabulary and laws of probability and present problems that illustrate how probabilities are determined. In later chapters, you will see how probability relates to statistical inference.

CONTENTS

STATISTICAL INSIGHT

LET'S MAKE A DEAL: THE MONTY HALL THREE-DOOR PROBLEM

For years, fans of the popular *Let's Make a Deal* television show watched Monty Hall give contestants the chance to choose, among three doors, the one that concealed the prize of the day. Behind two of the doors were gag gifts, but the other door concealed a valuable prize. After the contestant chose one of the doors, Monty opened one of the other doors with a gag gift. The contestant was then asked whether he or she wished to stay with the original choice or switch to the other closed door. What should the contestant do? Is it better to stay with the original choice or switch to the other closed door? Or does it really matter? The answer, of course, depends on whether contestants would improve their chances of winning by switching doors. In particular, what is the probability of winning by switching and what is the probability of winning by staying?

Columnist Marilyn vos Savant posed this question to her readers in the popular *Parade* magazine column "Ask Marilyn" (1990a). Thousands responded, including many Ph.D.s in mathematics and statistics (1990b, 1991a, 1991b). Most respondents, laypersons and experts alike, concluded that the probabilities of winning were the same for switching and for staying. Are they right?

In this chapter, we develop the tools necessary to understand this "three-door problem" as well as other interesting probability problems. See Computer Exercise 4.175 in Section 4.6 for a discussion of this problem.

4.1 BASIC PROBABILITY CONCEPTS

There is evidence that the concept of probability or chance has existed since the beginning of recorded history. Games of chance played by tossing bones have been documented from as early as 3000 B.C. They apparently evolved into our modern-day games involving the six-sided die. Gambling with dice has been popular since Roman times but was not studied mathematically until the 16th century.

As early as the 14th century, insurance companies insured the contents of cargo ships. To determine the premiums (12% to 15% of the value of the cargo), the insurance companies had to determine the probability that they might have to pay a claim. Another use of probability was in the recording of vital statistics such as births, deaths, and marriages. The *Bill of Mortality* in London in the early 1500s involved calculating the probability of death at a certain age. Such was the beginning of life insurance.

It is generally believed that the mathematical theory of probability was started by the French mathematicians, Blaise Pascal (1623–1662) and Pierre Fermat (1601–1665). Their solutions to certain gambling problems involving dice posed by the gambler and French nobleman, Chevalier de Méré, are considered to be among the first contributions to probability theory. As probability theory relates to statistics, significant contributors were Jacob Bernoulli (1654–1705), Abraham de Moivre (1667–1754), Pierre-Simon de Laplace (1749–1827), and Carl Friedrich Gauss (1777–1855). A significant contributor to modern-day probability theory is the Russian mathematician, A. N. Kolmogorov, who in 1933 presented a consistent axiomatic approach to the study of probability.

Today, probability is a useful tool in almost all fields of study, and its applications are much broader than the original gambling problems. Knowledge of probability and how it applies to statistical analysis is a necessity in areas such as astronomy, biology, computer science, criminal justice, economics, engineering, geography, geology, operations research, political science, psychology, and sociology.

One of the most important goals of statistics is to make inferences about a population from the information contained in a sample. Once an inference is made, however, we must measure the reliability of that inference. Probability is the tool that allows us to do so.

What Is Probability?

We all have some idea of what is meant by the term *probability* or *chance*. A student who "guesses" on a true/false problem has a 50–50 chance of getting it right. Or there is a 10% chance that a controversial bill will pass Congress. A baseball player has a .324 batting average; thus, each time at bat he has a 32.4% chance of getting a hit.

Intuitively we think of probability as a numerical value that is associated with some outcome and indicates how likely it is that the outcome will occur. We say that a small probability indicates that the outcome is not likely to occur and a large probability indicates that it is likely that the outcome will occur. The controversial bill has a 10% chance of passing; the 10% is small, which indicates that the bill has very little chance of passing. If the probability were 90%, the bill has a very good chance of passing.

To better understand probability, we need to introduce some terminology. An *experiment* is the process of making an observation, such as recording the results of the roll of a die, or taking a measurement, such as weighing a person. Other examples are recording voters' opinions on a certain issue, recording SAT scores for entering college students, measuring the dissolved oxygen content of a river, and observing the diameters of trees in a forest. We can say that any activity that results in outcomes is an experiment.

All experiments result in a certain collection of outcomes, and the set of all possible outcomes is called the *sample space* for the experiment.

An **experiment** is the process of making an observation or taking a measurement.

The collection of all possible outcomes of an experiment is called the **sample space, S.**

If the experiment is rolling a die, then the sample space is

$$S = \{1, 2, 3, 4, 5, 6\}$$

If the experiment is tossing a coin, the sample space is

$$S = \{\text{head, tail}\}$$

If the experiment is weighing a person, the sample space (assuming no one weighs more than 600 pounds) is

$$S = \{x \mid x \text{ is a real number between 0 and 600}\}$$

> Any subset of the sample space is called an **event**. An event is said to have **occurred** if any one of its elements is the outcome when the experiment is conducted.

In the die-rolling experiment, the event A that an odd number occurs is $A = \{1, 3, 5\}$. The event A occurs if, when the die is rolled, any one of the three possibilities 1, 3, or 5 comes up.

The coin-toss and die-rolling experiments are used extensively throughout our discussion of probability. The reason is not that we are interested in tossing coins or rolling dice, but rather that these experiments, simple as they may be, provide useful models for many real-world experiments. For example, diagnostic testing for such things as AIDS or prostate cancer in men can be modeled after the coin-toss experiment. Experiments with multiple outcomes like classifying a product as superior, average, or inferior are very similar to the die experiment. As these simple models are discussed, try to imagine how they could be modified to describe a particular experiment in your discipline.

EXAMPLE 4.1

According to the Bureau of the Census, 50% of all homes in Dallas are heated with gas (Statistical Brief SB/95-7, May 1995). For three randomly selected homes in Dallas, list the outcomes of the sample space corresponding to whether or not they are heated with gas. Also list the outcomes that make up the following events:

Event A: Exactly two of the three heat with gas

Event B: Only one heats with gas

Event C: All three heat with gas

Event D: At least one of the three heats with gas

SOLUTION

A tree diagram helps us list the possible outcomes. Each of the three homes will either heat with gas (g) or not (n). If the first selected home heats with gas, the second can either heat with gas or not, or if the first does not heat with gas, the second can again heat with gas or not. So, for two homes, there are four possibilities: gg, gn, ng, and nn. Then there are two possibilities, the third home heats with gas or not, for each of those four possibilities, resulting in eight different possible outcomes. The tree diagram is shown in Figure 4.1 on page 232. The sample space is

$$S = \{(g, g, g), (g, g, n), (g, n, g), (n, g, g), (g, n, n), (n, g, n), (n, n, g), (n, n, n)\}$$

where, for example, the outcome (g, n, g) represents that the first and third homes heat with gas and the second does not.

FIGURE 4.1

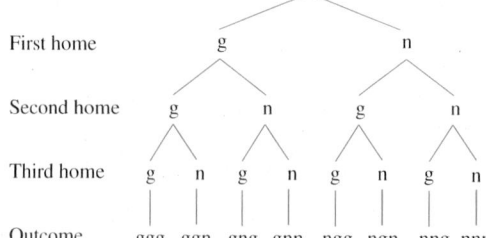

From the outcomes in the tree diagram, we can list the outcomes of the listed events:

$$A = \{(g, g, n), (g, n, g), (n, g, g)\}$$
$$B = \{(g, n, n), (n, g, n), (n, n, g)\}$$
$$C = \{(g, g, g)\}$$
$$D = \{(g, n, n), (n, g, n), (n, n, g), (g, g, n), (g, n, g), (n, g, g), (g, g, g)\}$$

Notice the similarity between this example and the experiment of tossing three coins. We could say that g means heads and n means tails. Also, note that the 50% figure is not used to list the outcomes in the sample space or events. Later we will need the 50% figure to assign probabilities to the outcomes.

EXAMPLE 4.2

In Example 4.1, suppose that the three homes selected are owned by the Jones, Brown, and Smith families, and it is determined that the Jones home is heated with gas, the Brown home is heated with gas, and the Smith home is not heated with gas. List which of the four events, A, B, C, and D, occurred and which did not occur.

SOLUTION

If we assume that the Jones home is the first selected, the Brown home is second, and the Smith home is the third selected, then the outcome we observed is (g, g, n), which is in events *A* and *D*. Thus, we can say that events *A* and *D* occurred but events *B* and *C* did not.

**Probability of
an Individual
Outcome**

The likelihood of the occurrence of an event depends on its *probability*. First, we look at the probability of individual outcomes.

The probability of an outcome of an experiment can be described as the *relative frequency* with which the outcome will occur if we repeat the experiment a large number of times. For example, if we toss a fair coin, we can say that the probability of getting a head on the coin is 50% or $\frac{1}{2}$ because half of the time the coin should land on a head and half of the time it should land on a tail. This does not mean that out of 10 tosses,

5 should be heads and 5 should be tails; rather, if we toss the coin 10,000 times, the number of heads should be near 5,000 (i.e., the relative frequency of heads should be near .5).

Earlier we used four different expressions for a probability: 50–50, 10%, .324, and $\frac{1}{2}$. Each can be transformed to any of the other forms. For example, 50–50 on the true/false question means half of the time the answer is right and half of the time it is wrong. This can be written as 50% or .50 or $\frac{1}{2}$, all meaning the same thing.

In any event, we can interpret the probability of an outcome as a number between 0 and 1. The closer it is to 0, the *less likely* the outcome is to occur, and the closer it is to 1, the *more likely* the outcome is to occur. In fact, if its probability is 0, then the outcome cannot happen, and if its probability is 1, then the outcome is certain to occur.

EXAMPLE 4.3

In a particular card game, all 52 cards are dealt to four players. Suppose you are dealt four aces. What is the probability that one of the other three players has an ace?

SOLUTION

Because there are only four aces in the deck and you have all four, no one else can have an ace. Therefore the probability is 0.

EXAMPLE 4.4

Ten people, four men and six women, are in a meeting and, at random, they select a person to preside. What is the probability that the person selected is a woman?

SOLUTION

Because six of the ten people are women and each person is equally likely to be selected, the probability that a woman is selected is six out of ten or .6.

Assigning Probabilities to Individual Outcomes

In assigning probabilities to the individual outcomes in a sample space, two conditions must be satisfied:

1. The probability of each outcome must be between 0 and 1, inclusive.
2. The probabilities of all outcomes in the sample space must sum to 1.

EXAMPLE 4.5

Suppose students who are registering for the fall semester are randomly assigned to one of six sections of freshman English. Because the students are randomly assigned to the sections, the six outcomes—{1}, {2}, {3}, {4}, {5}, {6}—are equally likely to occur.

Therefore, we assign a probability of $\frac{1}{6}$ to each because there is one chance out of six of getting any one of the sections. If we sum the probabilities of all outcomes, we get $\frac{6}{6} = 1$, which says that one of the outcomes will occur when the experiment is conducted. Notice the similarity between this example and the experiment of rolling a die.

EXAMPLE 4.6

Suppose that a student is twice as likely to be assigned to an even-numbered section than to an odd-numbered section. Assign probabilities to each possible outcome.

SOLUTION

There are still six outcomes—{1}, {2}, {3}, {4}, {5}, {6}. Suppose the probabilities of ending up in sections {1}, {3}, and {5} are each a value called p. Then the probabilities of {2}, {4}, and {6} are each $2p$. The sum of all the probabilities must be 1, so we have

$$1 = p + 2p + p + 2p + p + 2p = 9p$$

Because $9p = 1$, we have that $p = \frac{1}{9}$, so $2p = \frac{2}{9}$. That is, the probability of being assigned to each of the odd-numbered sections is $\frac{1}{9}$ and the probability of being assigned to each of the even-numbered sections is $\frac{2}{9}$.

C O M P U T E R   T I P

Random Data under the Calc menu in Minitab simulates data from a number of probability distributions, among which is the discrete uniform distribution. The discrete uniform distribution generates integer values (1, 2, 3, 4, etc.) such that all values have an equal chance of occurring. To simulate observations from the discrete uniform distribution, first select the **Calc** menu, choose **Random Data**, and then click on **Integer**. When the Integer window appears, choose the **number of observations** to generate, the **column to store the data**, and the **Minimum and Maximum values** for the data. For example, to simulate the roll of a die, the minimum value is 1 and the maximum value is 6; that is, roll a number between 1 and 6.

The same can be accomplished in the Session window. For example, to simulate 18 rolls of a fair die, execute the following commands at the MTB > prompt:

MTB > **Random 18 Observations, Put in C1;**
SUBC> **Integers Between 1 and 6.**

The Tally command can be used to give a frequency table for each column that summarizes how many times each value came up out of the 18 rolls. To execute a tally, first select the **Stat** menu, choose **Tables**, and then click on **Tally**. When the Tally window appears, **select all columns** and then click **OK**.

EXAMPLE 4.7

Simulate 18 rolls of a fair die and repeat it 10 times. Ideally, we would expect each of the six possible integers to turn up three times in 18 rolls. For each simulation, determine the frequency of occurrence of each possible outcome.

SOLUTION

In the Integer Distribution window, generate **18** observations, store in **C1-C10**, choose minimum value **1** and maximum value **6**, and then click **OK**. To summarize how many times each value came up in the ten simulations, at the Tally window, **select all 10 columns** and click **OK**. Exhibit 4.1(a) gives the simulation, and Exhibit 4.1(b) gives the summary.

EXHIBIT 4.1(A)

```
                                    MINITAB - Untitled Worksheet
 File   Edit   Manip   Calc   Stat   Graph   Editor   Window   Help
                                      Session
Worksheet size: 100000 cells


MTB > Random 18 c1-c10;
SUBC>   Integer 1 6.
MTB >
```

	C1	C2	C3	C4	C5	C6	C7	C8	C9	C10
1	3	3	3	2	3	6	6	3	4	3
2	5	5	3	3	4	5	4	5	6	2
3	3	5	1	4	2	4	1	5	5	5
4	1	5	1	3	1	3	5	5	3	6
5	5	6	2	6	3	6	6	5	4	1
6	1	4	5	5	2	4	3	5	3	6
7	2	4	6	3	3	2	1	1	1	5
8	5	4	2	3	1	1	2	3	4	4
9	6	2	6	3	5	5	5	6	4	4
10	1	3	3	6	1	3	3	6	2	6
11	3	2	1	6	3	6	5	3	5	1
12	3	2	3	1	5	2	6	1	3	1
13	4	3	5	4	4	3	5	6	6	3
14	6	5	3	2	2	2	4	5	6	1
15	3	4	6	3	5	1	5	2	1	2
16	1	4	1	6	4	1	5	6	5	4
17	2	3	4	4	1	4	2	5	1	1
18	1	3	2	4	6	6	2	1	3	6
19										

EXHIBIT 4.1(B)

```
MTB > Tally C1-C10;
SUBC>   Counts.
```

Summary Statistics for Discrete Variables

C1	Count		C2	Count		C3	Count		C4	Count
1	5		2	3		1	4		1	1
2	2		3	5		2	3		2	2
3	5		4	5		3	5		3	6
4	1		5	4		4	1		4	4
5	3		6	1		5	2		5	1
6	2		N=	18		6	3		6	4
N=	18					N=	18		N=	18

(continued)

EXHIBIT 4.1(B)
(continued)

C5	Count		C6	Count		C7	Count		C8	Count
1	4		1	3		1	2		1	3
2	3		2	3		2	3		2	1
3	4		3	3		3	2		3	3
4	3		4	3		4	2		5	7
5	3		5	2		5	6		6	4
6	1		6	4		6	3		N=	18
N=	18		N=	18		N=	18			

C9	Count		C10	Count
1	3		1	5
2	1		2	2
3	4		3	2
4	4		4	3
5	3		5	2
6	3		6	4
N=	18		N=	18

As stated, we would have expected each of the six possible values to come up three times each. We see in simulation number 8, however, that the 4 did not appear but the 5 appeared seven times. On the other hand, simulation number 6 is close to what one would expect.

Probability of an Event

Once probabilities are assigned to all outcomes of the sample space, we can find the probabilities of events.

> The **probability of an event** A is the sum of the probabilities of the outcomes in A. We write it as $P(A)$.

We saw in Example 4.5 that if students are randomly assigned to one of six sections of freshman English, then we can assign the probability $\frac{1}{6}$ to each of the six outcomes. Then the probability of event A, that a student is assigned to an odd-numbered section, is

$$P(A) = P(\{1\}) + P(\{3\}) + P(\{5\}) = \frac{1}{6} + \frac{1}{6} + \frac{1}{6} = \frac{1}{2}$$

On the other hand, for the situation in Example 4.6, where students are twice as likely to be assigned to an even-numbered section, the probability of being assigned to an odd-numbered section is

$$P(A) = P(\{1\}) + P(\{3\}) + P(\{5\}) = \frac{1}{9} + \frac{1}{9} + \frac{1}{9} = \frac{1}{3}$$

EXAMPLE **4.8**   In the game of craps, the "shooter" wins on the first roll of a pair of fair dice if the sum of the two dice is 7 or 11. Calculate the probability that the shooter wins on the first roll.

SOLUTION

The sample space for the roll of a pair of dice is

$$S = \{(1, 1), (1, 2), (1, 3), (1, 4), (1, 5), (1, 6),$$
$$(2, 1), (2, 2), (2, 3), (2, 4), (2, 5), (2, 6),$$
$$(3, 1), (3, 2), (3, 3), (3, 4), (3, 5), (3, 6),$$
$$(4, 1), (4, 2), (4, 3), (4, 4), (4, 5), (4, 6),$$
$$(5, 1), (5, 2), (5, 3), (5, 4), (5, 5), (5, 6),$$
$$(6, 1), (6, 2), (6, 3), (6, 4), (6, 5), (6, 6)\}$$

Assuming both dice are fair, we have 36 outcomes that are equally likely—that is, all have probability $\frac{1}{36}$. The event of interest is that a total of 7 or 11 turns up. In set notation, the event of interest is

$$A = \{(1, 6), (2, 5), (3, 4), (4, 3), (5, 2), (6, 1), (5, 6), (6, 5)\}$$

Adding the probabilities of the outcomes in A, we have

$$P(A) = 8 \left(\frac{1}{36} \right) = \frac{8}{36}$$

So the probability is $\frac{8}{36}$ or $\frac{2}{9}$ that the shooter wins on the first roll.

We can summarize the steps in calculating the probability of any event as follows:

Calculating the Probability of an Event

1. Define the experiment and list the outcomes in the sample space.
2. Assign probabilities to the outcomes such that each is between 0 and 1 and they sum to 1.
3. List the outcomes of the event of concern.
4. Sum the probabilities of the outcomes that are in the event of concern.

Thus far, listing the outcomes in a sample space has been fairly easy. Assigning probabilities has not been a difficult task either. However, many experiments are more difficult. In some situations, listing the sample space of outcomes may be straightforward but assigning probabilities may not be obvious. In other situations, the sample space may be so large that a listing is impractical. For example, we could not attempt to list all possible five-card poker hands. If we knew how many possible hands there were, however, we could assign an equal probability to each one (the hands are equally likely because you are just as likely to get any 5 of the 52 possible cards) and then find the probability of any type of hand. Combinatorial mathematics can be used to count the number of possible poker hands as well as to determine the number of outcomes in other sample spaces. In the next section, we present some laws of probability that will be helpful in determining the probability of events.

EXERCISES 4.1

4.1   If you roll a fair die, what is the probability that you will observe a number greater than 4?

4.2   There are 200 names in a bowl and one is drawn as the winner. What is the probability that you win, assuming that your name is in the bowl?

4.3   Suppose there are three doctors in a group of ten people. If we select a person at random, what is the probability that he or she is a doctor?

4.4   A card is drawn from an ordinary deck of 52 cards. What is the probability that it is red?

4.5   Two cards are drawn from a deck of cards one after the other without replacement (the first card is *not* placed back in the deck before the second draw). The first card is red. What is the probability that the second card is also red?

4.6   In the game of blackjack, the face cards (jack, queen, king) count 10 points, an ace counts as 1 or 11 (player's choice), and the remaining cards carry the point count that is on the card. To score blackjack, you must score 21 points with just two cards. If you score blackjack, what is the probability that you have an ace?

4.7   In a particular card game, all 52 cards are dealt to four players who are matched up as partners. Suppose you are dealt three aces. What is the probability that your partner has the other ace? Remember that there are three other players, one of whom is your partner.

4.8   There are three states on the West Coast: Washington, Oregon, and California.
       a. If we randomly choose one of the states, what is the probability that we select California?
       b. If the residents of the three states are grouped together and we randomly select a resident, is the probability that the person is a California resident the same as your answer in part **a**? Explain.

4.9   A basketball player has a 50% chance of making a free throw. Out of two tries, what is the probability that he misses both shots?

4.10   The accompanying table on local jail inmates and facilities, by size of facility, was produced by the Bureau of Justice Statistics. It appeared in the bulletin *Jails and Jail Inmates 1993–94*, April 1995.

	Inmates		Facilities	
Size of facility	Number	Percent	Number	Percent
Fewer than 50	34,332	7.5	1,874	56.7
50–99	37,135	8.1	545	16.5
100–149	31,293	6.8	253	7.7
150–249	41,472	9.0	218	6.6
250–499	73,938	16.1	209	6.3
500–999	90,481	19.7	129	3.9
1,000–1,499	44,000	9.6	35	1.1
1,500–1,999	30,764	6.7	18	.5
2,000 or more	76,389	16.6	23	.7

a. Do the percents sum to 100%?
b. What percent of the inmates are in facilities with fewer than 50 inmates?
c. What percent of the facilities have fewer than 50 inmates?
d. What percent of the inmates are housed in facilities that have fewer than 250 inmates?
e. What percent of the facilities have fewer than 250 inmates?
f. If we randomly select an inmate, what is the probability he or she is in a facility with fewer than 250 inmates?
g. If we randomly select a facility, what is the probability that it has fewer than 250 inmates?

4.11 Roulette is played by spinning a ball on a round table that is divided into 38 slots of equal size. The slots are numbered 00, 0, 1, 2, 3, . . . , 35, 36. The slots are also colored as follows:

Red: 1, 3, 5, 7, 9, 12, 14, 16, 18, 19, 21, 23, 25, 27, 30, 32, 34, 36

Black: 2, 4, 6, 8, 10, 11, 13, 15, 17, 20, 22, 24, 26, 28, 29, 31, 33, 35

Green: 00, 0

Find the probability of the ball falling into a black slot on a single spin. Find the probability that the ball falls into a slot that is neither red nor black on a single spin.

4.12 A group of people consists of two children under 12 years of age, three teenagers, and five adults. Suppose a person is selected at random. What is the probability that the person is an adult? What is the probability that the person selected is over 12 years old?

4.13 In a physical fitness study, subjects are randomly assigned to five different exercise groups. List the outcomes in the sample space and the following events:

A: Assigned to group 3

B: Assigned to one of the first three groups

C: Assigned to group 4 or 5

D: Assigned to a group between 2 and 5 exclusively

4.14 Assign probabilities to the outcomes in the sample space in Exercise 4.13 and then compute the probabilities of the different events.

4.15 A person conducting a poll randomly chooses one of two houses. He then randomly chooses either a man or a woman from the house. List the outcomes in the sample space of possibilities.

4.16 Two cards are drawn from a deck of cards with replacement and we are concerned only with the color. List the outcomes in the sample space of the experiment.

4.17 In Exercise 4.16 we are concerned with event R that both cards are red. One of the cards is the 2 of spades. Did event R occur?

4.18   In the roll of a pair of dice, we are concerned with the event W that a total of 7 is rolled. One of the dice turns up on an even number. What must the other die be for event W to occur?

4.19   Consider the sample space $S = \{s_1, s_2, s_3, s_4\}$ in which

$$P(s_1) = .4 \qquad P(s_2) = .2 \qquad P(s_3) = .1$$

What is $P(s_4)$?

4.20   A woman has a blue suit that she wears 30% of the time. Let B correspond to her wearing the blue suit and N mean that she does not. For 2 days, list the outcomes of the sample space. Are the outcomes equally likely?

4.21   Student loan applications are either approved or disapproved. If three students apply, list the sample space of possible outcomes. If 60% of the applications are approved, are the outcomes equally likely? If 50% of the applications are approved, are the outcomes equally likely? Assuming 50% are approved, assign probabilities to the outcomes.

4.22   There are three traffic lights on your way home. Assume that each light is either red (R) or green (G) and that it is green with probability .7. List the outcomes of the sample space. Are the outcomes equally likely?

4.23   Assume that 60% of the student body are women. Three students are selected at random. List the outcomes in the sample space of the possible genders of the three students. Are the outcomes equally likely? Assign probabilities to the outcomes if 50% of the student body are women.

4.24   Describe the basic difference between Exercises 4.21, 4.22, and 4.23.

4.25   For the game of roulette (see Exercise 4.11) assign the number 37 to 00 and the number 38 to 0 so that an integer from 1 to 38 is assigned to each of the 38 slots. Use the Random Data command in Minitab to simulate 20 rolls of the roulette wheel. Out of the 20 rolls, how many times did the ball fall into a black slot? How many times did it fall into a red slot? A green slot? Next, simulate 380 rolls and use the Tally command to summarize the results. Out of the 380 rolls, how often should we expect each value to turn up? How close are the actual results to what you would expect?

4.26   a. Roll a fair die. What is the probability that it turns up odd and greater than 4?
        b. A card is drawn from a standard deck of cards. What is the probability that the card is red and a face card?
        c. Roll a pair of fair dice. What is the probability that the sum of the two up faces is 7 or 11?

4.27   The big wheel on *The Price Is Right* game show is marked off in 20 equal increments from 5 cents to 1 dollar. The contestant gets two tries to spin a total as close as possible to 1 dollar without going over. Suppose a contestant gets 60 cents on the first spin.
        a. If she spins again, what is the probability she gets a total of 1 dollar?
        b. What is the probability she goes over 1 dollar?

4.28   There are eight books on a bookshelf, of which three are fiction.
        a. Suppose you select a book at random. What is the probability that it is nonfiction?
        b. A second book is selected without replacement of the first. If the first one was fiction, what is the probability the second is also fiction?

4.2   EVENT RELATIONSHIPS AND PROBABILITY LAWS

Events that are formed from intersections and unions of other events are called *compound events*. Most events of interest are compound events. This was true in Example 4.8 when we were interested in the probability of rolling a total of 7 *or* 11 on the toss of a pair of fair dice. We could have approached the problem as the combination of two events: One event, say *A*, is rolling a 7 and the other event, say *B*, is rolling an 11. The event of interest then is event (*A* or *B*). We could also speak of the event (*A* and *B*). Notice, however, that no outcomes are in event (*A* and *B*) because it is impossible to roll both a 7 and an 11 on a single toss. In this case, we say that event *A* and event *B* are disjoint or *mutually exclusive*.

Mutually Exclusive Events

> Events *A* and *B* are said to be **mutually exclusive** if they have no outcomes in common.

In other words, if the occurrence of one event precludes the possibility of the other event occurring, then the events are mutually exclusive. When events are mutually exclusive, this additive law applies.

The Additive Law for Mutually Exclusive Events

Let *A* and *B* be two mutually exclusive events. Then

$$P(A \text{ or } B) = P(A) + P(B)$$

Figure 4.2 is a *Venn diagram* that illustrates the additive law for mutually exclusive events. The shaded circle labeled *A* represents the event *A*, and the shaded circle labeled *B* represents the event *B*. The event (*A* or *B*), represented by both circles, can occur if either event *A* or event *B* occurs. Hence, the probability of (*A* or *B*) is given by $P(A) + P(B)$.

FIGURE 4.2

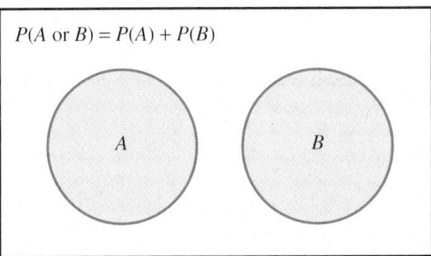

$P(A \text{ or } B) = P(A) + P(B)$

The additive rule for mutually exclusive events extends to multiple events. For example, the probability of rolling a total that exceeds 8 with a pair of dice is

$$P(\text{total} > 8) = P(9) + P(10) + P(11) + P(12)$$

(Note that it is impossible to roll a total greater than 12.)

Another useful probability law involves the *complement* of an event.

> The **complement of event** A is the collection of all outcomes in the sample space that are not in A. It is denoted as A^c.

Because A^c is all outcomes not in A, the two events make up the entire sample space. Accordingly, we have the complement law.

The Complement Law

Let A be an event with probability denoted by $P(A)$. Then

$$P(A^c) = 1 - P(A)$$

Thus, if the probability of an event is known, then the probability of its complement is also known.

Figure 4.3 illustrates the complement law. The shaded region outside the event A is the complement of A. Notice that $P(A)$ is subtracted from the probability of the whole sample space, which is 1.

FIGURE 4.3

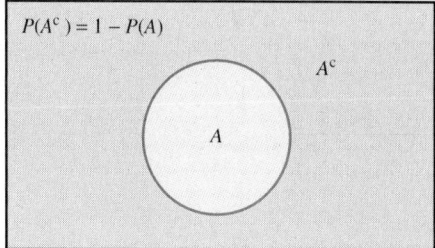

EXAMPLE 4.9

The Bureau of the Census states that 50% of all homes in Dallas are heated with gas. Out of six randomly selected homes in Dallas, what is the probability that at least one home is heated with gas?

SOLUTION

With three randomly selected homes, we saw in Example 4.1 that there are eight different outcomes. Recall that the sample space is

$$S = \{(g, g, g), (g, g, n), (g, n, g), (n, g, g), (g, n, n), (n, g, n), (n, n, g), (n, n, n)\}$$

If there are four homes, each outcome is of the form gggg, gggn, and so on, and there are 16 of them. The tree diagram in Figure 4.4 helps us list all 16. If there are five homes, there are 32 different outcomes of the form ggggg, ggggn, and so on. We see that the number of outcomes goes up by a power of 2 because there are two possibilities for each home: namely, they heat with gas (g) or they do not (n). Consequently for six homes there are 64 different outcomes of the form gggggg, gggggn, and so on. Because 50% heat with gas, each sequence of g's and n's has the same chance of occurring, and therefore all the outcomes are equally likely. Because there are 64 of them, each has probability $\frac{1}{64}$.

FIGURE 4.4

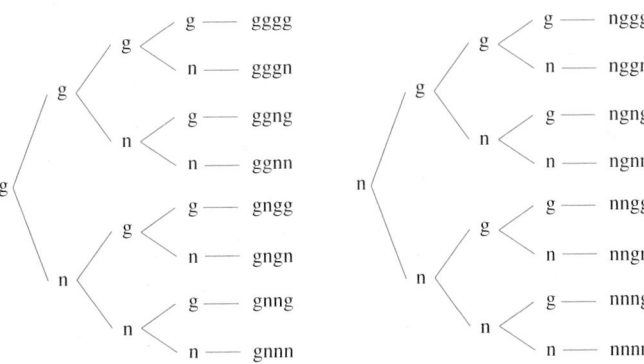

The event of interest, call it A, is that at least one of the six homes heats with gas. Event A contains 63 of the 64 outcomes in the sample space. Certainly we do not want to list all 63. However, the complement of A, A^c, contains only one outcome:

$$A^c = \{nnnnnn\} \quad \text{and} \quad P(A^c) = \frac{1}{64}$$

Using the complement law, we have

$$P(A) = 1 - P(A^c) = 1 - \frac{1}{64} = \frac{63}{64}$$

We see that there is a very high probability that at least one home heats with gas.

If two events are not mutually exclusive, then they must have outcomes in common. Figure 4.5 (page 244) illustrates the additive law for all events whether they are mutually exclusive or not.

The overlapping section between the two events represents the outcomes in common, which is event (A and B). When we add $P(B)$ to $P(A)$, we have added $P(A$ and $B)$ in twice, once with $P(A)$ and once with $P(B)$. Hence, to compute $P(A$ or $B)$, we must subtract $P(A$ and $B)$ one time from the sum of $P(A)$ and $P(B)$.

The Additive Law

Let A and B be any two events. Then

$$P(A \text{ or } B) = P(A) + P(B) - P(A \text{ and } B)$$

FIGURE 4.5

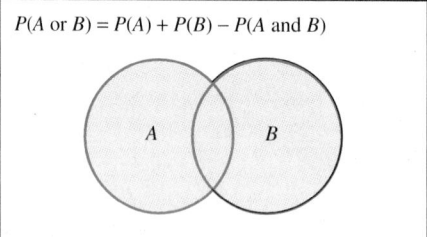

$P(A \text{ or } B) = P(A) + P(B) - P(A \text{ and } B)$

A B

EXAMPLE 4.10

In Example 4.5, students were randomly assigned to one of six sections of freshman English—section 1, 2, 3, 4, 5, or 6. Let events A and B be defined as follows:

    A: Susan is assigned to an even-numbered section

    B: Susan is assigned to a section numbered less than 3

Find the probability of A or B.

SOLUTION

Because students are randomly assigned to the sections, each assignment has probability $\frac{1}{6}$. Consequently,

$$P(A) = P(\{2\}) + P(\{4\}) + P(\{6\}) = \frac{1}{6} + \frac{1}{6} + \frac{1}{6} = \frac{3}{6}$$

and

$$P(B) = P(\{1\}) + P(\{2\}) = \frac{1}{6} + \frac{1}{6} = \frac{2}{6}$$

Recall that the event (A and B) contains outcomes that are common to both A and B; thus,

$$P(A \text{ and } B) = P(\{2\}) = \frac{1}{6}$$

Using the additive law, we have

$$P(A \text{ or } B) = P(A) + P(B) - P(A \text{ and } B) = \frac{3}{6} + \frac{2}{6} - \frac{1}{6} = \frac{4}{6} = \frac{2}{3}$$

Note that in adding $P(A)$ and $P(B)$, we are adding in $P(\{2\})$ twice. However, we have subtracted it out one time when we subtract $P(A \text{ and } B) = P(\{2\})$. In general, when finding $P(A \text{ or } B)$, we must subtract $P(A \text{ and } B)$ because it has been added in twice.

Independent Events

Mutually exclusive events A and B cannot both occur simultaneously because they have no outcomes in common. Let us now consider events that may occur simultaneously and if one occurs, it does not affect the probability of the other occurring. That is, the likelihood of event A occurring is unaffected by the occurrence of event B. In those cases, we say that events A and B are *independent*.

> Events A and B are said to be **independent** if the occurrence of one does not affect the probability of the occurrence of the other. Otherwise, A and B are **dependent**.

If we toss a quarter and a dime in the air, it is possible that both will land with heads facing up. The fact that the quarter landed with heads up, however, has no effect on what happens to the dime. Their outcomes are independent. That is to say, both have a $\frac{1}{2}$ chance of landing with heads up, and the probability remains $\frac{1}{2}$ for each coin regardless of what happens to the other coin. Because there are four equally likely outcomes in the sample space, $S = \{hh, ht, th, tt\}$, there is a $\frac{1}{4}$ chance that both land with heads up, $\{hh\}$. Notice that we get $\frac{1}{4}$ when we multiply $\frac{1}{2}$ by $\frac{1}{2}$. This is a result of the **multiplication law** for independent events.

The Multiplication Law for Independent Events

Let A and B be two independent events. Then

$$P(A \text{ and } B) = P(A)P(B)$$

EXAMPLE 4.11

Let us reconsider the problem of whether or not homes in Dallas are heated with gas. For three randomly selected homes, we found in Example 4.1 that the sample space of possible outcomes is

$$S = \{(g, g, g), (g, g, n), (g, n, g), (n, g, g), (g, n, n), (n, g, n), (n, n, g), (n, n, n)\}$$

Instead of assuming that 50% of all homes are heated with gas, suppose only 30% are heated with gas. Assign probabilities to the eight outcomes in the sample space.

SOLUTION

The outcome $\{(g, g, g)\}$ means that all three homes heat with gas, and each heats with gas with a 30% probability. One home heating with gas is independent of any other home heating with gas, so using an extension of the multiplication law, we have

$$P(\{g, g, g\}) = P(\text{1st heats with gas})P(\text{2nd heats with gas})P(\text{3rd heats with gas})$$

$$= (.3)(.3)(.3)$$

$$= .027$$

In a similar manner, outcome $\{(g, g, n)\}$ should be assigned probability $(.3)(.3)(.7) = .063$, as should outcomes $\{(g, n, g)\}$ and $\{(n, g, g)\}$, because each consists of two homes heating with gas and the other one not. Outcomes $\{(g, n, n)\}$, $\{(n, g, n)\}$, and $\{(n, n, g)\}$ should each be assigned probabilities $(.3)(.7)(.7) = .147$. Finally, outcome $\{(n, n, n)\}$ should be assigned probability $(.7)(.7)(.7) = .343$. Table 4.1 lists the probabilities associated with each outcome.

TABLE 4.1	Outcome	Probability
	$\{(g, g, g)\}$	.027
	$\{(g, g, n)\}$	.063
	$\{(g, n, g)\}$	.063
	$\{(n, g, g)\}$	.063
	$\{(g, n, n)\}$	.147
	$\{(n, g, n)\}$	.147
	$\{(n, n, g)\}$	.147
	$\{(n, n, n)\}$	.343
	Total	1.000

If we had used 50% instead of 30%, notice that $P(\{g, g, g\}) = (\frac{1}{2})(\frac{1}{2})(\frac{1}{2}) = \frac{1}{8}$. In fact, all outcomes would have probability $(\frac{1}{2})(\frac{1}{2})(\frac{1}{2}) = \frac{1}{8}$ because g and n both have probability $\frac{1}{2}$. In other words, the outcomes would be equally likely and have probability $\frac{1}{8}$ each.

EXAMPLE 4.12

In Example 4.1 events A, B, C, and D were defined as follows:

Event A: Exactly two of the three heat with gas

Event B: Only one heats with gas

Event C: All three heat with gas

Event D: At least one of the three heats with gas

If 30% of the homes heat with gas, determine the probabilities of events A, B, C, and D.

SOLUTION

Event A consists of outcomes

$$A = \{(g, g, n), (g, n, g), (n, g, g)\}$$

From Table 4.1, each outcome has probability .063. Using the additive law for mutually exclusive events, we have

$$P(A) = .063 + .063 + .063 = .189$$

Event B consists of

$$B = \{(g, n, n), (n, g, n), (n, n, g)\}$$

Each outcome has probability .147, so

$$P(B) = .147 + .147 + .147 = .441$$

Event C consists of only $C = \{(g, g, g)\}$. Thus,

$$P(C) = .027$$

Finally, event D consists of all outcomes *except* $\{(n, n, n)\}$, which has probability .343. Applying the complement rule, we have

$$P(D) = 1.000 - .343 = .657$$

The concepts of mutually exclusive events and independent events are often confused. The box shows an easy way to remember the difference between the two.

- If events A and B are mutually exclusive, then $P(A \text{ and } B) = 0$.
- If events A and B are independent, then $P(A \text{ and } B) = P(A)P(B)$.

Conditional Probability (Optional)

When events are dependent, the probability of one event is dependent on whether or not the other event has occurred. It is for this reason that we define the *conditional probability* of event A given that B has occurred. We denote it as $P(A \mid B)$ and read it as the probability of A given B.

The **conditional probability** of event A given that event B has occurred is found by dividing the probability that both A and B occur by the probability that B occurs. That is,

$$P(A \mid B) = \frac{P(A \text{ and } B)}{P(B)}$$

provided that $P(B)$ is not 0.

EXAMPLE 4.13

Suppose that at a large university 40% of the students take English, 25% take math, and 12% take both. If a student is randomly selected from an English class, what is the probability that he or she is also taking math?

SOLUTION

This is a conditional probability problem because we are given that the student is taking English and we are asked to find the probability he or she is taking math.

Let M be the event that the student is taking math, and let E be the event that the student is taking English. We have

$$P(M) = .25 \qquad P(E) = .40 \qquad P(M \text{ and } E) = .12$$

Therefore,

$$P(M \mid E) = \frac{P(M \text{ and } E)}{P(E)} = \frac{.12}{.40} = .30$$

Thus, if we know that a student is taking English, then there is a 30% chance that he or she is also taking math. If we do not know whether the student is taking English, then the probability that he or she is taking math is 25%.

The conditional probability formula can be rewritten so that

$$P(A \text{ and } B) = P(A \mid B)P(B)$$

In this form, the probability that both A and B happen is the product of the conditional probability of A given B and the unconditional probability of B. This leads to the multiplication law.

The Multiplication Law

Let A and B be two events with nonzero probabilities. Then

$$P(A \text{ and } B) = P(A \mid B)P(B)$$

Suppose there are two computer labs for a course you are taking this semester. Lab A has 10 Macintosh computers and 15 PC computers. Lab B has 18 Macintosh computers and 12 PC computers. You are randomly assigned to a lab, and then you randomly pick a computer in the lab.

a. If you are assigned to lab A, what is the probability that you end up with a Macintosh computer?

b. If you are assigned to lab B, what is the probability that you end up with a Macintosh computer?

c. If you do not know which lab you are assigned to, what is the probability that you end up with a Macintosh computer?

SOLUTION

Let M be the event that you end up with a Macintosh computer, A be the event that you are assigned to lab A, and B be the event that you are assigned to lab B.

Parts **a** and **b** ask for conditional probabilities. Knowing the number of Macintosh computers in each lab, we have:

$$P(M \mid A) = \frac{10}{25} = .4$$

$$P(M \mid B) = \frac{18}{30} = .6$$

Part **c** asks for the unconditional probability $P(M)$. To solve the problem, we must recognize that M can occur with A or with B but not with both; that is, you cannot be assigned to both labs. Notice that (M and A) is the event that you are assigned to lab A *and* you select a Macintosh computer, and (M and B) is the event that you are assigned to lab B *and* you select a Macintosh computer. Again, events (M and A) and (M and B) are mutually exclusive because you cannot be assigned to both labs. And event M occurs with compound event (M and A) *or* (M and B). Thus, from the additive law we have

$$P(M) = P(M \text{ and } A) + P(M \text{ and } B)$$

Using the multiplication law, we have

$$P(M) = P(M \mid A)P(A) + P(M \mid B)P(B)$$

Because the labs are assigned randomly, we have

$$P(A) = P(B) = \frac{1}{2}$$

Consequently,

$$P(M) = (.4)\left(\frac{1}{2}\right) + (.6)\left(\frac{1}{2}\right) = .5$$

Can you generalize this problem to a random assignment to one of three labs where lab C has 12 Macintoshes and 18 PCs?

We have considered probabilities involving dependent events. In most of the remaining applications, however, we will be dealing with independent events and using the multiplication rule for independent events.

EXERCISES 4.2

4.29   Suppose $S = \{2, 3, 4, 5, 6, 7, 8\}$, $A = \{3, 5, 7\}$, $B = \{4, 5, 6\}$, and $C = \{3, 4, 5, 6, 7\}$. List the outcomes in the following events:
    a. A or B
    b. A and B
    c. A and C
    d. C^c
    e. A^c and C
    f. A or B or C
    g. A and B and C
    h. What must the outcome be in order for event (A and B) to occur?
    i. Are A and B mutually exclusive? Why?

4.30   For events A and B, $P(A) = .5$, $P(B) = .4$, and $P(A$ and $B) = .3$.
    a. Find $P(A^c)$.
    b. Find $P(A$ or $B)$.
    c. Are A and B mutually exclusive? Why?

4.31   For events A and B, $P(A) = .5$, $P(B) = .4$, and $P(A$ or $B) = .8$.
    a. Find $P(B^c)$.
    b. Find $P(A$ and $B)$.
    c. Are A and B mutually exclusive? Why?

4.32   Suppose A and B are mutually exclusive. Explain why $P(A$ and $B) = 0$.

4.33   Two cards are drawn from a deck of cards with replacement and we are concerned only with the color. Assign probabilities to the outcomes in the sample space $S = \{rr, rb, br, bb\}$ found in Exercise 4.16 in Section 4.1. What is the probability of event R, that both cards are red?

4.34   Suppose a card is drawn from a standard deck of 52 cards. Find the probabilities of these events:
  a. The card is a club or a face card.
  b. The card is not a face card.
  c. Are the events of drawing a club and drawing a face card mutually exclusive?

4.35   A manufacturer of golf clubs has plants in Phoenix, Denver, and Memphis. The Memphis plant produces 60% of the company's clubs, and the remaining 40% are evenly divided between the Phoenix and Denver plants. A set of clubs is randomly selected from a Kmart in Amarillo, Texas.
  a. Are the events produced in Memphis, produced in Denver, and produced in Phoenix mutually exclusive?
  b. What is the probability that the clubs were produced by either the Memphis or Denver plant?
  c. What is the probability that the clubs were not produced by the Memphis plant?

4.36   A survey of teenagers 16 years of age and older showed that 35% think that the legal drinking age should be 18, 20% think that marijuana should be legalized, and 15% agree with both. What percent think that the legal drinking age should be 18 or that marijuana should be legalized?

4.37   For events A and B, $P(A) = .3$, $P(B) = .5$, and $P(A \text{ and } B) = .2$. Find the following probabilities:
  a. $P(A^c)$
  b. $P(A \text{ or } B)$
  c. $P(A \mid B)$
  d. Are A and B independent? Why?

4.38   For events A and B, $P(A) = .4$, $P(B) = .7$, and $P(A \text{ or } B) = .8$. Find these probabilities:
  a. $P(B^c)$
  b. $P(A \text{ and } B)$
  c. $P(B \mid A)$
  d. Are A and B independent? Why?

4.39   A woman wears her blue suit 30% of the time. For 2 days the sample space of possibilities is found from Exercise 4.20 of Section 4.1 to be $S = \{BB, BN, NB, NN\}$. Assign probabilities to the outcomes listed in the sample space. What is the probability that she wears the suit 2 days in a row?

4.40   Suppose 60% of all student loans are approved at your university. Assign probabilities to the outcomes in the sample space $S = \{aaa, aad, ada, daa, add, dad, dda, ddd\}$ found in Exercise 4.21 of Section 4.1.

4.41   In Exercise 4.40, what is the probability that at least two of three loans are approved?

4.42   The three traffic lights in Exercise 4.22 of Section 4.1 are green with probability .7 and red with probability .3. Assign probabilities to the outcomes listed in the sample space found in that exercise.

4.43   In Exercise 4.42, what is the probability that you are stopped no more than once by the three traffic lights?

4.44   Suppose 3 of the 12 bottles in a case of wine are bad. If you randomly select two bottles, what are the probabilities of the following outcomes?
  a. Both are good.
  b. Both are bad.
  c. One is good and one is bad.

4.45   Suppose two fair dice are rolled and we are concerned with these events:

  A: The sum on the two up faces is even

  B: The sum on the two up faces is less than 6

  Are events A and B independent? Are they mutually exclusive?

4.46   For events A and B, $P(A) = .4$ and $P(B) = .3$.
   a. If A and B are independent, find $P(A \text{ or } B)$.
   b. If A and B are mutually exclusive, find $P(A \text{ or } B)$.

4.47   Two rescue teams set out to find a lost hiker in the Grand Canyon. Team A has a 30% chance of finding the hiker, and team B has a 40% chance of finding the hiker. Assume the two rescue teams act independently. What is the probability that the hiker will be rescued?

4.48   In 1990, 25% of all American Indians lived on reservations, most in the Rocky Mountain states. The accompanying figure shows that of those American Indians with households on reservations, 17% live in new units (built in 1985 or later). For all U.S. households, the percentage is 10%.

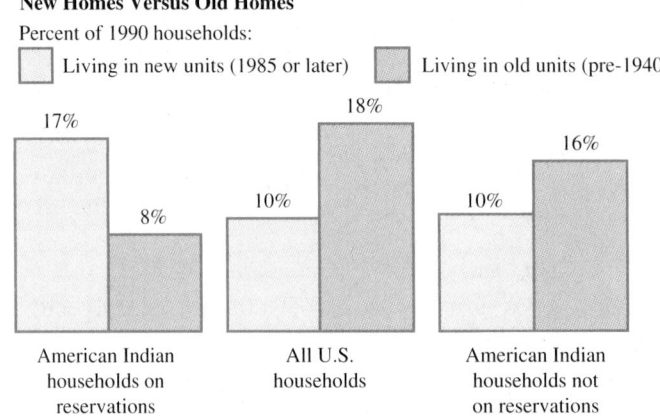

New Homes Versus Old Homes

Percent of 1990 households:

☐ Living in new units (1985 or later)   ▨ Living in old units (pre-1940)

Source: Bureau of the Census, U.S. Department of Commerce, *Housing of American Indians on Reservations,* Statistical Brief SB/95-10, April 1995.

   a. Are these percentages conditional probabilities?
   b. From the information provided, what percent of American Indians (reservation or not) live in new units?
   c. Of all U.S. households, what percent are neither new nor old?

4.49   In Statistical Brief SB/94-28 on health insurance coverage in 1993 produced by the Bureau of the Census (October 1994), it was reported that 39.7 million Americans (15.3%) were without health insurance coverage during the entire 1993 calendar year. The majority of Americans (70.2%) were covered by private insurance plans.
   a. Based on these figures, what was the estimated population size in 1993?
   b. How many Americans are covered by private insurance plans?
   c. What percent are covered by plans other than private plans?

4.50   Due to the weather during February, school is held only 80% of the time. Let H stand for school being held on a given day and N stand for school not being held.
   a. List the sample space of possibilities of school being held the first 3 days of February.
   b. Assign probabilities to each of the outcomes in part **a**.
   c. Suppose B is the event that school is held on exactly 1 day out of the first 3 days. List the outcomes that make up B.
   d. What is $P(B)$?

4.51   Three-fourths (75%) of the women who wear Andrea Acrylic Shield Nail Color have no significant chipping or peeling after 3 days. Let C denote chipping and N denote no significant chipping. Consider three women who try the nail color.

    a. List the sample space of possibilities of chipping and no chipping.

    b. Assign probabilities to the outcomes in the sample space.

    c. What is the probability that no more than one woman experiences chipping?

4.52 Suppose a man has a certain trait that will be passed on to his offspring with 30% probability. His wife has the same trait, and she will pass it to her offspring with 20% probability. What is the probability that their child will have the trait?

4.53 In the game of roulette (see Exercise 4.11 in Section 4.1), suppose you bet on black on three consecutive plays. What is the probability that you win on all three plays?

4.54 A trial lawyer claims that she wins 80% of her cases. Out of two independent cases, what is the probability that she wins both assuming that her claim is true? What is the probability that she loses both? What is the probability that she wins only one?

4.3   RANDOM VARIABLES

According to the Census Bureau, 50% of all homes in Dallas are heated with gas. In Example 4.9, we learned that if we randomly select six homes, the sample space of possibilities consists of 64 different outcomes of the form

$$(g, g, g, g, g, g), (g, g, g, g, g, n), (g, g, g, g, n, g), \ldots$$

It is not an easy matter to list all 64 possibilities; moreover, we may not even want to list them all. The outcome (g, n, n, g, n, g), for example, means that the first, fourth, and sixth homes heat with gas and the second, third, and fifth do not. Instead of concerning ourselves with distinct outcomes like this, it is much simpler to consider just the number of homes out of six that are heated with gas. That is, instead of considering the sample space of outcomes of the form $(g, g, g, g, g, g), (g, g, g, g, g, n), (g, g, g, g, n, g)$, and so on, we need concern ourselves only with the numerical values 0, 1, 2, 3, 4, 5, and 6, which is much simpler

What we need, it appears, is a rule that assigns a numerical quantity to each outcome in the sample space. This rule is called a *random variable*.

> A **random variable**, which is denoted by a letter such as x or y, is a rule that represents the possible numerical values associated with the outcomes of an experiment. The list of values is called the *sample space* or the *range* of the random variable.

In the preceding example, the random variable x denotes the number of homes out of six that are heated with gas. The sample space for x is

$$S_x = \{0, 1, 2, 3, 4, 5, 6\}$$

As another example, let x represent the sum of the two faces that show when we roll a pair of dice. The possible values that x can assume are

$$S_x = \{2, 3, 4, 5, 6, 7, 8, 9, 10, 11, 12\}$$

Thus, from the sample space, S, of an experiment, we define a random variable x (or y) that assumes a numerical value for each outcome in the experiment. The set of all possible numerical values of the random variable constitutes the sample space, S_x. In essence, the sample space for a random variable is analogous to the sample space for the experiment. That is, it lists all possible numerical values of the random variable, and their associated probabilities must sum to 1. Generally, it is much easier to apply the laws of probability to S_x than to the sample space of the experiment, S.

EXAMPLE **4.15**

Four students are asked whether they believe that marijuana should be legalized. Each responds with either yes or no. Define a random variable on the sample space of possible outcomes and list the sample space of the random variable.

SOLUTION

Because there are four students and each will respond with a yes (y) or a no (n), the sample space of the experiment consists of 16 possible outcomes of the form (y, y, y, y), (y, y, y, n), (y, y, n, y), and so on.

Several possible random variables could be defined on this sample space, but it seems reasonable that we are interested in only the number of students out of the four who favor the issue. Thus, we define the random variable as

y = number who believe marijuana should be legalized

The sample space of y is $S_y = \{0, 1, 2, 3, 4\}$. As you can see, it is much simpler than the sample space of outcomes for the experiment.

EXAMPLE **4.16**

An experiment consists of tossing a coin until a head appears. Define a random variable on this experiment.

SOLUTION

Because the head might appear on the first toss or perhaps the second toss or maybe not until the 17th toss, the sample space, S, will consists of an infinite number of outcomes:

$$S = \{h, th, tth, ttth, tttth, \ldots\}$$

where "..." means that the list continues forever. For this sample space, we let the random variable x be the number of tosses until the head appears, in which case the sample space for x is

$$S_x = \{1, 2, 3, 4, \ldots\}$$

Although this sample space is still infinite, it is much simpler than S because it consists of numbers instead of a collections of elements of the form h, th, tth, and so on.

Discrete and Continuous Random Variables

In the preceding examples, the sample space of each random variable consists of a finite number of values such as $\{0, 1, 2, 3, 4\}$, or at most a countable number of values such as $\{1, 2, 3, 4, \ldots\}$. Recall from Section 2.1 that variables of this type are called discrete. Thus, these are called *discrete random variables*.

Continuous random variables are those that can assume all the values in an interval of the number line.

EXAMPLE 4.17

In 1995, the average life expectancy in the United States reached an all-time high of 76.3 years. Obviously this does not mean that each of us can expect to live to the age of 76.3. In fact, if you are a woman, you can expect to live about 7 years longer than men (79.7 years for women and 72.8 years for men). Also, blacks have a significantly shorter life expectancy than whites (72.5 years for blacks and 77.0 for whites). Your occupation also has an effect on your life expectancy. Define a random variable related to life expectancy.

SOLUTION

Insurance companies that insure specific groups of people must study the life expectancy of each group. For example, companies that insure school teachers are very familiar with the random variable, call it y, that represents the life expectancy of teachers. Theoretically this random variable can assume any value greater than 0. Practically, however, the life expectancy of all teachers ranges from about 20 years to no more than 120 years. Most teachers can expect to live to about the age of 75, but a few will die very young and a few will live longer than 75 years. In any case, the random variable

$$y = \text{life expectancy of a school teacher}$$

can assume any value in the interval from about 20 years to 120 years and thus is a continuous random variable.

Typically we think of a discrete random variable as one that corresponds to a count or "the number of," and a continuous random variable as one that corresponds to some measurement process. Here are some examples of discrete random variables:

The number of penalties incurred by a football team in a single game ($x = 0, 1, 2, 3, \ldots$)

The number of students who pass a psychology test out of a class of 30 ($x = 0, 1, 2, 3, \ldots, 30$)

The number of students in the student body who favor beer being sold on campus ($x = 0, 1, 2, 3, \ldots, 10{,}000$, assuming the size of the student body is 10,000)

The number of violent crimes committed per month in a certain section of the city ($x = 0, 1, 2, 3, \ldots$)

The number of computer terminals that are in use during a certain period of the day ($x = 0, 1, 2, 3, \ldots, n$, where n is the number of terminals)

These are examples of continuous random variables:

The length of a prison term for possession of marijuana

The concentration of DDT in a sample of milk

The IQ of a randomly selected third-grader

The amount of a hospital bill for 2 weeks in coronary care

The amount of cola in a 12-ounce can

The weight of a box of oranges

The Probability Distribution of a Random Variable

Because a discrete random variable assumes only a discrete set of values, we next ask: What are the probabilities associated with the possible values of the random variable? For example, what is the probability of getting a 4 on the roll of a fair die, or what is the probability that two of the students believe that marijuana should be legalized?

> A table or function that lists all the possible values of a discrete random variable and their associated probabilities is called the **probability distribution** of the random variable.

Table 4.2 is the probability distribution for the random variable y associated with the outcomes of the roll of a fair die. The first column lists the possible values of y, and the second column gives the corresponding probabilities. The table lists all six possible values, and each occurs with a probability of $\frac{1}{6}$. Observe that the probabilities are all between 0 and 1. They sum to 1 (or 100%) because the table lists all possible values of the random variable. If any entry in the table falls outside the interval from 0 to 1 or if the probabilities do not sum to 1, then it is *not* a probability distribution. Again, this is a discrete random variable because it assumes only a discrete set of values.

TABLE 4.2

y	$P(y)$
1	$\frac{1}{6}$
2	$\frac{1}{6}$
3	$\frac{1}{6}$
4	$\frac{1}{6}$
5	$\frac{1}{6}$
6	$\frac{1}{6}$

Total 1.00

To illustrate how probabilities are *distributed* over the numerical values, we can graph the distribution of the random variable. Figure 4.6(a) is such a graph of the distribution, with line segments of height $\frac{1}{6}$ drawn at each value of the random variable. Figure 4.6(b) is the same distribution drawn in the form of a histogram. Both presentations are correct. Figure 4.6(a) illustrates the discrete nature of the random variable by showing that the probabilities are concentrated at the values 1, 2, 3, 4, 5, and 6. An advantage of the histogram form in Figure 4.6(b) is that the *areas* of the rectangles correspond to the probabilities associated with the numerical values. Because this concept of area carries over to continuous random variables, we shall graph distributions in the form of a histogram.

FIGURE 4.6

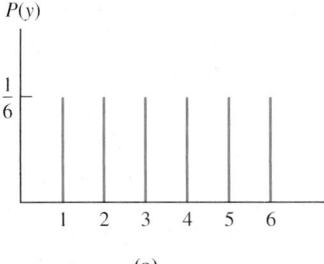

(a)

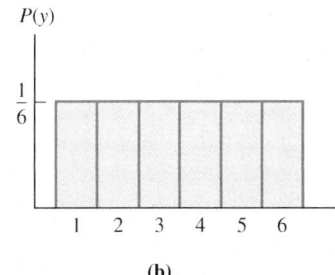

(b)

EXAMPLE 4.18

Consider families with three children. Let the variable w be the number of girls in a family. Find the probability distribution for w and graph the distribution.

SOLUTION

The possibilities for three children—that is, the sample space for the experiment—is

$$S = \{(g, g, g), (g, g, b), (g, b, g), (b, g, g), (b, b, g), (b, g, b), (g, b, b), (b, b, b)\}$$

We see that there are eight possibilities, and if we assume that a boy is just as likely as a girl, all of the eight possibilities are equally likely. Of the eight, only one, (b, b, b), corresponds to zero girls, in which case $w = 0$, and the probability of zero girls is $\frac{1}{8}$. Three of the eight possibilities correspond to $w = 1$ (one girl), and therefore the probability is $\frac{3}{8}$. Continuing in this fashion we get the following completed probability distribution of w:

w	0	1	2	3
$P(w)$	$\frac{1}{8}$	$\frac{3}{8}$	$\frac{3}{8}$	$\frac{1}{8}$

Observe that there could be 0, 1, 2, or 3 girls in the family, and the table gives the probability of each. The graph of this distribution is shown in Figure 4.7.

FIGURE 4.7

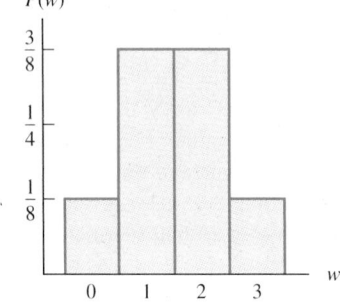

Note that the probabilities in Figure 4.7 sum to 1, which is always the case with a probability distribution table. Also, if you observe closely, this is exactly the same as the distribution for the random variable

$$x = \text{number of homes out of 3 that heat with gas in Dallas}$$

Recall from Example 4.1 that the sample space is

$$S = \{(g, g, g), (g, g, n), (g, n, g), (n, g, g), (g, n, n), (n, g, n), (n, n, g), (n, n, n)\}$$

which is analogous to the sample space given for the possibilities of three children. And because 50% of the homes heat with gas, just as we assumed a girl is born half the time, the probability distribution for x is exactly the same as the probability distribution for w.

You may also have noticed that the preceding two examples are analogous to the experiment of tossing three coins. In fact, we often use the coin-toss experiment as a model for many problems that confront us in the real world.

C O M P U T E R   T I P

To simulate experiments that are modeled with the coin-toss experiment, we use the Bernoulli distribution. You will learn in Section 4.4 that an experiment in which the outcome is classified as either a success or a failure is called a Bernoulli trial. The **Bernoulli** command will simulate the outcomes of a series of Bernoulli trials. In the Session window, the command is

MTB > **Random K Observations Into C, C,..., C;**
SUBC> **Bernoulli With P = K.**

Otherwise, select the **Calc** menu, choose **Random Data**, and then click on **Bernoulli**. When the Bernoulli Distribution window appears, specify the **number of observations** to generate, give the **column(s) to store** the data, give the **probability of success**, and then click **OK**. The outcomes are coded as a 0 or a 1, where the probability of a 1 on any given trial is P.

EXAMPLE **4.19**

Suppose a field-goal kicker has an 80% success rate inside the 35-yard line. Simulate eight field-goal kicks inside the 35 during a game. Determine the number of successes. Simulate eight kicks inside the 35 for ten consecutive games. Determine his success rate in each game and his success rate for the ten-game season.

SOLUTION

The following sequence simulates eight kicks with a success rate of 80%:

```
MTB >RANDOM 8 OBSERVATIONS IN C1;
SUBC>BERNOULLI TRIALS WITH P = 0.8000.
MTB >PRINT C1
      C1
      1    0    0    1    1    1    0    1
```

For a summary, we have

VALUE	FREQUENCY
0	3
1	5

Out of eight kicks, he missed three: the second, third, and seventh kicks.

Figure 4.8 on page 260 gives the simulations for the ten games with their summary: In game 1, the kicker missed the first two tries and made the rest, he made all eight kicks in game 2, and so on. Figure 4.9 (page 260) is a summary with the kicker's success rate in each game. We can see that his success rate for the ten-game season was 85%.

Figure 4.8

```
MTB >RANDOM 8 OBSERVATIONS INTO C1-C10;
SUBC>BERNOULLI WITH P = 0.8000.
MTB >PRINT C1-C10

        ROW   C1    C2    C3    C4    C5    C6    C7    C8    C9   C10

          1    0     1     1     1     1     1     1     1     0     1
          2    0     1     0     1     0     0     1     1     1     1
          3    1     1,    1     1     1     1     1     1     1     1
          4    1     1     1     1     0     1     1     0     1     0
          5    1     1     0     0     1     1     1     1     1     1
          6    1     1     1     1     1     1     1     0     1     1
          7    1     1     1     1     1     1     1     1     1     1
          8    1     1     1     1     1     1     1     1     1     1
```

Figure 4.9

Game	#hits	#misses	success rate
1	6	2	75%
2	8	0	100%
3	6	2	75%
4	7	1	87.5%
5	6	2	75%
6	7	1	87.5%
7	8	0	100%
8	6	2	75%
9	7	1	87.5%
10	7	1	87.5%
Total	68	12	85%

Notice in Example 4.19 that in no simulation did the kicker hit exactly 80% of his kicks (of course this is impossible with only eight kicks); rather, the success rate ranged from 75% to 100%. It is important to realize that, although the population rate is 80%, there is no guarantee that any sample will duplicate the population rate exactly. If we combine all 80 kicks, however, the overall success rate is 85%. What do you think the simulation would yield if we simulated 800 kicks? An important lesson is that the sample results will vary from sample to sample. This will become more apparent as we study sampling distributions in the next chapter.

The Mean and Standard Deviation of a Random Variable

The population mean, μ, and the population standard deviation, σ, were introduced in Section 2.4 as numerical summary measures of a population distribution. Because a population distribution is nothing more than a probability distribution generated by a random variable x, we refer to μ and σ as the mean and standard deviation of the population random variable x. To emphasize that the distribution is generated by a particular random variable x, we often refer to the mean and standard deviation as μ_x and σ_x, respectively.

The interpretation of the standard deviation of a sample, given in Section 2.5, applies to population variables as well. That is, one can determine the percent of the distribution that lies within one, two, or three standard deviations of the mean. Generally, for bell-shaped distributions, the empirical rule applies.

Empirical Rule Applied to Population Distributions

If the probability distribution is bell shaped, then:

1. Approximately 68% of the distribution falls within one standard deviation of the mean; that is, 68% of the distribution is between $\mu - \sigma$ and $\mu + \sigma$.
2. Approximately 95% of the distribution falls within two standard deviations of the mean; that is, 95% of the distribution is between $\mu - 2\sigma$ and $\mu + 2\sigma$.
3. Essentially all of the distribution falls within three standard deviations of the mean; that is, essentially all of the distribution is between $\mu - 3\sigma$ and $\mu + 3\sigma$.

For distributions that are not necessarily bell shaped—in fact, for any shape of distribution—we have Chebyshev's rule from the Russian mathematician P. L. Chebyshev (1821–1894).

Chebyshev's Rule

Regardless of the shape of the distribution:

1. At least $\frac{3}{4}$ of the distribution falls within two standard deviations of the mean—that is, from $\mu - 2\sigma$ to $\mu + 2\sigma$.
2. At least $\frac{8}{9}$ of the distribution falls within three standard deviations of the mean—that is, from $\mu - 3\sigma$ to $\mu + 3\sigma$.

EXAMPLE 4.20

It has been claimed that television violence leads to aggressiveness and violent behavior among children. Data have shown that in children's weekend shows, the average number of violent incidents has reached a record high of 30.3 per hour. If the standard deviation for the number of violent incidents per hour is 6.4, obtain an interval that would contain the number of violent incidents per hour for 95% of all weekend shows.

SOLUTION

If we can assume that the number of violent incidents has a bell-shaped distribution, then we can apply the empirical rule, which states that 95% of the distribution of the number of violent incidents per hour will be within two standard deviations of the mean.

This gives the limits

$$30.3 - 2(6.4) = 30.3 - 12.8 = 17.5$$

$$30.3 + 2(6.4) = 30.3 + 12.8 = 43.1$$

Thus, 95% of all weekend children's shows will have between 17.5 and 43.1 violent incidents per hour.

If we could not assume the bell-shaped distribution, then we would have to apply Chebyshev's rule, which states that at least 75% of the children's shows will have between 17.5 and 43.1 violent incidents per hour.

EXERCISES 4.3

4.55   An experiment consists of tossing five fair coins. Let x be the random variable that is the number of heads in the five tosses. List the sample space of values for x. Is x discrete or continuous?

4.56   The Bureau of Justice Statistics reported that in 1994, 10% of all jail inmates were women (*Jails and Jail Inmates 1993–94*, Bulletin NCJ-1511651, April 1995). Let the random variable y be the total number of women out of five randomly selected inmates. List the sample space of values for y. Is y discrete or continuous? How does this exercise differ from Exercise 4.55?

4.57   Let w be the random variable that gives the amount of time it takes a person to drive to work. List the sample space of values for w. Is w discrete or continuous?

4.58   The Red Cross is searching for blood donors who are HIV positive. It will continue to test subjects until they find one who is HIV positive. Let

y = number of blood donors tested until an HIV-positive donor is found

List the sample space of values for y. Is y discrete or continuous?

4.59   The Red Cross has 15 blood donors and is searching for type B blood. Let

b = number who have type B blood out of the 15

List the sample space of values for b. Is b discrete or continuous?

4.60   Let x be the SAT-math score of a student selected at random from the student body of your school. List the sample space of values for x. Is x discrete or continuous?

4.61   Classify the following as either discrete or continuous:
    a. The self-concept score on the Tennessee Self-Concept Scale
    b. The number of traffic accidents per week involving alcohol in your state
    c. The number of people who apply for food stamps per week in your city
    d. The length of time of a strike by steel workers
    e. The potential life expectancy of a 40-year old man taking out a life insurance policy

4.62   Suppose x is a discrete random variable with the following probability distribution table:

x	1	3	5	7	9
$P(x)$	1/15	2/15	3/15	4/15	5/15

a. What value of x is most likely to occur?

b. What is the probability that x is even?

c. What is the probability that x is less than 7?

4.63 Given the function

$$P(x) = \frac{x-1}{6} \qquad \text{for } x = 2, 3, 4$$

find $P(2)$, $P(3)$, and $P(4)$. Does the equation define a probability distribution? What is $P(1)$?

4.64 Given the function

$$P(x) = \frac{x-2}{10} \qquad \text{for } x = 3, 4, 5, 6$$

find $P(3)$, $P(4)$, $P(5)$, and $P(6)$. Does the equation define a probability distribution? What is $P(2)$?

4.65 Suppose y is a discrete random variable with the following probability distribution table:

y	1	2	3	4
$P(y)$	.111	.222	.333	.334

a. What value of y is most likely to occur?

b. What is the probability that y is even? Odd?

c. Graph this distribution.

4.66 Suppose y is a discrete random variable with the following probability distribution table:

y	1	3	5	7	9
$P(y)$	1/15	2/15	3/15	4/15	5/15

The mean and standard deviation of y are 6.3 and 2.5, respectively. Sketch a graph of the probability distribution and locate the interval $\mu \pm 2\sigma$ on the graph. Give an interpretation using either the empirical rule or Chebyshev's rule.

4.67 The probability distribution table for the random variable y is given here:

y	-1	0	1	2	3
$P(y)$	.05	.1	.15	.4	.3

The mean and standard deviation of y are 1.8 and 1.12, respectively. Sketch a graph of the probability distribution and locate the interval $\mu \pm 2\sigma$ on the graph. Give an interpretation using either the empirical rule or Chebyshev's rule.

4.68 Here is the probability distribution table for the number of daily requests for assistance from a poison control center:

Number of requests	0	1	2	3	4	5	6
Probability	.1	.1	.2	.3	.1	.1	.1

The mean number of requests and the standard deviation are 2.9 and 1.7, respectively. Sketch a graph of the probability distribution and locate the interval $\mu \pm 2\sigma$ on the graph. Give an interpretation using either the empirical rule or Chebyshev's rule.

4.69 An oil well drilling company generally drills four wells per month. The probability distribution table for the number of successful attempts out of the four wells is given at the top of page 264.

Number of successes	0	1	2	3	4
Probability	.1	.4	.3	.1	.1

The mean number of successful wells and the standard deviation are 1.7 and 1.1, respectively. Sketch a graph of the probability distribution and locate the interval $\mu \pm 2\sigma$ on the graph. Give an interpretation using either the empirical rule or Chebyshev's rule.

4.70   The mean of a discrete random variable y is given by the formula

$$\mu = \sum yP(y)$$

where the sum is over all possible values of y. Verify that the mean of y in Exercise 4.66 is 6.3.

4.71   Using the formula for the mean given in Exercise 4.70, verify that the mean of y in Exercise 4.67 is 1.8.

4.72   Using the formula for the mean given in Exercise 4.70, verify that the mean in Exercise 4.68 is 2.9.

4.73   Using the formula for the mean given in Exercise 4.70, verify that the mean in Exercise 4.69 is 1.7.

4.74   The standard deviation of a discrete random variable y is given by the formula

$$\sigma = \sqrt{\sum(y - \mu)^2 P(y)}$$

[an alternative formula is $\sigma = \sqrt{\sum y^2 P(y) - \mu^2}$] where the sum is over all possible values of y. Verify that the standard deviation of y in Exercise 4.66 is 2.5.

4.75   Using the formula given in Exercise 4.74, verify that the standard deviation of y in Exercise 4.67 is 1.12.

4.76   Using the alternative formula given in Exercise 4.74, verify that the standard deviation in Exercise 4.68 is 1.7.

4.77   Using the alternative formula given in Exercise 4.74, verify that the standard deviation in Exercise 4.69 is 1.1.

4.78   In Exercise 4.55, the experiment was tossing five fair coins and recording the number of heads. With the **Random Data** > **Bernoulli** command, simulate the toss of five coins and count how many times a head comes up. Repeat the simulation 32 times (32 different columns in the worksheet) and determine the number of heads in each of the 32 simulations. Out of the 32, how many times did you get no heads? One head? Two heads? Three heads? Four heads? Five heads? Construct a relative frequency table of the number of heads out of the 32 simulations.

4.79   In Exercise 4.56, the number of women out of five randomly selected inmates was recorded. With the **Random Data** > **Bernoulli** command, simulate the outcomes from 5 randomly selected inmates and count how many times a woman appears (remember only 10% are women). Repeat the simulation 32 times (32 different columns in the worksheet) and determine the number of women in each of the 32 simulations. Out of the 32, how many times did you get no women? One woman? Two women? Three women? Four women? Five women? Construct a relative frequency table of the number of women out of the 32 simulations.

4.80   Discrete distributions such as the ones given in Exercises 4.62–4.69 can be simulated with the **Random Data** > **Discrete** command in Minitab. First, store the values of the random variable in one column of the worksheet (C1) and the corresponding probabilities in an adjacent column (C2). Under the Calc menu, select **Random Data** and click on **Discrete**. When the Discrete Distribution window appears, fill in the **Generate** box with the desired number of simulated values. Specify the column to **Store** the values. Give the column containing the **Values** and the column containing the **Probabilities**, and click on **OK**. Simulate 20 days of the number of daily requests for assistance from the poison control center given in Exercise 4.68. Make a relative frequency table of the values. How closely does it agree with the theoretical distribution?

4.81 Use the simulation procedure outlined in Exercise 4.80 to simulate 24 months of drilling four oil wells each month by the oil well drilling company in Exercise 4.69. Make a relative frequency table of the values. What percent of the time were all four wells successful? What percent of the time were no wells successful? Do the simulated values agree with the theoretical values?

4.82 Suppose x is a discrete random variable with the following probability distribution table:

x	0	2	4	5
$P(x)$	1/10	2/10	3/10	4/10

a. What value of x is most likely to occur?
b. What is the probability that x is even?
c. What is the probability that x is less than 4?

4.83 A typical bar exam for lawyers might have someone on their fourth or fifth attempt at the exam. Suppose the following table is the distribution of the number of attempts at the bar exam for prospective lawyers:

Number of attempts	1	2	3	4	5
Probability	.4	.3	.1	.1	.1

Given that $\mu = 2.2$ and $\sigma = 1.33$, sketch a graph of the probability distribution and locate the interval $\mu \pm 2\sigma$ on the graph. Give an interpretation using either the empirical rule or Chebyshev's rule.

4.84 Suppose your skill at playing a particular video game is such that you stand a 40% chance of winning each time you play. Suppose you play three games and either win (W) or lose (L).
a. List the sample space of outcomes.
b. Assign probabilities to the outcomes.
c. Let x = number of games you win. What is the sample space for x?
d. Give the probability distribution table for x.
e. What is the probability that you win at least two of the three games?

4.4 THE BINOMIAL DISTRIBUTION

The distribution described in Example 4.18 in Section 4.3 is an example of the binomial distribution. At each stage of the experiment, the outcome is one of two possibilities— a girl or a boy. As we have seen, many experiments share this same characteristic; namely, the outcome is only one of two possibilities. This is what we commonly call a *Bernoulli population* after Jacob Bernoulli (1654–1705), the first of a series of gifted mathematicians in the Bernoulli family.

A **Bernoulli population** is a population in which each element is one of two possibilities. The two possibilities are usually designated as success and failure. A **Bernoulli trial** is observing one element in a Bernoulli population.

EXAMPLE 4.21

a. The toss of a coin results in a Bernoulli population because each toss results in a head (success) or a tail (failure).
b. The homes in Dallas either heat with gas (success) or do not (failure). Thus, we can treat the choice to heat with gas or not as a Bernoulli population.
c. In a local election, the voters either favor a candidate for mayor or do not. Thus, we can view the population of voter choices as a Bernoulli population, where a success is one who favors the candidate and a failure is one who opposes the candidate.

Binomial Experiments and Random Variables

We see that the choice to heat with gas and the opinion of voters on a specific issue are similar to tossing a coin, where the outcome is classified as either a success or a failure. A random sample from a Bernoulli population consists of n objects where each object is a success or a failure. The result, which is a sequence of Bernoulli trials, is called a *binomial experiment*.

A **binomial experiment** is an experiment that consists of n repeated independent Bernoulli trials in which the probability of success on each trial is π and the probability of failure on each trial is $1 - \pi$.

(*Note:* We use π as a symbol for the probability of a success in a binomial experiment. It should not be confused with the number 3.14159 that we usually associate with the symbol π.)

Thus, an experiment that consists of repeatedly drawing independently from a Bernoulli population is called a *binomial experiment*. The outcomes are labeled as success or failure, and normally one is interested in the number of successes out of the n trials.

The random variable x, which gives the number of successes in the n trials of a binomial experiment, is called a **binomial random variable**. The sample space of values of x is

$$S_x = \{0, 1, 2, \ldots, n\}$$

Because its sample space of values is a discrete set, the binomial random variable is a discrete random variable. From the examples of discrete random variables given in Section 4.3, the number of students who pass a psychology test out of a class of 30, the number of students out of 10,000 who favor beer being sold on campus, and the number of computer terminals in use out of a total of n terminals are binomial random variables. The random variable w = the number of girls in a family of three children, considered in Example 4.18, is also a binomial random variable.

EXAMPLE 4.22

Are the following random variables binomial random variables? Explain why or why not.

a. Forty percent of all airline pilots are over 40 years of age. Out of a random sample of 15 pilots, let the random variable x be the number of pilots who are over 40 years of age.

b. Suppose a salesperson makes sales to 20% of her customers. One day she counted her customers until she made a sale. Let x be the number of customers until her first sale.

c. A room contains six women and four men. Three people are selected to form a committee. Let the random variable x be the number of women on the committee.

SOLUTION

a. Each of the 15 pilots constitutes a trial, and each trial results in one of two outcomes: Success is being over 40 years of age and failure is not being over 40 years of age. The trials are independent because one pilot's age has no relationship to another pilot's age. The probability of a success is $\pi = .4$ and of a failure is $(1 - \pi) = .6$, and these probabilities remain the same from trial to trial because 40% of all pilots are over 40 years of age. Thus, the conditions of a binomial experiment are met. Because x is the number of successes, it is a binomial random variable.

b. The salesperson may never make a sale: hence, the sample space for x is

$$S_x = \{1, 2, 3, \ldots\}$$

which clearly is not the sample space of a binomial random variable. If x had been defined as the number of sales out of, say, 50 customers, then x would have been a binomial random variable because it would have been the number of successes out of 50 trials.

c. The probability of a woman being selected on the first pick is $\frac{6}{10}$. On the second pick, however, it is $\frac{5}{9}$ if a woman was selected on the first time or $\frac{6}{9}$ if a man was selected on the first time. Clearly, the trials are dependent, and hence the experiment does not satisfy the conditions of a binomial experiment. The random variable x is not a binomial random variable.

C O M P U T E R   T I P

In Section 4.3, we used **Random Data** under the Calc menu to simulate a Bernoulli experiment. For example, with the Bernoulli distribution command, we can simulate tossing a fair coin 20 times and then count the number of heads in the 20 tosses. But suppose we are interested in the number of heads in 20 tosses and we wish to conduct the experiment 30 times. We can repeat the Bernoulli command 30 times, each time counting the number of successes, or we can use the **Binomial distribution**. In the Session window, type

> MTB > **Random K Observations Into C,C,...,C;**
> SUBC> **Binomial With N = K, P = K.**

Otherwise, under the **Calc** menu, select **Random Data** and click on **Binomial**. At the Binomial Distribution window, specify the **number of observations** to generate, what **columns to store the data**, give the **number of trials** and **the probability of success**, and click on **OK**.

EXAMPLE 4.23

An experiment consists of rolling five dice and observing how many turn up on 6. The number of 6's that turn up is a binomial random variable with $n = 5$ and $\pi = \frac{1}{6} = .16666\ldots$. Simulate the experiment 20 times. How many times did all five dice turn up on 6? How many times did a 6 fail to turn up? Summarize the results.

FIGURE 4.10

```
MTB >RANDOM 20 OBSERVATIONS INTO C1;
SUBC>BINOMIAL WITH N = 5, P = .166667.

MTB >PRINT C1
 C1
   0    0    2    1    1    0    0    2    0    1
   2    1    0    0    0    1    0    1    2    2

MTB >TALLY C1
 SUMMARY
 C1   Count
  0      9
  1      6
  2      5
  3      0
  4      0
  5      0
 N=     20
```

SOLUTION

Rolling a 6 is called a success. Each die is treated as a trial that results in a success or a failure. Out of five dice, we wish to find the number of successes, where the probability of success is $\frac{1}{6}$. We repeat the binomial experiment 20 times and tally the results. From Figure 4.10 we see that on nine occasions none of the dice turned up on 6, on six occasions there was one 6 showing, and on five occasions there were two 6's showing. Not once did three or more of the five dice show 6's.

The Binomial Probability Distribution

Once a random variable is identified as being a binomial random variable, we can determine its probability distribution. Consider again Example 4.9, where we were concerned with the number of homes in Dallas that are heated with gas. Let

$$y = \text{number of homes out of six that are heated with gas}$$

The sample space of values of y is

$$S_y = \{0, 1, 2, 3, 4, 5, 6\}$$

To determine the probability distribution of y, we must be able to count the number of outcomes in the sample space that are associated with each value of y. As pointed out in Example 4.9, there are 64 equally likely outcomes of the form (gggggg), (gggggn), (ggggng), and so on in the sample space. Only one, (gggggg), corresponds to $y = 6$; hence, $p(6) = \frac{1}{64}$. Realizing that there are six outcomes with five g's, (gggggn), (gggggng), (gggngg), (ggnggg), (gngggg), and (ngggg), we have $p(5) = \frac{6}{64}$. But now, how many of the 64 outcomes have exactly four g's? The answer is found with the *combinations formula*.

> The number of **combinations** of k objects taken from n objects is given by
>
> $$\binom{n}{k} = \frac{n!}{k!\,(n-k)!}$$
>
> where $n!$ is read "n factorial" and is given by
>
> $$n! = n(n-1)(n-2)(n-3)\cdots(2)(1) \quad \text{and} \quad 0! = 1$$

Thus, there are

$$\binom{6}{4} = \frac{6!}{4!\,(6-4)!} = 15$$

combinations that have four g's. Continuing with the combinations formula, we come up with the following table, which is the probability distribution table for y:

y	0	1	2	3	4	5	6
$P(y)$	$\frac{1}{64}$	$\frac{6}{64}$	$\frac{15}{64}$	$\frac{20}{64}$	$\frac{15}{64}$	$\frac{6}{64}$	$\frac{1}{64}$

The preceding example is a special case where the outcomes in the sample space are equally likely. The outcomes are equally likely because the probability of a home being heated with gas is 50%; that is, $\pi = .5$. What adjustments must be made when $\pi \neq .5$? Recall that in Example 4.11, we changed the percentage from 50% to 30% and reduced the number of homes to three to see what would happen. Table 4.1, repeated here, gives the probabilities for the eight possible outcomes:

Outcome	Probability
{(g, g, g)}	.027
{(g, g, n)}	.063
{(g, n, g)}	.063
{(n, g, g)}	.063
{(g, n, n)}	.147
{(n, g, n)}	.147
{(n, n, g)}	.147
{(n, n, n)}	.343
Total	1.000

We redefine y to be

$$y = \text{number of homes out of three that are heated with gas}$$

The random variable y will assume the value 0 ($y = 0$) only if we observe the outcome (n, n, n) in the sample space. Therefore,

$$P(y = 0) = .343$$

Three favorable outcomes—(g, n, n), (n, g, n), and (n, n, g)—are associated with $y = 1$. Each of the three has probability $(.3)(.7)(.7) = .147$. Therefore,

$$P(y = 1) = 3(.147)$$

The probability distribution table for y is:

y	$P(y)$
0	1(.343)
1	3(.147)
2	3(.063)
3	1(.027)

To generalize, suppose there are n trials, with π being the probability of success on each trial. For there to be exactly k successes, there must be $n - k$ failures. Moreover, there are

$$\binom{n}{k}$$

different possibilities for the k successes. Because the probability of each success is π, the probability of a given arrangement of k successes and $n - k$ failures is

$$\pi^k(1 - \pi)^{n-k}$$

Thus, the probability of k successes in n trials is

$$P(k) = \binom{n}{k}\pi^k(1 - \pi)^{n-k}$$

Computing these probabilities can become rather tedious if n is very large. In fact, in most cases we use a computer to compute the probabilities or, for certain values of n and π, the probabilities are given in the binomial table (Table B.1 in Appendix B).

To aid our discussion of the binomial table, an excerpt is reproduced in Table 4.3. This is the section of the table corresponding to $n = 10$ independent trials. Down the left column we see values for k of 0, 1, 2, 3, 4, 5, 6, 7, 8, 9, and 10. These numbers correspond to k successes out of the ten trials. Along the top we see values for π of .10, .20, .30, and so on to .90. (Table B.1 also includes values for .01, .05, .95, and .99.) Remember that π stands for the probability of a single success in the Bernoulli population. The body of the table gives the probabilities corresponding to the particular values of k and π for $n = 10$. For example, the probability of four successes out of ten trials when $\pi = .7$ is .037.

TABLE 4.3 $n = 10$

$k \mid \pi$	.10	.20	.30	.40	.50	.60	.70	.80	.90
0	.349	.107	.028	.006	.001	.000+	.000+	.000+	.000+
1	.387	.268	.121	.040	.010	.002	.000+	.000+	.000+
2	.194	.302	.233	.121	.044	.011	.001	.000+	.000+
3	.057	.201	.267	.215	.117	.042	.009	.001	.000+
4	.011	.088	.200	.251	.205	.111	.037	.006	.000+
5	.001	.026	.103	.201	.246	.201	.103	.026	.001
6	.000+	.006	.037	.111	.205	.251	.200	.088	.011
7	.000+	.001	.009	.042	.117	.215	.267	.201	.057
8	.000+	.000+	.001	.011	.044	.121	.233	.302	.194
9	.000+	.000+	.000+	.002	.010	.040	.121	.268	.387
10	.000+	.000+	.000+	.000+	.001	.006	.028	.107	.349

Note: .000+ means that the probability rounded to three decimal points is 0 but is not exactly 0 because 0 is associated with an impossible event. The + means it is slightly greater than 0.

EXAMPLE 4.24   Suppose that the probability of successfully rehabilitating a convicted criminal in a penal institution is .4. If we let r represent the number successfully rehabilitated out of a random sample of ten convicted criminals, then r is a binomial random variable with $n = 10$ independent trials and $\pi = .4$. The probability distribution for r is given in the binomial probability distribution table.

    a. Find the probability that six of the ten prisoners are successfully rehabilitated.

    b. Find the probability that no more than two of the ten are successfully rehabilitated.

SOLUTION

    a. Focusing on the .4 column in Table 4.3, we see the probabilities associated with all the values of k when $n = 10$. Thus, the probability that six of the ten prisoners are rehabilitated is .111.

    b. No more than two implies that k is 0 or 1 or 2. Thus, to find the probability, we simply add those probabilities in the .4 column that correspond to the values 0, 1, and 2 in the k column:

$$.006 + .040 + .121 = .167$$

There is a 16.7% chance that two or fewer criminals are successfully rehabilitated.

If we plot all the probabilities for $n = 10$ and $\pi = .4$, we get a graph of this binomial distribution, shown in Figure 4.11.

FIGURE 4.11

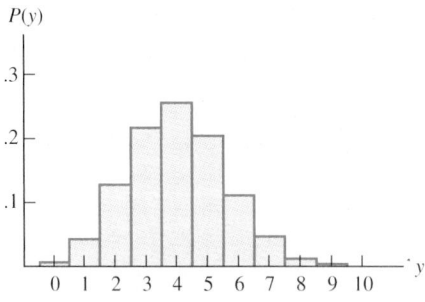

Binomial probability distribution
for $n = 10$ and $\pi = .4$

If π were only .1, we could go down the .10 column and get all the probabilities associated with the possible values of k. Figure 4.12(a) is a graph of this distribution. Note that for small values of π, the distribution is heavily concentrated on the lower values of k, which means that a smaller number of successes is more likely. Figure 4.12(b) shows that for large values of π (in this case $\pi = .9$), the binomial distribution is heavily concentrated on the higher values of k, which means that a greater number of successes is more likely. Note also that the distributions for $\pi = .1$ and $\pi = .9$ are mirror images of each other, as is the case for $\pi = .2$ and $\pi = .8$, and so on. If $\pi = .5$, the binomial distribution is symmetric, as shown in Figure 4.13.

FIGURE 4.12

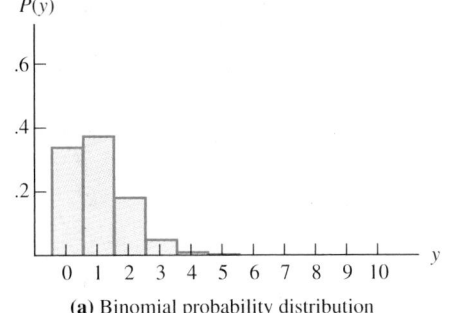

(a) Binomial probability distribution
for $n = 10$ and $\pi = .1$

(b) Binomial probability distribution
for $n = 10$ and $\pi = .9$

FIGURE 4.13

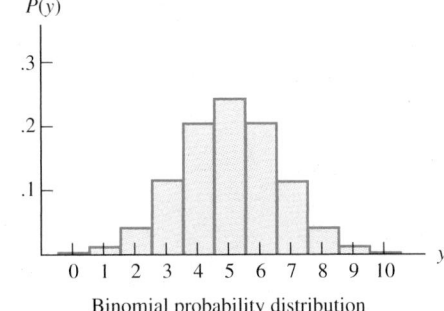

Binomial probability distribution
for $n = 10$ and $\pi = .5$

The Mean and Standard Deviation of the Binomial Distribution

It is apparent that the values of n and π are very important because they determine the various binomial distributions. They are the *parameters* for the binomial distribution. From the parameters we get other characteristics of the distribution, such as the mean and the standard deviation.

In the case of the binomial distribution, the mean is the product of n and π; that is,

$$\text{Mean} = n\pi$$

and the standard deviation is given by

$$\text{Standard deviation} = \sqrt{n\pi(1 - \pi)}$$

EXAMPLE 4.25

In Example 4.24, find the mean and the standard deviation of the variable r; that is, find the mean number of convicted criminals successfully rehabilitated out of the ten prisoners. Also, find the standard deviation.

SOLUTION

The parameters are $n = 10$ and $\pi = .4$. Therefore, the mean is

$$n\pi = 10(.4) = 4$$

and the standard deviation is

$$\sqrt{n\pi(1 - \pi)} = \sqrt{10(.4)(.6)} \approx 1.55$$

So we expect four of the ten prisoners to be successfully rehabilitated, with a standard deviation of 1.55 prisoners.

An interesting feature of the binomial distribution is what happens when n increases and π remains constant. Figure 4.14 gives the binomial distributions for $\pi = .2$ and $n = 5, 10, 20,$ and 30. Note that as n increases from 5 to 30, the distribution becomes more bell shaped. Recall that the empirical rule can be applied to bell-shaped distributions.

FIGURE 4.14

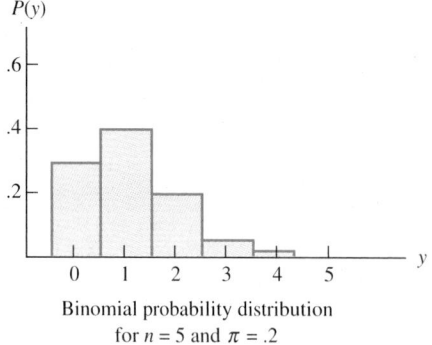

Binomial probability distribution
for $n = 5$ and $\pi = .2$

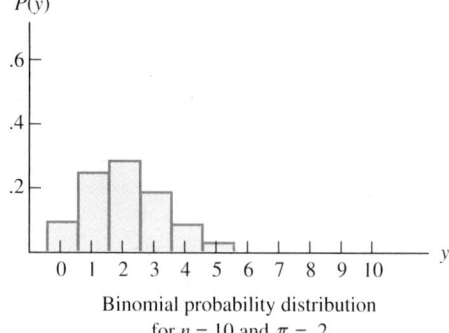

Binomial probability distribution
for $n = 10$ and $\pi = .2$

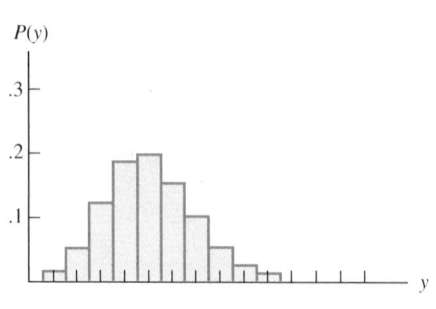

Binomial probability distribution
for $n = 20$ and $\pi = .2$

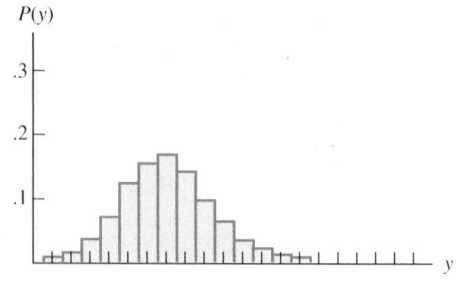

Binomial probability distribution
for $n = 30$ and $\pi = .2$

EXAMPLE **4.26**

A certain drug manufacturer claims that its vaccine is 80% effective; that is, each person who is vaccinated stands an 80% chance of developing immunity. Suppose 100 people are vaccinated. If we let v be the number who develop immunity, then v is a binomial random variable with parameters $n = 100$ and $\pi = .8$, and thus the mean is

$$n\pi = 100(.8) = 80$$

so we expect 80 to develop immunity. The standard deviation is

$$\sqrt{n\pi(1 - \pi)} = \sqrt{100(.8)(.2)} = \sqrt{16} = 4$$

So, using the empirical rule, we can say that approximately 95% of the distribution is between 72 and 88. That is, two standard deviations below the mean is

$$80 - 2(4) = 80 - 8 = 72$$

and two standard deviations above the mean is

$$80 + 2(4) = 80 + 8 = 88$$

So we know that if $\pi = .8$, then out of 100 people, we expect between 72 and 88 to develop immunity. Anything outside those limits is considered unusual.

To determine the probabilities associated with each of the possible values of a binomial random variable, we can work them out with the mathematical formula, look them up in the binomial tables (Table B.1 in the Appendix; remember only certain values of n and π are given), or use a computer.

C O M P U T E R   T I P

Suppose 68% of all requests for financial aid are approved by a university. Determine the probabilities associated with the number of approvals being 7, 8, 9, 10, 11, or 12 out of 20 requests.

First enter 7, 8, 9, 10, 11, and 12 in a column, say C1, in the Data worksheet. From the **Calc** menu, select **Probability Distributions** and then select **Binomial**. When the Binomial Distribution window appears, check **Probability**, key in the **Number of trials** = 20 and the **Probability of success** = .68, and specify C1 as the **Input Column**. If you wish to store the probabilities in, say, C2, specify it as **Optional Storage**. When you check **OK**, the probabilities will appear in the Session window or the optional storage column that you have specified.

(continued)

The contents of the Session window will be:

```
Probability Density Function

Binomial with n = 20 and p = 0.680000

        x           P( X = x)
     7.00            0.0019
     8.00            0.0066
     9.00            0.0188
    10.00            0.0440
    11.00            0.0849
    12.00            0.1354
```

EXERCISES 4.4

4.85 Suppose x is a binomial random variable. Compute the probabilities corresponding to the following values of n, π, and k:
   a. $n = 10$, $\pi = .8$, $k = 7$
   b. $n = 15$, $\pi = .3$, $k = 6$
   c. $n = 6$, $\pi = .5$, $k = 3$
   d. $n = 13$, $\pi = .2$, $k = 5$

4.86 For each of the following, graph the binomial distribution:
   a. $n = 10$, $\pi = .8$
   b. $n = 15$, $\pi = .3$
   c. $n = 6$, $\pi = .5$
   d. $n = 4$, $\pi = .9$

4.87 For each part in Exercise 4.86, find the mean and standard deviation.

4.88 If y is a binomial random variable with n and π given as follows, calculate the unknown probability:
   a. $n = 10$, $\pi = .3$, $P(y \leq 4)$
   b. $n = 6$, $\pi = .8$, $P(y > 3)$
   c. $n = 15$, $\pi = .9$, $P(y > 13)$
   d. $n = 5$, $\pi = .2$, $P(y < 3)$

4.89 How can the probabilities related to the binomial distribution be obtained if $n > 20$?

4.90 How can the probabilities related to the binomial distribution be obtained if $\pi = .68$?

4.91 Assume that 60% of a college's student loan applications are approved. Ten applications are chosen at random.
   a. What is the probability that eight or more are approved?
   b. How many applications are expected to be approved?
   c. What is the standard deviation of the number approved out of ten applications?

4.92   A new television show has a 20% chance of being successful. NBC will introduce eight new shows this season. What is the probability that fewer than three will be successful? How many are expected to succeed?

4.93   A presidential aide believes that half the members of Congress favor a particular action taken by the President. If this is true, what is the probability that out of a sample of 15 members of Congress, more than eight favor the action? Of the 15, how many are expected to favor the action?

4.94   A box contains six marbles, two of which are white. Three are drawn with replacement. What is the probability that two of the three are white?

4.95   Which of the following are binomial random variables?
a. The number of accidents per week involving alcohol in your state
b. The number of violent crimes per month committed in your city
c. The number of successful heart transplants out of five patients
d. The length of a prison term for possession of marijuana
e. The number of approved food stamp recipients out of 50 applications

4.96   According to the Census Bureau, 40% of the 1.6 million men with children whose mothers were absent were awarded child support in 1991 (Statistical Brief SB/95-16, June 1995). Out of 18 randomly selected men with children whose mothers were absent, what is the probability that less than half are awarded child support?

4.97   The population of a small town is 60% black. A jury is selected and consists of eight whites and four blacks. Was the jury randomly selected? (*Hint:*   Find the probability that four or fewer blacks would be selected.)

4.98   One-fourth of all pregnancies end in abortion (*Statistical Abstract of the United States*, 1994, p. 83). Suppose that three pregnant women are randomly selected. Let y be the number, out of the three, who get an abortion. Find the probability distribution of y. Find the mean and standard deviation of y. Sketch a graph of the probability distribution and locate $\mu \pm 2\sigma$ on the graph. Give an interpretation using either the empirical rule or Chebyshev's rule.

4.99   Hemophilia is a recessive sex-linked disease. A carrier is a person who has the recessive hemophilia gene but does not exhibit symptoms of the disease. If a woman carrier has a child with a non-hemophiliac man, the probability is $\frac{1}{4}$ that their child will be hemophiliac. Suppose the couple plans to have three children. Let y denote the number of hemophiliac children out of the three. Find the probability distribution of y. Find the expected number of hemophiliac children and the standard deviation. Sketch a graph of the probability distribution and locate the interval $\mu \pm 2\sigma$ on the graph. Give an interpretation using either the empirical rule or Chebyshev's rule.

4.100 According to the Census Bureau, 70% of all Americans have health coverage with a private insurance plan (Statistical Brief SB/94-28, October 1994). Out of a random sample of 15 Americans, what is the probability that 10 or more have health coverage with a private insurance plan?

4.101 Using the information in Exercise 4.100, simulate 30 random samples of 15 Americans regarding health insurance coverage. Out of the 30 samples, how many times did all 15 Americans have a private insurance plan? Out of 15, how many do you expect to have a private insurance plan? What number comes up most frequently in your simulation?

4.102 Of all persons who live in poverty areas, the Census Bureau has reported that 56% are white and 30% are black (Statistical Brief SB/95-13, June 1995). From a random sample of 50 persons who live in poverty areas, what is the probability that 40 or more are white? What is the probability that fewer than 10 are black?

4.103 Use the information in Exercise 4.102 to simulate 20 random samples of 50 persons who live in poverty areas. Out of a sample of 50, how many do you expect to be white? What number comes up most frequently in your simulation?

4.104   The mortality rate for a certain disease is 30%. Of ten patients who have the disease, what is the probability that more than half will die from the disease? Of the ten patients, how many are expected to die from the disease?

4.105   A drug, Nimodipine, holds considerable promise of providing relief for those people suffering from migraine headaches who have not responded to other drugs. Clinical trials have shown that 90% of the patients with severe migraines experience relief from their pain without suffering allergic reactions or side effects. Suppose 16 migraine patients try Nimodipine.

    a. What is the probability all 16 experience relief?

    b. What is the probability at least 14 experience relief?

    c. How many are expected to experience relief?

4.106   Jan Stenerud, who ended his career with the Minnesota Vikings, started kicking field goals in 1967 when he joined the Kansas City Chiefs. His season's best was in 1981 with the Green Bay Packers. That year he set an NFL record by making 22 of 24 attempts for a 91.7% success rate. His lifetime record is 66.2%. Assume that field goal attempts are independent. What are Stenerud's probabilities of the following outcomes out of three attempts?

    a. Hit all three

    b. Hit two of the three

    c. Missed all three

4.5   THE NORMAL DISTRIBUTION

The probability function for a discrete random variable gives the probabilities associated with the discrete set of values in the range of the random variable. Because a continuous random variable can assume all the values in a line interval, we are unable to assign probabilities as we do in the discrete case.

Probability Density Function

For continuous random variables, a function, called a *probability density function* or *density curve*, is defined over the interval of real numbers that the variable can assume, with the property that probabilities are represented by areas under the curve.

> A **probability density function**, $f(y)$, that describes the probability distribution for a continuous random variable y has the following properties:
>
> 1. $f(y) \geq 0$
> 2. The total area under the probability density curve is 1.00, which corresponds to 100%.
> 3. $P(a \leq y \leq b)$ = area under the probability density curve between a and b.

EXAMPLE 4.27

Let the variable y represent the annual earnings of year-round full-time plumbers. According to the Census Bureau, the mean annual income of full-time plumbers in 1995 was $29,942. Suppose that the probability density function given by the density curve in Figure 4.15 represents the distribution of annual earnings of full-time plumbers. The shaded area represents the percent of plumbers who have incomes between $22,000 and

$34,000. If the shaded area represents approximately 60% of the total area, we can write

$$P(22 \leq y \leq 34) \cong .60$$

FIGURE 4.15

Distribution of the incomes (in $ thousands) of full-time plumbers

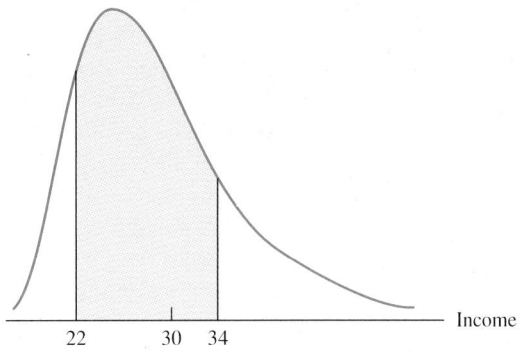

Because probabilities are represented by areas under the curve, it does not matter whether the endpoints of the interval are included. In other words, we have

$$P(a < y < b) = P(a \leq y \leq b)$$

Observe that the probability density curve in Example 4.27 is centered at about 30 (i.e., $30,000), is somewhat spread out, and tails off to the right. The shape of a probability density curve indicates what values of the random variable are most likely to occur in a sample. Two important characteristics of a random variable are its mean and its variance. The mean of a continuous random variable, like that of a discrete random variable, is the value of the random variable that balances the frequency curve. Both the mean and the variance of a continuous random variable can be calculated from the probability density function, but this is beyond the scope of this book.

The Normal Curve

One of the most commonly observed random variables is called the *normal random variable*. Its probability density curve, introduced in Section 2.6, is characterized by its mean and its variance. Recall that this distribution is *symmetric* (left half is mirror image of right half) and bell shaped, as shown in Figure 4.16.

FIGURE 4.16

A normal probability density curve

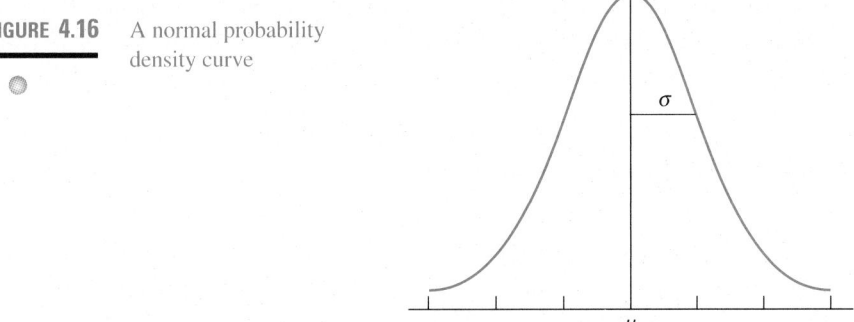

Characteristics of the Curve

The center of the normal distribution locates the mean and is denoted by the Greek letter μ. The amount of spread is determined by the standard deviation (square root of the variance) and is denoted by the Greek letter σ. Note in Figure 4.16 that at the peak of the probability density curve, the curvature is concave downward, whereas out toward the tails it is concave upward. The point of transition from concave downward to concave upward is called a *point of inflection*. For the normal curve, there is a point of inflection on each side of the mean. The distance from the mean, μ, to the point of inflection is one standard deviation, σ. Clearly, if the standard deviation were increased, the density curve would become more spread out. Also if the mean were changed, the density curve would shift so that it is centered at the mean. Thus, these two *parameters* change the appearance of the density curve, which is why we say that they characterize the normal distribution. Figure 4.17 shows three normal curves that have different means and standard deviations. We see that by changing the two parameters, μ and σ, we can describe any number of different normal distributions.

FIGURE 4.17

Probability density curves for three different normal distributions

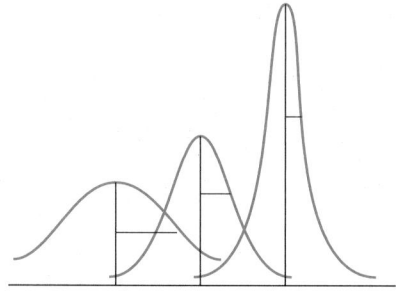

In addition to being symmetric about its mean, the normal density curve continues infinitely in both directions. However, most of the distribution lies within three standard deviations on either side of μ. In fact, the empirical rule, which we have already studied, was derived from the normal distribution. More precisely, for a normal distribution:

68.26% of the distribution is between $\mu - \sigma$ and $\mu + \sigma$

95.44% of the distribution is between $\mu - 2\sigma$ and $\mu + 2\sigma$

99.74% of the distribution is between $\mu - 3\sigma$ and $\mu + 3\sigma$

The normal distribution is important because it provides a model for many real-world distributions. Such measurements as aptitude test scores, physical measurements like heights and weights, and random error in production processes may follow the normal distribution.

Determining Areas Under the Curve

EXAMPLE 4.28

Scores on the Stanford-Binet IQ test are assumed to be normally distributed with a standardized mean of 100 and a standard deviation of 16. What percent of the population have IQs between 100 and 116?

SOLUTION

First, we suggest that a picture be drawn of the distribution and that the desired area be shaded. Figure 4.18 shows a normal curve centered at 100, with the desired area from 100 to 116 shaded. Observe that 116 is 16 points or exactly one standard deviation above the mean. From the empirical rule, we know that 68.26% of the distribution is within one standard deviation on *either* side of the mean, so the upper half would include one-half that; hence, 34.13% of the population have IQs between 100 and 116.

FIGURE 4.18

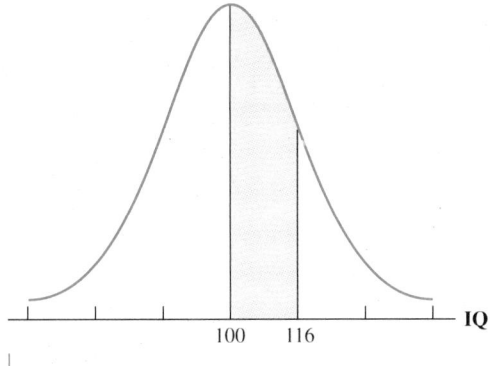

EXAMPLE 4.29

Assume that the time it takes a bank teller to serve a customer is normally distributed with a mean of 30 seconds and a standard deviation of 10 seconds. What percent of customers are served in less than 20 seconds?

SOLUTION

Figure 4.19 is a graph of the distribution in which the desired area (below 20) is shaded. Because of the symmetry of the curve, 50% of the area is below the mean of 30. Because 20 is 10 points, or one standard deviation, below 30, we know that 34.13% of the area

FIGURE 4.19

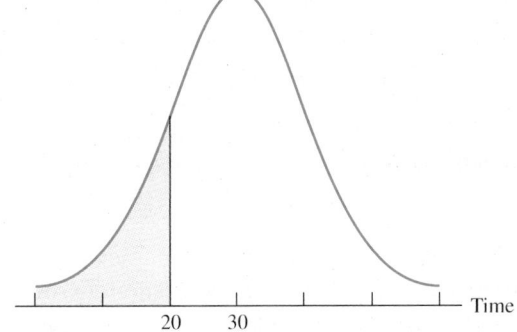

lies between 20 and 30. Thus,

$$P(t < 20) = .5000 - .3413 = .1587$$

Just under 16% of the customers are served in less than 20 seconds.

In the preceding two examples, the probability was determined by using the empirical rule, which gives the probabilities associated with one, two, and three standard deviations above or below the mean. But suppose we are not dealing with an integer number of standard deviations above or below the mean? For example, what percent of the population has an IQ (on the Stanford-Binet test) below 124? The probability is illustrated in Figure 4.20. Clearly, 124 is 24 points above the mean, which is not an exact multiple of the standard deviation of 16; thus, we cannot apply the empirical rule.

FIGURE 4.20

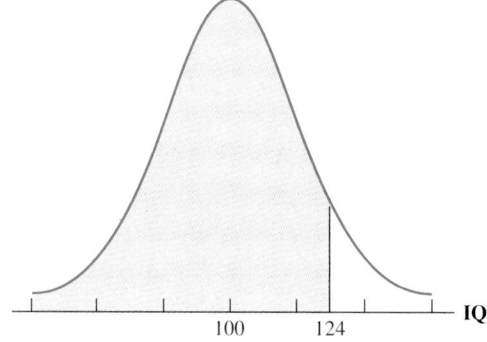

We know, however, that 24 points above μ represents $24/16 = 1.5$ standard deviations above μ. So the problem becomes: What percent of a normal distribution lies below a point that is 1.5 standard deviations above the mean?

Finding Probabilities Associated with z-scores

The *standard* normal probability distribution table in Appendix B (Table B.2) is constructed in a way that immediately answers this question. Each entry in Table B.2 corresponds to the area under the curve below a point that is z standard deviations from the mean. A negative z value corresponds to a score that is below the mean, and a positive z value corresponds to a score above the mean. A portion of Table B.2 corresponding to positive values is reproduced in Table 4.4 for our discussion.

To complete the problem, we simply find $+1.5$ in the z column and locate the associated probability of .9332. Thus, 93.32% of the area lies below a score that is 1.5 standard deviations above the mean. Because an IQ score of 124 is 1.5 standard deviations above the mean, we can say that 93.32% of the population have IQs below 124.

We can also say that 43.32% of the population have IQs between 100 and 124 because 50% of the total area is below 100. Subtracting 50% from 93.32% gives 43.32%. Furthermore, because of the symmetry of the normal distribution about the mean, the

TABLE 4.4

z	.00	.01	.02	.03	.04	.05	
.0	.5000	.5040	.5080	.5120	.5160	.5199	· · ·
+.1	.5398	.5438	.5478	.5517	.5557	.5596	· · ·
+.2	.5793	.5832	.5871	.5910	.5948	.5987	· · ·
⋮	⋮	⋮	⋮	⋮	⋮	⋮	
+1.4	.9192	.9207	.9222	.9236	.9251	.9265	· · ·
+1.5	.9332	.9345	.9357	.9370	.9382	.9394	· · ·
⋮	⋮	⋮	⋮	⋮	⋮	⋮	

area from 76 to 100 is the same as the area from 100 to 124. Therefore, 2(.4332) = .8664, or 86.64% have IQs between 76 and 124. Alternatively, we could find the area below 76, subtract it from the area below 124, and get the same answer.

To solve problems of this type, we must determine the number of standard deviations that a given score is from the mean. This is accomplished by calculating the *z-score*.

The **z-score**:

$$z = \frac{\text{score} - \mu}{\sigma}$$

gives the number of standard deviations that a score is from the mean. A negative z-score indicates that the score in question is below the mean, and a positive z-score indicates that the score is above the mean. The distribution of z has a mean of 0 and a standard deviation of 1 and is known as the *standard normal distribution*.

The probabilities associated with any normal distribution, standard or otherwise, can be found from the probabilities associated with the z-scores in the standard normal distribution table. The additional columns in Table B.2 are for z-scores carried out to two decimal places.

EXAMPLE 4.30

An important measurement in the textile industry is the tensile strength of a produced material. Suppose that the tensile strength (in pounds per square inch) of a roll of woven polypropylene is normally distributed with a mean of 89 and a standard deviation of 4. What is the probability that the tensile strength will be somewhere between 80 and 100 pounds per square inch?

SOLUTION

The probability we are seeking is represented by the shaded area in Figure 4.21 (page 284). We first find A_1, the area to the left of 100, and then A_2, the area to the left of 80. We then subtract A_2 from A_1 to find the desired probability. To find A_1 we

FIGURE 4.21

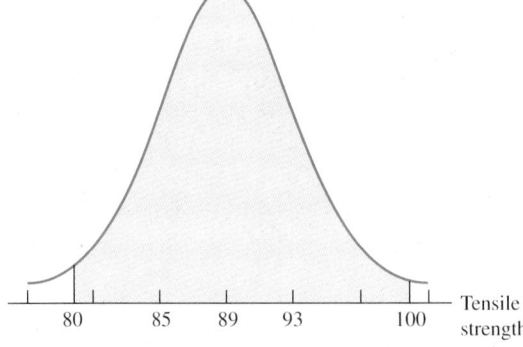

compute the z-score for 100 as follows:

$$z = \frac{100 - 89}{4} = +2.75$$

From Table B.2 we find that $A_1 = .9970$. The z-score for 80 is

$$z = \frac{80 - 89}{4} = -2.25$$

From Table B.2 we find that $A_2 = .0122$. So the probability that the tensile strength is between 80 and 100 is

$$A_1 - A_2 = .9970 + .0122 = .9848$$

By using the complement rule for probabilities, we can find areas in the upper tail of the normal distribution.

EXAMPLE 4.31   Use Example 4.30 to find the probability that the tensile strength of the woven polypropylene is greater than 95 pounds per square inch.

SOLUTION

The probability is the area above 95 in Figure 4.22. To find that area, we first find its complement, the area below 95. This area can be determined by finding the z-score for 95 and looking up the associated probability. Thus,

$$z = \frac{95 - 89}{4} = +1.5$$

which has a corresponding probability of .9332. Therefore, the desired area is

$$A = 1.0000 - .9332 = .0668$$

FIGURE 4.22

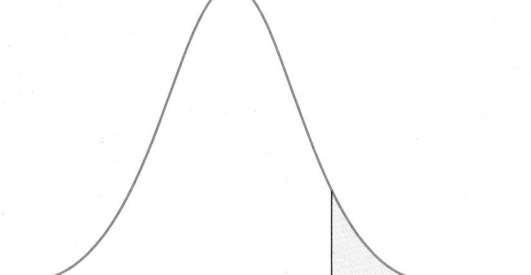

The probability is .0668 that the tensile strength of the material is greater than 95 pounds per square inch.

C O M P U T E R   T I P

It is extremely easy to find probabilities associated with a given z-score with Minitab. First store the z-score (or several z-scores) in a column of the worksheet. From the **Calc** menu, choose **Probability Distribution** followed by **Normal**. When the Normal Distribution window appears, click on **Cumulative probability** and then enter the **mean, standard deviation**, and the **input column** where you stored the z-scores. Click on **OK**, and the probability will appear in the Session window.

The following procedure will help when you work probability problems involving normally distributed populations.

To Work Probability Problems for Normally Distributed Populations

1. Draw a graph of a normal curve, label the mean, and shade the desired area.
2. Find the number of standard deviations the given score is from the mean by finding the z-score.
3. Find the associated probability for the z-score in the standard normal probability table.
4. Relate the result to the problem at hand.

Finding Percentiles

Often, instead of finding the probability associated with a certain score, we wish to find the score corresponding to a probability.

EXAMPLE 4.32

Consider again the tensile strength of a roll of woven polypropylene. In Example 4.30, we assumed that it is normally distributed with a mean of 89 and a standard deviation of 4. Find the value of b so that 80% of the distribution lies below it. In other words, if x is a normal random variable with mean 89 and standard deviation 4, find b so that $P(x \le b) = .80$.

SOLUTION

This is the reverse of the previous examples. Here we are given the probability and asked to find the associated score. We know that b is above the mean 89 because 50% of the distribution lies below the mean. Figure 4.23 illustrates that b is the value with .8000 area under the curve below it.

FIGURE 4.23

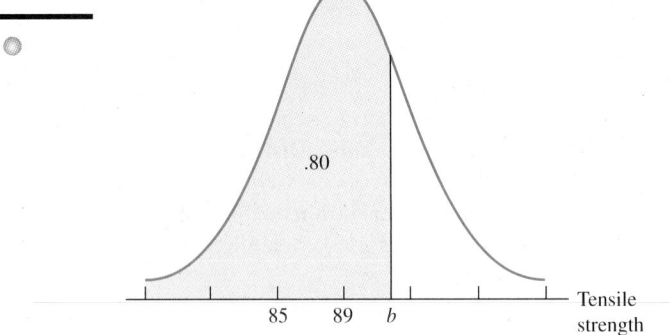

Looking in the z table, we find that a z-score of .84 corresponds to .7995 (the closest value to .8000). Hence, b is .84 standard deviation above 89. We have that

$$b = 89 + (.84)(4) = 89 + 3.36 = 92.36$$

We can say that 80% of the woven polypropylene will have a tensile strength less than 92.36 pounds per square inch. This value of b is called the 80th percentile.

The **pth percentile** is the value in the population such that $p\%$ of the distribution lies at or below that value.

EXAMPLE **4.33**

What IQ score on the Stanford-Binet corresponds to the 95th percentile?

SOLUTION

The 95th percentile is the IQ score such that 95% of all IQ scores are below it. From Figure 4.24 we see that we are looking for the score above the mean such that .95 of the area is below that score.

FIGURE **4.24**

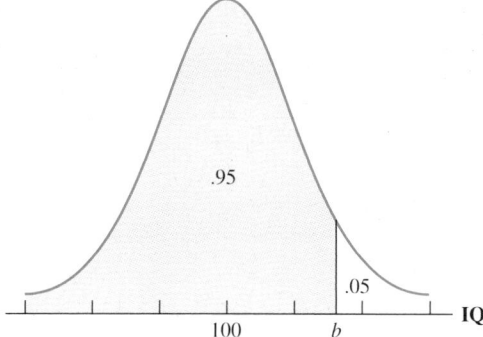

The z-score corresponding to an area of .9500 is approximately 1.64. Thus, we can say that the 95th percentile is 1.64 standard deviations above the mean. Since the standard deviation is 16, the 95th percentile score is

$$b = 100 + 1.64(16) = 100 + 26.24 = 126.24$$

Just as we used the computer to simulate Bernoulli and binomial distributions, we can simulate data drawn from a normally distributed population.

C O M P U T E R   T I P

Random with the subcommand **Normal** simulates a random sample from a normal population with a specified mean and standard deviation. The results are stored in a column to be printed out. In the Session window, the command is

MTB > **Random K Observations Into C,C,—,C;**
SUBC> **Normal With Mean = K, Sigma = K.**

Otherwise, under the **Calc** menu, select **Random Data** and click on **Normal**. At the Normal Distribution window, specify the **Number of observations** to generate, what **Columns to store the data**, give the **Mean** and the **Standard deviation**, and click on **OK**.

EXAMPLE 4.34

Assuming the mean IQ is 100 and the standard deviation is 16, simulate 80 IQs and construct a histogram of the results. Does the histogram resemble a normal distribution?

SOLUTION

See the Minitab commands in Figure 4.25. The histogram appears to approximate a normal distribution.

FIGURE 4.25

```
MTB > Random 80 Observations Put In C1;
SUBC> Normal With Mean = 100, Sigma = 16.

MTB > Histogram C1.
```

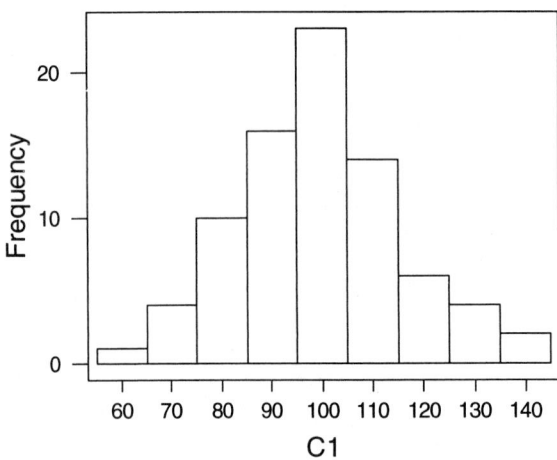

In Section 2.6, we looked at several of the more common distribution shapes that appear in the daily practice of statistics. Being able to detect the shape of the distribution from sample information is very important. As you will see later, many inferential procedures are based on the fact that the population that generates the sample is close to being normally distributed. Therefore, detecting normality is of utmost importance. We have shown how the midsummaries can tell us about symmetry and how the boxplot can tell us about the length of the tails of a distribution. Now we introduce the *normal probability plot*, a very sensitive graphical procedure for checking normality.

The normal probability plot is a graph of the sample data against the values we would expect if the sample had come from a normal population. If the sample is indeed from a normal population, then the graph should be close to a straight line. Deviations from a straight line indicate nonnormality. Judging whether or not the plot exhibits a linear pattern is somewhat subjective, and therefore one should not rule out normality unless the evidence is obvious. Figure 4.26 illustrates normal probability plots together with boxplots for data generated from several different theoretical distributions. The boxplots are given to show symmetry versus skewness and the length of the tails of the distributions.

FIGURE 4.26

(a) Sample of size 200 from a normally distributed population with mean 0 and standard deviation 1

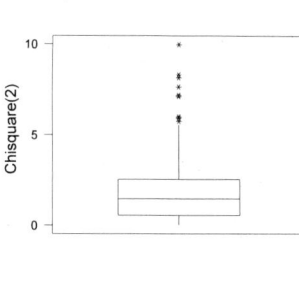

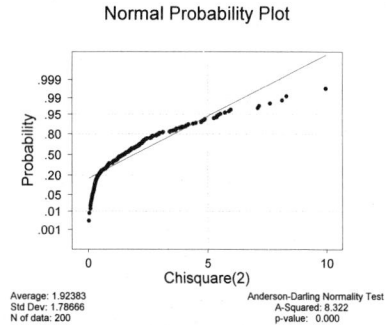

(b) Sample of size 200 from a skewed right distribution—Chi-square with 2 degrees of freedom

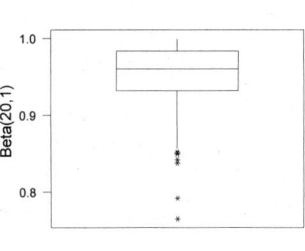

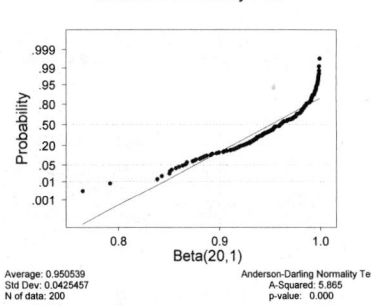

(c) Sample of size 200 from a skewed left distribution—Beta(20, 1)

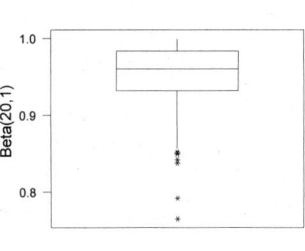

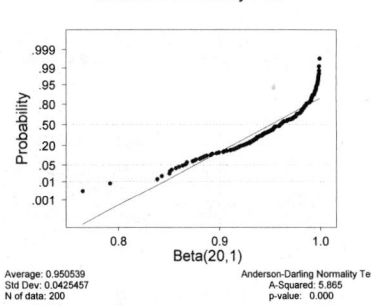

(*continued*)

FIGURE 4.26   (d) Sample of size 200 from a short-tailed distribution—Uniform(0, 1)
(continued)

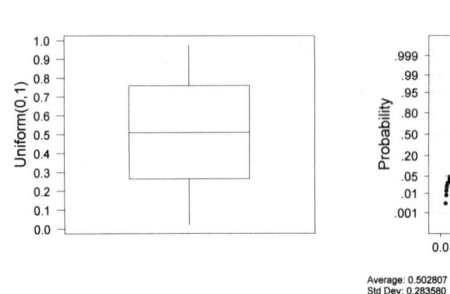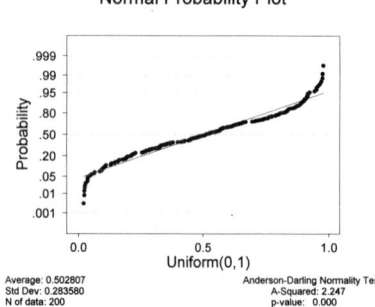

Normal Probability Plot

Average: 0.502807   Anderson-Darling Normality Test
Std Dev: 0.283580   A-Squared: 2.247
N of data: 200   p-value: 0.000

(e) Sample of size 200 from a long-tailed distribution—Cauchy

Normal Probability Plot

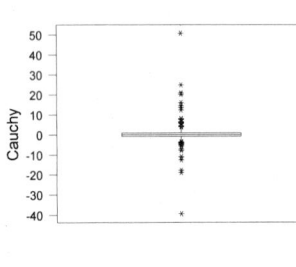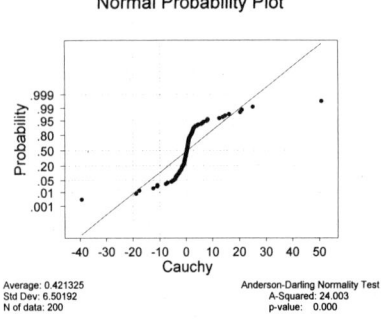

Average: 0.421325   Anderson-Darling Normality Test
Std Dev: 6.50192   A-Squared: 24.003
N of data: 200   p-value: 0.000

Notice first in Figure 4.26(a) that data from a normal population do exhibit a straight-line behavior. Figure 4.26(b) and (c) illustrate that the left tail of the probability plot curves downward for a skewed right distribution and the right tail curves upward for a skewed left distribution. Figure 4.26(d) shows the left tail curving down and the right tail curving up for a short-tailed distribution. Figure 4.26(e) exhibits the opposite behavior for a long-tailed distribution.

By hand, the computations to construct a probability plot are rather tedious and are not presented here. Minitab (or almost any other statistical package) makes the job feasible.

C O M P U T E R   T I P

First the data have been stored in a column of the Minitab worksheet. From the **Graph** menu, select **Normal Plot**. When the Normal Probability Plot window appears, **Select** the variable and click **OK**.

The construction of the normal probability plot is illustrated in the next example.

EXAMPLE **4.35**   Check the random data generated in Example 4.34 for normality with a normal probability plot.

SOLUTION

We have the data stored in C1 of the worksheet and named IQ. Figure 4.27 gives the resulting normal probability plot. The closeness of the plotted points to the straight line indicates that the data are reasonably close to being normally distributed.

FIGURE 4.27

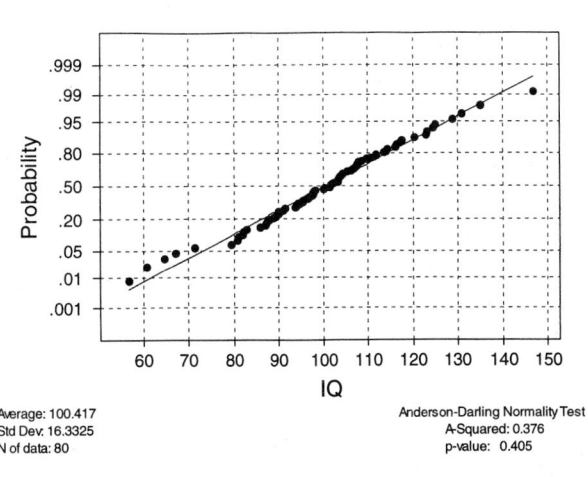

Normal Probability Plot

Average: 100.417
Std Dev. 16.3325
N of data: 80

Anderson-Darling Normality Test
A-Squared: 0.376
p-value: 0.405

The normal distribution is very important because many physical measurements and natural phenomena are closely approximated by it. However, a more fundamental reason for studying the normal distribution involves the theoretical properties of the *sample* mean, which allow us to make statistical inferences about the population mean. This will be discussed at length in Chapter 5.

EXERCISES 4.5

4.107   Suppose z has a standard normal distribution. Find the percent of the distribution in each case:
   a. Below $z = 2.0$
   b. Below $z = 2.6$
   c. Below $z = 1.36$
   d. Below $z = -2.0$
   e. Between $z = -1.42$ and $z = 1.25$
   f. Between $z = -2.82$ and $z = -.58$

4.108 Suppose z has a standard normal distribution. Find:
  a. $P(z < 1.64)$
  b. $P(z \geq 1.96)$
  c. $P(-1.35 \leq z \leq 1.35)$
  d. $P(1.22 \leq z \leq 2.47)$
  e. The value of z so that 5% of the area lies below it

4.109 Suppose x is a normally distributed random variable with a mean of 50 and a standard deviation of 10. Find:
  a. $P(40 \leq x \leq 56)$
  b. $P(x \geq 64)$

4.110 Suppose x is a normally distributed random variable with a mean of 8 and a standard deviation of 3. Find:
  a. $P(5 < x < 10)$
  b. $P(x > 9)$
  c. b such that $P(x \leq b) = .9$

4.111 What is the probability that it takes the bank teller in Example 4.29 longer than 1 minute to serve a customer? Recall that $\mu = 30$ and $\sigma = 10$.

4.112 Based on the Stanford-Binet IQ test, what percent of the population has an IQ above 120? Between 80 and 120? What is the probability of an IQ more than two standard deviations from the mean?

4.113 If a population of measurements can be assumed to be normally distributed, what two quantities distinguish it from another normal population?

4.114 How do you measure the relative magnitude of an arbitrary measurement that is taken from a population?

4.115 The figure shows a possible density curve for the variable w, which represents the weight of professional football players.

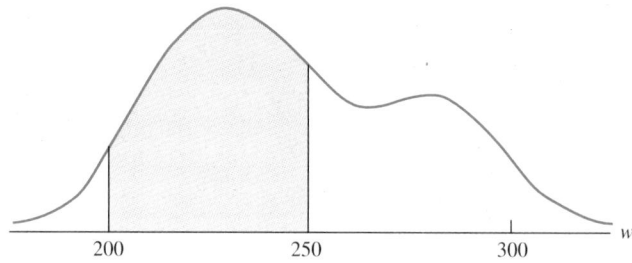

  a. Is w continuous or discrete?
  b. In general, can we apply the empirical rule to find probabilities associated with this distribution? Explain.
  c. Can the normal distribution tables be used to find probabilities associated with this distribution? Explain.
  d. Most of the professional players have weights between what values?
  e. Interpret the shaded area in the figure.
  f. Does the density curve seem reasonable to describe the weights of professional football players? (If you do not follow football, skip this part.)

4.116   Suppose a test of coordination for first-graders is scored so that the mean for all first-graders is 50 and the standard deviation is 15. If we assume further that the distribution is normal, what percent of the first-graders have the following scores?
    a. Below 30
    b. Between 40 and 70
    c. Above 75

4.117   Suppose that the test scores for a college entrance exam are normally distributed with a mean of 450 and a standard deviation of 100.
    a. What percent of those who take the exam score between 350 and 550?
    b. A student who scores above 400 is automatically admitted. What percent score above 400?
    c. The upper 5% receive scholarships. What score must they make on the exam to get a scholarship?

4.118   A job satisfaction index score for nurses is normally distributed with a mean of 50 and a standard deviation of 10. What is the probability that a nurse selected at random has an index score higher than 55?

4.119   The time it takes an ordinary mouse to run through a particular maze is assumed to be normally distributed with a mean of 15 seconds and a standard deviation of 3 seconds. The top 10% in speed will be selected for another experiment.
    a. What percent of the mice have times between 10 and 20 seconds?
    b. What time do they have to beat to be selected for the next experiment?
    c. What time corresponds to the slowest 10%?

4.120   Suppose the mean account in an investment firm is $8,000 and the standard deviation is $2,000. Assuming that the distribution of accounts is normal, what percent of the accounts are in the following ranges?
    a. Below $9,000
    b. Below $5,000
    c. Above $13,000
    d. Between $3,000 and $6,000

4.121   A public health department closes the beach when the contamination level index is in the 80th percentile. From research we know that the index level is normally distributed with a mean of 160 and a standard deviation of 20.
    a. What is the probability the index level exceeds 190?
    b. What is the probability the index level is between 150 and 170?
    c. What is the index level when the beach is closed?

4.122   Employees of a company are given a test that is distributed normally with mean 100 and variance 25. The top 5% will be awarded top positions with the company. What score is necessary to get one of the top positions?

4.123   Using the data in Exercise 2.109 in Section 2.6 on 1993 violent crime rates (worksheet: **Crime.mtw**), we have constructed the normal probability plot shown at the top of page 294. Use this result to comment on the shape of the distribution. Does your conclusion agree with your assessment in Exercise 2.109 in Section 2.6?

4.124   A normal probability plot of the radiocarbon ages studied in Examples 2.23 and 2.24 in Section 2.6 is shown on page 294. Recall that we concluded that the data appeared to be normally distributed but required further study. Judging from the appearance of this normal probability plot, is there evidence to suggest that the data are not normally distributed?

Normal probability plot for Exercise 4.123

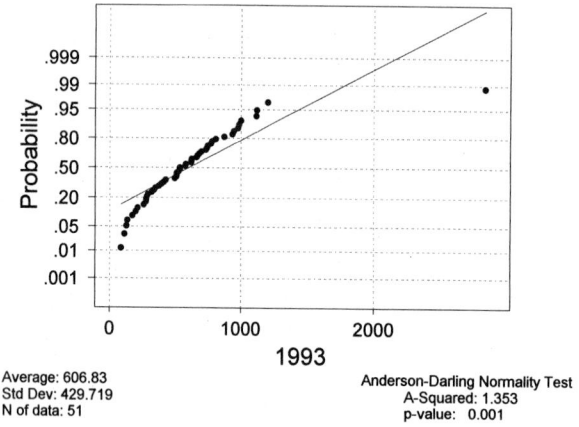

Normal probability plot for Exercise 4.124

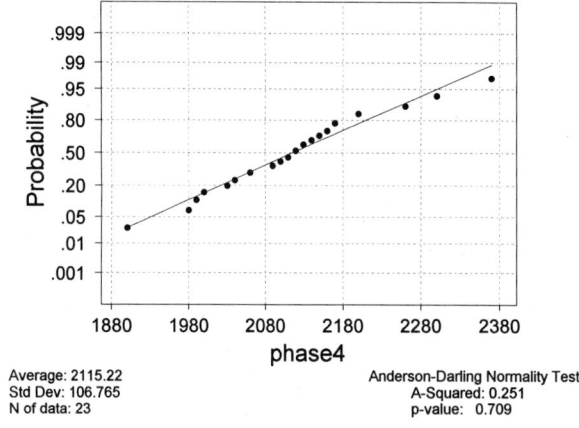

4.125   Using the data in Exercise 2.100 in Section 2.6 on the starting salaries of Ph.D. psychologists, (worksheet: **Salary.mtw**), construct a normal probability plot. Use the results to comment on the shape of the distribution. Does your conclusion agree with your assessment in Exercise 2.100 in Section 2.6?

4.126   Using the data in Exercise 2.102 in Section 2.6 on the reaction times of 30 senior citizens (worksheet: **Senior.mtw**), construct a normal probability plot. Use the results to comment on the shape of the distribution. Does your conclusion agree with your assessment in Exercise 2.102 in Section 2.6?

4.127   Using the data in Exercise 2.110 in Section 2.6 on the monthly payments to families on Aid to Families with Dependent Children (worksheet: **Aid.mtw**), construct a normal probability plot. Use the results to comment on the shape of the distribution. Does your conclusion agree with your assessment in Exercise 2.110 in Section 2.6?

4.128 Using the data in Exercise 2.55 in Section 2.3 on cholesterol values of subjects in the Framingham heart study (worksheet: **Framingh.mtw**), construct a normal probability plot. Use the results to comment on the shape of the distribution. Does your conclusion agree with your assessment in Exercise 2.55 in Section 2.3?

4.129 If a normal distribution has standard deviation σ, then show that

$$q\text{-spread} = 1.35\sigma \quad \text{and} \quad e\text{-spread} = 2.3\sigma$$

Lowercase q and e are used here because the spreads are for a population distribution.

4.6 SUMMARY AND REVIEW

KEY CONCEPTS

✓ An *experiment* is an activity of making an observation or taking a measurement that leads to a collection of outcomes. The set of all outcomes is called the *sample space*. Any subset of the sample space is called an *event*.

✓ *Probability* is a number between 0 and 1 that describes the likelihood with which an event is to occur. If probabilities can be assigned to the outcomes in a sample space, then the probability of any event A is the sum of the probabilities that have been assigned to the outcomes that are in event A.

✓ Several laws of probability apply to events. The *additive law* is

$$P(A \text{ or } B) = P(A) + P(B) - P(A \text{ and } B)$$

For *mutually exclusive* events A and B, $P(A \text{ and } B) = 0$.

✓ The *complement law* states that

$$P(A^c) = 1 - P(A)$$

✓ *Independent events* are such that if one event occurs, it does not affect the probability of the other event occurring.

✓ The *multiplicative law for independent events* states that

$$P(A \text{ and } B) = P(A)P(B)$$

✓ $P(A \mid B)$ is the *conditional probability of A given B* and is given by the formula:

$$P(A \mid B) = \frac{P(A \text{ and } B)}{P(B)}$$

✓ The terms *mutually exclusive* and *independent* are sometimes confused. If events A and B are mutually exclusive then $P(A \text{ and } B) = 0$. If events A and B are independent, then $P(A \text{ and } B) = P(A)P(B)$.

✓   A *random variable* is a rule that assigns numerical values to the outcomes of an experiment. If the sample space of values of the random variable is a discrete set of values, then the random variable is called *discrete*. If the sample space of values is a continuous set of values, then it is called a *continuous* random variable.

✓   The *probability distribution* of a discrete random variable is a table or function that lists the values of the variable and the probability with which it assumes those values. This table can be used to find the probabilities of events.

✓   The *mean* of a random variable is an average value of the random variable. It is the value that balances the distribution. The *standard deviation* is a measure of the variability associated with the random variable. The amount of variability of a random variable can be interpreted with either the *empirical rule* or *Chebyshev's rule*. The empirical rule applies to bell-shaped distributions, and Chebyshev's rule applies to any distribution.

✓   A *binomial experiment* consists of several independent trials in which each trial results in either a success or a failure. The *binomial random variable* is a discrete variable that represents the number of successes in a binomial experiment. Probabilities associated with the binomial random variable can be found in the binomial probability table (Table B.1) in the appendix.

✓   The *normal random variable* is a continuous random variable that is associated with the measurements of some numerical variable. The *normal probability density curve* (that corresponds to probabilities) is a bell-shaped curve that is symmetric. The mean and standard deviation are the parameters that distinguish normal random variables. Probabilities associated with the normal distribution can be found by determining how far a score is from the mean. To do this, we calculate the z-score and then find the associated probability in the standard normal table (z table, Table B.2) in the appendix.

LEARNING GOALS   Having completed this chapter, you should be able to:

1. Discuss the intuitive concepts of probability. *Section 4.1*
2. Calculate probabilities for simple experiments. *Section 4.1*
3. List the outcomes of the sample space for certain experiments. *Section 4.1*
4. Assign probabilities to outcomes and determine the probability of other events. *Section 4.2*
5. Be familiar with the laws of probability and use them to determine the probabilities of events. *Section 4.2*
6. Understand the concept of a random variable and be able to distinguish between a discrete and a continuous random variable. *Section 4.3*
7. Give the probability distribution of a random variable. *Section 4.3*
8. List the characteristics of the binomial experiment. *Section 4.4*
9. Find probabilities associated with the binomial distribution. *Section 4.4*
10. List the characteristics of the normal distribution. *Section 4.5*
11. Find probabilities associated with the normal distribution. *Section 4.5*
12. Use the normal probability plot to diagnose the shape of a distribution. *Section 4.5*

QUESTIONS
FOR REVIEW

Use the following problems to test your skills:

Multiple choice (4.130–4.135)

4.130   Suppose x is a binomial random variable with $n = 10$ and $\pi = .8$. Then

    a. The expected value of x is 8.
    b. The standard deviation of x is 1.6.
    c. x gives the number of trials until the first success.
    d. All of the above

4.131 Suppose x is a binomial random variable with $n = 10$ and $\pi = .8$. The probability that x is less than 7 is
    a. $.322^+$ b. $.201$ c. $.088$ d. $.121^+$

4.132 Suppose x is a normal random variable with $\mu = 25$ and $\sigma = 5$. The probability that x exceeds 32 is
    a. $.4192$ b. $.9192$ c. $.0808$ d. 1.4

4.133 Suppose there are five locations to drill wells and each location has a .6 chance of having water. What is the probability that all wells have water?
    a. $.6$ b. $.30$ c. $.5^6$ d. $.6^5$

4.134 Suppose a class contains five history majors, seven English majors, and eight psychology majors. If three students are to be selected, how many elements are there in the sample space of possible majors for the students?
    a. 20   b. 27   c. 3   d. 8

4.135 Suppose a class contains five history majors, seven English majors, and eight psychology majors. If two students are randomly selected, what is the approximate probability that both are psychology majors?
    a. $.15$ b. $.4$ c. $.33$ d. $.16$

4.136 A room has five women and five men. Two are selected at random. What is the probability that both are women?

4.137 Suppose 25% of all science majors return for an advanced degree within 5 years after graduation. A random sample of three science majors is selected.
    a. List the elements in the sample space of those who return for an advanced degree.
    b. Are the outcomes equally likely? Why or why not?
    c. What is the probability that none returns for an advanced degree within 5 years after graduation?
    d. What is the probability that at least two return within 5 years?

4.138 Suppose x has the following distribution:

x	1	2	3	4	5
$p(x)$	.2	.3	.3	.1	.1

    a. Find the probability that x is odd.
    b. Find the probability that x is greater than 3.
    c. What is the probability that x is less than or equal to 2?

4.139 A man has three keys, one of which will open a door. If he randomly picks one key after another until he finds the one to open the door, what is the probability that he will open the door on the first try? What is the probability that it will take more than three tries to open the door?

4.140 A tetrahedron (regular four-sided polyhedron) has four sides numbered 1, 2, 3, and 4. A pair of fair tetrahedra are tossed.
    a. How many outcomes are possible?
    b. Are they equally likely?
    c. A variable is defined to be the sum of the two numbers on the two tetrahedra. What are the possible outcomes of the variable?

    d. Are they equally likely?

    e. We win in a game if we roll a 5 or 7 on the first roll. What is the probability that we win on the first roll?

4.141 A circular game board is divided into five equal slots numbered 1, 2, 3, 4, and 5. The odd numbers are colored red, and the even numbers are colored black. A marble is spun around the board and randomly falls into one of the slots.

    a. What is the probability that the marble lands on red?

    b. Suppose the widths of the slots are such that an even number is twice as likely as an odd number. What is the probability that the marble lands on red?

4.142 Suppose 30% of the new employees hired by a computer firm are women. Three new employees are selected at random.

    a. List the sample space of the possible genders of the three employees.

    b. Assign a probability to each of the outcomes.

    c. Suppose A is the event that exactly two of the three selected are women. List the outcomes that make up A.

    d. What is $P(A)$?

4.143 The student council for a School of Science and Math has one representative from each of the five academic departments: biology (B), chemistry (C), mathematics (M), physics (P), and statistics (S). Two of these students are to be randomly selected for inclusion on a university-wide student committee.

    a. What are the ten possible outcomes?

    b. All outcomes are equally likely. What is the probability of each?

    c. What is the probability that one of the committee members is the statistics student?

    d. What is the probability that both members are from "laboratory science" departments?

4.144 Identify each random variable as either discrete or continuous:

    a. The amount of water pumped into a tank overnight

    b. The number of tickets drawn from a barrel until a lottery winner is found

    c. The cost of a week's groceries for a family of four

    d. The number of patients seen by the emergency room over a 24-hour period

    e. The length of life of a computer chip

4.145 A device to measure one's resistance to pain has a scale that is assumed to be normally distributed with a mean of 30 and a standard deviation of 5. What percent of those using the device score between 22 and 30? Above 34?

4.146 Sketch a graph of the distribution of a binomial variable when $n = 5$ and $\pi = .3$.

4.147 True or false? The following experiments satisfy the conditions of a binomial experiment.

    a. Select 20 random voters and record whether they favor reelection of the President.

    b. Select eight tickets from a barrel and record the names on the tickets.

    c. Draw three balls without replacement from an urn containing five red and seven white balls and record the color.

    d. Roll a pair of fair dice ten times and each time observe whether the total is seven.

    e. A drug to relieve pain is administered to ten patients. After 2 hours, each patient reports that he is better, worse, or not changed.

    f. The IQs of 20 college students are recorded.

4.148 It is believed that about 70% of convicted felons have a history of juvenile delinquency. Ten convicted felons are chosen at random.

    a. What is the probability that no more than five have a history of juvenile delinquency?

    b. What is the expected number to have a history of juvenile delinquency?

4.149 Suppose the grade point average of students at a university is normally distributed with a mean of 2.5 and a standard deviation of .5.

a. What percent of the students have grade point averages higher than 3.4?

b. If the top 10% in grade point average make the dean's list, what grade point average do they need?

4.150 Suppose 30% of the student body lives off campus. Five students are selected at random. Let the random variable x = the number of students who live off campus out of the five selected.

a. What kind of random variable is x?

b. Give the probability distribution table for x and graph it.

c. How many students do you expect to live off campus out of the five selected?

d. What is the probability that no more than two students live off campus?

4.151 Suppose a reading ability exam for 12-year-olds is normally distributed with a mean of 40 and a standard deviation of 6. What is the probability that a 12-year-old will score in the following ranges?

a. Above 60

b. Below 45

c. Between 50 and 70

4.152 A survey of students showed that 30% are in favor of a ban of alcohol on campus, 52% are in favor of eliminating hazing of pledges, and 20% favor both measures. What is the probability that a student selected at random favors one or the other? Are the events alcohol on campus and hazing of pledges mutually exclusive? Are they independent?

4.153 Suppose a variable, such as IQ or SAT, is distributed normally with a standardized mean of 50 and a standard deviation of 15.

a. What percent should score between 30 and 59?

b. What percent should get scores greater than 74?

c. What is the 90th percentile?

4.154 a. A test has two true/false questions on it. If a student guesses, what is the probability he gets both right?

b. A test has six multiple-choice problems, and each problem has four choices. If a student guesses, how many problems is she expected to get right?

4.155 In the game of craps, the "shooter" loses on the first roll of a pair of fair dice if the two dice are double 1's or double 6's, and wins on the first roll if the two dice total 7 or 11.

a. What is the probability that the shooter loses on the first roll?

b. What is the probability that the shooter neither loses nor wins on the first roll?

4.156 In the game "three strikes you're out" on *The Price Is Right* game show, seven circular chips are placed in a bag. Four of the chips are white and have numbers on them corresponding to the digits in the price of an automobile. The other three chips are red and have X's on them corresponding to strikes. If the contestant draws the four digits of the price of the car and places them in order before drawing the three strikes, then he wins the car. On the first draw, what is the probability that the contestant gets a white chip? If he get's a white chip and guesses, what is the probability that he selects the right position of the number in the price of the car?

4.157 An urn contains a red, a green, and a black marble. Two marbles are drawn from the urn with replacement. List the outcomes in the sample space and the following events:

A: Both marbles are red

B: None are red

C: One is red

4.158 Drivers can select any one of three pumps at a gas station. If two drivers enter the station at the same time (they obviously cannot use the same pump), list the sample space of possible selections for the two drivers.

4.159   A fair die is rolled. If it comes up even, the die is rolled again; if it comes up odd, a fair coin is tossed.
    a.   List the sample space and assign probabilities to each of the outcomes.
    b.   What is the probability of getting a head on the coin?

4.160   A beer drinker is asked to rank three unmarked glasses of beer according to taste. List the outcomes of the sample space. What must be true in order to assume that the outcomes are equally likely? Assuming that they are, assign probabilities to them.

4.161   A shopper wishes to buy two pairs of shoes but cannot decide among four different pairs. List the outcomes of the sample space of possibilities. If she randomly chooses the two pairs, assign probabilities to the outcomes.

4.162   Eighty students who are candidates for an honor society are classified according to gender and class:

	Men	Women	Total
Freshman	16	14	30
Sophomore	24	26	50
Total	40	40	80

Are the events sophomore and woman independent?

4.163   A professional football quarterback has a 60% completion record this season. Assume that pass attempts are independent. Consider his next ten passes.
    a.   What is the probability that he will complete from seven to nine passes?
    b.   How many is he expected to complete?
    c.   What is the standard deviation of the number he is to complete?

4.164   In 1982, 39% of all business/management bachelor's degrees were awarded to women. This compares with only 8.1% in 1971. Suppose that three business/management students are selected at random from the 1982 graduation class. What is the probability that all are women?

4.165   An estimated 2.3 million people are poisoned each year by dangerous chemicals and products found in the home. Sixty-four percent involve children under the age of 6. Out of the next four calls to the poison control center, what is the probability at least two are for children under the age of 6?

4.166   Ninety percent of the trees planted by a landscaping firm survive. What is the probability that 10 or more of the 14 trees just planted will survive?

4.167   Suppose it costs $1 to play a game in which you have a .01 chance of winning $10. Is it to your advantage to play the game? If you play 100 times, how much should you expect to gain?

4.168   Only a third of California homeowners carry earthquake insurance on their homes. A random sample of three homeowners is selected.
    a.   List the sample space of possibilities of those who have earthquake insurance. Code them S = have and F = do not have.
    b.   Assign probabilities to the outcomes.
    c.   What is the probability that at least two of the three will have insurance?
    d.   What is the probability that none of the three has insurance?
    e.   What is the expected number out of three that have earthquake insurance?

COMPUTER EXERCISES

4.169   Using the Random Data > Integer command, simulate the 200 rolls of a single die and store them in C1 of the worksheet. Repeat the process, and store the 200 results from a second die in C2 of the worksheet. Add C1 to C2 and store the sum in C3. The results in C3 should simulate 200 rolls of a pair of fair dice. Use the Tally command to summarize the results. From the simulation, what do you estimate the probability of rolling a total of 7 to be? How about a total of 11? Do these results appear to be close to the theoretical probabilities?

4.170   Open worksheet: **Dice.mtw**. In column 1 you will find the possibilities of the roll of a pair of dice. In column 2 you will find the associated probabilities. How do these percentages compare with your simulation in Exercise 4.169? Use the Random Data > Discrete command (see Exercise 4.80 in Section 4.3) to simulate the 200 rolls of a pair of dice. Use the Tally command to summarize the results. Compare this simulation with the one obtained in Exercise 4.169 and with the theoretical percentages in column 2.

4.171   In Exercise 4.141, a circular game board is divided into five equal slots numbered 1, 2, 3, 4, and 5. The odd numbers are colored red, and the even numbers are colored black. Simulate the game 200 times and tally the results. Are the outcomes fairly evenly distributed? Should they be? Out of the 200 times, what percent of the time did red come up? Is your result close to the correct percentage of 60%?

4.172   In Exercise 4.150, five students are randomly selected and the number who live off campus is recorded. The probability of living off campus is 30%. Simulate 100 realizations of the random variable x that gives the number out of five who live off campus. Tally the results. Does the simulation seem consistent with the probability distribution found in Exercise 4.150, part **b**?

4.173   In Exercise 4.151, you were given that scores on a reading exam for 12-year-old children are normally distributed with a mean of 40 and a standard deviation of 6. Simulate 200 reading scores and construct a histogram of the results. Does the histogram have a bell-shaped appearance? Check the normality of the data with a normal probability plot. Is there evidence that the data are not normally distributed?

4.174   Figure 4.14 (page 274) shows four binomial distributions, all with $\pi = .2$ but with $n = 5, 10, 20$, and 30. To determine the probabilities for $n = 5$ and $\pi = .2$, store the values 0, 1, 2, 3, 4, and 5 in column 1 of the worksheet. Using the Probability Distributions > Binomial command, generate the probabilities in column 2. Use the Plot command to graph the distribution. Is the graph similar to the histogram in Figure 4.14 when $n = 5$? Repeat the process for $n = 10$ (store values 0, 1, 2,..., 10 in column 1) and compare with the histogram in Figure 4.14 when $n = 10$. Repeat for $n = 20$ and $n = 30$. Do the graphs become more symmetric and bell shaped as n increases?

4.175   STATISTICAL INSIGHT REVISITED   To better understand the three-door problem presented in the Statistical Insight, conduct the following experiment.
   a. Randomly generate a door for the valuable prize by generating a 1, 2, or 3 with the Random Data–Integer command. Simulate 30 games by generating 30 random values and store the results in column 1.
   b. Randomly generate a guess by the contestant by again generating a 1, 2, or 3. Simulate 30 guesses to go along with the 30 winning doors in part a and store the results in column 2.
   c. Check the 30 games by comparing the corresponding values in columns 1 and 2. If the numbers match, Monty would show one of the other two doors that conceal gag gifts, and the winning strategy is to stay with the original choice. If, on the other hand, the numbers do not match—say, a 1 and a 3—then Monty would show door 2, which conceals a gag gift, and the winning strategy is to switch doors.
   d. Out of the 30 games, how many times was the winning strategy to switch and how may times was the winning strategy to stay?
   e. Based on this simulation, would you advise a contestant to switch or to stay?
   f. Simulate 300 games and see what happens.

4.176   STATISTICAL INSIGHT REVISITED   To develop the theoretical probability of winning by switching or by staying in the three-door problem, consider the following sample space analysis:

Location of prize	Possible guesses of the contestant	Monty opens door	Winning strategy
1	1	2 or 3	remain
1	2	3	switch
1	3	2	switch
2	1	3	switch
2	2	1 or 3	remain
2	3	1	switch
3	1	2	switch
3	2	1	switch
3	3	1 or 2	remain

a. In the table, have all possibilities of where the actual prize is located and all possible guesses been considered? How many are there?

b. Out of the total number of possibilities, how many win with the remain strategy? How many win with the switch strategy?

c. What is the probability of winning by remaining? What is the probability of winning by switching?

d. Do the results of Exercise 4.175 tend to follow these theoretical probabilities?

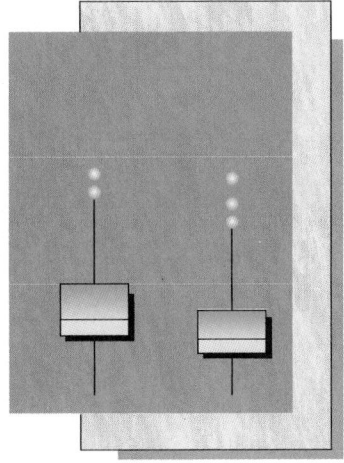

5

SAMPLING DISTRIBUTIONS

The computed value of a statistic depends on the sample that is observed. In other words, for each different sample, there is possibly a different value for the statistic. Thus, a statistic may be viewed as a random variable whose distribution corresponds to the different values that the statistic could assume in repeated sampling.

The distributions, called sampling distributions, of selected statistics are presented in this chapter. You will see that many can be approximated by the normal distribution. Also, through examples, you will see how sampling distributions can be used to evaluate generalizations about a population from a sample.

CONTENTS

STATISTICAL INSIGHT

POLLSTERS ANALYZE THEIR RESULTS

In the 1984 presidential election, Walter Mondale ran against Ronald Reagan. Of course, Reagan won the election. Prior to the election, however, several national political pollsters attempted to predict who would become president. Here are the results of the different polls that appeared in *USA Today* on November 7, 1984, along with the actual results:

Poll	Reagan	Mondale	Undecided	Margin
USA Today	60%	35%	5%	25%
NBC	58	34	8	24
Time	54	30	16	24
CBS/*NY Times*	58	37	5	21
Gallup	59	41	0	18
Actual results	59	41	—	18
Newsweek	57	40	3	17
ABC/*Washington Post*	54	40	6	14
NPR/Harris	56	44	0	12
Roper	52.5	42.5	5	10

Of nine polls, only the Gallup poll correctly predicted the actual results of the election. But can we say that the other polls were wrong? For instance, the *Newsweek* poll predicted 57% for Reagan and 40% for Mondale with 3% undecided. Suppose the 3% undecided vote were divided equally among the two candidates, as Gallup did. Then these would be the *Newsweek* figures:

Poll	Reagan	Mondale
Newsweek	58.5%	41.5%

In this case, the *Newsweek* poll is one-half of 1% off the actual results. Is this prediction incorrect?

We must realize that the *Newsweek* prediction was obtained from a sample of probably fewer than 2,000 voters. Yet the prediction was within .5% of the results from millions of voters. The prediction was quite accurate. Whenever an estimate is given from a sample, there is a certain amount of error, called the *margin of error* (not the same as the *margin* given in the table). In national polls of this type, the margin of error usually is from 3% to 5%. For example, even though the *Time* magazine poll appears to differ substantially from the actual results, we can split the undecided vote and the prediction becomes

Poll	Reagan	Mondale
Times	62%	38%

which is within 3% of the actual results.

We now list the results of the same polls with the undecided vote split equally among the two candidates (a common practice):

Poll	Reagan	Mondale
USA Today	62.5%	37.5%
NBC	62	38
Time	62	38
CBS/*NY Times*	60.5	39.5
Gallup	59	41
Actual results	59	41
Newsweek	58.5	41.5
ABC/*Washington Post*	57	43
NPR/Harris	56	44
Roper	55	45

All polls except the Roper and *USA Today* are within 3% of the actual results. Note that four of the polls overestimated the percent of voters favoring Reagan and four underestimated the percent favoring Reagan. This is certainly plausible because the margin of error can be either positive or negative. It is possible that the margin of error in the Roper and *USA Today* polls was ±4%, in which case their predictions were also within the margin of error. Also there might have been reason to divide the undecided in a way other than equally among the two candidates. After all, *USA Today* was within 1% of the actual results for Reagan.

The point of this discussion is that estimates vary from sample to sample, and they do not have to be the same to be valid estimates. The purpose of a poll of this type is to predict the next president of the United States, and *all* of these polls correctly predicted Ronald Reagan by a wide margin.

5.1 AN INTRODUCTION TO SAMPLING DISTRIBUTIONS

Parameters, such as μ and σ, that measure the characteristics of a population are usually unknown. To study the population, a sample must be selected and used to estimate the unknown parameters. We must decide which statistic, a numerical quantity calculated from the observations in a sample, best estimates the unknown parameter. These are some examples of statistics:

1. $\bar{y}$, the sample mean
2. $\bar{y}_T$, a trimmed sample mean
3. M, the sample median
4. s, the sample standard deviation
5. p, the sample proportion
6. r, the sample correlation coefficient
7. $\bar{y}_1 - \bar{y}_2$, the difference between two sample means
8. $M_1 - M_2$, the difference between two sample medians
9. $p_1 - p_2$, the difference between two sample proportions

Unlike the value of a population parameter, which is usually unknown and may never be known, the value of a statistic is known because it can be calculated from the collected sample. We must realize, however, that the value assumed by the statistic depends on the observed sample. The value of the statistic we observe is the value yielded by the sample we collect. A large number of possible samples can be chosen from a population, and each sample will yield its own value of the statistic. Thus, there is a distribution of potential values that the statistic can assume. This is called the *sampling distribution* of the statistic.

> The **sampling distribution of a sample statistic** is the probability distribution associated with the various values that the statistic can assume in repeated sampling.

Consider the problem of predicting the percent of registered voters who favor an incumbent mayor seeking reelection. The percent of *all* registered voters in favor of the mayor is the parameter π, which we wish to predict. The statistic p, which we will use to estimate the unknown value of π, is the percent of voters in a sample who are in favor of the mayor. Suppose that from a sample of 1,000 registered voters, we find that 380 favor the incumbent mayor for reelection. Then we say that the statistic *realized* the value

$$p = \frac{380}{1,000} = .38$$

Using this as our estimate of π, we are led to believe that approximately 38% of all registered voters favor the mayor for reelection.

If we had selected another sample of 1,000 registered voters, it is almost certain that there would *not* have been exactly 380 in favor of the mayor as before. The value of the statistic p will vary from sample to sample. Then we might ask: Is it reasonable to use the value of $p = .38$ to estimate the unknown value of π? The answer is yes, if the *sampling variability* of p is not too erratic. It is desirable that a statistic be *stable* from sample to sample. That is, the values it assumes from the different samples should be reasonably close to one another. If the statistic assumes widely varying values from sample to sample, then the conclusions we draw about the population parameter are less than reliable because we do not know which of the different values of the statistic to take as being close to the parameter. Thus, it is through the variability of the sampling distribution of the statistic that we can determine how precise an estimate of a parameter will be. The amount of variability associated with the sampling distribution of a statistic is measured by its *standard deviation*.

Suppose, for example, that the standard deviation of the statistic p used to estimate the percent of registered voters who favor the mayor is .015. We know from the empirical rule (assuming that the sampling distribution of p is bell shaped) that the estimated proportion of .38 should be within $2(.015) = .03$ of the true proportion of voters who favor the mayor. That is, the true proportion should be somewhere between .35 and .41. In other words, by knowing the standard deviation of the statistic p, we have narrowed down the possibilities for the true population proportion. In the next section we discuss how to compute the standard deviation of the statistic p. You will see that it depends on the sample size. As the sample size increases, the standard deviation becomes smaller, yielding a more stable estimate.

We close this section with an example that illustrates the general ideas about the sampling distribution of a statistic. The example presents the sampling distribution of the statistic $\bar{y}$ when the sample size is only $n = 2$.

EXAMPLE 5.1

Consider a population described by this probability distribution:

y	1	2	3	4	5
$P(y)$	.1	.2	.3	.3	.1

From the graph of the distribution in Figure 5.1, we see that the distribution is not exactly symmetric but is reasonably mound shaped.

FIGURE 5.1

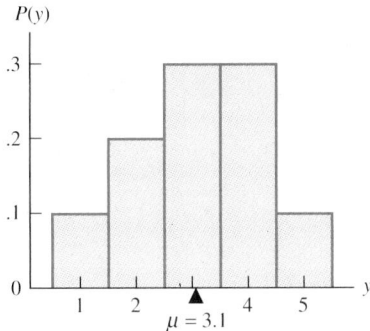

The mean of y is $\mu = 3.1$ (see Exercise 4.70 in Section 4.3 for a general method of calculating the mean of a discrete random variable). Suppose a sample of size 2 is chosen from this population. If we assume that the observations are independent, the table lists all possible samples of size 2 and their probabilities:

Sample	Prob.	Sample	Prob.	Sample	Prob.	Sample	Prob.	Sample	Prob.
(1, 1)	.01	(2, 1)	.02	(3, 1)	.03	(4, 1)	.03	(5, 1)	.01
(1, 2)	.02	(2, 2)	.04	(3, 2)	.06	(4, 2)	.06	(5, 2)	.02
(1, 3)	.03	(2, 3)	.06	(3, 3)	.09	(4, 3)	.09	(5, 3)	.03
(1, 4)	.03	(2, 4)	.06	(3, 4)	.09	(4, 4)	.09	(5, 4)	.03
(1, 5)	.01	(2, 5)	.02	(3, 5)	.03	(4, 5)	.03	(5, 5)	.01

The sample (4, 3), for example, means that the first observation is a 4 and the second is a 3. Because the observations are drawn independently, we can apply the multiplication rule for independent events to determine that the probability of (4, 3) is $P(4)P(3) = (.3)(.3) = .09$.

To find the sampling distribution of the sample mean, $\bar{y}$, we must list all the possibilities of $\bar{y}$ from the different samples and then determine their probabilities.

From the 25 different samples, we find the following possibilities for $\bar{y}$: 1, 1.5, 2, 2.5, 3, 3.5, 4, 4.5, and 5. The average 2.5, for example, is obtained from sample points (4, 1), (3, 2), (2, 3), and (1, 4). Thus, we have $P(\bar{y} = 2.5) = .03 + .06 + .06 + .03 = .18$. Continuing in the same manner, we find the following probability distribution of $\bar{y}$:

$\bar{y}$	1	1.5	2	2.5	3	3.5	4	4.5	5
$P(\bar{y})$	.01	.04	.10	.18	.23	.22	.15	.06	.01

Figure 5.2 (page 308) shows the center and shape of the distribution. Notice that the mean of $\bar{y}$ appears to be 3.1, the same as the mean of the population. In comparison to the population distribution in Figure 5.1, the distribution of $\bar{y}$ takes on a more bell-shaped appearance with less probability on the tails. In other words, the sampling distribution of $\bar{y}$ tends to be more tightly clustered about μ than the original population distribution.

All of this information is obtained by averaging just two observations. What would the sampling distribution look like if we considered all possible samples of size 3? It is more tedious to consider all samples of size 3, but the distribution of $\bar{y}$ using 3

FIGURE 5.2

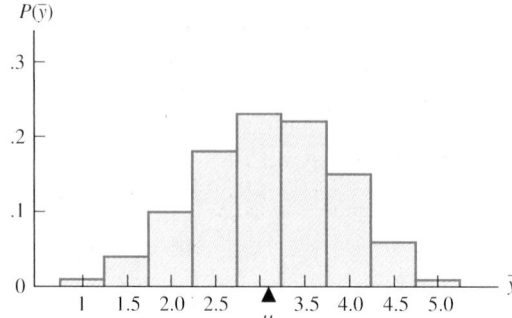

observations is also centered at 3.1, but it is not as spread out as the distribution graphed in Figure 5.2.

EXERCISES 5.1

In Exercises 5.1–5.6, identify each of the statistics from the list of nine statistics given at the beginning of this section.

5.1 The median salary of 50 workers in a plant is $320 per week.

5.2 In a 1994 *Times Mirror Center* nationwide survey of 3,800 adults, 66% said that elected officials do not care what people think about the issues.

5.3 The median monthly income for households in western states is $245 higher than for those in southern states.

5.4 From a sample of size 2,000, it was found that 60% of the teenagers killed in auto accidents were drinking before their accidents.

5.5 The average weight loss for patients on a certain diet plan was 4.6 pounds.

5.6 The difference between the percents of seniors and juniors who have cars on campus is 9%.

5.7 According to a survey of 6,000 service stations in 1995, the average price per gallon of gasoline was $1.22, which was 3 cents less than in 1994. Is the $1.22 a parameter or a statistic?

5.8 A population is described by this probability distribution:

y	0	2	5	7
$P(y)$	.2	.3	.4	.1

a. List all possible random samples of size 2 and calculate their probabilities as in Example 5.1.
b. Find the sampling distribution of $\bar{y}$ that is calculated from random samples of size 2.
c. The mean of the population is $\mu = 3.3$. Graph the population distribution and illustrate the position of μ. Graph the distribution of $\bar{y}$ and illustrate the position of μ as in Figure 5.2.
d. Compare the sampling distribution of $\bar{y}$ with the distribution of the population. Is the sampling distribution of $\bar{y}$ more tightly concentrated about μ?

5.9   Consider a population described by this probability distribution:

y	1	2	6
$P(y)$	$\frac{1}{3}$	$\frac{1}{3}$	$\frac{1}{3}$

The mean of y is 3.
- a. Find the sampling distribution of $\bar{y}$ that is calculated from a random sample of size 3.
- b. Find the sampling distribution of M, the sample median, that is found from a sample of size 3.
- c. Graph the population distribution and the sampling distributions of $\bar{y}$ and M.
- d. Are both sampling distributions centered at 3, the mean of y? Which distribution is spread out more?
- e. Which statistic, $\bar{y}$ or M, do you think is the better measure of the center of this distribution? Explain.

5.10   A game board in a booth at the county fair has a spinner that randomly lands on one of the areas (see the accompanying figure). There is a 50% chance that it lands on a 1, a 40% chance it lands on a 3, and a 10% chance it lands on a 5.

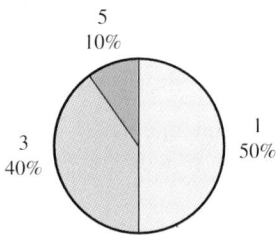

A player pays $2 and gets two independent spins. The player is then paid the average of the two spins. For example, a player who spins a 1 and a 3 is paid $2; a player who spins a 3 and a 5 is paid $4; and a player who spins a 5 and a 5 is paid $5.
- a. List the possible payoffs.
- b. Determine the probabilities of the payoffs.
- c. Graph the probability distribution obtained from parts **a** and **b**.
- d. What is the probability that a player is paid $3 or more in a single play?
- e. Is it to your advantage to play the game?

5.11   A *USA Today*/CNN/Gallup nationwide poll of 1,022 adults conducted on September 6–7, 1994, showed that 34% approved of President Clinton's handling of foreign policy. The article stated that the prediction was accurate to within ±3 percentage points.
- a. How many of the 1,022 adults approved of the President's handling of foreign policy?
- b. From the list of statistics at the beginning of the section, what statistic does the 34% represent?
- c. What was the standard deviation of the statistic?
- d. Is it possible that a second sample of 1,022 adults would yield a different percent approving the President's foreign policy?
- e. Most samples of 1,022 adults would give a percent somewhere between what two values?

5.12   In *Career Opportunity News* (1993, Vol. 10, No. 2), it was reported that college students at 4-year private liberal arts colleges spend an average of $496 per year for books.
    a. Is the $496 figure a parameter or a statistic?
    b. For it to be a parameter, data from what students at what colleges must be collected in order to compute the average?
    c. If it is a parameter, is a standard deviation associated with it? Explain.

5.13   Half of all Americans are involved in alcohol-related accidents in their lifetimes. Describe the population of interest. Is the one-half (50%) a parameter or a statistic?

5.2   THE SAMPLING DISTRIBUTION OF $\bar{y}$

In Section 5.1, we observed that the value of a statistic varies from sample to sample. This sampling variability is described by the sampling distribution of the statistic. In this section, we take a closer look at the sampling distribution of the sample mean $\bar{y}$. In Section 5.3, we consider the sampling distribution of the sample proportion p.

Sampling Distribution of $\bar{y}$ from a Normal Population

Consider, for the moment, a population that is normally distributed with a mean of 50 and a standard deviation of 15. The distribution curve is shown in Figure 5.3.

FIGURE 5.3   Population distribution with mean 50 and standard deviation 15

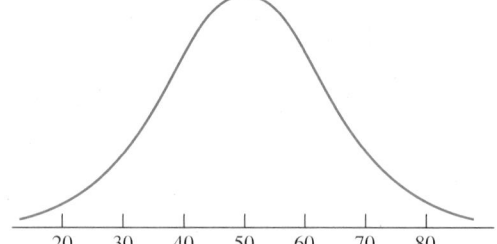

If we take a random sample from this population, most observations will be between 20 and 80 because 20 is two standard deviations below the mean of 50 and 80 is two standard deviations above the mean of 50.

Figure 5.4 is a histogram of a sample of 100 observations selected at random from the population, with the frequency distribution curve superimposed over the histogram. Most observations do indeed fall between 20 and 80, with the shape of the histogram similar to the shape of the frequency distribution of the population, as expected.

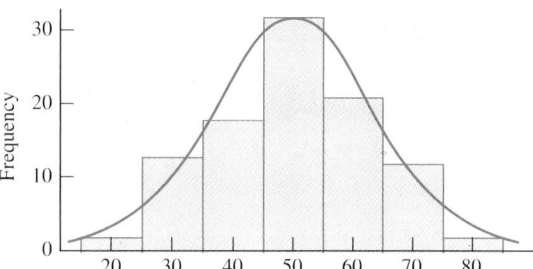

Population distribution with mean 50 and standard deviation 15 superimposed over a histogram of 100 observations

Descriptive Statistics

Variable	N	Mean	Median	TrMean	StDev	SEMean
C1	100	49.75	49.78	49.57	12.80	1.28

Variable	Min	Max	Q1	Q3
C1	23.27	81.16	39.09	57.35

From the descriptive statistics for the sample, we see that the average of this sample is 49.75, which is very close to 50, the mean of the population. Certainly another sample of 100 observations will not yield the same histogram or the same average, but we would expect to see something similar. In fact, from the following descriptive statistics for a second sample, we see that the average is 50.89, which again is close to 50:

Descriptive Statistics

Variable	N	Mean	Median	TrMean	StDev	SEMean
C2	100	50.89	52.36	50.87	15.89	1.59

Variable	Min	Max	Q1	Q3
C2	15.36	91.10	39.19	63.11

Notice that in the second sample, the standard deviation is greater than in the first sample. The greater variability in the second sample is also apparent from the values of Min and Max in both samples. Other samples will yield still different sample averages, but again the averages should be close to the population mean, μ.

If we consider the distribution of all potential values that $\bar{y}$ could assume in repeated sampling, we might ask:

What is the general shape of the distribution of potential values that $\bar{y}$ can assume?

Are the potential values of $\bar{y}$ centered about a certain quantity?

How much variability is associated with the potential values of $\bar{y}$?

These questions are answered with a description of the sampling distribution of $\bar{y}$.

Prior to sampling, the statistic can be thought of as a random variable because different samples can lead to different values of the statistic. Its sampling distribution is the probability distribution of that random variable. Knowing the probability distribution of a random variable, we are able to find the mean, standard deviation, and probabilities associated with the random variable. Consequently, if we know the sampling distribution of $\bar{y}$, we can find its mean, standard deviation, and the probabilities associated with the various values that it can assume. That is, we can answer the three questions posed above.

Returning to the problem at hand, recall that the first random sample of size 100 had an average of 49.75 and a second random sample of 100 had an average of 50.89. We have two *realizations* of the random variable $\bar{y}$. Figure 5.5 gives a histogram and a description of 200 realizations of the random variable $\bar{y}$ obtained from 200 different random samples, each of size 100. The histogram of these 200 $\bar{y}$'s gives the general appearance of the theoretical sampling distribution of $\bar{y}$.

FIGURE 5.5   Histogram of 200 $\bar{y}$s

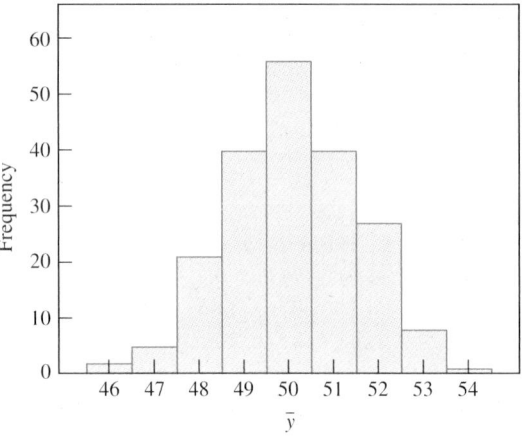

```
Descriptive Statistics

Variable        N      Mean    Median    TrMean     StDev    SEMean
ȳ             200    50.050    50.018    50.051     1.459     0.103

Variable       Min      Max       Q1        Q3
ȳ           45.839   53.815   48.993    51.147
```

Notice that the values of $\bar{y}$ tend to mound up around 50, the mean of the population. If we think about it, the 100 observations in each sample are nearly all somewhere between 20 and 80, so each average should be around 50, the mean of the population. Of course, some samples will give averages that are below or above 50, but all will cluster around 50. Also observe that the standard deviation of the population is 15, but the standard deviation of the sampling distribution of $\bar{y}$ does not tend to be so great. In fact, most realizations of $\bar{y}$ lie between 47 and 53. If 95% of the $\bar{y}$'s range from 47 to 53, as they appear to, then the standard deviation is 1.5. (Working backward, we know that if the standard deviation is 1.5, then *two* standard deviations on each side of 50 will create a range from 47 to 53.) We show later that the standard deviation of the sampling distribution of $\bar{y}$ is given by $\sigma/\sqrt{n} = 15/\sqrt{100} = 1.5$.

From the descriptive statistics in Figure 5.5, we see that the 200 simulated values have a mean of 50.05 (very close to 50) and a standard deviation (which is an estimate of the standard deviation of $\bar{y}$) of 1.459, which is close to 1.5. This suggests that approximately 95% of the sample averages are between 47 and 53. Furthermore, the smallest realization of $\bar{y}$ is 45.839 and the largest is 53.815.

Finally, the normal probability plot of the 200 simulated values in Figure 5.6 indicates that the sampling distribution is normally distributed.

FIGURE 5.6

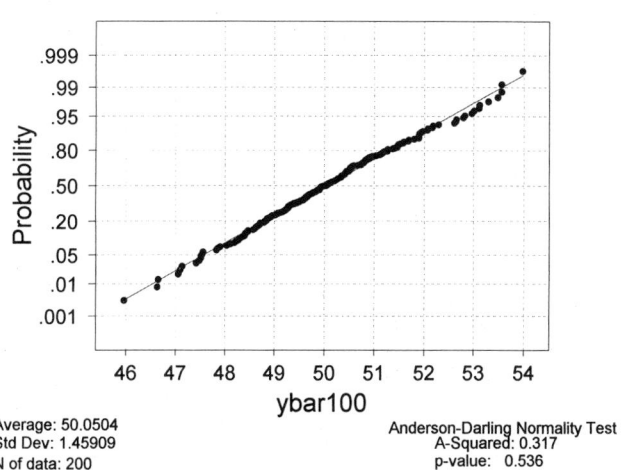

Normal Probability Plot

Average: 50.0504
Std Dev: 1.45909
N of data: 200

Anderson-Darling Normality Test
A-Squared: 0.317
p-value: 0.536

We are now in a position to answer the three questions about the sampling distribution of $\bar{y}$ when sampling is from a normally distributed population with a mean of 50 and a standard deviation of 15:

1. The general shape is bell-shaped; in fact, it is normally distributed.
2. The distribution of potential values of $\bar{y}$ is centered at 50, the mean of the population.
3. The standard deviation is

$$\sigma/\sqrt{n} = 15/\sqrt{100} = 1.5,$$

where σ is the standard deviation of the population.

Caution: We are referring to two different standard deviations in this discussion. First, we have the standard deviation of the population, σ, and second, the standard deviation of the sampling distribution of $\bar{y}$, which is given by $\sigma/\sqrt{n}$.

The preceding observations are based on a specific example. In particular, the population is normally distributed and centered at 50, with a standard deviation of 15 and a sample size of 100. However, the results can be generalized for any sample size and arbitrary means and standard deviations.

Sampling Distribution of $\bar{y}$ When Sampling from a Normally Distributed Population

Let $\bar{y}$ be the mean of a sample of size n from a normally distributed population that has mean μ and standard deviation σ. For all sample sizes n, the sampling distribution of $\bar{y}$:

1. Is exactly normally distributed
2. Is centered at μ, the mean of the population
3. Has a standard deviation of $\sigma/\sqrt{n}$, where σ is the standard deviation of the population

---

Sampling Distribution of $\bar{y}$ from a Nonnormal Population

We next ask: What is the sampling distribution of $\bar{y}$ when we sample from a population that is not normally distributed? Figure 5.7 shows the distribution of a population that is skewed right (notably not normally distributed) and has mean $\mu = 5$ and standard deviation $\sigma = \sqrt{10}$.

We will examine the sampling distribution of $\bar{y}$ when samples of size $n = 100$ are randomly selected from this population. As before, we randomly generate 200 realizations of $\bar{y}$ to simulate its sampling distribution. Figure 5.8 is a histogram of the 200 values of $\bar{y}$.

FIGURE 5.7

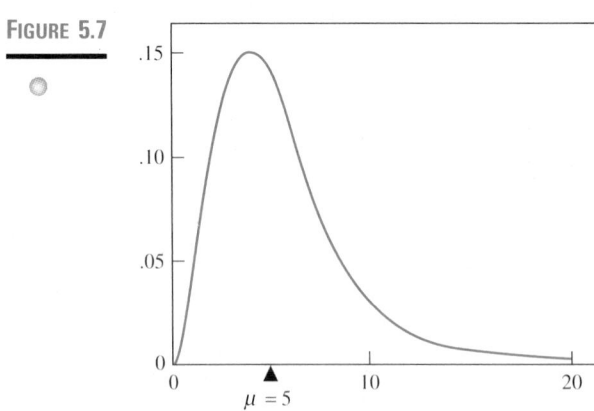

FIGURE 5.8

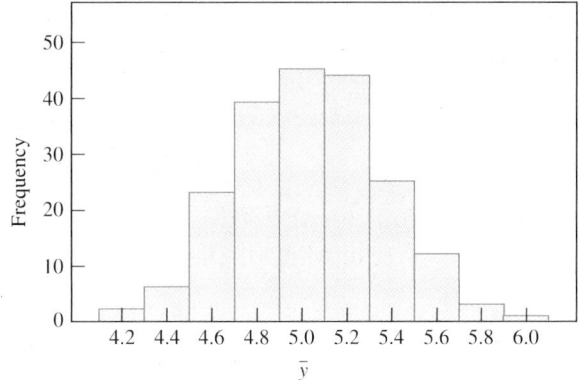

```
Descriptive Statistics

Variable          N      Mean    Median    TrMean     StDev    SEMean
ȳ               200    5.0336    5.0311    5.0313    0.3158    0.0223

Variable        Min       Max        Q1        Q3
ȳ            4.2494    6.0166    4.8149    5.2040
```

Although the population is skewed right, the simulated sampling distribution is almost symmetric, is centered at 5, and is not nearly so spread out as the population. The population assumes values from 0 to 20, whereas most of the values of $\bar{y}$ are somewhere between 4.4 and 5.6. Recall that the standard deviation of the population is $\sigma = \sqrt{10} \approx 3.16$. The standard deviation of the simulated values is .3158, which is almost exactly $\sigma/\sqrt{n} = 3.16/\sqrt{100} = .316$. It appears that the sampling distribution of $\bar{y}$ is close to being normally distributed with a mean the same as the population mean and a standard deviation of $\sigma/\sqrt{n}$.

The Central Limit Theorem

What we have observed in this simulation is a result of one of the most important theorems in statistics: the Central Limit Theorem. What was stated previously about the sampling distribution of $\bar{y}$ when sampling is from normally distributed populations generalizes to almost any population. The only difference is that the sampling distribution is not exactly normal but is only approximately normally distributed and the sample size must be rather large. The larger the sample size, the more closely the distribution is approximated by a normal distribution.

Central Limit Theorem

Let $\bar{y}$ be the mean of a sample of size n from a population with an unknown distribution. When n is relatively large, the sampling distribution of $\bar{y}$ is approximately normally distributed. The approximation becomes better as the sample size increases.

Generally speaking, the more the population deviates from a normal distribution, the larger the sample size must be to conclude that the sampling distribution of $\bar{y}$ is approximately normally distributed. In most cases, however, a sample size of 30 or more is adequate.

Sampling Distribution of $\bar{y}$ When Sampling from a General Population Distribution

Let $\bar{y}$ be the mean of a sample of size n from a population that has mean μ and standard deviation σ. When the sample size, n, is sufficiently large,[*] the sampling distribution of $\bar{y}$:

1. Is approximately normally distributed
2. Is centered at μ, the mean of the population
3. Has a standard deviation of $\sigma/\sqrt{n}$, where σ is the standard deviation of the population

[*] In most cases, a sample size of 30 or more is sufficient.

The Effect of Population Shape and Sample Size

To see how the shape of the sampling distribution is affected by the shape of the population and the sample size, we will simulate the sampling distribution of $\bar{y}$ based on three different sample sizes (5, 10, and 30) from three different distributions. Each simulation is based on 200 realizations of $\bar{y}$, as before, computed from 200 different random samples.

Figure 5.9(a) gives the three different population distributions. The first is the normal distribution with mean 50 and standard deviation 15 studied previously. Second is the symmetric uniform distribution in which all values are equally probable. The last is a highly skewed distribution called the lognormal. Figure 5.9(b) shows histograms of the 200 realizations of $\bar{y}$ when the sample sizes are 5. For the normal and uniform distributions, both of which are symmetric, we see that the sampling distribution is also symmetric but has less variability. In the case of the skewed lognormal distribution, the sampling distribution is less skewed and less variable. Remember that the standard deviation of $\bar{y}$ is $\sigma/\sqrt{n}$, which gets smaller as n increases because n is in the denominator. In Figure 5.9(c), we see the sampling distributions when the sample sizes are 10. Again, when the distribution is symmetric, the sampling distribution is symmetric, and even in the nonsymmetric lognormal case, the sampling distribution is becoming more symmetric as the sample size increases. In all three cases, the variability of the sampling distribution is decreasing as the sample size is increasing. In Figure 5.9(d), we see that when the sample size is 30, the sampling distribution is more bell shaped with a smaller standard deviation. With the aid of boxplots in Figure 5.9(e), we see how the sampling distributions become more symmetric with smaller standard deviations as the sample size increases from 5 to 10 to 30. The normal probability plots for the sampling distributions when $n = 30$ in Figure 5.9(f) show that the distributions are approximately normally distributed in the normal and uniform cases. Because of the severe skewness of the lognormal distribution, however, it may take a sample size as large as $n = 100$

to remove the skewness and get a reasonable normal approximation of the sampling distribution. As guaranteed by the Central Limit Theorem, however, there is a sample size n for which each of the three sampling distributions can be reasonably approximated with a normal curve centered at μ with a standard deviation of $\sigma/\sqrt{n}$.

Again, we point out that in the first case, when the population is normally distributed, the sampling distribution of $\bar{y}$ is *exactly* normally distributed for *all* sample sizes. It is only approximately normal when sampling is from a nonnormal population. The approximation becomes better as the sample size becomes larger. The more the population deviates from normality, the larger the sample size must be in order to get a reasonable approximation. In most cases, a sample size of 30 or more is sufficient.

FIGURE 5.9

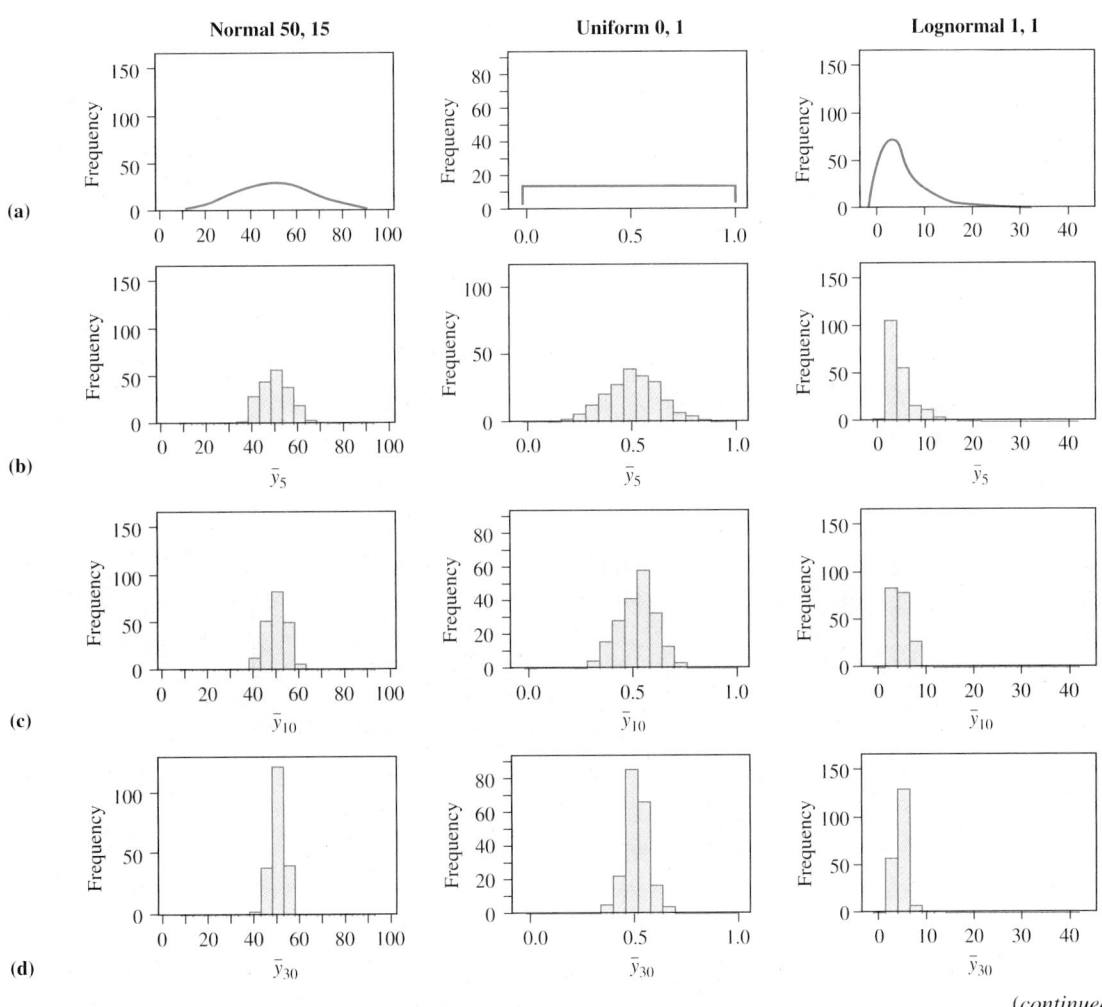

(*continued*)

FIGURE 5.9
(continued)

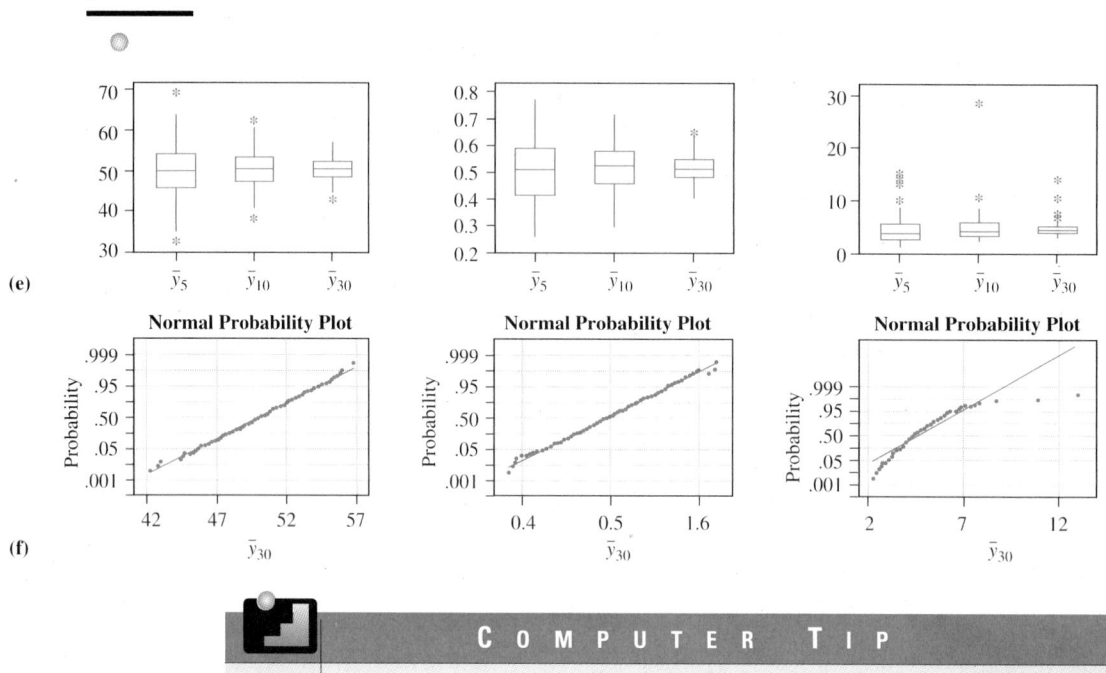

(e)

(f)

COMPUTER TIP

The above simulations were obtained by using the Random Data command in Minitab. Under the **Calc** menu, select **Random Data** and then choose the distribution. When the Distribution window appears, fill in the **Generate** box with the desired number of rows of data. It is possible to store random data in several columns at the same time by specifying the different columns in the **Store in column(s)** window. To generate the data, click **OK**.

For these simulations it is best to consider the rows as the different random samples and then use the row averages as the different realizations of $\bar{y}$. For example, to generate 200 random samples each of size 30, we generate 200 rows of data and store in C1–C30. To compute the 200 realizations of $\bar{y}$, under the **Calc** menu, select **Row Statistics**. In the Row Statistics window, click on **Mean** and then specify the **Input Variables** as C1–C30. **Store** the **results** in a column—say, C100—and click **OK**. We are then able to construct a histogram or a boxplot of the 200 different values of $\bar{y}$ in C100 to get a simulation of the sampling distribution of $\bar{y}$ when the sample size is 30.

Applications of the Central Limit Theorem

EXAMPLE 5.2   The federal program, Aid to Families with Dependent Children (AFDC), made an average monthly payment of $546 to families in New York in 1993 (U.S. Department of Health and Human Services). A social worker in upstate New York would like to know something about the monthly payments to welfare recipients in her region of the state. The average monthly payment to all welfare recipients in her region is the unknown population parameter, μ, which she wants to estimate. She decides to take a random sample of 100 welfare recipients from her region, calculate the sample mean $\bar{y}$, and use it as an estimate of μ. The sample average turns out to be $528, so she estimates the mean payment to all welfare recipients to be $528. If the standard deviation of all welfare recipients is $200—that is, $\sigma = 200$—then how close will her estimate be to the true mean payment to all welfare recipients in her region?

SOLUTION

Because $\sigma = 200$ and $n = 100$, the standard deviation of $\bar{y}$ is

$$\frac{\sigma}{\sqrt{n}} = \frac{200}{\sqrt{100}} = 20$$

Thus, we can be reasonably assured that the observed $\bar{y}$ will be within $40 [two standard deviations, or 2(20)] of the population mean. That is, the mean monthly payment for all welfare recipients in her region is most likely between $488 and $568. Notice that the average monthly payment of $546 paid to welfare recipients in all of New York state is within the interval of values.

EXAMPLE 5.3

Suppose in Example 5.2 that the social worker had a random sample of size 400 instead of 100. How close then would her estimate be to the true mean income level of all welfare recipients in her region?

SOLUTION

With a sample size of 400, the standard deviation of $\bar{y}$ is

$$\frac{\sigma}{\sqrt{n}} = \frac{200}{\sqrt{400}} = 10$$

Now 95% of the $\bar{y}$'s, and in particular her estimate, are within $20 (instead of $40) of the population mean. We see that by increasing the sample size fourfold, the standard deviation is halved. The accuracy of the estimate improves as the sample size increases.

Knowing the sampling distribution of the sample mean, $\bar{y}$, we are able to evaluate how close a certain $\bar{y}$ is to μ once it is calculated from the sample.

EXAMPLE 5.4

Census data from the National Center for Education Statistics (U.S. Department of Education, 1993) indicate that the distribution of annual income for school teachers in the United States has a mean of $34,100 and a standard deviation of $4,000. This means that the majority of school teachers earn between $26,100 and $42,100 (two standard deviations below and above the mean). Now suppose we randomly select 64 teachers from a certain state, and we find that their average income is $32,600. Can anything be said about the incomes of teachers in that state?

SOLUTION

Suppose the salaries in this state are comparable to the salaries across the United States. Then, from the Central Limit Theorem, the sampling distribution of $\bar{y}$ (obtained from samples of size 64 from this state) is approximately normally distributed with

$$\text{mean} = 34,100 \qquad \text{standard deviation} = \frac{4,000}{\sqrt{64}} = 500$$

Thus, 95% of all potential $\bar{y}$'s should fall between 33,100 and 35,100 (within two standard deviations of 34,100). We observed $\bar{y} = 32,600$, a full three standard deviations below the mean. This is unusual if the mean salary in the state is $34,100. We are led to believe that the mean of the distribution of teacher's salaries in this state is not $34,100 but is somewhat below the national average.

Probability problems involving $\bar{y}$ can be worked if we know its sampling distribution, as indicated in the next example.

EXAMPLE 5.5

The average length of stay in a certain Alcoholics Anonymous clinic is 17 days, and the standard deviation of the length of stay is 3 days. A random sample of 36 patients is chosen.

a. Find the probability that the average stay is more than 18.5 days.

b. What is the probability that the average stay is between 16 and 19 days?

SOLUTION

We have from the Central Limit Theorem that $\bar{y}$ is approximately normally distributed with

$$\text{mean} = 17 \qquad \text{standard deviation} = \frac{3}{\sqrt{36}} = .5$$

a. To find the probability that $\bar{y} > 18.5$, we find the z-score associated with 18.5 as follows:

$$z = \frac{18.5 - 17.0}{.5} = 3.0$$

From the standard normal probability table, we find that .9987 of the area lies to the left of a point that is three standard deviations above the mean. Using the complement rule, we have that the probability of observing $\bar{y} > 18.5$ is

$$1.0000 - .9987 = .0013$$

The probability that the average stay is longer than 18.5 days is .0013, extremely small.

b. To find the probability that $\bar{y}$ lies somewhere between 16 and 19, we must work two problems: First find the area below 16 and then find the area below 19. Finally, we subtract the first from the second. The z-score associated with 16 is

$$z = \frac{16.0 - 17.0}{.5} = -2.0$$

From the normal probability table, we find an area of .0228 associated with a z-score of -2. The z-score associated with 19 is

$$z = \frac{19.0 - 17.0}{.5} = 4.0$$

From the normal probability table, we find an area of .99997 associated with a z-score of 4.0. Consequently, the desired probability is

$$.99997 - .0228 = .97717$$

which is very large. It is very likely that the average length of stay is between 16 and 19 days. In fact, with approximately 95% probability, the average stay is somewhere between 16 and 18 days (two standard deviations on either side of 17).

The sampling distribution also allows us to determine which of the potential values of the statistic are reasonable and which are unlikely. Thus, when we obtain a value for a particular statistic, knowing the sampling distribution of that statistic allows us to determine immediately whether that value is in agreement with our ideas about the population.

EXAMPLE 5.6

Returning to the welfare problem in Example 5.2, let us suppose a conjecture is made that the mean monthly payment to the welfare recipients in the social worker's region is significantly less than $546, the mean amount paid to all New York welfare recipients. What is your opinion of this conjecture?

SOLUTION

The mean amount paid to all New York recipients is $546. Based on the sample mean, $\bar{y}$ = $528, of the 100 recipients randomly selected in her region, we determined in Example 5.2 that the mean payment to all recipients in the social worker's region is between $488 and $568. Because the mean amount paid in her region could be as high as $568 and the mean amount paid to all New York recipients is $546, we cannot support the conjecture that the recipients in her region are paid less.

The test of hypothesis procedures in Chapter 8 will give a formal treatment to problems of this type.

EXERCISES 5.2

5.14   A population has mean 500 and standard deviation 100. A sample of size 200 is randomly selected from the population. Describe the sampling distribution of the sample mean $\bar{y}$.

5.15   A population has mean μ and we wish to estimate it. We take a sample, calculate the sample mean $\bar{y}$, and use it to estimate μ. What is the accuracy of that estimate; that is, in general, how close will $\bar{y}$ be to μ?

5.16   A random sample of size n is selected from a population with a mean of 25 and a standard deviation of 8. For each value of n, give the mean and standard deviation of the sampling distribution of $\bar{y}$:

    a. $n = 10$ b. $n = 16$ c. $n = 30$ d. $n = 100$

In which of these can we be assured that the sampling distribution will be adequately approximated by a normal distribution?

5.17   The California Achievement Test was given to 84,361 third-grade students in North Carolina in 1992. The mean score for the total battery was 690.8 (North Carolina Department of Public Instruction, *Report of Student Performance, 1986–1992*, July 1992). A random sample of 200 third-grade students is chosen.

    a. Would you expect to see a $\bar{y}$ in excess of 710 if $\sigma = 100$?
    b. Would you expect to see a $\bar{y}$ in excess of 710 if $\sigma = 200$?
    c. With $\sigma = 100$, what is the largest $\bar{y}$ you would expect to observe from a random sample of 200 students?

5.18   The average number of days spent in a North Carolina hospital for a coronary bypass in 1992 was 9 days and the standard deviation was 4 days (North Carolina Medical Database Commission, *Consumer's Guide to Hospitalization Charges in North Carolina Hospitals*, August 1994). What is the probability that a random sample of 30 patients will have an average stay longer than 9.5 days?

5.19   The average length of a field goal in the National Football League is 38.2 yards, and the standard deviation is 6.4 yards. Suppose a typical kicker kicks 41 times in one season. What is the probability that his kicks average less than 37 yards?

5.20   It was reported in an Atlanta newspaper that for the 1994 Christmas holidays, the average price per gallon of self-service gasoline in the southeastern states was $1.18 and the standard deviation was 6 cents. A check of 36 randomly selected stations revealed an average price of $1.21. Did this average exceed $1.18 purely by chance, or is there statistical evidence that the $1.18 is low?

5.21   Annual incomes for intracity social workers are assumed to be normally distributed with a mean of \$28,500 and a standard deviation of \$2,400.
    a.   What percent of the workers receive an income greater than \$30,000?
    b.   What is the probability that 36 workers will have an average income in excess of \$30,000?

5.22   The national norm of a science test for tenth-graders has a mean of 75 and a standard deviation of 20. A sample of 100 tenth-grade students from the New York City public school system had an average of 72 on the test. Obtain bounds that would include the mean score of almost all New York City tenth-graders.

5.23   Suppose the starting salary for new Ph.D. psychologists is normally distributed with a mean of \$48,000 and a standard deviation of \$4,000.
    a.   What percent have starting salaries greater than \$53,000?
    b.   Suppose a certain university has 16 Ph.D. graduates and their average starting salary is \$53,000. Is this unusual?

5.24   The Bayley Scale of Infant Development, which has a mean of 100 and a standard deviation of 16, was given to a group of 34 infants who were exposed to a new educational procedure developed at a university. Their average score was 104.6. Is there evidence that this group of children scored unusually high on the test?

5.25   An insurance company's records show that the mean payout for all automobile claims is \$1,800 and the standard deviation is \$400. Suppose 90 claims are filed in 1 week. What is the probability they average more than \$1,900?

5.26   The average length of a felony sentence imposed by state courts for trafficking in drugs in 1992 was 72 months (Bureau of Justice Statistics, *Felony Sentences in State Courts, 1992*, Bulletin NCJ-151167, January 1995). If the standard deviation is 10 months, is it unusual for a random sample of 50 subjects accused of trafficking in drugs to have an average sentence longer than 75 months? To answer this question, find the probability that a sample of size 50 will have a sample mean above 75.

5.27   It takes General Motors 31 hours of labor to build a car (*Term*, Vol. 4, No. 1, January 1995). Suppose that the duration of time is normally distributed with $\sigma = 2$ hours.
    a.   What is the probability that a single car, selected at random, will take between 28 and 34 hours?
    b.   What is the probability that the mean of the times for a sample of 25 cars will be between 28 and 34 hours?

5.28   A standardized social studies test was given to 83,621 sixth-grade students in North Carolina in 1992. The mean was 39.6 (North Carolina Department of Public Instruction, *Report of Student Performance, 1986–1992*, July 1992).
    a.   If we assume that the distribution is normal with a standard deviation of 10, what percent of the students scored somewhere between 35 and 45?
    b.   What is the probability that 36 students will have an average score between 35 and 45?

5.29   A standardized science test has a national mean of 250 and a standard deviation of 50. The test is to be given to a group of prospective science majors. What is the standard deviation of the estimate, $\bar{y}$, calculated from a sample of size 64? From a sample of size 400?

5.30   An Asheville, North Carolina, newspaper reported that a man waited 20 minutes for an ambulance while coughing up blood. During the next 3 months, the ambulance response times were recorded on 236 calls. The average response time was 12.63 minutes. If the standard deviation of all response times is 5 minutes, give an interval that you are reasonably sure (95% sure) contains the population ambulance response time for Asheville.

5.31   The incubation temperature to hatch ostrich eggs for a Type SR-50 incubator is set at 99° F and is allowed to vary with a standard deviation of 2° F. Each hour the temperature is recorded at 50 randomly selected times. If the average of the 50 measurements is less than 98.5 or greater than 99.5, an alarm goes off. What is the probability of a false alarm? That is, what is the probability that the average is not between 98.5 and 99.5 when the actual temperature is 99°F?

5.32   Simulate 200 random samples of size 16 from a normally distributed population with a mean of 30 and a standard deviation of 8.

    a. Determine the sample mean of each of the 200 samples.

    b. Construct a histogram of the 200 sample means. Does it appear to be normally distributed?

    c. Construct a normal probability plot of the 200 sample means. Is normality plausible?

    d. Compute descriptive statistics for the 200 sample means.

    e. What is the mean of the 200 sample means? Is it close to the population mean? Should it be?

    f. What is the standard deviation of the 200 sample means? Is it close to the population standard deviation? Should it be?

5.33   Repeat Exercise 5.32, but this time each random sample should be of size 64. After answering parts **a–e**, compare these results to those found in Exercise 5.32. What general observations can you make?

5.34   The exponential distribution is a skewed right distribution that is often used to model lifetimes or waiting times. It has a single parameter, which is the reciprocal of the mean. That is, if the parameter is $\frac{1}{2}$, then the mean is 2. Simulate 200 random samples of size 16 from an exponential distribution with a mean of 5.

    a. Determine the sample mean of each of the 200 samples.

    b. Construct a histogram of the 200 sample means. Does it appear to be normally distributed?

    c. Construct a normal probability plot of the 200 sample means. Is normality plausible?

    d. Compute descriptive statistics for the 200 sample means.

    e. What is the mean of the 200 sample means? Is it close to the population mean? Should it be?

5.35   Repeat Exercise 5.34, but this time each random sample should be of size 64. After answering parts **a-e**, compare these results to those found in Exercise 5.34. What general observations can you make?

5.36   The sampling distribution of $\bar{y}$ depends on the sample size and on the distribution of the population (as the sample size increases, the dependence on the population distribution becomes less and less). The simulations of the sampling distribution of $\bar{y}$ conducted in this section involved a large number of random samples taken from a specific population distribution, such as a normal or uniform distribution. In some cases, however, we may have a random sample of size n from a population with an unknown distribution, and yet we still wish to simulate the sampling distribution of $\bar{y}$. This can be accomplished by a method called a *bootstrap*. We simply treat the random sample of size n as the population and resample the sample, *with replacement*, a large number of times. A histogram of the $\bar{y}$'s calculated from each of the many samples is an estimate of the sampling distribution of $\bar{y}$, and the standard deviation of the simulated $\bar{y}$'s gives an approximation of the standard deviation of $\bar{y}$.

    Worksheet: **ddmc-dat.mtw** contains 206 measurements of the lead content of soil samples taken by the Environmental Protection Agency in the vicinity of an old smelter located in Dallas, Texas (data provided by Donald E. Myers). The bootstrap method was used to take 500 samples of size 206, which have been stored in rows 1–500 and columns C2–C207 of the worksheet: **Bootstrp.mtw**. Column C1 contains the original lead measurements.

    a. Using the row statistic mean, compute the sample mean of the 206 observations for each of the 500 samples. Store the means in C210.

    b. Construct histograms of the original data in C1 and the 500 sample means in C210.

c. Compute descriptive statistics of the original data and the 500 sample means and make comparisons.

d. The standard deviation of the simulated means gives an approximation of the standard deviation of $\bar{y}$ when the sample size is 206. What is the standard deviation of the simulated means?

5.3   SAMPLING DISTRIBUTION OF THE SAMPLE PROPORTION p

Recall from Chapter 4 that a Bernoulli population is one in which each element is either a success or a failure. We are generally interested in the proportion of successes, denoted by π. If we wish to estimate π from a sample, it seems reasonable to calculate the sample proportion, p. In order to evaluate how well p estimates π, we need to study its sampling distribution.

In Section 5.1, we saw that a sample of 1,000 registered voters revealed that 380 favored the mayor seeking reelection. Thus, we obtained a sample proportion of

$$p = \frac{380}{1,000} = .38 = 38\%$$

We also observed that a second sample of size 1,000 most likely would not result in exactly 380 positive responses. Just as $\bar{y}$ varies from sample to sample, so does the sample proportion p. This variability is described by its sampling distribution. As with $\bar{y}$, prior to sampling, p is a random variable and thus has a probability distribution. We call it a *sampling distribution* because it is arrived at by repeated sampling.

The three questions we posed with regard to the sampling distribution of $\bar{y}$ may also be asked of the sampling distribution of p:

1. What is the general shape of the distribution of potential values that p can assume?
2. Are the potential values of p centered about a certain quantity?
3. How much variability is associated with the potential values of p?

Central Limit Theorem   The following version of the Central Limit Theorem applies to the sample proportion:

Central Limit Theorem Applied to the Sample Proportion p

If the sample size, n, is sufficiently large, then the sampling distribution of p:

1. Is approximately normally distributed
2. Is centered at π, the true proportion of successes in the Bernoulli population
3. Has a standard deviation of

$$\sqrt{\pi(1 - \pi)/n}$$

The next example illustrates the Central Limit Theorem as it applies to p.

EXAMPLE 5.7

Suppose the proportion of voters who favor the mayor seeking reelection is $\pi = .4$. With a computer, simulate a random sample of 1,000 voters and calculate p, the proportion of the 1,000 who favor the mayor seeking reelection. Repeat the process 200 times; that is, simulate 200 random samples each of size 1,000.

SOLUTION

A histogram of the 200 realizations of p should resemble the sampling distribution of p when $n = 1,000$. Figure 5.10 shows the results of the simulation.

FIGURE 5.10

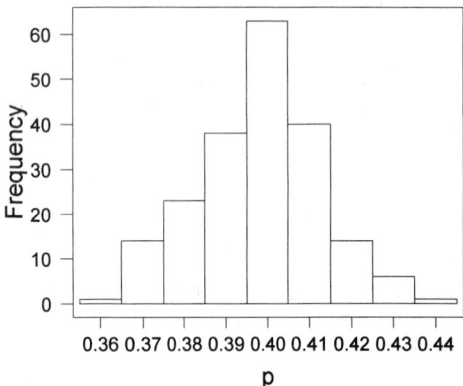

Observe that the histogram appears to be approximately normally distributed (a normal probability plot would verify normality), is centered around .4, the true proportion in the population, and exhibits little variability. In fact, all the potential values of p are somewhere between .36 and .44, which is very close to π. Knowing that $\pi = .4$, we can calculate the standard deviation of the sampling distribution of p to be

$$\sqrt{\frac{\pi(1 - \pi)}{n}} = \sqrt{\frac{(.4)(.6)}{1,000}} = .0155$$

Two standard deviations on either side of .4 give the interval (.37, .43). We see from the simulation that at least 95% of the potential values of p are indeed between .37 and .43.

To perform the simulation in Example 5.7, we can use the Random Data > Bernoulli command in Minitab. It would require, however, a worksheet that has 200 rows and 1,000 columns (each row represents one of the 200 samples that has a size of 1,000). An alternative is to use the following relationship between the proportion of successes and the number of successes:

*Relationship
Between the
Proportion of
Successes and
a Binomial
Random Variable*

If p represents the proportion of successes in a random sample of size n from a Bernoulli population with parameter π, then the number of successes

$$x = np$$

is a binomial random variable with parameters n and π.

This follows from the fact that if

$$p = \frac{\text{number of successes}}{n}$$

then we can multiply both sides by n to get

$$np = \text{number of successes}$$

And recall that the number of successes in a random sample of size n from a Bernoulli population with parameter π was defined in Chapter 4 to be a binomial random variable with parameters n and π. So instead of simulating 1,000 Bernoulli values and calculating the proportion of successes as described previously, we can simulate one binomial variable with $n = 1,000$ and then divide by n to get a realization of p.

C O M P U T E R   T I P

To simulate a binomial random variable, under the **Calc** menu select **Random Data** followed by **Binomial**. Once the Binomial Distribution window appears, specify the number of rows of data in the **Generate** box and specify a column for the data in the **Store in column(s)** box. Fill in the **Number of trials** (n) and the **Probability of success** (π) and click on **OK**.

The Effect of Sample Size

EXAMPLE 5.8

To illustrate how the sampling distribution of p is affected by the sample size, simulate 200 realizations of p when $\pi = .2$ and $n = 50, 100, 200, 400,$ and $1,000$. Construct a histogram for each of the sample sizes and then compare them with boxplots.

SOLUTION

Figure 5.11 on page 328 shows the simulated results.

FIGURE 5.11

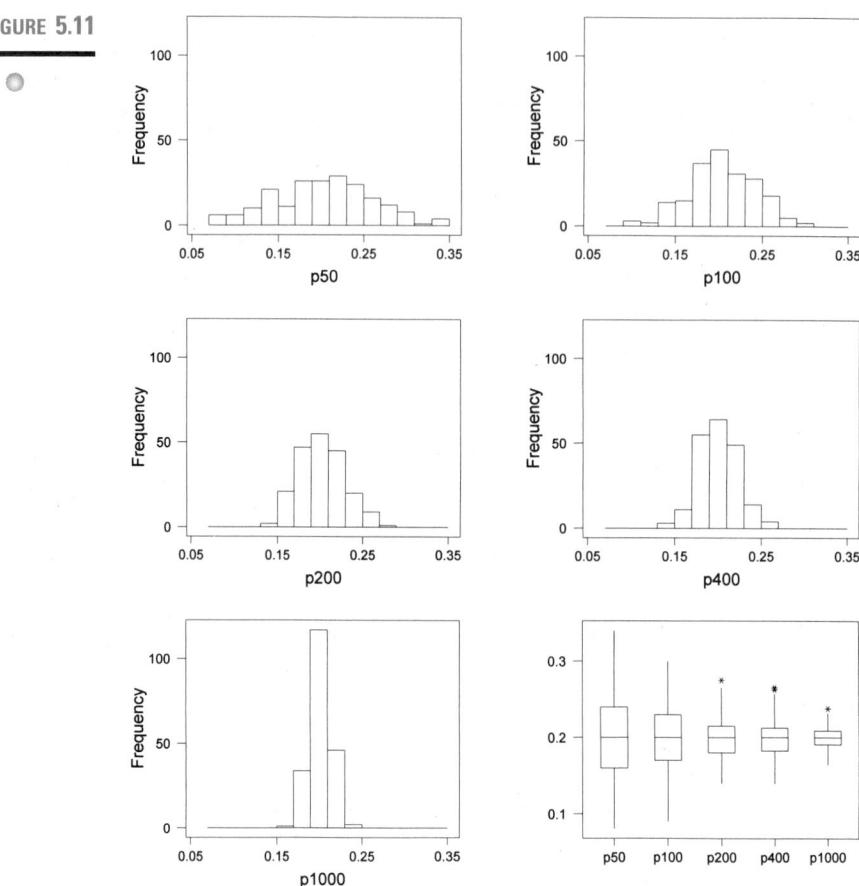

Notice that in all five simulations the sampling distribution of p is centered at $\pi = .2$. As n increases from 50 to 1,000, the sampling distribution becomes more bell shaped and the standard deviation decreases. The boxplots clearly illustrate that the sampling distributions become less variable as n increases and collapses down on $\pi = .2$.

The reason that the standard deviation decreases as the sample size increases, as in the case of the standard deviation of $\bar{y}$, is that n appears in the denominator. That is, the standard deviation of p is

$$\sqrt{\frac{\pi(1 - \pi)}{n}}$$

and clearly for fixed values of π, it will decrease as n increases. In fact, if n is increased fourfold, the standard deviation will be halved.

The Effect of π on the Standard Deviation

Next let us fix n and see how different values of π affect the standard deviation. Consider the following table of values of the standard deviation for various values of π when $n = 1,600$:

π	$\sqrt{\dfrac{\pi(1-\pi)}{1,600}}$
.2	.01
.3	.0115
.4	.0122
.5	.0125
.6	.0122
.7	.0115
.8	.01

We see that the standard deviation increases until we get to $\pi = .5$ and then it decreases. The *maximum* standard deviation occurs when $\pi = .5$. Notice that the sample size $n = 1,600$ was arbitrary; it is easy to see that this scheme is true for any sample size. Also, note the symmetry in the table; the standard deviation is the same for $\pi = .2$ and .8, for $\pi = .3$ and .7, and the same for $\pi = .4$ and .6.

EXAMPLE 5.9

A Statistical Brief (SB/95-7) issued in May 1995 by the Bureau of the Census reported the percent of housing units in major metropolitan areas that are heated with gas. The percents ranged from a low of 6.6% in Fort Lauderdale to a high of 91.9% in Salt Lake City. Figures for Tucson, Arizona, were not given. Suppose a random sample of $n = 400$ housing units in Tucson gives a sample proportion of 37% being heated with gas. If we let π denote the Bernoulli proportion of *all* housing units in Tucson that are heated with gas, how close do you think the 37% is to π?

SOLUTION

Because a random sample of size 400 is large, the Central Limit Theorem says that the sampling distribution of p tends to mound up around π. Moreover, the standard deviation is maximum when $\pi = .5$, so we know that the standard deviation is no more than

$$\sqrt{\frac{(.5)(.5)}{400}} = .025$$

Thus, we feel confident that 95% of all possible values for p will lie within $2(.025) = .05$ of the true value of π. So we believe that the value of $p = .37$ should be within $\pm.05$ of π. That is, we estimate that π is somewhere within the interval $(.32, .42)$.

EXAMPLE 5.10

Recent studies have shown that 40% of the adult population in the United States believe that abortion is permissible under any circumstances. In a random sample of 1,000 adults, 450 said abortion is permissible under any circumstance. Does this cast doubt on the initial claim?

SOLUTION

If the true proportion of adults who believe that abortion is permissible is 40%, then the Central Limit Theorem says that the sampling distribution of the random variable p is approximately normal with mean .4 and standard deviation $\sqrt{(.4)(.6)/1,000} = .0155$. So we would expect approximately 95% of the potential p's to be between

$$.4 - 2(.0155) \quad \text{and} \quad .4 + 2(.0155)$$

which reduces to the interval

$$(.369, .431)$$

However, the p we obtained from our sample is

$$p = \frac{450}{1,000} = .45$$

which clearly is not in the interval; it is more than two standard deviations above .4. Consequently, we have sufficient evidence to question the validity of the claim that $\pi = .40$.

Because of the relationship between the number of successes, x, the proportion of successes, p (recall $x = np$), and the Central Limit Theorem, we have the next rule.

Normal Approximation of the Binomial Distribution

When n becomes large, the binomial distribution can be reasonably approximated with a normal distribution that has a mean of $n\pi$ and a standard deviation of $\sqrt{n\pi(1 - \pi)}$.

For a good approximation, n should be large enough that both $n\pi > 5$ and $n(1 - \pi) > 5$.

EXAMPLE 5.11

It has been reported that 46% of all U.S. homes have more than one television set. Out of a random sample of 500 homes, what is the probability that less than half have more than one television set?

SOLUTION

If we let x be the number of homes that have more than one television set, then the probability that less than half of the homes in a sample of 500 have more than one television is written as

$$P(x < 250)$$

Because the distribution of x is approximately normal with a mean of $n\pi$ and a standard deviation of $\sqrt{n\pi(1 - \pi)}$, we will determine the probability by finding the z-score corresponding to 250 and then looking up the associated probability in the standard normal probability table. We have

$$z = \frac{x - n\pi}{\sqrt{n\pi(1 - \pi)}} = \frac{250 - 500(.46)}{\sqrt{500(.46)(.54)}} = 1.79$$

From Table B.2 we find the associated probability to be .9633, so

$$P(x < 250) = .9633$$

It is highly probable that fewer than half the homes in the sample have more than one television set.

An alternative solution to this problem is to consider the *proportion* of homes with more than one television set instead of the *number* of homes with more than one television set. A proportion, p, calculated from a sample of 500 homes would have a sampling distribution that is approximately normal with a mean of .46 and a standard deviation of $\sqrt{(.46)(.54)/500} = .0223$. The probability that less than half of the homes in a sample of 500 have more than one television can be written as

$$P(p < .5)$$

As before, we find the z-score, but this time the form is

$$z = \frac{p - \pi}{\sqrt{\pi(1 - \pi)/n}}$$

Substituting in values, we get the same z-score:

$$z = \frac{.5 - .46}{\sqrt{(.46)(.54)/500}} = 1.79$$

and hence the same answer. In other words, we have

$$P(p < .5) = P(x < 250) = .9633$$

This section and the previous section described the sampling distributions of two often used statistics. Several examples illustrated different applications. What we have learned will be useful as we proceed to statistical inferences about the two population parameters, μ and π.

EXERCISES 5.3

5.37 Briefly describe the characteristics of a Bernoulli population and give two examples.

5.38 A sample of 400 observations is randomly selected from a Bernoulli population with parameter $\pi = .8$. Describe the sampling distribution of the sample proportion p.

5.39 A random sample of size n is selected from a Bernoulli population with 70% successes. For each value of n, give the mean and standard deviation of the sampling distribution of p:

    a. $n = 100$ b. $n = 400$ c. $n = 1,000$ d. $n = 1,600$

5.40 A random sample of 1600 is selected from a Bernoulli population. For each value of π, find the mean and standard deviation of the sampling distribution of p:

    a. $\pi = .1$ c. $\pi = .5$ e. $\pi = .9$

    b. $\pi = .3$ d. $\pi = .7$ f. $\pi = .575$

5.41 Determine the maximum standard deviation of p for the following sample sizes:

    a. 25   c. 200   e. 1,000

    b. 100   d. 500   f. 2,000

5.42 A new drug is proposed as a treatment for lung cancer. A sample of 100 patients is to be tested. What is the maximum standard deviation of the proportion of successfully treated patients? What would be the maximum standard deviation of the estimate if the sample size were 900?

5.43 Of all persons who live in poverty areas in the United States, 87% are white (Bureau of the Census, *Poverty Areas*, Statistical Brief SB/95-13, June 1995). A random sample of $n = 400$ is selected from poverty areas in the United States. Approximate the probability that the sample proportion, p, falls in these ranges:

    a. Greater than 88%

    b. Less than 85%

    c. Between .80 and .85

5.44 Only 30% of all single parents own their own homes (Bureau of the Census, *Single-Parent Families*, Statistical Brief SB/94-6, November 1994). Consider a random sample of size $n = 100$ from this Bernoulli population (they either own their home or they do not) with $\pi = .3$.

    a. Would you expect to see a sample proportion, p, in excess of .4?

    b. What is the largest p you would expect to see from the sample?

5.45 A random sample of 1,000 adults in a regional survey revealed that 23% of the residents in western states dine out on a credit card. How close is the 23% to the true proportion of westerners who dine out on a credit card?

5.46 AIDS-related deaths accounted for one-third of all deaths of state prison inmates during 1993 (Bureau of Justice Statistics, Bulletin NCJ-152765, August 1995). There were 121 deaths from all causes in state prisons in New Jersey in 1993. What is the probability that 45 or more of those deaths were AIDS-related?

5.47 The Census Bureau reported that 70% of all Americans have hospitalization coverage by a private insurance plan (Statistical Brief SB94/28, October 1994). In a large hospital there were 200 patients admitted one week. What is the probability that more than 65% have a private insurance plan?

5.48   The poverty cutoff line for a family of four was $14,335 in 1992. Fifty-three percent of all families in Tunica, Mississippi, are below that poverty line. Is it unusual that a sample of 100 families in Tunica will have more than 60 families below the poverty line?

5.49   Forty-one percent of those who took the California bar exam in 1993 passed. Consider the sample proportion, p, from a random sample of 50 who took the exam.
   a. What is the value of π?
   b. What is the standard deviation of p?
   c. If the sample size is changed to 200, what effect will it have on the standard deviation of p?
   d. From the sample of 50, what is the probability that more than half passed?
   e. From a sample of 200, what is the probability that more than half passed?
   f. Explain why there is a difference in the answers to parts **d** and **e**.

5.50   A survey of 40,000 working women by *McCall's* magazine revealed that 80% of them are married. We randomly select 1,000 of the women and calculate the proportion, p, who are married.
   a. What is the standard deviation of p?
   b. What value did you use for π?
   c. Is the 40,000 figure used in your calculations?
   d. In the random sample of 1,000, what is the probability that fewer than 750 are married?

5.51   President Reagan got 59% of the popular vote in the November 1984 election, for a landslide victory.
   a. From a random sample of 400 voters, what is the probability that more than 240 voted for Reagan?
   b. In 1992, President Clinton got 43% of the popular vote. From a random sample of 400 voters, what is the probability that more than 240 voted for Clinton?
   c. Explain why there is such a big difference in the answers to parts **a** and **b**.

5.52   Suppose 40% of the adult residents in North Dakota favor the death penalty. In a simple random sample of 100 North Dakota adult residents, what is the probability that more than 50% of them favor the death penalty?

5.53   A report from the President's office stated that more than half of the nation agrees with the current foreign policy. However, a CNN/Gallup poll of 1,008 Americans found only 44% support the President's foreign policy. Did the 44% simply happen by chance or is there statistical evidence that the report is in error?

5.54   One-third of all ex-convicts return to jail within 3 years. One state is trying a new rehabilitation system for prisoners. After 200 prisoners who had participated in the new system were released, only 48 returned within 3 years. Is there statistical evidence that the new system is working to reduce the number of repeat offenders?

5.55   According to *Racing Update*, three out of four racehorses lose money. A racing partnership had 11 of its 36 horses show a profit. Are the results of the partnership substantially better than what would normally be expected from 36 randomly selected horses?

5.4   SAMPLING DISTRIBUTIONS OF OTHER USEFUL STATISTICS

Sampling Distributions of the Median and the Midrange

Because the Central Limit Theorem says that the sampling distribution of $\bar{y}$ is centered at the population mean μ, it seems natural that we should use $\bar{y}$ to estimate μ in any given situation. There are several statistics, however, that can be used to estimate the center of a population distribution. Each has its own sampling distribution generated through repeated sampling. Some produce estimates that are consistently too large or too small, and some are more tightly clustered about the population center than others. We now compare the sampling distribution of $\overline{y}$ to the sampling distributions of two statistics—the median and the midrange—that conceivably could be used to estimate the center of a distribution.

As previously observed, the sampling distribution of a statistic depends on the shape of the population distribution. We consider three familiar cases: a normal population with mean 50 and standard deviation 15, a uniformly distributed population, and a lognormal population.

Figure 5.12 gives simulations of the sampling distributions of $\overline{y}$, the median, and the midrange for the three different population distributions. Each simulation is based on 500 realizations (500 samples randomly selected from the population) of the statistic with a sample size of $n = 100$.

FIGURE 5.12

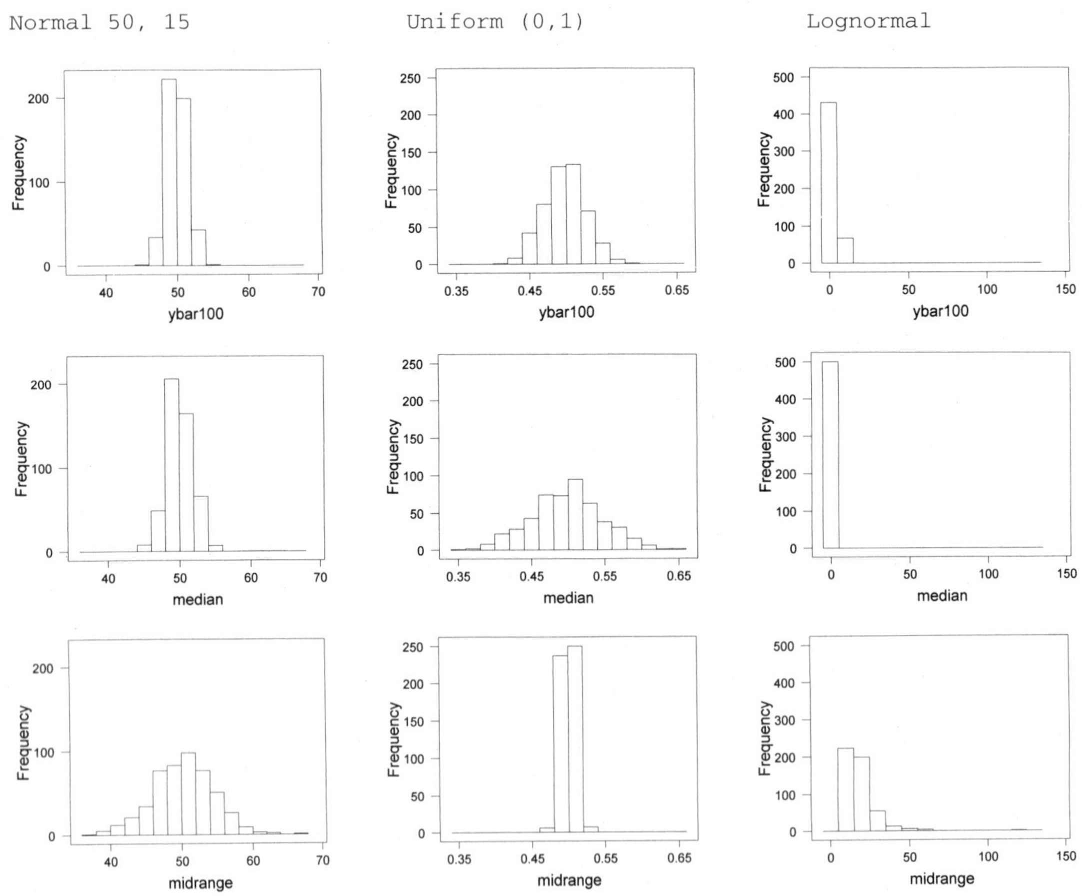

Normal 50, 15   Uniform (0,1)   Lognormal

(continued)

FIGURE 5.12
(continued)

Normal 50, 15

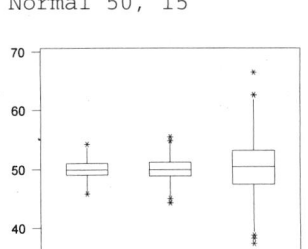

Uniform (0,1)

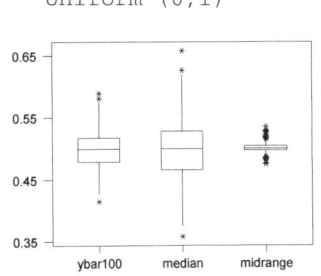

Lognormal

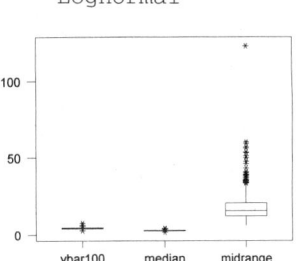

The first column in the figure gives the sampling distributions of the three statistics when the population distribution is normal. Notice that all are centered at 50, the population mean. Of the three, the sample mean, $\bar{y}$, has the smallest variation, which says that its potential values are more tightly clustered about the population mean, μ. We conclude that when we are sampling from a normally distributed population, the sample mean has values that are centered at the population mean and produces more accurate estimates.

The second column gives the sampling distributions when the population distribution is uniform. Recall that a uniform distribution is rectangular with all values equally probable. Here we see that all three sampling distributions are symmetrically distributed about the mean of the population, with the midrange having the least variability. From the boxplots, we clearly see that the values of the midrange are tightly clustered about the mean.

In the case of the highly skewed lognormal distribution, shown in the third column, the mean and median are reasonably symmetric but the midrange is severely skewed. Moreover, the center of the midrange sampling distribution is much higher than the mean of the lognormal distribution (for this particular lognormal distribution, $\mu = 4.48$). It is apparent that the midrange is a poor estimate of the center of a lognormal distribution. The scale used for the mean and median makes it difficult to compare their sampling distributions. Figure 5.13 on page 336 compares the sampling distributions of the mean and median on a more refined scale with their corresponding descriptive statistics. Both are reasonably symmetric, with the distribution of the mean centered at 4.4326 and the distribution of the median centered at 2.7333. Given that the population mean is 4.48, we can say that the possible values of the sample mean are centered near the population mean but the median tends to underestimate the population mean. On the other hand, the standard deviation of the distribution of the median is less than that of the mean. Its values are more tightly clustered but around a value below the population mean. Which one provides the best estimate of the center of the population? We will address this question in the next chapter.

FIGURE 5.13

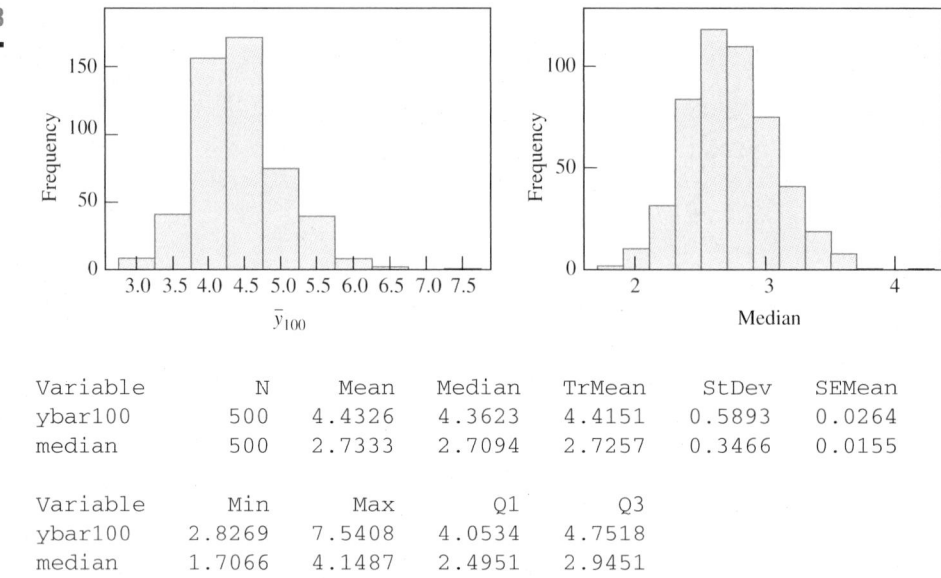

Variable	N	Mean	Median	TrMean	StDev	SEMean
ybar100	500	4.4326	4.3623	4.4151	0.5893	0.0264
median	500	2.7333	2.7094	2.7257	0.3466	0.0155

Variable	Min	Max	Q1	Q3
ybar100	2.8269	7.5408	4.0534	4.7518
median	1.7066	4.1487	2.4951	2.9451

Sampling Distribution of the Difference Between Two Statistics

In the chapters that follow, we will see the importance of being able to compare the difference between two parameters. For example, we might wish to investigate the difference in the mean grade point averages of men and women college students. Because we are unable to conduct a census, we would have to investigate the difference between two sample means: the mean GPA for a random sample of men and the mean GPA for a random sample of women. Generally, to compare two population means we need to compare the sample means that are obtained from two independent random samples taken from the two respective populations. To assess how close the difference between the two sample means is to the difference between the two population means, we need to investigate the sampling distribution of

$$\overline{y}_1 - \overline{y}_2$$

where $\overline{y}_1$ is the mean of a random sample from population 1 and $\overline{y}_2$ is the mean of an independent random sample from population 2.

Suppose population 1 has mean μ_1 and standard deviation σ_1 and population 2 has mean μ_2 and standard deviation σ_2. Furthermore, suppose that independent random samples of size n_1 and n_2 are taken, respectively, from the two populations. If n_1 and n_2 are sufficiently large ($n_1 > 30$ and $n_2 > 30$), we have from the Central Limit Theorem that the sampling distributions of $\overline{y}_1$ and $\overline{y}_2$ are both approximately normally distributed. Because the samples are assumed to be independent, it can be shown that the difference $\overline{y}_1 - \overline{y}_2$ is approximately normally distributed. Furthermore, if the mean of the sampling distribution of $\overline{y}_1$ is μ_1 and the mean of the sampling distribution of $\overline{y}_2$ is μ_2, it follows that the mean of the sampling distribution of $\overline{y}_1 - \overline{y}_2$ is $\mu_1 - \mu_2$.

The standard deviations of $\overline{y}_1$ and $\overline{y}_2$ are $\sigma_1/\sqrt{n_1}$ and $\sigma_2/\sqrt{n_2}$, respectively. Again assuming independence of the two samples, we can show that the standard deviation of

$\bar{y}_1 - \bar{y}_2$ is

$$\sqrt{\frac{\sigma_1^2}{n_1} + \frac{\sigma_2^2}{n_2}}$$

(This follows because the variance of the difference between two independent random variables is the sum of the individual variances.)

These results are summarized in the box.

Sampling Distribution of $\bar{y}_1 - \bar{y}_2$

Independent random samples of size n_1 and n_2 are selected from two populations that have means μ_1 and μ_2 and standard deviations σ_1 and σ_2, respectively. If n_1 and n_2 are each larger than 30, then the sampling distribution of the difference between the sample means, $\bar{y}_1 - \bar{y}_2$:

1. Is approximately normally distributed
2. Is centered at $\mu_1 - \mu_2$
3. Has a standard deviation of $\sqrt{\sigma_1^2/n_1 + \sigma_2^2/n_2}$

Note: If the two populations are normally distributed, then the sampling distribution of $\bar{y}_1 - \bar{y}_2$ is also normally distributed regardless of the sample sizes. It is only when the population distributions deviate from normality that the sample sizes must be large.

EXAMPLE 5.12

A manufacturer of a new gasoline additive claims that his product will increase gas mileage by 3 miles per gallon. Suppose μ_1 denotes the mean miles per gallon for a particular type of automobile with the additive, and μ_2 denotes the mean miles per gallon without the additive. Suppose also that $\sigma_1 = \sigma_2 = 8$ miles per gallon. If samples of sizes 50 and 54 are taken with and without the additive and the average miles per gallon are computed, describe the sampling distribution of $\bar{y}_1 - \bar{y}_2$.

SOLUTION

The claim is that the additive will increase the mileage by 3 miles per gallon; therefore, $\mu_1 - \mu_2 = 3$. Furthermore, we have

$$\sqrt{\frac{\sigma_1^2}{n_1} + \frac{\sigma_2^2}{n_2}} = \sqrt{\frac{64}{50} + \frac{64}{54}} = 1.57$$

Now we have that the sampling distribution of $\bar{y}_1 - \bar{y}_2$ is approximately normally distributed with a mean of 3 and a standard deviation of 1.57.

Sampling Distribution of the Difference Between Two Sample Proportions

We also wish to investigate the sampling distribution of the difference between two population proportions. Assume Bernoulli population 1 has π_1 proportion of successes and Bernoulli population 2 has π_2 proportion of successes. Independent random samples of sizes n_1 and n_2 from the two populations yield two sample proportions, p_1 and p_2.

The sampling distribution of the difference between sample proportions is summarized in the box.

Sampling Distribution of $p_1 - p_2$

If n_1 and n_2 are sufficiently large, then the sampling distribution of $p_1 - p_2$:

1. Is approximately normally distributed
2. Is centered at $\pi_1 - \pi_2$
3. Has a standard deviation of $\sqrt{[\pi_1(1 - \pi_1)/n_1] + [\pi_2(1 - \pi_2)/n_2]}$

EXAMPLE 5.13

Simmons Market Research Bureau claims that 36% of adult men and 26% of adult women read magazines in the bathroom. If these figures are accurate, describe the sampling distribution of the difference between sample proportions from samples of sizes 200 men and 180 women.

SOLUTION

The claim is that $\pi_1 = .36$ and $\pi_2 = .26$, so that $\pi_1 - \pi_2 = .10$. We have

$$\sqrt{\frac{\pi_1(1 - \pi_1)}{n_1} + \frac{\pi_2(1 - \pi_2)}{n_2}} = \sqrt{\frac{(.36)(.64)}{200} + \frac{(.26)(.74)}{180}} = .047$$

So the sampling distribution of $p_1 - p_2$ is approximately normally distributed with a mean of .10 and a standard deviation of .047.

EXERCISES 5.4

5.56 Refer to Figure 5.12 (column 1) where the population is normally distributed.
   a. Which of the three statistics has a sampling distribution that appears to be centered around the population mean?
   b. Make a general statement about where the sampling distribution of these statistics will be centered when the population is normally distributed.

     c. Which sampling distribution has the greatest standard deviation? The smallest standard deviation?

     d. Overall, which statistic provides the best estimate of the population mean?

5.57   Refer to Figure 5.12 (column 2) where the population is uniformly distributed.

     a. Which of the three statistics has a sampling distribution that appears to be centered around the population mean, which is .5?

     b. Make a general statement about where the sampling distribution of these statistics will be centered when the population is uniformly distributed.

     c. Which sampling distribution has the greatest standard deviation? The smallest standard deviation?

     d. Overall, which statistic provides the best estimate of the population mean?

5.58   Refer to Figure 5.12 (column 3) where the population has a lognormal distribution.

     a. Which of the three statistics has a sampling distribution that appears to be centered around the population mean, which is 4.48?

     b. Make a general statement about where the sampling distribution of these statistics will be centered when the population has a lognormal distribution.

     c. Which sampling distribution has the greatest standard deviation? The smallest standard deviation?

     d. Overall, which statistic provides the best estimate of the population mean?

5.59   A 14-ounce box of cornflakes may contain slightly more or slightly less than 14 ounces of actual cornflakes. Suppose μ_1 denotes the mean weight of a batch of 14-ounce boxes produced on shift 1, and μ_2 denotes the mean weight of a batch of 14-ounce boxes produced on shift 2. Suppose also that $\sigma_1 = \sigma_2 = .4$ ounce. If samples of sizes 40 and 45 are taken from the two shifts and their average weights are computed, what is the standard deviation of the sampling distribution of $\bar{y}_1 - \bar{y}_2$? Where should the sampling distribution be centered? What is the approximate distribution of $\bar{y}_1 - \bar{y}_2$?

5.60   Suppose in Exercise 5.59 that $\bar{y}_1 = 13.76$ and $\bar{y}_2 = 13.92$. What is the difference between the two sample means? Based on the results of Exercise 5.59, is this difference to be expected? Based on these sample means, are you inclined to say that there is a statistical difference between the amounts of cornflakes packaged by the two shifts? Explain your answer.

5.61   According to the Bureau of the Census, in 1990, 36 million or 15% of Americans participated for at least 1 month in some type of government assistance program (food stamps, Medicaid, rent assistance, public housing). In 1991, the figure grew to 38 million or 16%. Consider the sample proportions, p_1 and p_2, obtained from samples of sizes 250 and 300, respectively, from 1990 and 1991. What is the standard deviation of the difference between the sample proportions? Where is the sampling distribution of the difference centered? What is the approximate distribution of the sampling distribution of the difference between the sample proportions?

5.5 SUMMARY AND REVIEW

KEY CONCEPTS   ✓ The *sampling distribution* of a sample statistic describes the variability associated with the statistic in repeated sampling. If the sampling distribution of a statistic is somewhat stable from sample to sample, then it is possible to evaluate the statistic once it is found.

✓   As with any distribution, describing the sampling distribution of a statistic involves three tasks:

1. Describe the general shape of the distribution.
2. Give a measure of where the distribution is centered.
3. Give a measure of the amount of variability in the distribution.

✓   The variability in the sampling distribution of a statistic is measured by the *standard deviation* of the sampling distribution of the statistic.

✓   The *Central Limit Theorem* is a very important theorem in statistics that describes the sampling distribution of the sample mean, $\bar{y}$. It says that regardless of the shape of the population distribution, the sampling distribution of $\bar{y}$ is approximately normal for sufficiently large samples. In addition, the mean of the sampling distribution is the same as the mean of the population distribution, and the standard deviation is $\sigma/\sqrt{n}$.

✓   The Central Limit Theorem also applies to the sampling distribution of the sample proportion, p. For sufficiently large sample sizes, its sampling distribution is approximately normally distributed with a mean of π and a standard deviation of $\sqrt{\pi(1 - \pi)/n}$.

✓   The sampling distributions of the *median* and the *midrange* suggest that in some cases they may be preferred over the mean as an estimate of the center of a distribution.

✓   The sampling distribution of the difference between two $\bar{y}$'s or the difference between two sample proportions is also approximately normally distributed for sufficiently large sample sizes.

LEARNING GOALS   Having completed this chapter, you should be able to:

1. Understand the concept of a sampling distribution. *Section 5.1*
2. Understand how the standard deviation of a statistic affects the different possibilities of the statistic. *Section 5.1*
3. State and understand the Central Limit Theorem. *Section 5.2*
4. Describe the sampling distribution of $\bar{y}$. *Section 5.2*
5. Describe the sampling distribution of p. *Section 5.3*
6. Describe the differences between the sampling distributions of the mean, median, and midrange. *Section 5.4*
7. Describe the sampling distribution of the difference between two sample means or two sample proportions. *Section 5.4*

QUESTIONS
FOR REVIEW   Use the following problems to test your skills:

5.62   In 1968 Richard Nixon won 43.4% of the popular vote and Hubert Humphrey won 42.7% of the popular vote. Are these figures parameters or statistics?

5.63   According to the U.S. Department of Health and Human Services, the death rate due to automobile accidents in 1993 was 15.9 per 100,000. Is 15.9 a parameter or a statistic?

5.64   A Gallup poll in September 1968 found that 43% favored Nixon and 28% favored Humphrey. Are these figures parameters or statistics?

5.65   According to 1995 figures, 17.6% of U.S. households earn more than $50,000. Is it unusual that a random sample of 200 households would have more than 40 with incomes in excess of $50,000? Calculate the probability that more than 40 would have incomes in excess of $50,000.

5.66   Suppose the mean of a population is 65 and the standard deviation is 15. For samples of size 100, give the mean and standard deviation of the sampling distribution of $\bar{y}$. Will the sampling distribution be adequately approximated by a normal distribution? Why?

5.67   A random sample of size 100 is selected from a population with mean μ and standard deviation σ. For each value of μ and σ, find the mean and standard deviation of the sampling distribution of $\bar{y}$:

   a. $\mu = 10$, $\sigma = 2$ c. $\mu = 50$, $\sigma = 12$
   b. $\mu = 10$, $\sigma = 4$ d. $\mu = 50$, $\sigma = 24$

5.68   Which of the following does the sampling distribution of a statistic describe?
   a. The shape of the distribution of all potential values of the statistic in repeated sampling
   b. The center of all potential values of the statistic in repeated sampling
   c. The amount of variability associated with all potential values of the statistic in repeated sampling
   d. All of the above

5.69   Suppose, over a period of time, we took a large number of samples, each of size 100, from a population and calculated the average, $\bar{y}$, of each sample. If the $\bar{y}$'s were concentrated around 8,000 and varied from 7,600 to 8,400, what would be an approximate value of the standard deviation of $\bar{y}$?

5.70   Jerry Rice is considered by many to be the best wide receiver in National Football League history. He has averaged 75 receptions each year of his NFL career, and he has averaged 17 yards per reception. If the standard deviation of the yards gained for each reception is $\sigma = 8$ yards, is it reasonable to expect him to average 20 yards for a season in which he catches 75 passes?

5.71   A large garage has determined that the average time it takes to install new brakes on an automobile is 56.7 minutes and the standard deviation is 9.3 minutes. What is the probability that a random sample of 36 installations has an average exceeding 1 hour?

5.72   Scores on a standardized test of mathematical ability have a mean of 70 and a standard deviation of 15. Is it unusual that a random sample of 40 scores will average more than 75? Calculate the probability that a random sample of 40 scores will average more than 75.

5.73   A sample of n observations is randomly selected from a Bernoulli population with parameter π. Describe the sampling distribution of the sample proportion p by stating the Central Limit Theorem.

5.74   The cure rate for colon/rectum cancer is 70% if it is detected early enough. Out of a random sample of 40 patients, what is the probability that more than 30 are cured?

5.75   What is the maximum standard deviation of the sample proportion when the sample size is 200?

5.76   The *Corporate Travel/RIT* reports that traveling employees spend an average of $167.21 a day on lodging, rental car, and food. A company has 60 employees who spent an average of $148.75 on travel. If the standard deviation of the daily amount spent traveling is $50, what can be said about the average amount spent by the 60 employees?

5.77   In 1992 the average sentence for homicide was 149 months (Bureau of Justice Statistics, *Prison Sentences and Time Served for Violence*, NCJ-153858, April 1995). A random sample of size 100 is selected from the population of homicide convictions. With a standard deviation of 40 months, approximate the probability that the sample mean sentence, $\bar{y}$, falls in each range:
   a. Greater than 145
   b. Less than 155
   c. Between 151.2 and 159.6

5.78   It is not known what percent of the student body favors the construction of a new building on campus. However, if 60% are in favor, what is the probability of each of the following when we have a random sample of 30 students:

     a.   More than half are in favor.

     b.   Twenty-five or more are in favor.

     c.   All are in favor.

5.79   Suppose the scores on a management test are normally distributed with a mean of 75 and a standard deviation of 15.

     a.   What percent of the scores are above 80?

     b.   What percent are between 84 and 95?

     c.   What is the 80th percentile?

     d.   What is the probability that the average of 25 scores is above 80?

5.80   Two independent random samples are taken from two populations with means of 100 and 150 and standard deviations of 20 and 30, respectively. If the sample sizes are 40 and 45, what is the standard deviation of the sampling distribution of the difference between the sample means?

5.81   A zoologist is interested in estimating the life span of the white-tailed deer. Assuming the standard deviation in the life span is 2.5 years, how large a sample is needed so that the standard deviation of the estimate is .4 year?

5.82   Suppose a bimodal population has a mean of 20 and a standard deviation of 4. A sample of size 100 is to be randomly selected from the population, and the sample mean is to be computed. Prior to selection of the sample, tell how the potential values of the sample mean will be distributed. Give the mean and standard deviation of the sampling distribution of the sample mean.

5.83   An index score is given to new nurses after they have been on the job for 2 months. From past records, the mean index score is 45 and the standard deviation is 8 points. What is the probability that a group of 35 new nurses will have an average index score less than 43?

5.84   An airline company has historical data that suggest that the mean number of passengers on its flights is 212 and the standard deviation is 42 passengers. What is the probability that the next 50 flights will average less than 200 passengers?

5.85   In the eastern states, 12% of those who dine out use a credit card to charge their dinner. Out of a random sample of 160 people from eastern states, what is the probability that fewer than 15 will use a credit card to pay for their dinner?

5.86   We wish to estimate the proportion of students who have smoked marijuana on at least three different occasions. What will be the maximum standard deviation of our estimate if we randomly select 200 students? How large a sample is necessary for the maximum standard deviation to be no more than 5%?

5.87   According to the National Crime Victimization Survey, there were 43.6 million criminal victimizations in 1993. Of the victims of these violent crimes, 29% stated that a firearm was involved in the crime. From a random sample of 200 criminal victimizations, is it conceivable that more than 65 involve firearms? Explain your answer.

5.88   In 1989, 52% of all Ph.D.s awarded by U.S. colleges went to foreign students. Of the 32 Ph.D.s awarded by the university system of a particular state, 14 were awarded to U.S. students. Was this an unusually low percentage being awarded to U.S. students.? (*Hint*:   Calculate the probability that 14 or fewer in a sample of 32 Ph.D.s are awarded to U.S. students.)

5.89   The owner of a local ski resort would like to estimate the mean daily amount of money spent at the resort by the guests. Assuming that $15 is a reasonable estimate of the standard deviation of the amounts spent by guests, how large a sample is needed so that the standard deviation of the estimated mean daily amount of money spent is $2?

5.90   The state highway patrol would like to estimate the average speed of motorists on a certain section of interstate. How large a sample is necessary so that the standard deviation of their estimate is 1 mile per hour? Assume the standard deviation of the speeds of the motorist is 12 miles per hour.

5.91   A study showed that typical college students study an average of 8 hours per week with a standard deviation of 3 hours. From a random sample of 100 students, it was found that they studied on average 9.2 hours per week. Do these students study significantly more than typical college students, or did the average of 9.2 happen purely by chance? Explain your answer.

5.92   To illustrate the variability associated with a statistic, conduct the following project: From the classified section of your local Sunday newspaper, randomly select 20 houses that are for sale. Calculate the average price of the 20 houses. From the same newspaper, randomly select another 20 houses and again calculate the average. Do this several times and then compare the averages you found. Remember that the average cost of a house in your city for a given week is not variable; it is a fixed quantity, μ. However, the $\bar{y}$'s you calculate will vary; each one is a separate estimate of μ. The variability associated with the $\bar{y}$'s is the main idea of the sampling distribution of $\bar{y}$.

COMPUTER
EXERCISES

5.93   The amount of dye dispensed in a gallon of blue paint at the auto paint store is normally distributed with a mean of 2 ounces and a standard deviation of .5 ounce. Simulate 200 random samples, each of size 5, from the population and calculate the sample mean and sample median of each sample.
   a. Construct histograms with a common scale for the sampling distributions of the sample mean and the sample median.
   b. Where are the two sampling distributions centered?
   c. Describe the general shape of the two distributions.
   d. Which distribution is more variable?
   e. Can you think of a reason one distribution is more variable?

5.94   Repeat Exercise 5.93 except this time each random sample should be of size 30. Compare your results with those from Exercise 5.93. What basic difference did the increase in sample size cause?

5.95   The amount of time one has to wait for a bus at a particular bus stop is exponentially distributed with a mean of 5 minutes. Simulate 200 random samples, each of size 5, from the population and calculate the sample mean and sample median of each sample.
   a. Construct histograms with a common scale for the sampling distributions of the sample mean and the sample median.
   b. Where are the two sampling distributions centered?
   c. Describe the general shape of the two distributions.
   d. Which distribution is more variable?
   e. Can you think of a reason why the distributions are skewed right?

5.96   Repeat Exercise 5.95 except this time each random sample should be of size 30. Compare your results with those from Exercise 5.95. What basic difference did the increase in sample size cause?

5.97   A survey of 10,000 youths by the National Center for Health Statistics (NCHS) in 1993 showed that 16% of boys and girls aged 12 to 18 smoke. Over half said they expected to quit within a year. Worksheet: **Kidsmoke.mtw** contains the results of a similar survey of 1,000 youths. Column 1 indicates the gender of the child (0 = female, 1 = male), and column 2 indicates whether they smoke (0 = no, 1 = yes).
   a. Find the percent of kids who smoke.
   b. What is the percent of girls who smoke?
   c. What is the percent of boys who smoke?
   d. Are there an equal number of boys and girls in the study?
   e. Which group has the higher smoking rate?
   f. Are these results similar to those from the NCHS survey?

5.98   STATISTICAL INSIGHT REVISITED   Nine different polls attempted to predict the percent of the popular vote that Ronald Reagan would get in the 1984 presidential election. Use the Random Data > Bernoulli command to simulate the results of nine polls in C1–C9 where each poll has a sample size of 1,000. Use the actual percent of .59 for the probability of success. What is the standard deviation of the sample proportion when $\pi = .59$ and $n = 1,000$? How many of your nine polls have a sample proportion that falls within one standard deviation of .59? How many are within two standard deviations of .59? Are your results similar to those of the actual pollsters in the Statistical Insight?

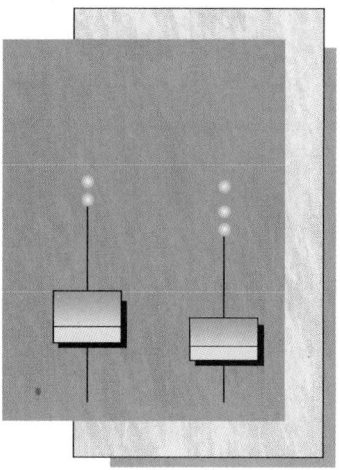

6

DESCRIBING
DISTRIBUTIONS

n the practice of statistics, we are concerned with the task
of extracting information from data. By properly analyzing data,
we detect patterns and recognize relationships. The preceding
chapters introduced the tools that are commonly used to explore
data. Graphical displays such as histograms, boxplots, and
scatterplots help us look at the overall pattern produced by the
data. Numerical summaries such as the mean, median, standard
deviation, and correlation coefficient are used to identify specific
characteristics of the data. We introduced these exploratory
tools in a rather disjointed fashion and used them in a purely
descriptive sense. These same tools, however, can be used as a
preliminary to formal statistical inference. Thus, as we begin our
study of inferential statistics, we will use these tools collectively
to help develop a strategy for analyzing data and conducting
formal inferences. Our first task is to infer a general shape for the
population distribution.

CONTENTS

Statistical Insight

Scholastic Assessment Test Scores

Scholastic Assessment Test (SAT) scores are often thought of as a barometer that measures how well high school seniors will do in college. The nationwide average was the highest in 1963, when the average SAT math score was 502 and the average SAT verbal score was 478. The averages began a downward trend that persisted through the 1970s and 1980s. The average SAT verbal score is still near its all-time low of 422 recorded in 1991. The average SAT math score fell to 466 in 1980 but has posted a modest increase to 479 over the last 15 years.

Boys generally perform better than girls on the SAT, and some states have significantly higher averages than other states. Many reasons have been given for the low scores, such as the permissive attitude of our society and the state funding of education. The gap between scores of boys and girls has narrowed to only 4 points on the verbal section but is some 40 points on the math section. Some argue that boys outperform girls on math because they generally take more

math and science courses in high school. The differences among the states has caused some state legislatures to commission studies of the problem, such as the one conducted by the General Assembly of Georgia. Their study suggested that the decline was linked directly to the per-student funding by the states; they claimed that in states where funding was high, the scores were high, and where funding was low, the scores were low. Critics of the study suggest that it is not fair to compare state averages because in some states as many as 80% of the students take the exam and in other states only 4% take it. The 4% represent the so-called "cream of the crop," and one would expect their average to be high. So the critics may have a valid argument.

The table lists the 1994 SAT averages by state, together with the percent of students taking the exam and the state funding per student. You will analyze these scores in Exercise 6.73 of the Computer Exercises in Section 6.4.

Worksheet: Sat.mtw

State	1994 Verbal	Math	Total	Percent	Expend
Alabama	482	529	1,011	8	$3,616
Alaska	434	477	911	49	8,450
Arizona	443	496	939	26	4,381
Arkansas	417	518	935	6	4,031
California	413	482	895	46	4,746
Colorado	456	513	969	28	5,172
Connecticut	426	472	898	80	8,017
Delaware	428	464	892	68	6,093
District of Columbia	406	443	849	53	9,549
Florida	413	466	879	49	5,243
Georgia	398	446	844	65	4,375
Hawaii	401	480	881	58	5,420
Idaho	461	508	969	16	3,556
Illinois	478	546	1,024	14	5,670
Indiana	410	466	876	60	5,074
Iowa	506	574	1,080	5	5,096

(continued)

1994 State	Verbal	Math	Total	Percent	Expend
Kansas	494	550	1,044	10	5,007
Kentucky	474	523	997	11	4,719
Louisiana	481	530	1,011	9	4,354
Maine	420	463	883	68	5,652
Maryland	429	479	908	64	6,679
Massachusetts	426	475	901	79	6,408
Michigan	472	537	1,009	11	6,228
Minnesota	495	562	1,057	9	5,409
Mississippi	485	528	1,013	4	3,245
Missouri	485	532	1,017	10	4,830
Montana	463	523	986	21	5,423
Nebraska	482	543	1,025	9	5,263
Nevada	429	484	913	30	4,926
New Hampshire	438	486	924	69	5,790
New Jersey	418	475	893	71	9,317
New Mexico	475	528	1,003	12	3,765
New York	416	472	888	76	8,527
North Carolina	405	455	860	60	4,555
North Dakota	497	559	1,056	5	4,441
Ohio	456	510	966	24	5,694
Oklahoma	482	537	1,019	9	4,078
Oregon	436	491	927	53	5,913
Pennsylvania	417	462	879	70	6,613
Rhode Island	420	462	882	68	6,546
South Carolina	395	443	838	60	4,436
South Dakota	483	548	1,031	5	4,173
Tennessee	488	535	1,023	12	3,692
Texas	412	474	886	48	4,632
Utah	509	558	1,067	4	3,040
Vermont	427	472	899	68	6,944
Virginia	424	469	893	65	4,880
Washington	434	488	922	49	5,271
West Virginia	439	482	921	17	5,109
Wisconsin	487	557	1,044	9	6,139
Wyoming	459	521	980	12	5,812

SOURCE: *World Almanac and Book of Facts* (New York: Funk & Wagnalls, 1995).

6.1   DATA ANALYSIS

Data analysis, referred to as *exploratory data analysis* (EDA) by some, is designed to provide us with a strategy for investigating statistical questions. It is more than just a collection of summary measures and graphic tools that are used to look at data; it is a philosophy by which we attempt to extract knowledge from data. When we begin a statistical problem, we may have specific questions in mind that we wish to answer by investigating the data. More frequently than not, however, our questions are not fully developed, and thus we look to the data with the idea that they will suggest our course of action. Even when we have a well-structured path of analysis, a careful exploration of the data can turn up unexpected features that suggest new and better questions to improve our analysis of the data.

Formal statistical inference is based on specific mathematical models and assumptions. When the model is inaccurate or the assumptions are violated, the inference procedures can lead to false conclusions. Data analysis gives us diagnostic tools that are used to assess the adequacy of the model and to check assumptions. It is important that we use these tools to carefully analyze the data prior to making any formal inferences.

We begin the analysis of a univariate data set by attempting to describe the *parent distribution*.

> The distribution of the measurements in the original population is called the **underlying** or **parent distribution.**

To describe the parent distribution we are concerned with three characteristics:

1. The general *shape* of the distribution
2. A measure of the *center* of the distribution
3. A measure of the *variability* of the distribution

Determining the Shape of the Distribution

The choice of parameters to summarize the center and the amount of variability depends on the shape of the parent distribution. Thus, our first task is to examine the shape of the parent distribution. In previous chapters, we introduced several diagnostic tools for this purpose:

- The stem and leaf plot and the histogram give the general appearance of the collected data, which, assuming we have a representative sample, resembles the shape of the parent distribution.
- The boxplot shows skewness and symmetry and identifies outliers.
- The midsummary analysis confirms the skewness or the symmetry.
- The normal probability plot checks for normality as well as identifies other unusual behavior in the data.

Symmetric or Skewed

For a unimodal (single-peaked) distribution, we first attempt to classify the shape as either symmetric or skewed.

EXAMPLE **6.1**

Challenger Accident

The U.S. space shuttle *Challenger* accident (see Examples 2.12 and 2.15 in Section 2.4) occurred because of failure of the O-rings that seal the field joints of the rocket boosters. It was well known at the time that the rubber O-rings, which are designed to prevent the release of hot gases produced during combustion, are quick to recover their shape after compression in warm weather but not in cold weather. The stem and leaf plot in Example 2.15 for the space shuttle launch temperatures is repeated here. Comment on the distributional shape after you analyze the midsummaries, a boxplot, and a normal probability plot.

Worksheet: Challeng.mtw

```
3 | 1
3 |
4 |
4 |
5 | 3
5 | 7 8
6 | 3
6 | 6 7 7 7 8 9
7 | 0 0 0 0 2 3
7 | 5 5 6 6 8 9
8 | 0 1
```

Solution

Figure 6.1 contains a letter value display, a boxplot, and a normal probability plot. With the exception of the midQ, the midsummaries get progressively smaller as we scan down from the median to the midR, which suggests that the frequency distribution is skewed left. The boxplot also identifies two outliers in the left tail, and that supports the conjecture that the distribution is skewed left. The normal probability plot on page 350 has the general appearance of a skewed left distribution.

FIGURE 6.1   *Challenger* data including 31°

Letter Value Display

	DEPTH	LOWER	UPPER	MID	SPREAD
N=	25				
M	13.0	70.000		70.000	
H	7.0	67.000	75.000	71.000	8.000
E	4.0	58.000	78.000	68.000	20.000
D	2.5	55.000	79.500	67.250	24.500
C	1.5	42.000	80.500	61.250	38.500
	1	31.000	81.000	56.000	50.000

Character Boxplot

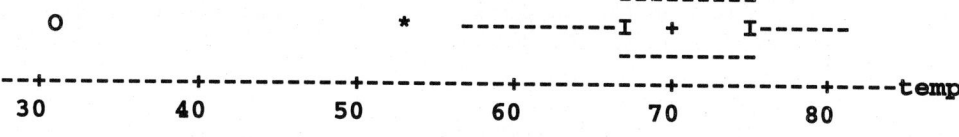

(continued)

FIGURE 6.1
(continued)

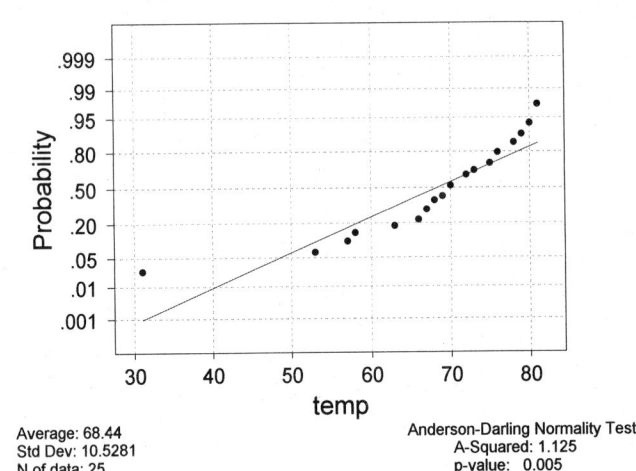

Normal Probability Plot

Average: 68.44
Std Dev: 10.5281
N of data: 25

Anderson-Darling Normality Test
A-Squared: 1.125
p-value: 0.005

On a closer investigation, however, we see in Figure 6.2 that without the 31° temperature, the data appear somewhat symmetric. The midsummaries, with the exception of the midQ, still decrease but not nearly so dramatically as before. In fact, we could say that they are reasonably close in value, which suggests symmetry.

FIGURE 6.2   *Challenger* data without 31°

Letter Value Display

	DEPTH	LOWER	UPPER	MID	SPREAD
N=	24				
M	12.5	70.000		70.000	
H	6.5	67.000	75.500	71.250	8.500
E	3.5	60.500	78.500	69.500	18.000
D	2.0	57.000	80.000	68.500	23.000
	1	53.000	81.000	67.000	28.000

Character Boxplot

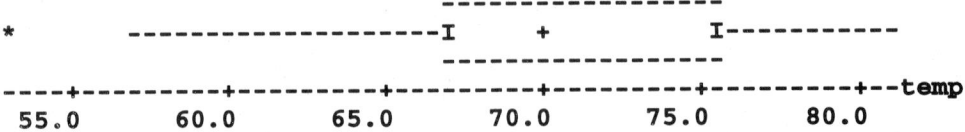

(continued)

FIGURE 6.2
(continued)

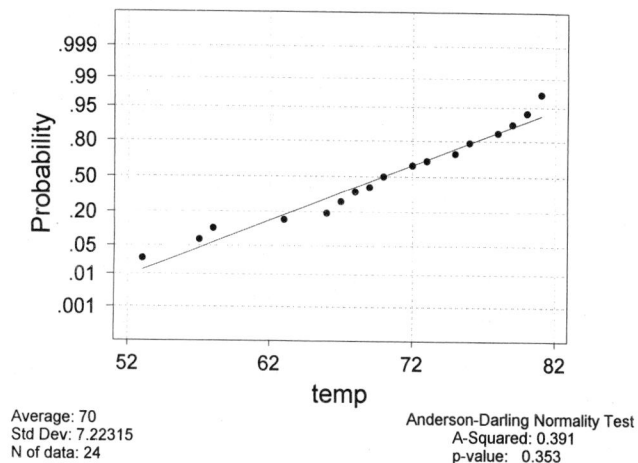

Normal Probability Plot

Average: 70
Std Dev: 7.22315
N of data: 24

Anderson-Darling Normality Test
A-Squared: 0.391
p-value: 0.353

The boxplot does not show the same degree of skewness as before, and the normal probability plot suggests that the distribution may even be normally distributed. The point is this: By carefully examining the data, we see that without the 31° temperature, the data are close to being normally distributed and temperatures below about 55° are extreme temperatures. The 31° temperature on the day of the launch of *Challenger* is most certainly an anomaly.

As we examine data, one of the main features of our strategy should be flexibility. No two data sets are exactly alike, and therefore, no standard procedure works best all the time. We must be able to adapt the analysis to the problem. As we examine the data for structure, we must be able to respond to different patterns that are suggested by our analysis. We must approach the analysis with the goal of answering our questions but at the same time be open to the possibility of unexpected features that will prompt new questions.

Because of the ease with which we can analyze data with a computer, we must be willing to look at our data in any number of ways. For example, if the managers in charge of the launch of the *Challenger* had not ignored relevant data, it is generally agreed that the fatal launch would have been averted.

EXAMPLE **6.2**

Challenger Accident

In seven previous flights of the space shuttle *Challenger*, there was damage to at least one O-ring, as documented in these data:

Worksheet: Challeng.mtw

Flight	Date	Temp	Failures
1	4/12/81	66	0
2	11/12/81	70	1
3	3/22/82	69	0
4	6/27/82	80	(hardware lost at sea)
5	11/11/82	68	0
6	4/4/83	67	0
7	6/18/83	72	0
8	8/30/83	73	0
9	11/28/83	70	0
41-b	2/3/84	57	1
41-c	4/6/84	63	1
41-d	8/30/84	70	1
41-g	10/5/84	78	0
51-a	11/8/84	67	0
51-c	1/24/85	53	3
51-d	4/12/85	67	0
51-b	4/29/85	75	0
51-g	6/17/85	70	0
51-f	7/29/85	81	0
51-i	8/27/85	76	0
51-j	10/3/85	79	0
61-a	10/30/85	75	2
61-b	11/26/85	76	0
61-c	1/12/86	58	1
61-i	1/28/86	31	(*Challenger* accident)

SOURCE: S. R. Dalal, E. B. Fowlkes, and B. Hoadley, "Risk Analysis of the Space Shuttle: Pre-Challenger Prediction of Failure," *Journal of the American Statistical Association* 84: 408 (1989): 945–957.

The scatterplot in Figure 6.3 of the number of failures versus the temperature at launch for the seven flights with at least one O-ring failure reveals no clear linear trend. The correlation between temperature and number of failures is only $r = -.263$. Based on

FIGURE 6.3   Plot where there was damage to at least one O-ring

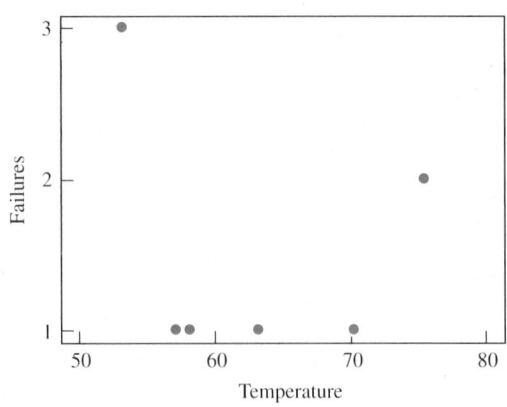

this graph and the small correlation coefficient, the managers recommended to proceed with the launch, although the forecasted temperature for the day was 31°F. The coldest temperature for any previous launch was 53°F and then there were three failures.

The fatal mistake was the failure to consider the 16 flights where there was *no* O-ring damage. With the exception of one data point, (75, 2), the scatterplot of all the data in Figure 6.4 shows that there is a definite downward linear trend between the number of damaged O-rings and the temperature. The correlation between temperature and number of failures using all the data is $r = -.561$. If we exclude the one outlier, (75, 2), the correlation is $r = -.714$. Had the managers considered this scatterplot and observed these correlations, surely they would have canceled the launch.

FIGURE 6.4   Plot with all data points

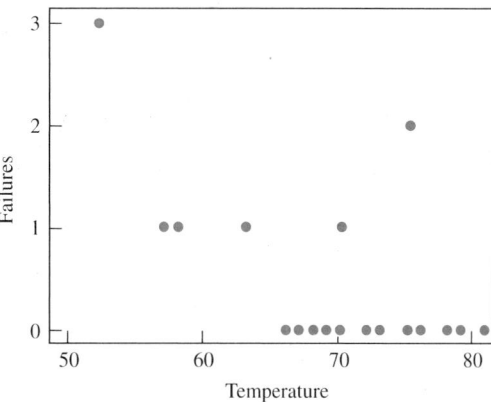

Normal

As stated earlier, formal statistical inference is based on certain mathematical models and assumptions. In many cases, we must assume that the parent distribution is normally distributed. Thus, it is important that we be able to determine whether or not data are coming from a normally distributed population.

EXAMPLE 6.3   **R. A. Fisher Data on *Iris setosa***

A historic data set studied by R. A. Fisher (one of the founding fathers of modern-day statistics) is the measurements of four flower parts on 50 specimens of each of three species of irises. Here are the sepal lengths (in centimeters) of the species *Iris setosa*:

Worksheet: Irises.mtw Column 1

5.1 4.9 4.7 4.6 5.0 5.4 4.6 5.0 4.4 4.9
5.4 4.8 4.8 4.3 5.8 5.7 5.4 5.1 5.7 5.1
5.4 5.1 4.6 5.1 4.8 5.0 5.0 5.2 5.2 4.7
4.8 5.4 5.2 5.5 4.9 5.0 5.5 4.9 4.4 5.1
5.0 4.5 4.4 5.0 5.1 4.8 5.1 4.6 5.3 5.0

SOURCE: R. A. Fisher, "The Use of Multiple Measurements in Taxonomic Problems," *Annals of Eugenics*, 7 (1936): 179–184.

Analyze these data and determine whether the parent distribution is normal.

SOLUTION

The analysis is presented in Figure 6.5. We first conduct a midsummary analysis. Notice that the midsummaries in the letter value display produced by Minitab are all very close in value, which suggests symmetry. Next we check a histogram and a boxplot of the data. The histogram appears very symmetric and resembles a normal distribution. The boxplot also appears very symmetric with no outliers, which is what we expect from a normal distribution. Finally, the normal probability plot indicates that the data are normally distributed. In conclusion, all of the exploratory evidence lead us to classify the distribution of sepal lengths of *Iris setosa* as normally distributed.

FIGURE 6.5   Descriptive Statistics

Variable	N	Mean	Median	TrMean	StDev	SEMean
sepalL1	50	5.0060	5.0000	5.0000	0.3525	0.0498

Variable	Min	Max	Q1	Q3
sepalL1	4.3000	5.8000	4.8000	5.2000

Letter Value Display

	DEPTH	LOWER	UPPER	MID	SPREAD
N=	50				
M	25.5		5.000	5.000	
H	13.0	4.800	5.200	5.000	0.400
E	7.0	4.600	5.400	5.000	0.800
D	4.0	4.400	5.500	4.950	1.100
C	2.5	4.400	5.700	5.050	1.300
B	1.5	4.350	5.750	5.050	1.400
	1	4.300	5.800	5.050	1.500

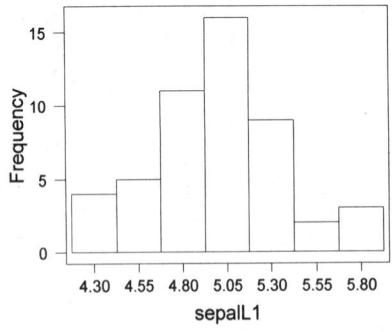

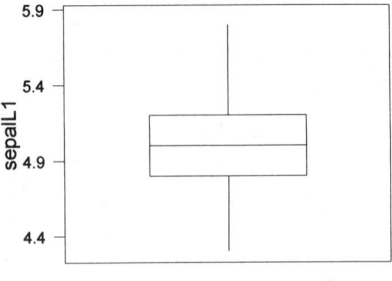

(continued)

FIGURE 6.5
(continued)

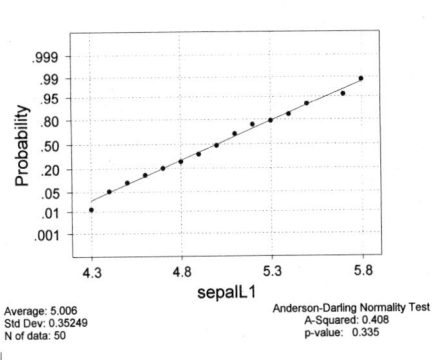

Normal Probability Plot

Average: 5.006
Std Dev: 0.35249
N of data: 50

Anderson-Darling Normality Test
A-Squared: 0.408
p-value: 0.335

Multimodal

Most often the distributions we encounter in practice are either symmetric or skewed. There are other possibilities, however, such as multiple modes or gaps in the data, and these issues can be addressed with a proper analysis of the data.

EXAMPLE 6.4   **Clustering of Galaxies**

A popular theory in astronomy is that after the Big Bang, matter expanded at a tremendous rate, and because of the local attraction of matter, the galaxies were formed. It is believed that gravitational pull would cause some clustering of galaxies, but would the gravitational pull be great enough to form superclusters of galaxies surrounded by large voids? To help answer this question, we need to measure the distances between galaxies—not a trivial task. If the universe is indeed expanding, then points more distant from our galaxy are moving at greater velocities; that is, we can estimate distances by measuring velocities. The next data are the velocities (measured in kilometers per second) of 82 galaxies found in the Corona Borealis region. Is there evidence in the data to suggest that there are superclusters of galaxies?

Worksheet: Galaxie.mtw

9,172	9,558	10,406	18,419	18,972	19,330	19,440	19,541	19,846	19,914
19,989	20,179	20,221	20,795	20,875	21,492	21,921	22,209	22,314	
22,746	22,914	23,263	23,542	23,711	24,289	24,990	26,995	34,279	
9,350	9,775	16,084	18,552	19,052	19,343	19,473	19,547	19,856	
19,918	20,166	20,196	20,415	20,821	20,986	21,701	21,960	22,242	
22,374	22,747	23,206	23,484	23,666	24,129	24,366	25,633	32,065	
9,483	10,227	16,170	18,600	19,070	19,349	19,529	19,663	19,863	
19,973	20,175	20,215	20,629	20,846	21,137	21,814	22,185	22,249	
22,495	22,888	23,241	23,538	23,706	24,285	24,717	26,960	32,789	

SOURCE: K. Roeder, "Density Estimation with Confidence Sets Exemplified by Superclusters and Voids in the Galaxies," *Journal of the American Statistical Association*, 85(1990): 617–624.

SOLUTION

With a computer we construct the boxplot and normal probability plot shown in Figure 6.6. From the appearance of the boxplot and the normal probability plot, we should classify the shape of the distribution as a long-tailed distribution. The boxplot clearly shows numerous outliers on each tail, and the normal probability plot gives the appearance of a long-tailed distribution. This means that the distances between galaxies tend to mound up around a common value with some galaxies far removed from the center.

FIGURE 6.6

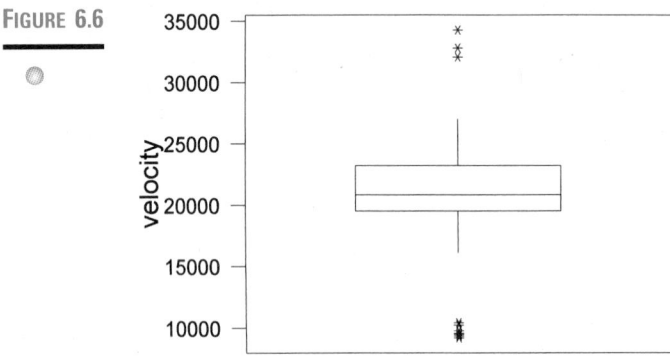

Normal Probability Plot

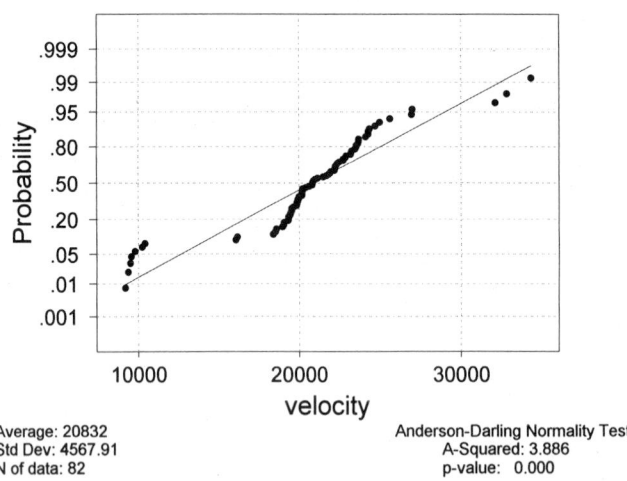

Average: 20832
Std Dev: 4567.91
N of data: 82

Anderson-Darling Normality Test
A-Squared: 3.886
p-value: 0.000

The histogram and dotplot shown in Figure 6.7, however, reveal something overlooked by the boxplot and the normal probability plot. The histogram reveals the possibility of four different modes. The dotplot shows as many as seven different modes. If the distribution is multimodal, then the galaxies are gathering in different clusters. Each mode represents a cluster of galaxies. The question then is whether or not the modes are "real" or just random variations in the data.

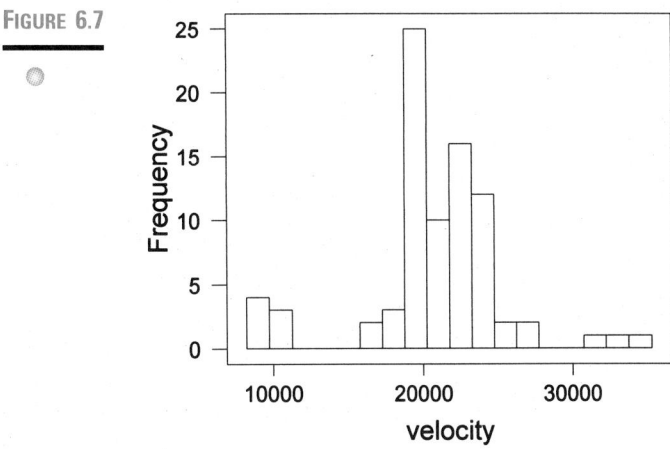

FIGURE 6.7

Character Dotplot

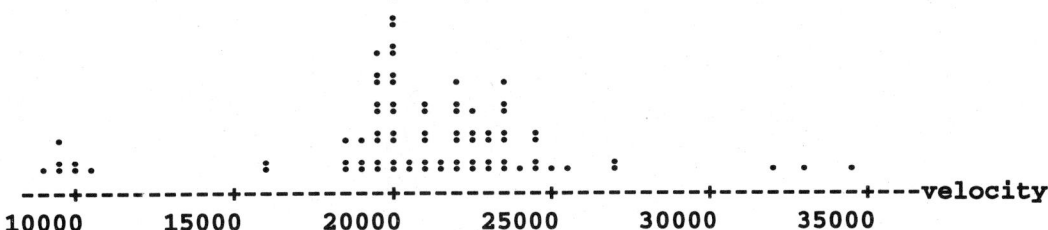

Very sophisticated data analysis procedures, beyond the scope of this book, can be used to study the significance of the modes. But even with simple exploratory tools we can study what is perhaps an important scientific discovery in astronomy. Incidentally, Roeder concluded that there are somewhere between three and seven significant modes in these data.

Determining Anomalies in the Data

Other peculiarities in data can be revealed with proper data analysis, as in the next example.

EXAMPLE 6.5   Simpson's Paradox

The athletic director was told by the registrar of a university that of the men and women who compete in sports, men have a higher overall grade point average than women. Having some doubts about this conclusion, the athletic director randomly selected 50 male and 50 female student athletes who participate in basketball, soccer, and track. The listing gives the grade point average, the sport they participate in (1 = basketball, 2 = track, and 3 = soccer), and the gender (1 = male, 2 = female) of the students.

Worksheet: Simpson.mtw

Male (1)

2.78	3	2.80	3	2.39	2	2.71	3	2.46	2
2.64	3	2.74	3	2.29	2	3.10	3	1.95	1
2.77	3	2.75	3	1.96	1	2.57	2	3.02	3
2.86	3	2.92	3	2.00	1	1.99	1	2.32	2
2.84	3	1.83	1	2.90	3	2.41	2	2.85	3
2.43	2	1.93	1	2.29	2	2.76	3	2.93	3
2.32	2	2.76	3	2.39	2	1.99	1	2.24	2
2.03	1	1.88	1	2.10	1	2.70	3	2.69	3
2.74	3	2.67	3	3.02	3	2.30	2	2.34	2
2.67	3	2.82	3	2.42	2	2.63	3	2.43	2

Female (2)

2.08	1	2.56	2	2.37	2	2.46	2	3.12	3
2.30	2	2.47	2	3.05	3	2.50	2	2.63	2
3.01	3	2.10	1	2.57	2	2.47	2	2.38	2
2.12	1	2.46	2	2.37	1	2.21	1	3.06	3
2.97	3	2.87	3	2.11	1	2.16	1	2.35	1
2.53	2	3.14	3	2.98	3	2.55	2	2.19	1
2.49	2	2.29	1	2.47	2	2.98	3	2.50	1
2.61	2	2.25	1	2.38	2	2.22	1	2.52	2
2.42	2	2.25	1	3.07	3	2.32	1	2.19	1
2.29	1	3.05	3	3.00	3	3.10	3	2.43	2

Analyze these data by comparing the grade point averages of the male and female student athletes.

SOLUTION

Figure 6.8 shows stem and leaf plots of the two genders along with side-by-side boxplots. The stem and leaf plots of the grade point averages of the two genders of athletes do not reveal any big differences between the men and women. But the boxplots (page 360) do show that the median grade point average for men is higher than the median grade point average for women. On the surface it appears that the registrar is correct.

Taking a closer look, however, we see that the stem and leaf plot for men appears to be trimodal, and the stem and leaf plot for women appears to be bimodal, which indicates an extraneous variable confounded with the data. Investigating further, we decide to look at the grade point averages of the athletes in the different sports.

FIGURE 6.8

```
Character Stem-and-Leaf Display

Stem-and-leaf of gpa
Leaf Unit = 0.010
gender = 1       N  = 50

     2     18 38
     7     19 35699
     9     20 03
    10     21 0
    13     22 499
    19     23 022499
    24     24 12336
    25     25 7
    25     26 34779
    20     27 014456678
    11     28 02456
     6     29 023
     3     30 22
     1     31 0


Stem-and-leaf of gpa
Leaf Unit = 0.010
gender = 2       N  = 50

     1     20 8
     7     21 012699
    13     22 125599
    20     23 0257788
   (8)     24 23667779
    22     25 0023567
    15     26 13
    13     27
    13     28 7
    12     29 788
     9     30 015567
     3     31 024
```

(continued)

FIGURE 6.8
(continued)

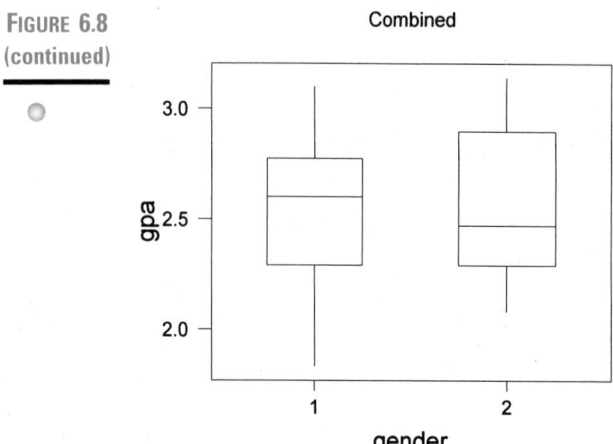

Combined

Figure 6.9 gives side-by-side boxplots of the grade point averages for males and females for the three different sports.

FIGURE 6.9

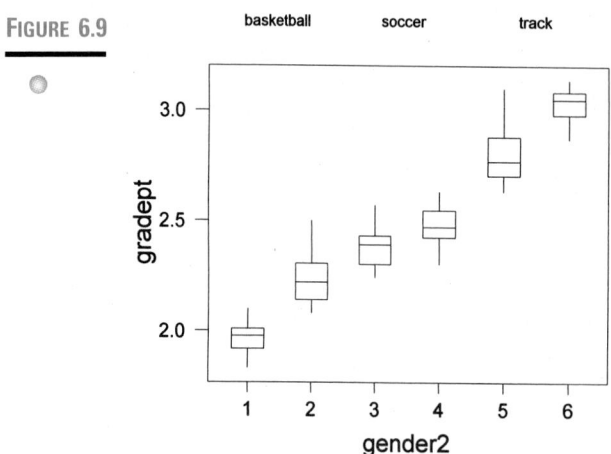

Gender2 code: 1, 3, and 5 are male athletes, and 2, 4, and 6 are female athletes.

We observe from the boxplots that in each sport the female athletes appear to have higher grade point averages than the male athletes. In fact, in basketball women's grade point averages completely dominate those of the men. When all the athletes from the different sports are combined, as in Figure 6.8, the median male grade point average exceeds the median female grade point average. When we look at each individual sport, however, just the opposite happens. This phenomenon is know as *Simpson's paradox*. Can you explain how this can happen?

In this section, we have presented several different examples that illustrate the general philosophy of data analysis. In summary, we hope to extract as much information from the data as possible. In those cases where inference is the goal, we first describe the general shape of the parent distribution. In the remaining sections of this chapter, we identify specific parameters that characterize the distribution and then learn what statistics best estimate them.

EXERCISES 6.1

6.1   The table gives the 1995 EPA fuel-efficiency ratings for the ten highest rated cars and the ten highest rated trucks. Based on the midsummaries in the letter value display, how do you classify the shape of the distribution?

Worksheet: Fuel.mtw

Cars		Trucks	
Honda Civic HB	56	Suzuki Samurai	29
Honda Civic VX	51	GMC Sonoma	29
Geo Metro	49	Chevrolet S10	29
Suzuki Swift	49	Suzuki Sidekick	27
Honda Civic SL	46	Geo Tracker Convert	27
Honda Civic	45	Dodge Dakota	27
Geo Metro	43	Suzuki Sidekick 2dr	26
Suzuki Swift	43	Suzuki Sidekick 4x4	26
Ford Aspire	42	Suzuki Sidekick 4dr	26
Honda Civic Del Sol	41	Suzuki Sidekick 4dx4	26

SOURCE: Environmental Protection Agency.

```
Letter Value Display

        DEPTH      LOWER       UPPER         MID        SPREAD
   N=   20
   M    10.5           35.000             35.000
   H    5.5        27.000      45.500     36.250       18.500
   E    3.0        26.000      49.000     37.500       23.000
   D    2.0        26.000      51.000     38.500       25.000
        1          26.000      56.000     41.000       30.000
```

6.2   From this boxplot of the EPA fuel-efficiency ratings listed in Exercise 6.1, comment on the shape of the distribution. Does this agree with your answer to Exercise 6.1?

Character Boxplot

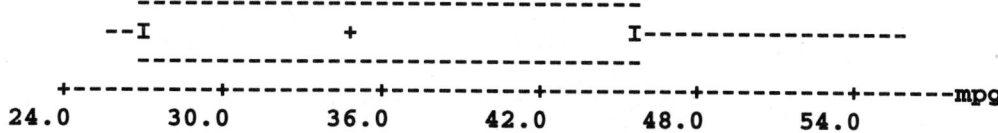

6.3   Here is a dotplot of the EPA fuel-efficiency ratings given in Exercise 6.1.

Character Dotplot

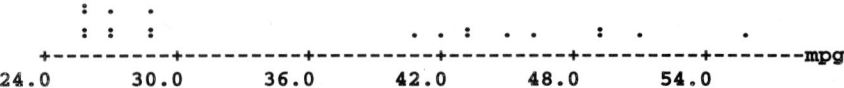

a. What information can you gather from the dotplot that was not available in either the letter value display or the boxplot given in Exercises 6.1 and 6.2?

b. Give an explanation for the different modes.

c. Calculate the average rating from the data in Exercise 6.1. Does this average give a representative measure of the fuel efficiency of all the vehicles?

d. What are your recommendations for summarizing these data?

6.4   The visual acuity of a group of subjects is tested under different doses of a drug. The visual acuity measurements for drug dose equal to 5 units are listed here:

Worksheet: Visual.mtw

.36	.41	.55	.33	.28	.25	.17	.14	.23
.27	.35	.37	.25	.24	.13	.19	.25	.34

a. Construct an ordered stem and leaf plot of the data.

b. Construct a five-number summary diagram including the midsummaries and spreads.

c. Perform a midsummary analysis to check for skewness.

d. Construct a boxplot.

e. Based on your results, comment on the shape of the parent distribution.

6.5   A group of 25 college students applying for graduate school take the Miller Personality Test for admission purposes and get these scores:

Worksheet: Miller.mtw

21	18	20	25	23	19	30	24	29	14	25	22	35
26	23	16	33	25	22	18	34	22	31	27	25	

a. Construct an ordered stem and leaf plot of the data.

b. Construct a five-number summary diagram including the midsummaries and spreads.

c. Perform a midsummary analysis to check for skewness.

d. Construct a boxplot.

e. Based on your results, comment on the shape of the parent distribution.

6.6   A social scientist examined the records at several hospitals and recorded the following ages of women at the birth of their first child:

Worksheet: Firstchi.mtw

30	18	35	22	23	22	36	24	23	28	19	23
25	24	33	21	24	19	33	23	19	32	21	18
36	21	25	17	21	24	39	22	23	18	22	28
18	15	25	21	23	26	38	24	20	36	27	21
28	26	22	28	33	18	17	21	15	20	16	21
23	15	20	38	16	24	42	22	24	24	20	17
26	39	22	21	28	20	29	14	25	20	19	17
21	24	26									

From these data, you were asked in Exercise 2.88 of Section 2.5 to construct a five-number summary diagram. The midsummaries are

Median	23
MidQ	23
MidE	25.5
MidRange	28

Based on these midsummaries, do you think the parent distribution is skewed? What additional information would be helpful in assessing the shape of the parent distribution?

6.7   The figure shows a normal probability plot of the Miller Personality Test scores given in Exercise 6.5. Does the plot indicate that the data are from a normal population? Does this agree with your conclusion in Exercise 6.5?

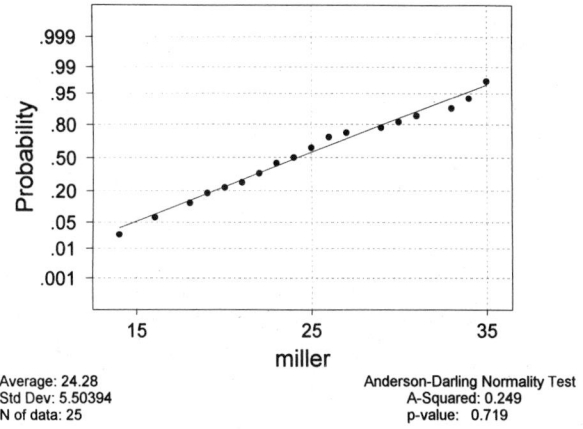

Normal Probability Plot

Average: 24.28
Std Dev: 5.50394
N of data: 25

Anderson-Darling Normality Test
A-Squared: 0.249
p-value: 0.719

6.8   A boxplot of the data from Exercise 6.6 is shown here. Based on the appearance of this boxplot, how do you classify the shape of the parent distribution? Does this agree with the midsummary analysis in Exercise 6.6? Is it reasonable to suggest that the parent distribution is normally distributed? Why or why not?

Character Boxplot

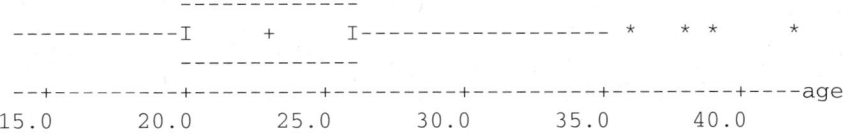

6.9   A normal probability plot of the data from Exercise 6.6 is shown at the top of page 364. Based on the appearance of this normal probability plot, how do you classify the shape of the parent distribution? Does this agree with your assessment in Exercises 6.6 and 6.8?

Normal Probability Plot

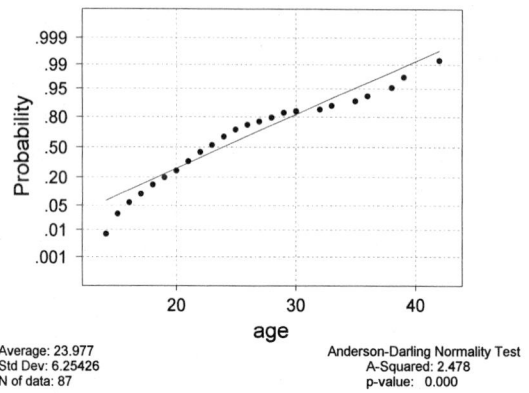

Average: 23.977
Std Dev: 6.25426
N of data: 87

Anderson-Darling Normality Test
A-Squared: 2.478
p-value: 0.000

6.10   A tort is a claim arising from personal injury or property damage caused by the negligent or intentional act of another person or business. Over half (60%) of all tort cases are automobile accident cases. Other tort cases include injury caused by dangerous conditions of residential or commercial property, medical malpractice, injury caused by defective products, and injury caused by toxic substances. The data presented here are from court files from a sample of 45 of the nation's 75 largest counties. They include the county, the average number of months to process a tort, the population of the county, the number of torts, and the rate per 100,000 residents. We want to analyze the number of months to process a tort.

Worksheet: Tort.mtw

County	Months	Populat	Torts	Rate
Maricopa, AZ	11	2,209,567	9,914	449
Pima, AZ	17	690,202	3,346	485
Alameda, CA	22	1,307,572	3,258	249
Contra Costa, CA	15	840,585	2,469	294
Fresno, CA	42	705,613	2,364	335
Los Angeles, CA	12	3,485,398	21,954	630
Orange, CA	21	2,484,789	18,297	736
San Bernadino, CA	30	1,534,343	7,583	494
San Francisco, CA	21	728,921	3,467	476
Santa Clara, CA	20	1,528,527	5,148	337
Ventura, CA	22	686,560	1,917	279
Fairfield, CT	23	267,099	2,303	862
Hartford, CT	19	858,831	4,184	487
Dade, FL	13	2,007,972	7,122	355
Orange, FL	15	714,579	1,700	238
Palm Beach, FL	14	900,655	4,194	466
Fulton, GA	13	665,765	1,470	221
Honolulu, HI	16	863,117	1,795	208
Cook, IL	34	5,139,341	20,573	400
Dupage, IL	16	816,116	1,884	231
Marion, IN	14	812,835	1,725	212
Jefferson, KY	14	670,837	1,543	230

(continued)

County	Months	Populat	Torts	Rate
Essex, MA	28	669,984	1,843	275
Middlesex, MA	27	1,394,408	5,735	411
Norfolk, MA	19	620,957	1,733	279
Suffolk, MA	22	639,192	5,023	786
Worcester, MA	24	708,164	1,673	236
Oakland, MI	12	1,118,611	5,279	472
Wayne, MI	12	2,096,179	16,739	799
Hennepin, MN	13	1,041,332	2,485	239
St. Louis, MO	21	1,000,690	1,684	168
Bergen, NJ	20	834,983	6,830	818
Essex, NJ	21	773,420	10,544	1,363
Middlesex, NJ	19	684,456	7,080	1,034
New York, NY	25	1,489,066	9,150	614
Cuyahoga, OH	12	1,411,209	9,589	679
Franklin, OH	16	992,095	3,951	398
Allegheny, PA	20	1,334,396	5,430	407
Philadelphia, PA	24	1,552,572	18,283	1,178
Bexar, TX	18	1,233,096	3,087	250
Dallas, TX	13	1,913,395	6,411	335
Harris, TX	23	2,971,755	11,483	386
Fairfax, VA	16	877,531	3,370	384
King, WA	13	1,557,537	5,057	325
Milwaukee, WI	10	951,884	3,234	340

SOURCE: U.S. Department of Justice, *Tort Cases in Large Counties*, Bureau of Justice Statistics Special Report, April 1995.

a. Create a histogram and boxplot of the months variable. What is the general shape of the distribution?

b. Generate descriptive statistics for the months variable. Compare the mean and median. If you classified the distribution as being skewed right in part **a**, the mean should be greater than the median. Is it? Why not?

c. Create a dotplot of the data. Do you see two or maybe three modes? Should we classify this distribution as skewed right or multimodal? Can you explain the different modes in these data?

d. How would this multimodal distribution cause the mean to be smaller than the median in an otherwise skewed right distribution?

6.11   Regarding the tort cases in Exercise 6.10, consider whether the time required to process a tort is related to the number of torts to be processed or to the size of the county.

a. Construct a scatterplot of the number of months to process a tort versus the number of torts. Do you detect a relationship between the two variables?

b. Construct a scatterplot of the number of months to process a tort versus the population of the county. Is there a relationship between these two variables?

c. Construct a scatterplot of the number of months to process a tort versus the rate per 100,000 residents. Is there a relationship between these two variables?

d. Based on the scatterplots, is the size of the county or the number of torts processed in any way related to the time required to process a tort?

6.12   It is suspected that schizophrenia causes changes in the activity of a substance called dopamine in the central nervous system. Twenty-five patients with schizophrenia were classified as psychotic or nonpsychotic after being treated with an antipsychotic drug. Cerebrospinal fluid was taken from each patient and assayed for dopamine b-hydroxylase (DBH) activity. Compare the DBH activity [units are nmol/(ml)(h)/mg of protein] for the two groups. For ease of calculation, the data values in the worksheet may be multiplied by 10,000, so that .0104, for example, becomes 104.

Worksheet: Dopamine.mtw

Nonpsych (group 1)

.0104	.0105	.0112	.0116	.0130	.0145	.0154	.0156
.0170	.0180	.0200	.0200	.0210	.0230	.0252	

Psychotic (group 2)

.0150	.0204	.0208	.0222	.0226	.0245	.0270
.0275	.0306	.0320				

SOURCE: D. E. Sternberg, D. P. Van Kammen, and W. E. Bunney, "Schizophrenia: Dopamine b-Hydroxylase Activity and Treatment Response," *Science* 216 (1982): 1423–1425.

a. Create boxplots of the two samples. Do the samples appear to be symmetric or skewed?
b. Construct normal probability plots of the two groups. Does each group appear to be normally distributed?
c. Calculate descriptive statistics for the two groups. Is there much difference between the means? Between the medians?
d. From the appearance of the boxplots and the descriptive statistics, does one group tend to have higher DBH activity? Which group? Was the antipsychotic drug effective?

6.13 Old Faithful is a geyser in Yellowstone National Park in Wyoming. It is a hot-water spring that becomes unstable and spews hot water and steam into the air. It is called Old Faithful because the eruptions follow a rather predictable pattern. Data on the wait times between successive eruptions of Old Faithful were obtained between August 1 and August 15, 1985, and are stored in the worksheet: **Faithful.mtw**. The histogram is for 299 wait times (in minutes) associated with the 300 observed eruptions.

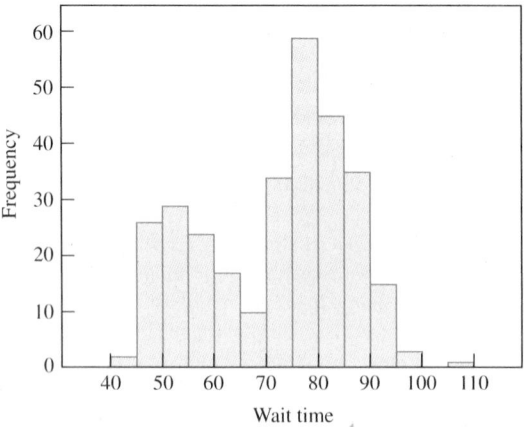

SOURCE: A. Azzalini and A. Bowman, "A Look at Some Data on the Old Faithful Geyser," *Journal of the Royal Statistical Society*, Series C, 39 (1990): 357–366.

What observations can you make from this histogram? Certainly, Old Faithful is not 100% predictable. There is going to be some variability in the wait times between eruptions, but is the histogram what one would expect? Shouldn't the wait times be bell shaped? If visitors to the park came up to you and asked how long they should expect to wait for an eruption, what would you tell them? Here are the descriptive statistics associated with the data:

```
Descriptive Statistics

Variable        N     Mean    Median    TrMean    StDev    SEMean
waitime        299   72.314   76.000    72.439   13.890    0.803


Variable      Min      Max       Q1        Q3
waitime     43.000   108.000   59.000    83.000
```

Would you tell them that they should expect to wait about 72 minutes (72.314 to be exact) from the last eruption? Is 72 minutes a reasonable measure of the expected wait time? We can see clearly from the histogram that 72 minutes is not very likely to occur.

Geophysicists have studied Old Faithful and have determined that there are two types of eruptions: those of short duration and those of long duration. The short eruptions have a duration just less than 3 minutes, and the long eruptions have a duration in excess of 3 minutes. A long-duration eruption tends to clear out the geyser tube of hot water, which causes a longer wait time (the time it takes to heat the new water to the boiling point) before the next eruption. Thus, a long wait time follows a long eruption, and a short wait time follows a short eruption. So, if the last eruption lasted more than 3 minutes, then visitors should expect to have a long wait time before the next eruption. Here are descriptive statistics for the wait times associated with short and long eruptions:

```
Descriptive Statistics

Variable  eruption    N    Mean   Median   TrMean   StDev   SEMean
waitime   short 1    110  57.109  55.000   56.612   8.148   0.777
          long  2    189  81.164  81.000   81.263   7.303   0.531


Variable  eruption   Min      Max       Q1       Q3
waitime   short 1   43.000   77.000   50.000   61.250
          long  2   59.000  108.000   77.000   87.000
```

Following a short eruption, how long should a visitor expect to wait before the next eruption? Following a long eruption, how long should a visitor expect to wait before the next eruption? Do these values seem consistent with the histogram?

6.2 PARAMETERS AND ESTIMATORS

Using the procedures outlined in the previous section, we are able to use sample data to predict the shape of a population distribution. To further describe the population distribution, we need to look at the numerical characteristics of the population. Recall from Chapter 2 that these descriptive measures are called *parameters*.

**Choosing a
Parameter**

In an inference problem, we must decide which parameter best describes the population characteristic of interest. For Bernoulli populations, the task is straightforward; we are concerned with π, the proportion of successes. For bivariate data, we are generally concerned with estimating ρ, the population correlation coefficient, and the regression equation that relates the two variables. For a univariate measurement distribution, we are interested in where the distribution is centered and a measure of its variability. Deciding which parameters best describe the center and the variability is not quite so simple. For example, to measure the center of a distribution, which do you prefer: the population mean, μ, or the population median, θ? The answer depends on the shape of the population distribution.

*Symmetric
Distributions*

For symmetric distributions, all natural measures of the center, including the mean, median, and trimmed means, coincide. Any one of the three is a good measure of the center. To avoid confusion, however, we identify the population mean, μ, as the measure of center associated with a symmetric distribution.

The most common parameter used to measure variability is the population standard deviation, σ. Because it is a measure of variation about the population mean, μ, it should be used when μ is the parameter that measures the center of a distribution. That is, σ is used most often to measure population variability when the distribution is symmetric.

EXAMPLE 6.6   **R. A. Fisher Data on *Iris setosa***

In Example 6.3, we concluded that the sepal length of *Iris setosa* is most likely normally distributed. Therefore, the mean sepal length, μ, and the standard deviation, σ, are the two parameters that should be identified as measures of the center and variability of the distribution.

*Skewed
Distributions*

For skewed distributions, the choice of a parameter that measures the center depends on the situation. For example, an insurance company concerned with losses incurred by its clients' claims may be more interested in the "mean payout" than in the "median payout" because premiums are determined by the mean payouts. In fact, it is conceivable that the insurance company will be interested in estimating a quantity even beyond the mean payout in order to protect its profit. More generally, however, in disciplines such as economics, education, and the sciences, the population median is the preferred measure of the center of a skewed distribution.

To measure variability in skewed distributions, recall that the population median, θ, divides the population frequency distribution in half, and the population quartiles, θ_1 and θ_3, further subdivide it into quarters. The distance between the first and third quartiles, the q-spread $= \theta_3 - \theta_1$, is the preferred measure of dispersion for skewed populations.

EXAMPLE 6.7

Challenger Accident

The analysis of the *Challenger* temperatures in Example 6.1 gave us two interpretations. When the 31° temperature the day of the *Challenger* accident was included, we concluded that the distribution is skewed left. Based on this conclusion, we can identify the population median, θ, as the parameter that best represents the center of the distribution. If we choose to exclude the crucial 31° temperature, then the distribution of temperatures is close to being symmetric. Then either the mean or the median suffices as a measure of the center of the distribution. Because the median, θ, is a resistant measure, it is probably a more reasonable measure of a "typical" temperature for the launch of a space shuttle.

Choosing an Estimator

Point Estimator

Having chosen a parameter to describe a particular characteristic of a population distribution, we must choose a statistic that will be reasonably close to the unknown value of the parameter. Which sample statistic to use in a given situation is an important decision. Obviously we want to choose the one that generally gives the best estimate of the parameter. The statistic that we decide to use to estimate the parameter is called a *point estimator* (or just *estimator*) of the parameter.

> A **point estimator** of a parameter is a *statistic* whose values should be close to the true value of the parameter. The actual numerical value that the point estimator assumes from the collected data (the sample) is called the **point estimate**.

EXAMPLE 6.8

Here is a sample of starting salaries of 50 chemistry majors selected at random from a large midwestern university:

Worksheet: Chemist.mtw

29,400	29,720	27,600	29,500	30,520	30,230	28,100	30,400	23,200	29,950
30,200	32,330	32,410	30,530	34,200	32,670	30,300	26,950	29,760	31,460
27,950	31,500	36,250	29,510	27,980	30,350	29,280	28,300	33,100	26,930
30,890	32,840	27,760	26,950	31,250	29,970	28,960	30,680	25,480	29,940
31,760	28,640	24,100	31,320	28,130	28,600	27,720	33,550	31,380	29,900

A histogram and a normal probability plot will demonstrate that the distribution is symmetric and very close to being normally distributed. Therefore, we are interested in the population mean as a measure of the center of the distribution and a typical salary. Several possible estimators of the mean salary along with the resulting estimates are given here:

If the estimator is $\bar{y}$, then the estimate is 29,808.

If the estimator is M, then the estimate is 29,945.

If the estimator is a 10% trimmed mean $\bar{y}_{T.10}$, then the estimate is 29,845.

If the estimator is the midrange, then the estimate is 29,725.

In the preceding example, we must decide which estimator is the most reliable for estimating the mean salary. That is, which estimator gives consistently closer values to the unknown parameter? By studying the sampling distribution of an estimator, we can investigate the various values it can assume and possibly determine whether its potential values are "close" to the value of the parameter.

Unbiased Estimator

Figure 6.10 shows the sampling distributions of two estimators, A and B. We see that the sampling distribution of estimator A is centered at α, and the sampling distribution of estimator B is centered at β. Because the potential values of A are clustered about α, we would expect A to provide a good estimate of α. By the same token, we would not expect A to provide a good estimate of β. It would tend to always underestimate β. On the other hand, estimator B would provide a good estimate of β and not of α. When the sampling distribution of an estimator is centered at the parameter being estimated, we say that it is an *unbiased estimator*.

FIGURE 6.10

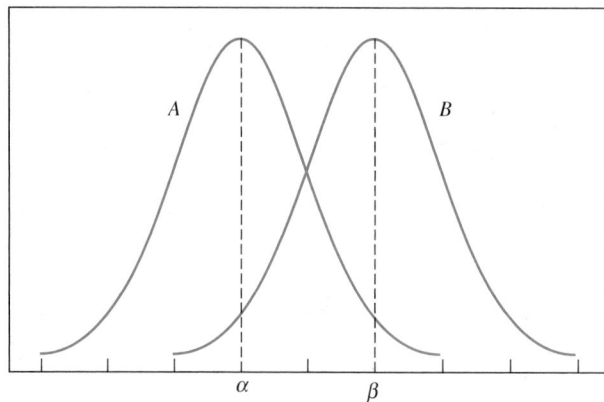

If the sampling distribution of an estimator has a mean equal to the parameter being estimated, then it is called an **unbiased estimator** of the parameter. If the mean of the sampling distribution is not equal to the parameter, the estimator is said to be **biased**.

EXAMPLE 6.9   *p* Is an Unbiased Estimator of π

Consider a Bernoulli population with unknown parameter π. Let p be the proportion of successes in a sample of size n. The Central Limit Theorem says that the sampling distribution of p is approximately normal. Furthermore, it has a mean of π and a standard deviation of $\sqrt{\pi(1 - \pi)/n}$. Figure 6.11 illustrates the sampling distribution of p, and we see that the potential values of p are centered at π, the mean of the parent distribution. Because the sampling distribution of p has mean π, we say that p is an unbiased estimator of π.

FIGURE 6.11   Sampling distribution of p

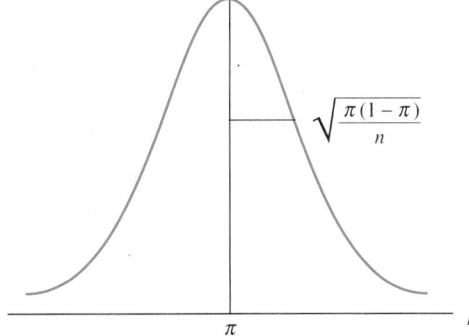

There are several characteristics that an estimator can have when estimating a parameter. Unbiasedness is one of the more important characteristics used for choosing an estimator.

EXAMPLE 6.10

$\bar{y}$ Is an Unbiased Estimator of μ

Suppose that a population distribution has unknown mean μ and standard deviation σ. A random sample, $y_1, y_2, y_3, \ldots, y_n$, is selected from the population. We noted in Chapter 5 that the mean of the sampling distribution of $\bar{y}$ is μ and the standard deviation is $\sigma/\sqrt{n}$. Thus, if $\bar{y}$ is the chosen estimator of μ, then it is an unbiased estimator because its sampling distribution has mean μ.

In repeated sampling, an unbiased estimator will average out to equal the parameter in question. Using it will not result in a systematic overestimate or underestimate of the unknown parameter, as is the case with a biased estimator.

EXAMPLE 6.11

s^2 Is an Unbiased Estimator of σ^2

Suppose that we wish to estimate the population variance σ^2. A reasonable estimator of σ^2 is the sample variance, which is given by the formula

$$s^2 = \sum \frac{(y_i - \bar{y})^2}{n - 1}$$

It can be shown with more advanced statistical theory that the sampling distribution of s^2 has mean σ^2; that is, s^2, as defined above, is an unbiased estimator of σ^2.

When we first learned about s^2 in Section 2.5, it was tempting to divide by n instead of by the degrees of freedom, $n - 1$. Suppose the sample variance had been defined as

$$V = \sum \frac{(y_i - \bar{y})^2}{n}$$

with a denominator of n instead of $n - 1$. In this case, it can be shown that the sampling distribution of V has a mean of

$$\left(\frac{n - 1}{n} \right) \sigma^2$$

Because V does not have mean σ^2, it is a biased estimator of σ^2 and will tend to underestimate it. For example, if $n = 10$, then V will, on average, estimate $\frac{9}{10}$ of σ^2.

Example 6.10 shows that $\bar{y}$ is always an unbiased estimator of μ. If, in addition, we know that the parent distribution is continuous and symmetric, then the sample median and the trimmed means are also unbiased estimators of μ. Thus, deciding which to use to estimate μ in a given situation requires that we look further at the sampling distribution of the possible estimators.

Variability of Estimators

The possible values of an unbiased estimator are centered at the parameter being estimated. But unless the standard deviation of the estimator is small, there is no guarantee that its value from a particular sample will be close to the parameter being estimated. For example, estimator A in Figure 6.12 is an unbiased estimator of μ but has a high probability of considerable deviation from μ. This is because its standard deviation, which measures the variability in its sampling distribution, is large. On the other hand, unbiased estimator B in Figure 6.12 has a sampling distribution with a smaller standard deviation. Consequently, the potential values (from the various samples) of estimator B have a higher probability of being closer to μ.

FIGURE 6.12   Sampling distribution of two unbiased estimators

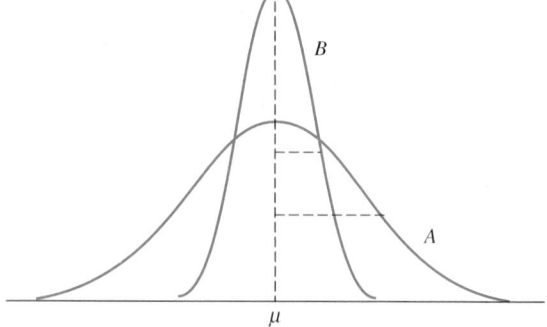

Restricting ourselves to unbiased estimators, we can say that the smaller the standard deviation, the more tightly clustered will be the potential values of the estimator about the unknown parameter. If the standard deviation of an unbiased estimator were zero, then every sample would yield the same value for the estimator, which would coincide with the value of the parameter. That is, we could estimate the unknown parameter with 100% accuracy. However, no estimators (except in trivial cases) have zero standard deviation. What we strive for is an unbiased estimator that has a standard deviation smaller than the standard deviation of any other unbiased estimator.

EXAMPLE 6.12

Standard Deviation of $\bar{y}$ When the Parent Distribution Is Normal

If the parent distribution is *normal* with mean μ and standard deviation σ, then the sample median and the sample mean are both unbiased estimators of μ. Which is better?

SOLUTION

We know that the standard deviation of the sample mean is always

$$\text{Stdev}(\bar{y}) = \frac{\sigma}{\sqrt{n}}$$

Using the normality condition, we can show that, for large samples, the standard deviation of the sample median is approximately 1.2533 times greater than the standard deviation of the sample mean; that is,

$$\text{Stdev}(M) \approx 1.2533 \left(\frac{\sigma}{\sqrt{n}} \right)$$

Because $\text{Stdev}(\bar{y})$ is less than $\text{Stdev}(M)$, the potential values of $\bar{y}$ are more tightly clustered about μ than are the potential values of the sample median. This was illustrated with the simulations in Figure 5.12 of Section 5.4, which showed that the sampling distribution of $\bar{y}$ has a smaller amount of variation than either the median or the midrange when the parent population is normally distributed. In fact, under the normality condition, it can be shown that $\bar{y}$ has a smaller standard deviation than any other unbiased estimator of μ.

If the normality assumption in Example 6.12 is dropped, it is possible that the sample median might have a smaller standard deviation than the sample mean. We cannot say for sure because it depends on the distribution of the parent population.

The Laplace distribution (a discussion of its characteristics can be found in a more advanced book) is a symmetric distribution with *long tails*. In this case, it can be shown that for large samples, the sample median has an approximate standard deviation given by

$$\text{Stdev}(M) \approx .7071 \left(\frac{\sigma}{\sqrt{n}} \right)$$

Because $\text{Stdev}(\bar{y}) = \sigma/\sqrt{n}$, we have, in the case of this long-tailed distribution, that the median is a better estimator of μ than is $\bar{y}$.

Robust Estimator   In some applications we may not know whether we are sampling from a normal, a Laplace, or some other type of distribution. So we cannot say for sure which is the better estimator. In those cases, we need a *robust estimator*.

> Remember that the normal probability plot and the boxplot can be used to give some indication of the lengths of the tails of the parent distribution.

> Statistical procedures that are insensitive to departures from assumptions are called *robust*.

A robust estimator is one that works well in a wide variety of population distributions. One robust estimator of the center of a distribution is the trimmed mean, $\bar{y}_T$. A long-tailed distribution tends to produce outliers that have an adverse effect on $\bar{y}$. Trimming eliminates the outliers, and hence $\bar{y}_T$ is not adversely affected by them.

The choice of an estimator to estimate a particular parameter depends on the shape of the underlying parent distribution and the properties of the estimator. We seek an estimator that is unbiased, has a small standard deviation, and is reasonably robust. In the next section, we look at some of the more important estimation problems.

EXERCISES 6.2

6.14   The average IQ is **100** as measured by the Stanford-Binet IQ test. What parameter is this: π, μ, θ, or σ?

6.15   The salaries of teachers in our state do not vary much. More than half are within $\pm\$3,000$ of the mean. What parameter is this: π, μ, θ, σ, or ρ?

6.16   A score of **20** on a test divides the upper 50% from the lower 50%. What parameter is this: π, μ, θ, σ, or ρ?

6.17   The leading cause of death among 15- to 19-year-olds is car accidents, which account for **45%** of teenage deaths. What parameter is this: π, μ, θ, σ, or ρ?

6.18   Why is p an unbiased estimator of π?

6.19   If the parent distribution is symmetric with tails that are not excessively long, then which parameter best represents the center of the distribution?
   a. Median   c. Trimmed mean
   b. Mean   d. Midrange

6.20   Suppose the parent population is *not* normal and has mean μ.
   a. Does $\bar{y}$ have to be an unbiased estimator of μ?
   b. Is there possibly an estimator of μ with a smaller standard deviation than $\bar{y}$?

6.21 Sketch a picture of the sampling distribution of a biased estimator of μ.

6.22 Should $\bar{y}_T$ always be used to estimate μ in a nonnormal population?

6.23 Suppose that the sample size is 100 and we wish to estimate the population variance σ^2 with the estimator V given in Example 6.11. Will V underestimate or overestimate σ^2? In general, what percent of σ^2 will V estimate? Is the bias in using V to estimate σ^2 a serious problem when the sample size is large, as it is here?

6.24 Assume the bull's-eye in the targets shown here represents a parameter that is to be estimated. Each shot represents a value of the estimator that is to be used to estimate the parameter. To illustrate the concepts of bias and variability of estimators, classify each of the four targets according to bias (high or low) and variability (high or low).

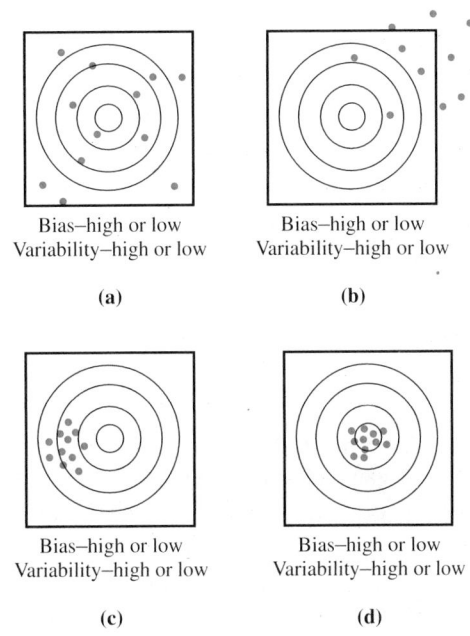

Bias–high or low
Variability–high or low

(a)

Bias–high or low
Variability–high or low

(b)

Bias–high or low
Variability–high or low

(c)

Bias–high or low
Variability–high or low

(d)

6.25 The manufacturer of a computerized airline ticket reservation system has contracted to deliver a system that will check reservations in a mean time of 40 seconds. Identify the parameter of interest. Give two possible statistics that might be used to estimate this parameter. From the following sample of 20 reservations, calculate the values of your two statistics. Which do you prefer?

Worksheet: Ticket.mtw

| 40 | 45 | 29 | 58 | 48 | 39 | 30 | 50 | 33 | 42 |
| 63 | 31 | 27 | 47 | 58 | 32 | 51 | 34 | 39 | 51 |

6.26 The time it takes a subway to travel from the airport to downtown was recorded on 30 different trips. Identify a parameter of interest. From the sample on page 376, calculate the values of two statistics that might be used to estimate the parameter. Which do you prefer?

Worksheet: Subway.mtw

34.4	41.6	39.9	40.1	40.4	39.7	37.6	39.2	40.1	38.8
41.2	37.9	40.3	42.5	38.9	39.5	39.9	40.4	42.5	41.1
39.4	37.7	40.2	40.1	39.3	39.6	40.0	41.3	40.4	38.3

6.27   An animal behaviorist is studying the time it takes an animal to perform a certain behavioral task. The variable measured is the time from the beginning of the experiment until each individual animal has performed. The researcher is interested in the mean time of performance. Some animals are slow to respond and, in fact, some may never respond properly, which makes the calculation of the mean impossible. What parameter is a meaningful measure of the "typical" time of performance?

6.28   When researchers study the response time to a drug or poison, some subjects will react very quickly whereas others do not react for a long time. What is a meaningful measure of the average response time?

6.29   Consider the starting salaries of the chemistry majors in Example 6.8. From worksheet: **Chemist.mtw**, construct a histogram, boxplot, and normal probability plot. Is the parent population normally distributed? Of the four estimators given in Example 6.8, which is the best estimator of the mean salary?

6.30   In medical studies on the survival times of patients, it is often necessary to *censor* the data. Consider the problem of studying the survival times of cancer patients who have received a certain treatment. At some point, researchers must stop the study and evaluate the results. Those patients who continue to live beyond the time the study is terminated will have a survival time longer than what was observed. For example, a patient who has survived 765 days and continues to live after the study is terminated has the survival time recorded as 765+, meaning that the patient has survived beyond 765 days.

Patients with small cell lung cancer (SCLC) generally receive a standard treatment of a combination of etoposide (E) and cisplatin (P); however, the optimal sequencing of the two has not been established. The following data were obtained from a study that compared two regimens: arm A—P followed by E and arm B—E followed by P. Days give the survival time in days, and the age variable is the entry age of the patient at the beginning of the study.

Worksheet: Censored.mtw

Arm A

Days	Age	Days	Age	Days	Age	Days	Age
730	56	288	66	311	52	490	53
1,980+	70	1,123+	57	1,843+	51	755	62
260	56	442	67	455	68	1,008	64
1,883+	54	1,133+	56	315	59	525	62
1,194	74	1,204+	57	624	50	220	59
1,624+	65	429	49	473	69	464	65
967	60	470	74	354	71	1,102+	58
1,779+	66	667	65	893	55	938	72
643	74	1,110+	62	577	64	597	63
1,645+	63	622	66	441	55	476	53
749	39	980+	68	478	69	251	69
882	64	935	57	1,433+	57	539	71
164	65	152	79	1,043	64	746+	52
1,221	71	552	55	465	47	835+	54
523	47	256	52	524	58		
201	75	988	52	529	55		

(continued)

Arm B

Days	Age	Days	Age	Days	Age	Days	Age
1,225	72	199	60	1,820+	60	329	69
556	55	426	62	728	68	1,120+	56
170	68	340	59	613	70	181	61
174	60	488	68	352	36	490	67
219	58	292	66	343	51	285	72
241	62	426	59	1,232	57	1,043+	59
394	72	305	54	232	65	435	65
731	64	1,005+	68	428	68	897+	58
395	72	382	63	1,573+	42	440	79
687	58	325	77	1,457+	68	251	71
230	67	916+	52	398	65	254	63
209	75	172	73	166	70		
703	55	339	78	364	56		
799	72	371	65	789	72		
1,315	58	511	60	83	63		
265	72	372	44	757	45		

SOURCE: Z. Ying, S. Jung, and L. Wei, "Survival Analysis with Median Regression Models," *Journal of the American Statistical Association* 90 (1995): 178–184.

a. For the survival time variable, create stem and leaf plots for each of the two samples. What is the general shape of the two distributions?

b. Use the Describe command to compute descriptive statistics for the two samples. Is there much difference between the mean survival time and the median survival time? Why is there such a big difference between the two? Which do you think is a better measure of how long a typical cancer patient survives?

c. Create side-by-side boxplots of the two samples. Are the two distributions similar in shape? Does there appear to be a difference in the locations of the two distributions? Does one of the regimens, A or B, appear to be better than the other?

d. Notice in arm A there are several censored values. If the study were not terminated and continued on until all patients died, what effect would this have on the median survival time? For example, if the second patient who has survived at least 1,980 days, lived another 3,000 days, would the median survival time change? Would the mean survival time change?

6.3   ESTIMATING CERTAIN POPULATION PARAMETERS

For a population with a continuous distribution, we are concerned with measures of the center of the distribution (mean and median) and measures of dispersion (standard deviation). For Bernoulli populations, we are concerned with the proportion of successes. In this section, we estimate these parameters.

Measures of a Distribution's Center

Population Mean   In Section 6.2, we pointed out that all natural measures of the center of a symmetric distribution coincide. Thus, the problem of estimating the center of a symmetric distribution becomes estimating the point of symmetry. For simplicity, we estimate the population mean, μ.

Many different point estimators might be used to estimate μ. Among the possible choices are the sample mean, the sample median, the midrange, and any of the trimmed means. The appropriate choice depends on the underlying parent distribution.

Normal Parent Distribution   If the parent distribution is normal, then there is no better estimator of μ than the sample mean, $\bar{y}$. There are two reasons for this:

1. $\bar{y}$ is an unbiased estimator of μ, regardless of the parent distribution (Example 6.10).
2. Under the normality condition, $\bar{y}$ has a smaller standard deviation than any other unbiased estimator of μ (Example 6.12).

Thus, $\bar{y}$ is the best choice for estimating μ when we are convinced that the parent distribution is normal or near normal.

EXAMPLE 6.13   **R. A. Fisher Data on *Iris setosa***

In Example 6.3, we concluded that the sepal length of *Iris setosa* is most likely normally distributed. In Example 6.6, we determined that the mean sepal length, μ, is the parameter that we should use to measure the center of the distribution. Given that the distribution is normal, the sample mean, $\bar{y}$, is the best estimator of μ. From the descriptive statistics given in Figure 6.5, and repeated here, we see that $\bar{y} = 5.006$. Notice that the sample median and the trimmed mean are both equal to 5.000, which is very close to $\bar{y}$. This is generally the case when the parent distribution is normally distributed.

```
Descriptive Statistics
```

Variable	N	Mean	Median	TrMean	StDev	SEMean
sepalL1	50	5.0060	5.0000	5.0000	0.3525	0.0498

Variable	Min	Max	Q1	Q3
sepalL1	4.3000	5.8000	4.8000	5.2000

Symmetric Long-Tailed Distribution   If the parent distribution is symmetric and has long tails (compared with the normal distribution), $\bar{y}$ can be a terrible estimator of μ because it is very sensitive to outliers. And with a long-tailed distribution, it is likely that outliers will appear in any sample. As pointed out in Section 6.2, the trimmed mean is a robust estimator that is not adversely affected by outliers. Therefore, we should use $\bar{y}_T$ to estimate μ when there is evidence that the parent distribution is symmetric with long tails. How much to trim is a matter of debate, but normally a 10% or 20% trimmed mean is suggested. Trimming should be symmetric and enough to remove outliers.

EXAMPLE 6.14

Estimating the Mean of a Long-Tailed Distribution

A kilowatt-hour of electricity is the amount of energy required to burn ten 100-watt light bulbs for 1 hour. The list gives the basic rates per kilowatt-hour for each of the 50 states and the District of Columbia. Construct a stem and leaf plot, a letter value display, a boxplot, and a normal probability plot. Comment on the distribution's shape. Choose an appropriate measure of central tendency, and use the data to give an estimate.

Worksheet: Kilowatt.mtw

Ala	6.34	Ky	5.63	ND	5.61
Alaska	8.38	La	6.25	Ohio	7.34
Ariz	7.35	Maine	6.98	Okla	6.11
Ark	6.68	Md	6.84	Ore	3.88
Calif	6.74	Mass	8.44	Pa	7.36
Colo	6.14	Mich	6.71	RI	9.02
Conn	9.12	Minn	6.21	SC	6.00
Del	9.17	Miss	6.08	SD	6.07
D.C.	6.31	Mo	6.01	Tenn	4.80
Fla	7.26	Mont	4.21	Texas	7.17
Ga	5.94	Neb	5.91	Utah	6.88
Hawaii	11.29	Nev	6.02	Vt	6.33
Idaho	3.58	NH	8.97	Va	6.57
Ill	8.47	NJ	10.01	Wash	3.38
Ind	6.11	NM	7.34	W.Va	5.50
Iowa	6.72	NY	10.40	Wis	6.70
Kan	7.10	NC	6.19	Wyo	5.20

SOURCE: The Energy Information Administration.

SOLUTION

A complete analysis is shown in Figure 6.13 on pages 380–381. With the exception of the midE, the midsummaries are increasing slightly, which suggests that the distribution might be slightly skewed right, if not symmetric. The appearance of the normal probability plot rules out normality. The boxplot strongly suggests a long-tailed distribution, which is consistent with the normal probability plot. Even though the midsummaries are slightly increasing, the boxplot evidence is stronger, and therefore we classify this distribution as a long-tailed distribution. Therefore, a trimmed mean is a meaningful measure of the center. If indeed the population distribution is symmetric, then the trimmed mean coincides with the population mean, in which case the problem is to estimate the population mean. Note that $\bar{y} = 6.761$ is close to $\bar{y}_{T.05} = 6.717$. Either is a reasonable estimate of the population mean; however, due to the long-tailed nature of the distribution, we are inclined to go with the trimmed mean.

FIGURE 6.13

```
Character Stem-and-Leaf Display

Stem-and-leaf of rate      N  = 51
Leaf Unit = 0.10

     1      3 3
     3      3 58
     4      4 2
     5      4 8
     6      5 2
    11      5 56699
    25      6 00000111122333
    (9)     6 567777889
    17      7 1123333
    10      7
    10      8 344
     7      8 9
     6      9 011
     3      9
     3     10 04
     1     10
     1     11 2
```

```
Letter Value Display

        DEPTH       LOWER        UPPER         MID        SPREAD
N=      51
M       26.0              6.570               6.570
H       13.5        6.015        7.340        6.678        1.325
E        7.0        5.500        8.970      * 7.235        3.470
D        4.0        4.210        9.170        6.690        4.960
C        2.5        3.730       10.205        6.967        6.475
B        1.5        3.480       10.845        7.162        7.365
         1          3.380       11.290        7.335        7.910
```

(continued)

Descriptive Statistics

Variable	N	Mean	Median	TrMean	StDev	SEMean
rate	51	6.761	6.570	6.717	1.610	0.225

Variable	Min	Max	Q1	Q3
rate	3.380	11.290	6.010	7.340

Character Boxplot

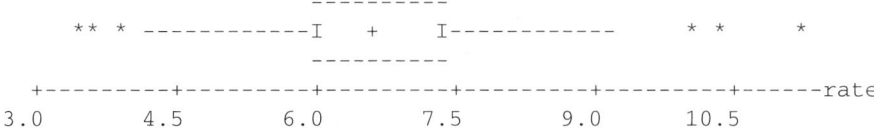

```
                      ----------
         ** *  -------------I    +    I------------      *  *          *
                      ----------

     +---------+---------+---------+---------+---------+------rate
    3.0       4.5       6.0       7.5       9.0      10.5
```

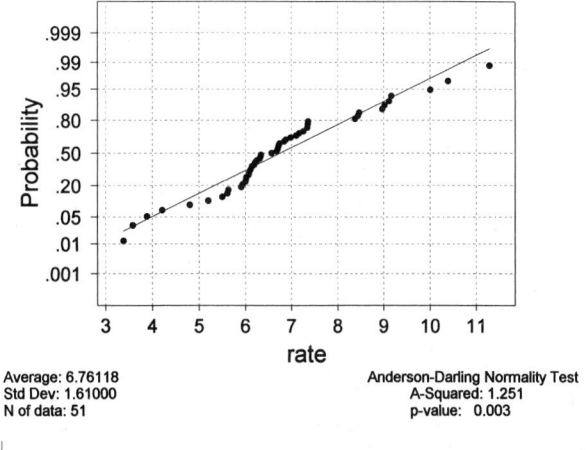

Normal Probability Plot

Average: 6.76118
Std Dev: 1.61000
N of data: 51

Anderson-Darling Normality Test
A-Squared: 1.251
p-value: 0.003

Other Symmetric Distributions   It is difficult to determine the exact shape of a parent distribution. If we are able to determine that the parent distribution is symmetric, however, recent statistical research has shown that $\bar{y}$ and the trimmed means are reasonably good estimators of μ. Generally speaking, the longer the tails of the distribution, the more one should trim. If, on the other hand, there are no tails, as in the distribution in column 2 of Figure 5.12, then the best estimator of μ is the midrange because its sampling distribution is centered at μ and has less variability than the sampling distribution of either $\bar{y}$ or M. However, $\bar{y}$ does perform reasonably well in this case, and because it is a more common measure of the center of a distribution, we choose to use it to estimate the center of short-tailed distributions.

EXAMPLE **6.15**

Estimating the Mean of a Short-Tailed Distribution

An engineer for NASA wishes to investigate the heat resistance of a particular type of electronic component to be used in a space probe. He measures the temperatures at which 30 such components fail and analyzes the results in the stem and leaf plot, boxplot, and normal probability plot in Figure 6.14. Comment on the shape of the underlying parent distribution, and estimate the mean temperature of failure of this particular type of component.

FIGURE **6.14**

Worksheet: NASA.mtw

```
Character Stem-and-Leaf Display

Stem-and-leaf of temp      N   = 30
Leaf Unit = 10

     5     2 12222
   (11)    2 55567888899
    14     3 011112444
     5     3 88899
```

```
Descriptive Statistics

Variable         N      Mean    Median    TrMean     StDev    SEMean
temp            30     300.5     298.0     300.0      54.9      10.0

Variable       Min       Max        Q1        Q3
temp         214.0     390.0     254.5     347.2
```

```
Letter Value Display

         DEPTH        LOWER        UPPER           MID        SPREAD
    N=    30
    M     15.5                   298.000                   298.000
    H      8.0        255.000    347.000       301.000      92.000
    E      4.5        226.000    386.000       306.000     160.000
    D      2.5        222.500    389.500       306.000     167.000
    C      1.5        218.000    390.000       304.000     172.000
           1          214.000    390.000       302.000     176.000
```

(*continued*)

Character Boxplot

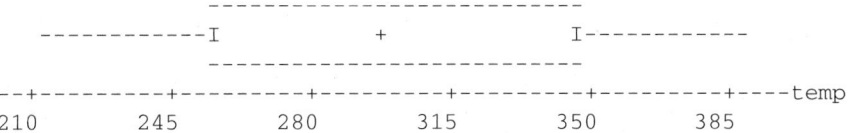

```
                        ---------------------------
        ------------I             +            I------------
                        ---------------------------
 --+---------+---------+---------+---------+---------+----temp
   210        245       280       315       350       385
```

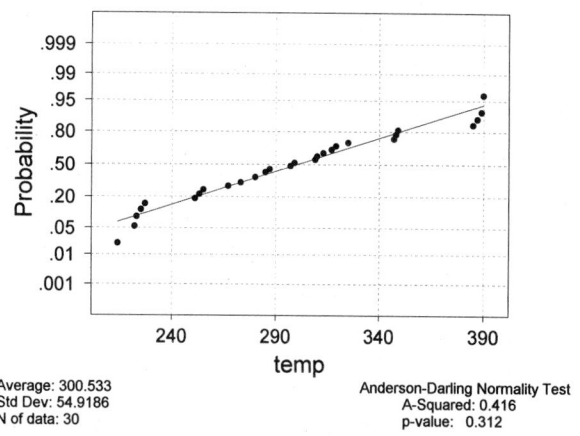

Normal Probability Plot

Average: 300.533
Std Dev: 54.9186
N of data: 30

Anderson-Darling Normality Test
A-Squared: 0.416
p-value: 0.312

SOLUTION

From the appearance of the boxplot and the normal probability plot, it is reasonable to assume that the parent distribution is symmetric with short tails. We observe that the midrange, $\bar{y}$, and $\bar{y}_{T.05}$ are all very close and any is a reasonable estimate of the mean temperature of failure. We choose to use $\bar{y} = 300.5$ as our estimate.

The performance of the estimator $\bar{y}$ is not nearly so adversely affected by a short-tailed distribution as it is by a long-tailed distribution. Although the midrange has a smaller standard deviation in the case of a distribution with no tails, $\bar{y}$ is still a reasonable estimator of the center of a distribution with tails that are shorter than normal. In fact, we recommend using $\bar{y}$ to estimate the mean of any symmetric distribution with tails that are not excessively long. But when a boxplot or other descriptive tools indicate long tails, we suggest using a trimmed mean to estimate μ.

Nonsymmetric Distributions   For nonsymmetric parent distributions, we consider parameters other than the population mean.

Population
Median

If the parent distribution is skewed, the median may be a better indicator of the center of the distribution. In such cases, the sample median should be used to estimate the population median.

EXAMPLE 6.16

Estimating the Center of a Skewed Distribution

A psychological test for measuring racial prejudice is given to a random sample of 25 high school students. Analyze the midsummaries, a boxplot, and a normal probability plot and comment on the shape of the parent distribution. Identify a parameter that best describes a "typical" score, and estimate it from the data.

Worksheet: Prejudic.mtw

59	54	41	51	87	42	65	42	44
46	74	41	58	83	58	47	62	
48	48	45	72	79	52	61	48	

SOLUTION

The letter value display, boxplot, and probability plot in Figure 6.15 are produced by Minitab. The midsummaries show a significant increase as we scan down from the median to the midrange, which indicates a skewed right distribution. The boxplot and the normal probability plot also indicate that the parent distribution is skewed right. Because the distribution is skewed, the median is a more representative measure of the center of the distribution and a typical score. From the letter value display, we find the estimate of a typical score to be the sample median, $M = 52$.

FIGURE 6.15   Letter Value Display

```
          DEPTH      LOWER        UPPER          MID        SPREAD
   N=     25
   M      13.0                52.000            52.000
   H       7.0      46.000      62.000          54.000       16.000
   E       4.0      42.000      74.000          58.000       32.000
   D       2.5      41.500      81.000          61.250       39.500
   C       1.5      41.000      85.000          63.000       44.000
           1        41.000      87.000          64.000       46.000
```

Character Boxplot

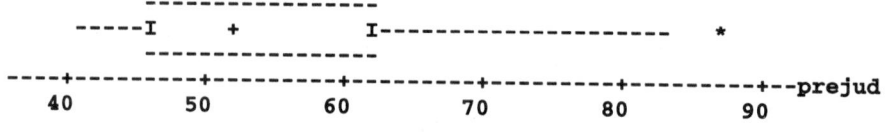

(continued)

FIGURE 6.15
(continued)

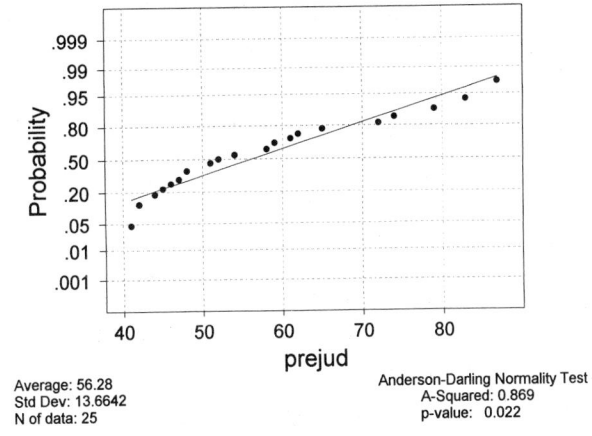

Normal Probability Plot

Average: 56.28
Std Dev: 13.6642
N of data: 25

Anderson-Darling Normality Test
A-Squared: 0.869
p-value: 0.022

There are numerous other estimators of the center of a distribution. In most cases, however, we can simplify the problem to three choices.

Estimating the Center of a Distribution

1. The sample mean, $\bar{y}$, is the recommended estimator of the population mean when the parent distribution is normal, near normal, or symmetric with tails that are *not* excessively long.

2. A trimmed sample mean, $\bar{y}_T$, is the recommended estimator of the population mean when the parent distribution is symmetric with long tails.

3. The sample median, M, is the recommended estimator of the population center when the parent distribution is skewed in either direction.

Population Proportion

To estimate π, the proportion of successes in a Bernoulli population, use p, the proportion of successes in the sample. In Example 6.9, we showed that p is an unbiased estimator of π. Because its standard deviation, $\sqrt{\pi(1-\pi)/n}$, has n in the denominator, the variability of p can be made very small by increasing the sample size. In fact, its standard deviation is smaller than the standard deviation of any other unbiased estimator of π.

EXAMPLE 6.17   **Estimating a Population Proportion**

What percent of the population think they have to wait too long when they visit their family physician? A random sample of 400 patients revealed that 280 thought they had to sit in the waiting room too long. Give an estimate of π, the proportion of the population who believe they have to wait too long.

SOLUTION

Because 280 of the 400 indicated that they had excessive waiting periods, the point estimate of π is

$$p = \frac{280}{400} = .70$$

or 70%.

Measures of a Distribution's Dispersion

Population Variance

We observed in Section 6.2 (Example 6.11) that s^2 is an unbiased estimator of the population variance, σ^2. Although the standard deviation of the sampling distribution of s^2 is not as small as the standard deviation of some other obscure estimators of σ^2, it does get arbitrarily small as the sample size increases and it is the most widely used estimator of the population variance.

Estimating the Population Variance

The sample variance, s^2, is an unbiased estimator of the population variance, σ^2.

Standard Deviation

Just because s^2 is an unbiased estimator of σ^2, it does not follow that s is an unbiased estimator of σ. The bias in using s to estimate σ is slight, however, and becomes negligible for large samples. For lack of a better estimator, we use the standard deviation, s, to estimate the population standard deviation, σ.

> The Q-spread estimates population variability and not specifically the population standard deviation.

Because the computation of s involves squared deviations, it is affected by outliers even more than $\bar{y}$. The Q-spread is a robust estimator of spread that is not as affected by outliers. Hence, when there are outliers, as in a long-tailed distribution, the Q-spread seems to be a reasonable estimator of the population variability.

The empirical rule can be used to give a quick estimate of the standard deviation. Recall that for distributions that are somewhat normal in shape, the empirical rule states that most (95%) of the distribution lies within two standard deviations of the mean. Thus, the range should cover roughly four standard deviations. Consequently, dividing the range by 4 should give an approximate estimate of the standard deviation. If we believe that the range covers the entire distribution, then we should probably divide by 6 instead of 4 because the empirical rule states that virtually all of the distribution (99.7%) is within three standard deviations of the mean.

> ## Estimating the Standard Deviation Based on the Range
>
> 1. If we believe that the range covers the middle 95% of the distribution, then an estimate of the standard deviation is Range/4.
> 2. If we believe that the range covers all of the distribution, then an estimate of the standard deviation is Range/6.

EXAMPLE **6.18**   **Using the Range to Estimate Standard Deviation**

The minimum SAT verbal score is 200 and the maximum is 800. Thus, the range is

$$R = 800 - 200 = 600$$

Dividing by 6, we have as an estimate of the standard deviation:

$$\frac{R}{6} = \frac{600}{6} = 100$$

which happens to be the actual standard deviation of SAT scores.

Remember that Range/4 or Range /6, whichever you choose to use, is a crude estimate of the population standard deviation and should not serve as a replacement for s. It should be used only when no data are available for calculating an estimate such as s. For example, in later sections, we will be confronted with the problem of finding the required sample size for a statistical study. Because the sample is unavailable (we are finding the sample size), we will have to use a method such as this to estimate the population standard deviation when it is unknown.

E X E R C I S E S   6 . 3

6.31   The survival time of a skin graft is extremely important to burn victims. The list gives survival times (in days) of closely and poorly matched skin grafts on the same burn patient. Of interest to the statistician is the difference between the survival times.

Worksheet: Skin.mtw

Patient	1	2	3	4	5	6	7	8	9	10	11
close	37	19	57	93	16	22	20	18	63	29	60
poor	29	13	15	26	11	17	26	21	43	15	40

SOURCE: R. F. Woolson and P. A. Lachenbruch, "Rank Tests for Censored Matched Pairs," *Biometrika* 67 (1980): 597–606.

a. Examine the following stem and leaf plot of the difference scores. How do you classify the general shape?

```
Character Stem-and-Leaf Display

Stem-and-leaf of differ    N  = 11
Leaf Unit = 1.0


    2     -0 63
   (4)     0 5568
    5      1 4
    4      2 00
    2      3
    2      4 2
    1      5
    1      6 7
```

b. Examine the boxplot and normal probability plot. Based on these graphs, how do you classify the general shape? Does this agree with your answer in part **a**?

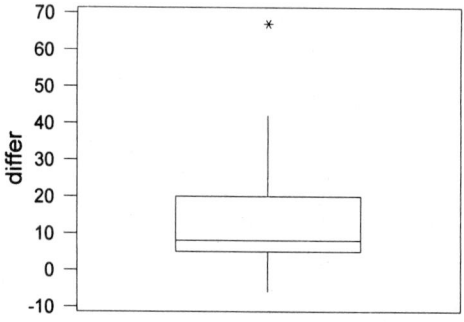
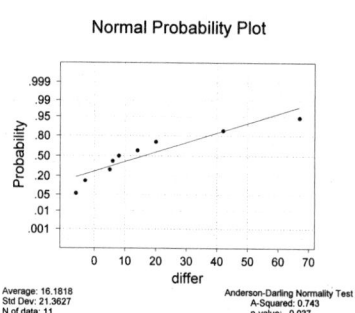

c. Based on the results obtained in parts **a** and **b**, from the following descriptive statistics, what is your estimate of the center of the distribution of difference scores? What is your estimate of the amount of variability in the difference scores?

```
Descriptive Statistics
Variable       N     Mean   Median    TrMean     StDev    SEMean
differ         11    16.18    8.00     13.00     21.36      6.44


Variable    Min      Max        Q1        Q3
differ     -6.00    67.00      5.00     20.00
```

6.32   A survey of 400 randomly selected homes in a large community reveals that a television set is turned on an average of 6.2 hours per day. Does this mean that all sets are on 6.2 hours per day? Does it mean that the average number of hours is 6.2 in all large communities? What about small or rural communities?

6.33   A survey of 182,370 freshmen at 345 schools by the American Council on Education reported that 68% enter college to make more money. Along those lines the report gave the following distribution of incomes of the parents of students who attend universities:

Income	Percent
Up to $9,999	6%
$10,000 to 19,999	13
$20,000 to 29,999	18
$30,000 to 39,999	20
$40,000 and up	43

Construct a relative frequency histogram of the data and comment on the distribution's shape. What measure of central tendency seems appropriate? Is it possible to calculate an estimate of your measure of central tendency?

6.34   The report in Exercise 6.33 also gave this distribution of incomes of the parents of students who attend predominantly black schools:

Income	Percent
Up to $9,999	25
$10,000 to 19,999	28
$20,000 to 29,999	19
$30,000 to 39,999	13
$40,000 and up	15

Construct a relative frequency histogram of the data and comment on the distributional shape. What measure of central tendency seems appropriate? Is there enough information to compute an estimate?

6.35   Surface-water salinity measurements were taken in a bottom-sampling project in Whitewater Bay, Florida. Geographic considerations lead geologists to believe that the salinity variation should be normally distributed. If this is true, it means there is free mixing and interchange between open marine water and fresh water entering the bay.

Worksheet: Salinity.mtw

46	37	62	59	40	53	58	49	60	56	58	46
47	52	51	60	46	36	34	51	60	47	40	40
35	49	48	39	36	47	59	42	61	67	53	48
50	43	44	49	46	63	53	40	50	78	48	42

SOURCE: J. Davis, *Statistics and Data Analysis in Geology*, 2nd ed. (New York: Wiley, 1986).

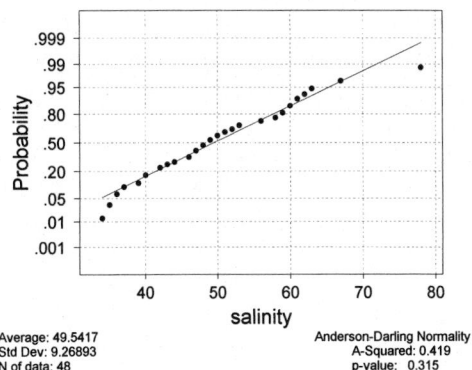

Normal Probability Plot

Average: 49.5417
Std Dev: 9.26893
N of data: 48

Anderson-Darling Normality Test
A-Squared: 0.419
p-value: 0.315

a. Does the normal probability plot rule out normality?

b. Construct a stem and leaf plot of the data. Does it appear bell shaped?

c. Construct a five-number summary diagram and examine the midsummaries. Are they all about the same?

d. If the above analysis suggests that the salinity level is approximately normally distributed, what estimator do you recommend for the mean salinity level? What is the estimate?

6.36   Suppose that in Example 6.8 we wish to investigate the amount of variability in the salaries of the chemistry majors. What statistic from the sample should we calculate to estimate this variability? Give two possible choices along with the estimate obtained from the data. Of those that you list, which do you think is the more reliable estimator of the variability?

6.37   A psychologist has developed an IQ test that should not discriminate against any ethnic group. To obtain a norm for the test, she gives the test to a random sample of 100 citizens in her community. Presented here is an analysis of the data (worksheet: **Iq.mtw**) including a stem and leaf plot, letter value display, descriptive statistics, boxplot, and normal probability plot.

```
Character Stem-and-Leaf Display

Stem-and-leaf of iq          N  = 100
Leaf Unit = 1.0

    1      7 2
    4      7 577
    9      8 12344
   18      8 667888899
   28      9 0112233344
   43      9 556666667777889
  (18)    10 000000111112223344
   39     10 55666778888999
   25     11 001123444
   16     11 55666778
    8     12 22334
    3     12 55
    1     13 0
```

```
Letter Value Display
```

	DEPTH	LOWER	UPPER	MID	SPREAD
N=	100				
M	50.5		101.000	101.000	
H	25.5	93.000	109.500	101.250	16.500
E	13.0	88.000	116.000	102.000	28.000
D	7.0	83.000	122.000	102.500	39.000
C	4.0	77.000	124.000	100.500	47.000
B	2.5	76.000	125.000	100.500	49.000
A	1.5	73.500	127.500	100.500	54.000
	1	72.000	130.000	101.000	58.000

(*continued*)

Descriptive Statistics

Variable	N	Mean	Median	TrMean	StDev	SEMean
iq	100	101.35	101.00	101.40	12.30	1.23

Variable	Min	Max	Q1	Q3
iq	72.00	130.00	93.00	109.75

Character Boxplot

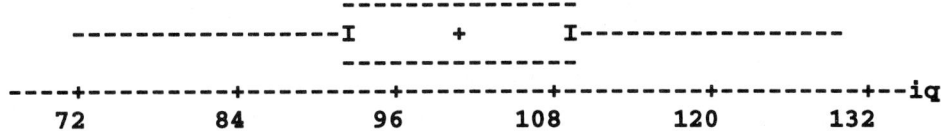

```
                                 ---------------
            ------------------I        +        I-----------------
                                 ---------------
    ----+---------+---------+---------+---------+---------+---------+--iq
        72        84        96       108       120       132
```

Normal Probability Plot

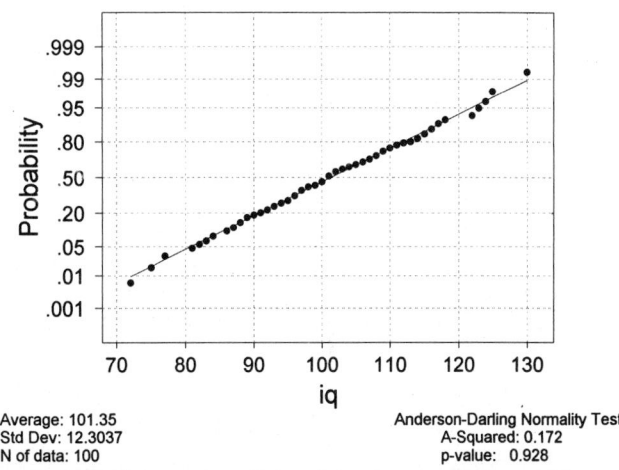

Average: 101.35
Std Dev: 12.3037
N of data: 100

Anderson-Darling Normality Test
A-Squared: 0.172
p-value: 0.928

a. What do the midsummaries tell you about the general shape of the parent distribution?
b. Does the boxplot seem consistent with the stem and leaf plot and the midsummary analysis?
c. What does the normal probability plot tell you about the parent distribution?
d. Comment on the shape of the underlying parent distribution, and estimate the mean score on the IQ test.

6.38   Here are the educational levels (in years) of a sample of 40 auto workers in a plant in Detroit:

Worksheet: Detroit.mtw

22	16	11	11	8	12	21	8	12	12
17	14	12	5	10	8	6	12	8	12
11	10	9	14	13	12	10	10	12	1
11	13	14	10	12	12	12	12	9	12

Based on the analysis presented on pages 392–393, comment on the shape of the underlying parent distribution of the educational levels, and estimate the mean of the distribution.

Descriptive Statistics

Variable	N	Mean	Median	TrMean	StDev	SEMean
educ	40	11.400	12.000	11.306	3.699	0.585

Variable	Min	Max	Q1	Q3
educ	1.000	22.000	10.000	12.000

Letter Value Display

	DEPTH	LOWER	UPPER	MID	SPREAD
N=	40				
M	20.5	12.000		12.000	
H	10.5	10.000	12.000	11.000	2.000
E	5.5	8.000	14.000	11.000	6.000
D	3.0	6.000	17.000	11.500	11.000
C	2.0	5.000	21.000	13.000	16.000
	1	1.000	22.000	11.500	21.000

Character Stem-and-Leaf Display

Stem-and-leaf of educ   N = 40
Leaf Unit = 1.0

```
     1     0 1
     1     0
     2     0 5
     3     0 6
     9     0 888899
    18     1 000001111
   (15)    1 222222222222233
     7     1 444
     4     1 67
     2     1
     2     2 1
     1     2 2
```

Character Boxplot

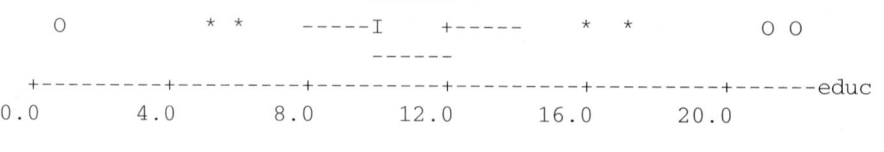

(continued)

Normal Probability Plot

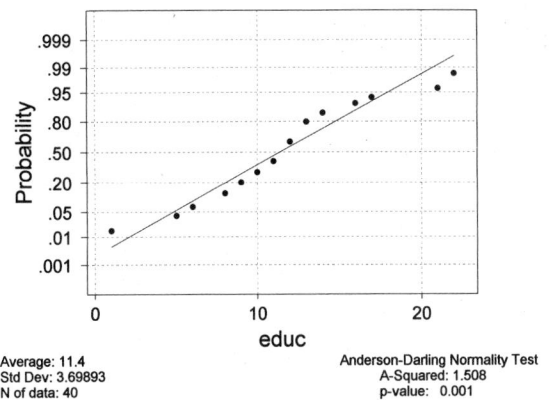

Average: 11.4
Std Dev: 3.69893
N of data: 40

Anderson-Darling Normality Test
A-Squared: 1.508
p-value: 0.001

6.39   To study the distance a professional golfer can drive a golf ball, the following distances in yards were recorded by 20 pros:

Worksheet: Golf.mtw

259	270	248	262	271	255	261	242	251	238
273	271	265	268	251	273	265	241	239	254

a. Construct a stem and leaf plot and a boxplot. What is your general impression about the shape of the underlying parent distribution?

b. Does the following normal probability plot suggest that the parent distribution might be short-tailed? Does this agree with your analysis in part **a**?

Normal Probability Plot

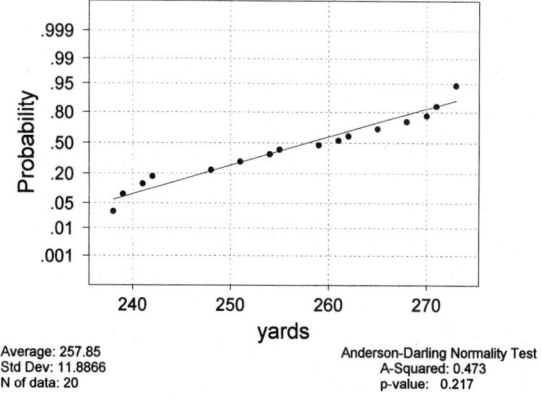

Average: 257.85
Std Dev: 11.8866
N of data: 20

Anderson-Darling Normality Test
A-Squared: 0.473
p-value: 0.217

c. From the descriptive statistics on page 394, calculate the midrange. Does it differ from the other measures of central tendency?

Descriptive Statistics

Variable	N	Mean	Median	TrMean	StDev	SEMean
yards	20	257.85	260.00	258.11	11.89	2.66

Variable	Min	Max	Q1	Q3
yards	238.00	273.00	248.75	269.50

d. Comment on the general shape of the underlying parent distribution, and estimate the center of the distribution.

6.40 A survey of freshmen was conducted to find out how many hours per week they studied for their courses. The sample of 50 freshmen revealed these scores:

Worksheet: Study.mtw

30	34	32	40	33	25	15	29	37	30
28	42	28	34	60	35	20	35	44	5
32	25	45	39	40	40	41	38	50	30
20	35	28	43	30	35	41	35	38	45
10	39	30	20	40	36	25	37	35	65

a. Construct a stem and leaf plot and a boxplot. What is your general impression about the shape of the underlying parent distribution?

b. Does the following normal probability plot suggest that the parent distribution might be long-tailed? Does this agree with your analysis in part **a**?

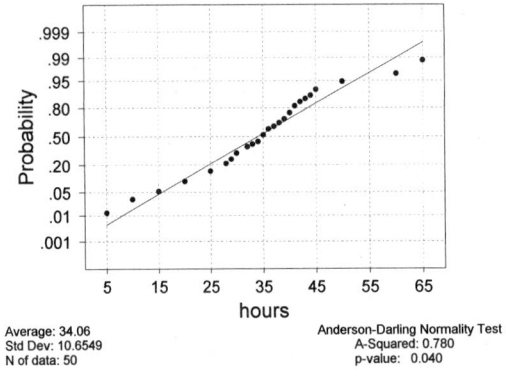

Normal Probability Plot

Average: 34.06
Std Dev: 10.6549
N of data: 50

Anderson-Darling Normality Test
A-Squared: 0.780
p-value: 0.040

c. Examine the descriptive statistics. Calculate the midrange. Why are the median and the midrange the same? Why are the mean and trimmed mean the same?

Descriptive Statistics

Variable	N	Mean	Median	TrMean	StDev	SEMean
hours	50	34.06	35.00	34.05	10.65	1.51

Variable	Min	Max	Q1	Q3
hours	5.00	65.00	28.75	40.00

d. Which statistic best estimates the center of the distribution?

6.41 A standardized exam for math competency has a national mean of 70 and a standard deviation of 12. The exam is given to 31 entering freshmen at a small community college with the results listed here:

Worksheet: Mathcomp.mtw

61	67	73	68	76	82	90	83	75	55	53
70	92	75	100	61	46	65	90	77	95	
73	85	68	75	87	70	69	65	50	70	

a. Construct a stem and leaf plot and a boxplot. What is your general impression about the shape of the underlying parent distribution?

b. Does the following normal probability plot suggest that the parent population is normally distributed? Does this agree with your analysis in part **a**?

Normal Probability Plot

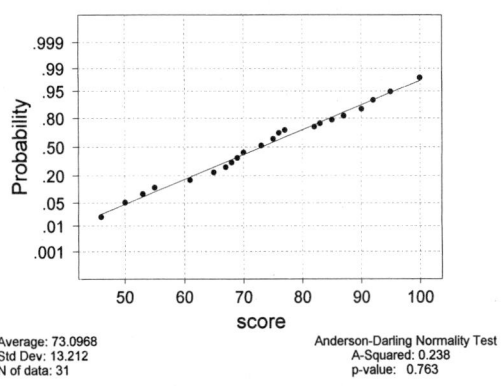

Average: 73.0968
Std Dev: 13.212
N of data: 31

Anderson-Darling Normality Test
A-Squared: 0.238
p-value: 0.763

c. Comment on the general shape of the underlying parent distribution and estimate the center of the distribution.

6.42 A sociologist who is studying ethnic problems randomly selects 50 Puerto Rican families in Miami. She records these weekly family incomes:

Worksheet: Puerto.mtw

150	280	175	190	305	380	290	300	170	315
280	255	335	180	200	210	350	360	550	225
245	175	180	260	320	345	180	225	275	280
325	350	190	220	265	270	335	310	270	240
235	355	370	260	280	350	390	250	375	250

a. Construct a stem and leaf plot, a boxplot, and a normal probability plot.

b. Are there any outliers in the data?

c. Excluding outliers, what is the general shape of the distribution?

d. Calculate descriptive statistics. Compare the mean, median, trimmed mean, and midrange. Why is the midrange so different from the other measures of central tendency?

e. Of the descriptive statistics in part **d**, which one do you think best represents the center of the distribution?

6.43 The sociologist in Exercise 6.42 has a random sample of 225 names from the registered voter roll in Miami and finds that 45 are Puerto Rican. Estimate the percent of registered voters in Miami who are Puerto Rican.

6.44   Here is a list of 25 scores on the first test in a statistics class taught at a small midwestern university:

Worksheet: Test1.mtw

61	38	64	70	81	11	61	92	77
47	98	76	83	58	97	78	14	
80	41	65	90	81	98	72	78	

From the stem and leaf plot, letter value display, descriptive statistics, boxplot, and normal probability plot of the data, classify the shape of the parent distribution and give an estimate of the typical score.

```
Character Stem-and-Leaf Display

Stem-and-leaf of test1    N  = 25
Leaf Unit = 1.0


     2    1 14
     2    2
     3    3 8
     5    4 17
     6    5 8
    10    6 1145
    (6)   7 026788
     9    8 0113
     5    9 02788
```

```
Letter Value Display

        DEPTH      LOWER        UPPER          MID         SPREAD
  N=    25
  M     13.0                 76.000           76.000
  H      7.0    61.000       81.000           71.000       20.000
  E      4.0    41.000       92.000           66.500       51.000
  D      2.5    26.000       97.500           61.750       71.500
  C      1.5    12.500       98.000           55.250       85.500
         1      11.000       98.000           54.500       87.000
```

```
Descriptive Statistics


Variable        N      Mean    Median    TrMean     StDev    SEMean
test1          25     68.44     76.00     69.65     23.48      4.70


Variable       Min       Max        Q1        Q3
test1        11.00     98.00     59.50     82.00
```

(continued)

Character Boxplot

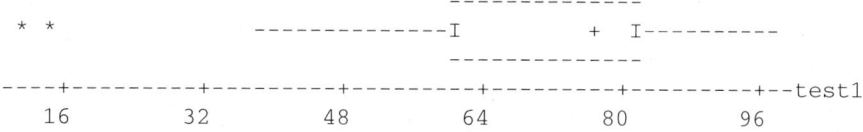

```
                                        --------------
  *  *                    --------------I        +   I----------
                                        --------------
  ----+---------+---------+---------+---------+---------+--test1
     16        32        48        64        80        96
```

Normal Probability Plot

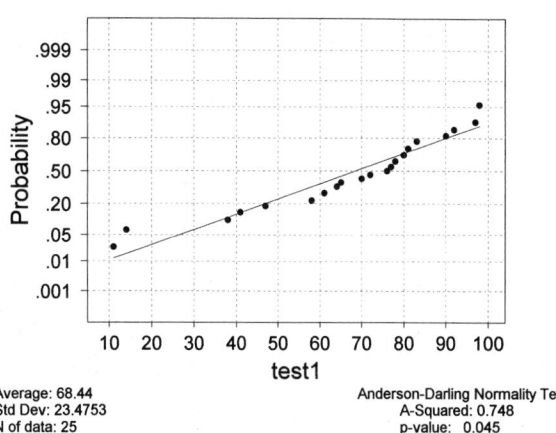

Average: 68.44
Std Dev: 23.4753
N of data: 25

Anderson-Darling Normality Test
A-Squared: 0.748
p-value: 0.045

6.4 SUMMARY AND REVIEW

KEY CONCEPTS

✓ This chapter is aimed at describing a population distribution from the collected data. The description involves three characteristics: the shape of the distribution, measures of location, and measures of variability. *Distributional shapes* are classified as near normal, skewed, symmetric short-tailed, symmetric long-tailed, and multimodal.

✓ The *stem and leaf plot, histogram*, and *boxplot* can be used to describe the general shape of the distribution. The *midsummaries* can tell us about skewness. The *normal probability plot* can be used to compare the distribution with the normal distribution.

✓ Once the shape of a distribution is determined, we can decide which parameter best describes the characteristic of interest. For example, is it better to examine the population mean or the population median?

✓ Once the parameter is determined, we are in a position to decide which statistic best estimates the parameter. In making this decision, we seek an *estimator* (the statistic used to estimate the parameter) that is unbiased, has a small standard deviation, and is robust.

✓   For symmetric distributions, the mean, the median, and all the trimmed means coincide; therefore, to estimate a typical score, we simply estimate the mean. The estimator used to estimate the mean is determined by the type of symmetric distribution we have observed.

✓   For skewed distributions, the population median is usually the measure of center used to represent a typical score. The sample median is used to estimate the population median.

LEARNING GOALS   Having completed this chapter, you should be able to:

1. Identify the various population distribution shapes. *Section 6.1*
2. Use the midsummaries to diagnose shape. *Section 6.1*
3. Interpret the results of a boxplot. *Section 6.1*
4. Use a normal probability plot to describe the shape of a distribution. *Section 6.1*
5. Choose a parameter that measures the population characteristic of interest. *Section 6.2*
6. Understand the important properties of an estimator. *Section 6.2*
7. Determine what statistic best estimates the parameter of concern. *Section 6.3*

QUESTIONS FOR REVIEW   Use the following problems to test your skills:

6.45   What statistic would you use to estimate the center of a distribution if a boxplot and a normal probability plot of the sample suggest each of these parent populations?
a. Approximately normal in shape
b. Significantly longer tails than the normal distribution
c. Significantly shorter tails than the normal distribution
d. Highly skewed left
e. Highly skewed right

6.46   If the parent distribution is symmetric with tails that are not excessively long, then which estimator best estimates the center of the distribution?
a. Median
b. Mean
c. Trimmed mean
d. Midrange

6.47   What kind of estimator works well in a wide variety of population distributions?

6.48   According to Ford Motor Company, 5 years ago, fewer than 20% of the functions in its cars were handled by computers. Now **82%** are computer-controlled. What parameter is this: π, μ, θ, σ, or ρ?

6.49   General auto mechanics earn **$30,000** a year to start, and those with top diagnostic skills sometimes earn as much as $100,000. What parameter is this: π, μ, θ, σ, or ρ?

6.50   According to the Census Bureau (Statistical Brief SB/95-20, August 1995), half of all people stay in poverty longer than **4.3** months. What parameter is this: π, μ, θ, σ, or ρ?

6.51   **Eighty percent** of the adult population drives cars. What parameter is this: π, μ, θ, σ, or ρ?

6.52   Why would you select an unbiased estimator over one that is biased? If an estimator is unbiased, are we assured that the estimate will be close to the parameter being estimated? Explain.

6.53   A lottery agent sells lottery tickets in states that have legalized lotteries. The amount the agent earns depends on the number of tickets he or she sells. Here are the average annual sales commissions paid lottery agents in 17 states and the District of Columbia:

Worksheet: Lottery.mtw

State	Number of agents	Average sales commission
Arizona	2,100	$2,228
Colorado	2,500	4,880
Connecticut	3,000	3,200
District of Columbia	850	3,986
Delaware	324	5,247
Illinois	7,710	3,320
Maine	1,629	490
Maryland	1,239	20,399
Massachusetts	3,600	5,279
Michigan	7,293	4,758
New Hampshire	1,150	574
New Jersey	4,000	12,000
New York	12,449	4,281
Ohio	5,400	3,704
Pennsylvania	7,700	8,013
Rhode Island	883	7,667
Vermont	750	307
Washington	4,711	1,804

SOURCE: Gaming and wagering business.

a. Construct a stem and leaf plot of the average sales commissions.
b. From the following letter value display, examine the midsummaries for skewness. If the distribution appears to be skewed, is it worthwhile to check for normality? Explain.

Letter Value Display

	DEPTH	LOWER	UPPER	MID	SPREAD
N=	18				
M	9.5	4133.500		4133.500	
H	5.0	2228.000	5279.000	3753.500	3051.000
E	3.0	574.000	8013.000	4293.500	7439.000
D	2.0	490.000	12000.000	6245.000	11510.000
	1	307.000	20399.000	10353.000	20092.000

6.54   The boxplot shows the average annual sales commissions paid to lottery agents in the 18 lotteries listed in Exercise 6.53. Make a general statement about the shape of the distribution, and identify a parameter that best measures the center of the distribution. Give an estimate of the parameter.

Character Boxplot

6.55   The list on page 400 gives porosity measurements (percent) on 20 samples of Tensleep Sandstone, Pennsylvanian, from Wind River Basin and Bighorn Basin in Wyoming. Construct a stem and leaf plot and a five-number summary diagram for these data. Calculate the midsummaries and use them to evaluate possible skewness in the distribution.

Worksheet: Porosity.mtw

15	10	15	23	18	26	24	18	19	21
13	17	15	23	27	29	18	27	20	24

SOURCE: J. C. Davis, *Statistics and Data Analysis in Geology*, 2nd ed. (New York: John Wiley, 1986), pp. 63–65.

6.56 Use the information in Exercise 6.55 to construct a boxplot. From the boxplot make a general statement about the shape of the distribution. What parameter best describes a "typical" score? What is your estimate of the parameter?

6.57 Here is a normal probability plot of the porosity measurements given in Exercise 6.55. Does the plot indicate that the data are from a normal population? If not, how do you classify the shape of the distribution? Does this agree with your conclusion in Exercises 6.55 and 6.56?

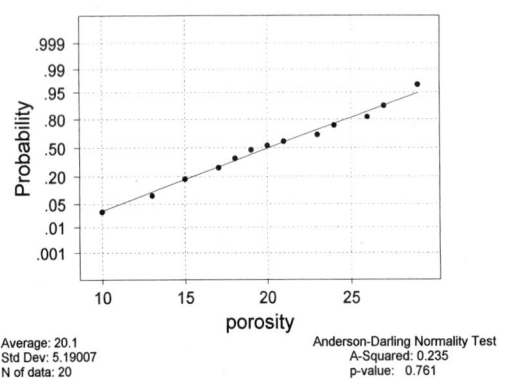

Normal Probability Plot

Average: 20.1
Std Dev: 5.19007
N of data: 20

Anderson-Darling Normality Test
A-Squared: 0.235
p-value: 0.761

6.58 Explain why the estimator V in Example 6.11 is a biased estimator of σ^2. Explain what happens when n becomes very large.

6.59 The table gives the median charges for coronary bypass surgery at 17 hospitals in North Carolina:

Worksheet: Bypass.mtw

Hospital	Charge	Hospital	Charge
Carolinas Medical Center	38,578	New Hanover Regional	35,831
Duke Medical Center	31,935	Pitt County Memorial	34,001
Durham Regional	34,465	Presbyterian	27,445
Forsyth Memorial	24,810	Rex	31,419
Frye Regional	35,144	Moses Cone Memorial	34,695
High Point Region	29,245	N.C. Baptist	29,154
Memorial Mission	29,473	Univ. of North Carolina	36,745
Mercy	34,376	Wake County	31,163
Moore Regional	32,428		

SOURCE: North Carolina Medical Database Commission, *Consumer's Guide to Hospitalization Charges in North Carolina Hospitals*, Raleigh, August 1994.

a. Construct a stem and leaf plot of the charges.
b. From the following letter value display, evaluate the skewness of the distribution.

```
Letter Value Display

        DEPTH          LOWER            UPPER            MID            SPREAD
N=   17
M     9.0               32428.000                        32428.000
H     5.0          29473.000        34695.000            32084.000        5222.000
E     3.0          29154.000        35831.000            32492.500        6677.000
D     2.0          27445.000        36745.000            32095.000        9300.000
      1            24810.000        38578.000            31694.000       13768.000
```

c. If the distribution appears symmetric, is it worthwhile to check for normality? Explain.

6.60   The boxplot shows the medical charges for a coronary bypass listed in Exercise 6.59. Use it along with the results from Exercise 6.59 to make a general statement about the shape of the distribution and identify a parameter that best measures the center of the distribution. Give an estimate of the parameter.

Character Boxplot

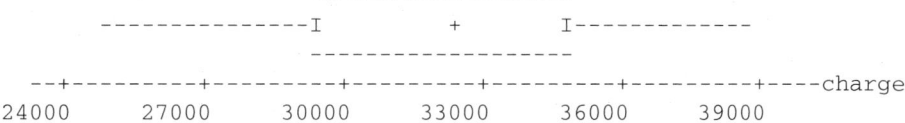

```
--+---------+---------+---------+---------+---------+---------+----charge
24000     27000     30000     33000     36000     39000
```

6.61   Here is a normal probability plot of the coronary bypass data given in Exercise 6.59. Does the plot indicate that the data are from a normal population? If not, how do you classify the shape of the distribution? Does this agree with your conclusion in Exercise 6.60?

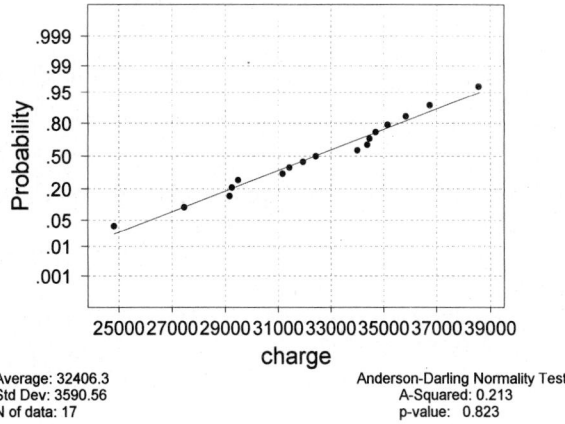

Normal Probability Plot

Average: 32406.3
Std Dev: 3590.56
N of data: 17

Anderson-Darling Normality Test
A-Squared: 0.213
p-value: 0.823

6.62   Annual food expenditures for a random sample of 40 single households in Ohio are listed at the top of page 402.

Worksheet: Food.mtw

$2,845	3,170	2,352	4,978	3,820	2,475	3,160	5,780	2,175	2,648
2,872	4,250	3,970	2,534	6,870	2,734	2,847	4,670	5,176	3,640
2,765	1,180	3,679	3,320	7,580	2,416	3,743	2,830	3,127	3,249
2,648	1,976	2,784	3,869	2,086	5,587	3,420	2,645	8,147	4,367

SOURCE: Bureau of Labor Statistics.

a. Construct a stem and leaf plot and a letter value display. Examine the midsummaries and evaluate the skewness of the distribution.

b. If the distribution appears to be skewed, is it worthwhile to check for normality? Explain.

6.63   Construct a boxplot of the data in Exercise 6.22. Use it along with the results of Exercise 6.62 to make a general statement about the shape of the distribution and identify a parameter that best measures the center of the distribution. Give an estimate of the parameter.

6.64   Shown here is a normal probability plot of the food expenditures listed in Exercise 6.62. Does the plot indicate that the data are from a normal population? If not, how do you classify the shape of the distribution? Does this agree with your conclusion in Exercise 6.63?

Normal Probability Plot

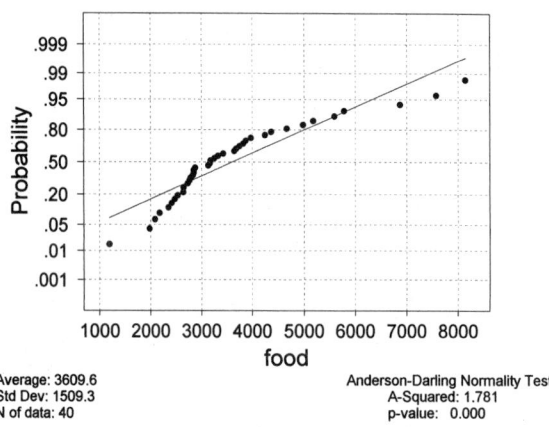

Average: 3609.6
Std Dev: 1509.3
N of data: 40

Anderson-Darling Normality Test
A-Squared: 1.781
p-value: 0.000

6.65   Generally speaking, what parameter best represents the center of a distribution in which some of the data are censored?

6.66   The student government association at a large university is interested in investigating the amount spent on a date by a typical male university student. A poll of 50 men was conducted and found these amounts (in dollars) spent on a typical weekend date:

Worksheet: Date.mtw

$22	17	5	25	20	10	8	5	40	15
0	10	15	25	0	18	15	25	12	30
0	15	25	35	30	18	22	35	20	10
0	12	23	15	10	0	5	10	50	20
15	25	15	55	25	20	17	13	30	25

Study the distribution of the amounts spent by completing a midsummary analysis and constructing a boxplot.

6.67   These are the 1994 annual salaries of the state governors:

Worksheet: Governor.mtw

Ala	81,151	Neb	65,000
Alaska	81,648	Nev	90,000
Ariz	75,000	NH	86,235
Ark	60,000	NJ	85,000
Calif	120,000	NM	90,000
Colo	70,000	NY	130,000
Conn	78,000	NC	123,300
Del	95,000	ND	68,280
Fla	101,764	Ohio	115,752
Ga	91,092	Okla	70,000
Hawaii	94,780	Ore	80,000
Idaho	75,000	Pa	105,000
Ill	103,097	RI	69,900
Ind	77,200	SC	101,959
Iowa	76,700	SD	74,649
Kan	76,476	Tenn	85,000
Ky	86,352	Texas	99,122
La	73,440	Utah	70,000
Maine	70,000	Va	110,000
Mass	75,000	Wash	121,000
Mich	112,025	W.Va	90,000
Minn	109,053	Wis	92,283
Miss	75,600	Wyo	70,000
Mo	90,312		
Mont	55,502		

SOURCE: *The World Almanac and Book of Facts*, 1995.

Complete an analysis of the data, comment on the shape of the distribution of governors' salaries, and give a measure of the center of the data.

6.68   According to the National Association of Realtors, the median price of a single-family home in the United States was $73,600 in 1984. In 1993, the median price had risen to $104,200. Listed here are the median home prices for 1984 and 1993 in 37 U.S. markets:

Worksheet: Housing.mtw

	1984	1993		1984	1993
Albany, NY	$ 52,400	109,900	Detroit	48,000	92,200
Anaheim, CA	134,900	222,200	Fort Lauderdale	76,000	99,500
Atlanta	64,600	94,000	Houston	79,600	78,000
Baltimore	65,200	113,200	Indianapolis	54,200	83,800
Birmingham, AL	66,600	89,000	Kansas City, MO	57,600	79,100
Boston	102,000	165,200	Los Angeles	115,300	199,700
Chicago	77,500	131,300	Louisville, KY	49,500	69,900
Cincinnati	59,600	85,600	Memphis, TN	65,200	84,200
Cleveland	65,600	89,200	Miami	84,400	98,000
Columbus, OH	60,400	89,300	Milwaukee	69,600	98,400
Dallas–Fort Worth	83,400	89,600	Minneapolis–St. Paul	75,100	96,400
Denver	85,000	96,300	Nashville, TN	64,000	87,000

(continued)

	1984	1993		1984	1993
New York metro area	106,900	168,000	San Antonio, TX	71,600	72,600
Oklahoma City	63,600	61,100	San Diego	102,900	175,500
Philadelphia	59,500	108,900	San Francisco	132,600	249,300
Providence, RI	61,400	112,500	San Jose, CA	121,600	297,100
Rochester, NY	62,100	83,300	Tampa, FL	60,800	69,900
St. Louis	64,400	80,900	Washington, DC	92,900	153,500
Salt Lake City	67,600	79,000			

SOURCE: National Association of Realtors.

To compare the cost of housing in 1984 and 1993, use a computer to construct summary diagrams such as stem and leaf plots and dotplots of the two samples. Calculate the midsummaries, and conduct a midsummary analysis. If the distributions are skewed, is a check for normality useful? Construct normal probability plots to shed light on the length of the tails of the two distributions. Finally construct a side-by-side boxplot. From the information, classify the shapes of the distributions and give the parameter that you think is best to compare the centers of the two distributions. Give your two estimates and compare the cost of housing in 1993 to what it was in 1984.

6.69   These are the average teachers' salaries across the states for school years 1973–74, 1983–84, and 1993–94:

Worksheet: Teacher.mtw

State	Average salary 1973–74	1983–84	1993–94	State	Average salary 1973–74	1983–84	1993–94
Ala	$ 9,226	$18,000	$27,000	Mont	$9,429	$20,657	$27,600
Alaska	15,667	36,564	44,700	Neb	9,174	18,785	27,200
Ariz	10,414	21,605	31,200	Nev	11,549	23,000	33,900
Ark	7,820	16,929	26,600	NH	9,613	17,376	33,200
Calif	13,113	26,403	40,200	NJ	11,920	23,044	41,000
Colo	10,131	22,895	33,100	NM	9,100	20,760	26,700
Conn	11,030	22,624	47,000	NY	13,371	26,750	43,300
Del	11,304	20,925	34,500	NC	10,223	18,014	29,200
D.C.	na	27,659	41,300	ND	8,493	20,363	24,500
Fla	10,018	19,545	31,100	Ohio	10,107	21,421	33,300
Ga	9,392	18,505	29,500	Okla	8,238	18,490	25,300
Hawaii	11,112	24,357	34,500	Ore	10,180	22,833	34,100
Idaho	8,383	18,640	26,300	Pa	10,921	22,800	38,700
Ill	11,871	23,345	36,500	RI	11,407	24,641	36,000
Ind	10,508	21,587	34,800	SC	8,654	17,500	28,300
Iowa	9,863	20,140	29,200	SD	8,150	16,480	23,300
Kan	8,894	19,598	30,700	Tenn	8,840	17,900	28,600
Ky	8,295	19,780	30,900	Texas	8,920	20,100	29,000
La	9,166	19,100	27,000	Utah	9,146	20,256	26,500
Maine	9,238	17,328	30,100	Vt	8,932	17,931	33,600
Md	11,741	24,095	39,500	Va	9,919	19,867	31,900
Mass	11,121	22,500	37,300	Wash	11,295	24,780	34,800
Mich	12,545	28,877	41,100	W.Va	8,467	17,482	27,400
Minn	11,122	24,480	33,700	Wis	10,830	23,000	35,200
Miss	7,604	15,895	24,400	Wyo	9,668	24,500	30,400
Mo	9,530	19,300	28,900	Average	10,778	22,019	34,100

SOURCE: National Education Association.

Clearly, the average salary for 1973–74 is lower than the average for 1983–84, which in turn is lower than for 1993–94, but other than that, are the distributions similar? How variable are the salaries? What are the shapes of the three distributions? To investigate these questions, compare side-by-side dotplots and descriptive statistics for the three groups. Construct side-by-side boxplots and normal probability plots to shed light on the length of the tails of the three distributions. From this information, compare the shapes of the three distributions.

6.70   The list gives the national track records for women in the 100-meter, 200-meter, and 400-meter races. (The source also gives the men's records.) Study the distributions of the three races by constructing dotplots and boxplots of each sample. Do the distributions appear to be normally distributed? Are all three similar in shape? Construct normal probability plots for each sample. Calculate descriptive statistics for each of the three samples. Are the means, medians, and trimmed means about the same for each of the three samples? Is the amount of variability about the same for each of the three races? Are the standard deviations about what you would expect? Write a short paragraph explaining what you found out by analyzing these data.

Worksheet: Track.mtw

Country	100 m	200 m	400 m
Argentina	11.61	22.94	54.50
Australia	11.20	22.35	51.08
Austria	11.43	23.09	50.62
Belgium	11.41	23.04	52.00
Bermuda	11.46	23.05	53.30
Brazil	11.31	23.17	52.80
Burma	12.14	24.47	55.00
Canada	11.00	22.25	50.06
Chile	12.00	24.52	54.90
China	11.95	24.41	54.97
Colombia	11.60	24.00	53.26
Cook Islands	12.90	27.10	60.40
Costa Rica	11.96	24.60	58.25
Czech	11.09	21.97	47.99
Denmark	11.42	23.52	53.60
Dominican Republic	11.79	24.05	56.05
Finland	11.13	22.39	50.14
France	11.15	22.59	51.73
GDR	10.81	21.71	48.16
FRG	11.01	22.39	49.75
Gbni	11.00	22.13	50.46
Greece	11.79	24.08	54.93
Guatemala	11.84	24.54	56.09
Hungary	11.45	23.06	51.50
India	11.95	24.28	53.60
Indonesia	11.85	24.24	55.34
Ireland	11.43	23.51	53.24
Israel	11.45	23.57	54.90
Italy	11.29	23.00	52.01
Japan	11.73	24.00	53.73
Kenya	11.73	23.88	52.70
Korea	11.96	24.49	55.70
DPR Korea	12.25	25.78	51.20

(*continued*)

Country	100 m	200 m	400 m
Luxembourg	12.03	24.96	56.10
Malaysia	12.23	24.21	55.09
Mauritius	11.76	25.08	58.10
Mexico	11.89	23.62	53.76
Netherlands	11.25	22.81	52.38
New Zealand	11.55	23.13	51.60
Norway	11.58	23.31	53.12
PNG	12.25	25.07	56.96
Philippines	11.76	23.54	54.60
Poland	11.13	22.21	49.29
Portugal	11.81	24.22	54.30
Rumania	11.44	23.46	51.20
Singapore	12.30	25.00	55.08
Spain	11.80	23.98	53.59
Sweden	11.16	22.82	51.79
Switzerland	11.45	23.31	53.11
Taipei	11.22	22.62	52.50
Thailand	11.75	24.46	55.80
Turkey	11.98	24.44	56.45
USA	10.79	21.83	50.62
USSR	11.06	22.19	49.19
W Samoa	12.74	25.85	58.73

SOURCE: B. Dawkins, "Multivariate Analysis of National Track Records," *The American Statistician* 43:2 (1989): 110–115.

6.71   The *uniform distribution* is a very common distribution used to model statistical data. It gets its name from the fact that all values seem to have the same chance of occurring. A good example of the discrete uniform distribution is the roll of a die. We know that each of the six sides has the same chance ($\frac{1}{6}$) of coming up. The continuous uniform distribution is similar except that there is an uncountable number of possible outcomes. To get an idea about the shape of the uniform distribution, simulate 1,000 observations and store them in column 1 of the computer worksheet. In Minitab, from the Calc menu, select **Random Data** and click on **Uniform**. When the Uniform Distribution window appears, key **1000** in the **Generate** box and **Store** results in **C1**. We can change the lower and upper endpoints of the uniform distribution, but for now leave the lower endpoint at 0.0 and the upper endpoint at 1.0. Construct a histogram, boxplot, and normal probability plot of the data in C1. Does the distribution appear to be symmetric? Are there any outliers? Do all values appear to have the same chance of occurring? Do you classify the shape of the distribution as short-tailed or long-tailed? Calculate the descriptive statistics for the data. What is the mean? Median? Are they what you would expect? If we had generated values between 1 and 5 instead of between 0 and 1, what would the mean be? Generate 1,000 observations between 1 and 5 and store them in column C2. Is the mean what you would expect? Repeat the same analysis as above. Did you get similar results except that the numbers are now between 1 and 5?

6.72   Blood pressure is a complex quantitative human trait that is affected by environmental as well as genetic factors. Red blood cell sodium–lithium countertransport (SLC) is correlated with blood pressure and is of interest to geneticists because it is easier to study than blood pressure. They study the general distribution shape of SLC and a measure of the center of the distribution. The list gives the SLC activity measured on 190 individuals from six large English kindred:

Worksheet: SLC.mtw

.467	.430	.192	.192	.293	.160	.164	.126	.328	.202
.282	.328	.247	.132	.138	.224	.512	.221	.252	.193
.263	.186	.346	.219	.177	.349	.272	.245	.213	.197
.229	.245	.210	.281	.175	.273	.439	.471	.451	.237
.313	.136	.245	.391	.349	.158	.252	.416	.232	.183
.254	.195	.141	.151	.073	.300	.231	.075	.208	.267
.187	.244	.245	.231	.167	.337	.251	.209	.181	.411
.191	.288	.280	.119	.394	.443	.423	.534	.393	.273
.149	.225	.159	.170	.329	.183	.262	.250	.179	.329
.253	.270	.310	.321	.333	.284	.380	.222	.178	.265
.289	.199	.309	.279	.194	.203	.139	.162	.251	.619
.343	.155	.340	.332	.412	.218	.304	.261	.206	.231
.182	.267	.198	.191	.258	.179	.197	.188	.202	.150
.201	.255	.293	.255	.189	.414	.292	.253	.168	.295
.215	.213	.267	.216	.264	.138	.239	.288	.311	.414
.462	.361	.623	.199	.215	.321	.273	.259	.206	.376
.228	.155	.186	.097	.179	.174	.386	.393	.198	.243
.326	.250	.590	.461	.361	.321	.236	.139	.316	.313
.263	.180	.184	.354	.264	.269	.171	.359	.338	.163

SOURCE: K. Roeder, "A Graphical Technique for Determining the Number of Components in a Mixture of Normals," *Journal of the American Statistical Association*, 89 (1994): 487–495.

a. Construct a stem and leaf plot and a boxplot of the data. Based on these two graphs, how do you describe the shape of the distribution?
b. Construct a normal probability plot. Can you reject normality based on this plot? How do you describe the shape? Does it agree with your answer to part **a**?
c. Does the mean or the median best describe the center of the distribution? Explain your reasoning.
d. Give your estimate of the center.
e. How do you measure the variability in the distribution? What is your estimate?

6.73   STATISTICAL INSIGHT REVISITED   The SAT data set described in the Statistical Insight at the beginning of the chapter is stored in worksheet: **Sat.mtw**. Retrieve Sat.mtw and analyze the data. As with any data set, first construct an ordered stem and leaf plot (or histogram). From the stem and leaf plot, do you classify the distribution's shape as bimodal? In Section 2.6, it was pointed out that bimodal data indicate that some nonhomogeneous factor (extraneous variable) is acting on the data. If possible, we would like to identify that variable. Is it possible to associate the SAT scores in the data set with the percent of the student population in each state who took the test? Is it possible that lower SAT scores come from states where the percent of the student population who took the test is greater than 40%? Separate the SAT scores into two groups: those where fewer than 40% took the test and those where more than 40% took the test. This is easily done by creating a code variable. Let it be 1 if the percent is less than .4, and let it be 2 if the percent is greater than .4. Now create side-by-side dotplots of the two groups. Do you see a clear separation between the two groups? Next create side-by-side boxplots to further illustrate the difference between the two groups. Do the boxplots clearly demonstrate that there are two distinct groups? What measure of center do you recommend for the two distributions? Calculate descriptive statistics for the two groups. Compare the means, medians and standard deviations of the two groups. A scatterplot of the original data will also show a definite relationship between the SAT scores and the percent taking the test. Create the scatterplot and calculate the correlation coefficient.

The laws of probability are designed to aid in the computation of probabilities for complicated events. Care must be exercised, however, in applying the laws of probability, as is illustrated in the following example.

PROBABILITY
IN THE
COURTROOM

An elderly woman was robbed in the suburb of a large city. A black man and his blond female companion were convicted of a crime simply because they were together in a yellow Trans Am in the vicinity of a crime scene. The prosecutor convinced the jury that the chance of another couple of that description in a yellow Trans Am was so unlikely that they had to be the two who committed the crime. The prosecution used the multiplication law of probability on the following events:

Event	Assumed Probability
Driving a yellow Trans Am	1/10
Interracial couple	1/1,000
Blond woman	1/4
Woman wears ponytail	1/10
Man has beard	1/10
Black man	1/3

These probabilities were multiplied to find the probability that a couple of this description would be in the vicinity of the crime. The claimed probability was 1/12,000,000. The jury convicted the defendants because the likelihood of another couple with that description was so small. We recall, however, that to use the multiplication law, the events must be independent. Clearly, for example, interracial couple and black man are not independent. The defense lawyer pointed this out in the appeal and, consequently, the supreme court of the state reversed the conviction (*Time* magazine, April 26, 1968, p. 41).

When we introduced the probability of an event in Chapter 4, we defined it as the long-term relative frequency of occurrence of the event. For example, the probability of a 1 showing when we roll a fair die is $\frac{1}{6}$ because the long-term relative frequency of occurrence of a 1 is $\frac{1}{6}$. In some situations, however, it is difficult to imagine a repetition of the experiment a large number of times. In those cases, we may assign a *subjective probability* to the event. A subjective probability is our personal opinion of the likelihood of the occurrence of the event.

This intuitive notion of probability can sometimes be misleading. For instance, it seems that the event that two people in your class have the same birthday would be a "rare event," one for which the probability is extremely small. Yet, if there are at least 23 people, the chances are greater than 50% that at least two will have the same birthday. In this case, our intuition gives us an incorrect conclusion.

The following news article shows that rare events do actually happen:

Delivery Room Birthdays

Quincy, Mass. (AP)

When Justine Lee Mitchell was born last week, she wasn't the only one in the operating room celebrating a birthday. Her mother turned 18, the obstetrician turned 37, and the nurse turned 32, all on Saturday. "It's mind boggling," the obstetrician, Dr. Richard R. Adams, said in an interview published today.

While Lauralyn Mitchell was delivering Justine at Quincy City Hospital, Adams said he boasted to registered nurse Beth MacLeod, Ms. Mitchell was "going to have a great kid because we have the same birthday." Ms. McLeod responded that it was her birthday, too, and then the new mother announced it was her birthday. Adams said he demanded a look at her driver's license to be sure, and it was confirmed, she was telling the truth.

Justine was born at 5:30 A.M. and weighed 8 pounds, 3 ounces.

SOURCE: Associated Press wire release, July 11, 1984.

As one might expect, the chance of having four randomly selected people (random as far as their birthdays are concerned) with the same birthday in a room is extremely rare, approximately 1 chance out of 50 million! Yet it did happen. Even two randomly selected people in a room having the same birthday is somewhat rare, with less than 3 chances out of 1,000. But, as stated above, the chances improve as the number of people in the room increases. In fact, if there are 50 people in a room, it is almost certain that at least two will have the same birthday.

BIRTHDAY PROBLEM

We now take a closer look at the *birthday problem*. First, suppose there are two people in a room. To determine the probability that both have the same birthday, let b_1 denote the birthday of one of the two and b_2 denote the birthday of the other. Both b_1 and b_2 can be any one of 365 possible days of the year (assuming it is not leap year). The sample space of possibilities is

$$S = \{(b_1, b_2) \mid b_1 = 1, 2, 3, \ldots, 365 \text{ and } b_2 = 1, 2, 3, \ldots, 365\}$$

Clearly, the number of outcomes in S is

$$(365)(365) = 365^2$$

and they are equally likely.

Let B be the event that $b_1 = b_2$; that is, the two people have the same birthday. How many elements of the sample space are in the event B? There is no restriction on b_1, so it can be any one of 365 days. However, b_2 must equal b_1, so there is only one possibility for b_2. Hence, there are $(365)(1) = 365$ outcomes in event B. Listing the elements of B, we have

$$B = \{(1, 1), (2, 2), (3, 3), \ldots, (365, 365)\}$$

and we see that there are indeed 365 outcomes. Because the outcomes in B are equally likely, we have

$$P(B) = \frac{365}{(365)^2} = \frac{1}{365} = .00274$$

Next suppose there are three people in a room. The sample space of possible birthdays is

$$S = \{(b_1, b_2, b_3) \mid b_i = 1, 2, 3, \ldots, 365; i = 1, 2, 3\}$$

The number of outcomes in S is $(365)^3$, and they are equally likely. Let B be the event that $b_1 = b_2 = b_3$ (the three have the same birthday). Then we have

$$B = \{(1, 1, 1), (2, 2, 2), (3, 3, 3), \ldots, (365, 365, 365)\}$$

Then, as before, the number of outcomes in B is 365, so

$$P(B) = \frac{365}{(365)^3} = \frac{1}{(365)^2} = .0000075$$

Now consider the delivery room problem. There were four people (one was the newborn) in the room. The sample space of possible birthdays is

$$S = \{(b_1, b_2, b_3, b_4) \mid b_i = 1, 2, 3, \ldots, 365; i = 1, 2, 3, 4\}$$

As before, the number of outcomes in S is $(365)^4$ and they are equally likely. The event B that they all have the same birthday is the event $b_1 = b_2 = b_3 = b_4$. That is, we have

$$B = \{(1, 1, 1, 1), (2, 2, 2, 2), (3, 3, 3, 3), \ldots, (365, 365, 365, 365)\}$$

and the number of outcomes in B is 365. Because the outcomes are equally likely, we have

$$P(B) = \frac{365}{(365)^4}$$

$$= \frac{1}{(365)^3}$$

$$= .00000002056$$

$$= 1 \text{ out of } 48,627,125$$

or about 1 out of 50 million.

Let's now look at a slightly different problem. Suppose there are ten people in a room. What is the probability that at least two have the same birthday? Let $b_1, b_2, b_3, \ldots, b_{10}$ denote the birthdays of the ten people. Then the sample space of possible birthdays is

$$S = \{(b_1, b_2, b_3, \ldots, b_{10}) \mid b_i = 1, 2, 3, \ldots, 365; i = 1, 2, \ldots, 10\}$$

As before, the number of outcomes in S is $(365)^{10}$. We are interested in the event A that at least two of the ten have the same birthday. However, the complement of A, A^c, is easier to work with. A^c is the event that no two have the same birthday; that is,

$$b_1 \neq b_2 \neq b_3 \neq \cdots \neq b_{10}$$

There are 365 possibilities for b_1, but because $b_2 \neq b_1$, there are only 364 possibilities for b_2. Similarly, there are 363 possibilities for b_3, and so on. We can see that the number of elements in A^c is $(365)(364)(363) \cdots (356)$. Therefore, we have

$$P(A^c) = \frac{(365)(364)(363) \cdots (356)}{(365)^{10}} = .883$$

Using the complement law, we have

$$P(\text{at least two have the same birthday}) = P(A)$$
$$= 1 - P(A^c)$$
$$= 1 - .883$$
$$= .117$$

The following table gives the probability that at least two people will have the same birthday when the number of people ranges from 10 to 50:

Number of people	Probability that at least two have the same birthday
10	.117
15	.253
20	.411
21	.444
22	.476
23	.507
24	.538
25	.569
30	.706
40	.891
50	.970

In a group of 23 people the chance is greater than 50% that at least two will have the same birthday. Among 50 people, it is almost certain that at least two will have the same birthday.

USES AND MISUSES OF THE CENTRAL LIMIT THEOREM

When one is asked to describe the Central Limit Theorem, a common response is: "When the sample size is large, it becomes normal." What is meant by this statement? What does the "it" refer to? A misconception is that "it" refers to the sample. The perception is that the larger the sample size, the closer the distribution of the sample comes to the normal distribution. Remember that the sample should be representative of the parent distribution. If the parent distribution is skewed left, for example, then the sample should be skewed left, regardless of the sample size. The Central Limit Theorem does nothing to alter the distribution of the sample. The issue is: What are the possibilities for the *average* of the sample? The distribution of all possible values of the average is the distribution that becomes more normally distributed as the sample size becomes larger. It is called a *sampling distribution* because it is arrived at by considering all possible values of the statistic from all the different possible samples. All statistics have a sampling distribution of some sort, and the Central Limit Theorem, in some cases, can be applied to show that the sampling distribution becomes more normally distributed as the sample size increases. We have studied in particular the sampling distributions of the sample mean (average), $\bar{y}$, and the sample proportion of successes, p, in a Bernoulli population. Understanding the sampling variability of these two statistics is of the utmost importance if we are to understand the concepts of statistical inference that are covered in the remainder of this text.

Although we studied the sampling distributions of $\bar{y}$ and p separately, they are actually closely related. In particular, if we code the values in the Bernoulli population as a 1 for a success and a 0 for a failure, then the sum of the elements in a sample of size n is simply the sum of all the 1's in the sample. Then the sample mean, $\bar{y} = \sum y_i/n$, gives the proportion of 1's in the sample; that is, $\bar{y} = p$. Thus, using this coding technique, we see that the Central Limit Theorem applied to proportions is a special case of the Central Limit Theorem applied to $\bar{y}$.

Another useful feature of the Central Limit Theorem is to recognize that if we multiply $\bar{y}$ by n, we get the sum of the elements in the random sample; that is, $n\bar{y} = \sum y_i$. Applying the Central Limit Theorem, we have, for a sufficiently large sample size, that the sampling distribution of $\sum y_i$ is approximately normally distributed with a mean of $n\mu$ and a standard deviation of $(n)(\sigma/\sqrt{n}) = \sigma\sqrt{n}$. Consequently, if a particular measurement is the sum of a large number of independent quantities, then the distribution of the measurement will be closely approximated by a normal distribution. For example, variables such as temperature, wind speed, IQ scores, and weights are generally made up of sums of a large number of smaller contributions and thus are approximately normally distributed.

We now give an application of the Central Limit Theorem to proportions: The Christmas tree growers in the Blue Ridge Mountains of North Carolina pride themselves on the fact that they grow one of the prettiest trees of all, the Fraser fir. Their biggest competitor as a Christmas tree, strangely enough, is the artificial tree. So a problem of interest to the growers in this region and to all Christmas tree growers in general is: What percent of the potential buyers of Christmas trees purchase artificial trees?

To formulate this problem, we can view the population of buyers of Christmas trees as a Bernoulli population because each member of the population will purchase either a live tree or an artificial tree. The Bernoulli proportion, π, in this problem is the proportion of buyers of artificial trees in the population. We wish to estimate π.

Suppose we choose a random sample of 500 potential Christmas tree buyers and find that 135 intend to purchase artificial trees. Will a second sample of 500 buyers give the exact same results? Certainly not; there are thousands of potential buyers of Christmas trees and we know what only 500 intend to buy. We can say, however, that $135/500 = .27$ is *one* of the possible values of the sample proportion, p. What are some of the other possible values of p?

The Central Limit Theorem governs the possible values of a sample proportion, p. In fact, when the sample size is large enough, the theorem guarantees that all the potential values of p will be symmetrically distributed about π in *normal* fashion. Thus, the observed value of .27 from our sample is just one of the values that should be close to the actual value of π. Before we go any further, we should ask, Why did we get .27 in the first place? This is where the actual value of π enters the picture. We do not know what it is (that is the point of the problem), but whatever it is, each representative sample will give a value of p somewhere around π. Some samples will give values of p below π, and some samples will give values of p above π. We do not really know whether our $p = .27$ is above or below π. All we know is that it is *close* to π. How close? Again, the Central Limit Theorem helps out. Closeness is measured by the standard deviation of the statistic. In the case of p, we learned that its standard deviation is $\sqrt{\pi(1 - \pi)/n}$. Using the empirical rule, we can say that p should be within $\pm 2\sqrt{\pi(1 - \pi)/n}$ of π, 95% of the time.

Remember also that the standard deviation is a maximum when $\pi = .5$, so we are reasonably assured (95% sure) that our value of $p = .27$ is within

$$\pm 2\sqrt{\frac{\pi(1 - \pi)}{n}} = \pm 2\sqrt{\frac{(.5)(1 - .5)}{500}} = \pm.045$$

units of π. In other words, π is probably somewhere between

$$.27 - .045 \quad \text{and} \quad .27 + .045$$

or between .225 and .315. This interval is reasonably wide, leaving some doubt as to what is the actual value of π. If we wish to narrow the width of the interval, we will see later that we need a sample larger than 500.

UNIT REVIEW

Exercises 1–4 pertain to the Christmas tree growers' example.

1. We are interested in the percent of potential buyers who will purchase an artificial tree.
   a. Describe the parameter that we are interested in predicting.
   b. To predict the value of the parameter, should we conduct a census or take a random sample?
   c. If we choose to take a random sample, do you think that 500 potential Christmas tree buyers is large enough?

2. Describe how the buyers of Christmas trees can be viewed as a Bernoulli population. (*Note:* People in the sample who choose not to buy a tree at all are discarded because they are not in the population of interest.)

3. Recall that π represents the proportion of successes in the Bernoulli population.
   a. What does π represent in the Christmas tree problem?
   b. Suppose in a random sample of 500 buyers, we find that 135 intend to purchase artificial trees. Is the ratio $135/500 = .27$ the value of π or one of the possible values for p?
   c. Can we say that 27% of all buyers will purchase artificial trees?

4. Given that one sample of size 500 yielded .27, is it conceivable that a second sample of 500 will yield, say, .87? Explain your answer.

5. Suppose 40% of the employees of a company are women. Three employees are selected at random to participate in a discussion of salaries.

a. List the sample space of possible genders of the three employees.
b. Assign probabilities to each outcome in the sample space.
c. Suppose A is the event that exactly 2 of the 3 selected are women. List the elements that make up A.
d. What is $P(A)$?

6. At the end of the television season, a show is either renewed or canceled. One producer has three shows on the air.
a. List the sample space of possibilities of his shows being canceled or renewed.
b. If the probability that a show is canceled is 60%, assign probabilities to the outcomes in the sample space.
c. What is the probability at least two of his shows will be renewed?
d. How does this exercise differ from Exercise 5?

7. According to the state Center for Health Statistics, 116,564 pregnancies were reported in North Carolina in 1983. Of those 116,564 pregnancies, 31,892 ended in abortions. It was reported that it was the first abortion for 23,217 of the women, the second for 6,680 women, the third for 1,561 women, and the fourth or higher for 434 women.
a. What percent of the pregnancies ended in abortion?
b. Of those who had an abortion, what percent had previously had an abortion?

8. A box contains a nickel, a dime, and a quarter. Two coins are selected without replacement. List the elements of the sample space and the elements of the following events:

A: The selection has a quarter

B: The selection totals less than 25 cents

9. There are three checkout lanes in a department store. Suppose that three customers randomly pick a lane. Assuming that all three can choose the same lane, list the elements of the sample space. Assign probabilities to the outcomes.

10. What is the probability that the next four babies born at a hospital will be girls? Assume a boy and a girl are equally likely.

11. A company packages pictures of 40 former U.S. presidents in its product. Assuming an equal number of pictures of each president are distributed, what is the probability that out of 5 packages, you will get 5 different pictures?

12. Compute the probability that, out of five people, at least two have the same birth month.

13. In a local clinic, 30% of the patients have high blood pressure, 40% are overweight, and 10% are both. Fifteen patients are chosen at random.
a. What is the probability that no more than three have high blood pressure?
b. What is the probability that no more than three are overweight?
c. What is the probability that no more than three are both?
d. How many are expected to have high blood pressure?
e. How many are expected to be overweight?
f. How many are expected to be both?

14. A recent article reports that 40% of all cancer patients are cured. Out of ten patients, what are the chances that more than half are cured? Out of 100 patients, what are the chances that more than half are cured?

15. In a large metropolitan area, 70% of the families with four or more children are on welfare. Out of 8 randomly selected families (with four or more children), what is the probability that all are on welfare?

16. Suppose 70% of the students live in dorms. Ten students are selected at random. What is the probability that more than half of them live in dorms? Of the 10, how many are expected to live in dorms? Out of 100 students, what is the probability that more than 60% live in dorms?

17. Suppose a basketball player has a lifetime average of .80 at the free-throw line. He starts shooting until he misses. Let x be the number of shots until he misses. List the range of values of x. Is x a discrete or continuous random variable?

18. Suppose scores on a National Merit Scholarship exam are normally distributed with a mean of 70 and a standard deviation of 8. What percent of those taking the exam score in these ranges?
    a. Between 60 and 80
    b. Between 75 and 90
    c. Below 67
    d. What is the 80th percentile?

19. A physical education instructor told members of his class that they would earn an A for the triple jump if they could jump farther than 24 feet.
    a. If the distance jumped is normally distributed with a mean of 22 feet and a standard deviation of 3, what percent of the class will earn A's?
    b. If the instructor wants 70% of his class to make a C or better, how far will they have to jump?
    c. The instructor has 32 students in his class. What is the probability that their average exceeds 23 feet?

20. Scores on the STEP science test are assumed to be normally distributed with a mean of 20 and a standard deviation of 5. What percent of scores will be in each range?
    a. Above 30
    b. Below 17
    c. Between 18 and 23

21. Students who get the top 25% of scores on the STEP science test described in Exercise 20 receive scholarships. What score must a student get to receive a scholarship?

22. Suppose you know that the amount of time your watch gains or loses (i.e., gains negatively) in seconds per day is normally distributed with a mean of 0 and a standard deviation of 1; that is, it is a standard normal distribution. What are the probabilities of these events?
    a. It gains no more than 1.5 seconds in a day.
    b. It gains at least .5 second in a day.
    c. It loses .5 second in a day.
    d. It gains between .7 and 1.4 seconds in a day.
    e. The average time gained during the month of September is no more than .5 second.

23. Approximately 40% of all business/management degrees are awarded to women. Determine the probability that out of 26 business/management graduates, more than half are women.

Multiple Choice (24–29)

24. Recent events in California indicate that the chance of major destruction from an earthquake at a given location depends on the distance from the epicenter of the earthquake as well as the type of geological conditions at the location. Suppose it has been determined that the probabilities of major destruction from a 7.0 earthquake at two different locations are .4 and .3, respectively. If destructions at the two locations are independent, what is the probability of major destruction at both locations?
    a. .7   b. .58   c. .12   d. 0

25. Suppose a manufacturer of an item claims that 80% of all items are not defective. If you buy three of the items, what is the probability that at least one is defective?

    a. .488   b. .2   c. .6   d. .008

26. Suppose x is a binomial random variable with $n = 12$ and $\pi = .7$. Then
    a. the expected value of x is 8.
    b. the variance of x is 1.6.
    c. x gives the number of successes in 12 trials.
    d. All of the above

27. Suppose an experiment consists of choosing six digits (each number from 0 to 9) for the winning lottery number. The number of elements in the sample space is

    a. 6   b. 60   c. 10   d. 1 million

28. The probabilities assigned to the outcomes in the sample space of an experiment must
    a. be greater than 0.
    b. be less than 1.
    c. sum to 1.
    d. All of the above

29. The sampling distribution of a statistic describes
    a. the shape of the distribution of all potential values of the statistic in repeated sampling.
    b. the center of all potential values of the statistic in repeated sampling.
    c. the amount of variability associated with all potential values of the statistic in repeated sampling.
    d. All of the above

30. Suppose 60% of the student body favors the construction of a new building on campus. What are the probabilities of the following in a random sample of 15 students?
    a. More than half are in favor.
    b. Five or fewer are in favor.
    c. All are in favor.

31. One-third of all science majors return for an advanced degree within 5 years after graduation. From a random sample of three science majors, what are the probabilities of the following events?
    a. None return for an advanced degree within 5 years after graduation.
    b. At least two return within 5 years.
    c. All return within 5 years.

32. Three books in the library are checked out 20% of the time. Suppose they are independently checked out (C) or not (N) on any given day.
    a. List the sample space of possibilities for any randomly selected day.
    b. Assign probabilities to the outcomes.
    c. What is the probability that at least two are checked out on any randomly selected day?

33. The useful lifetime of a dishwasher is assumed to be normally distributed with a mean of 12 years and a standard deviation of 18 months.
    a. What percent will last more than 15 years?
    b. What percent will last between 8 and 11 years?
    c. Ninety percent of all dishwashers will expire within how many years?

34. A chemist is analyzing the concentration of copper in hair samples taken from individuals who have copper plumbing in their homes. Identify the parent population, a possible parameter of interest, and an associated statistic that estimates the parameter.

35. A psychologist designs an experiment to study the effect of electroshock treatment on the time it takes a subject to complete a difficult task. Here are the numbers of trials necessary to complete the task:

Worksheet: Electro.mtw

6	11	14	3	9	15	4	7	12
8	14	16	6	10	15	3	9	
18	31	26	21	25	20	48	23	

a. Construct a stem and leaf plot.

b. Construct a five-number summary diagram including midsummaries and spreads.

c. From the midsummaries, comment on the distribution's shape.

d. Compute $\bar{y}$ and s using your calculator.

e. Construct a boxplot and comment on the distribution.

f. Identify parameters that measure the center and the variability of the distribution. Give estimates of the two parameters.

36. These data represent the educational levels attained by employees in three different industries in a medium-sized community:

Worksheet: Edlevel.mtw

Industry		
A	B	C
8	8	12
8	9	12
1	9	8
12	8	9
14	7	12
9	6	9
12	12	18
14	16	12
12	2	14

Does it appear that the educational levels for the three industries differ? Does there appear to be a large variation in educational levels? Construct side-by-side boxplots and compare the distributions.

COMPUTER EXERCISES

37. Use the Random Data > Integer command to simulate the roll of a die 60 times. How often would one expect each of the six integers to turn up? Determine the frequency of occurrence of each possible outcome in the 60 rolls. Repeat the process five times and compare the results.

38. The Random Data > Integer command can be used to simulate birthdays if we associate each birthday with its corresponding day of the year. For example, January 23 is day 23, February 19 is day 50 ($31 + 19 = 50$), until we get to December 31, which is day 365. Earlier in the unit summary, we saw that in a sample of 23 people, the chance of two having the same birthday is greater than 50%. Simulate 23 birthdays with the Random Data > Integer command and see whether there are any matches. In a sample of 50 people, it is almost certain that two will have the same birthday. Simulate 50 birthdays and see whether there are any matches.

39. Approximately 55% of entering freshmen end up graduating from a 4-year college. Use the Random Data > Bernoulli command to simulate the success or failure to graduate for 50 entering freshmen. How many graduated? Does the number seem reasonable?

40. A component for a spacecraft launch fails 5% of the time. Suppose five such components as connected in series for a launch. Simulate the success or failure of the five components in a launch. Did all work? Simulate ten different launches. How many got off the ground?

41. Approximately 13% of all medical school seniors go into internal medicine. If 30 seniors are completing medical school at a university, determine the probability distribution of the

number who plan to go into internal medicine. What is the probability that none of the 30 goes into internal medicine? How many are expected to go into internal medicine? What is the standard deviation of the distribution?

42. Suppose 25 universities plan to graduate 30 medical students each. Simulate the number of students who plan to go into internal medicine from the 25 universities (see Exercise 41). Out of all 25 universities, how many students plan to go into internal medicine?

43. Use the Random Data > Normal command to simulate the drawing of a random sample of size 50 from a normal population that has a mean of 70 and a standard deviation of 8. Construct a histogram of the sample.

44. Suppose the average cost for 1 hour of consultation with a lawyer is $88.64 and the standard deviation is $5.23. Assuming that the distribution of the costs for various lawyers across the nation is normally distributed, simulate the fees of 20 different lawyers. Construct a stem and leaf plot of the generated data.

45. The fertility rate is the number of births a woman can expect in her child-bearing years. An average rate of 2.12 is needed to keep the population constant. Here are the fertility rates in all states and the District of Columbia:

Worksheet: Fertilit.Mtw

Ala	1.9	Mont	2.1
Alaska	2.3	Neb	2.0
Ariz	2.1	Nev	1.8
Ark	2.0	NH	1.7
Calif	1.9	NJ	1.6
Colo	1.8	NM	2.2
Conn	1.5	NY	1.6
Del	1.8	NC	1.6
D.C.	1.5	ND	2.1
Fla	1.7	Ohio	1.8
Ga	1.9	Okla	2.0
Hawaii	2.1	Ore	1.8
Idaho	2.5	Pa	1.6
Ill	1.9	RI	1.5
Ind	1.8	SC	1.8
Iowa	2.0	SD	2.4
Kan	2.0	Tenn	1.7
Ky	1.9	Texas	2.1
La	2.2	Utah	3.2
Maine	1.7	Vt	1.7
Md	1.6	Va	1.6
Mass	1.5	Wash	1.8
Mich	1.8	W.Va	1.8
Minn	1.9	Wis	1.9
Miss	2.2	Wyo	2.4
Mo	1.9		

SOURCE: Population Reference Bureau.

Construct a stem and leaf plot and a letter value display. Calculate the midsummaries and spreads. Use the midsummaries to evaluate the skewness of the distribution.

46. Construct a boxplot and a normal probability plot of the data in Exercise 45. Make a general statement about the shape of the distribution and identify a parameter that best measures the center of the distribution. Give an estimate of the parameter.

DISCUSSION QUESTIONS

47. Ninety percent of heroin addicts confess to using marijuana prior to the use of heroin. Does this mean that if you smoke marijuana, you most likely will use heroin? Explain.

48. Determine whether two people in your class have the same birthday. Give the number of students in the class and compare the results with the table of probabilities given in our earlier discussion.

49. It is claimed that scores on a college placement exam are normally distributed with a mean of 50 and a standard deviation of 10. A group of 25 students from a private school all scored above 55. What is the probability of scoring above 55? Do you think it is unusual that all 25 scored above 55? Explain.

50. Four out of five accidents in Watauga County, North Carolina, were alcohol-related, whereas nationally only 50% are. The police chief is concerned. Is this unusual?

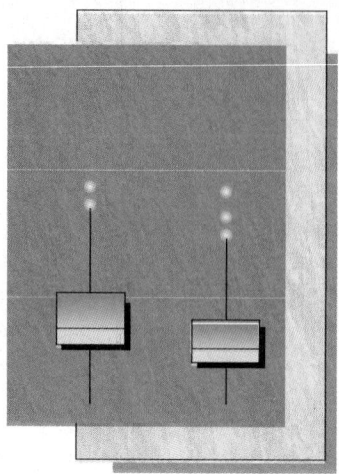

7

CONFIDENCE INTERVAL ESTIMATION

In the preceding chapter, you learned how to examine a data set intelligently and how to describe some of its more important characteristics. We introduced the basic idea of statistical inference—drawing generalizations about a population from the information in a sample. Now we use the information gained thus far to expand our knowledge of statistical inference. Two main areas of inferential statistics that we consider are estimation and hypothesis testing. This chapter deals with confidence interval estimation.

STATISTICAL INSIGHT

THE COST OF A NEW HOME
IN THE UNITED STATES

A home is the most expensive item most of us will ever buy. In fact, many may never be able to afford one. The cost of a new home depends on several factors, including labor, material, land, and financing. In 1949, financing accounted for only 5% of the total cost of a new home. Today, it accounts for close to 20% of the total cost. Because of labor, material, and land costs, the price of a new home can vary significantly from one region of the country to another. For example, in 1994, the mean cost of a new home in the West was $154,750; in the Northeast, $123,536; in the Midwest, $90,335; and in the South, $82,791. The mean price difference between the West and South was almost $72,000.

Instead of reporting the mean cost of a new home, the National Association of Realtors reports the median cost of a new home. It reported that in 1994 the median cost of a new home in the West was $119,250; in the Northeast, $116,800; in the Midwest, $84,900; and in the South, $82,900.

So we see that not only do the costs vary across regions, but also the choice of the statis-tic used to estimate the cost varies. Which is the better measure of the price one will have to pay for a new home—the mean or the median? For example, is the mean, $154,750, or the median, $119,250, more representative of the purchase price of a new home in the West? We know from Chapter 6 that the shape of the parent distribution is a determining factor in choosing a parameter to represent the center of the distribution.

Furthermore, after deciding on a parameter to represent the center, should we be satisfied with just a single number as an estimate of what one will have to pay for a new home? For those who plan to purchase a new home, it might be more appealing to give a range of values for the center of the distribution. In this chapter, we take a closer look at these issues.

The list gives the 1994 median prices of single-family homes in 65 metropolitan statistical areas, as defined by the U.S. Office of Management and Budget, across the United States. In Exercises 7.110 and 7.111 in Section 7.6, you are asked to analyze these data further.

Worksheet: Homes.mtw

City	Price	City	Price
Akron, OH	$81,600	Cleveland, OH	$94,200
Albuquerque, NM	103,100	Columbia, SC	82,900
Anaheim, CA	209,500	Columbus, OH	92,800
Atlanta, GA	93,200	Corpus Christi, TX	71,700
Baltimore, MD	115,700	Dallas, TX	95,100
Baton Rouge, LA	78,400	Daytona Beach, FL	66,200
Birmingham, AL	99,500	Denver, CO	111,200
Boston, MA	170,600	Des Moines, IA	77,400
Bradenton, FL	86,400	Detroit, MI	84,500
Buffalo, NY	82,400	El Paso, TX	73,600
Charleston, SC	91,300	Grand Rapids, MI	76,600
Chicago, IL	135,500	Hartford, CT	132,900
Cincinnati, OH	93,600	Honolulu, HI	355,000

(*continued*)

City	Price	City	Price
Houston, TX	84,800	Orlando, FL	89,900
Indianapolis, IN	90,500	Philadelphia, PA	116,800
Jacksonville, FL	79,700	Phoenix, AZ	89,200
Kansas City, MO	84,900	Pittsburgh, PA	80,000
Knoxville, TN	88,600	Portland, OR	111,200
Las Vegas, NV	110,400	Providence, RI	115,600
Los Angeles, CA	188,500	Sacramento, CA	127,300
Louisville, KY	77,400	St. Louis, MO	83,100
Madison, WI	111,500	Salt Lake City, UT	92,800
Memphis, TN	85,600	San Antonio, TX	76,800
Miami, FL	105,000	San Diego, CA	177,800
Milwaukee, WI	106,500	San Francisco, CA	246,900
Minneapolis, MN	100,000	Seattle, WA	152,900
Mobile, AL	69,500	Spokane, WA	90,700
Nashville, TN	95,200	Syracuse, NY	82,100
New Haven, CT	137,600	Tampa, FL	74,300
New Orleans, LA	75,400	Toledo, OH	72,800
New York, NY	170,300	Tulsa, OK	73,500
Oklahoma City, OK	67,600	Washington, DC	154,900
Omaha, NE	72,800		

SOURCE: National Association of Realtors.

7.1 CONFIDENCE INTERVAL FOR A POPULATION PROPORTION

A point estimator of a population parameter produces a single numerical value from the information in a random sample. That value is an estimate of the parameter. You learned in Chapter 6 that a point estimator is subject to sample variability and thus can assume values that are somewhat removed from the true value of the parameter. Instead of giving a single number as the estimate of an unknown parameter, a *confidence interval* gives an interval of values that should contain the true value of the parameter with a high degree of confidence.

> A **confidence interval** for a population parameter is an interval of possible values for the unknown parameter. The interval is computed from sample data in such a way that we have a high degree of confidence that the interval contains the true value of the parameter. The confidence, stated as a percent, is the **confidence level.**

In practice, estimates of unknown parameters are usually given in the form

$$\text{estimate} \pm \text{margin of error}$$

The *margin of error* is the maximum error that one would expect in the estimate for a specified confidence level. For example, the Bureau of Labor Statistics estimates the number of unemployed in a certain area to be $1,500 \pm 300$. The estimate of the number unemployed is 1,500, yet it can be off by as much as ± 300. The Bureau is rather confident that the actual number is somewhere between 1,200 and 1,800. If a degree of confidence can be attached to this interval, we have a confidence interval estimate for the unknown parameter.

Three determinations must be made to develop a confidence interval:

1. A good point estimator of the parameter
2. The sampling distribution (or approximate sampling distribution) of the point estimator
3. The desired confidence level, usually stated as a percent

Knowledge of the sampling distribution of the estimator allows us to make a probability statement that involves the estimator and the parameter with the desired confidence level.

Confidence Interval for π

A *USA Today*/CNN/Gallup nationwide telephone poll of 1,022 adults reported on August 14, 1994, that 42% approved of President Clinton's handling of the economy. Because the respondents either approve or do not, the population can be considered a Bernoulli population consisting of a collection of successes and failures. If we let π be the proportion of successes (approvals), then 42% is an estimate of its numerical value. The actual value of π is the proportion of *all* adults who approve of the President's handling of the economy.

In Section 6.3, you learned that the sample proportion, p, is the point estimator of π. As stated above, to develop the confidence interval, we need to know the sampling distribution of the estimator. In most situations when we are estimating a proportion, the sample size is large; for example, the opinion poll consisted of 1,022 adults. When n is large, the Central Limit Theorem applies and says that the sample proportion, p, has a sampling distribution that is approximately normal. Furthermore, the mean of the sampling distribution is π and the standard deviation is $\sqrt{\pi(1-\pi)/n}$.

The normal probability table shows that a normally distributed variable will be within ± 1.96 standard deviations of its mean with 95% probability. Because p is approximately normally distributed and its standard deviation is $\sqrt{\pi(1-\pi)/n}$, it will be within

$$\pm 1.96\sqrt{\frac{\pi(1-\pi)}{n}}$$

of its mean, π, with 95% probability.

Notice, however, that the standard deviation, $\sqrt{\pi(1-\pi)/n}$, depends on the unknown π. That being the case, we are unable to compute numerical values for the confidence interval. Because the sample proportion p estimates π, however, the standard deviation can be estimated by replacing π with p. The result

$$\text{SE}(p) = \sqrt{\frac{p(1-p)}{n}}$$

is called the *standard error of the estimate p.*

> The **standard error of a statistic** is the standard deviation of its sampling distribution when all unknown population parameters have been estimated.

[*Note:* Some statisticians call $\sqrt{\pi(1-\pi)/n}$ the standard error and then refer to $\sqrt{p(1-p)/n}$ as the *estimated* standard error.]

The margin of error in estimating π with 95% probability is then found to be

$$\pm 1.96\, \mathrm{SE}(p) = \pm 1.96\sqrt{\frac{p(1-p)}{n}}$$

Adding and subtracting the margin of error to the estimator p, we have the 95% confidence interval for π:

$$p \pm 1.96\sqrt{\frac{p(1-p)}{n}}$$

Substituting in the values from the *USA Today*/CNN/Gallup poll, we have

$$p = .42 \qquad \text{Margin of error} = \pm 1.96\sqrt{\frac{(.42)(.58)}{1,022}} = \pm .0303$$

(The poll reported the margin of error as ± 3 percentage points.)

The result says that we are 95% confident that the true proportion of adults who approve of the President's handling of the economy is somewhere between $.42 - .03$ and $.42 + .03$—that is, between 39% and 45%.

Suppose we desire a confidence level other than 95%. For example, suppose we desire 98% confidence, or maybe only 80% confidence. The point estimator is still p, but what about the margin of error?

We see from Figure 7.1 that z^* denotes the value of a standard normal variable chosen so that $\alpha/2$ of the probability lies above it, or $(1-\alpha)100\%$ of the probability lies between $-z^*$ and $+z^*$. By choosing different values for α, we can specify different confidence levels. For example, if $\alpha = .02$, then $1 - \alpha = .98$, so that $(1-\alpha)100\% = 98\%$. That is, we are interested in a 98% confidence interval when we choose $\alpha = .02$.

FIGURE 7.1   Definition of z^*

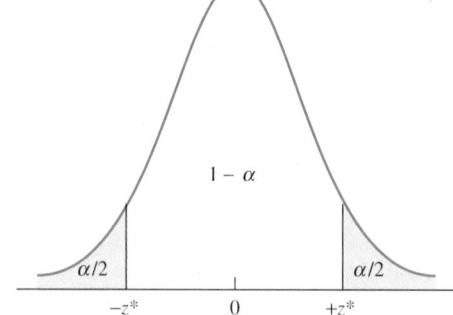

We call z^* the *upper critical value* for the standard normal distribution. Here are a few values of z^* obtained from the standard normal probability table:

Upper critical values for z

$1 - \alpha$	.80	.85	.90	.95	.98	.99
z^*	1.28	1.44	1.645	1.96	2.33	2.58

Substituting z^* for 1.96, we have that the margin of error in using p to estimate π with $(1 - \alpha)100\%$ confidence is

$$\pm z^* \sqrt{\frac{p(1 - p)}{n}}$$

That is, the margin of error is found by multiplying the z^* critical value by the standard error of the estimator.

We have the following general statement:

A large-sample $(1 - \alpha)100\%$ confidence interval for π based on p is given by the limits

$$p \pm z^* \sqrt{\frac{p(1 - p)}{n}}$$

EXAMPLE 7.1

To investigate the proportion of smokers who believe that smoking should be banned from public buildings, a researcher randomly samples 500 smokers from all walks of life and finds that 155 believe smoking should be banned from public buildings. Obtain a 99% confidence interval for the true proportion of smokers who believe that smoking should be banned from public buildings.

SOLUTION

The form of a 99% confidence interval for π, the true population proportion, is

$$p \pm 2.58 \sqrt{\frac{p(1 - p)}{n}}$$

The sample reveals that

$$p = \frac{155}{500} = .31$$

and thus

$$1 - p = .69$$

so the interval for π becomes

$$.31 \pm 2.58\sqrt{\frac{(.31)(.69)}{500}} = .31 \pm .053$$

The interval is from .257 to .363, and we are 99% confident that this interval does contain the proportion of smokers who believe that smoking should be banned from public buildings.

Interpreting Confidence Intervals

Recall from Chapter 4 that the probability that is assigned to an event is relevant only before the experiment is conducted. For example, you may have a 1/10,000 chance of winning a new car in a raffle, but after the winning number is chosen, you either won the car or you did not. The same can be said about confidence intervals. The probability that is attached to an interval is relevant only prior to sampling. After we sample and calculate the endpoints of the interval, the probability statements no longer apply. That is, in Example 7.1, it is meaningless to say that π lies between .257 and .363 with 99% probability. Either π is between the two values or it is not; there is no probability associated with it. What is meant by a 99% confidence interval is that if 100 different intervals are obtained from 100 different samples, then it is likely that 99 of those intervals will contain π and 1 will not. To illustrate, we will use the computer to simulate 100 confidence intervals and then count how many actually contain the value of π and how many do not.

EXAMPLE 7.2

Suppose in Example 7.1 we know that the true value of π is .3. That is, somehow we know that 30% of all smokers believe that smoking should be banned from public buildings. Let us now simulate 100 random samples, each of size 500, and then calculate, for each sample, the proportion of the 500 who believe smoking should be banned from public buildings. From each simulated proportion, we can then construct a 99% confidence interval for π. Knowing that $\pi = .3$, we can determine how many of the 100 intervals successfully contain π. To simulate the number of successes (banning smoking in public buildings) out of 500 Bernoulli trials, we generate a value of a binomial random variable with $n = 500$ and $\pi = .3$. Dividing the results by 500 gives a realization of p, from which we obtain a confidence interval. For each of the 100 realizations of p, we constructed the 99% confidence intervals for π listed in Table 7.1. From the table, we find that two of the intervals do not contain $\pi = .3$; consequently, 98 of the 100 do contain $\pi = .3$. Notice that we did not get 99 out of 100. Just because an event has a 99% probability, there is no guarantee that the event will occur *exactly* 99 times out of every 100. In the next simulation of 100 intervals, for example, all 100 intervals may contain $\pi = .3$.

To display a confidence interval graphically, we mark off the endpoints on a line graph and put a circle at the location of the point estimate of the parameter. The first

TABLE 7.1

(.257, .363)	(.268, .376)	(.270, .378)	(.238, .342)
(.268, .376)	(.262, .370)	(.234, .338)	(.257, .363)
(.230, .334)	(.247, .353)	(.259, .365)	(.249, .355)
(.276, .384)	(.228, .332)	(.260, .368)	(.243, .349)
(.240, .344)	(.259, .365)	(.260, .368)	(.255, .361)
(.257, .363)	(.232, .336)	(.247, .353)	(.253, .359)
(.247, .353)	(.249, .355)	(.253, .359)	(.213, .315)
(.245, .351)	(.264, .372)	(.268, .376)	(.266, .374)
(.247, .353)	(.249, .355)	(.238, .342)	(.204, .304)
(.274, .382)	(.264, .372)	(.240, .344)	(.245, .351)
(.234, .338)	(.241, .347)	(.234, .338)	(.283, .393)
(.245, .351)	(.240, .344)	(.282, .390)	(.247, .353)
(.278, .386)	(.238, .342)	(.234, .338)	(.191, .289) X
(.234, .338)	(.249, .355)	(.243, .349)	(.240, .344)
(.230, .334)	(.251, .357)	(.236, .340)	(.215, .317)
(.249, .355)	(.259, .365)	(.276, .384)	(.272, .380)
(.228, .332)	(.274, .382)	(.208, .308)	(.257, .363)
(.255, .361)	(.223, .325)	(.226, .330)	(.230, .334)
(.259, .365)	(.260, .368)	(.224, .328)	(.243, .349)
(.266, .374)	(.285, .395)	(.238, .342)	(.270, .378)
(.211, .313)	(.208, .308)	(.266, .374)	(.262, .370)
(.274, .382)	(.226, .330)	(.240, .344)	(.240, .344)
(.260, .368)	(.234, .338)	(.219, .321)	(.305, .415) X
(.274, .382)	(.268, .376)	(.245, .351)	(.253, .359)
(.255, .361)	(.228, .332)	(.282, .390)	(.247, .353)

seven intervals listed in Table 7.1 are graphed in Figure 7.2. Notice that the intervals are approximately the same length but are centered in different locations. All seven intervals do cover the true value $\pi = .3$.

FIGURE 7.2

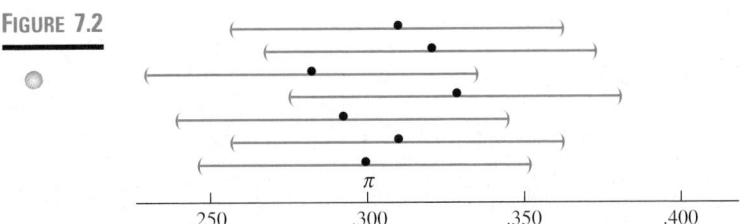

From the simulation experiment in Example 7.2, do not be led to believe that in practice we calculate 100 different intervals and find which ones contain the parameter. To the contrary, we calculate only *one* interval from the *one* random sample that we have chosen. After it is found, however, we know from the above simulation that there is a very good chance that the interval does indeed contain π. Unfortunately, there is always the chance that it does not, but, as we have seen, this is rare.

To execute the simulation in Example 7.2, from the **Calc** menu select **Random Data > Binomial**. Once the Binomial Distribution window appears, **Generate** 100 rows of data and **Store in column** C1. In the **Number of trials** box, type 500 and in the **Probability of success** box, type .3 and then click **OK**. Divide C1 by 500 and store the resulting values of p in C2. To compute the endpoints of the 100 confidence intervals, in the Session window, at the MTB prompt, type

MTB > **let c3 = c2 − 2.58*sqrt(c2*(1 − c2)/500)**
MTB > **let c4 = c2 + 2.58*sqrt(c2*(1 − c2)/500)**
MTB > **let c3 = round(c3*1000)/1000**
MTB > **let c4 = round(c4*1000)/1000**
MTB > **print c3 c4**

(The third and fourth lines are to round the results to three decimal places.)

Another important point about confidence intervals is that once an interval is obtained, no preferential treatment is given to any value in the interval. True, p is the center of the confidence interval for π and is the point estimate of π, but for interpretation purposes, it is just one of the values in the interval.

Validity and Precision of Confidence Intervals

The usefulness of a confidence interval is judged by its *validity* and *precision*. Validity is measured by the confidence level, which is the probability that the interval will contain the true value of the parameter. Precision is measured by the length of the interval. Because the length is twice the margin of error (the interval is usually estimator ± margin of error), the smaller the margin of error, the more precise the interval.

In general, these two characteristics, validity and precision, compete with each other. Because validity is measured by the confidence level, it would be desirable to have a 100% confidence level—that is, a guarantee that the parameter is somewhere in the interval. There would be *no* precision then, however, because the interval would have to contain all possible values; we could say with 100% confidence that π is somewhere between 0 and 1. Because this is unreasonable, the validity is fixed at a level—say, 95%—and then, given the assumptions, we search for an interval that has a small margin of error.

Recall that the margin of error of p is

$$z^* \sqrt{\frac{p(1 - p)}{n}}$$

Clearly for a fixed sample size n, the only way to reduce the margin of error (increase precision) is to decrease the value of z^*, which in effect reduces validity. On the other hand, to raise the validity—say from 95% confidence to 99% confidence—the z^* critical value increases from 1.96 to 2.58, which increases the margin of error and decreases precision.

It appears that it is not possible to increase precision while maintaining a fixed validity level. Because the sample size, n, appears in the denominator, however, it is possible to decrease the margin of error and hence increase precision by increasing n. In fact, we can have as much precision as we desire if we can afford the increased sample size. Remember, though, that large samples are usually costly.

To determine the sample size for a desired confidence level and a specific bound on the margin of error, we set the equation for the margin of error equal to the specified bound and then solve for n.

Sample Size Determination for Estimating π

In general, to estimate π with a $(1 - \alpha)100\%$ confidence interval so that the bound on the margin of error is B, we solve this equation for n:

$$B = z^* \sqrt{\frac{p(1 - p)}{n}}$$

The solution is

$$n = \left[\frac{z^*}{B} \sqrt{p(1 - p)} \right]^2$$

(If the computation yields a decimal expression, always round the result up to the next whole number in order not to exceed the desired bound on the margin of error.)

Notice that the solution depends on the value of the sample proportion, p. At this point, however, we do not have a sample; we are trying to find an appropriate sample size. This being the case, we must guess a value for p. One approach is to use a value of p that may have been obtained in a similar study. If that is unavailable, we recall that the standard deviation of p, $\sqrt{\pi(1 - \pi)/n}$, is maximized when we choose $\pi = .5$. The same applies to the margin of error; that is, the margin of error, $z^* \sqrt{p(1 - p)/n}$, is maximized when $p = .5$. By choosing $p = .5$, we guarantee that the margin of error will not exceed the specified bound B.

EXAMPLE 7.3

The estimated margin of error for the 99% confidence interval in Example 7.1 was .053. Find the sample size necessary to reduce that margin to .03.

SOLUTION

To retain the same validity (99%), the margin of error is

$$2.58 \sqrt{\frac{p(1 - p)}{n}}$$

which now must also equal .03. We have the formula

$$.03 = 2.58\sqrt{\frac{p(1-p)}{n}}$$

which must be solved for n.

To solve the equation we need a value for p. We can take the conservative approach and use $p = .5$ because it maximizes the standard error, or, in this case, we can use $p = .31$, which is the estimate obtained from the sample in Example 7.1. Because we have an estimate of π ($p = .31$) from a previous study, we will use it. The equation becomes

$$.03 = 2.58\sqrt{\frac{(.31)(.69)}{n}}$$

So

$$\sqrt{n} = \frac{2.58\sqrt{(.31)(.69)}}{.03} = 39.77$$

$$n = (39.77)^2 = 1{,}582.0044$$

Thus, to estimate π with a 99% confidence interval so that the margin of error is no greater than .03, we need a sample size of 1,583.

Using the conservative approach with $p = .5$, we get

$$\sqrt{n} = \frac{2.58\sqrt{(.5)(.5)}}{.03} = 43$$

$$n = 1{,}849$$

The results of this section are summarized in the boxes:

Large-Sample Confidence Interval for π

Application: A random sample from a Bernoulli population

Estimator: p, the sample proportion

Standard error of the estimator:

$$SE(p) = \sqrt{\frac{p(1-p)}{n}}$$

A $(1 - \pi)100\%$ confidence interval for π is given by the limits

$$p \pm z^*\sqrt{\frac{p(1-p)}{n}}$$

For a given bound B on the margin of error, the sample size required for estimating a population proportion is

$$n = \left[\frac{z^*}{B} \sqrt{p(1-p)} \right]^2$$

E X E R C I S E S   7.1

7.1   What will be the effect on a 95% confidence interval for π if the confidence level is changed from 95% to 98%?

7.2   What will be the effect on the margin of error in a 95% confidence interval for π if the sample size is changed from 100 to 400?

7.3   To learn whether financial aid is a determining factor in a student's choice of a college, the College Board conducted a survey of 1,183 students. They determined that 722 high school graduates entered their first choice of college regardless of financial aid offers. Find a 99% confidence interval for the proportion who did not use financial aid as a determining factor in their college choice. Is a 95% confidence interval wider or narrower than the interval you found? Explain your answer.

7.4   What is the margin of error in the interval estimate of Exercise 7.3? What sample size will reduce the margin of error to no greater than .03?

7.5   Of 900 people treated with a new drug, 180 showed an allergic reaction. Estimate with a 90% confidence interval the proportion of the population who would show an allergic reaction. How large a sample is necessary to ensure that the margin of error is no greater than .03?

7.6   The state of Massachusetts matched bank statements with welfare recipients and found that of 1,000 welfare recipients, 200 had accounts ranging from $20,000 to $50,000, 50 had accounts ranging from $50,000 to $100,000, and 10 had accounts in excess of $100,000. The state says that if a person has a bank account in excess of $20,000, he or she should not receive welfare benefits. Estimate with a 99% confidence interval the percent of welfare recipients who should not receive welfare benefits. How large a sample is necessary to ensure that the margin of error is no greater than .03?

7.7   A nationwide poll of 1,000 students by the Gordon S. Black Corporation for *USA Today* in 1996 showed that 880 believed that education is very important to their future (www.usatoday.com, May 13, 1996). Find a 90% confidence interval for the proportion of all students who believe that education is very important to their future.

7.8   In 1993, the Census Bureau estimated that 15.3% of all Americans had no health insurance coverage (Statistical Brief SB/94-28, October 1994). The data were obtained from the March 1994 supplement to the Current Population Survey (CPS). The CPS is a nationwide monthly survey of about 60,000 households conducted by the Census Bureau. In the report, the estimates and standard errors at the top of page 432 were given for the percents not covered by health insurance in several states:

State	Percent	Standard Error
Arizona	19.7	1.4
Colorado	12.6	1.3
Florida	19.6	.7
Montana	15.3	1.3
New Jersey	13.7	.7
New Mexico	22.0	1.4

SOURCE: U.S. Census Bureau.

Find individual 95% confidence intervals for the percent without health insurance in each of the states. Graph the intervals on a common number line. Describe any differences you observe among the percents without health coverage in the different states.

7.9   A Roper poll of 2,000 American adults showed that 1,440 thought that chemical dumps are among the most serious environmental problems. Estimate with a 98% confidence interval the proportion of the population who consider chemical dumps among the most serious environmental problems. After obtaining the interval, can we say that there is a 98% chance that the population proportion is inside the interval? Explain your answer.

7.10   A national survey of 6th- through 12th-grade students was conducted in the spring of 1993 as part of the 1993 National Household Educational Survey by the U.S. Department of Education. Based on responses from 6,504 students, these percents of students stated that they had witnessed a crime or threat at school:

Student's Race/Ethnicity	Percent	Standard Error
White, non-Hispanic	57	2.6
Black, non-Hispanic	56	3.4
Hispanic	51	2.3
Other races	48	4.8

SOURCE: U.S. Department of Education.

Construct 99% confidence intervals for the percent in each race/ethnicity category that had witnessed a crime or threat. Graph the intervals on a common number line. Which group had the highest percent witnessing a crime or threat? Can you detect any differences among the groups?

7.11   A 1996 poll of 1,200 African American adults conducted by Yankelovich Partners Inc. for *The New Yorker* found that 708 think that the American dream has become impossible to achieve. Construct a 90% confidence interval for the proportion of all African American adults who feel this way. What is the maximum margin of error for the survey? How large a sample will yield a maximum margin of error no greater than .03?

7.12   It is believed that 30% of all adults are overweight. How large a sample is necessary to estimate the true proportion of adults who are overweight with a 95% confidence interval so that the margin of error of the estimate is no greater than 2 percentage points?

7.13   The nationwide poll of 1,000 students by the Gordon S. Black Corporation for *USA Today* reported in Exercise 7.7 found that 60% of the students think that the behavior of students on school buses is a serious problem. How many students need to be surveyed so that the margin of error of a 98% confidence interval estimate of all students who feel this way is no greater than 3%?

7.14   Suppose one wishes to estimate the percent of American citizens who receive some sort of government aid. How large a sample is needed so that the margin of error of a 95% confidence interval estimate is no greater than 2%?

7.15   On September 22, 1993, President Clinton presented his health care plan to Congress. The next day a poll of 500 adult Americans was taken for TIME/CNN by Yankelovich Partners Inc. They reported that 36% thought increased federal involvement in the U.S. health care system would make the system worse and 33% foresaw improvement (given the margin of sampling error, that amounts to a statistical tie).

What is meant by the statement in parentheses: "given the margin of sampling error, that amounts to a statistical tie"? The article in *Time* (November 8, 1993) gave the sampling error as ±4.5%. How was that figure reached? Is it the correct value for the sampling error? Assuming that it is the correct value, give your interpretation of it.

7.16   The list gives the results of a simulation of 100 different 95% confidence intervals for π when it is known that the true value of π is .3. Mark the intervals that do not contain .3. How many of the intervals contain .3? Is this a reasonable number given that these are 95% confidence intervals?

(.295, .377)	(.239, .317)	(.266, .346)
(.277, .359)	(.279, .361)	(.250, .330)
(.256, .336)	(.281, .363)	(.244, .324)
(.268, .348)	(.250, .330)	(.285, .367)
(.248, .328)	(.256, .336)	(.300, .384)
(.244, .324)	(.291, .373)	(.275, .357)
(.254, .334)	(.281, .363)	(.254, .334)
(.225, .303)	(.295, .377)	(.271, .353)
(.283, .365)	(.246, .326)	(.250, .330)
(.254, .334)	(.250, .330)	(.271, .353)
(.269, .351)	(.243, .321)	(.291, .373)
(.275, .357)	(.243, .321)	(.248, .328)
(.223, .301)	(.277, .359)	(.262, .342)
(.258, .338)	(.262, .342)	(.314, .398)
(.235, .313)	(.262, .342)	(.208, .284)
(.260, .340)	(.237, .315)	(.233, .311)
(.298, .382)	(.269, .351)	(.223, .301)
(.237, .315)	(.273, .355)	(.264, .344)
(.254, .334)	(.266, .346)	(.243, .321)
(.269, .351)	(.266, .346)	(.239, .317)
(.248, .328)	(.254, .334)	(.266, .346)
(.279, .361)	(.250, .330)	(.216, .292)
(.262, .342)	(.254, .334)	(.266, .346)
(.285, .367)	(.268, .348)	(.289, .371)
(.285, .367)	(.273, .355)	(.235, .313)
(.268, .348)	(.254, .334)	(.283, .365)
(.244, .324)	(.250, .330)	(.283, .365)
(.271, .353)	(.273, .355)	(.262, .342)
(.241, .319)	(.231, .309)	(.281, .363)
(.262, .342)	(.269, .351)	(.293, .375)
(.212, .288)	(.295, .377)	(.285, .367)
(.287, .369)	(.306, .390)	(.273, .355)
(.252, .332)	(.244, .324)	
(.316, .400)	(.250, .330)	

7.17   Assume that π = .7. With a computer, simulate 100 different 98% confidence intervals for π and determine how many actually contain .7.

7.2   CONFIDENCE INTERVAL FOR μ BASED ON $\bar{y}$: THE z-INTERVAL

Having developed a confidence interval for the Bernoulli proportion π in the preceding section, we turn our attention to developing confidence intervals for the center of a measurement distribution. You learned in Chapter 6 that the choice of the parameter to measure the center depends on the shape of the parent distribution. When the parent distribution is symmetric, most natural measures of the center coincide and we use the population mean μ to measure the center. When the parent distribution is skewed, we usually use the population median θ to measure the center. In Section 7.4, we consider confidence intervals for the median θ, but now we consider confidence intervals for μ.

There are numerous estimators of the population mean μ; however, as discussed in Section 6.3, if the parent distribution is symmetric with tails that are not unusually long, then the sample mean $\bar{y}$ is the most logical estimator to use. If, on the other hand, the parent distribution is symmetric with long tails (a long-tailed distribution), then a trimmed mean $\bar{y}_T$ is a better estimator of μ. We present a confidence interval estimate of μ based on $\bar{y}$ in this section and a confidence interval estimate for μ based on $\bar{y}_T$ in Section 7.5.

Confidence Interval for μ Based on $\bar{y}$ When σ Is Known

Suppose, for the moment, that the parent population is normally (or nearly normally) distributed and its standard deviation, σ, is a known quantity. Under these conditions, the sampling distribution of $\bar{y}$ is also normally distributed with a mean μ and a standard deviation given by $\sigma/\sqrt{n}$. Because σ is known, we also have

$$\text{SE}(\bar{y}) = \frac{\sigma}{\sqrt{n}}$$

The normal probability table shows that a normal variable will be within ± 1.96 standard deviations of its mean with 95% probability. Because $\bar{y}$ is normally distributed with standard deviation $\sigma/\sqrt{n}$, the margin of error in using $\bar{y}$ to estimate μ with 95% probability is

$$\pm 1.96\text{SE}(\bar{y}) = \pm 1.96\frac{\sigma}{\sqrt{n}}$$

Adding and subtracting the margin of error to the estimator, we have that a 95% confidence interval estimate of μ is

$$\bar{y} \pm 1.96\frac{\sigma}{\sqrt{n}}$$

For confidence intervals with confidence levels other than 95%, we replace 1.96 with the appropriate z^* value from the z-tables. Thus, we have the z-interval:

> ## The z-Interval
>
> When the population standard deviation σ is known, a $(1 - \alpha)100\%$ confidence interval for μ based on $\bar{y}$ is given by the limits
>
> $$\bar{y} \pm z^* \frac{\sigma}{\sqrt{n}}$$

EXAMPLE 7.4

Tensile strength is the ability of a material to resist rupture when pressure is applied under specified conditions to one of its sides by an instrument. In the manufacture of a certain woven polypropylene, the process is operating properly when the standard deviation of the tensile strength is $\sigma = 4$ pounds per inch. Measurements of the tensile strength on a random sample of 40 rolls of woven polypropylene produced a mean of 87.3 pounds (information from Dr. Fred Morgan, J. P. Stevens & Co., Inc.). Find a 98% confidence interval for μ, the mean tensile strength of the material.

SOLUTION

Because a 98% confidence interval is desired, we have .01 probability on each tail, which corresponds to $z^* = 2.33$. The desired confidence interval becomes

$$\bar{y} \pm 2.33 \frac{\sigma}{\sqrt{n}}$$

Substituting in the numerical values, we have

$$87.3 \pm (2.33)\frac{4}{\sqrt{40}} = 87.3 \pm 1.474$$

which is 85.826 to 88.774. We are 98% confident that this interval contains the mean tensile strength of the woven polypropylene.

In the preceding discussion, we assumed that the parent population is normally distributed. Because of the Central Limit Theorem, the results apply to any shaped distribution as long as the sample size is large. Recall, however, that in a severely skewed distribution, the sample size must be very large for the Central Limit Theorem to overcome the skewness. Furthermore, the population median θ, not μ, is the preferred parameter to represent the center of most skewed distributions. As previously stated, confidence intervals for θ are presented in Section 7.4. Additionally, we know that $\bar{y}$ performs badly as an estimator of μ when the parent distribution has extremely long tails. Here again, we do not recommend the confidence interval based on $\bar{y}$ but rather the interval based on $\bar{y}_T$. For these reasons, you should restrict the use of the above confidence interval for μ based on $\bar{y}$ for those cases where the parent distribution is symmetric with tails that are not unusually long. Furthermore, the more the parent distribution deviates from normality, the larger the sample size should be. In most cases, a sample size in excess of 30 is adequate.

Determining Sample Size

Using the formula for the margin of error, we can determine the sample size necessary to guarantee a given margin of error for a given confidence level.

EXAMPLE 7.5

The precision of the 98% confidence interval obtained in Example 7.4 is quantified by its margin of error, which is 1.474. Suppose we wish to maintain the same validity—namely, 98%—but increase the precision so that the margin of error is only 1.2 pounds per inch. How large should the sample size be?

SOLUTION

The margin of error for a 98% confidence interval is

$$2.33 \frac{\sigma}{\sqrt{n}}$$

which now must equal 1.2. Knowing that $\sigma = 4$ (which was given in the problem), we can solve this equation for n:

$$1.2 = (2.33) \frac{4}{\sqrt{n}}$$

We have

$$\sqrt{n} = (2.33) \frac{4}{1.2} = 7.77$$

$$n = (7.77)^2 = 60.37$$

Rounding up to the next whole number (to maintain the specified margin of error), we need a sample size of 61.

Sample Size Determination for Estimating μ

In general, to estimate μ with a $(1-\alpha)100\%$ confidence interval so that the bound on the margin of error is B, solve this equation for n:

$$B = z^* \frac{\sigma}{\sqrt{n}}$$

The solution is

$$n = \left(z^* \frac{\sigma}{B}\right)^2$$

In Example 7.5, we were able to determine the sample size because the value of σ was known. In most cases, however, the value of σ is not available and must be estimated. Normally, the sample standard deviation, s, is used to estimate σ. Here, however, we are trying to determine the sample size for the random sample, in which case s is not available to us. We must find some other means of estimating σ. Perhaps a previous study gives a sample standard deviation that can be used, or, as was pointed out in Section 6.3, range/4 (or range/6) provides a crude estimate of σ.

EXAMPLE 7.6

Each February in Indianapolis, prospective National Football League players meet to try out for the NFL draft. A determining factor is a player's time to run the 40-yard dash. In 1996, the fastest time recorded was 4.36 seconds by Leeland McElroy of Texas A&M. The slowest time was 5.6 seconds (www.usatoday.com). A scout for an NFL team would like to know how many measurements for a particular player should be used to estimate the player's mean time in the dash to within .15 second with a 90% confidence interval.

SOLUTION

Because a 90% confidence interval is desired, we have

$$z^* = 1.645 \quad \text{Margin of error} = 1.645 \frac{\sigma}{\sqrt{n}}$$

From the 1996 data (which are typical for recent years), the range is $5.6 - 4.36 = 1.24$. An estimate of σ is

$$\frac{\text{Range}}{4} = \frac{1.24}{4} = .31$$

(Because these data are from one year and hence the true range may be greater, we use range/4 instead of range/6.) Substituting .15 second for the margin of error and .31 for

σ, we have

$$.15 = (1.645)\frac{.31}{\sqrt{n}}$$

Solving for n, or using the formula developed previously, we have

$$n = \left[(1.645)\frac{.31}{.15}\right]^2 = 11.56$$

Thus, he will need a sample of 12 measurements of the times from each player in the 40-yard dash.

The results of this section are summarized in the box:

Confidence Interval for μ Based on $\bar{y}$ When σ Is Known: The z-Interval

Application: Symmetric population distributions with tails that are not excessively long. If the population distribution deviates substantially from normality, then the sample size should be larger than 30.

Estimator: $\bar{y}$, the sample mean

Standard error of the estimator:

$$SE(\bar{y}) = \frac{\sigma}{\sqrt{n}}$$

A $(1 - \alpha)100\%$ confidence interval for μ is given by the limits

$$\bar{y} \pm z^* \frac{\sigma}{\sqrt{n}}$$

For a given bound, B, on the margin of error, the sample size is

$$n = \left(z^* \frac{\sigma}{B}\right)^2$$

EXAMPLE 7.7

From a random sample of 12 measurements of a certain player's time in the 40-yard dash, the NFL scout referred to in Example 7.6 found the average time to be 4.56 seconds. Using $\sigma = .31$ second, find a 90% confidence interval for this player's mean time in the 40-yard dash.

SOLUTION

If we assume the times have a symmetric distribution close to the normal distribution, then the confidence interval will be based on $\bar{y}$. The margin of error for a 90% confidence interval is

$$\pm 1.645 \frac{\sigma}{\sqrt{n}} = \pm(1.645)\frac{.31}{\sqrt{12}} = \pm.147$$

From the collected data we have $\bar{y} = 4.56$ seconds. Thus, a 90% confidence interval for the mean time is

$$4.56 \pm .147$$

which gives the limits 4.413 to 4.707. Based on the times collected in 12 trials, we are 90% confident that the mean time for this player is somewhere between 4.413 and 4.707 seconds.

Figure 7.3 gives a graphical illustration of the confidence interval. We can clearly see the length of the interval and thus evaluate its precision. If we had data from another player, we could graph a confidence interval for his time on the same scale and then compare the results.

FIGURE 7.3

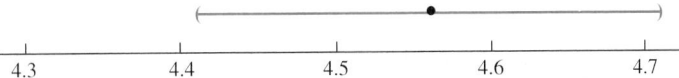

| 4.3 | 4.4 | 4.5 | 4.6 | 4.7 |

C O M P U T E R   T I P

After we have stored data in the worksheet, it is a simple procedure to construct a z-interval. From the **Stat** menu, choose **Basic Statistics** and then select **1-Sample z**. When the 1-Sample z window appears, **Select** the **Variable** containing the data and click on **Confidence interval** and specify the desired **Level**. For a z-interval, you must give the value for the population standard deviation in the **Sigma** box. To execute the command, click **OK**.

E X E R C I S E S 7.2

7.18   A random sample of size n is selected from a population with unknown mean μ and standard deviation $\sigma = 20$. Calculate a 95% confidence interval for μ based on $\bar{y}$ for each of these samples:
   a. $n = 30, \bar{y} = 94.3$ b. $n = 45, \bar{y} = 96.4$
   c. $n = 100, \bar{y} = 95.6$ d. $n = 200, \bar{y} = 95.82$

7.19   A random sample of 60 observations selected from a population with unknown mean μ and standard deviation $\sigma = 50$ yielded $\bar{y} = 465.8$.
   a. Find a 95% confidence interval for μ based on $\bar{y}$.
   b. Find a 99% confidence interval for μ based on $\bar{y}$.

7.20   Find the following confidence intervals for μ based on $\bar{y}$ from the given information:
   a. $n = 36, \bar{y} = 72.4, \sigma = 11.2$, confidence level = .95
   b. $n = 64, \bar{y} = 128.3, \sigma = 32.4$, confidence level = .98
   c. $n = 100, \bar{y} = 465, \sigma = 112$, confidence level = .99

7.21   The illustration shows 95% and 99% confidence intervals for μ based on $\bar{y}$ using the same data. Which one (A or B) is the 95% confidence interval? Explain your answer.

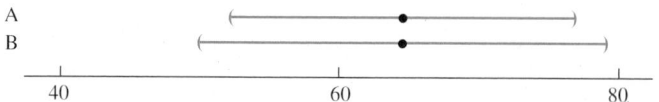

7.22   Two 95% confidence intervals for μ based on $\bar{y}$ are shown here. One has sample size $n = 80$ and the other has sample size $n = 120$. If we assume the standard deviations of the two populations are close to each other, which interval (A or B) has $n = 120$? Explain your answer. Why are the intervals centered in different locations?

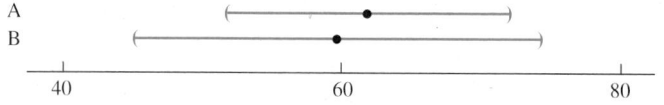

7.23   A city planner randomly samples 100 apartments in the inner city to estimate the mean living area per apartment. Find a 95% confidence interval based on $\bar{y}$ for the mean living area if the sample yielded an average of 1,325 square feet. Assume that the standard deviation σ is 42 square feet. Having found the interval, can you say that there is a 95% chance that the mean living area is within the interval? Explain your answer.

7.24   To determine the diameter of Venus, an astronomer makes 36 measurements of the diameter and finds $\bar{y} = 7,848$ miles. Assuming $\sigma = 310$ miles, find a 95% confidence interval estimate based on $\bar{y}$ of the diameter of Venus. What sample size is required so that the margin of error in determining the diameter of Venus is only 50 miles?

7.25   A sample of 22 public high school students spent an average of 6.5 hours studying per week. Assuming the distribution of hours spent studying is near normal and the standard deviation is 2.3 hours, find a 95% confidence interval for the mean number of hours spent studying by public high school students. Give your interpretation of the interval.

7.26   A new anesthetic was tested on ten patients who were to undergo surgery. These are their recovery times (in hours):

Worksheet: Anesthet.mtw

2.6   3.0   2.8   3.1   3.5   2.9   3.1   2.7   2.9   3.3

Normal Probability Plot

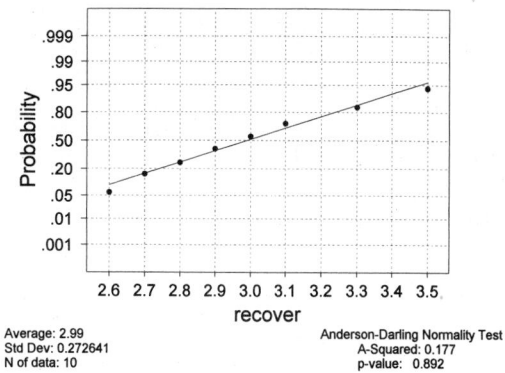

Average: 2.99
Std Dev: 0.272641
N of data: 10

Anderson-Darling Normality Test
A-Squared: 0.177
p-value: 0.892

From the accompanying normal probability plot, can we assume that the recovery time is normally distributed? Is this condition required for a z-interval? Assuming $\sigma = .3$, construct a 90% confidence interval for the mean recovery time for the new drug. Give your interpretation of the interval.

7.27   A group of students who frequent the student union are concerned with the amount of coffee dispensed from a certain coffee machine. To estimate the mean number of ounces of coffee in a cup, they take a random sample of 40 cups and find that $\bar{y} = 5.2$ ounces. Assuming $\sigma = .24$ ounce, find a 99% confidence interval estimate based on $\bar{y}$ of the mean number of ounces dispensed per cup. If the students had taken a random sample of 100 cups, what would be the effect on the length of the interval? Explain.

7.28   A consumer research group sampled 100 hand-held video games, all of the same make and model. The sample mean life (hours of operation before failure) was 560 hours. Assuming the standard deviation σ is 35 hours, construct a 90% confidence interval estimate based on $\bar{y}$ of the true mean life span of the video games. Is a 95% confidence interval wider or narrower than the interval you got? Explain your answer.

7.29   To study the birth weights of infants whose mothers smoke, a physician records the weights of 100 newborns whose mothers smoke. Find a 98% confidence interval based on $\bar{y}$ for the mean birth weight of children of smoking mothers if $\bar{y}$ was found to be 6.1 pounds. Use $\sigma = 2.1$ pounds. What sample size is required so that the margin of error in determining the mean birth weight is only .3 pound?

7.30   An ad for Philips Longer Life Outdoor floodlight gave an average lifetime of 2,500 hours and a standard deviation of 325 hours. A random sample of 100 floodlights gave an average life of 2,418 hours. Find a 99% confidence interval for the mean lifetime of the floodlights. Does the interval contain 2,500 hours? What does this mean about the advertiser's claim?

7.31   A banker would like to estimate the average amount of the loans made to farmers in his community for the past growing season. He randomly selects 40 accounts and finds that the average loan is $78,460. Assuming that the standard deviation is $22,000, he proceeds to find a 90% confidence interval for the mean loan amount. Should he have checked the underlying parent distribution for skewness or long tails before finding the confidence interval? Explain.

7.3 CONFIDENCE INTERVAL FOR μ BASED ON $\bar{y}$: THE t-INTERVAL

The development of the z-interval in the preceding section relied on the assumption that the population standard deviation σ is a known quantity. It is more realistic, however, that σ is unknown to the experimenter and must be estimated with the sample data.

As before, we begin our discussion by assuming that the parent distribution is normally distributed with mean μ and standard deviation σ. The only difference now is that we do not know the numerical value of σ. Again, under these assumptions, the sampling distribution of $\bar{y}$ is exactly normally distributed with mean μ and standard deviation of $\sigma/\sqrt{n}$.

Computing the standardized version of $\bar{y}$ (the z-score for $\bar{y}$), we have that

$$z = \frac{\bar{y} - \mu}{\sigma/\sqrt{n}}$$

has a *standard normal distribution*. If we replace σ with its estimate s in the above formula, what is the effect on the distribution?

Student's
t Distribution

This is precisely the question posed by the chemist W. S. Gosset in 1908. In particular, Gosset, who worked for an Irish brewing company, was concerned with the distribution of

$$t = \frac{\bar{y} - \mu}{s/\sqrt{n}}$$

We will use the letter t to distinguish it from z. Notice that the only difference between t and z is that s has replaced σ. Because his employer did not allow its chemists to publish their own research, Gosset published his work under the pseudonym "Student." Thus, the distribution became known as *Student's t distribution*.

To better understand Gosset's problem and the effect that replacing σ with s has on the distribution of t, we will use the computer to simulate the sampling distributions of both z and t. We first generate 2,000 different random samples of size 10 from a normally distributed population that has mean $\mu = 50$ and standard deviation $\sigma = 15$. For each of the 2,000 random samples we compute $\bar{y}$ and s and then compute z and t. In computing z, we use $\sigma = 15$ for all 2,000 cases. In computing t, we use the sample standard deviation s, which may change from sample to sample.

Figure 7.4 gives histograms of the 2,000 realizations of z and of t. Notice that the simulated distribution of z looks very much like the standard normal distribution. It is symmetrically distributed about zero and spreads out from about -3 to $+3$. The simulated distribution of t is also symmetrically distributed about zero but is spread out slightly more than the distribution of z. The side-by-side boxplots and normal probability plots in Figure 7.5 also illustrate the difference. In the boxplots, we see the symmetry about zero and the greater variance in the t distribution. In the normal probability plots, we see that z is indeed normally distributed but the distribution of t has tails that are longer than those of a normal distribution. The long tails are illustrated by the many outliers exhibited in the boxplot.

FIGURE 7.4

FIGURE 7.5

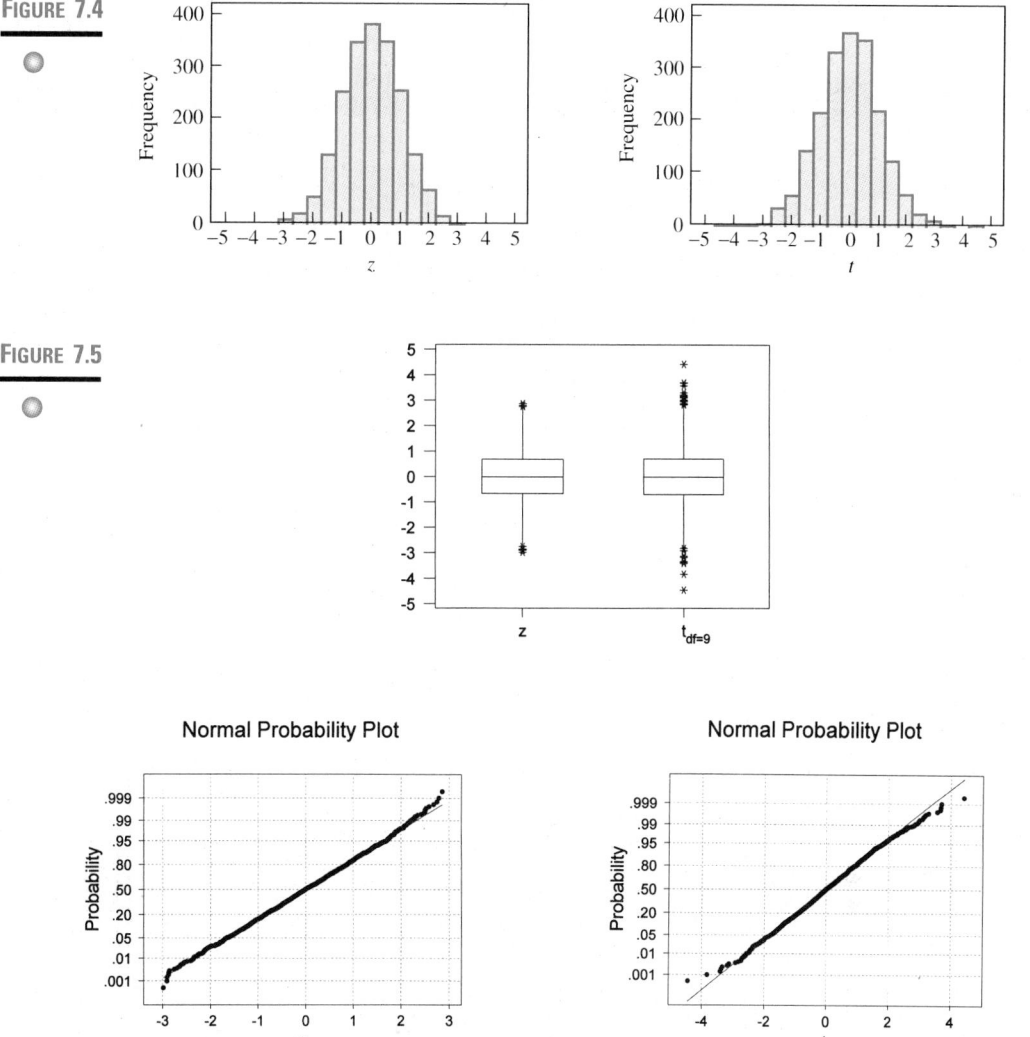

Gosset, of course, did not have a computer to draw normal probability plots, nor did he have boxplots to illustrate variability. He observed the difference, however, between z and t and noticed that the shape of the t distribution changes when the sample size changes. Later it was proved that, for $n > 3$, the standard deviation of the t distribution is given by

$$\text{Standard deviation of } t = \sqrt{\frac{n-1}{n-3}}$$

Notice that as n gets larger, this standard deviation gets closer to 1, which is the standard deviation of z. In fact, as n becomes infinitely large, the t distribution approaches the standard normal distribution. If n is small, however, the t distribution has thicker and longer tails than does the standard normal distribution.

Recall from Section 2.5 that the standard deviation, s, has $n - 1$ degrees of freedom. Because s is in the denominator of t, we say that the t distribution also has $n - 1$ degrees of freedom. In place of saying that the shape of the t distribution depends on the sample size, it is more common to say that its shape depends on the degrees of freedom.

Side-by-side boxplots of simulated t distributions in Figure 7.6 show how the spread of the t distribution decreases as the degrees of freedom increase from 4 to 29. Notice that if the degrees of freedom are 30 or more, there is not much difference in the spreads of the t and the z distributions.

FIGURE 7.6

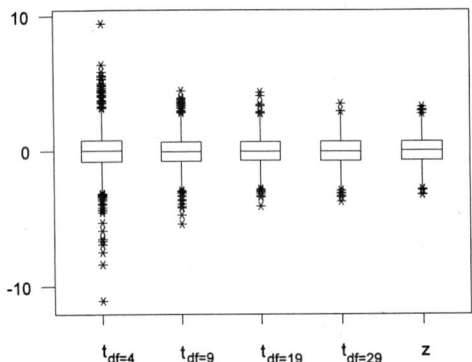

Just as z^* is the critical value of the standard normal variable such that $\alpha/2$ of the area lies on the tail above it, t^* denotes the critical value of the t variable with $n - 1$ degrees of freedom such that $\alpha/2$ of the area lies on the tail above it. Because of the symmetry about zero, the limits that include the middle $(1 - \alpha)100\%$ are $-t^*$ to $+t^*$ (see Figure 7.7).

FIGURE 7.7   A t distribution with $n - 1$ degrees of freedom

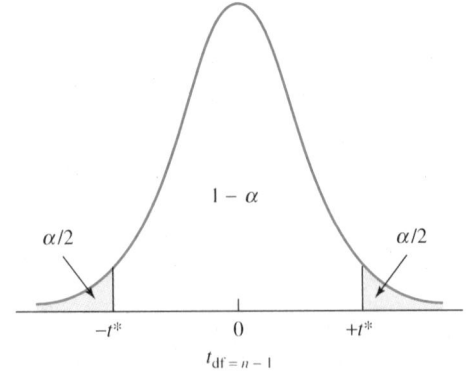

Because the t distribution has longer tails than the standard normal distribution, the limits $\pm t^*$ that include the middle $(1 - \alpha)100\%$ will naturally be larger than the limits $\pm z^*$ found in the normal tables.

Table B.3 gives the values of t^* for different tail probabilities and different values of the degrees of freedom (abbreviated df). For example, for df $= 10$, $t^* = 2.228$ has .025 probability above it. Because the t distribution is symmetric about zero, the middle 95% will be between -2.228 and $+2.228$. Recalling that the middle 95% of a standard normal distribution is between -1.96 and $+1.96$, we see that the t distribution is indeed spread out more than the standard normal.

EXAMPLE 7.8

Using Table B.3, find t^* that has .05 probability above it for a t distribution with 15 degrees of freedom.

SOLUTION

With 15 degrees of freedom, the upper 5% critical value is found to be $t^* = 1.753$. Because the distribution is symmetric about zero, the lower 5% critical value is $-t^* = -1.753$. We then have that the middle 90% of the distribution lies between -1.753 and $+1.753$.

Confidence Interval for μ Based on $\bar{y}$ When σ Is Unknown

To develop the confidence interval for μ based on $\bar{y}$ when σ is unknown, we must replace the z^* critical value found in the z-tables with a t^* critical value found in Table B.3. Consequently, the $(1 - \alpha)100\%$ confidence interval for μ becomes

$$\bar{y} \pm t^* \frac{s}{\sqrt{n}}$$

When to Use the t-Interval

The derivation of the t distribution assumed that the parent population is normally distributed. In practice, however, perfect normally distributed populations never happen. Fortunately, statistical research has shown that the t-interval is remarkably *robust* (validity and precision are unaffected by moderate departures from normality) except when the parent distribution is severely skewed or has long tails. In other words, moderate departures from normality are permissible when applying the t-interval.

Prior to using the t-interval, therefore, we should check whether or not there are unusual violations of the normality assumption. This is done quite easily with the data analysis procedures given in Section 6.1. If a normal probability plot (or some other procedure) indicates the presence of a large number of outliers or severe skewness, we should avoid using the t-interval.

For nonnormal populations, recall that when the sample size is reasonably large, the Central Limit Theorem guarantees that the sampling distribution of $\bar{y}$ is approximately normally distributed. This, along with the fact that when the sample size is large, the sample standard deviation s is a close approximation to σ, assures the validity of the *t*-interval. In other words, for large sample sizes, we need not be unusually cautious about the nonnormality of the population.

Follow these rules of thumb when deciding whether to use the *t*-interval procedure:

- If the sample size is less than 15, the population distribution should be close to normal.
- If the sample size is between 15 and 30, the *t*-interval can be used when the population distribution is reasonably symmetric and has no outliers.
- If the sample size is greater than 30, the *t*-interval procedure is permissible except when there are extreme outliers or strong skewness.

For those situations when the *t*-interval is not appropriate, you should consider the procedures described in Sections 7.4 and 7.5.

Confidence Interval for μ Based on $\bar{y}$ When σ Is Unknown: The *t*-Interval

Application: Symmetric population distributions with tails that are not excessively long. If the sample size is small, then the population distribution should not deviate substantially from normality.

Estimator: $\bar{y}$, the sample mean

Standard error of the estimator:

$$\text{SE}(\bar{y}) = \frac{s}{\sqrt{n}}$$

A $(1 - \alpha)100\%$ confidence interval for μ is given by the limits

$$\bar{y} \pm t^* \frac{s}{\sqrt{n}}$$

where t^* is the upper $\alpha/2$ critical value found in the *t*-table with degrees of freedom $n - 1$.

EXAMPLE 7.9

In Example 2.23 in Section 2.6, radiocarbon dates were given for a sample of 23 observations (worksheet: **Archaeo.mtw column 4**) taken from an archaeological excavation of the Danebury iron-age hill fort (Cunliffe, 1984). In Examples 2.23 and 2.24, it was concluded that the distribution of radiocarbon ages (recorded as years before 1983) is symmetric and possibly even normally distributed. In Figure 7.8, the normal probability plot of the data shows no significant departures from normality. Find a 99% confidence interval for the mean age of the archaeological site.

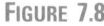

FIGURE 7.8

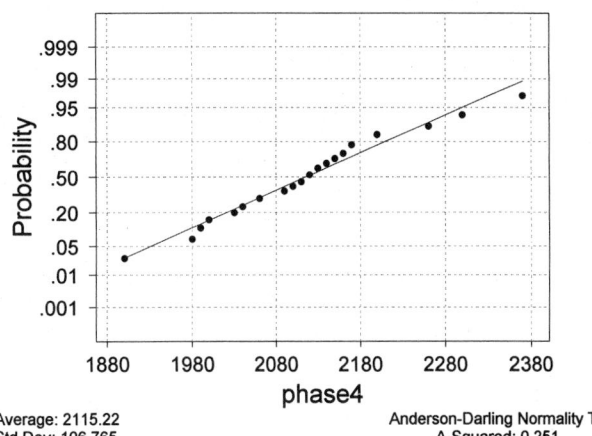

Normal Probability Plot
Probability
.999
.99
.95
.80
.50
.20
.05
.01
.001
1880
1980
2080
2180
2280
2380
phase4
Average: 2115.22
Std Dev: 106.765
N of data: 23
Anderson-Darling Normality Test
A-Squared: 0.251
p-value: 0.709

FIGURE 7.9

```
Stem-and-leaf of phase 4    N = 23
Leaf Unit = 10

      1      19  0
      3      19  89
      6      20  034
      9      20  669
    (6)      21  012234
      8      21  5677
      4      22  0
      3      22  6
      2      23  0
      1      23  7
```

SOLUTION

A stem and leaf plot of the data is presented in Figure 7.9 for convenience. Because the data are reasonably close to being normally distributed, we can find a confidence interval based on the Student's t distribution. It takes the form

$$\bar{y} \pm t^* \frac{s}{\sqrt{n}}$$

Because a 99% confidence interval is requested, we have .005 on the upper tail of the t distribution. With degrees of freedom $= n - 1 = 22$, we find

$$t^* = 2.819$$

From the data, we find

$$\bar{y} = 2{,}115.2 \qquad s = 106.8$$

Thus, the 99% confidence interval for μ becomes

$$2{,}115.2 \pm (2.819)\frac{106.8}{\sqrt{23}} = 2{,}115.2 \pm 62.8$$

This gives the interval from 2,052.4 to 2,178.0. Because the "ages" were recorded as years before 1983, we are 99% confident that the mean age of the Danebury archaeological site is somewhere between 195 and 69.4 B.C.

C O M P U T E R   T I P

After the data are stored in the worksheet, the t-interval is executed by selecting the **Stat** menu and choosing **Basic Statistics** and then selecting **1-Sample t**. In the 1-Sample t window, **Select** the **Variable** where the data are stored, click on **Confidence interval**, and specify the desired **Level** and click **OK**.

Figure 7.10 is the Minitab output from the archaeological data in worksheet: **Archaeo.mtw column 4**. Note that there is a slight difference between the interval we obtained in Example 7.9 and the one found by Minitab. The difference is due to round-off error and should be ignored.

FIGURE 7.10　Confidence Intervals

Variable	N	Mean	StDev	SE Mean	99.0 % C.I.	
phase4	23	2115.2	106.8	22.3	(2052.5,	2178.0)

E X E R C I S E S   7.3

7.32　For the following sample sizes, find t^* when the upper-tail probability is p:

　　a. $n = 14, p = .05$　　　　　　　　　　b. $n = 26, p = .05$
　　c. $n = 10, p = .10$　　　　　　　　　　d. $n = 20, p = .10$
　　e. $n = 22, p = .01$　　　　　　　　　　f. $n = 40, p = .01$

7.33　On August 20, 1984, a skull and almost complete skeleton of a 12-year-old male *Homo erectus* who died some 1.6 million years ago was found on the west side of Lake Turkana in northern Kenya. It is believed that the 5 foot 6 inch youth would have matured to 6 feet had he lived (*USA Today*, October 19, 1984). This is contrary to the assumption that early humans were shorter than modern man. *Homo erectus* has been found in Java, China, and various African sites. Suppose we were able to determine the adult heights of five *Homo erectus* skeletons to be 64, 70, 73, 69, and 70 inches. Assume that the parent distribution of heights is near normal. Find a 95%

confidence interval for the mean adult height of *Homo erectus*. Does the interval contain 6 feet? Is the normality condition required for constructing the t-interval?

7.34   To estimate the number of visitors per month to the John F. Kennedy Library, the following numbers of visitors were recorded for 12 randomly selected months:

Worksheet: Kennedy.mtw

198,286	249,821	294,653	255,728	267,475	231,759
275,641	191,374	228,391	307,613	250,834	259,540

From the normal probability plot, can we assume that the parent distribution is near normal?

Normal Probability Plot

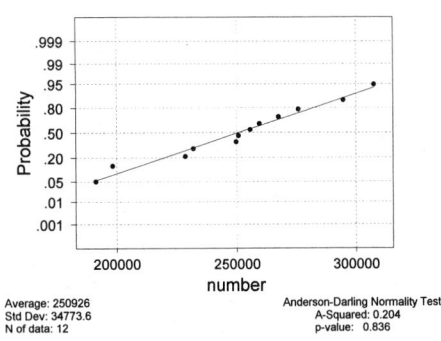

Average: 250926
Std Dev: 34773.6
N of data: 12

Anderson-Darling Normality Test
A-Squared: 0.204
p-value: 0.836

To construct a t-interval, is it required that the parent distribution be near normal? Assuming that the conditions for the t-interval are met, estimate with a 90% confidence interval the mean number of visitors per month.

7.35   An air pollution index for 15 randomly selected days during the summer months was recorded for a major western city:

Worksheet: Pollutio.mtw

57.6	61.2	59.4	65.6	58.3	44.7	63.2	48.8
55.7	59.0	64.3	43.2	59.7	71.2	45.6	

Based on the midsummaries in the accompanying letter value display, can we assume that the parent distribution of the index is symmetric? Should we check the distribution for normality?

```
Letter Value Display
```

	DEPTH	LOWER	UPPER	MID	SPREAD
N=	15				
M	8.0	59.000		59.000	
H	4.5	52.250	62.200	57.225	9.950
E	2.5	45.150	64.950	55.050	19.800
D	1.5	43.950	68.400	56.175	24.450
	1	43.200	71.200	57.200	28.000

(continued)

```
Descriptive Statistics

Variable          N     Mean   Median   TrMean    StDev   SEMean
index            15    57.17    59.00    57.16     8.21     2.12

Variable        Min      Max       Q1       Q3
index         43.20    71.20    48.80    63.20
```

Assuming that the conditions for a *t*-interval are met, use the information in the descriptive statistics to estimate the mean air pollution index with a 98% confidence interval.

7.36   These are the cholesterol values from a sample of 62 subjects in the Framingham Heart Study presented in Exercise 2.55 in Section 2.3:

Worksheet: Framingh.mtw

393	353	334	336	327	300	300	308	283	285	270
270	272	278	278	263	264	267	267	267	268	254
254	254	256	256	258	240	243	246	247	248	
230	230	230	230	231	232	232	232	234	234	
236	236	238	220	225	225	226	210	211	212	
215	216	217	218	200	202	192	198	184	167	

SOURCE: R. D'Agostino et al., "A Suggestion for Using Powerful and Informative Tests of Normality," *The American Statistician* 44 (1990): 316–321.

Normal Probability Plot

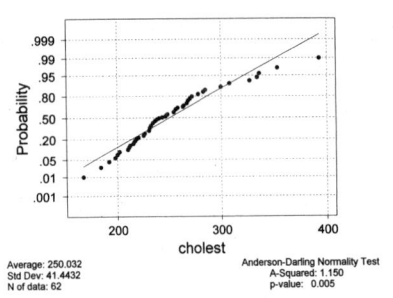
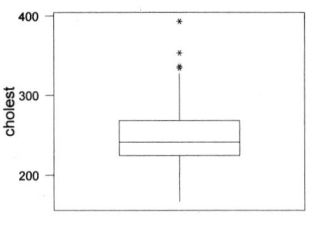

```
Average: 250.032              Anderson-Darling Normality Test
Std Dev: 41.4432                  A-Squared: 1.150
N of data: 62                     p-value: 0.005
```

Notice that the accompanying normal probability plot has a slight curvature and thus does not fit the straight line. What does this mean about the shape of the parent distribution? Does the boxplot give the same conclusion? Considering the apparent shape of the parent distribution, should we use the population mean or median to represent a typical cholesterol value?

7.37   These descriptive statistics are for the cholesterol levels given in Exercise 7.36. Suppose we decide to estimate the population mean. From the descriptive statistics, calculate a 95% confidence interval for μ based on $\bar{y}$.

```
Descriptive Statistics

Variable          N     Mean   Median   TrMean    StDev   SEMean
cholest          62   250.03   241.50   247.80    41.44     5.26
```

(continued)

Variable	Min	Max	Q1	Q3
cholest	167.00	393.00	225.00	268.50

7.38   In Exercise 2.56 in Section 2.3, the survival times in weeks were given for 20 male rats that were exposed to a high level of radiation:

Worksheet: Rat.mtw

152	152	115	109	137	88	94	77	160	165
125	40	128	123	136	101	62	153	83	69

SOURCE: J. Lawless, *Statistical Models and Methods for Lifetime Data* (New York: Wiley, 1982).

Based on the accompanying normal probability plot, is it reasonable to assume that the survival time is normally distributed? Should the data be normally distributed in order to construct a t-interval?

Normal Probability Plot

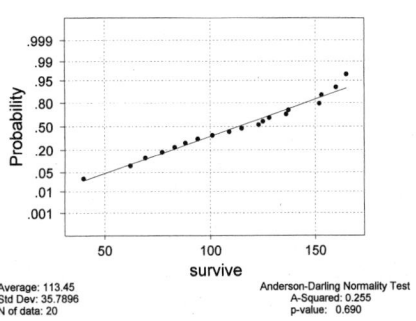

Assuming that the conditions for a t-interval are met, use the information in the descriptive statistics to estimate the mean survival time with a 95% confidence interval.

Descriptive Statistics

Variable	N	Mean	Median	TrMean	StDev	SEMean
survive	20	113.45	119.00	114.67	35.79	8.00

Variable	Min	Max	Q1	Q3
survive	40.00	165.00	84.25	148.25

7.39   Surface-water salinity measurements were taken in a bottom-sampling project in Whitewater Bay, Florida, and analyzed in Exercise 6.35 of Section 6.3:

Worksheet: Salinity.mtw

46	37	62	59	40	53	58	49	60	56
58	46	47	52	51	60	46	36	34	51
60	47	40	40	35	49	48	39	36	47
59	42	61	67	53	48	50	43	44	
49	46	63	53	40	50	78	48	42	

SOURCE: J. Davis, *Statistics and Data Analysis in Geology*, 2nd ed. (New York: Wiley, 1986).

In the analysis of the data given previously, we concluded that the salinity variation should be approximately normally distributed. Is this a required condition in order to find a confidence interval for the mean salinity? Based on these descriptive statistics, find a 99% confidence interval for the mean salinity level.

```
Descriptive Statistics

Variable          N      Mean    Median    TrMean      StDev    SEMean
salinity         48     49.54     48.50     49.18       9.27      1.34

Variable        Min       Max        Q1        Q3
salinity      34.00     78.00     42.25     57.50
```

7.40   The weekly family incomes of a random sample of 50 Puerto Rican families in Miami were analyzed in Exercise 6.42 of Section 6.3:

Worksheet: Puerto.mtw

150	280	175	190	305	380	290	300	170	315
280	255	335	180	200	210	350	360	550	225
245	175	180	260	320	345	180	225	275	280
325	350	190	220	265	270	335	310	270	240
235	355	370	260	280	350	390	250	375	250

In the previous analysis, after one outlier (550) was removed, the remainder of the data appeared to be short-tailed. If we remove the outlier, are there any serious consequences in finding a confidence interval for μ based on $\bar{y}$? Does the fact that the sample size is 50 have any affect on your decision? Remove the outlier and construct a 90% confidence interval estimate for the mean weekly income for the Puerto Rican families of Miami.

7.41   Exercise 2.57 in Section 2.3 described a special diet that was mixed with a drug compound designed to reduce low-density lipoprotein (LDL) cholesterol and was fed to a treatment group of quail. A placebo group of quail were fed the same special diet for the same period of time but without the drug compound.

Worksheet: Quail.mtw

Placebo

64	49	54	64	97	66	76	44	71	89
70	72	71	55	60	62	46	77	86	71

Treatment

40	31	50	48	152	44	74	38	81	64

SOURCE: J. McKean and T. Vidmar, "A Comparison of Two Rank-Based Methods for the Analysis of Linear Models," *The American Statistician* 48 (1994): 220–229.

The normal probability plots are of the plasma LDL levels for the two groups. Does it appear that the parent population of the placebo group is normally distributed? What about the treatment group?

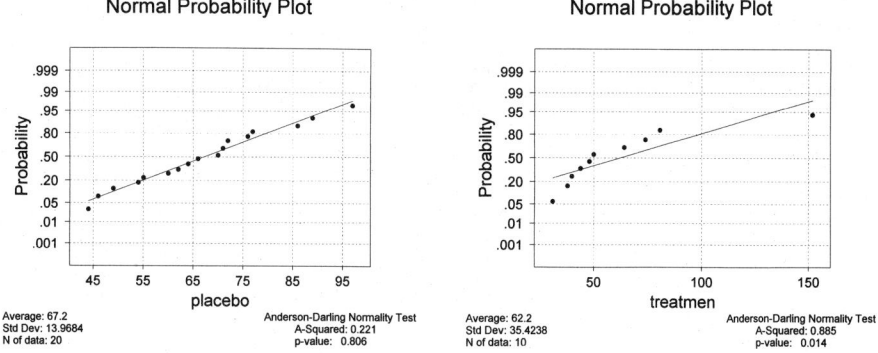

It was pointed out that there is one very large value in the treatment group. The researchers stated that this was typical with data in the study. If we remove that one observation, we get the following normal probability plot for the treatment group:

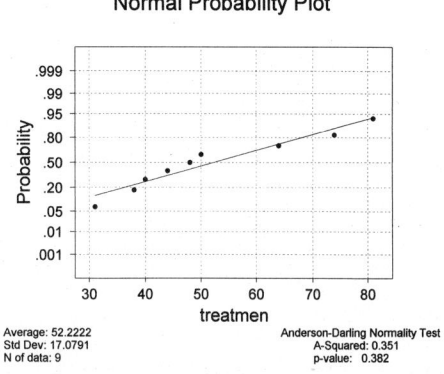

By removing the one observation, can we assume that the parent population of the treatment group is normally distributed? If we wish to construct confidence intervals for the placebo and treatment groups, is it advisable to remove the one extreme outlier in the treatment group and find t-intervals?

7.42   The accompanying individual t-intervals are for the quail data in Exercise 7.41. The first set of intervals is a comparison of the placebo and treatment groups when no data are removed from the data sets. The second set is the same comparison with the one extreme outlier in the treatment group removed. Without removing the outlier, do you detect a difference between the placebo and treatment groups? After removing the outlier, do you detect a difference between the two groups? What conclusion should we draw from these data?

```
Individual 95% CIs For Mean

  Level      N      Mean     StDev   ---+---------+---------+---------+---------+---------+------
  placebo    20     67.20    13.97                              (------*------)
  treatmen   10     62.20    35.42   (---------------------------*------------------------------)
                                     ---+---------+---------+---------+---------+---------+----------
                                        40        50        60        70        80
```

(continued)

```
Individual 95% CIs For Mean

Level      N    Mean    StDev   ---+---------+---------+---------+---------+------
placebo    20   67.20   13.97                                (------*------)
treatmen   9    52.22   17.08   (------------*------------)
                                ---+---------+---------+---------+---------+------
                                   40        50        60        70        80
```

7.43 Investing in coins can be very profitable as well as risky. Had one invested $10,000 in coins in 1980, in 5 years the investment would have grown to $23,602. Over this 5-year period, coins were the best investment, with more than a 27% per year return. Investing in the New York Stock Exchange would have yielded an average rate of return of 10.7% per year. Suppose that an investor calculates the yearly return on each of 12 possible investments and gets these results:

Worksheet: Coins.mtw

12.6	9.8	13.2	11.6	12.1	10.7
14.6	10.4	18.4	11.2	27.0	10.7

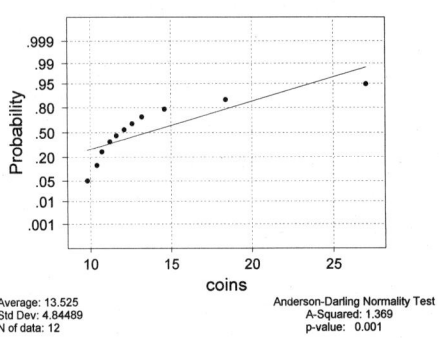

Normal Probability Plot

Average: 13.525
Std Dev: 4.84489
N of data: 12

Anderson-Darling Normality Test
A-Squared: 1.369
p-value: 0.001

What does the normal probability plot say about the shape of the parent distribution? In choosing a parameter to measure a "typical" return on an investment, what do you recommend: the mean or the median? Why? What condition must be satisfied in order to calculate a t-interval? Do you think that condition is met with these data?

7.44 Average home electric bills in the winter can vary as much as 600% depending on where you live. Here are the costs of some 3-month bills (1,500 kilowatt-hours) in 15 randomly selected cities on the West Coast:

Worksheet: Electric.mtw

$78.26	184.28	124.36	35.28	144.70	197.75	20.60	110.18
131.49	148.14	94.50	129.34	101.26	35.70	110.90	

Construct a normal probability plot of the data. Does it appear that the parent population is normally distributed? Assume that the conditions for the t-interval are met and construct a 90% confidence interval for the mean 3-month bill for cities on the West Coast.

7.45 Ozone, a prevalent photochemical oxidant, has been linked to forest decline and severe crop loss. Column 3 of worksheet: **Tablrock.mtw** contains ozone concentrations that were obtained on Mt. Mitchell in Yancey County, North Carolina, in September 1992 (data provided by Dr. Thomas C. Rhyne). We provide the descriptive statistics, boxplot, and normal probability plot of the ozone concentration here.

```
Descriptive Statistics

Variable        N       N*     Mean    Median   TrMean    StDev    SEMean
03             629       90   23.693   20.000   22.183   19.276    0.769

Variable      Min      Max      Q1       Q3
03          0.000  254.000   11.000   33.000
```

Normal Probability Plot

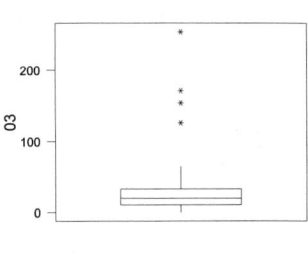 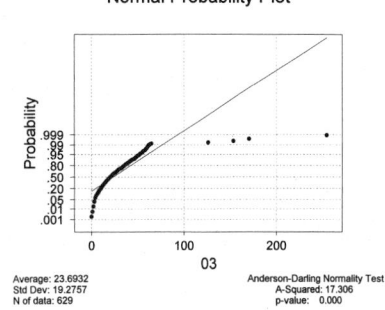

a. What do the boxplot and normal probability plot tell about the distribution of the data?
b. What is the sample size? What effect does the sample size have on the decision to find a confidence interval for the mean ozone concentration based on the t distribution?
c. Using the descriptive statistics, find a 99% confidence interval for the mean ozone concentration.
d. Do you think the validity of the confidence interval is affected by the shape of the population distribution?

7.4 CONFIDENCE INTERVAL FOR THE POPULATION MEDIAN

The z-interval and the t-interval, covered in the two preceding sections, are used to find confidence intervals for the center of a population that has a reasonably symmetric distribution without long tails. In cases where the sample size is rather small, the population distribution should not deviate substantially from the normal distribution. Furthermore, the confidence intervals are for the population mean, μ.

In many situations, however, the population median, θ, is the parameter to estimate. In particular, it was suggested in Section 6.3 that if the parent distribution is significantly skewed, then the median is a better indicator of the center of the distribution than is the mean. Also in Section 6.2, you saw that in the case of at least one symmetric long-tailed distribution (the Laplace), the sample median has a smaller standard deviation than the sample mean. This section shows how to construct confidence intervals for the population median given small and large samples.

Small-Sample Confidence Interval for the Population Median

The interval we give is based on the binomial distribution and is commonly referred to as the confidence interval obtained from the *sign test* (refer to a nonparametric text for an explanation). To obtain a confidence interval for the population median, θ, we first need to order the sample (ordered stem and leaf plot). Then the confidence interval is formed from two of the sample observations, C_L and C_H. C_L is located by counting up from the low end of the ordered data, and C_H is located by counting down from the high end of the ordered data. The distance we count in from each end (location of C) is found from the binomial table with $\pi = \frac{1}{2}$, or more quickly from Table 7.2. If, for example, we desire a 95% confidence interval (C.I.) for the population median and the sample size is $n = 15$, then Table 7.2 gives:

$$\text{Location of } C: 4$$

Thus, C_L is the fourth observation from the low end of the ordered data and C_H is the fourth observation from the high end of the ordered data.

TABLE 7.2	Location of C			
⊙	n	90% C.I.	95% C.I.	99% C.I.
	4	1		
	5	1		
	6	2	1	
	7	2	1	
	8	3	2	1
	9	3	2	1
	10	3	2	1
	11	3	2	1
	12	4	3	2
	13	4	3	2
	14	5	4	3
	15	5	4	3
	16	5	4	3
	17	6	5	4
	18	6	5	4
	19	6	5	4
	20	7	6	5

As stated, the construction of a confidence interval for the population median is based on the binomial distribution. Because binomial probabilities with $\pi = .5$ can be accurately approximated with the normal distribution when $n > 20$, we use a sample size of 20 to distinguish between large and small samples when finding a confidence interval for the population median.

EXAMPLE 7.10

A study was undertaken to investigate the number of days to recover after being treated with a new drug for patients who suffer from a certain type of viral disease. From the data on recovery time, in days, recorded in the stem and leaf plot, obtain a 95% confidence interval for the median number of days to recovery.

```
0 | 5  6  7  8  8  9
1 | 1  2  2  3
1 | 7  8
2 |
2 | 8
3 | 2
3 | 7
```

SOLUTION

From the stem and leaf plot, it appears that the data are skewed right. Therefore, we choose to find a confidence interval for the median number of days to recovery. From Table 7.2, with 15 observations and 95% confidence, we find:

$$\text{Location of } C: 4$$

Thus, C_L, the fourth value from the low side, is 8 and C_H, the fourth value from the high side, is 18. Therefore, the 95% confidence interval for the median number of days is given by the limits (8, 18).

We should point out that the confidence level given in Table 7.2 is only approximate because of the discrete nature of the binomial distribution. For example, the exact confidence level for the interval obtained in Example 7.10 is found, using the binomial probability table, to be 96.48%, which is reasonably close to 95%. It is as close to 95% as we can get because we must count to the fourth observation; we cannot count to a fraction of an observation.

Large-Sample Confidence Interval for the Population Median

We now construct a large-sample confidence interval for the population median, θ. Assuming that we have a large sample, we can use the Central Limit Theorem and approximate the binomial distribution with the normal distribution.

To construct the interval, we proceed as in the small-sample case by first ordering the sample from smallest to largest values. Then the $(1 - \alpha)100\%$ confidence interval is formed (as in the small-sample case) by two of the sample observations, C_L and C_H. C_L is located by counting up from the low end of the ordered data, and C_H is located by counting down from the high end of the ordered data (the same way we found C_L and

C_H in the small-sample case). The distance we count in from each end, the location of C, is approximated by the normal distribution.

To find a $(1 - \alpha)100\%$ confidence interval for the population median when the sample size is greater than 20, use the following formula to find the location of the endpoints of the interval:

$$\text{Location of } C: \frac{n - z^* \sqrt{n}}{2}$$

where z^* is the upper $\alpha/2$ critical value found in the standard normal distribution table.

Normally the location of C will not be a whole number, so we round up to the next whole number.

EXAMPLE **7.11**

In Example 6.16, it was determined that the parent distribution of the scores on the psychological test used for measuring racial prejudice (worksheet: **Prejudic.mtw**) is skewed right and that the population median best represents the center of the distribution. Find a 90% confidence interval for the population median, θ.

SOLUTION

The sample size in Example 6.16 was given to be $n = 25$. For a 90% confidence interval, we have $z^* = 1.645$. Substituting these values in the formula for the location

FIGURE **7.11**   Character Stem-and-Leaf Display

Stem-and-leaf of prejud   N = 25
Leaf Unit = 1.0

```
     5      4  11224
    11      4  567888
    (3)     5  124
    11      5  889
     8      6  12
     6      6  5
     5      7  24
     3      7  9
     2      8  3
     1      8  7
```

of C, we find:

$$\text{Location of } C: \frac{n - z^* \sqrt{n}}{2} = \frac{25 - 1.645 \sqrt{25}}{2}$$

$$= 8.4$$

$$\cong 9$$

From the worksheet, we construct the ordered stem and leaf plot shown in Figure 7.11. Notice the skewness in the data. $C_L = 48$ (the ninth observation from the low end) and $C_H = 59$ (the ninth observation from the high end). Thus, we are 90% confident that the population median is somewhere inside the interval from 48 to 59.

C O M P U T E R   T I P

After the data are stored in the worksheet, the confidence interval for the median is found by selecting the **Stat** menu, choosing **Nonparametrics**, and then selecting **1-Sample Sign**. In the 1-Sample Sign window, **Select** the **Variable** where the data are stored, click on **Confidence interval**, specify the desired **Level**, and click **OK**.

Figure 7.12 gives the computer results for the data in Example 7.11. Minitab prints three confidence intervals that have a confidence level close to the requested 90%. The first confidence interval has an actual confidence level below 90% (.8922) and the third confidence interval has an actual confidence level above 90% (.9567). The middle confidence interval with an achieved confidence level of the requested 90% is obtained by nonlinear interpolation (NLI) between the first and third intervals. Notice that the first interval is the one we got in Example 7.11.

FIGURE 7.12   `Sign Confidence Interval`

`Sign confidence interval for median`

	N	MEDIAN	ACHIEVED CONFIDENCE	CONFIDENCE INTERVAL		POSITION
prejud	25	52.00	0.8922	(48.00,	59.00)	9
			0.9000	(47.94,	59.12)	NLI
			0.9567	(47.00,	61.00)	8

Although the confidence interval for the population median presented in this section is suggested when the parent distribution is skewed, it is appropriate for estimating the population median regardless of the shape of the parent distribution. In particular, if the distribution is symmetric and has numerous outliers (a long-tailed distribution), this interval is recommended over the *t*-interval. For large samples, however, another alternative is the robust confidence interval based on trimmed means given in the next section.

EXERCISES 7.4

7.46   For the following sample sizes, determine the location of the two observations in an ordered stem and leaf plot that form a 95% confidence interval for the population median, θ.
   a. $n = 10$ b. $n = 12$ c. $n = 15$ d. $n = 19$

7.47   For a sample size of 18, determine the location of the two observations in an ordered stem and leaf plot that form a confidence interval for the population median, θ, with each confidence level:
   a. 90%   b. 95%   c. 99%

7.48   From the ordered stem and leaf plots, construct 95% confidence intervals for the population median:

a. 4	0 3 5		b. 4	0		c. 5	0 2 9
5	0 2 3 5 5 7 9		5	0 2 5 9		6	1 4 5 8
6	1 2 5		6	1 2 4 5 5 8		7	4 9
7	4		7	4 7 9		8	1
8	1		8	1		9	2
9	2		9	2 5		10	1 4
			10	1			

7.49   Shoplifting and employee theft cost retailers more than $2 billion annually. These are the values of merchandise found in the possession of shoplifters apprehended in a department store over a busy weekend:

$38   20   12   100   22   45   5   75   150   20   25   19

Construct a 90% confidence interval for the median value of shoplifted merchandise.

7.50   Medical malpractice awards often exceed $1 million. Because some malpractice awards are extremely large, it may be appropriate to report the median award. From these data (recorded in $1,000s) construct a 90% confidence interval for the median malpractice award:

Worksheet: Malpract.mtw

760	380	125	250	2,800	450	100	150	2,000
180	650	275	850	1,700	1,500	3,000	390	

Also construct a 90% confidence interval based on $\bar{y}$ and plot the two intervals on the same line graph. Compare the two intervals. Which do you think better represents the amount of malpractice awards?

7.51   An experiment was designed to examine the distribution of the times (in minutes) to breakdown of an insulating fluid under various levels of voltage stress (E. Soofi, N. Ebrahimi, and M. Habibullah 1995). Here are the fluid breakdown times at 34 kilovolts:

Worksheet: Fluid.mtw

.19	.78	.96	1.31	2.78	3.16	4.15	4.67	4.85	6.50
7.35	8.01	8.27	12.06	31.75	32.52	33.91	36.71	72.89	

a. Construct a stem and leaf plot and conduct a midsummary analysis to verify that the data are skewed.

b. Construct a 95% confidence interval for the median breakdown time.

7.52 The motor vehicle traffic death rate varies greatly among cities with more than 1 million people. The table gives the reported rates for six major cities. Use the rates to construct a 90% confidence interval for the median traffic death rate.

City	Rate per 100,000
Houston	20.3
Los Angeles	13.8
Detroit	8.7
Chicago	8.1
New York	7.2
Philadelphia	7.1

SOURCE: National Safety Council, 1995.

7.53 For the following sample sizes, determine the location in an ordered stem and leaf plot of the two observations that form a 95% confidence interval for the population median, θ:

   a. $n = 25$ b. $n = 50$ c. $n = 100$ d. $n = 200$

7.54 For a sample size of 100, determine the location in an ordered stem and leaf plot of the two observations that form a confidence interval for the population median, θ, with each confidence level:

   a. 90%   b. 98%   c. 95%   d. 99%

7.55 From the ordered stem and leaf plots, construct a 95% confidence interval for the population median, θ:

a.	0	5		b.	0	0 0 1 5		c.	0	0 0 0 1 5 7 9
	1	2 8			1	2 4 5 5 8 9 9			1	2 5 8 9
	2	1 2 3 5 7 8			2	1 2 3 5			2	1 3 5
	3	1 3 4 5 5 6 9			3	3 4 6			3	3 4 6
	4	2 5 8 8 9			4	2 5			4	2 5 7
	5	1 4			5	1 4			5	1
	6	3			6	3			6	3
	7	4			7	4			7	4
					8	1			8	1

7.56 The time to repair the body of an automobile involved in a wreck can vary from reasonably short to rather long when special parts must be ordered. It is reasonable to expect that the distribution of times for repairs is skewed positively. From the following recorded times (in hours) for repairing 22 automobiles, construct a 98% confidence interval for the median time of a repair:

Worksheet: Repair.mtw

| 10.3 | 4.1 | 8.6 | 2.1 | 3.7 | 5.6 | 11.4 | 5.7 | 3.8 | 4.5 | 2.7 |
| 22.6 | 5.5 | 9.6 | 4.8 | 3.9 | 10.4 | 4.8 | 5.9 | 6.7 | 7.6 | 12.0 |

7.57 Top executives are in demand by the major corporations. From the random sample of advertised salaries listed at the top of page 462, construct a 95% confidence interval for the median salary offered general managers in 1995:

Worksheet: Manager.mtw

$95,000	85,000	65,500	98,000	150,000	75,000	60,000
78,000	100,000	75,500	90,000	85,000	150,000	120,000
55,000	70,000	85,500	120,000	85,000	99,900	
110,000	150,000	95,000	60,000	185,000	75,000	

7.58 The normal probability plot in Exercise 7.43 in Section 7.3 indicates that the distribution of the returns on the investments studied by the investor is skewed right. It is recommended that the "typical" return be measured with the population median. From the stem and leaf plot, construct a 90% confidence interval for the median return on investment.

```
Character Stem-and-Leaf Display

Stem-and-leaf of coins      N  = 12
Leaf Unit = 1.0

     1      0 9
     6      1 00011
     6      1 223
     3      1 4
     2      1
     2      1 8
     1      2
     1      2
     1      2
     1      2 7
```

7.59 The female alcoholic generally begins drinking at a later age than does the male alcoholic. Here are the ages at which 14 female alcoholics began drinking:

Worksheet: Alcohol.mtw

18	16	24	14	19	22	28
16	19	21	22	35	20	15

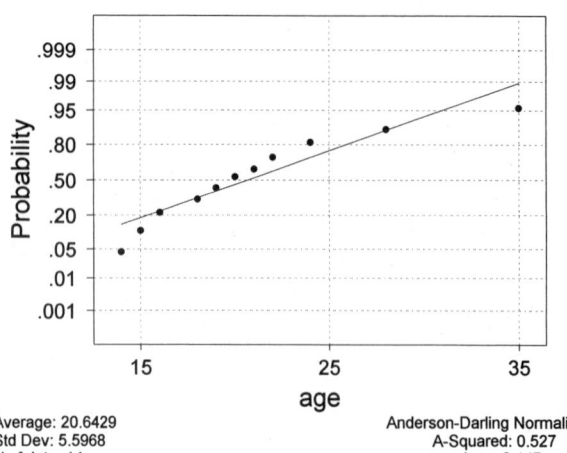

Normal Probability Plot

Average: 20.6429
Std Dev: 5.5968
N of data: 14

Anderson-Darling Normality Test
A-Squared: 0.527
p-value: 0.147

Notice the curvature in the normal probability plot. What does this suggest about the shape of the parent distribution? Under these conditions it might be best to estimate the population median instead of the population mean. Find a 99% confidence interval for the median age.

7.60   In Computer Exercise 6.68 in Section 6.4, the following median prices were given for single-family homes in 37 markets across the United States:

Worksheet: Housing.mtw

City	1984	1993	City	1984	1993
Albany, NY	$52,400	$109,900	Memphis, TN	$65,200	$84,200
Anaheim, CA	134,900	222,200	Miami	84,400	98,000
Atlanta	64,600	94,000	Milwaukee	69,600	98,400
Baltimore	65,200	113,200	Minneapolis–St. Paul	75,100	96,400
Birmingham, AL	66,600	89,000	Nashville, TN	64,000	87,000
Boston	102,000	165,200	New York metro area	106,900	168,000
Chicago	77,500	131,300	Oklahoma City	63,600	61,100
Cincinnati	59,600	85,600	Philadelphia	59,500	108,900
Cleveland	65,600	89,200	Providence, RI	61,400	112,500
Columbus, OH	60,400	89,300	Rochester, NY	62,100	83,300
Dallas–Fort Worth	83,400	89,600	St. Louis	64,400	80,900
Denver	85,000	96,300	Salt Lake City	67,600	79,000
Detroit	48,000	92,200	San Antonio, TX	71,600	72,600
Fort Lauderdale	76,000	99,500	San Diego	102,900	175,500
Houston	79,600	78,000	San Francisco	132,600	249,300
Indianapolis	54,200	83,800	San Jose, CA	121,600	297,100
Kansas City, MO	57,600	79,100	Tampa, FL	60,800	69,900
Los Angeles	115,300	199,700	Washington, DC	92,900	153,500
Louisville, KY	49,500	69,900			

SOURCE: National Association of Realtors.

We determined that both distributions are skewed right. Find 98% confidence intervals for the median costs in 1984 and in 1993. Graph both intervals on a common number line. Discuss the differences between the two intervals.

7.61   For fishing in the ocean, the trawl gear consists of a cone-shaped net towed behind a vessel at a speed similar to a walking pace. Fish that are caught in the trawl work themselves to the end of the net to a portion called the codend. The size and shape of the codend determine what size fish will be caught and what size fish will escape the net. The list gives the size and the number of haddock caught with a 35-mm diamond-mesh (small-mesh) codend and an 87-mm diamond-mesh (large-mesh) codend. The data are in frequency table form, with the first column listing the length of the fish and the second and third columns giving the number of fish of the various lengths caught.

Worksheet: Fish.mtw

Length (cm)	Small mesh	Large mesh
24	1	0
25	1	0
26	3	0

(continued)

Length (cm)	Small mesh	Large mesh
27	14	1
28	30	5
29	49	19
30	60	29
31	50	51
32	70	71
33	108	120
34	88	118
35	84	107
36	68	78
37	37	52
38	33	40
39	12	17
40	5	17
41	6	14
42	10	10
43	1	4
44	6	6
45	2	2
46	1	5
47	0	1

SOURCE: R. Millar, "Estimating the Size-Selectivity of Fishing Gear by Conditioning on the Total Catch," *Journal of the American Statistical Association* 87 (1992): 962–968.

a. What is the median length fish caught by the two sizes of codend?
b. What is the *q*-spread for the two distributions?
c. Does it appear that the lengths of fish caught by the two codends are different?
d. Calculate individual 99% confidence intervals for the median length fish caught by the two different codends. Describe any differences between the two confidence intervals.

7.5   ROBUST CONFIDENCE INTERVAL FOR THE CENTER OF A DISTRIBUTION (OPTIONAL)

One goal of interval estimation is to come up with an interval estimate that gives good results in a large variety of cases. When we estimate the population mean μ, the confidence interval based on $\bar{y}$ is robust and should be used in most cases. It has been pointed out, however, that the precision of the confidence interval based on $\bar{y}$ is significantly reduced when the parent distribution has long tails. In Section 6.3, we suggested that a trimmed mean (a robust estimator) be used as the estimator of the population mean when the parent distribution is symmetric with long tails. We now construct a confidence interval for μ based on a trimmed mean.

Large-Sample Confidence Interval for μ based on $\bar{y}_T$

Assume we have a large sample from a symmetric population that has mean μ and standard deviation σ. For large samples, the sampling distribution of $\bar{y}_T$ is approximately normal with a mean μ and a standard error that we will estimate presently.

Recall that the standard error of $\bar{y}$ is $s/\sqrt{n}$. To estimate the standard error of $\bar{y}_T$, it is tempting simply to calculate the ordinary standard deviation of the trimmed sample and divide by $\sqrt{k}$, where k is the size of the trimmed sample. It has been shown, however, that a better estimator of the standard error of $\bar{y}_T$ is obtained by calculating the standard deviation of the trimmed sample, call it s_T, based on the so-called Winsorized sample.

> The **Winsorized sample** is obtained by replacing the trimmed values with the values that were next in line for trimming.

EXAMPLE 7.12

The boxplot of the educational levels of auto workers in Detroit (worksheet: **Detroit.mtw**), given in Figure 7.13, exhibits a long-tailed distribution. From the data in the stem and leaf plot, construct the Winsorized sample based on a 10% trimming.

FIGURE 7.13

Character Boxplot

```
                                    ------
      O             *  *    -----I   +-----     *   *          O  O
                                    ------
      +---------+---------+---------+---------+---------+------educ
     0.0       4.0       8.0      12.0      16.0      20.0
```

```
0 | 1
0 |
0 | 5
0 | 6
0 | 8  8  8  8  9  9
1 | 0  0  0  0  0  1  1  1  1
1 | 2  2  2  2  2  2  2  2  2  2  2  2  2  3  3
1 | 4  4  4
1 | 6  7
1 |
2 | 1
2 | 2
```

SOLUTION

The sample size is 40, so a 10% trimming trims four observations from each end. They are underlined in the stem and leaf plot above. The Winsorized sample is obtained by replacing the trimmed values with the values that are next in line for trimming. Those are

8 on the low end and 14 on the upper end. Here is a stem and leaf plot of the Winsorized sample:

```
0 | 8  8  8  8  8  8  8  9  9
1 | 0  0  0  0  0  1  1  1  1
1 | 2  2  2  2  2  2  2  2  2  2  2  2  2  3  3
1 | 4  4  4  4  4  4  4
```

> The **standard deviation of a trimmed sample** based on the Winsorized sample is given by
>
> $$s_T = s_W \sqrt{\frac{n-1}{k-1}}$$
>
> where s_W is the ordinary sample standard deviation of the Winsorized sample and k is the size of the trimmed sample.

EXAMPLE 7.13   Calculate s_T for the Winsorized sample in Example 7.12.

SOLUTION

To find s_T we first find the ordinary sample standard deviation s_W of the Winsorized sample. From the data in the stem and leaf plot, we find $s_W = 2.015$. Thus,

$$s_T = s_W \sqrt{\frac{n-1}{k-1}} = 2.015 \sqrt{\frac{39}{31}} = 2.26$$

We now give the standard error of the trimmed mean, $\overline{y}_T$.

> The **standard error of the trimmed mean**, $\overline{y}_T$, is given by
>
> $$SE(\overline{y}_T) = \frac{s_T}{\sqrt{k}}$$
>
> where s_T is the standard deviation of the trimmed sample and k is the size of the trimmed sample.

As stated earlier, for a sufficiently large sample, the sampling distribution of $\overline{y}_T$ is approximately normal. Therefore, $\overline{y}_T$ should not deviate more than ± 1.96 standard errors from μ with 95% probability. That is, the estimate of μ is $\overline{y}_T$, which has a margin of error of

$$\pm 1.96 \frac{s_T}{\sqrt{k}}$$

Replacing 1.96 with z^*, we have this summary:

Large-Sample Confidence Interval for μ Based on $\overline{y}_T$

Application: Symmetric population distribution with long tails

Assumption: $n > 30$

Estimator: $\overline{y}_T$, the trimmed sample mean

Standard error of the estimator:

$$\text{SE}(\overline{y}_T) = \frac{s_T}{\sqrt{k}}$$

A $(1 - \alpha)100\%$ confidence interval for μ is given by the limits

$$\overline{y}_T \pm z^* \frac{s_T}{\sqrt{k}}$$

where s_T is the standard deviation of the trimmed sample based on the Winsorized sample and k is the size of the trimmed sample.

EXAMPLE 7.14

Calculate a 98% confidence interval based on $\overline{y}_T$ for the mean educational level of the auto workers described in Example 7.12. Show that it is more precise than an interval based on $\overline{y}$.

SOLUTION

From Example 7.12, we find $\overline{y}_T = 11.25$, and from Example 7.13, we have $s_T = 2.26$. The 98% confidence interval for μ takes the form

$$\overline{y}_T \pm 2.33 \frac{s_T}{\sqrt{k}}$$

which is

$$11.25 \pm (2.33) \frac{2.26}{\sqrt{32}} = 11.25 \pm .931$$

So the 98% confidence interval for μ based on $\bar{y}_T$ is (10.319, 12.181). In other words, we are 98% confident that the mean educational level of the auto workers is somewhere between 10.319 and 12.181 years.

A 98% confidence interval for μ based on $\bar{y}$ takes the form

$$\bar{y} \pm 2.33 \frac{s}{\sqrt{n}}$$

and thus, using the data in Example 7.12, we have

$$11.4 \pm (2.33)\frac{3.699}{\sqrt{40}} = 11.4 \pm 1.363$$

or 10.037 to 12.763.

Figure 7.14 compares the two intervals for μ. The interval based on $\bar{y}_T$ is more precise and equally valid.

FIGURE 7.14

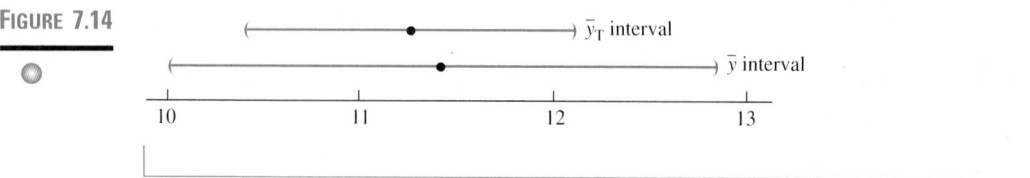

Many hand-held calculators are capable of simplifying the calculations of trimmed means and standard deviations. Any calculator that has statistical functions such as a standard deviation allows one to enter data and also to remove data. One can enter all the sample data, calculate the mean and standard deviation, and then remove the observations that are to be trimmed. Recalculating the sample mean gives the trimmed mean. One can then enter the additional observations for the Winsorized sample and calculate the standard deviation, which is the standard deviation based on the Winsorized sample. With a little practice, the calculations become quite easy.

Small-Sample Robust Confidence Interval

The large-sample robust confidence interval based on the trimmed mean, $\bar{y}_T$, is recommended for those situations when the parent distribution is symmetric with long tails. Suppose now that we have a small sample. What is the recommended procedure for finding a small-sample confidence interval for the center of a symmetric long-tailed distribution? It has been shown, for example, that the standardized version of the trimmed mean follows Student's t distribution when the parent population is near normal and thus we could develop a t-interval based on the trimmed mean. But our parent distribution is not normal; it has long tails. This problem has received much attention in the statistical literature but no definitive answer. Many argue that the t-interval based on $\bar{y}$ is robust and can be used in situations where the normality assumption is not met. This is sound advice except when the parent distribution is skewed or has long tails. In the case of a skewed parent distribution, it was recommended in Section 7.4 that we find a confidence interval for the population median. Moreover, the development of the confidence interval for the median did not depend on the shape of the parent distribution. Thus, in this small-sample case, perhaps we could find a confidence interval for the median when the parent distribution is symmetric with long tails. Recall that the mean and median coincide for all symmetric distributions, so estimating the center of the distribution can be interpreted as estimating the median rather than the mean. This will be our approach; that is, if we desire a small-sample confidence interval for the center of a symmetric long-tailed distribution, we will construct the confidence interval for the population median that was developed in Section 7.4.

EXAMPLE 7.15

A consumer group was interested in the medical costs associated with patients in Kentucky who had black lung disease. A sample of 16 coal miners who had recovered from black lung recorded the medical expenses associated with their treatment. From the data in the ordered stem and leaf plot, construct a 95% confidence interval for the median cost. (The costs are given in units of $100.)

Worksheet: Blklung.mtw

```
 0 | 2
 1 |
 2 | 4  6
 3 | 4  5  7  8  8
 4 | 2  5  5  6
 5 | 2  4
Hi | 90   100
```

SOLUTION

The boxplot in Figure 7.15 indicates a long-tailed distribution. Given that the sample size is small, we will construct a 95% confidence interval for the population median. Using Table 7.2 for a 95% confidence interval with $n = 16$, we find that the

Location of C: 4

FIGURE 7.15   Character Boxplot

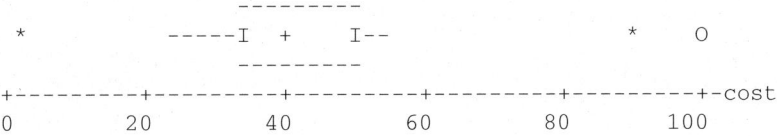

Returning to the stem and leaf plot, we find that the 95% confidence interval for the median cost of medical care is (34, 52).

In contrast, if we ignore the nonnormality of the parent distribution and find a *t* confidence interval, we get the interval (31.78, 56.72), as demonstrated in the Minitab output in Figure 7.16. Note that the standard deviation is really inflated, which is always the case in the presence of extreme outliers.

FIGURE 7.16   Confidence Intervals

Variable	N	Mean	StDev	SE Mean	95.0 % C.I.
cost	16	44.25	23.40	5.85	(31.78, 56.72)

The graph in Figure 7.17 points out that the interval for the median has greater precision than the *t*-interval based on $\bar{y}$.

FIGURE 7.17

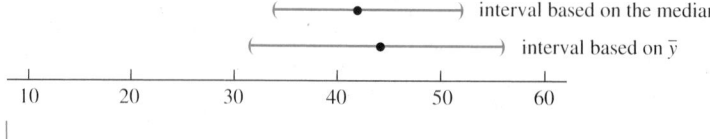

EXERCISES 7.5

7.62   We present a normal probability plot of the educational levels of the auto workers considered in Example 7.12. What evidence in the plot suggests that the parent population is a long-tailed distribution?

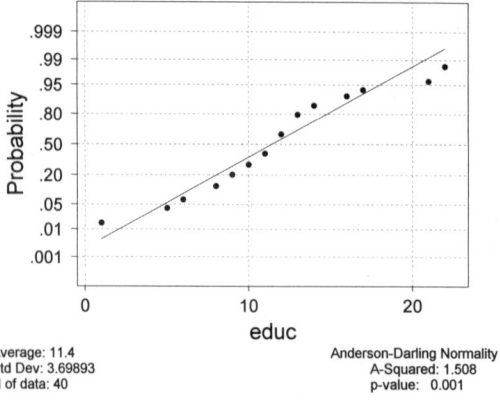

Normal Probability Plot

7.63 In Exercise 6.40 in Section 6.3, we used a random sample of 50 freshmen to investigate the number of hours per week that freshmen study. This boxplot was constructed from worksheet: **Study.mtw**. Does the parent distribution have symmetric long tails?

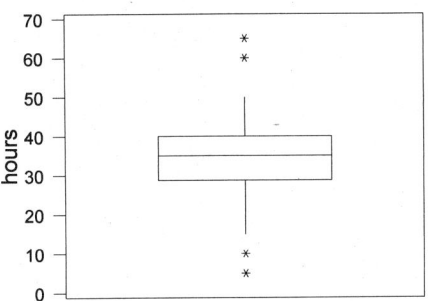

Based on a 10% trimming, we find

$$\bar{y}_{T.10} = 34.2 \qquad s_T = 8.373$$

Construct a 98% confidence interval for the mean study time for all freshmen.

7.64 A college coach is interested in estimating the mean amount spent on recruiting a high school football player. He tabulates the amounts spent on recruiting his last 35 players in the following stem and leaf plot:

Worksheet: Coach.mtw

0	00	00	00	00	00			
1	00	20						
2	10	50						
3	00	25	40	50	50	70	90	
4	00	25	50	50	75	90	95	
5	00	20	30	50	50	65	75	80
HI	1,000	1,050	1,500	2,000				

Find a 90% confidence interval for the mean amount spent per player.

7.65 A new pain reliever is to be tested on hospital patients. The drug is administered to a random sample of 50 patients and the time to relief is recorded. From the given times (recorded in minutes), construct a stem and leaf plot and decide whether $\bar{y}$ or $\bar{y}_T$ is the better estimator of μ, the mean time to relief. Construct a 95% confidence interval for μ based on your choice of estimator.

Worksheet: Pain.mtw

4.1	3.8	4.7	5.9	4.6	5.4	4.5	4.6	5.6	5.1
4.9	5.6	6.4	5.3	4.8	3.5	5.2	3.6	5.7	6.8
6.4	8.3	5.8	4.7	5.4	6.3	7.4	4.6	5.8	5.3
5.6	5.2	5.9	6.7	8.4	3.1	4.5	7.1	4.8	5.9
4.0	3.4	5.7	8.8	3.3	4.9	5.0	4.6	5.9	4.8

7.66 A university is thinking of underwriting its own health insurance. To get a feeling for the size for a typical claim, a random sample of 30 claims submitted under the old plan is chosen for analysis. Arrange the claim amounts at the top of page 472 in a stem and leaf plot and construct a 90% confidence interval for the median claim amount:

Worksheet: Health.mtw

$126	201	910	540	350	225	155	260	110	456
1,122	350	275	375	560	1,025	150	94	280	375
460	650	385	120	450	720	275	85	428	360

7.67 A realtor is studying the cost of single-family homes in a small city in a southern state. From these costs of homes (in $100s), construct a stem and leaf plot and a 99% confidence interval for the median cost of a home:

Worksheet: Realtor.mtw

455	610	543	722	645	316	700	580	615
1,280	743	515	689	575	445	842	1,100	
758	665	1,350	780	475	680	714	1,000	

7.68 Here are the daily receipts in a small hardware store for 31 working days. Construct a boxplot of the data to determine whether $\bar{y}$ or $\bar{y}_T$ is the better estimator of μ. Construct a 95% confidence interval for μ based on your choice of estimator.

Worksheet: Hardware.mtw

$98.50	195.60	73.60	156.80	184.70	210.50	375.60	164.00	55.90
45.00	251.70	195.00	155.70	173.50	287.50	151.40	423.50	35.00
98.70	165.40	172.30	185.70	168.90	214.70	100.60	216.20	196.30
257.80	195.70	233.70	528.50					

7.69 Suppose we want to estimate the mean weekly weight for all refuse that enters a landfill. The weight of the refuse entering the dump was measured on 36 randomly selected weeks. The data indicate that there were some very slow weeks where little refuse entered the landfill and some very heavy weeks with a lot of activity. The analyst chose to calculate a 10% trimmed mean and found:

$$\bar{y}_{T.10} = 153 \text{ tons} \qquad s_T = 21.5 \text{ tons}$$

Based on these results, estimate the mean weekly tonnage with a 95% confidence interval.

7.70 The ages of nine randomly selected people attending a PG-rated movie were

Worksheet: PGMovie.mtw

19	18	26	17	22	16	25	20	17

Construct a stem and leaf plot and a 99% confidence interval for the median age of people attending the movie. Comment on the precision of the interval.

7.71 The distribution of monthly salaries of unskilled construction workers is assumed to be long-tailed. A random sample of 64 workers yielded this information:

$$\bar{y}_{T.10} = \$1,530 \qquad s_T = \$285$$

Find a 98% confidence interval based on $\bar{y}_{T.10}$ for the mean monthly salary for all unskilled construction workers.

7.72 A sample of 32 commuters were asked how many miles they travel to work each day. Here are the data:

Worksheet: Travel.mtw

2.4	4.6	3.1	.2	3.8	5.9	4.2	5.7	27.8	6.4	
3.8	6.4	5.8	3.9	.1	4.6	3.6	2.7	.1	5.2	6.7
7.8	3.4	4.9	5.8	4.7	31.0	4.7	2.8	3.9	.3	5.5

Construct a boxplot of the data. The boxplot should identify two extreme outliers. What effect will the outliers have on a confidence interval for μ based on $\bar{y}$? What effect will the outliers have on a confidence interval for μ based on $\bar{y}_{T.10}$? How should we deal with the outliers? Construct a 95% confidence interval for the mean miles traveled.

7.6 SUMMARY AND REVIEW

KEY CONCEPTS

✓ The large-sample confidence interval for a population proportion, π, is given by

$$p \pm z^* \sqrt{\frac{p(1-p)}{n}}$$

✓ If the distribution is normal, near normal, or symmetric with tails that are not excessively long, then the confidence interval for the mean should be based on $\bar{y}$. When the population standard deviation, σ, is known, the confidence interval (z-interval) takes the form

$$\bar{y} \pm z^* \frac{\sigma}{\sqrt{n}}$$

✓ If the population standard deviation is unknown, the confidence interval (t-interval) takes the form

$$\bar{y} \pm t^* \frac{s}{\sqrt{n}}$$

✓ If the distribution is skewed, the best measure of the center is the median and therefore the confidence interval should be for the population median. The confidence interval is formed from two sample observations, (C_L, C_H), where C_L is found by counting up from the low end of an ordered stem and leaf plot and C_H is found by counting down from the high end of the ordered stem and leaf plot. The distance we count is given by the location of C. For large samples,

$$\text{Location of } C: \frac{n - z^* \sqrt{n}}{2}$$

For small samples, the location of C is found in Table 7.2 on page 456.

✓ If the distribution is symmetric with long tails and the sample size is large, then the confidence interval for the mean, μ, should be based on $\bar{y}_T$. It takes the form

$$\bar{y}_T \pm z^* \frac{s_T}{\sqrt{k}}$$

✓ If the distribution is symmetric with long tails and the sample size is small, it is recommended that the confidence interval be for the population median.

LEARNING GOALS   Having completed this chapter, you should be able to:

1. Construct a confidence interval for a population proportion. *Section 7.1*
2. Interpret a confidence interval. *Section 7.1*
3. Understand the concepts of validity and precision of confidence intervals. *Section 7.1*
4. Determine sample size for a given margin of error and confidence level. *Sections 7.1 and 7.2*
5. Construct a z-interval for the population mean based on $\bar{y}$ when σ is known. *Section 7.2*
6. Construct a t-interval for the population mean based on $\bar{y}$ when σ is unknown. *Section 7.3*
7. Construct a small-sample confidence interval for the population median. *Section 7.4*
8. Construct a large-sample confidence interval for the population median. *Section 7.4*
9. Construct a large-sample confidence interval for the population mean based on $\bar{y}_T$. *Section 7.5*

QUESTIONS FOR REVIEW   Use the following problems to test your skills:

7.73   Suppose $n = 16$. To construct confidence intervals, find the following values when the upper-tail probability is p:
    a. z^* when $p = .05$ b. z^* when $p = .01$
    c. t^* when $p = .05$ d. t^* when $p = .01$

7.74   We wish to estimate with a 95% confidence interval the proportion of television viewers who would watch a 1-hour national news program. How large a sample is necessary so our precision of estimation will be ± 3 percentage points?

7.75   We wish to estimate with a confidence interval the center of a population distribution using the random sample presented in this stem and leaf plot:

```
0 | 1   4   6
1 | 6   8   9   3   9   7
2 | 8   6   9   4   2
3 | 4   8   6
4 | 5   2
5 | 3   6
6 | 4
7 | 3   8   1
9 | 4
```

    a. Is the distribution symmetric with long tails, symmetric with short tails, normal, skewed right, or skewed left?
    b. What statistic would you base your confidence interval on: $\bar{y}$, $\bar{y}_T$, the median, or p?
    c. Construct a 99% confidence interval for the center of the distribution based on your answer to part **b**.

7.76   Suppose the parent population is normal with mean μ and standard deviation σ. A random sample $y_1, y_2, \ldots, y_n$ is selected. What statistic would you use to estimate μ? Explain why.

7.77   You are studying the percent of convicted felons who have a history of juvenile delinquency.
    a. How large a sample do you need to estimate the percent to within .03 with a 99% confidence interval?
    b. A sample of 1,600 convicted felons revealed that 1,120 had a history of juvenile delinquency. Find a 99% confidence interval for the true percent. Interpret the results.

7.78   In Exercises 6.55, 6.56, and 6.57 of Section 6.4, it was determined that the following porosity measurements (in percent) obtained from Wind River Basin and Bighorn Basin in Wyoming come from a normally distributed population:

Worksheet: Porosity.mtw

15	10	15	23	18	26	24	18	19	21
13	17	15	23	27	29	18	27	20	24

SOURCE: J. C. Davis, *Statistics and Data Analysis in Geology*, 2nd ed. (New York: Wiley, 1986), pp. 63–65.

Construct a 90% confidence interval for the mean porosity percent in these two basins.

7.79   Are bounced check fees too high? The California Supreme Court thinks so. The court unanimously ordered a lower court to decide on the fairness of bounced-check fees in California. The court said that California banks collect more than $200 million each year from bounced-check fees when the actual cost of processing a bounced check is around $1.00. These are the fees (in dollars) charged per check at some randomly selected banks across the United States:

Worksheet: Checkfee.mtw

20	15	20	25	20	10	30	10	35
10	5	20	35	25	20	15	30	
25	20	5	30	20	30	15	25	

a. Construct a stem and leaf plot of the data.
b. Do the data appear to violate the normality assumption that is required for a *t*-interval?
c. Construct a 90% confidence interval for the mean bounced-check fee.

7.80   A random sample of 64 homes in a small community found that an average of 160 gallons of heating oil were consumed over a given period of time. If it is known that the amount of heating oil consumed is close to normally distributed with a standard deviation of 32 gallons, find a 95% confidence interval based on $\bar{y}$ for the mean number of gallons of heating oil consumed over this period by all residents of the community.

7.81   From a random sample of 2,400 college students, 960 believe that the penalties for the use of marijuana should be reduced.
a. Find a 99% confidence interval for the proportion of all college students who believe the penalties should be reduced.
b. How large a sample is needed so that the margin of error of the estimate is no greater than 2%?

7.82   To estimate the self-concept of a group of college administrators, a psychologist administers the Tennessee Self-Concept Scale exam to 16 administrators and finds that their average self-concept score is 23. Assume that the distribution of scores is close to normal and the standard deviation is 3.
a. Find a 95% confidence interval for the mean self-concept.
b. How might precision be increased while maintaining the same validity?

7.83   A random sample of 100 cups of coffee from a coffee vending machine contained an average of 6.7 ounces in a 7-ounce cup. Assuming the distribution of the amount of coffee is near normal and the standard deviation is .2 ounce, find a 96% confidence interval based on $\bar{y}$ for the mean amount of coffee served by the machine.

7.84   When the flow characteristics of oil through a valve are determined, the inlet oil temperature is measured in degrees Fahrenheit. Here is a sample of 12 readings:

Worksheet: Inletoil.mtw

93	99	97	99	94	91
93	90	89	92	90	93

Construct a 95% confidence interval based on $\bar{y}$ for the mean temperature. What assumption is required for the interval to be 95% valid?

7.85   A survey of 1,564 daily newspapers by the National Federation of Press Women found that of all key editor jobs, 873 were held by women and 3,628 were held by men (*USA Today*, April 19, 1993). At the position of editor, women held 144 and men held 879 positions. At the position of associate/assistant editor, the numbers were 49 women and 148 men. For news editor, 278 were held by women and 854 were held by men. Find individual 99% confidence intervals for the percents of women in the three editor positions. Graph the three confidence intervals on a common number line. Discuss the differences among the confidence intervals. Finally, construct a 99% confidence interval for all key editor jobs held by women.

7.86   The number of hazardous toxic waste sites in each of the 50 states is recorded in worksheet: **Toxic.mtw**. Shown here is a normal probability plot of the number of hazardous sites. Notice the curvature in the plot and that it does not fit the straight line very well. What does this say about the shape of the parent distribution? Do you recommend investigating the mean number of sites or the median number of sites?

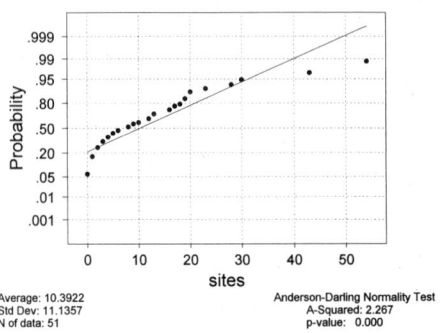

Normal Probability Plot

Average: 10.3922
Std Dev: 11.1357
N of data: 51

Anderson-Darling Normality Test
A-Squared: 2.267
p-value: 0.000

7.87   The stem and leaf plot gives the number of hazardous waste sites discussed in Exercise 7.86 and given in worksheet: **Toxic.mtw**. Based on these data, construct a 98% confidence interval for the median number of hazardous waste sites.

```
Character Stem-and-Leaf Display

Stem-and-leaf of sites     N  = 51
Leaf Unit = 1.0

    21     0 000000111112222333444
   (8)     0 55668889
    22     1 0222233
    15     1 666789999
     6     2 03
     4     2 8
     3     3 0
     2     3
     2     4 3
     1     4
     1     5 4
```

7.88   In Exercise 2.58 in Section 2.3, we learned that the value of a rare stamp is affected by the thickness of the paper that the stamp was printed on. Recorded in worksheet: **Stamp.mtw** are the thicknesses of the *1872 Hidalgo* that was issued in Mexico. From the descriptive statistics, construct a 99% confidence interval for the mean thickness of the *1872 Hidalgo*.

Descriptive Statistics

Variable	N	Mean	Median	TrMean	StDev	SEMean
thick	485	0.08602	0.08000	0.08500	0.01496	0.00068

Variable	Min	Max	Q1	Q3
thick	0.06000	0.13100	0.07500	0.09800

7.89   The Old Faithful geyser, located in Yellowstone National Park in Wyoming, has two types of eruptions: a short eruption somewhat less than 3 minutes and a long eruption with a duration in excess of 3 minutes (see Exercise 6.13 in Section 6.1). The waiting time between eruptions consequently has a bimodal distribution. In worksheet: **Faithful.mtw**, the wait times are classified into two categories: a short eruption and a long eruption. From the descriptive statistics, construct individual 95% confidence intervals for the mean wait time between eruptions for the two types of eruptions. Graph the two intervals on a common number line. Is the precision of the two intervals about the same? If they are different, explain why.

Descriptive Statistics

Variable	eruption	N	Mean	Median	TrMean	StDev	SEMean
waitime	1	110	57.109	55.000	56.612	8.148	0.777
	2	189	81.164	81.000	81.263	7.303	0.531

Variable	eruption	Min	Max	Q1	Q3
waitime	1	43.000	77.000	50.000	61.250
	2	59.000	108.000	77.000	87.000

7.90   In a study to determine whether alcoholism has a genetic basis, several genetic markers were observed on a group of 50 Caucasian alcoholics. For 10 of the 50, the antigen B15 was present. Estimate with a 90% confidence interval the proportion who have this antigen in the population of Caucasian alcoholics.

7.91   We wish to give a personality test to a certain group of people, and we know that their scores will range from 0 to 20. How large a sample is needed to estimate the mean test score with a 95% confidence interval if the margin of error is no greater than .5?

7.92   How large a sample is needed to estimate the average hospital costs to within $30 with a 90% confidence interval if the population distribution is near normal with a standard deviation of $120?

7.93   A reading teacher would like to estimate the mean reading speed of students in fifth grade. A sample of 16 students had an average reading speed of 285 words per minute and a standard deviation of 48. To construct a confidence interval for the mean reading speed based on $\bar{y}$, what assumption must be made about the parent population? Assuming that the assumption is satisfied, find a 95% confidence interval for the mean reading speed.

7.94   A pollster believes that approximately 40% of the registered voters will favor a particular issue; however, he wishes to estimate the true proportion with a 98% confidence interval.
   a. How many people will he need to interview so that the error of estimation is no greater than 4%?
   b. The pollster found 260 in favor of the issue. What is the 98% confidence interval for the true proportion?

7.95   A sample of 35 hotels in Miami yielded an average daily rate of $82.40 and a standard deviation of $23.50. Find a 90% confidence interval based on $\bar{y}$ for the mean hotel rate in Miami.

7.96   In the early 1980s, samples of Tylenol were found to be tampered with and contained traces of poison. The manufacturers of Tylenol designed a new safety bottle and launched a mass advertising campaign to convince their customers that their product was safe. After the advertising campaign, 320 out of a random sample of 400 users of Tylenol said they would continue to use the product. Find a 98% confidence interval for the proportion of all Tylenol users who would continue to use the product.

7.97   An instrument used to measure anxiety has a standardized mean of 50 and a standard deviation of 15. The instrument will be used to estimate the average anxiety level of emotionally disturbed children. How large a sample is needed so that the standard error of the estimate is no greater than 2 points?

7.98   A company wishes to estimate the proportion of its accounts that are paid on time.
   a. How large a sample is needed to estimate the true proportion to within 3% with a 95% confidence interval?
   b. Construct a 95% confidence interval if 300 out of 400 accounts were paid on time.

7.99   A survey of 17,000 seniors in approximately 140 public and private high schools nationwide (by the University of Michigan's Institute for Social Research) showed that 10,540 had tried an illegal drug. Construct a 99% confidence interval for the proportion of all high school seniors who have tried an illegal drug.

7.100   In 1983, a record 64,877 applicants took bar exams in the United States. The table gives the number of takers and the number who passed in three states. Construct separate 95% confidence intervals for the pass rates in the three states. Graphically display all three confidence intervals on the same line axis.

State	Test takers	Number who passed
California	12,499	5,112
Iowa	379	344
Ohio	2,054	1,660

7.101   Suppose you wish to estimate the mean income of new Ph.D. psychologists. A random sample of 16 new Ph.D. psychologists had an average income of $42,300 and a standard deviation of $4,000.
   a. Obtain a 98% confidence interval for the mean income of new Ph.D. psychologists.
   b. What should the parent distribution look like in order for the precision of the interval to be high?
   c. How might you increase precision and still maintain the same validity?

7.102   A Michigan State University survey of 549 employers found that 220 indicated they sought students who had practical work experience such as a co-op or an internship. Construct a 90% confidence interval for the percent of all employers who prefer work experience prior to graduation.

7.103   A survey of 49 randomly selected graduates from the class of 1995 with bachelor's degrees in chemical engineering found the average starting salary was $38,394 and the standard deviation was $8,482. Based on this random sample, find a 99% confidence interval for the mean starting salary of all chemical engineers.

7.104   Have your classmates determine the amount of money they spent in the community last weekend (make sure that it was a normal weekend that tends to be representative of all weekends). Using your collected data, construct a 95% confidence interval for the mean amount that students spend in the community on a weekend.

COMPUTER   **7.105**   The Census Bureau maintains housing data on 335 metropolitan areas across the United States.
EXERCISES   Here are the monthly rental costs in those metro areas with 1 million or more population:

Worksheet: Metrent.mtw

790	529	490	646	656	426	491	367	406	421
456	431	455	575	428	406	413	425	626	
493	680	447	479	778	397	503	583	480	
642	524	516	465	366	437	562	466	531	
379	380	611	709	773	516	415	448	667	

SOURCE: U.S. Bureau of the Census, *Housing in Metropolitan Areas*, Statistical Brief SB/94/19, September 1994.

a. Construct a boxplot of the data. Does the data appear to violate the normality assumption?
b. To construct a confidence interval for the mean rental cost, is the normality assumption required?
c. Construct a 99% confidence interval for the mean rental cost.

7.106 Using the Random Data command with the Normal Distribution, randomly generate 20 samples of 50 observations each from a normal population with mean 70 and standard deviation 12. Construct 20 separate 90% confidence intervals for μ using $\sigma = 12$ and the Zinterval command.
a. How many of the 20 intervals contain μ, which is specified to be 70?
b. How many intervals would you expect to contain μ?
c. If the confidence level was 99% instead of 90%, would the interval be longer or shorter?
d. Would you expect more of the intervals to contain μ if the confidence level was 99%?

7.107 Into the Minitab worksheet, recall worksheet: **Fish.mtw**, which contains the lengths and numbers of fish caught with a small-mesh and a large-mesh codend (the base of a fishing net). From the Graph menu, select the Plot command. Using the length of the fish as the X variable and the frequency count as the Y variable, graph the two distributions. Under Display, select Connect so that the points in the graph are connected. What are the general shapes of the two distributions? To estimate the centers of the two distributions with confidence intervals, what statistic should we base the interval on? Are the confidence intervals for the medians (see Exercise 7.61 in Section 7.4) of the two distributions appropriate or would you recommend confidence intervals for the means of the two distributions? Calculate individual 99% confidence intervals for the two means and compare the results to the results found in Exercise 7.61 in Section 7.4.

7.108 The 1993 violent crime rates for all 50 states and the District of Columbia are given in worksheet: **Crime.mtw**. An analysis of these data in Exercise 2.109 of Section 2.6 determined that there was one outlier, the rate for the District of Columbia. Recall worksheet: **Crime.mtw** and construct 95% confidence intervals for the mean with and without the one outlier. What are the basic differences between the two intervals? Which one is the better representation of a 95% confidence interval for the mean? Explain why.

7.109 Worksheet: **Archaeo.mtw** contains radiocarbon dates of pottery samples taken from an archaeological excavation of the Danebury iron-age hill fort (see Example 2.23 in Section 2.6). The data are generally accepted by archaeologists as being classified into one of four phases called Ceramic Phases 1–4. Verify that all four samples are from populations that are reasonably close to being normally distributed. Construct individual 95% t-intervals from the four samples and graph them on a common line. Does it appear from the four confidence intervals that the phases correspond to four nonoverlapping periods of time?

7.110 STATISTICAL INSIGHT REVISITED   Enter the housing cost data at the beginning of the chapter into the computer (worksheet: **Homes.mtw**).
a. Construct a 95% confidence interval for the population mean using the Tinterval command.
b. Construct a normal probability plot of the data and describe their general shape.
c. Construct a stem and leaf plot and a boxplot of the data.
d. Based on your analysis in parts **b** and **c**, comment on the shape of the distribution.

e. Given your answer to part **d**, do you think the *t*-interval found in part **a** is a reasonable estimate of the typical cost of a new home in 1994?

f. What alternative interval procedure can you suggest for these data?

g. Construct the interval suggested in part **f** and compare it with the *t*-interval found in part **a**.

7.111   STATISTICAL INSIGHT REVISITED   The housing costs in worksheet: **Homes.mtw** are classified into four regions of the country (Northeast—1, Midwest—2, South—3, and West—4).

a. Construct side-by-side boxplots of the housing costs for the four regions.

b. Compare the costs of housing in the four regions by examining where the boxplots are centered and the amount of variability they have.

c. Calculate descriptive statistics for the four regions. Calculate individual 98% confidence intervals for the housing costs in the four regions.

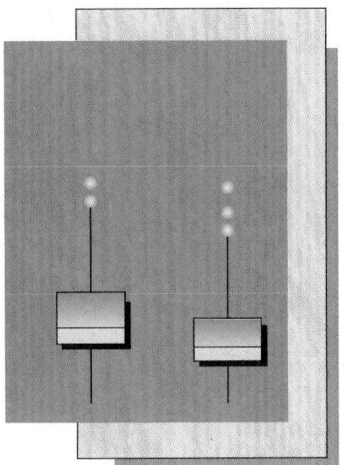

8 HYPOTHESIS TESTING

Many of our statistical questions can be answered by estimation procedures; however, others require a verification that can be done only with a test of hypothesis.

Hypothesis testing is a means by which statistical decisions are made. For example, a new medication will be marketed if it is more than 70% effective. This statement can be formulated in a hypothesis and evaluated with data from a random sample. If the data support the hypothesis, then the decision will be made to market the product. In this chapter, we investigate hypothesis-testing problems that involve a single population. In Chapter 9, we investigate those situations that involve two populations. We address several different statistical problems that can be solved with a test of hypothesis.

CONTENTS

STATISTICAL INSIGHT

EPA MILEAGE CHARTS

How reliable is the EPA mileage chart? Can someone who buys a Honda Civic expect to get 49 miles per gallon in city driving, as suggested by the Environmental Protection Agency's annual mileage ratings?

One obvious way to answer this question is to drive the car and check the mileage. Just as the EPA cautions, however, a motorist's actual mileage can vary, depending on such factors as the amount of traffic, the weather, and car maintenance. Therefore, one should check the mileage on several different occasions under a variety of conditions.

Suppose that the mileage of a Honda Civic was checked on 35 different occasions with these miles per gallon recorded:

Worksheet: Honda.mtw

48.3	49.8	39.6	43.5	46.8	49.4	52.6	45.3	49.7	45.3	40.7	45.6
48.2	50.3	48.2	40.9	43.5	54.2	50.0	51.2	50.8	45.5	49.2	40.5
45.3	50.2	44.8	47.3	48.7	49.2	47.4	51.4	49.6	50.3	48.5	

Is there evidence in these data to suggest that the average mileage for the Honda Civic is less than 49 miles per gallon? Exercise 8.125 in Section 8.6 considers this question.

8.1   INTRODUCTION TO HYPOTHESIS TESTING

Null and Alternative Hypotheses

Hypothesis testing is an area of statistical inference in which we evaluate a conjecture (which we will call a *hypothesis*) about some characteristic of the parent population. Usually the hypothesis concerns an unknown parameter of the population. Consider, for example, the hypothesis that the mean per capita income of all residents in a certain rural county is $15,000 per year. This hypothesis is concerned with the population mean and can be written as $\mu = 15,000$, where μ denotes the true mean per capita income of all residents of the rural county.

> The statement being tested in a test of hypothesis is called the **null hypothesis** and is denoted as H_0. Because it usually is stated as an equality ($\mu = 15,000$), the null hypothesis is referred to as the hypothesis of "no difference," meaning that the difference between the parameter and the value being tested is zero, that is, $\mu - 15,000 = 0$.

In a test of the null hypothesis, data are collected and analyzed to assess the strength of evidence against the null hypothesis. If the evidence casts doubt on the truth of the null hypothesis, it will be rejected in favor of the *alternative hypothesis*.

> The **alternative hypothesis**, denoted as H_a, is what is believed to be true if the null hypothesis is false. Usually the person conducting the research wishes to establish that there is a difference between the parameter and the value being tested, and thus the alternative is also called the **research hypothesis.**

Suppose, in the preceding example, that the researcher believes the mean per capita income of the county residents is greater than \$15,000. This is formulated as $\mu > 15,000$, and because this is what the researcher wishes to support, it is a statement of the alternative hypothesis. Thus, the null hypothesis is

$$H_0: \mu = 15,000$$

and the alternative hypothesis is

$$H_a: \mu > 15,000$$

EXAMPLE 8.1

A conjecture is made that the mean starting salary for computer science graduates is \$30,000 per year. You believe it is less than \$30,000. Formulate the null and alternative hypotheses to evaluate the claim.

SOLUTION

The parameter about which the conjecture is made is the *mean* starting salary, μ. The statement "you believe it is less than \$30,000" is formulated as $\mu < 30,000$. Because this is the statement you wish to support, it is the alternative (research) hypothesis. Consequently, we have

$$H_0: \mu = 30,000 \quad \text{versus} \quad H_a: \mu < 30,000$$

EXAMPLE 8.2

The standard medication for a certain disease is effective in 60% of all cases. A pharmaceutical company believes that its new drug is more effective than the old treatment. Formulate the null and alternative hypotheses to test whether there is statistical evidence to support the new drug.

SOLUTION

Let π denote the proportion of cases for which the new drug is effective. The pharmaceutical company thinks that the new drug is more effective than the standard medication; therefore, the research hypothesis is that $\pi > .60$. We have

$$H_0: \pi = .60 \quad \text{versus} \quad H_a: \pi > .60$$

The alternative hypotheses in the preceding examples are *one-sided alternatives* because the researcher is interested in deviations from the null hypothesis only in a specified direction. If deviations from the null hypothesis in either direction are of concern, then it is called a *two-sided alternative*.

EXAMPLE 8.3   Researcher Alfredo Morabia of University Hospital in Geneva, Switzerland, was interested in comparing the proportions of smokers among women with breast cancer and women without breast cancer (www.usatoday.com/life/health/lhs502.htm). Morabia's report, in the May 1996 issue of *The American Journal of Epidemiology*, is based on a study of 244 women with breast cancer and 1,032 women without the disease. Set up null and alternative hypotheses to evaluate the claim that there is no difference between the two proportions.

SOLUTION

Let π_1 represent the proportion who smoke among women with breast cancer, and let π_2 represent the proportion who smoke among women without breast cancer. The null hypothesis says that there is no difference between the two proportions, and the alternative hypothesis says that there is a difference (π_1 could be greater than or less than π_2). Thus, we have

$$H_0: \pi_1 = \pi_2 \quad \text{versus} \quad H_a: \pi_1 \neq \pi_2$$

(This is a two-sided alternative.)

The Test Statistic

Having formulated the null and alternative hypotheses, we next develop the test procedure to assess the evidence against H_0. If the evidence is attributable to factors other than chance, then H_0 will be rejected in favor of the alternative, H_a.

An important step in developing the test procedure is to determine an appropriate *test statistic*—that is, a statistic from the sample that seems appropriate for testing the null hypothesis.

> A **test statistic** is a statistic, calculated from the sample data, that is used to test the null hypothesis.

The choice of the test statistic depends on the parameter being tested and the underlying parent distribution. Just as in confidence interval estimation, one must use exploratory data analysis techniques to describe the underlying distribution so that an appropriate test statistic can be chosen. For example, in Section 6.3, you learned that if the underlying parent distribution is near normal, then the statistic $\overline{y}$ is the best estimator of μ. Similarly, under the normality assumption, using $\overline{y}$ as a test statistic results in the best test procedure for testing the population mean μ. If the parent distribution is not normal, however, then $\bar{y}$ may not be the best test statistic for testing μ. For example, if the parent distribution is symmetric with long tails, a better test statistic is one of the trimmed sample means. General speaking, the test statistic we will use is the point estimator of the parameter that was found in Chapter 6.

Let us return to the example of the mean per capita income of the residents of the rural county. As stated, the parameter of concern is the population mean, and we wish to test

$$H_0: \mu = 15,000 \quad \text{versus} \quad H_a: \mu > 15,000$$

Assuming that the parent distribution is nearly normal, we will use $\bar{y}$ as the test statistic. Suppose that a random sample of 100 residents yielded an average per capita income of $\bar{y} = \$25,000$. We recall that sample means vary from the population mean, but if $\mu = 15,000$, is it reasonable to observe a $\bar{y}$ this large? Because of the large difference between μ and $\bar{y}$ in this case, we might suspect that there is something wrong with our hypothesis. If, on the other hand, the random sample gave an average of $\bar{y} = \$15,100$, we could not say for sure that $\mu = 15,000$, but it is doubtful that there is sufficient evidence to reject it. Thus, an important element in testing this hypothesis is the difference between the test statistic $\bar{y}$ and the hypothesized value of μ.

The Sampling Distribution of the Test Statistic

The strategy we will use to evaluate the difference is to assume that the null hypothesis is true and then from the random sample determine whether the observed $\bar{y}$ is different from what we generally would expect given the hypothesized value of μ. To accomplish this, we must determine the sampling distribution of the test statistic under the assumption that H_0 is true. With the sampling distribution, it is possible to determine what values of the test statistic seem reasonable for rejection of the null hypothesis. These values are collectively known as the *rejection region*.

> The **rejection region** consists of those values of the test statistic that will lead to the rejection of the null hypothesis.

For the example in which the alternative hypothesis is $\mu > 15,000$, we should not reject the null hypothesis for any value of $\bar{y}$ that is smaller than 15,000. In fact, the rejection region should consist of only those values of $\bar{y}$ that are considerably greater than 15,000.

Figure 8.1 on page 486 depicts the rejection region on the right tail of the sampling distribution of the test statistic, $\bar{y}$. When the parent distribution is near normal or the sample size is large, recall that the sampling distribution of $\bar{y}$ is approximately normally distributed with a mean equal to the mean of the parent population and a standard deviation equal to $\sigma/\sqrt{n}$. Of course, the more the parent distribution deviates from normality, the larger the sample size should be. Assuming H_0 is true, we have that the mean of the parent population is 15,000, and thus the sampling distribution is drawn as a normal curve centered at 15,000. This test with a one-sided alternative is called a *right-tailed test* because the rejection region is on the right tail of the sampling distribution.

FIGURE 8.1 Sampling distribution of $\bar{y}$ with rejection region

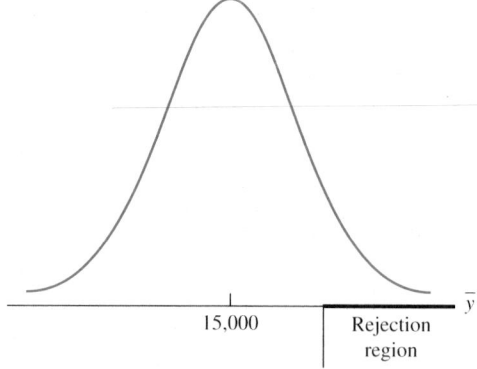

The *p*-Value

To continue the example, suppose the researcher selects a random sample of 100 county residents and finds that their average per capita income is $\bar{y}_{\text{obs}} = \$16,200$. (We use $\bar{y}_{\text{obs}}$ to denote the value of $\bar{y}$ that was observed from the random sample.) Is this evidence sufficient to reject H_0 and suggest that $\mu > 15,000$? To assess the strength of evidence against the null hypothesis and determine whether the test statistic falls in the rejection region, we will calculate the probability of observing a value of the test statistic at least as extreme as the actual observed outcome from the sample. This probability, called the *p-value*, is computed under the assumption that the null hypothesis is true. Thus, a large probability will support the truth of the null hypothesis, and a small probability provides evidence that the null hypothesis is false and should be rejected.

> The **p-value** is the probability (computed when H_0 is assumed to be true) of observing a value of the test statistic at least as extreme as that given by the actual observed data. The smaller the *p*-value, the stronger is the evidence against the null hypothesis.

Figure 8.2 illustrates the *p*-value, which is given by

$$p\text{-value} = P(\bar{y} \geq \bar{y}_{\text{obs}}) = P(\bar{y} \geq 16,200)$$

FIGURE 8.2   Sampling distribution of $\bar{y}$ and $p\text{-value} = P(\bar{y} \geq \bar{y}_{\text{obs}})$

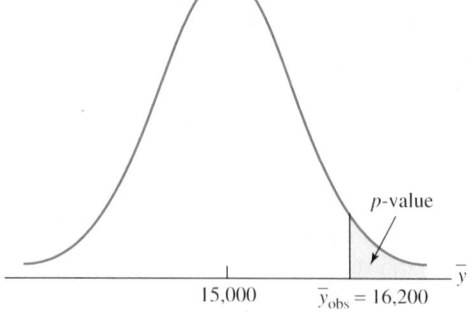

Knowing that the sampling distribution of $\bar{y}$ is normal with mean 15,000 and standard deviation $\sigma/\sqrt{n}$, we can find the p-value in the normal probability table by determining the distance between $\bar{y}_{obs} = 16,200$ and 15,000 with a z-score.

The z-Test

The z-score,

$$z = \frac{\bar{y} - 15,000}{\sigma/\sqrt{n}}$$

is called the *standardized test statistic*, and the p-value is given by

$$p\text{-value} = P(\bar{y} \geq \bar{y}_{obs}) = P(z \geq z_{obs})$$

Calculating the observed z-score for this z-test requires that we know the population standard deviation σ. In most applications, σ is not known and must be estimated from sample data. We address this issue in Section 8.3, but for the time being, suppose we know that $\sigma = \$4,000$. Substituting values, we have

$$z_{obs} = \frac{\bar{y}_{obs} - 15,000}{\sigma/\sqrt{n}} = \frac{16,200 - 15,000}{4,000/\sqrt{100}} = 3$$

Already the evidence suggests that the null hypothesis should be rejected. A z-score of 3 means that 16,200 is three standard deviations above 15,000, which is highly unlikely. In fact, looking up a z-score of 3 in the normal table, we find that .9987 of the area lies below 3, so

$$p\text{-value} = 1.0 - .9987 = .0013$$

which is a very small probability. The farther $\bar{y}_{obs}$ is from 15,000, the larger will be z_{obs} and consequently, the smaller will be the p-value. Thus, the smaller the p-value, the stronger is the evidence that H_0 should be rejected.

Level of Significance

Just how small should a p-value be before we reject H_0? As observed, a small p-value indicates that the test statistic is far removed from the hypothesized value of the parameter

FIGURE 8.3   Sampling distribution of $\bar{y}$ with rejection region and level of significance

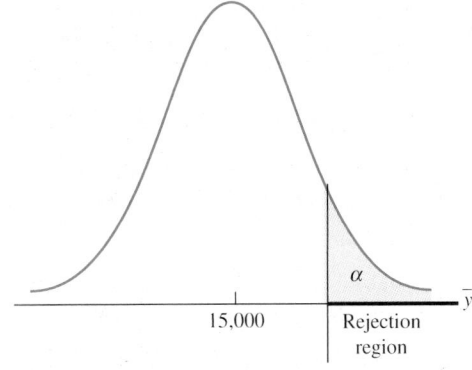

15,000

α

Rejection region

$\bar{y}$

and consequently is farther into the rejection region. Therefore, the *size* of the rejection region, called the *level of significance* (denoted by the Greek letter α), determines how small the *p*-value should be before we reject the null hypothesis. Figure 8.3 shows that the area above the rejection region is the level of significance α.

If $\bar{y}_{obs}$ falls in the rejection region, as shown in Figure 8.4, then the *p*-value is less than α and we reject the hypothesis at the α level of significance. In practice, we set the level of significance at a small probability, such as $\alpha = .05$. Then if *p*-value < .05, we say, "The results are *statistically significant* ($P < .05$)." If *p*-value > .05, we say, "The results are *insignificant*." In some applications, the criterion for rejection may be as high as $\alpha = .10$; in others, as low as $\alpha = .01$.

FIGURE 8.4

Sampling distribution of $\bar{y}$ with $\bar{y}_{obs}$ in the rejection region

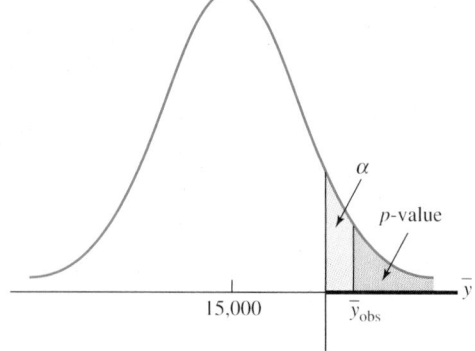

Setting a specified value for α prior to testing makes sense if we must make a clear-cut decision about the hypotheses. For example, a pharmaceutical company may decide to market a new drug if the test results are statistically significant at the $\alpha = .01$ level. Or, in the area of *acceptance sampling*, a manufacturer of log homes accepts logs from a milling company if the mean moisture content of the logs is less than a specified amount. The manufacturer decides to reject the logs if the test results of the moisture content are statistically significant at the $\alpha = .05$ level.

If, on the other hand, we are testing simply to assess the strength of the statistical evidence, then reporting the *p*-value will suffice. In this context, the evaluation of a hypothesis is referred to as a *test of significance*.

Interpreting the Results

In most situations, the following criteria may be used for rejecting the null hypothesis:

Criteria for Rejection of H_0

1. If *p*-value > .10, then fail to reject H_0 and declare the results insignificant.
2. If .05 < *p*-value ≤ .10, you may reject H_0, but the results are only mildly significant.
3. If .01 < *p*-value ≤ .05, then reject H_0 and declare the results significant.
4. If *p*-value ≤ .01, then reject H_0 and declare the results highly significant.

Because p-value $= .0013$ in the per capita income example, we conclude by saying, "On the basis of a random sample of 100 residents of the county, there is *highly significant* evidence that the mean per capita income of all residents of the county is greater than $15,000."

Testing hypotheses, then, is a problem of deciding between the null and alternative hypotheses based on the information contained in a random sample. Because the alternative is the hypothesis that the researcher believes to be true, the goal is to reject H_0 in favor of H_a. To summarize, there are five basic steps in the hypothesis-testing procedure:

Procedure for Testing Hypotheses

1. Formulate the null and alternative hypotheses.
2. Decide on an appropriate test statistic.
3. Determine the sampling distribution of the test statistic under the assumption that the null hypothesis is true.
4. Calculate the p-value and determine whether it is sufficiently small to reject the null hypothesis.
5. Interpret the results in a way that a nonstatistician could understand.

EXAMPLE 8.4

The scores on a college placement exam in mathematics are assumed to be normally distributed with a mean of 70 and a standard deviation of 18. The exam is given to a random sample of 50 high school seniors who have been admitted to college. Their average score on the exam was 67. If this is a true random sample, is the evidence sufficient to suggest that the population mean score is lower than 70?

SOLUTION

Step 1   Let μ denote the true population mean of the placement exam. We wish to see whether there is evidence that $\mu < 70$. This is the research hypothesis. The null and alternative hypotheses are

$$H_0: \mu = 70 \quad \text{versus} \quad H_a: \mu < 70$$

Step 2   Because the parent distribution is assumed to be normal, the best test statistic is the sample mean, $\bar{y}$.

Step 3   Given that we know the numerical value of σ and we assume that $\mu = 70$, the standardized form of the test statistic

$$z = \frac{\bar{y} - 70}{\sigma/\sqrt{n}}$$

has a standard normal distribution.

Step 4   Because the alternative hypothesis is $\mu < 70$, the rejection region consists of values on the left tail of the sampling distribution of $\bar{y}$, as illustrated in Figure 8.5. This test with a one-sided alternative is called a *left-tailed test*.

FIGURE 8.5   A left-tailed test

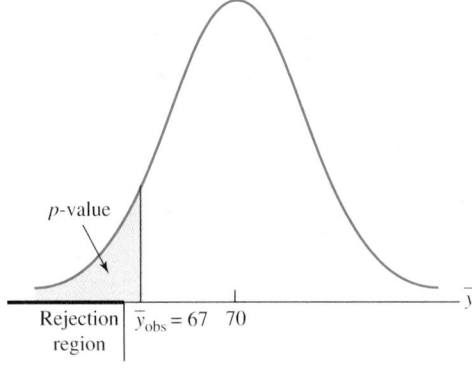

The *p*-value (also shown in Figure 8.5) is the probability of observing a value of $\bar{y}$ less than or equal to $\bar{y}_{obs}$ when the true population mean is indeed 70; that is, *p*-value $= P(\bar{y} \leq 67)$. Substituting in values, we have

$$z_{obs} = \frac{67 - 70}{18/\sqrt{50}} = -1.18$$

and thus,

$$p\text{-value} = P(\bar{y} \leq 67) = P(z \leq -1.18) = .1190$$

A *p*-value this large exceeds any reasonable level of significance α, and thus we fail to reject H_0.

Step 5   Conclusion: Based on the results of a random sample of 50 high school seniors, there is insufficient evidence to say that the mean score on the college placement exam should be lower than 70.

Type I and Type II Errors

Because sample results vary from sample to sample, the evidence provided by a particular sample can be misleading. Even with a truly random sample, a test of hypothesis could result in an incorrect decision. To understand this, we can draw an analogy between hypothesis testing and a trial by jury. The null hypothesis says "The defendant is not guilty," and the jury must decide whether the presented evidence (observed sample) is convincing enough to warrant conviction (or rejection) of the defendant. Even though we hope that the jury reaches the correct decision in all cases, we are aware that an innocent person can be convicted or a guilty person freed. Similarly, two errors are possible when testing hypotheses. The first type of error occurs when a null hypothesis is rejected (convicted) when, in fact, it is true (innocent). The second type of error occurs when the null hypothesis is not rejected (not convicted) when, in fact, it is false (guilty).

> A **Type I error** is rejecting a null hypothesis that is true. The probability of committing a Type I error is denoted as α.
>
> A **Type II error** is failing to reject a null hypothesis that is false. The probability of committing a Type II error is denoted as β.

If we are testing a hypothesis for the purpose of decision making, Table 8.1 illustrates four possibilities, two of which result in incorrect decisions. If H_0 is true, then either our decision is correct or we make a Type I error. If H_0 is false, then either our decision is correct or we make a Type II error. Because H_0 is either true or false, it is impossible to make both errors, and we hope we will make neither.

TABLE 8.1

		Decision	
		Reject H_0	**Fail to Reject H_0**
Null Hypothesis	True	Type I error	Correct decision
	False	Correct decision	Type II error

Rarely would we set the level of significance at a value greater than .10 because seldom do we want the chance of a Type I error to exceed 10%.

An important concern here is the probabilities of the two types of error. Recall that the size of the rejection region of a test is the level of significance, α. As illustrated in Figure 8.3, α is the probability that the test statistic falls in the rejection region when the null hypothesis is true. In other words, α is the probability of a Type I error. Because we wish not to make this error, we choose a small value, such as $\alpha = .05$, for the level of significance. This means that the test procedure, if used over and over again with different samples, will reject a true null hypothesis about 5 times out of 100. Why don't we make α extremely small—say, .00001—by reducing the size of the rejection region? The result would be a very small rejection region, and consequently we would accept H_0 more frequently, thus increasing our chances of making a Type II error. That is, decreasing α automatically increases β, the probability of a Type II error. Also, increasing α automatically decreases β. Because of this relationship between α and β, it is not possible to make both values arbitrarily small. Thus, in decision making we fix α at a reasonable level and then attempt to find a test procedure that minimizes β. The level set for α depends on the seriousness of the two types of errors. If making a Type I error results in a serious consequence, then α should be set at a small level. If making a Type II error results in a more serious consequence, then α should be set at a high level. Thus, we can, in a sense, control both probabilities by setting α at a specified level.

Suppose the decision is whether we should accept or reject a parachute for a sky dive. The null hypothesis might be

$$H_0: \text{The parachute will open}$$

and the alternative is then

$$H_a: \text{The parachute will not open}$$

A Type I error is committed if we reject H_0 when it is true; that is, we say the parachute will not open when in fact it will. On the other hand, a Type II error is committed if we accept H_0 when it is false; that is, we say the parachute will open when in fact it will not. Clearly, in this case, we do not want to make a Type II error! We want the chance of the Type II error to be extremely small, so we would set α at a very high level. In other situations, the opposite might be true, and the Type I error is the more serious error.

EXAMPLE 8.5

A manufacturer of log homes accepts logs from a milling company if the mean moisture content of the logs does not exceed 10 on her moisture reading equipment. The manufacturer randomly selects a sample of the logs, examines them, and rejects the shipment if the test results do not meet her standard. Formulate the null and alternative hypotheses to determine whether the manufacturer should reject the shipment. Interpret Type I and Type II errors and determine which is more serious.

SOLUTION

Let μ denote the mean moisture content of the shipment. The null hypothesis says the shipment meets the standard established by the manufacturer, and the alternative hypothesis says the shipment does not meet the standard. Therefore,

$$H_0: \mu = 10 \quad \text{versus} \quad H_a: \mu > 10$$

Rejecting the null hypothesis amounts to rejecting the shipment. Rejecting it when it is true, or committing a Type I error, means rejecting the shipment when the moisture content is acceptable. A Type II error accepts the shipment when the moisture content does not meet the standard. Which is more serious: rejecting a good shipment of logs or accepting a bad shipment of logs? The answer depends on whether you are the manufacturer or the milling company that supplies the logs. To the manufacturer, accepting a bad shipment of logs (a Type II error) is clearly more serious. To the supplier of the logs, however, a Type I error means that a shipment of good logs is rejected.

EXERCISES 8.1

8.1   In each case, formulate the null and alternative hypotheses involving the population mean, μ.
   a. You wish to show that μ exceeds 50.
   b. You wish to show that μ is less than 50.
   c. You wish to find evidence against the claim that μ is 50 when you believe it is greater than 50.
   d. You wish to find evidence against the claim that μ is 50 when you believe it is less than 50.
   e. You wish to test the claim that μ is 50.

8.2   In each case, formulate the null and alternative hypotheses involving the Bernoulli proportion, π.
   a. You wish to show that π exceeds 70%.
   b. You wish to show that π is less than 70%.
   c. You wish to find evidence against the claim that π is 70% when you believe it is greater than 70%.
   d. You wish to find evidence against the claim that π is 70% when you believe it is less than 70%.
   e. You wish to test the claim that π is 70%.

8.3   Draw a normal curve and shade in the rejection region as in Figure 8.1 (page 486) for these test of hypothesis problems. Identify each as being a right-tailed or a left-tailed test.
   a. $H_0: \mu \geq 15$ b. $H_0: \mu \leq 650$ c. $H_0: \mu \geq 6{,}000$
      $H_a: \mu < 15$ $H_a: \mu > 650$ $H_a: \mu < 6{,}000$

8.4   A sociologist is interested in the percent of blacks in the inner city who are unemployed. She has pleaded with local industries to employ more blacks. She thinks that more than 30% of inner-city blacks are unemployed. How should she set up the null and alternative hypotheses to support her point?

8.5   A psychologist suspects that more than 10% of the adult population is illiterate. She takes a random sample of 1,600 adults and gives them the Wechsler Adult Intelligence Scale. State the null and alternative hypotheses to evaluate the psychologist's claim.

8.6   A claim is made that 60% of the adult population feels that there is too much violence on television. Assuming you think this figure is too high, set up null and alternative hypotheses to evaluate the claim.

8.7   It has been claimed that by the year 2010, 20% of all American adults will be drawing retirement benefits. If you believe that the number is greater than 20%, which is the correct way to formulate the hypotheses?
   a. $H_0: p = .20$ versus $H_a: p > .20$
   b. $H_0: p \geq .20$ versus $H_a: p < .20$
   c. $H_0: \pi = .20$ versus $H_a: \pi > .20$
   d. $H_0: \pi \geq .20$ versus $H_a: \pi < .20$

8.8   If the null hypothesis is rejected, does this mean that the alternative is true? Explain.

8.9   If the null hypothesis is not rejected, does this mean that it is true? Explain.

8.10   Suppose a testing procedure leads to the rejection of the null hypothesis. Is it possible that a Type I error was committed? Is it possible that a Type II error was committed? Explain.

8.11   Does $\alpha + \beta = 1$? Why or why not?

8.12   If the null hypothesis is rejected at the 1% level of significance, will it also be rejected at the 5% level? Explain.

8.13   During a group study session, one student says, "I know what a p-value is. It is the probability that the null hypothesis is true." A second student says, "No, it is the probability that the null hypothesis is rejected." Discuss these two interpretations of the p-value.

8.14   Suppose that the null hypothesis says, "The patient will recover." Interpret the Type I and Type II errors. Which is more serious?

8.15   Suppose that the null hypothesis says, "The fire is out." Interpret the Type I and Type II errors. Which is more serious?

8.16   In the oil industry, a wildcat is an experimental oil well that is drilled in an area not known to be successful. The wildcat success rate is about 10% in the United States. For this reason, the oil industry considers the consequences of failing to drill in locations where there is oil to be more serious than drilling dry holes. Suppose the null hypothesis states that the wildcat will not produce oil and the alternative states that the wildcat will be productive. Interpret the Type I and Type II errors. Which would oil industry people consider the more serious?

8.17   Suppose a geologist believes that the mean porosity measurement of Tensleep Sandstone from the Bighorn Basin in Wyoming exceeds 18%. To find statistical evidence to support his theory, he computes the porosity percents on ten core samples of Tensleep Sandstone. State the null and alternative hypotheses required to evaluate his theory. Explain the situation if the geologist makes a Type II error.

8.18   Draw a normal curve and shade in the rejection region and the p-value area for these test of hypothesis problems. Calculate the p-values.

  a. $H_0: \mu \geq 15$
     $H_a: \mu < 15$
     $n = 48$, $\bar{y} = 14.2$, $\sigma = 4.1$
  b. $H_0: \mu = 120$
     $H_a: \mu \neq 120$
     $n = 100$, $\bar{y} = 124.6$, $\sigma = 16.3$
  c. $H_0: \mu \leq 650$
     $H_a: \mu > 650$
     $n = 250$, $\bar{y} = 694$, $\sigma = 235.7$

8.19   Here is the computer printout of a test of hypothesis about the mean age of food stamp recipients.

```
Z-Test

Test of mu = 40.000 vs mu not = 40.000
The assumed sigma = 5.00
```

Variable	N	Mean	StDev	SE Mean	Z	P-Value
C1	37	38.378	4.609	0.822	-1.97	0.049

  a. State the null and alternative hypotheses.
  b. Was the population standard deviation given or was it estimated?
  c. How large was the sample?
  d. What was the average of the sample? Was it significantly different from the hypothesized mean age?
  e. What was the p-value? Should the null hypothesis be rejected?
  f. State your conclusion.

8.20   Suppose the mean entrance exam score for incoming freshmen at a large university is 550 and the standard deviation is 120. A sample of 90 students from this year's freshman class had an average score of 582.6. Is this unusual? Use the computer printout to perform a test of significance.

```
Z-Test

Test of mu = 550.0 vs mu > 550.0
The assumed sigma = 120
```

Variable	N	Mean	StDev	SE Mean	Z	P-Value
C1	90	582.6	112.3	12.6	2.57	0.0051

  a. State the null and alternative hypotheses.
  b. Was the population standard deviation given or was it estimated?
  c. What was the difference between the sample mean and the hypothesized value of the population mean? Was the difference significant?

d. What was the *p*-value? Should the null hypothesis be rejected?

e. State your conclusion.

8.21 An instructor gives his class an examination that, as he knows from years of experience, yields $\mu = 78$ and $\sigma = 7$. His present class of 35 students scores an average of 81.4. Is he correct in assuming that this is a superior class? Is it necessary that the parent distribution be approximately normally distributed?

8.2 TESTING A POPULATION PROPORTION

We often wish to evaluate the proportion of successes in a Bernoulli population. Recall that a Bernoulli population is one in which each outcome is classified as either a success or a failure, and we are concerned with π, the proportion of successes. In this section, conjectures about π are investigated with a test of hypothesis. For example, is it true that 69% of all American drivers wear seat belts? Is it true that no more than 40% of school-aged children receive regular dental care? These conjectures can be evaluated with a test of hypothesis on the Bernoulli proportion π.

Testing the Proportion π

Let us evaluate the conjecture made by the pharmaceutical company in Example 8.2. We wish to test

$$H_0: \pi = .60 \quad \text{versus} \quad H_a: \pi > .60$$

As outlined in Section 8.1, after formulating the null and alternative hypotheses, we must determine an appropriate test statistic for testing the hypothesis. As in confidence interval estimation, the best estimator of π is the sample proportion p; hence, p is the test statistic.

Because of the alternative hypothesis, this is a right-tailed test, and consequently the rejection region consists of those values of p that are considerably greater than .60. Figure 8.6 illustrates the rejection region on the right tail of the sampling distribution of the test statistic p. The graph is drawn as a normal curve because, with a large sample, the

FIGURE 8.6 Sampling distribution of p with a right-tailed rejection region

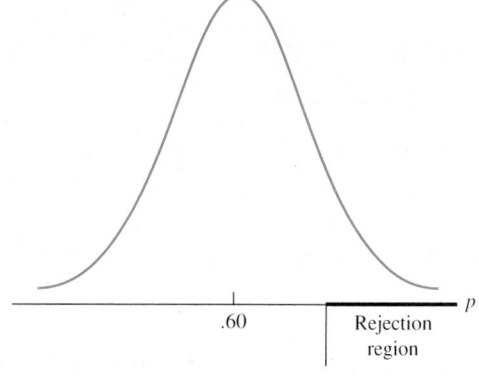

Central Limit Theorem guarantees that the sampling distribution of p is approximately normal. Furthermore, the mean and standard deviation of the sampling distribution are π and $\sqrt{\pi(1 - \pi)/n}$, respectively. Under the assumption that H_0 is true, the sampling distribution is centered at .60.

Suppose that the drug company investigates 200 cases and finds that the new drug is effective in 134 cases ($p_{obs} = 134/200 = .67$). Is this enough evidence to reject H_0 and say that $\pi > .60$? That is, does $p_{obs} = .67$ fall in the rejection region?

To assess the evidence against H_0 and determine whether p_{obs} falls in the rejection region, we calculate the p-value, which is illustrated in Figure 8.7 and given by

$$p\text{-value} = P(p \geq p_{obs}) = P(p \geq .67)$$

FIGURE 8.7 Sampling distribution of p and p-value $= P(p \geq .67)$

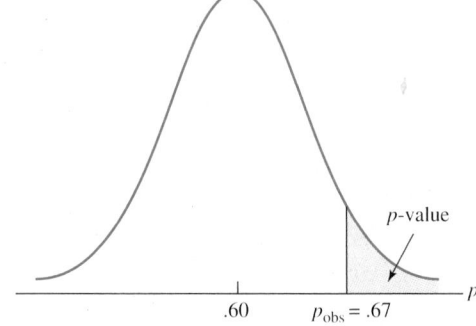

Because the sampling distribution of p is approximately normal with mean π and standard deviation $\sqrt{\pi(1 - \pi)/n}$, the standardized test statistic is

$$z = \frac{p - \pi_0}{\sqrt{\pi_0(1 - \pi_0)/n}}$$

and the p-value is given by

$$p\text{-value} = P(p \geq p_{obs}) = P(z \geq z_{obs})$$

[*Note:* Under the assumption that the null hypothesis H_0 is true, the standard deviation becomes $\sqrt{\pi_0(1 - \pi_0)/n}$, where π_0 denotes the value of π being tested.] Substituting values, we have

$$z_{obs} = \frac{p_{obs} - \pi_0}{\sqrt{\pi_0(1 - \pi_0)/n}} = \frac{.67 - .60}{\sqrt{(.60)(.40)/200}} = 2.02$$

Looking up a z-score of 2.02 in the normal probability table, we find that .9783 of the area lies below $+2.02$, so

$$p\text{-value} = 1.0 - .9783 = .0217$$

Thus, we reject H_0 ($\pi = .60$) at any level of significance greater than .0217. This is strong evidence that the null hypothesis is false. In conclusion, based on a random sample of size 200, there is statistical evidence that the new drug is more effective than the standard medication.

Testing a More General H_0

Suppose that in the preceding example the hypotheses had been stated as

$$H_0: \pi \leq .60 \quad \text{versus} \quad H_a: \pi > .60$$

instead of

$$H_0: \pi = .60 \quad \text{versus} \quad H_a: \pi > .60$$

Recall that the p-value is calculated under the assumption that the null hypothesis is true. When H_0 states that $\pi = .60$, the procedure is straightforward because we can use the specified value of $\pi = .60$ in the z-score and proceed to find the p-value. If H_0 states that $\pi \leq .60$, however, which value of π do we use to calculate the z-score? Should we, perhaps, choose .50 in the calculation of the z-score or just any value of π that is less than .60? In Figure 8.8, we see two sampling distributions of p: one assuming that $\pi = .60$ and the other assuming that π is some value $m < .60$. Because the p-value is the tail probability beyond the observed test statistic, it is clear that the p-value is greatest when we assume that $\pi = .60$.

FIGURE 8.8

The p-value assuming that $\pi = .60$ or a value $m < .60$

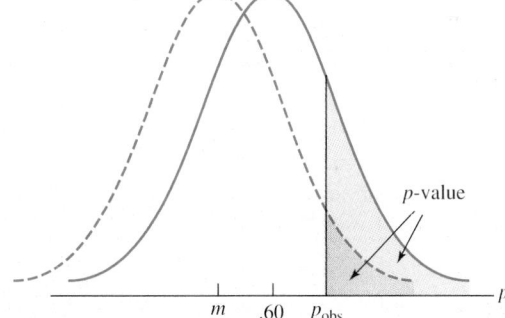

Thus, when testing the more general null hypothesis

$$H_0: \pi \leq \pi_0$$

we will compute the p-value under the assumption

$$H_0: \pi = \pi_0$$

The important
inequality is in
the statement of
the alternative
hypothesis H_a
because that
determines the
direction of the
test.

so as to obtain its maximum value. In this sense, testing

$$H_0: \pi \leq \pi_0 \quad \text{versus} \quad H_a: \pi > \pi_0$$

is equivalent to testing

$$H_0: \pi = \pi_0 \quad \text{versus} \quad H_a: \pi > \pi_0$$

Unless stated otherwise, we will give the null hypothesis in the more general form.

Two-Tailed Test

To this point we have considered only tests with one-sided alternatives. We now look at the *two-tailed test*. For the Bernoulli proportion π, the hypotheses are

$$H_0: \pi = \pi_0 \quad \text{versus} \quad H_a: \pi \neq \pi_0$$

In light of the two-sided alternative, we could conceivably reject the null hypothesis in favor of the alternative if p_{obs} is either significantly greater than π_0 or significantly less than π_0. That is, the rejection region lies on both tails. Clearly, p_{obs} cannot fall on both tails of the sampling distribution, so let us suppose that it falls on the upper tail. If this were a one-sided alternative, the p-value would be the area under the curve on the upper tail as in a right-tailed test. But because this is a two-tailed test, we must account for both tails by doubling the probability. If

$$p_{obs} > \pi_0$$

we calculate the p-value as in a right-tailed test and then double it. In a similar fashion, if

$$p_{obs} < \pi_0$$

we calculate the p-value as in a left-tailed test and then double it.

Figure 8.9 illustrates the rejection region and calculation of the p-value for the two-tailed test.

FIGURE 8.9   The two-tailed test

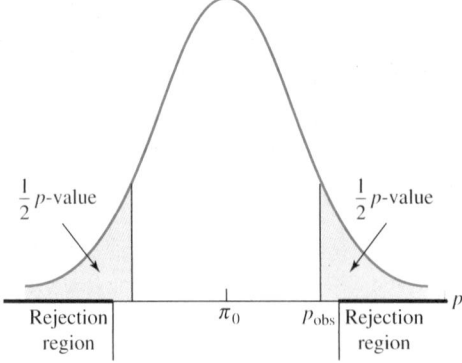

EXAMPLE **8.6**

A recent report claimed that 20% of all college graduates find a job in their chosen field of study. A survey of a random sample of 500 graduates found that 110 obtained work in their field. Is there statistical evidence to refute the claim?

SOLUTION

If we let π denote the percent of college graduates who find a job in their field of study, then the null and alternative hypotheses are

$$H_0: \pi = .20 \quad \text{versus} \quad H_a: \pi \neq .20$$

The test statistic is p, the proportion of successes in the sample. The observed value of p is found to be

$$p_{obs} = \frac{110}{500} = .22$$

The observed value of the standardized test statistic is

$$z_{obs} = \frac{p_{obs} - .20}{\sqrt{(.20)(.80)/500}} = \frac{.22 - .20}{.018} = 1.11$$

If this were a one-sided alternative, the p-value would be

$$P(z \geq 1.11) = 1.0 - .8665 = .1335$$

But because this is a two-sided alternative, we have

$$p\text{-value} = 2(.1335) = .267$$

Clearly, this is insignificant evidence to reject the null hypothesis. On the basis of a random sample of 500 college graduates, there is no evidence to refute the claim that 20% find work in their field of study.

Testing with Confidence Intervals

In the notation for confidence intervals and tests of hypotheses, it is no coincidence that we use that same Greek letter α in a $(1 - \alpha)100\%$ confidence interval and for the level of significance in a test of hypothesis. In fact, it is possible to carry out a two-tailed test of hypothesis by constructing a confidence interval for the unknown parameter.

Equivalence of Confidence Intervals and Two-tailed Tests

The null hypothesis H_0: $\pi = \pi_0$ versus the alternative hypothesis H_a: $\pi \neq \pi_0$ is rejected at an α level of significance if and only if the hypothesized value π_0 falls outside a $(1 - \alpha)100\%$ confidence interval for π.

EXAMPLE 8.7

A news report in a major eastern city stated that 80% of all violent crimes in that city involve firearms. A survey of all violent crimes in the city for the past 2 years revealed that of 283 violent crimes, 240 involved firearms. Determine with a confidence interval whether the news report is correct.

SOLUTION

The claim made by the news report is that the percent is 80; therefore, we set up the hypothesis as a two-tailed test:

$$H_0\text{: }\pi = .80 \quad \text{versus} \quad H_a\text{: }\pi \neq .80$$

Instead of finding the p-value, however, we will find a 95% confidence interval for π. Recall that the form of a 95% confidence interval for π is

$$p \pm 1.96\sqrt{\frac{p(1 - p)}{n}}$$

The sample reveals that

$$p_{\text{obs}} = \frac{240}{283} = .848$$

so the interval for π becomes

$$.848 \pm 1.96\sqrt{\frac{(.848)(.152)}{283}} = .848 \pm .042$$

Thus, the 95% confidence interval for π is $(.806, .890)$. Notice that the hypothesized value $\pi_0 = .80$ is outside the interval. Therefore, the null hypothesis is rejected at the 5% level of significance.

Will it also be rejected at the 1% level of significance? To find out, we construct a 99% confidence interval for π. The interval becomes

$$.848 \pm 2.58\sqrt{\frac{(.848)(.152)}{283}} = .848 \pm .055$$

Thus, the 99% confidence interval for π is $(.793, .903)$. This time the hypothesized value $\pi_0 = .80$ is inside the confidence interval, which means that the null hypothesis cannot be rejected at the 1% level of significance. Figure 8.10 shows that $\pi_0 = .80$ is covered by the 99% confidence interval but not by the 95% confidence interval.

In conclusion, we can say that there is moderately significant evidence that the percent of violent crimes involving firearms is different from 80%.

FIGURE 8.10

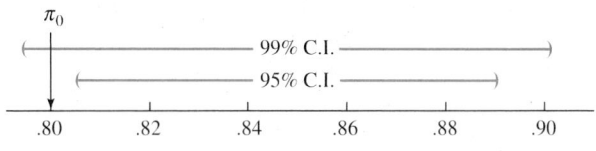

The test of a population proportion is summarized in the box:

Large-Sample Test of a Population Proportion

Application: Bernoulli Population

Assumption: $n > 30$

Left-Tailed Test	Right-Tailed Test	Two-Tailed Test
$H_0: \pi \geq \pi_0$	$H_0: \pi \leq \pi_0$	$H_0: \pi = \pi_0$
$H_a: \pi < \pi_0$	$H_a: \pi > \pi_0$	$H_a: \pi \neq \pi_0$

Standardized test statistic:
$$z = \frac{p - \pi_0}{\sqrt{\pi_0(1 - \pi_0)/n}}$$

From the z-table (Table B.2), find:

p-value $= P(z < z_{obs})$ p-value $= P(z > z_{obs})$ p-value $= 2P(z < z_{obs})$
                                                                                if $p_{obs} < \pi_0$

                                                                      or p-value $= 2P(z > z_{obs})$
                                                                                if $p_{obs} > \pi_0$

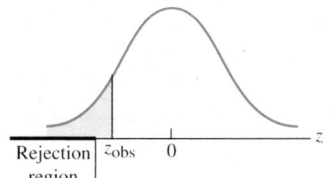
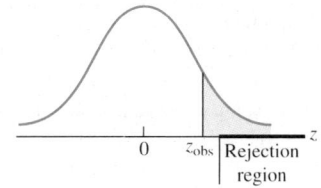
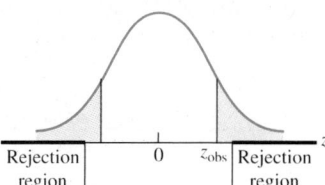

If a level of significance α is specified, reject H_0 if p-value $< \alpha$.

EXERCISES 8.2

8.22   In Exercise 8.5 of Section 8.1, a psychologist suspected that more than 10% of the adult population was illiterate. The hypotheses should be stated as

$$H_0: \pi = .10 \quad \text{versus} \quad H_a: \pi > .10$$

a. Is this a left-tailed, right-tailed, or two-tailed test?
b. What is the correct test statistic to use to test H_0?
c. If the test statistic falls on the left tail of the sampling distribution, is it possible to reject H_0? Explain your answer.

8.23   Following up on Exercise 8.22, the psychologist gave the Wechsler Adult Intelligence Scale to a sample of 1,600 adults. Based on the results, she classified 180 of the 1,600 as illiterate. Is there sufficient evidence to verify her claim?

8.24   In Exercise 8.6 of Section 8.1, a claim was made that 60% of the adult population thinks that there is too much violence on television. You thought that figure was too high and were asked to set up null and alternative hypotheses to evaluate the claim. The hypotheses should be stated as

$$H_0: \pi = .60 \quad \text{versus} \quad H_a: \pi < .60$$

a. Is this a left-tailed, right-tailed, or two-tailed test?
b. What is the correct test statistic to use to test H_0?
c. If the test statistic falls on the right tail of the sampling distribution, is it possible to reject H_0? Explain your answer.

8.25   As follow-up to Exercise 8.24, a random sample of 200 adults found that 110 thought that there is too much violence on television. Is this enough evidence to reject the claim?

8.26   The government believes that no more than 25% of all college students would favor reducing the penalties for the use of marijuana. A sample of 2,400 college students revealed that 750 favor reducing the penalties.
a. Set up null and alternative hypotheses to evaluate the government's claim.
b. Give the form of the standardized test statistic and calculate its observed value.
c. Compute the p-value and determine whether there is sufficient statistical evidence to reject the government's claim.
d. State your conclusion so that a nonstatistician could understand it.

8.27   In Exercise 8.26, if you use the government's estimate of 25%, how large a sample is needed to estimate the percent to within 1%, using a 95% confidence interval?

8.28   We wish to determine whether a greater proportion of students than local citizens favor the sale of beer in the county. We know that half of the local citizens oppose the sale of beer. In a random sample of 400 students, we find that 230 favor the sale of beer. Is there sufficient evidence to suggest that a greater proportion of students than local citizens favor the sale of beer in the county?
a. Set up null and alternative hypotheses to determine whether a greater proportion of students than local citizens favor the sale of beer in the county.
b. Give the form of the standardized test statistic and calculate its observed value.
c. Compute the p-value and determine whether there is sufficient statistical evidence to support the alternative hypothesis.
d. State your conclusion so that a nonstatistician could understand it.

8.29 In 1992, juries in the 75 largest U.S. counties disposed of 12,026 tort, contract, and real property cases (Bureau of Justice Statistics, *Civil Jury Cases and Verdicts in Large Counties*, Special Report NCJ-154346, July 1995). Of all those cases, 33% were classified as automobile accident suits. In Cook County, Illinois there were 600 jury cases. With a test of significance determine whether the percent of automobile accident suits in Cook County exceeds the national average if there were 210 automobile accident suits out of the 600 cases. State the null and alternative hypotheses and give the z statistic and the p-value. State your conclusion so that a person with no statistical training could understand it.

8.30 In Special Report NCJ-154346 (see Exercise 8.29), the Bureau of Justice Statistics reported that 1,362 medical malpractice suits were decided by a jury. In 413 of those cases, the plaintiff won. In this sample, only 30.3% of the cases were won by the plaintiff. Is the evidence from this sample strong enough to conclude that fewer than one-third of all medical malpractice suits are won by the plaintiff? State the null and alternative hypotheses and give the z statistic and the p-value. State your conclusion so that a person with no statistical training could understand it.

8.31 In a *USA Today*/CNN/Gallup nationwide telephone poll of 1,022 randomly selected adults, 470 said that they were pleased with the way the president was dealing with the economy (*USA Today*, September 9, 1994). A presidential consultant stated that the president has the support of the majority of all American adults. Is that claim valid?
   a. State the null and alternative hypotheses to test the consultant's claim.
   b. Calculate the standardized test statistic, z.
   c. A 10% level of significance is generally accepted as the dividing line between statistical significance and insignificance. What must the standard z-test statistic be in order for there to be insignificant evidence to reject the consultant's claim?
   d. Is the z statistic found in part **b** in the rejection region?
   e. Calculate the p-value. Is it less than a 10% level of significance?
   f. Should the null hypothesis be rejected?
   g. State your conclusion.

8.32 A manufacturer of automobiles purchases machine bolts from a supplier who claims that no more than 5% of his bolts are defective. From a random sample of 400 bolts, it is found that 28 are defective. Is there sufficient evidence to reject the supplier's claim?

8.33 A psychologist has developed a new aptitude test and believes that 80% of the public should score above 50 on the test. From a sample of 200 people, 164 scored above 50. Is there statistical evidence that the claim made by the psychologist is not valid? For the results to be significant at the 5% level of significance, how many out of 200 will have to score above 50 on the aptitude test?

8.34 A presidential aide said that at least half of the nation agreed with the U.S. invasion of Grenada in 1984. However, a Roper poll of 2,000 Americans found that only 34% supported the invasion. Did the 34% happen simply by chance, or is there statistical evidence that the aide was in error?

8.35 One-third of all ex-convicts return to jail within 3 years. One state is trying a new rehabilitation system on its prisoners. After the release of 200 prisoners who had participated in the new system, only 48 returned within 3 years. Is there statistical evidence that the new system is reducing the proportion of repeat offenders? What is the smallest level of significance at which these results will still be significant?

8.36 A doctor claims that at least 80% of his patients never have significant arthritic pain after a 6-month stay in his Caribbean treatment center. Out of a random sample of 75 of his former patients, 45 said that his treatment was ineffective and that they now have arthritic pain. Is there statistical evidence to refute the doctor's claim?

8.3 TESTING A POPULATION MEAN

In Section 8.1, we introduced hypothesis-testing problems that involve the mean of a population. The z-test was presented to test H_0: $\mu = \mu_0$. Recall, however, that the z-score

$$z = \frac{\bar{y} - \mu_0}{\sigma/\sqrt{n}}$$

depends on the population standard deviation σ. Therefore, to find the p-value, we need to know the numerical value of σ. In most applications, it is unreasonable to assume that we would know the population standard deviation when we are conducting inferences about the population mean.

Testing μ
When σ Is
Unknown: The
t-test

To develop the procedure for testing H_0: $\mu = \mu_0$ when the population standard deviation σ is unknown, we will, as in the development of the confidence interval for μ, modify the z-score by estimating σ with the sample standard deviation s.

As in the z-test, the evidence against H_0 is measured by the distance the test statistic $\bar{y}$ is from μ_0, the value of μ being tested. This distance is evaluated by calculating the observed value of the standardized test statistic

$$t = \frac{\bar{y} - \mu_0}{s/\sqrt{n}}$$

Notice that σ in the z-score has been replaced by s.

When the parent population is *normally distributed*, recall that t is distributed as Student's t with $n - 1$ degrees of freedom (see Section 7.3). Therefore, the p-value associated with the observed value of t is found in the t-table (Table B.3) with $n - 1$ degrees of freedom. Most computer programs print the p-value; however, if you must look it up in the table, you should be cautioned that the exact p-value cannot be found as with a z-table. With the t-table, you will only be able to find bounds for the p-value as described later in the summary and illustrated in Examples 8.8 and 8.9.

Like the t-interval, the t-test is robust against violations of the normality assumption; that is, modest departures from normality are permissible. We must, however, guard against severely skewed distributions or symmetric distributions with long tails. Again, normal probability plots and boxplots should be used to identify outliers and departures from normality. The same rules of thumb for using the t-interval procedure applies to using the t-test:

- If the sample size is less than 15, the population distribution should be close to normal.
- If the sample size is between 15 and 30, the t-interval can be used when the population distribution is reasonably symmetric and has no outliers.
- If the sample size is greater than 30, the t-interval procedure is permissible except when there are extreme outliers or strong skewness.

Sections 8.4 and 8.5 give alternatives to the t-test.

The *t*-test is summarized in the box:

Test of μ Based on $\bar{y}$ When σ Is Unknown

Application: Symmetric population distributions with tails that are not unusually long. If the sample size is small, then the parent distribution should not deviate substantially from normality.

Assumption: σ is unknown

Left-Tailed Test

$H_0: \mu \geq \mu_0$
$H_a: \mu < \mu_0$

Right-Tailed Test

$H_0: \mu \leq \mu_0$
$H_a: \mu > \mu_0$

Two-Tailed Test

$H_0: \mu = \mu_0$
$H_a: \mu \neq \mu_0$

Standardized test statistic:
$$t = \frac{\bar{y} - \mu_0}{s/\sqrt{n}}$$

p-value = $P(t \leq t_{\text{obs}})$

p-value = $P(t \geq t_{\text{obs}})$

p-value is same as for one-tailed and doubled to account for both tails.

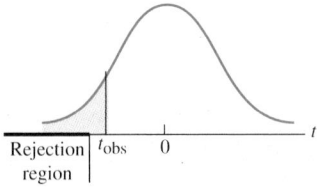
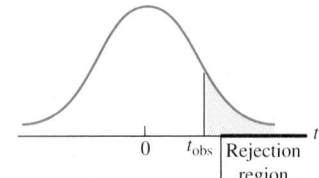
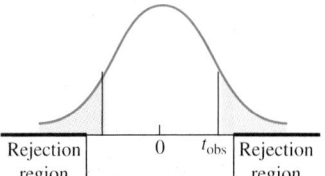

From the *t*-table (Table B.3), with $n - 1$ degrees of freedom, find the *p*-value most closely associated with t_{obs}. If t_{obs} falls between two table values, give the two associated probabilities as bounds for the *p*-value. The *p*-value for the two-tailed test is calculated as in a one-tailed test and then doubled to account for both tails.

If a level of significance α is specified, reject H_0 if *p*-value $< \alpha$.

EXAMPLE 8.8

To justify raising its rates, an insurance company claims that the mean medical expense for all middle-class families is at least $700 per year. A survey of 100 randomly selected middle-class families found that their mean medical expense for the year was $670 and the standard deviation was $140. Assuming that the tails of the distribution of medical expenses are not unusually long, is there evidence that the insurance company is misinformed?

SOLUTION

The insurance company claims that the mean medical expense is at least \$700. We doubt that, however, and believe that it is less than \$700. This becomes the research hypothesis. Denoting the mean medical expense for all middle-class families as μ, we can state the null and alternative hypotheses as

$$H_0: \mu \geq 700 \quad \text{versus} \quad H_a: \mu < 700$$

Because the underlying distribution of medical expenses does not have excessively long tails, the choice for a test statistic is $\bar{y}$. The standardized test statistic is

$$t = \frac{\bar{y} - 700}{s/\sqrt{n}}$$

and has 99 degrees of freedom ($n = 100$). The observed value of t becomes

$$t_{obs} = \frac{670 - 700}{140/\sqrt{100}} = -2.14$$

The degrees of freedom in the t-table jump from 80 to 100 because there is very little difference between the corresponding critical values. Because the degrees of freedom here are 99, we use the table values associated with 100 degrees of freedom. The absolute value of t_{obs}, 2.14, falls between table values 2.081 (upper-tail probability .02) and 2.364 (upper-tail probability .01). Therefore, we have

$$.01 < p\text{-value} < .02$$

which suggests that the null hypothesis should be rejected. There is significant evidence to indicate that the insurance company's claim is inaccurate. We might suggest that the insurance company check its sources.

Comparing the z-test and the t-test

Because of the large number of degrees of freedom (df $= 99$) in Example 8.8, there is very little difference between the z distribution and the t distribution. For example, If we had used the z distribution to find the p-value in the example, we would have found

$$p\text{-value} = P(z \leq -2.14) = 1.0 - .9838 = .0162$$

This is in complete agreement with the t-test, which found

$$.01 < p\text{-value} < .02$$

The point here is that there is very little difference between a z-test and a t-test when the sample size is relatively large—say, $n > 30$. When the sample size is small, however, care must be exercised in applying the t-test.

EXAMPLE **8.9**

A chemist measures the haptoglobin concentration (in grams per liter) in the blood serum taken from a random sample of eight healthy adults. These are the values:

Worksheet: Haptoglo.mtw

1.82   3.32   1.07   1.27   .49   3.79   .15   1.98

SOURCE: J. C. Miller and J. N. Miller, *Statistics for Analytical Chemistry*, 2nd ed. (New York: Halsted Press, 1988).

Is there statistical evidence that the mean haptoglobin concentration in adults is less than 2 grams per liter?

SOLUTION

Let μ denote the population mean haptoglobin concentration in adults. The research hypothesis is that the concentration is less than 2; therefore, we have

$$H_0: \mu \geq 2 \quad \text{versus} \quad H_a: \mu < 2$$

The normal probability plot of the data in Figure 8.11 does not rule out normality; therefore, we will assume that the parent distribution is normally distributed and use $\bar{y}$ as the test statistic. Because we are sampling from a normally distributed population, the standardized form of the test statistic is

$$t = \frac{\bar{y} - 2}{s/\sqrt{n}}$$

and has a t distribution with $n - 1 = 7$ degrees of freedom. The observed value is

$$t_{obs} = \frac{1.736 - 2}{1.283/\sqrt{8}} = -.58$$

FIGURE 8.11

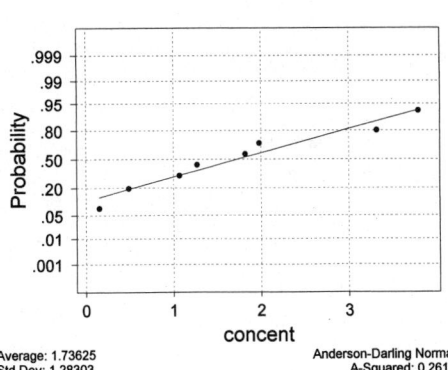

Normal Probability Plot

Average: 1.73625
Std Dev: 1.28303
N of data: 8

Anderson-Darling Normality Test
A-Squared: 0.261
p-value: 0.599

Thus, we have

$$p\text{-value} = P(t \leq -.58)$$

Going to the t-table with 7 degrees of freedom, we see that a tail probability of .10 (10%) corresponds to a t-score of -1.415 and the value of t_{obs} is only $-.58$. Consequently, we can say that

$$p\text{-value} > .10$$

which means that we do not have evidence to reject the null hypothesis. In conclusion, we can say that there is insufficient evidence that the mean haptoglobin concentration in adults is less than 2 grams per liter.

COMPUTER TIP

Just as in the one-sample confidence interval procedures, there are two alternatives for the one-sample test of hypothesis about a population mean. They are the one-sample z-test and the one-sample t-test. To execute either one, from the **Stat** menu select **Basic Statistics** and click on either **1-Sample z** or **1-Sample t**. When the next window appears, **Select** the **Variable** containing the data. Click on the **Test Mean** box and type in the value for the mean being tested. Next choose the **Alternative** depending on whether it is a two-tailed or one-tailed test. For the one-sample z-test, you must give a value for the population standard deviation, σ, in the **Sigma** box. Finally click on **OK**.

EXAMPLE 8.10

Psychologists wish to investigate the learning ability of schizophrenic people after they have taken a specified dose of a tranquilizer. Thirteen patients were given the drug, and 1 hour later they were given a standardized exam. Their scores are listed here:

Worksheet: Schizoph.mtw

15	20	30	27	24	22	22
17	21	25	23	27	25	

Generally patients score around 20 on the exam. Is there statistical evidence that taking the tranquilizer has affected their scores?

SOLUTION

Let μ denote the mean score on the standardized exam for all schizophrenics who could be administered the tranquilizer. The null and alternative hypotheses are:

$$H_0: \mu = 20 \quad \text{versus} \quad H_a: \mu \neq 20$$

The dotplot in Figure 8.12 shows that only three patients scored 20 or below. Is it conceivable that we would observe this pattern of variability if the population mean were indeed 20?

FIGURE 8.12

Character Dotplot

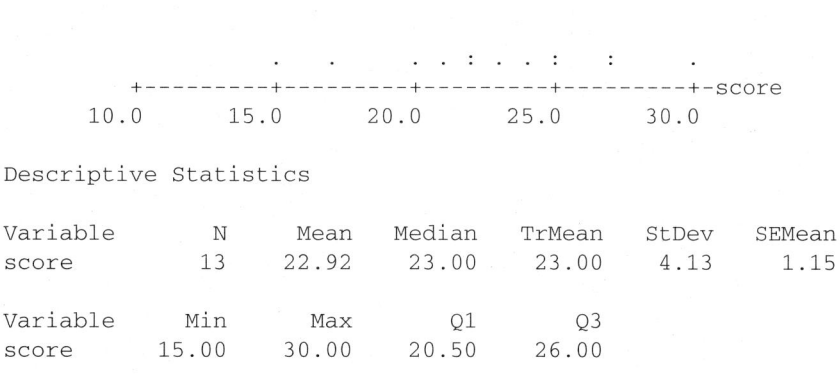

```
                        .     .    . :   . . :    :      .
          +---------+---------+---------+---------+---------+-score
        10.0      15.0      20.0      25.0      30.0
```

Descriptive Statistics

Variable	N	Mean	Median	TrMean	StDev	SEMean
score	13	22.92	23.00	23.00	4.13	1.15

Variable	Min	Max	Q1	Q3
score	15.00	30.00	20.50	26.00

From the descriptive statistics in Figure 8.12, we see that the sample mean is 22.92. Is the difference between $\bar{y} = 22.92$ and $\mu_0 = 20$ statistically significant? To answer this, we must first verify that conditions for the t-test are met. The boxplot and the normal probability plot of the data in Figure 8.13 reveal no outliers or significant departures from normality. There appears to be no unusual behavior in the data that would prohibit using a t-test.

FIGURE 8.13

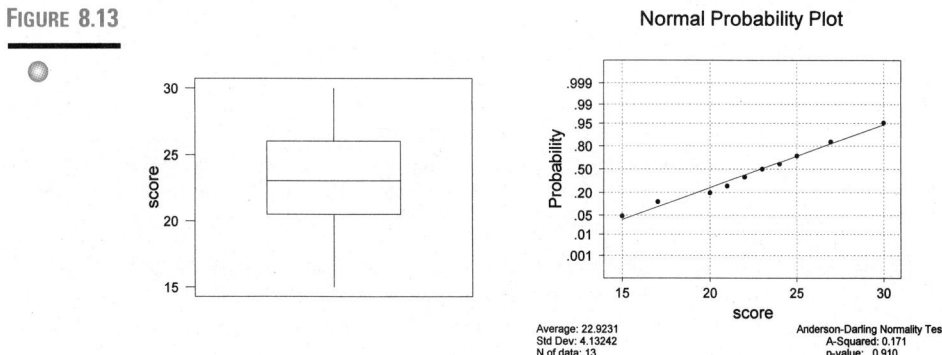

Normal Probability Plot

Average: 22.9231
Std Dev: 4.13242
N of data: 13

Anderson-Darling Normality Test
A-Squared: 0.171
p-value: 0.910

Figure 8.14 (page 510) gives the results of the t-test applied to the data in worksheet: **Schizoph.mtw**. The p-value $= .025$ is computed exactly by the computer instead of finding bounds as we did using the t-table in the previous example. In conclusion, there is significant evidence to indicate that the mean score is not 20 after the tranquilizer has been taken.

FIGURE 8.14   T-Test of the Mean

Test of mu = 20.00 vs mu not = 20.00

Variable	N	Mean	StDev	SE Mean	T	P-Value
score	13	22.92	4.13	1.15	2.55	0.025

EXERCISES 8.3

8.37   Identify each of the following as a left-, right-, or two-tailed test. For each, describe the characteristics that the parent distribution should have to conduct a t-test. Assuming the conditions are met, use the summary data to complete the test. Be sure to calculate the p-value and interpret the results.

    a. H_0: $\mu \geq 2.5$
       H_a: $\mu < 2.5$
       $n = 25, \bar{y} = 2.33, s = .67$
    b. H_0: $\mu = 50$
       H_a: $\mu \neq 50$
       $n = 70, \bar{y} = 51.6, s = 5.9$
    c. H_0: $\mu \leq 100$
       H_a: $\mu > 100$
       $n = 14, \bar{y} = 102.3, s = 2.45$

8.38   The Chamber of Commerce of a particular city claims that the mean carbon dioxide level of air pollution is no greater than 4.9 parts per million (ppm). A random sample of 16 readings resulted in $\bar{y} = 5.6$ ppm and $s = 2.1$ ppm. If we assume that the carbon dioxide level is normally distributed, is there sufficient evidence against the Chamber of Commerce's claim? State null and alternative hypotheses to evaluate the conjecture. Calculate the t statistic and its p-value and state your conclusion to the test. Why is it important to assume that the parent distribution is normal?

8.39   A car manufacturer claims that its cars use, on the average, no more than 5.5 gallons of gas for each 100 miles. A consumer group tests 40 of the cars and finds an average consumption of 5.65 gallons per 100 miles and a standard deviation of 1.52 gallons. Do these results cast doubt on the claim made by the car manufacturer? Is it necessary that the parent distribution be approximately normally distributed?

8.40   According to a recent report, after 5 years on the job, American workers get an average of 24 days of paid holidays and vacation leave each year. These are the numbers of days of paid holidays and vacation leave taken by a random sample of 35 workers in the American textile industry.

    Worksheet: Vacation.mtw

23	12	10	34	25	16	27	18	28
13	14	20	8	21	23	33	30	13
16	14	38	19	6	11	15	21	10
39	42	25	12	17	19	26	20	

    a. Construct a boxplot of the data. Are there any unusual observations in the data set?

  b. To test the population mean, should the test be based on $\bar{y}$, or can you think of a better test statistic?

  c. Test the null hypothesis that the mean vacation leave for textile workers is no different from the mean for all American workers.

8.41   Here are the daily price returns (in pence) of Abbey National shares between July 31 and October 8, 1991:

Worksheet: Abbey.mtw

296	296	300	302	300	304	303	299	293	294	294	293	295
287	288	297	305	307	307	304	303	304	304	309	309	309
307	306	304	300	296	301	298	295	295	293	292	297	
293	306	303	301	303	308	305	302	301	297	299	294	

SOURCE: D. Buckle, "Bayesian Inference for Stable Distributions," *Journal of the American Statistical Association* 90 (1995): 605–613.

  a. From the histogram and boxplot of the data is there any reason that a test of the population mean should be based on any statistic other than $\bar{y}$?

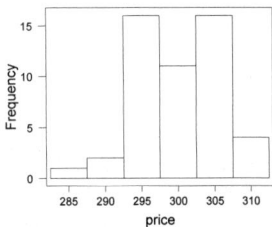 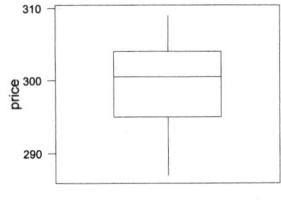

  b. Using these descriptive statistics determine with a test of significance whether the mean price differs from 300 pence.

Descriptive Statistics

Variable	N	Mean	Median	TrMean	StDev	SEMean
price	50	299.96	300.50	300.09	5.61	0.79

Variable	Min	Max	Q1	Q3
price	287.00	309.00	295.00	304.00

8.42   Water pH levels for 75 water samples in the Great Smoky Mountains were collected by Kaufman and colleagues (1988) as part of an Environmental Protection Agency (EPA) study. The data listed in worksheet: **Smokyph.mtw** also appear in Schmoyer (1994).

  a. From the stem and leaf plot of the water pH levels (at the top of page 512), do you detect any unusual behavior in the data?

  b. Would $\bar{y}$ be a good test statistic to use to test a hypothesis about the population mean?

  c. State the null and alternative hypotheses necessary for determining whether the mean pH level is 7.

  d. Test the hypothesis stated in part **c**. Be sure to calculate the t statistic and the p-value. Give your conclusion so that a nonstatistican could understand it.

```
Character Stem-and-Leaf Display

Stem-and-leaf of waterph    N  = 75
Leaf Unit = 0.10

     1      6 3
     4      6 555
    11      6 6677777
    32      6 88888888899999999999
   (19)     7 0000000000000001111
    24      7 22222223333
    13      7 444
    10      7 6
     9      7 899
     6      8 01
     4      8 223
     1      8 4
```

```
Descriptive Statistics
```

Variable	N	Mean	Median	TrMean	StDev	SEMean
waterph	75	7.1396	7.0300	7.1052	0.4394	0.0507

Variable	Min	Max	Q1	Q3
waterph	6.3800	8.4300	6.8700	7.2400

8.43   The average power consumption of 25 randomly selected families in a community for a given period is 125.6 kilowatt-hours and the standard deviation is 20.3 kilowatt-hours. If we assume that kilowatt usage is normally distributed, is there evidence that the mean usage for the whole community exceeds 120 kilowatt-hours? State null and alternative hypotheses to evaluate the conjecture. Calculate the t statistic and its p-value and state your conclusion to the test.

8.44   The average length of time required to complete a certain aptitude test is claimed to be 80 minutes. A sample of 25 students yielded an average of 86.5 minutes and a standard deviation of 15.4 minutes. If we assume normality of the parent distribution, is there evidence to support the claim? State null and alternative hypotheses to evaluate the conjecture. Calculate the t statistic and its p-value and state your conclusion to the test. Why is it important to assume that the parent distribution is normal?

8.45   Past experience indicates that scores on the first exam in a beginning history class are normally distributed with a mean of 72. This semester's class of 16 students had an average of 76 and a standard deviation of 14.4. Is this class "better than usual"? Evaluate this claim with a test of significance. Why is it important to assume that the scores in the past have been normally distributed?

8.46   To test the claim that the average home in a certain town is within 5.5 miles of the nearest fire department, an insurance company measured the distances from 25 randomly selected homes to the nearest fire department and found $\bar{y} = 5.8$ miles and $s = 2.4$ miles. Determine what the insurance company found out with a test of significance. What assumption did you make about the underlying parent distribution?

8.47 An instructor gave his class an examination that he knows from years of experience has a mean of 78. His present class of 20 students got an average score of 82 with a standard deviation of 7. Is he correct in assuming that this is a superior class? What assumption did you make about the underlying parent distribution?

8.48 A class of 50 eighth-graders took a standardized reading test. Their scores had a mean of 107.5 and a standard deviation of 10.5. The national mean score on the test is 100. Do we have sufficient statistical evidence to conclude that this class is superior in reading ability?

8.49 A study was undertaken to determine the number of trials necessary for a person to master a task under the influence of a drug. A sample of nine people had these results:

    6   8   14   9   10   11   7   5   12

  a. Is there evidence in the boxplot and normal probability plot to suggest that the parent distribution is not close to being normally distributed?

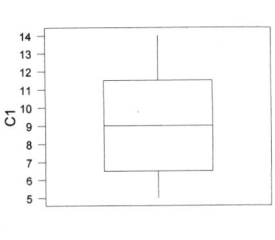
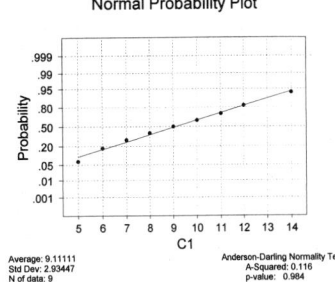

Normal Probability Plot

Average: 9.11111
Std Dev: 2.93447
N of data: 9

Anderson-Darling Normality Test
A-Squared: 0.116
p-value: 0.984

  b. Does the computer printout provide evidence against the claim that the mean number of trials is 8? State the hypotheses and give a conclusion.

```
T-Test of the Mean

Test of mu = 8.000 vs mu not = 8.000

Variable    N     Mean    StDev    SE Mean      T    P-Value
C1          9    9.111    2.934      0.978   1.14       0.29
```

8.50 An instructor believes that students who score below 80 on their first test most likely will fail developmental mathematics. In order to determine whether this is reasonable, she records the following scores on the first test for all students who failed developmental mathematics in the fall semester of 1995:

Worksheet: Devmath.mtw

84	88	96	87	65	98	41	92	78	70
93	77	39	73	62	74	51	88	100	89
100	79	69	74	69	84	49	65	77	48
61	86	68	90	68	76	67	40	84	77

SOURCE: Data provided by Dr. Anita Kitchens.

Is there statistical evidence that the mean test score for the students who failed developmental mathematics is less than 80?

8.51 To determine the flow characteristics of oil through a valve, the inlet oil temperature is measured in degrees Fahrenheit. Here is a sample of 12 readings:

Worksheet: Inletoil.mtw

93	99	97	99	94	91
93	90	89	92	90	93

We want to know whether there is evidence that the mean inlet oil temperature is significantly less than 98 degrees.

    a. Based on the normal probability plot, can we assume that the parent distribution of oil temperatures is normally distributed?

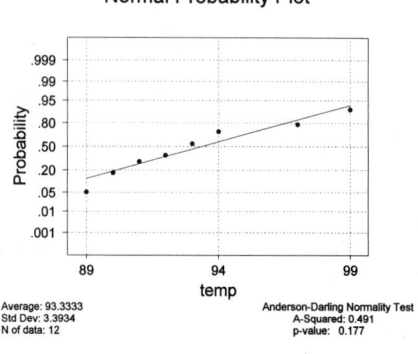

    b. What are the null and alternative hypotheses to evaluate the claim that the mean inlet oil temperature is significantly less than 98 degrees?

    c. Perform the test of significance suggested in part **b**. Be sure to give the standardized test statistic, its *p*-value, and a conclusion.

8.52 In Chesapeake Bay, complex changes in salinity are caused by the mixture of fresh water and sea water during the diurnal tidal cycle. The fresh water from the Chesapeake River floats across the denser brine in the bay, and during low tide it travels farther down the estuary. There is a counterflow, however, along the bottom that carries the dense marine water up the bay during the waning tide. Here are the surface salinity measurements (in parts per thousand) taken at station 11, offshore from Annapolis, Maryland, on July 3–4, 1927:

Worksheet: Chesapea.mtw

6.97	6.20	5.93	6.32	6.36	6.72	6.80	6.90
7.14	6.91	6.76	6.74	7.20	7.45	7.47	7.47

SOURCE: J. Davis, *Statistics and Data Analysis in Geology*, 2nd ed. (New York: Wiley, 1986).

From the normal probability plot at the top of page 515, is there evidence that the normality assumption is violated by the data? Is it appropriate to test the population mean using the *t*-test?

Normal Probability Plot

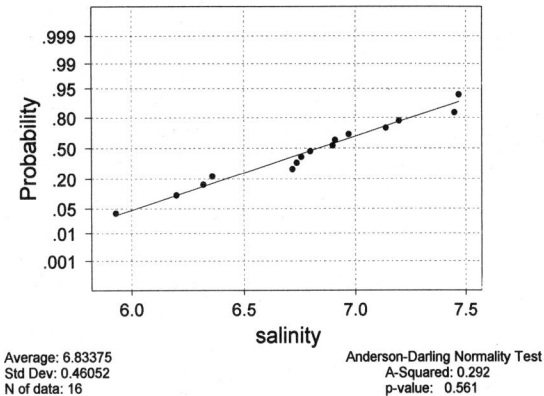

Average: 6.83375
Std Dev: 0.46052
N of data: 16

Anderson-Darling Normality Test
A-Squared: 0.292
p-value: 0.561

8.53   The computer printout is for a test of hypothesis about the mean surface salinity levels at station 11 offshore from Annapolis, Maryland (see Exercise 8.52).

```
T-Test of the Mean

Test of mu = 7.000 vs mu not = 7.000

Variable     N     Mean    StDev    SE Mean       T    P-Value
salinity    16    6.834    0.461      0.115    -1.44       0.17
```

    a. State the null and alternative hypotheses.
    b. Is it a right-tailed, left-tailed, or two-tailed test?
    c. Is there statistical evidence to reject the null hypothesis?
    d. State your conclusion for the test.

8.54   In the manufacture of integrated circuits (chips) used in computers, approximately 200 chips are typically processed as part of a wafer, which is a thin disk about 20 centimeters in diameter. The production of wafers (which can take months) is handled in lots of approximately 20 wafers. In one part of the process, a thin layer of silicon oxide is placed on the surface of a wafer. The thickness of the oxide layer is critical to the performance of the resulting chips. Two wafers are randomly selected from each of 30 lots. Four measurements of the thickness of the oxide layer on each wafer are then taken. The data set gives the averages of the eight measurements obtained from each set of two wafers:

Worksheet: Chipavg.mtw

962.50	1,047.50	966.25	998.75	951.25
927.50	1,096.25	1,076.25	1,052.50	992.50
1,035.00	1,058.75	1,033.75	1,050.00	986.25
1,001.25	1,101.25	966.25	1,001.25	1,003.75
1,086.25	1,067.50	1,026.25	1,058.75	1,030.00
1,035.00	1,038.75	967.50	865.00	1,006.25

SOURCE: E. Yashchin, "Likelihood Ratio Methods for Monitoring Parameters of a Nested Random Effect Model," *Journal of the American Statistical Association* 90 (1995): 729–738.

a. From the normal probability plot, is there evidence that the normality assumption is violated by the data? Is it appropriate to test the population mean using the *t*-test?

Normal Probability Plot

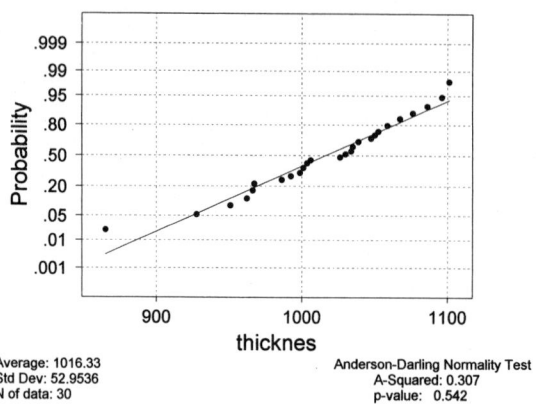

Average: 1016.33
Std Dev: 52.9536
N of data: 30

Anderson-Darling Normality Test
A-Squared: 0.307
p-value: 0.542

b. Is there evidence in the computer printout against the claim that the mean thickness is 1,000? State the hypotheses and give a conclusion.

```
T-Test of the Mean

Test of mu = 1000.00 vs mu not = 1000.00

Variable     N      Mean     StDev    SE Mean       T    P-Value
thicknes    30   1016.33     52.95       9.67    1.69       0.10
```

8.55   The charges for a coronary bypass at 17 hospitals in North Carolina are given in worksheet: **Bypass.mtw**. The normal probability plot in Exercise 6.61 of Section 6.4 strongly supports the normality of the data. Use these descriptive statistics to test the hypothesis that the mean charge for a coronary bypass in North Carolina is $35,000.

```
Descriptive Statistics

Variable          N      Mean    Median    TrMean     StDev    SEMean
charge           17     32406     32428     32501      3591       871

Variable        Min       Max        Q1        Q3
charge        24810     38578     29359     34919
```

8.56   These data are the survival times in weeks for 20 male rats that were exposed to a high level of radiation (see Exercise 2.56 in Section 2.3):

Worksheet: Rat.mtw

152	152	115	109	137	88	94	77	160	165
125	40	128	123	136	101	62	153	83	69

SOURCE: J. Lawless, *Statistical Models and Methods for Lifetime Data* (New York: Wiley, 1982).

a. Is there evidence in the boxplot and normal probability plot that the parent distribution is not close to being normally distributed?

Normal Probability Plot

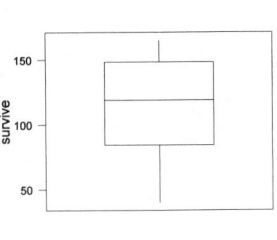 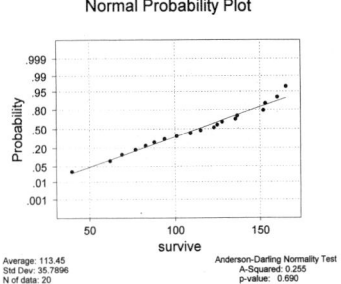

Average: 113.45
Std Dev: 35.7896
N of data: 20

Anderson-Darling Normality Test
A-Squared: 0.255
p-value: 0.690

b. From the computer printout from the *t*-test, state the null and alternative hypotheses.

```
T-Test of the Mean

Test of mu = 100.00 vs mu not = 100.00

Variable    N      Mean     StDev    SE Mean       T    P-Value
survive    20    113.45     35.79       8.00    1.68       0.11
```

c. What is the *t* statistic? What is its *p*-value?
d. Is there evidence to reject the null hypothesis?
e. State your conclusion to this test of significance.

8.57 The Eocene lake deposits of the Rocky Mountains consist of thinly laminated dolomitic oil shales hundreds of feet thick. It is generally accepted that the laminations are varves, or layered deposits caused by seasonal climatic changes in the lake basins. By measuring the thickness of these laminations, scientists record annual changes in the rate of deposition throughout the lake's history. These data are the thicknesses (in millimeters) of a varved section of the Green River oil shale deposit near the western shore of one of the major lakes:

Worksheet: Grnriv2.mtw

6.0	7.2	7.1	7.1	7.2	7.4	8.0	8.6	10.0	11.4	12.0
11.0	9.6	8.7	7.6	7.2	7.2	7.8	8.1	7.8	7.1	
7.2	7.1	7.0	7.0	7.7	8.6	9.0	12.0	13.7	14.0	
13.6	12.1	12.9	12.8	11.1	9.0	7.5	7.5	8.4	8.4	
7.9	7.0	6.7	6.8	7.3	7.3	7.2	8.1	9.8	11.0	
10.8	9.5	8.1	7.2	7.1	6.8	7.0	7.1	5.6	3.8	
3.4	4.2	4.8	4.5	3.6	3.0	2.8	4.1	6.8	8.1	
7.8	6.4	4.6	3.7	4.0	4.2	4.5	5.9	7.3	7.3	
6.7	6.0	5.8	5.7	6.5	8.2	10.2	12.3	13.2	13.2	
12.4	9.7	9.2	9.3	8.3	6.0	5.7	6.1	6.3	6.3	

SOURCE: J. Davis, *Statistics and Data Analysis in Geology*, 2nd ed. (New York: Wiley, 1986).

a. Does the stem and leaf plot on page 518 show any unusual characteristics in the data?

```
Character Stem-and-Leaf Display

Stem-and-leaf of thick N = 101
Leaf Unit = 0.10

    1      2 8
    6      3 04678
   14      4 01225568
   19      5 67789
   32      6 0001334577888
  (30)     7 000011111122222223333455678889
   39      8 011112344667
   27      9 00235678
   19     10 028
   16     11 0014
   12     12 0013489
    5     13 2267
    1     14 0
```

```
Descriptive Statistics

Variable       N     Mean    Median   TrMean    StDev    SEMean
thick        101    7.831     7.300    7.766    2.587    0.257

Variable      Min      Max       Q1       Q3
thick       2.800   14.000    6.350    9.100
```

b. Does a test of the population mean using $\bar{y}$ as the test statistic seem appropriate?

c. Using the descriptive statistics, perform a test of significance to determine whether the mean thickness of the varves is less than 8 millimeters. Be sure to state the null and alternative hypotheses and calculate the t statistic and the p-value. State your conclusion so that a geologist without much statistical training could understand it.

8.4 TESTING A POPULATION MEDIAN

In confidence interval estimation, if the parent population distribution is highly skewed, then the population median is a better indicator of the center of the distribution than is the population mean. In this section, we describe a procedure commonly known as the *sign test* for testing the population median. Recall that the population median is denoted by the symbol θ. Not unlike the population mean, the null and alternative hypotheses can take one of three forms:

Left-tailed—$H_0: \theta \geq \theta_0$ versus $H_a: \theta < \theta_0$

Right-tailed—$H_0: \theta \leq \theta_0$ versus $H_a: \theta > \theta_0$

Two-tailed—$H_0: \theta = \theta_0$ versus $H_a: \theta \neq \theta_0$

where θ_0 represents the value of θ that we wish to test.

**Small-Sample
Test of the
Population
Median**

Suppose we are interested in investigating the study habits of students in a beginning college math class. We suspect that the number of hours of study outside class each week is skewed right. Therefore, the parameter of concern is the median number of hours of study outside class each week. We believe that the median number of hours is more than 5, and we decide to test

$$H_0: \theta \le 5 \quad \text{versus} \quad H_a: \theta > 5$$

If the median is 5 hours, one would expect approximately half of the sample observations to be below 5 and half above 5. On the other hand, if the population median is more than 5 (the research hypothesis claim), one would expect an abundance of observations greater than 5. Thus, it seems reasonable to reject the null hypothesis if a "large" number of observations are more than 5. If the alternative had been

$$H_a: \theta < 5$$

that is, a left-tailed test, we would reject the null hypothesis if a very "small" number of observations were more than 5.

The test statistic is

$$T = \text{number of observations} > 5$$

Clearly, the range of values for T is

$$\{0, 1, 2, \ldots, n\}$$

where n is the sample size. Moreover, if the null hypothesis is true, then the proportion of successes (observations greater than 5) should be 50%. The sampling distribution of T is a binomial distribution with parameters n and $\pi = .5$.

Suppose that from a sample of 15 students, we find that 9 study more than 5 hours per week; that is, the observed value of T is $T_{obs} = 9$. Is this evidence sufficient to reject H_0 and conclude that the median exceeds 5 hours per week? The p-value is the probability of observing a value of the test statistic at least as large as T_{obs} when the null hypothesis is true—namely,

$$p\text{-value} = P(T \ge T_{obs}) = P(T \ge 9)$$

where T is binomial with $n = 15$ and $\pi = .5$.

Going to the binomial table (Table B.1), we find

$$p\text{-value} = P(T \ge 9)$$
$$= 1 - P(T \le 8)$$
$$= 1 - .696$$
$$= .304$$

Clearly, this *p*-value is insignificant, and therefore there is insufficient evidence to indicate that the true median is more than 5 hours per week. We see from the binomial table that it would have taken 12 or more observations (out of 15) greater than 5 to reject H_0 with a *p*-value of .014 + .003 = .017.

One problem that may occur when this test is conducted is that some observations may equal the value of the median being tested. Those observations that are "tied" with the median value are discarded, and the sample size *n* is reduced by that amount. For example, in the preceding example, if 2 of the 15 observations were equal to 5, then the sample size would be reduced to 13.

The complete small-sample ($n \leq 20$) test of the population median is summarized in the box:

Small-Sample Test of the Population Median

Application: Mainly skewed distributions but can be used to test the median of any continuous distribution

Assumption: Parent distribution is continuous and $n \leq 20$

Left-Tailed Test

$H_0: \theta \geq \theta_0$

$H_a: \theta < \theta_0$

Right-Tailed Test

$H_0: \theta \leq \theta_0$

$H_a: \theta > \theta_0$

Two-Tailed Test

$H_0: \theta = \theta_0$

$H_a: \theta \neq \theta_0$

Test statistic: T = number of observations $> \theta_0$

From the binomial probability table (Table B.1) with *n* and $\pi = .5$, we find

p-value = $P(T \leq T_{obs})$ *p*-value = $P(T \geq T_{obs})$ *p*-value = $2P(T \leq T_{obs})$
if $T_{obs} < n/2$

or *p*-value = $2P(T \geq T_{obs})$
if $T_{obs} > n/2$

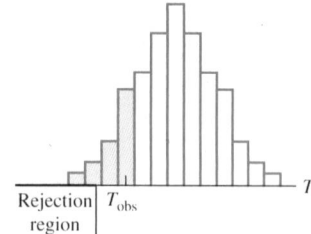

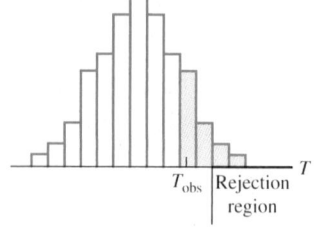

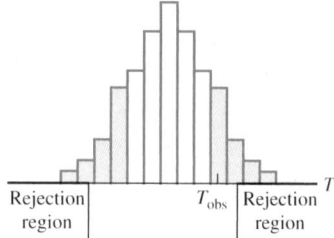

If a level of significance is specified, reject H_0 if *p*-value $< \alpha$.

EXAMPLE 8.11

A psychologist measures the threshold reaction times (in seconds) for persons subjected to emotional stress and obtains these results:

Worksheet: Reaction.mtw

14.3	13.7	15.4	14.7	12.4	13.1
9.2	14.2	14.4	15.8	11.3	15.0

Is there evidence that the median threshold reaction time is less than 15 seconds?

SOLUTION

The stem and leaf plot of the data in Figure 8.15 indicates that the distribution is skewed left, and therefore the median, θ, is the appropriate measure to investigate. We wish to obtain evidence that θ is less than 15 (the research hypothesis). The null and the alternative hypotheses are

$$H_0: \theta \geq 15 \quad \text{versus} \quad H_a: \theta < 15$$

FIGURE 8.15

⊙

Unit value = .10

```
 9 | 2
10 |
11 | 3
12 | 4
13 | 1  7
14 | 2  3  4  7
15 | 0  4  8
```

From the ordered stem and leaf plot, we find that the test statistic, T = number of observations > 15, has an observed value of

$$T_{obs} = 2$$

FIGURE 8.16

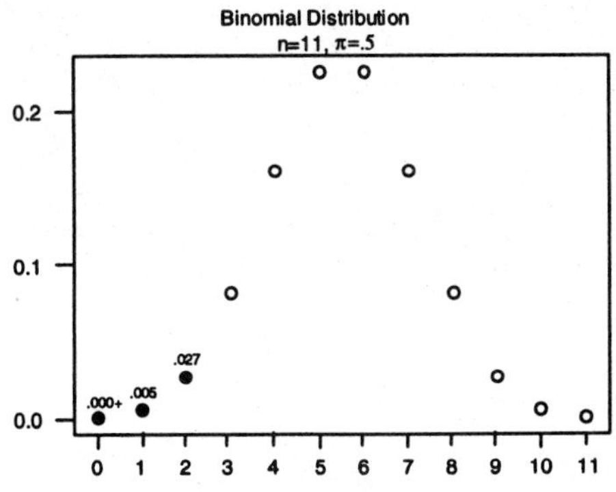

⊙

Binomial Distribution
n=11, π=.5

One observation is 15, the value being tested, so we must reduce the sample size by one. Going to the binomial table with $n = 11$ and $\pi = .5$, we find (see Figure 8.16)

$$p\text{-value} = P(T \leq 2) = .032^+$$

We have sufficient evidence to reject the null hypothesis and conclude that the median threshold reaction time is less than 15 seconds.

Large-Sample Test of the Population Median

For large samples, we can no longer use the binomial table, so we use a normal approximation of the binomial distribution. Recall from Chapter 4 that the mean of the binomial distribution is $n\pi$ and the standard deviation is $\sqrt{n\pi(1 - \pi)}$. Therefore, the standardized form of the test statistic T becomes

$$z = \frac{T - n\pi}{\sqrt{n\pi(1 - \pi)}}$$

It can be shown that the sampling distribution of z is approximately a standard normal distribution. Therefore, the p-value can be found in the normal table (Table B.2). For the median test, π is always equal to .5. Thus, the test statistic, z, simplifies to

$$z = \frac{T - n/2}{\sqrt{n/4}} = \frac{2T - n}{\sqrt{n}}$$

The accompanying boxes contain a summary of this test:

Large-Sample Test of the Population Median

Application: Mainly skewed distributions but can be used to test the median of any continuous distribution

Assumption: Parent distribution is continuous and $n > 20$.

Left-Tailed Test	Right-Tailed Test	Two-Tailed Test
$H_0: \theta \geq \theta_0$	$H_0: \theta \leq \theta_0$	$H_0: \theta = \theta_0$
$H_a: \theta < \theta_0$	$H_a: \theta > \theta_0$	$H_a: \theta \neq \theta_0$

Standardized test statistic: $z = \dfrac{2T - n}{\sqrt{n}}$

where T = number of observations $> \theta_0$.

From the z-table (Table B.2), we find

p-value $= P(z < z_{obs})$ p-value $= P(z > z_{obs})$ p-value $= 2P(z < z_{obs})$
                                                                        if $T_{obs} < n/2$

                                                    or p-value $= 2P(z > z_{obs})$
                                                        if $T_{obs} > n/2$

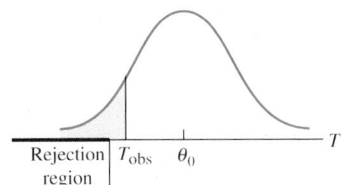 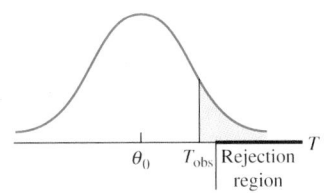 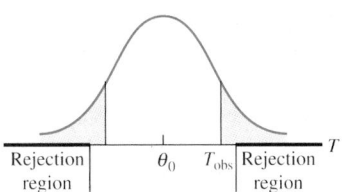

If a level of significance α is specified, reject H_0 if p-value $< \alpha$.

EXAMPLE 8.12

An experimental lethal drug is injected into mice, and the survival times (in seconds) are recorded. The researcher suspects that the median survival time is less than 45 seconds. From these data, see whether there is evidence to support her theory:

Worksheet: Lethal.mtw

32	37	44	41	65	27	35	29	54	42
38	57	40	24	39	43	82	52	74	25
34	25	47	34	24	38	91	30	35	31

SOLUTION

The stem and leaf plot and the boxplot of the data in Figure 8.17 (page 524) point out that the data are skewed right. Thus, the investigator will test the population median. The null and alternative hypotheses are

$$H_0: \theta \geq 45 \quad \text{versus} \quad H_a: \theta < 45$$

The sample is large, and consequently the test statistic is

$$z = \frac{2T - n}{\sqrt{n}}$$

The number of observations greater than 45 is 8, so the observed value of z is

$$z_{obs} = \frac{(2)(8) - 30}{\sqrt{30}} = -2.56$$

From the normal table (Table B.2) we have

$$p\text{-value} = P(z \leq -2.56) = .0052$$

FIGURE 8.17

```
Stem-and-leaf of survival   N  = 30
Leaf Unit = 1.0

       2     2 44
       6     2 5579
      11     3 01244
     (6)     3 557889
      13     4 01234
       8     4 7
       7     5 24
       5     5 7
       4     6
       4     6 5
       3     7 4
       2     7
       2     8 2
       1     8
       1     9 1

Character Boxplot
```

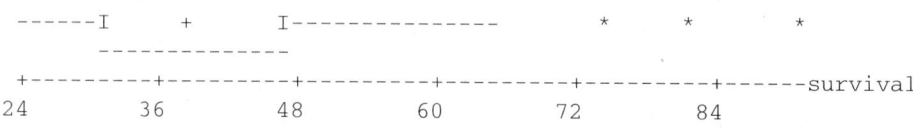

```
              ---------------
    ------I     +        I---------------            *       *         *
              ---------------
    +---------+---------+---------+---------+---------+------survival
    24        36        48        60        72        84
```

which is highly significant. The researcher has statistical evidence that the median survival time is less than 45 seconds.

The test procedures for the population median described in this section are intended for skewed distributions. They are appropriate, however, for testing the median of any shape of continuous distribution. If the median best describes the population characteristic of concern, then these procedures (large or small) should be used. In fact, if the sample size is small and the normality requirement for the t-test described in Section 8.3 is not met, then this test is recommended.

COMPUTER TIP

To perform the Minitab test for a population median, from the **Stat** menu choose **Nonparametrics** and click on **1-Sample Sign** test. When the Sign Test window appears, **Select** the variable to be tested, click on **Test median**, and type in the value of the median being tested in the null hypothesis. Choose an **Alternative** depending on whether it is a left-, right-, or two-tailed test. Then click on **OK.**

Figure 8.18 shows the results obtained by applying the sign test to the data in Example 8.11. Notice that the p-value is slightly different from the value we obtained in Example 8.11. This is because the binomial probabilities in Table B.1 have been rounded to three decimal places. If more significant decimal places are taken, then we would obtain the same p-value as the computer. Also, notice that the computer gives the number of observations that are below the value being tested and the number of observations equal to the value of the median being tested.

FIGURE 8.18

Sign Test for Median

Sign test of median = 15.00 versus   L.T.   15.00

	N	BELOW	EQUAL	ABOVE	P-VALUE	MEDIAN
time	12	9	1	2	0.0327	14.25

E X E R C I S E S   8 . 4

8.58   Test the hypotheses using the sample information. Be sure to find the p-value and interpret the results.
   a. $H_0: \theta \geq 50$
      $H_a: \theta < 50$
      $n = 14, T =$ number of observations > 50
         $= 5$
   b. $H_0: \theta \leq 3.5$
      $H_a: \theta > 3.5$
      $n = 8, T =$ number of observations > 3.5
         $= 7$
   c. $H_0: \theta = 100$
      $H_a: \theta \neq 100$
      $n = 18, T =$ number of observations > 100
         $= 4$

8.59   Test the hypotheses using the sample information. Be sure to find the p-value and interpret the results.
   a. $H_0: \theta \geq 400$
      $H_a: \theta < 400$
      $n = 50, T =$ number of observations > 400
         $= 14$
   b. $H_0: \theta \leq 12.5$
      $H_a: \theta > 12.5$
      $n = 100, T =$ number of observations > 12.5
         $= 56$

c. $H_0: \theta = 75$
   $H_a: \theta \neq 75$
   $n = 200$, $T = $ number of observations > 75
   $\qquad = 78$

8.60   A study of the length of long-distance phone calls for a small business firm found this random sample of times (in minutes):

Worksheet: Phone.mtw

12.8	3.5	2.9	9.4	8.7	3.5	4.8	7.7	5.9	6.2
2.8	4.7	1.7	2.6	1.9	6.5	2.8	3.1	7.2	15.8

The normal probability plot indicates a degree of skewness in the data, and therefore we test the population median.

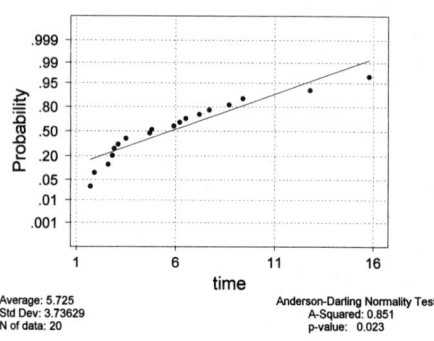

Normal Probability Plot

Average: 5.725
Std Dev: 3.73629
N of data: 20

Anderson-Darling Normality Test
A-Squared: 0.851
p-value: 0.023

```
Sign Test for Median

Sign test of median = 5.000 versus  G.T.  5.000

               N     BELOW   EQUAL   ABOVE   P-VALUE     MEDIAN
time          20        11       0       9    0.7483      4.750
```

Based on the computer printout:
   a. State the null and alternative hypotheses.
   b. Is it a one-tailed or a two-tailed test?
   c. What is the value of the test statistic, T?
   d. Is this a large-sample or a small-sample test?
   e. Why is the p-value so large?
   f. Is there evidence to reject the null hypothesis?
   g. State your conclusion to the test.

8.61   What does it cost to operate a state law enforcement agency? Listed are the operating expenditures (dollars) per resident per year for each of the states in 1993:

Worksheet: Statelaw.mtw

State	Cost	State	Cost	State	Cost
Alabama	5	Louisiana	4	Ohio	13
Alaska	4	Maine	2	Oklahoma	6
Arizona	8	Maryland	16	Oregon	7
Arkansas	3	Massachusetts	11	Pennsylvania	32
California	59	Michigan	24	Rhode Island	2
Colorado	4	Minnesota	4	South Carolina	3
Connecticut	9	Mississippi	*	South Dakota	1
Delaware	3	Missouri	8	Tennessee	8
Florida	11	Montana	1	Texas	33
Georgia	8	Nebraska	3	Utah	*
Hawaii	*	Nevada	3	Vermont	2
Idaho	1	New Hampshire	*	Virginia	11
Illinois	17	New Jersey	11	Washington	11
Indiana	7	New Mexico	1	West Virginia	3
Iowa	5	New York	23	Wisconsin	3
Kansas	3	North Carolina	9	Wyoming	1
Kentucky	8	North Dakota	1		

SOURCE: U.S. Bureau of Justice Statistics, *Law Enforcement Management and Administrative Statistics, 1993*, Report NCJ-148825, September 1995, p. 84.

a. Notice in the descriptive statistics that the mean cost is \$8.96 per resident and the median cost is \$5.50 per resident. Why is there such a difference between the two measures of the center of the distribution?

```
Descriptive Statistics

Variable     N     N*   Mean   Median   TrMean   StDev   SEMean
cost        46     4    8.96   5.50     7.57     10.70   1.58


Variable    Min    Max    Q1     Q3
cost        1.00   59.00  3.00   11.00
```

b. Construct a stem and leaf plot of the data and see whether it explains the difference between the mean and median costs of operating the agencies.
c. To conduct a test of hypothesis, should we test the population mean or the population median? Explain your answer.
d. From the computer printout, state the null and alternative hypotheses being tested.

```
Sign Test for Median

Sign test of median = 8.000 versus  L.T.  8.000


            N     N*    BELOW   EQUAL   ABOVE   P-VALUE   MEDIAN
cost        46    4     26      5       15      0.0586    5.500
```

e. Is there statistical evidence to reject the null hypothesis? Write a conclusion to this test of significance.

8.62 Double Nobel Prize winner Linus Pauling strongly advocated the use of vitamin C as a treatment for cancer. The survival times (in days) of terminal stomach cancer patients who were treated with vitamin C are given at the top of page 528.

Worksheet: Cancer.mtw column 1

| 124 | 42 | 25 | 45 | 412 | 51 | 1,112 |
| 46 | 103 | 876 | 146 | 340 | 396 | |

SOURCE: E. Cameron and L. Pauling, "Supplemental Ascorbate in the Supportive Treatment of Cancer," *Proceedings of the National Academy of Science USA*, 75 (1978): 4538–4542.

a. What do the boxplot and normal probability plot of the data tell about the underlying parent distribution?

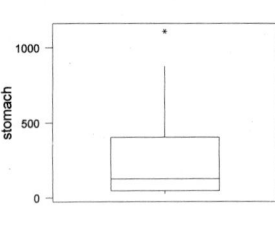

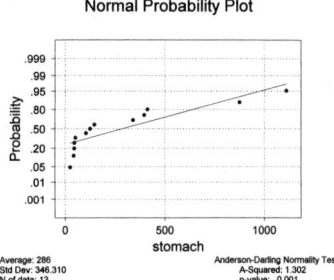

b. Does the computer output indicate that the median survival time exceeds 100 days? Explain by setting up the null and alternative hypotheses and performing a test of significance. Be sure to write a conclusion.

```
Sign Test for Median

Sign test of median = 100.0 versus  G.T.  100.0

              N   BELOW   EQUAL   ABOVE   P-VALUE    MEDIAN
stomach      13      5       0       8    0.2905     124.0
```

8.63 According to test theory, the median mental age for the 16 girls with scores listed below should be 100. Does the median mental age differ significantly from 100? Set up null and alternative hypotheses and perform a test of significance. Be sure to calculate the test statistic and its *p*-value and state your conclusion.

Worksheet: Mental.mtw

| 87 | 89 | 93 | 93 | 93 | 95 | 95 | 99 |
| 99 | 102 | 108 | 108 | 113 | 113 | 114 | 114 |

8.64 To assess the predictive validity of Klopfer's Prognostic Rating Scale (PRS) for people who have received behavior modification therapy, these data were obtained following psychotherapy:

Worksheet: Prognost.mtw

| 11.9 | 8.2 | 6.9 | 11.7 | 7.4 | 6.5 | 9.5 | 7.4 |
| 6.3 | 9.4 | 7.3 | 6.8 | 8.7 | 7.1 | 4.9 | |

SOURCE: C. Newmark et al., "Predictive Validity of the Rorschach Prognostic Rating Scale with Behavior Modification Techniques," *Journal of Clinical Psychology* 29 (1973): 246–248.

We want to know whether the mean or the median rating score differs significantly from a standard score of 9.

    a. Construct a stem and leaf plot and a boxplot of the data. Does it appear that the assumptions for the t-test are satisfied, or should we test the population median?

    b. State the hypotheses for the test.

    c. Perform the test and state your conclusion.

8.65 A retail merchant is considering opening a new department store that caters to college students in the community. One concern is the age of students in the freshmen class. She obtained the following ages from a random sample of 30 college freshmen:

Worksheet: Freshman.mtw

19	18	19	22	18	21	20	19	19	28
19	20	19	18	20	15	33	19	19	20
20	19	18	20	19	20	19	26	21	19

Test the hypothesis that the median age of all college freshmen at the university is 19.

8.66 A supplier of storm windows claims that the median leakage from a 50-miles-per-hour wind is not more than 12.5% (.125). A sample of nine windows yields these results:

Worksheet: Window.mtw

.13	.17	.13	.18	.14	.12	.11	.14	.20

Does the sample support the supplier's claim? Conduct a test of significance and state your conclusion so that the supplier will understand it.

8.67 It is reported that the median annual income for social workers with less than 5 years of experience is $27,500. A random sample of 25 social workers from North Carolina (with less than 5 years of experience) had these incomes:

Worksheet: Social.mtw

25,200	26,500	26,700	27,900	28,000	25,500	22,200	24,700	27,000
26,500	27,700	21,500	26,000	23,500	28,500	27,000	24,500	
27,600	28,100	26,400	27,900	25,900	27,600	26,100	27,400	

Perform a test of significance to determine whether there is evidence that the North Carolina social workers are underpaid.

8.68 Column 2 of worksheet: **Cancer.mtw** contains survival times of terminal bronchus cancer patients. Perform an analysis on these data similar to that given in Exercise 8.62. With a test of significance determine whether the median survival time for the bronchus patients exceeds 100 days.

8.69 The stem and leaf plot at the top of page 530 shows the varve thicknesses from a sequence through an Eocene lake deposit in the Rocky Mountains. The deposit, part of the Green River Formation, consists of thinly laminated dolomitic oil shales hundreds of feet thick. In Exercise 8.57 in Section 8.3, you analyzed a similar oil shale deposit that is only 10 miles from this deposit. In that exercise, you learned that the median varve thickness was 7.3 millimeters. Is there statistical evidence that the median varve thickness for this deposit exceeds 7.3 millimeters? Set up hypotheses, perform the test of significance, calculate the test statistic and p-value, and state your conclusion.

    Is the apparent skewness in the data also demonstrated in the boxplot and the normal probability plot? Explain what you see in the two graphs. Is the population median the correct parameter to investigate in this problem, or should we conduct a test of the population mean?

Worksheet: Greenriv.mtw

```
Character Stem-and-Leaf Display

Stem-and-leaf of thick      N  = 37
Leaf Unit = 0.10

     1      8 9
     8      9 0012244
    11      9 899
    17     10 001233
   (5)     10 67888
    15     11 0013
    11     11 7
    10     12 03
     8     12 6
     7     13 04
     5     13 5
     4     14 2
     3     14 6
     2     15 0
     1     15 6
```

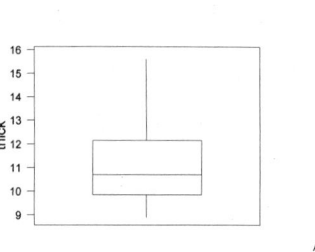

Normal Probability Plot

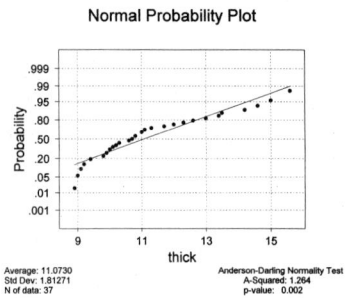

Average: 11.0730   Anderson-Darling Normality Test
Std Dev: 1.81271   A-Squared: 1.264
N of data: 37   p-value: 0.002

8.70   Exercise 8.42 of Section 8.3 looked at the pH levels of water samples collected in the Great Smoky
Mountains. You performed a test of hypothesis on the population mean using $\bar{y}$ as the test statistic.
Do the boxplot and the normal probability plot of the pH levels support the decision to test the
population mean, or would a test of the population median be more appropriate? Explain why.

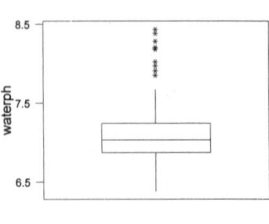

Normal Probability Plot

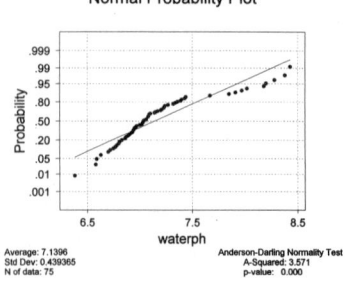

Average: 7.1396   Anderson-Darling Normality Test
Std Dev: 0.439365   A-Squared: 3.571
N of data: 75   p-value: 0.000

8.71   These are the results of a test of the median pH levels of the water samples collected in the Great Smoky Mountains (worksheet: **Smokyph.mtw**):

```
Sign Test for Median

Sign test of median = 7.000 versus  N.E.  7.000

              N     BELOW   EQUAL   ABOVE   P-VALUE    MEDIAN
waterph      75        32       1      42    0.2955     7.030
```

    a. State the null and alternative hypotheses.
    b. Is the test a one-tailed or a two-tailed test?
    c. Compute the value of the standardized test statistic.
    d. What is the p-value? Should the null hypothesis be rejected?
    e. State your conclusion.

8.72   Of critical concern in AIDS research is the "incubation" time of the HIV virus—that is, the time from HIV infection to the clinical manifestation of full-blown AIDS. In column 1 of worksheet: **Aids.mtw**, the incubation times (duration in months) for 295 patients who were thought to be infected with HIV by a blood transfusion are presented (Kalbfleisch and Lawless, 1989).
    a. Examine the boxplot and the normal probability plot and comment on the general shape of the distribution of the incubation time.
    b. Should the investigator draw inferences about the mean or the median incubation time?

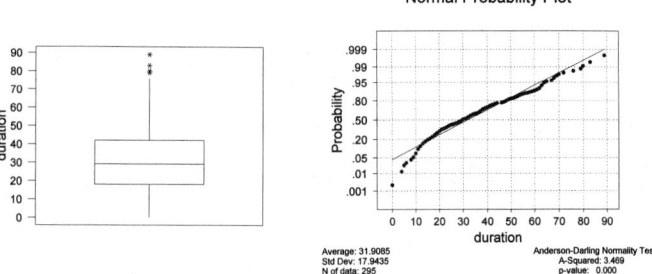

These data are analyzed further in Computer Exercises 8.121 in Section 8.6.

8.5 A ROBUST, LARGE-SAMPLE TEST OF A POPULATION MEAN (OPTIONAL)

We now consider the case where the underlying parent population is assumed to be symmetric with long tails. Under this assumption, a test of μ based on a trimmed mean performs better than the test based on $\bar{y}$ presented in Section 8.3. For long-tailed distributions, we present the test of μ based on $\bar{y}_T$.

As before, the null and alternative hypotheses can take one of three forms:

Left-tailed—$H_0: \mu \geq \mu_0$ versus $H_a: \mu < \mu_0$

Right-tailed—$H_0: \mu \leq \mu_0$ versus $H_a: \mu > \mu_0$

Two-tailed —-$H_0: \mu = \mu_0$ versus $H_a: \mu \neq \mu_0$

where μ_0 represents the value of μ that we wish to test.

Section 7.5 explained that for a sufficiently large sample size, the sampling distribution of $\bar{y}_T$ is approximately normal with a mean μ and a standard error of

$$SE(\bar{y}_T) = \frac{s_T}{\sqrt{k}}$$

where s_T is the standard deviation of the trimmed sample based on a Winsorized sample and k is the size of the trimmed sample. Therefore, the standardized test statistic is

$$z = \frac{\bar{y}_T - \mu_0}{s_T/\sqrt{k}}$$

and we get the p-value from the observed value of z, z_{obs}, by inserting the observed values of $\bar{y}_T$ and s_T in the preceding formula for z.

The complete test is summarized in the boxes:

Large-Sample Test of μ Based on $\bar{y}_T$

Application: Symmetric distribution with long tails

Assumption: $n > 30$

Left-Tailed Test	Right-Tailed Test	Two-Tailed Test
$H_0: \mu \geq \mu_0$	$H_0: \mu \leq \mu_0$	$H_0: \mu = \mu_0$
$H_a: \mu < \mu_0$	$H_a: \mu > \mu_0$	$H_a: \mu \neq \mu_0$

Standardized test statistic: $z = \dfrac{\bar{y}_T - \mu_0}{s_T/\sqrt{k}}$

where s_T is the standard deviation of the trimmed sample based on the Winsorized sample and k is the size of the trimmed sample.

From the z-table (Table B.2), we find

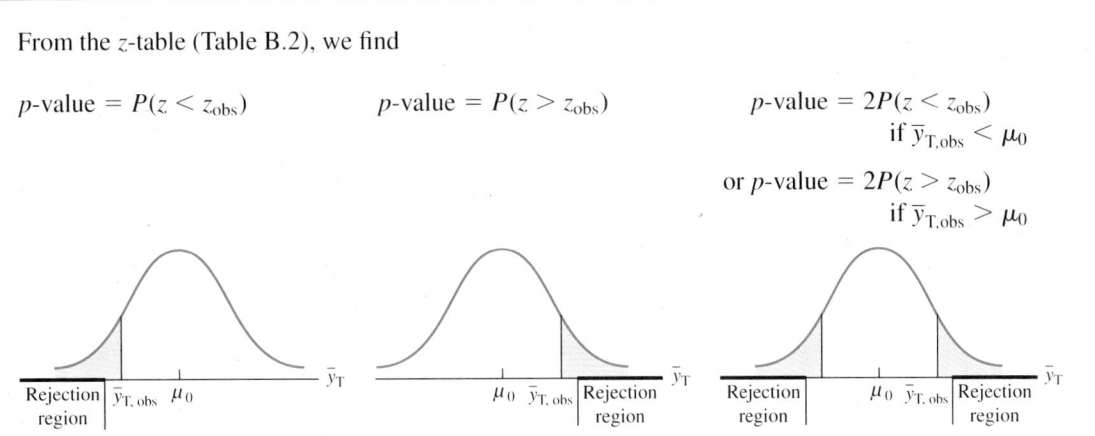

$$p\text{-value} = P(z < z_{obs})$$

$$p\text{-value} = P(z > z_{obs})$$

$$p\text{-value} = 2P(z < z_{obs})$$
$$\text{if } \bar{y}_{T, obs} < \mu_0$$

$$\text{or } p\text{-value} = 2P(z > z_{obs})$$
$$\text{if } \bar{y}_{T, obs} > \mu_0$$

If a level of significance α is specified, reject H_0 if p-value $< \alpha$.

EXAMPLE 8.13

A recruiter for a university marketing department suggests that the mean starting salary for master's level marketing majors is at least $25,000. Here is a sample of 65 starting salaries reported to the nearest $100:

Worksheet: Market.mtw

26,000	25,100	25,300	24,000	22,900	23,400	27,000	27,500
22,800	21,500	14,600	25,500	24,700	23,000	24,000	26,500
23,700	35,000	25,100	25,000	24,900	22,000	24,500	
20,000	15,500	24,400	23,500	20,000	28,000	23,900	
25,000	23,000	26,400	26,800	25,200	26,500	36,000	
28,500	25,000	22,500	21,800	18,000	26,000	24,000	
26,800	27,000	28,500	21,500	20,000	24,100	25,500	
24,800	23,500	22,000	27,500	28,000	25,500	24,500	
24,900	21,000	22,700	24,000	25,900	27,000	24,200	

Is there evidence to deny the claim?

SOLUTION

If we denote the mean starting salary as μ, the null and alternative hypotheses are

$$H_0: \mu \geq 25,000 \quad \text{versus} \quad H_a: \mu < 25,000$$

which is a left-tailed test.

An ordered stem and leaf plot and a boxplot of the data are given in Figure 8.19 (page 534). They indicate that the data are symmetric with long tails. Therefore, the test statistic is $\bar{y}_T$, and the standardized form is

$$z = \frac{\bar{y}_T - 25,000}{s_T / \sqrt{k}}$$

FIGURE 8.19

Low	146, 155, 180

```
20 | 0 0 0
21 | 0 5 5 8
22 | 0 0 5 7 8 9
23 | 0 0 4 5 5 7 9
24 | 0 0 0 0 1 2 4 5 5 7 8 9 9
25 | 0 0 0 1 1 2 3 5 5 5 9
26 | 0 0 4 5 5 8 8
27 | 0 0 0 5 5
28 | 0 0 5 5                    (observations × 100)
```

High	350, 360

Character Boxplot

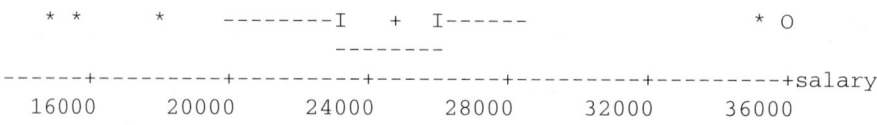

```
                         --------
    *  *        *     --------I   +   I------                * O
                         --------
------+---------+---------+---------+---------+---------+salary
   16000     20000     24000     28000     32000     36000
```

To find the *p*-value, we must find the observed value of *z*, which means we need $\bar{y}_T$ and s_T. A 10% trimming would trim seven observations from each end, which seems unnecessary; therefore, we will trim only 5% or four observations from each end. Figure 8.20 is the stem and leaf plot of the Winsorized sample. Using a calculator, we find

$$\bar{y}_{T.05} = 24{,}505.26$$

$$s_T = s_W \sqrt{\frac{n-1}{k-1}} = (2{,}301.29)\sqrt{\frac{64}{56}} = 2{,}460.18$$

Thus, the observed value of *z* is

$$z_{obs} = \frac{24{,}505.26 - 25{,}000}{2{,}460.18/\sqrt{57}} = -1.52$$

and so

$$p\text{-value} = P(z \le -1.52) = .0643$$

FIGURE 8.20

```
20 | 0 0 0 0 0 0
21 | 0 5 5 8
22 | 0 0 5 7 8 9
23 | 0 0 4 5 5 7 9
24 | 0 0 0 0 1 2 4 5 5 7 8 9 9
25 | 0 0 0 1 1 2 3 5 5 5 9 9
26 | 0 0 4 5 5 8 8
27 | 0 0 0 5 5
28 | 0 0 0 0 0 0                (observations × 100)
```

This p-value indicates that the difference between the test statistic $\bar{y}_{T.05}$ and $\mu_0 = 25{,}000$ is moderately significant. On the basis of a random sample of 65 starting salaries for new marketing graduates, there is mildly significant evidence that the starting salaries for all marketing graduates is less than \$25,000. There is some indication that the university recruiter's claim is in error.

From the computer printout in Figure 8.21 for the data in Example 8.13, the ordinary t-test (which assumes that the parent population is nearly normally distributed) does not detect a significant difference (p-value $= .12$). When the parent distribution is symmetric with long tails, the robust procedure based on a trimmed mean may find significant differences that might otherwise go undetected with the ordinary t-test.

FIGURE 8.21

```
T-Test of the Mean

Test of mu = 25000 vs mu L.T. 25000

Variable     N      Mean    StDev   SE Mean      T    P-Value
salary      65     24506     3366       418   -1.18     0.12
```

EXERCISES 8.5

8.73   A sample of 120 observations yielded $\bar{y}_{T.10} = 19.4$ and $s_T = 4.3$. With $\bar{y}_T$ as the test statistic, use this information to test the hypotheses:

$$H_0: \mu \geq 20 \quad \text{versus} \quad H_a: \mu < 20$$

Be sure to calculate the p-value and interpret the results.

8.74   A sample of 50 observations yielded $\bar{y}_{T.10} = 57.8$ and $s_T = 6.4$. With $\bar{y}_T$ as the test statistic, use this information to test the hypotheses:

$$H_0: \mu \geq 60 \quad \text{versus} \quad H_a: \mu < 60$$

Be sure to calculate the p-value and interpret the results.

8.75   A sample of 100 students from a large high school took a college entrance exam that has a mean score of 450. After an examination of the data, it appeared that a trimmed mean might better represent the data than the ordinary sample mean. The 10% trimmed mean and trimmed standard deviation turned out to be 474.6 and 127.3, respectively. With a test of significance, determine whether the trimmed average is unusually high? Be sure to state your conclusion so that those who have not had a course in statistics could understand it.

8.76   A psychology test is given to 36 students in a beginning psychology class. From these scores, we want to know whether there is evidence to suggest that the mean test score for all beginning psychology students is anything other than 50.

Worksheet: Psychol.mtw

45	36	48	55	32	43	35	57	62
42	58	46	21	42	58	84	30	6
45	37	49	54	59	7	33	41	47
91	50	44	41	36	48	56	52	39

a. Based on the boxplot of the data, how would you classify the shape of the distribution?

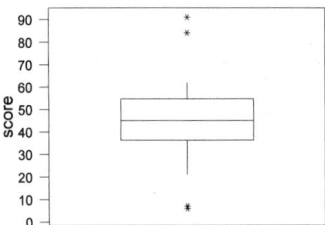

b. In light of your answer to part **a**, should the test of significance be based on $\bar{y}$ or $\bar{y}_T$?

c. Complete the test you recommended in part **b**. Be sure to give the p-value and state your conclusion.

8.77 The mean cost of an evening meal at a typical restaurant in a small city is claimed to be $8.50. Suppose a random sample of 49 tickets had a symmetric distribution with long tails and had a 10% trimmed mean of $9.23 and a trimmed standard deviation of $1.69. Is there statistical evidence to refute the claim about the mean cost of an evening meal?

8.78 A survey of 100 parents of first- and second-grade children revealed that the number of hours per week their children watch television has a distribution that is symmetric with long tails. The data have a 5% trimmed mean of 25.8 hours and a trimmed standard deviation of 4.4 hours. Determine whether there is statistical evidence to conclude that μ exceeds 25 hours.

8.79 A city official claims that the average city tax is very close to $120 per year. From a random sample of 100 residents, we found that the distribution is symmetric with long tails. A 20% trimmed mean was required to remove outliers. The trimmed mean was $134.60 and the trimmed standard deviation was $37.40. Does this information cast doubt on the official's claim?

8.80 The distribution of the daily wages in a particular industry is generally symmetric with long tails and has a mean of $55. To determine whether a particular company in this industry is paying inferior wages, a random sample of 36 workers was selected and their wages recorded. The 10% trimmed mean was $52 and the trimmed standard deviation was $12.50. Can we conclude that the company is paying inferior wages?

8.81 A manufacturer of television sets claims that the average life of its picture tubes is at least 10 years. A sample of 100 picture tubes revealed a long-tailed distribution with a 20% trimmed mean of 9.6 years and a trimmed standard deviation of 2.6 years. Is there evidence to suggest that the manufacturer is in error?

8.82 A researcher claims that the average self-criticism score on the Tennessee Self Concept Scale is at least 30 for all gifted high school students at one school. A sample of 20 students had these scores:

Worksheet: Tenness.mtw

29.8	30.4	27.5	29.8	31.0	30.2	29.5	29.0	27.0	35.8
25.4	34.2	30.7	31.5	28.2	28.9	24.2	32.4	29.2	28.5

Notice in the histogram and boxplot that the parent distribution may be a long-tailed distribution.

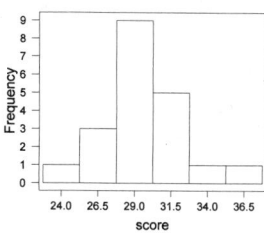
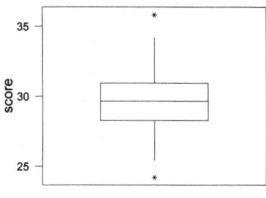

a. Do the long tails create a problem with the *t*-test in the computer printout? Is there evidence to refute the researcher's claim? Be sure to state the null and alternative hypotheses and give a conclusion.

```
T-Test of the Mean

Test of mu = 30.000 vs mu < 30.000

Variable    N      Mean    StDev    SE Mean       T    P-Value
C1         20    29.660    2.689     0.601    -0.57       0.29
```

b. The researcher proceeded to find a 5% trimmed mean and a trimmed standard deviation:

$$\bar{y}_{T.05} = 29.622 \quad s_T = 2.528$$

Using these results, test the above claim using $\bar{y}_T$ as the test statistic. Do the results differ from those of the standard *t*-test?

c. Because of the sample size, do you think that a test of the population median might be better? Remember that a test of the median does not require the parent distribution to be normally distributed. State the hypotheses and give a conclusion based on this printout. Do these results differ from the results in parts **a** and **b**?

```
Sign Test for Median

Sign test of median = 30.00 versus  L.T. 30.00

                N    BELOW  EQUAL  ABOVE   P-VALUE    MEDIAN
    score      20      12      0      8    0.2517     29.65
```

8.83 The reading scores of the 30 fifth-grade students in Example 2.25 were shown to have a symmetric distribution with long tails. The data are recorded in column 1 of worksheet: **Reading.mtw**. The 10% trimmed sample is in column 2, and the Winsorized sample is in column 3. From column 2 calculate the 10% trimmed mean, and from column 3 calculate the Winsorized standard deviation. With these summary statistics, test the hypothesis that the mean reading score is 95 using $\bar{y}_{T.10}$ as your test statistic.

8.84   The parallax of the sun is the angle subtended by the earth's radius, as if viewed from the surface of the sun. Knowing this angle and the radius of the earth, one can determine the distance between the earth and the sun. Astronomers in the 18th century were concerned with the absolute dimensions of the solar system and were eager to determine the parallax of the sun. Astronomer Edmund Halley (1656–1742) suggested that the parallax of the sun could be determined by observing a "transit of Venus," which is the time it takes the planet Venus to travel across the face of the sun, as viewed from the Earth. Transits of Venus are quite rare. James Short (1763) used the one in 1761 to make several determinations of the parallax of the sun. His measurements are stored in worksheet: **Short.mtw**. Columns C1 through C8 in the worksheet are Short's calculations from different pairs of observations of the transit. Column C9 gives all 158 measurements made by Short, and C10 indicates the sample number.*

   a. From the histogram of Short's 158 measurements, do you observe any unusual observations?

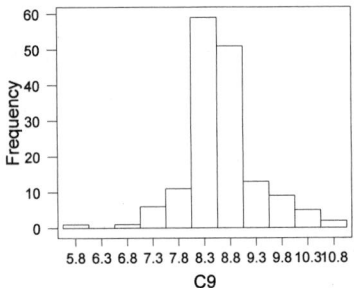

   b. Would you classify the distribution as being normally distributed or symmetric with long tails?
   c. What parameter would you suggest as a measure of the "true" parallax of the sun?

8.85   Here are a boxplot and a normal probability plot of Short's 158 measurements of the parallax of the sun (see Exercise 8.84). Notice the S shape in the normal probability plot and the outliers in the boxplot.

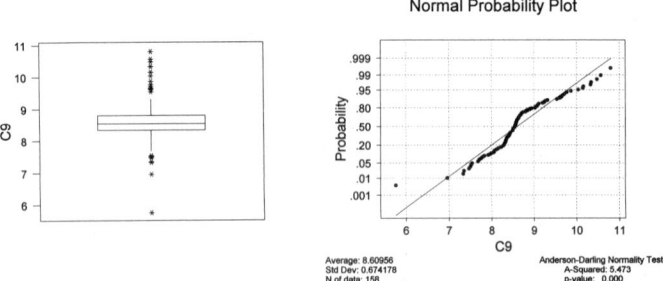

*SOURCE: S. Stigler, "Do Robust Estimators Work with Real Data?" *The Annals of Statistics* 5 (1977): 1055–1078; and J. Short, "Second Paper Concerning the Parallax of the Sun Determined from the Observations of the Late Transit of Venus," *Philos. Trans. Roy. Soc. London*: 53 (1763): 300–345.

a. Based on the boxplot and the normal probability plot, would you classify the distribution as being close to normal or symmetric with long tails? Does this agree with your analysis in Exercise 8.84?

b. What estimator of the "true" parallax of the sun do you recommend, $\bar{y}$ or $\bar{y}_T$?

8.86   The "true" parallax of the sun has been determined to be 8.798 seconds of a degree (see Exercise 8.84). Short's 158 measurements are stored in column 9 of worksheet **Short.mtw**. The data are sorted from smallest to largest in column 11. Column 12 contains the data that remain after trimming 10%, and column 13 contains the Winsorized sample. Averaging C12, we find $\bar{y}_{T.10} = 8.573$, and from the Winsorized sample, we find $s_W = .4561$ and $s_T = .5111$. Using the summary information and a test based on the 10% trimmed mean, test the hypothesis that Short's measurements were consistent with the "true" parallax of the sun.

8.6   SUMMARY AND REVIEW

KEY CONCEPTS

✓   The *null hypothesis* is the statement being tested in a *test of significance*. The *alternative hypothesis*, what is believed by the researcher to be true, is also called the *research hypothesis*.

✓   A *Type I error* is rejecting a true null hypothesis. A *Type II error* is failing to reject a false null hypothesis. The probability of a Type I error is denoted by α and is called the *level of significance*. The probability of a Type II error is denoted by β.

✓   A *test statistic* is the statistic that is used to test the null hypothesis. The null hypothesis will be rejected if the test statistic falls in the rejection region.

✓   A *right-tailed test* is a test of hypothesis where the rejection region is on the right tail of the sampling distribution of the test statistic. A *left-tailed test* is a test where the rejection region is on the left tail of the sampling distribution of the test statistic. A *two-tailed test* is one where the rejection region is on both tails of the sampling distribution of the test statistic.

✓   The *p-value* is the probability of observing a value of the test statistic at least as extreme as that given by the observed data under the assumption that the null hypothesis is true. The test statistic will fall in the rejection region if and only if the *p*-value is smaller than the level of significance. In that case, the null hypothesis is rejected.

✓   These are the steps to follow in testing hypotheses:

1. Formulate the null and alternative hypotheses.
2. Decide on an appropriate test statistic.
3. Determine the sampling distribution of the test statistic.
4. Collect the data and calculate the *p*-value.
5. Interpret the results.

✓   The test of a population mean can be based on either $\bar{y}$ or $\bar{y}_T$. The test is based on $\bar{y}$ if the tails of the parent distribution are not excessively long. It is based on $\bar{y}_T$ if the parent distribution is symmetric with long tails.

✓   If the parent distribution is skewed, the inference should be concerned with the population median and the test based on the sample median.

✓   The test of μ based on $\bar{y}$ is either a z-test or a t-test depending on whether we know the population standard deviation.

✓   The test of a population proportion is a z-test and is similar to the test of the population mean based on $\bar{y}$.

LEARNING GOALS   Having completed this chapter, you should be able to:

1. Formulate null and alternative hypotheses. *Section 8.1*
2. Interpret Type I and Type II errors. *Section 8.1*
3. Identify a right-tailed, a left-tailed, and a two-tailed alternative. *Section 8.1*
4. Test a population mean with a z-test based on $\bar{y}$. *Section 8.1*
5. Determine which test statistic to use in a testing procedure. *Sections 8.1–8.5*
6. Understand the concept of and the calculation of the p-value. *Sections 8.1–8.5*
7. Test a population proportion with a large sample. *Section 8.2*
8. Test a population mean with a t-test based on $\bar{y}$. *Section 8.3*
9. Test a population median with a large or small sample. *Section 8.4*
10. Test a population mean with a large sample based on $\bar{y}_T$. *Section 8.5*

QUESTIONS FOR REVIEW   Use the following problems to test your skills:
Multiple Choice (Exercises 8.87–8.93)

8.87   In a test of significance,
  a. the null hypothesis is believed to be true and we are trying to find evidence to support it.
  b. the alternative hypothesis is believed to be true and we are trying to find evidence to support it.
  c. Neither of the above

8.88   If the null hypothesis is rejected, then
  a. only a Type I error is possible.
  b. only a Type II error is possible.
  c. both Type I and Type II errors are possible.
  d. neither a Type I nor a Type II error is possible.

8.89   If the probability of a Type I error is α, then the probability of a Type II error is
  a. also α b. $1 - \alpha$ c. 0   d. unknown

8.90   The null hypothesis should be rejected if
  a. the p-value is smaller than the level of significance.
  b. the p-value is greater than the level of significance.
  c. the p-value is greater than 90%.
  d. the p-value is greater than 10%.

8.91   In testing $H_0: \mu = 50$ versus $H_a: \mu > 50$ with $n = 100$, $\bar{y} = 52.4$, which value of s leads to the rejection of H_0?
  a. 200   b. 20
  c. 2   d. None of the above

8.92   If the parent distribution is normal, then the choice of estimator for the population mean is
  a. $\bar{y}$ b. $\bar{y}_T$ c. M d. s

8.93 In a test of significance,
  a. it is possible to prove the alternative hypothesis.
  b. it is possible to find evidence to support the alternative.
  c. it is not possible to find evidence to support the alternative.
  d. None of the above

8.94 Why is the alternative hypothesis also called the research hypothesis?

8.95 A study showed that more than 70% of the women who undergo breast biopsies to remove benign lumps face no unusual risk of later developing breast cancer. Formulate null and alternative hypotheses to test the validity of this claim.

8.96 The pH factor measures the acidity or alkalinity of a substance. A soil pH of 5.5 is recommended for raising Christmas trees. Formulate the null and alternative hypotheses to test that the soil pH is at the desired level.

8.97 A sample of 42 observations yielded $\bar{y} = 642$ and $s = 135$. Use this information to test these hypotheses, with $\bar{y}$ being the test statistic: $H_0: \mu = 600$ versus $H_a: \mu \neq 600$. Be sure to calculate the p-value and interpret the results.

8.98 A sample of 40 observations yielded $\bar{y}_{T.10} = 655.4$ and $s_T = 241.7$. With $\bar{y}_T$ as the test statistic, use this information to test these hypotheses: $H_0: \mu = 700$ versus $H_a: \mu \neq 700$. Be sure to calculate the p-value and interpret the results.

8.99 The county agent believes that the farm land in his county needs more than 2,000 pounds of lime per acre. Out of 60 random soil samples, it was found that the average required lime per acre was 2,080 pounds and the standard deviation was 240 pounds. Set up the null and alternative hypotheses to determine whether the data support the agent's claim. Test the hypotheses by determining the value of the test statistic and its p-value and then state your conclusion.

8.100 The mean self-concept score on a standardized test is 15. A group of 24 politicians were given the exam and had an average score of 17.8 with a standard deviation of 6.2. Generally scores on standardized tests are close to normally distributed. This is a reasonable assumption in this case. With a test of significance, determine whether the politicians scored significantly above the norm.

8.101 It is believed that the percent of convicted felons who have a history of juvenile delinquency is 70%. Is there evidence to contradict this claim if out of 200 convicted felons, we find that 154 have a history of juvenile delinquency?

8.102 Do these data on STEP science test scores for a class of ability-grouped students provide sufficient evidence to conclude that they fall significantly below the national median of 80? First verify that the data are close to normal with a normal probability plot.

Worksheet: Step.mtw

58	60	82	80	67	70
65	73	75	77	82	68

8.103 In 1989, 52% of all Ph.D. degrees awarded by U.S. colleges went to foreign students. Of the 32 Ph.D.s awarded by the university system of a particular state, 14 were awarded to U.S. students. Was this an unusually low percentage awarded to U.S. students? Evaluate with a test of significance.

8.104 A manufacturer claims that the average lifespan of its washing machines is at least 4 years. It is assumed that the distribution of the lifespans is approximately normal. A random sample of 32 of these machines had an average lifespan of 3.6 years and a standard deviation of 1.5 years. Is there statistical evidence to dispute the manufacturer's claim?

8.105 A psychologist wished to evaluate a new technique that he has developed to improve rote memorization. Subjects are asked to memorize 50 word phrases using his technique. He thinks his technique is successful if the subjects average more than 40 correct phrases. How should he set up the null and alternative hypotheses?

8.106   Interpret the Type I and II errors for Exercise 8.105.

8.107   The city government claims that the average residence tax is no higher than $800 per year. We wish to test this claim; therefore, we randomly select 45 property owners and determine the average residence tax to be $848. Can we reject the claim made by city government if it is known that the distribution of residence taxes is near normal with a standard deviation of $200?

8.108   According to the norms published for a certain intelligence test, the average score for college freshmen is expected to be 65 points and the standard deviation is 10 points. A sample of 81 freshmen at State U. averaged 68.2 points on the test. Is there sufficient evidence to say that State U. freshmen are more intelligent, as measured by this test, than the average college freshman?

8.109   To study the effect of birth control pills on exercise capacity, a physiologist measures the maximum oxygen uptake during a treadmill session. It is known that the average maximum oxygen uptake for women not on the pill is 36 milliliters per kilogram of body weight. A random sample of 36 women on the pill had an average maximum oxygen uptake of 33.6 ml/kg of body weight with a standard deviation of 4.8 ml/kg. Is there statistical evidence that women on the pill have a significantly smaller maximum oxygen uptake than women not on the pill?

8.110   A group of nine slow-ability-grouped children was given a standardized exam that has a national mean score of 50. Their mean was 42 and the standard deviation was 15. Are these children really slow-ability children as measured by this test? What assumption must be made about the distribution of the scores on the standardized exam?

8.111   The height of adults in a certain town has a mean of 65.42 inches. A sample of 144 adults living in a depressed section of town is found to have a mean height of 64.82 inches and a standard deviation of 2.32 inches. Does this indicate that the adult height of residents in the depressed area is significantly less than that of all residents of the town?

8.112   A utility company claims that its customers' heating bills average no more than $250 during the winter period of December, January, and February. A consumer group sampled 100 accounts for the 3-month period and found an average cost of $263 and a standard deviation of $72. Is there statistical evidence to cast doubt on the utility company's claim?

8.113   A chemistry student found a way to manufacture imitation diamonds. From the process, he determines that he can make a profit if the produced stones have a median weight of more than .5 carat. From his sample he found these weights:

Worksheet: Carat.mtw

.46   .61   .52   .48   .57   .54

Should he go into the diamond-making business or stay in school?

8.114   A manufacturer of string has established from several years of experience that the breaking strength of his string has a mean of 15.9 pounds and a standard deviation of 2.4 pounds. A change is made in the manufacturing process, after which a sample of 64 pieces is taken, with a mean breaking strength of 16.2 pounds. Assuming breaking strength is normally distributed and the new process has the same standard deviation as the old, can we say that the average breaking strength of the string from the new process is greater than the average from the old process? Is the normality assumption really required in this exercise? Explain.

8.115   A spokesperson for a statewide organization for raising the legal drinking age claims that at least 70% of the state population thinks that the legal drinking age should be raised from 18 to at least 20. A random sample of 200 people were asked whether they approve of raising the drinking age to at least 20. Of the 200, 132 favored raising the limit. Is there statistical evidence to doubt the spokesperson's claim?

8.116   A factory makes a certain computer part that, according to specifications, must have a mean length of 1.5 centimeters. In a random sample of 16 parts from a shipment, the average length was found to be 1.56 centimeters and the standard deviation was .09 centimeter. Should this shipment be rejected?

8.117   Suppose you own a large racing stable and are contemplating the purchase of a 2-year-old stallion. Based on past experience, you think a horse of this age should be able to run the mile in less than 98.5 seconds. You race the horse and obtain these times (in seconds):

Worksheet: Stable.mtw

| 104.6 | 98.8 | 101.4 | 98.2 | 99.7 | 102.5 | 103.6 | 98.7 | 101.5 |

You decide that you will reject the horse if you feel that his lifetime median time on the mile will exceed 98.5 seconds. Based on the data, should you reject the horse? Three years later it is determined that his lifetime median time is 100 seconds. Did you make an error? If so, what type?

8.118   Suppose a preadmission algebra exam has a standardized mean of 200 and a standard deviation of 50. A sample of 100 students from a large high school take the college admission exam. The average score turns out to be 215. Is this unusual? Explain with a test of significance.

8.119   A standardized psychology exam has a mean score of 70. A research psychologist wished to see whether a counseling process has an effect on performance on the exam. She administered the exam to 18 volunteers who had participated in the counseling process, and they got these scores:

Worksheet: Counsel.mtw

| 68 | 71 | 75 | 65 | 61 | 70 | 70 | 64 | 71 |
| 73 | 62 | 78 | 70 | 69 | 76 | 67 | 69 | 72 |

From the stem and leaf plot, boxplot, and normal probability plot of the data, assess the assumptions required for the t-test.

```
6 | 1
6 | 2
6 | 4 5
6 | 7
6 | 8 9 9
7 | 0 0 0 1 1
7 | 2 3
7 | 5
7 | 6
7 | 8
```

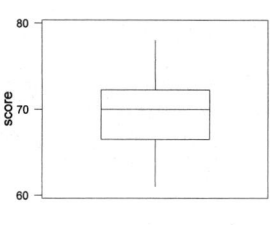

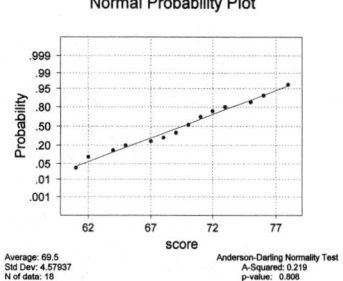

Use the summary statistics to determine whether there is evidence that the counseling process reduces one's score on the exam.

```
Descriptive Statistics

Variable        N      Mean    Median    TrMean    StDev    SEMean
score          18     69.50     70.00     69.50     4.58      1.08


Variable      Min       Max       Q1        Q3
score       61.00     78.00     66.50     72.25
```

COMPUTER **8.120**
EXERCISES

The table gives a 100-year history (1770–1869) of worldwide seismic activity based on the annual incidence of severe earthquakes. Examine these data to determine whether the mean severity exceeds 100.

Worksheet: Earthqk.mtw

Year	Severity	Year	Severity	Year	Severity	Year	Severity
1770	66	1795	78	1820	90	1845	86
1771	62	1796	110	1821	86	1846	127
1772	66	1797	79	1822	119	1847	201
1773	197	1798	85	1823	82	1848	76
1774	63	1799	113	1824	79	1849	64
1775	0	1800	59	1825	111	1850	31
1776	121	1801	86	1826	60	1851	·138
1777	0	1802	199	1827	118	1852	163
1778	113	1803	53	1828	206	1853	98
1779	27	1804	81	1829	122	1854	70
1780	107	1805	81	1830	134	1855	155
1781	50	1806	156	1831	131	1856	97
1782	122	1807	27	1832	84	1857	82
1783	127	1808	81	1833	100	1858	90
1784	152	1809	107	1834	99	1859	122
1785	216	1810	152	1835	99	1860	70
1786	171	1811	99	1836	69	1861	96
1787	70	1812	177	1837	67	1862	111
1788	141	1813	48	1838	26	1863	42
1789	69	1814	70	1839	106	1864	97
1790	160	1815	158	1840	108	1865	91
1791	92	1816	22	1841	155	1866	64
1792	70	1817	43	1842	40	1867	81
1793	46	1818	102	1843	75	1868	162
1794	96	1819	111	1844	99	1869	137

SOURCE: M. H. Quenouille, *Associated Measurements* (London: Butterworth, 1952), p. 279.

a. Construct a histogram of the severity measurements. What is the general shape of the histogram?
b. Construct a boxplot and a normal probability plot of the data. Are they consistent with the shape given in part **a**?
c. Do you think it is more important to study the mean severity or the median severity?
d. Using $\bar{y}$ as your test statistic, test the hypothesis that the mean severity over this time exceeds 100. Be sure to state the hypotheses, give the *p*-value, and form a conclusion.

8.121 A study of AIDS cases in which patients were thought to have been infected with HIV by a blood transfusion was conducted by the Centers of Disease Control in Atlanta, Georgia (Kalbfleisch and Lawless, 1989). The durations of the incubation period (in months) for a sample of 295 patients are given in column 1 of worksheet: **Aids.mtw**.

    a. Construct a histogram of the duration times. What is the general shape of the histogram?

    b. Construct a boxplot and a normal probability plot of the data. Are they consistent with the shape given in part **a**?

    c. Do you think it is more important to study the mean incubation period or the median incubation period?

    d. Using $\bar{y}$ as your test statistic, test the hypothesis that the mean incubation period exceeds 30 months. Be sure to state the hypotheses, give the p-value, and form a conclusion.

    e. Perform a test of significance to determine whether the median incubation time exceeds 2 years.

    f. Which test of significance, part **d** or **e**, do you recommend as the appropriate analysis of the data?

8.122 According to a recent report, after 5 years on the job, American workers get an average of 24 days of paid holidays and vacation leave per year. Here are the numbers of days of paid holidays and vacation leave taken by a random sample of 35 workers in the American textile industry:

Worksheet: Vacation.mtw

23	12	10	34	25	16	27	18	28
13	14	20	8	21	23	33	30	13
50	14	38	19	6	11	15	21	10
39	42	25	12	17	49	26	20	

    a. Construct a histogram of the data. What is the general shape of the histogram?

    b. Construct a boxplot and a normal probability plot of the data. Are they consistent with the shape given in part **a**?

    c. In Exercise 8.40 we completed a test of the population mean based on $\bar{y}$. It was determined that there is a significant difference between the mean vacation leave for textile workers and the mean of all American workers. Judging from your answers to parts **a** and **b**, would a test of the population median be more appropriate?

    d. Test the hypothesis that the median number of days of vacation and leave in the textile industry is 24 days. Be sure to state the hypotheses, give the p-value, and form a conclusion. Compare these results to those found in Exercise 8.40.

8.123 As a follow-up on Example 8.10, a psychologist examines the amount of learning exhibited by patients with schizophrenia after they take a specified dose of a tranquilizer. One hour after taking the drug, 17 patients were given a second exam on which schizophrenics normally score an average of 22. Their scores are listed here:

Worksheet: Schizop2.mtw

36	29	30	32	37	15	34	23	32
5	24	10	25	34	13	30	33	

    a. Construct a histogram of the data. What is the general shape of the histogram?

    b. Construct a boxplot and a normal probability plot of the data. Are they consistent with the shape given in part **a**?

    c. For a typical or representative measure, should the psychologist investigate the mean exam score or the median exam score?

    d. Based on your answer to part **c**, formulate the null and alternative hypotheses to determine whether the tranquilizer significantly improved the amount of learning exhibited by all schizophrenics.

e. Test the hypothesis stated in part **d**. Give the test statistic and its *p*-value and state your conclusion.

8.124   In most states, police departments have a certain percentage of unmarked police cars. These are the percentages of marked cars in 65 police departments in Florida:

Worksheet: Marked.mtw

61	74	49	65	62	62	73	71	37	54
79	64	54	75	92	63	74	63	55	58
56	78	63	74	65	52	53	55	57	
79	56	69	51	54	56	74	51	68	
61	61	60	64	58	56	58	51	72	
51	63	62	62	50	58	60	51	50	
68	73	56	60	52	41	51	75	52	

SOURCE: Bureau of Justice Statistics, *Law Enforcement Management and Administrative Statistics, 1993*, Report NCJ-148825, September 1995, pp. 147–148.

a. Construct a histogram of the data. What is the general shape of the histogram?
b. Construct a boxplot and a normal probability plot of the data. Are they consistent with the shape given in part **a**? Are there any outliers in the data?
c. A criminal justice expert is interested in whether or not the mean for Florida exceeds 60%. Formulate the null and alternative hypotheses for evaluating the question.
d. Based on the results found in part **b**, should the test statistic be $\bar{y}$ or $\bar{y}_T$?
e. Test the hypothesis stated in part **c**. Give the test statistic and its *p*-value and state your conclusion.

8.125   STATISTICAL INSIGHT REVISITED   The Statistical Insight given at the beginning of this chapter was about the EPA gas mileage estimates for a Honda Civic. The EPA says that a Civic should average at least 49 miles per gallon in city driving. Based on the sample of 35 different occasions in worksheet: **Honda.mtw**, we want to know whether there is statistical evidence to deny the EPA claim.

a. Construct a histogram of the data. What is the general shape of the histogram?
b. Construct a boxplot and a normal probability plot of the data. Are they consistent with the shape given in part **a**? Are there any outliers in the data?
c. Formulate the null and alternative hypotheses for evaluating the EPA prediction.
d. Based on the results found in part **b**, should the test statistic be $\bar{y}$, $\bar{y}_T$, or M?
e. Test the hypothesis stated in part **c**. Give the test statistic and its *p*-value and state your conclusion.

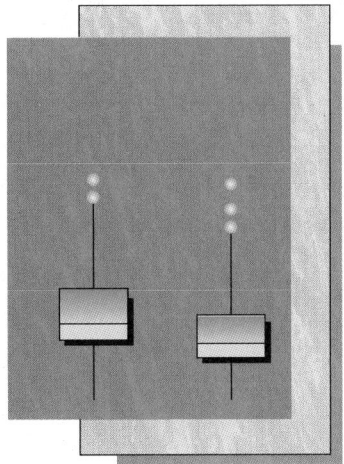

9

INFERENCES ABOUT THE DIFFERENCE BETWEEN TWO PARAMETERS

In the preceding two chapters, we studied inference procedures for a single population. Now we investigate problems that involve two populations. Samples from each are used to make inferences about two population parameters. The inferences from the two samples may take the form of an interval estimate or a test of hypothesis.

Tests that compare two population means are among the most widely used statistical procedures. In addition to comparing two population means, we study the problem of comparing two population proportions. You will see that a simple extension of the concepts developed for the single population can enormously expand the realm of statistical analysis.

CONTENTS

STATISTICAL INSIGHT

ARE VITAMINS HELPFUL IN WARDING OFF CANCER, HEART DISEASE, AND THE RAVAGES OF AGING?

In the early 1970s, Nobel Prize winner Linus Pauling was ridiculed for his claim that large doses of vitamin C could cure everything from the common cold to cancer. A Canadian study, however, suggests that large doses of vitamins C and E appear to reduce the risk of cataracts by at least 50%. Other studies have linked increased doses of vitamin C with reduced risks of cancer and heart disease.

In other studies, beta carotene, which is turned into vitamin A by the body, has appeared to reduce the risks of heart disease, stroke, and certain cancers. Doctors at Harvard Medical School studied 22,000 male physicians over a 10-year period and discovered that men with a history of cardiac disease who were given beta carotene supplements of 50 milligrams every other day suffered half as many heart attacks, strokes, and deaths as those who took placebo pills. A University of Arizona Cancer Center study found that daily beta carotene pills taken for 3 to 6 months dramatically reduced precancerous mouth lesions in 70% of patients.

All evidence is not positive, however. A Mayo Clinic study found that 51 terminally ill cancer patients given massive doses of vitamin C for up to 1 year fared no better than 49 similar patients who were given useless placebo pills. About equal percentages of both groups died within 1 year: 49% of the vitamin C group and 47% of the placebo group. And 96 of the total 100 patients, all of whom had advanced colorectal cancer, experienced a worsening of the disease during the study.

The common theme in each of these studies is the attempt to compare an experimental group with a control group. In some studies we compare mean measurements, and in others we compare percentages. In the Mayo Clinic study, for example, 49% of the terminally ill cancer patients in the vitamin C group, as compared with 47% in the placebo group, died within 1 year. This is a comparison of Bernoulli proportions. Here the data speak for themselves; the vitamin C group did not fare any better than the placebo group, and thus there is no need for a formal analysis of the data. Other studies, however, require a formal test of significance.

SOURCE: "The New Scoop on Vitamins," *Time,* April 6, 1992.

9.1 INTRODUCTION

A widely used method of statistical analysis is the comparison of two populations. Often we wish to compare a new procedure with an old established procedure, or compare one product with another, or one treatment with another treatment. The key word is *compare*; we wish to compare the characteristics of one population with those of another population. To compare the two populations, random samples from each are chosen and then used to make inferences about the parameters of the two populations.

Often the observations in the two samples come from an experimental design, as discussed in Chapter 1. Recall that in a comparative experiment, we randomly divide the subjects into two equivalent groups, with one group receiving the treatment and the other group serving as the control. Also it is possible to compare two treatments by having each serve as the control for the other.

The procedure used to choose and assign subjects to the groups is the experimental design. We consider two types of designs that are appropriate for a comparative experiment. In the first design, the sample assigned to group 1 is selected independently of the sample assigned to group 2. In the second design, the two samples are related because the subjects assigned to the two groups are matched or paired in some way. The subjects are chosen in pairs or are matched in such a way that, without a treatment, the subjects in each pair will perform almost identically. One member of each pair receives treatment 1, and the other member receives treatment 2. Any observed differences in the responses are attributed to the treatments. As pointed out in Chapter 1, this design, often referred to as the matched pairs design, is a special case of the randomized block design.

EXAMPLE 9.1

A study published in the *Journal of the American Medical Association* by Swedish doctors at Sahlgren's Hospital in Goteborg suggests that nicotine-laced gum helps smokers to stop smoking. The study showed that 29 of 106 smokers who chewed nicotine gum while trying to quit remained smoke free for 1 year. In a similar group, 16 of 100 smokers who chewed regular gum remained smoke free for 1 year. Both groups participated in counseling groups to help them quit.

In this example, we assume that the two groups of smokers are independent of each other, and we are interested in comparing two Bernoulli proportions. For instance, we can let π_1 denote the proportion of smokers who use nicotine-laced gum and remain smoke free for 1 year, and let π_2 denote the proportion of smokers who chew regular gum and remain smoke free for 1 year. The problem then is to compare π_1 and π_2. The quantity $\pi_1 - \pi_2$ represents the difference between the proportions of the two groups. An inference problem can involve finding a confidence interval estimate of $\pi_1 - \pi_2$ or a test of hypothesis about the difference between the two proportions. If the research hypothesis is that the nicotine-laced gum is beneficial in helping smokers to remain smoke free, then the null and alternative hypotheses are

$$H_0: \pi_1 = \pi_2 \quad \text{versus} \quad H_a: \pi_1 > \pi_2$$

The analysis for this type of problem is explained in Section 9.2.

EXAMPLE 9.2

A team of physicians would like to compare the recovery times of two different techniques after a certain operation. Subjects are to be randomly and independently assigned to the two techniques, and after the operation their recovery times are to be recorded. If μ_1 denotes the mean recovery time for one technique and μ_2 denotes the mean recovery time for the other technique, then $\mu_1 - \mu_2$ represents the difference in the mean recovery times for the two techniques. An inference can be one of finding a confidence interval estimate of $\mu_1 - \mu_2$ or a test of hypothesis about the difference between the mean recovery times for the two techniques. If the null hypothesis says that there is no difference between the two techniques, then the hypotheses are

$$H_0: \mu_1 = \mu_2 \quad \text{versus} \quad H_a: \mu_1 \neq \mu_2$$

The computations for the confidence interval and the hypothesis testing procedures are presented in Section 9.3 when information about the population variances is unknown, and in Section 9.4 if we can assume that the population variances are equal.

EXAMPLE 9.3

To examine the effects of jogging on the resting pulse rate, a researcher divided a sample of 100 men and women into an experimental and a control group, with each member of one group paired with a member of the other group according to gender, age, weight, and height. All subjects were asked to continue their everyday activities except that the members of the experimental group were asked to jog 4 days a week. They began jogging a quarter-mile each day and built up, over a 6-month period, to 2 miles a day. After 6 months, the resting pulse rate was measured for both members of each pair of subjects.

The pulse rate of a member of the experimental group is related to the pulse rate of the corresponding member of the control group because the subjects were paired according to gender, age, weight, and height. If we let μ_d denote the mean of the population of differences between the experimental and control groups, then the test of significance that there is no difference between the groups can be stated as

$$H_0: \mu_d = 0 \quad \text{versus} \quad H_a: \mu_d \neq 0$$

The analysis of the matched pairs experiment is described in Section 9.5.

In each example, the aim of the study is to compare the two groups by evaluating the difference between specific parameters in the respective populations. In the following sections, we develop confidence interval estimates and test of hypotheses procedures for comparing the difference between these parameters.

EXERCISES 9.1

9.1   In each situation, identify the parameters under study and state the null and alternative hypotheses.
   a. An economist for the Census Bureau wishes to compare the costs associated with renting a home in the West versus the East. Realizing that the distribution of rents is typically skewed, she focuses on the median gross rents for the two regions. She has no information to suggest that one region is generally more expensive than the other.
   b. The British bioscience company Zeneca Group PLC has harnessed microorganisms that ingest corn sugar to make plastic (*Wall Street Journal*, March 23, 1994). The goal is to make a plastic product that is biodegradable. Mean biodegradation rates for the plastic and

ordinary paper grocery bags in a compost pile were compared. The experimenter wishes to show that the plastic biodegrades faster than the paper.

c. Has the percent of persons not covered by health insurance changed from last year? One researcher believes that it has increased.

9.2 Does Example 9.2 describe a controlled experiment or an observational study?

9.3 Does Example 9.3 describe a controlled experiment or an observational study?

9.4 In Example 9.2, suppose that the results showed a statistically significant difference (p-value < .05) between the mean recovery times for the two techniques. Explain what this means using terminology that a nonstatistician could understand. Do not use statistical language such as "statistically significant."

9.5 In Example 9.3, suppose that the statistical analysis of data produced a p-value of .014. What does this mean in terms of a test of significance? Should H_0 be rejected? If so, what does this mean? Explain your conclusion so that someone who is not knowledgeable about statistics could understand.

9.6 Example 9.1 does not specifically state whether the study is a controlled experiment or an observational study. It appears on the surface that there is a significant difference between the two proportions of smokers who remained smoke free for 1 year. If this is an observational study and the difference is statistically significant, can we say that the nicotine-laced gum caused the smokers to remain smoke free for 1 year? Explain your answer.

9.7 Is in-line skating more dangerous than skateboarding? According to the Consumer Product Safety Commission, 37,000 injuries were caused by in-line skating in 1993 and 27,700 injuries were caused by skateboarding (*USA Today*, June 10, 1994). What other information is needed in order to compare the relative danger of the two sports? Define the parameters of interest and state the null and alternative hypotheses to evaluate whether in-line skating is more dangerous than skateboarding.

9.8 An 18-month Duke University study of 413 patients in a rural North Carolina clinic suggests that family stress is linked to increased health problems (AP wire release, March 15, 1995). The report states that those patients with high family stress had more follow-up visits to the clinic, more referrals to specialists, more hospitalizations, and more severe illnesses. Can we conclude from this study that family stress causes increased health problems? Is this an observational study or a designed experiment?

9.9 A study in the *Journal of the National Cancer Institute* showed a possible link between exposure to electricity and breast cancer (*USA Today*, June 15, 1994). The researchers compared the death records of 267 female electrical workers and 138,564 other working women. They found that the electrical workers were 38% more likely to die of breast cancer. The report also stated that the link was strongest for telephone installers and line workers but that there was no link among telephone operators, computer operators, and others in traditionally female jobs with potential electrical exposure. Does this mean that unusually high doses of electricity cause breast cancer? Is this an observational study or a designed experiment?

9.10 Water pH levels and elevations (in miles) for 75 water samples in the Great Smoky Mountains were collected by Kaufman et al. (1988) as part of an Environmental Protection Agency (EPA) study. The data in worksheet: **Smokyph.mtw** also appear in Schmoyer (1994). The elevations are coded as 0 (low) for elevations below .60 mile and 1 (high) for elevations above or equal to .60 mile. State the null and alternative hypotheses necessary for determining whether the pH level is greater at the lower elevations. From the data, construct side-by-side dotplots of the water pH levels for the low and high elevations. Comment on the shapes of the two distributions. Are they similar? Are they symmetric? Does either distribution appear to be long-tailed? Would normal probability plots of the two samples indicate anything about the lengths of the tails of the distributions?

9.11   Worksheet: **Indy500.mtw** gives the qualifying speeds and the numbers of previous starts for drivers in the 79th Indianapolis 500 car race held on May 28, 1995. In an effort to determine whether the qualifying speed is related to the number of previous starts, the drivers were separated into two groups: those with fewer than four previous Indy 500 starts and those with five or more previous Indy 500 starts. State the null and alternative hypotheses for determining whether there is a difference between the qualifying speeds for experienced and inexperienced drivers. From the data, construct side-by-side dotplots and boxplots of the two samples. Comment on the shapes of the two distributions. Are the two samples equally variable? Does normality seem reasonable in either distribution? Does it appear that the more experienced drivers had higher qualifying speeds?

9.2   INFERENCE ABOUT THE DIFFERENCE BETWEEN TWO POPULATION PROPORTIONS

In Chapters 7 and 8, you learned how to statistically evaluate a single Bernoulli proportion with a confidence interval estimate (Section 7.1) and with a test of hypothesis (Section 8.2). For example, when a CNN/Gallup nationwide telephone poll of 1,022 adults reports a 54% approval rate for the president's health care policy, we know how to interpret this information.

Now suppose the results of the poll indicate that 56% of Democrats and 47% of Republicans approve of the president's health care policy. How do we deal with two proportions? How do we compare the results? Are the percentages significantly different? In this section, you will learn how to statistically compare two Bernoulli proportions. Recall that a Bernoulli proportion is the probability of a success in a situation where the outcome is classified as either a success or a failure. For example, if the respondent favors the president's health care policy, we classify the outcome as a success; otherwise, it is classified as a failure. In the example, let π_1 denote the Bernoulli proportion of Democrats who approve of the president's health care policy and let π_2 denote the Bernoulli proportion of Republicans who approve of the president's health care policy. The statistical problem then is to compare π_1 and π_2. The procedures developed previously to estimate and test hypotheses about a single Bernoulli proportion can be modified to apply to two Bernoulli proportions. In this section we develop the inference procedures for the *difference* between two Bernoulli proportions.

As suggested, let π_1 and π_2 denote the proportions of successes in the two Bernoulli populations. Assume that random samples of size n_1 and n_2 are drawn *independently* from the two populations. Let p_1 and p_2 denote the two sample proportions of success. From the two samples, we can make inferences about the difference between the two proportions, $\pi_1 - \pi_2$.

Confidence Interval for $\pi_1 - \pi_2$

The confidence interval for $\pi_1 - \pi_2$ is constructed in the same way that the one-sample confidence interval for π was constructed in Section 7.1. We first need an estimator of the parameter $\pi_1 - \pi_2$. We also need to know the sampling distribution of the estimator and the desired level of confidence.

The statistic $p_1 - p_2$ is an unbiased estimator of the parameter $\pi_1 - \pi_2$. In Section 5.4, we pointed out that if n_1 and n_2 are large, then the sampling distribution of $p_1 - p_2$ is approximately normally distributed with a mean of $\pi_1 - \pi_2$ and a standard deviation of

$$\sqrt{\frac{\pi_1(1 - \pi_1)}{n_1} + \frac{\pi_2(1 - \pi_2)}{n_2}}$$

Estimating π_1 and π_2, we obtain the standard error:

$$SE(p_1 - p_2) = \sqrt{\frac{p_1(1 - p_1)}{n_1} + \frac{p_2(1 - p_2)}{n_2}}$$

As in the one-sample procedure, the confidence interval is constructed like this:

$$\text{Estimator} \pm (\text{critical value})(\text{standard error})$$

Because of the normality of the sampling distribution, the critical value for a confidence interval is z^*, found in the standard normal probability table.

The form of the confidence interval for $\pi_1 - \pi_2$ is summarized in the box:

Confidence Interval for $\pi_1 - \pi_2$

Application: Bernoulli populations

Assumptions: Independent samples; $n_1 > 30$ and $n_2 > 30$

A $(1 - \alpha)100\%$ confidence interval for $\pi_1 - \pi_2$ is given by the limits

$$(p_1 - p_2) \pm z^* \sqrt{\frac{p_1(1 - p_1)}{n_1} + \frac{p_2(1 - p_2)}{n_2}}$$

where z^* is the upper $\alpha/2$ critical value found in the standard normal probability table.

EXAMPLE 9.4

Example 9.1 reported that 29 out of 106 smokers who chewed nicotine-laced gum remained smoke free for 1 year. Of the 100 smokers who chewed regular gum, only 16 remained smoke free for 1 year. Use this information to find a 98% confidence interval for the difference between the proportions of smokers who successfully use nicotine-laced gum and those who successfully use regular gum.

SOLUTION

Following the suggestion in Example 9.1, we let π_1 denote the proportion of smokers who use nicotine-laced gum and remain smoke free for 1 year, and let π_2 denote the proportion of smokers who chew regular gum and remain smoke free for 1 year. The quantity $\pi_1 - \pi_2$ then represents the difference between the proportions from the two groups. We wish to find a 98% confidence interval for $\pi_1 - \pi_2$.

From the sample data, we have

$$p_1 = \frac{29}{106} = .274 \qquad p_2 = \frac{16}{100} = .16$$

and we get

$$SE(p_1 - p_2) = \sqrt{\frac{(.274)(.726)}{106} + \frac{(.16)(.84)}{100}}$$
$$= .0568$$

For a 98% confidence interval, we have $z^* = 2.33$. The confidence interval for $\pi_1 - \pi_2$ becomes

$$(.274 - .16) \pm 2.33(.0568) \quad \text{or} \quad .114 \pm .132$$

so the interval is $(-.018, .246)$.

Note that the confidence interval for the difference between the proportions contains both positive and negative values. This indicates that a test of hypothesis would not find a significant difference between the two proportions when the level of significance is 2%. Thus, although the percents (27.4% and 16%) appear to be significantly different, it is likely that the perceived difference between these two samples happened purely by chance. Repeating the experiment may yield percents in the opposite direction.

Large-Sample Hypothesis Test of $\pi_1 - \pi_2$

When we compare two proportions, as in Example 9.4, it may be that the sample proportions are considerably different and yet there is no difference between the population proportions. The difference between the two sample proportions may be due to the particular samples that we happened to observe. With a test of hypothesis, we can determine how likely it is that we will see a difference in the sample proportions when there is no difference in the population proportions.

Following the procedures outlined in the one-sample theory, the null and alternative hypotheses to evaluate claims involving π_1 and π_2 can take one of three forms:

Left-tailed—H_0: $\pi_1 - \pi_2 \geq 0$ versus H_a: $\pi_1 - \pi_2 < 0$

Right-tailed—H_0: $\pi_1 - \pi_2 \leq 0$ versus H_a: $\pi_1 - \pi_2 > 0$

Two-tailed—H_0: $\pi_1 - \pi_2 = 0$ versus H_a: $\pi_1 - \pi_2 \neq 0$

Recall that in the development of the hypothesis test, we assume that the null hypothesis is true and then attempt to find statistical evidence to the contrary. More specifically, when calculating the p-value, we assume the equality part of the null hypothesis so as to maximize the p-value; that is, we assume that $\pi_1 - \pi_2 = 0$. Under this assumption, the standard deviation of $p_1 - p_2$ is

$$\sqrt{\pi(1 - \pi)}\sqrt{\frac{1}{n_1} + \frac{1}{n_2}}$$

where π is the common value of π_1 and π_2.

Pooling together the information in the two samples, we obtain p, a pooled estimate of π:

$$p = \frac{x_1 + x_2}{n_1 + n_2}$$

where x_1 is the number of successes in sample 1 and x_2 is the number of successes in sample 2. Using this estimate of π, we have for the standard error:

$$\text{SE}(p_1 - p_2) = \sqrt{p(1 - p)}\sqrt{\frac{1}{n_1} + \frac{1}{n_2}}$$

Because H_0 is assumed to be true (until evidence to the contrary is found), the standardized test statistic is

$$z = \frac{p_1 - p_2}{\sqrt{p(1 - p)}\sqrt{1/n_1 + 1/n_2}}$$

which has an approximate *standard* normal sampling distribution when n_1 and n_2 are sufficiently large. At this point, it is easy to see that the remainder of the test is just like the z-test for a single proportion presented in Section 8.2.

Large-Sample Hypothesis Test of $\pi_1 - \pi_2$

Application: Bernoulli populations

Assumptions: Independent samples, $n_1 > 30$ and $n_2 > 30$

Left-Tailed Test	Right-Tailed Test	Two-Tailed Test
H_0: $\pi_1 - \pi_2 \geq 0$	H_0: $\pi_1 - \pi_2 \leq 0$	H_0: $\pi_1 - \pi_2 = 0$
H_a: $\pi_1 - \pi_2 < 0$	H_a: $\pi_1 - \pi_2 > 0$	H_a: $\pi_1 - \pi_2 \neq 0$

Standardized test statistic: $z = \dfrac{p_1 - p_2}{\sqrt{p(1 - p)}\sqrt{1/n_1 + 1/n_2}}$

We are given that

$$p_1 = \frac{x_1}{n_1} \qquad p_2 = \frac{x_2}{n_2} \qquad p = \frac{x_1 + x_2}{n_1 + n_2}$$

where x_1 is the number of successes in sample 1 and x_2 is the number of successes in sample 2.

(continued)

From the z-table we find

Left-Tailed Test

p-value $= P(z < z_{\text{obs}})$

Right-Tailed Test

p-value $= P(z > z_{\text{obs}})$

Two-Tailed Test

p-value $= 2P(z < z_{\text{obs}})$ if $z_{\text{obs}} < 0$
or p-value $= 2P(z > z_{\text{obs}})$ if $z_{\text{obs}} > 0$

If a level of significance α is specified, then reject H_0 if p-value $< \alpha$.

EXAMPLE 9.5

The campaign manager for a presidential candidate wishes to test the claim that the proportion of Ohio voters who favor the candidate is at least as large as the proportion of California voters who favor the candidate. Given these data, test the manager's claim at a 5% level of significance:

	n	Number in favor
Ohio	100	36
California	200	84

SOLUTION

Let π_1 denote the proportion of Ohio voters who favor the candidate, and let π_2 denote the proportion of California voters who favor the candidate. The claim of the campaign manager is that $\pi_1 \geq \pi_2$. To test this claim, the null and alternative hypotheses are

$$H_0: \pi_1 \geq \pi_2 \quad \text{versus} \quad H_a: \pi_1 < \pi_2$$

Pooling the sample information, we find

$$p = \frac{36 + 84}{100 + 200} = \frac{120}{300} = .4$$

Thus, we have

$$SE(p_1 - p_2) = \sqrt{(.4)(.6)}\sqrt{\frac{1}{100} + \frac{1}{200}}$$
$$= (.4899)(.12247) = .06$$

The observed value of the test statistic is

$$z_{\text{obs}} = \frac{.36 - .42}{.06} = -1.0$$

From the z-table we find

$$p\text{-value} = .1587 > \alpha \ (= .05)$$

Consequently, there is insufficient evidence to reject the null hypothesis. There is no statistical information to refute the campaign manager's claim.

Occasionally the evidence in a statistical study is such that a formal test of significance is not required. A case in point is the Mayo Clinic study described in the Statistical Insight at the beginning of this chapter. The objective of the study was to determine whether large doses of vitamin C prolonged the lives of patients who had cancer. The hypotheses of interest are

$$H_0: \pi_1 \geq \pi_2 \quad \text{versus} \quad H_a: \pi_1 < \pi_2$$

where π_1 is the proportion of patients in the experimental group who die within 1 year and π_2 is the proportion of patients in the control group who die within 1 year. The sample results of the study showed that a greater percentage of the experimental group died within 1 year. Without completing a test, we realize that the null hypothesis is supported by the sample data: There is no statistical evidence to support the research hypothesis. In fact, the p-value of the test would exceed .50 because the test is left-tailed and the test statistic is on the right tail of the sampling distribution. Only in those cases where the evidence is in question is a formal test of significance required.

EXERCISES 9.2

9.12   Construct a 95% confidence interval for $\pi_1 - \pi_2$ in each case:
   a. $n_1 = 240$, $x_1 = 160$, $n_2 = 250$, $x_2 = 145$
   b. $n_1 = 1{,}200$, $x_1 = 65$, $n_2 = 1{,}000$, $x_2 = 70$
   c. $n_1 = 50$, $x_1 = 38$, $n_2 = 50$, $x_2 = 26$
   d. Of the three intervals constructed in parts **a–c**, which one has the smallest standard error? Explain why it is the most precise (is the narrowest).
   e. What can be said about the validity of the three intervals?

9.13   Independent random samples of size 200 each were selected from two Bernoulli populations. The numbers of successes in sample 1 and sample 2 were 65 and 74, respectively.
   a. Calculate individual estimates of the two Bernoulli proportions, π_1 and π_2. If $\pi_1 = \pi_2$, what is the pooled estimate of π, the common value of π_1 and π_2?
   b. Calculate the standard error of $p_1 - p_2$ if we cannot assume that $\pi_1 = \pi_2$.
   c. Calculate the standard error of $p_1 - p_2$ if we can assume that $\pi_1 = \pi_2$.
   d. To test the null hypothesis that $\pi_1 = \pi_2$ versus the alternative that $\pi_1 \neq \pi_2$, should the standard error of $p_1 - p_2$ be calculated as in part **b** or part **c**?
   e. Test the null hypothesis stated in part **d**. Be sure to perform all calculations, including the z statistic and the p-value.
   f. Suppose the two Bernoulli proportions represent the percentages of female and male students who have cars registered on campus. How does this information affect your analysis of the data? How does it affect your conclusion to the test of significance?

9.14   In a survey taken to help understand emotions, 22 of 70 persons under 18 years of age expressed a fear of meeting people and 23 of 90 persons 18 years old and over expressed the same fear.

a. Formulate the null and alternative hypotheses necessary to determine whether there is a statistical difference between the proportions of those who fear meeting people in the two age groups.

b. What test statistic is used to test the hypotheses in part **a**?

c. To test the hypotheses, should the standard error of the test statistic be computed with a pooled estimate of π?

d. Test the hypotheses. Be sure to compute the test statistic and its p-value and state your conclusion.

9.15   Suppose 250 castings produced in mold A contained 19 defectives and 300 castings produced in mold B contained 27 defectives.

a. Calculate individual estimates of the proportions of defectives produced in mold A and mold B.

b. To find a 99% confidence interval for the difference between the proportions of defectives produced by the two molds, should the two sample proportions be pooled together to calculate the standard error of $p_1 - p_2$?

c. Calculate the confidence interval from part **b**.

d. Does the interval contain both positive and negative values? What does this mean about the two proportions of defectives?

9.16   The FDA requires that the mean antibody strength for a measles vaccine exceed 1.6 before it can be placed on the market. In a sample of size 55, firm A's vaccine exceeded the requirement 37 times. In a sample of size 46, firm B's vaccine exceeded the requirement 33 times. Is there a significant difference between the success rates for the two firms?

a. Formulate null and alternative hypotheses to evaluate the difference.

b. What test statistic is used to test the hypotheses?

c. To test the hypotheses, should the standard error of the test statistic be computed with a pooled estimate of π?

d. Test the hypotheses. Be sure to compute the test statistic and its p-value and state your conclusion.

9.17   In hypothesis testing, why do we calculate a pooled estimate of π when we calculate the standard error of $p_1 - p_2$? Why do we not pool in confidence interval estimation?

9.18   In a particular occupation, a random sample of 210 men found that 65 smoked. In a random sample of 240 women, 87 smoked. Are the percents of men and women in this occupation who smoke significantly different? Formulate null and alternative hypotheses to evaluate the difference. Test the hypotheses and be sure to give the p-value of the test statistic as well as a summary.

9.19   To test the effectiveness of a vaccine against a certain disease, 120 experimental animals were given the vaccine and 180 were not. All 300 animals were then infected with the disease. Among those vaccinated, 15 died as a result of the disease. In the control group, 36 died. Can we conclude that the vaccine was effective in reducing the mortality rate? Formulate the null and alternative hypotheses if we believe that the vaccine was effective in reducing the mortality rate. Test the hypotheses and be sure to give the p-value of the test statistic as well as a summary. Is this a controlled experiment or an observational study?

9.20   A sample of 500 adults ranging in age from 24 through 54 was divided randomly into two groups. Those in group 1 consumed 1,000 milligrams of vitamin C each day for 6 months. Those in group 2 consumed a placebo each day for the same 6 months. Of the 238 in group 1 (12 dropped out of the experiment for various reasons), 42 had at least one cold during the period. Of the 241 in group 2, 61 had at least one cold during the period. Is there a significant difference between the percents that had at least one cold in each group? Formulate the hypotheses and conduct the test. Is this a controlled experiment or an observational study?

9.21   To be judged proficient in reading at a certain university, a student must score 85 or above on the Nelson–Denny Reading test. Of the 1,250 entering freshmen, 190 were randomly selected and given a special 2-week course in reading. After the course, 146 of the 190 were judged proficient in reading. Of the remaining 1,060 entering freshmen, 762 were judged proficient in reading. Did the reading course increase reading scores? To evaluate with a test of significance, should the test be a one-tailed or a two-tailed test? Formulate the hypotheses and conduct the test. Be sure to give the p-value and state your conclusion.

9.22   To study the effectiveness of the drug AZT for preventing AIDS among their offspring, 164 pregnant HIV-positive women were randomly assigned to receive AZT and a like group of 160 were randomly assigned to receive a placebo (*Newsweek*, March 7, 1994). Among the 164 births in the AZT group, 13 children were diagnosed as HIV-positive, and among the 160 births in the placebo group, 40 children were diagnosed as HIV-positive. Is there statistical evidence that AZT reduces the chances of children of HIV-positive mothers developing AIDS? Evaluate with a test of significance. Should the test be a one-tailed or a two-tailed test? Formulate the hypotheses and conduct the test. Be sure to give the p-value and state your conclusion.

9.23   Do you worry about being a victim of crime? Do people who live in cities feel more vulnerable than those who live in the suburbs? A telephone poll of 500 adult Americans conducted for Time/CNN on August 12, 1993, found that 59% of those living in cities versus 57% of those living in suburbs were worried about being a victim of crime. Are these figures statistically significantly different? Assume there were 250 adults in each sample; that is, 250 lived in cities and 250 lived in suburbs. Test the hypothesis that there is no difference in the population proportions.

9.24   It has been reported that at least 10% of the adult male population suffers from kidney stones. To compare the percentages of men and women who suffer from kidney stones, random samples of 900 men and 400 women were selected and tested. From the two samples, it was found that 99 men and 36 women suffer from kidney stones. Is there statistical evidence, at the .05 level of significance, that the percent of male kidney stone sufferers is greater than the percent of females?

9.25   In 1992, there were 534,543 male and 118,519 female physicians in the United States according to the American Medical Association. Of the 133,718 doctors under the age of 35, 40,431 were women and of the 198,257 between the ages of 35 and 44, 44,336 were women. Is there statistical evidence that more women are entering the medical profession now than in the past?

9.26   A survey by the American Association of University Women reported that 64% of third-grade girls and 66% of third-grade boys said they were good in math. Assume that the survey involved 150 girls and 200 boys. Are the percentages statistically different? Formulate null and alternative hypotheses to test the equality of the two proportions. Conduct a test of significance to evaluate the findings. Report the p-value and write a summary of the results.

9.27   In the same study reported in Exercise 9.26, the American Association of University Women determined that only 48% of 11th-grade girls versus 60% of 11th-grade boys felt they were good in math. Assume again that the sample sizes were 150 and 200, respectively. Conduct the same test of significance for the 11th-graders as you did for the third-graders in Exercise 9.26. Are the results different? What can you conclude about girls' and boys' feelings about math as they get older?

9.28   It is reported by the Census Bureau that 42% of single-parent homes are in the central cities versus 37% in the suburbs. Can the methods presented in this section be used to statistically compare these figures? Remember that the confidence interval and test of significance presented here assume that we have two samples drawn independently from two different Bernoulli populations.

9.3 INFERENCE ABOUT THE DIFFERENCE BETWEEN TWO POPULATION MEANS

For a certain medical operation, patients are randomly and independently assigned to two different techniques to determine whether the recovery times differ for the two methods (see Example 9.2). A manufacturer of plastic bags for grocery stores wishes to compare the mean tensile strengths of bags produced by two different manufacturing processes. A commercial fishery compares the haddock catches from a small-mesh net versus a large-mesh net. Is the mean pH level of water in the Great Smoky Mountains the same at high and low elevations? In each of these problems, we wish to compare the means of two independent groups that are treated as independent samples selected from two different populations. Initially, we should examine the two samples from dotplots that have a common scale and side-by-side boxplots. If the graphical information indicates that the distributions are somewhat symmetric with tails that are not excessively long, then we can proceed to make inferences about the means of the two populations much like the single-population inference procedures given in Section 8.3. As in the inference procedures for proportions given in the preceding section, the procedures used to estimate and test hypotheses about a single population mean are modified to apply to the problem of comparing two population means.

Two-Sample
***t*-Procedures**

Let μ_1 and σ_1 denote the mean and standard deviation of population 1 and let μ_2 and σ_2 denote the mean and standard deviation of population 2, respectively. Suppose we have two *independent* samples from the two populations, with n_1 and n_2 denoting the two respective sample sizes. From the two samples we can make inferences about the difference, $\mu_1 - \mu_2$, between the two population means.

We first need a good point estimator for the parameter $\mu_1 - \mu_2$. As in single-parameter estimation, the choice of an estimator depends on the distribution of the underlying parent populations. Initially we will assume that both populations are normally distributed. Because of the robustness of the inference procedures, however, we can relax the condition and apply the results to situations where the two parent populations are symmetric with tails that are not excessively long. In later sections, we consider skewed and long-tailed populations.

Following the single-parameter estimation problem, a reasonable estimator of $\mu_1 - \mu_2$ is the difference between the two sample means, $\bar{y}_1 - \bar{y}_2$. In Section 5.4, we pointed out that when the parent distributions are normally distributed or if the sample sizes are large, the sampling distribution of $\bar{y}_1 - \bar{y}_2$ is approximately normal with a mean of $\mu_1 - \mu_2$ and a standard deviation of $\sqrt{\sigma_1^2/n_1 + \sigma_2^2/n_2}$. Thus, the standardized form

$$z = \frac{(\bar{y}_1 - \bar{y}_2) - (\mu_1 - \mu_2)}{\sqrt{\sigma_1^2/n_1 + \sigma_2^2/n_2}}$$

has an approximate *standard* normal sampling distribution.

In practice, however, we rarely ever know the numerical values of σ_1^2 and σ_2^2. Consequently, they are replaced with their respective estimators, s_1^2 and s_2^2. Then we have the standard error:

$$SE(\bar{y}_1 - \bar{y}_2) = \sqrt{\frac{s_1^2}{n_1} + \frac{s_2^2}{n_2}}$$

Because we have estimated population variances with sample variances, the sampling distribution of

$$t = \frac{(\bar{y}_1 - \bar{y}_2) - (\mu_1 - \mu_2)}{\sqrt{s_1^2/n_1 + s_2^2/n_2}}$$

is no longer standard normal. The distribution is, however, approximated with a t distribution that has degrees of freedom given by this formula:

$$df = \frac{(s_1^2/n_1 + s_2^2/n_2)^2}{\frac{(s_1^2/n_1)^2}{n_1 - 1} + \frac{(s_2^2/n_2)^2}{n_2 - 1}}$$

If the formula yields a decimal number, round it down to the nearest whole number so that the stated confidence level and/or p-value is accurate.

Confidence Interval for $\mu_1 - \mu_2$ Based on $\bar{y}_1 - \bar{y}_2$

As before, a confidence interval is constructed like this:

Estimator $\pm$ (critical value)(standard error)

For a $(1 - \alpha)100\%$ confidence interval for $\mu_1 - \mu_2$, the estimator is $\bar{y}_1 - \bar{y}_2$, the standard error is $\sqrt{s_1^2/n_1 + s_2^2/n_2}$, and the critical value, t^*, is found in the t-table with degrees of freedom given by the above formula.

The confidence interval procedure is summarized in the box:

Confidence Interval for $\mu_1 - \mu_2$ Based on $\bar{y}_1 - \bar{y}_2$

Application: Symmetric distributions with tails that are not excessively long. If the parent distributions deviate substantially from normality, then the sample sizes should be larger than 30.

Assumption: The two samples are independent of each other.

Estimator: $\bar{y}_1 - \bar{y}_2$

Standard error:

$$SE(\bar{y}_1 - \bar{y}_2) = \sqrt{\frac{s_1^2}{n_1} + \frac{s_2^2}{n_2}}$$

(continued)

A $(1 - \alpha)100\%$ confidence interval for $\mu_1 - \mu_2$ is given by the limits

$$(\bar{y}_1 - \bar{y}_2) \pm t^* \sqrt{\frac{s_1^2}{n_1} + \frac{s_2^2}{n_2}}$$

where t^* is the upper $\alpha/2$ critical value found in the t-table with degrees of freedom given by

$$df = \frac{(s_1^2/n_1 + s_2^2/n_2)^2}{\dfrac{(s_1^2/n_1)^2}{n_1 - 1} + \dfrac{(s_2^2/n_2)^2}{n_2 - 1}} \qquad \text{(rounded down to the nearest whole number)}$$

EXAMPLE 9.6

Wind speed data were gathered during January and July at the site proposed for a wind generator to determine whether the production of electricity by the wind generator will be different in the two months. Assume that the distribution of wind speeds (in knots per hour) at this location is symmetric and does not have long tails. From the summary data, construct a 99% confidence interval for the difference between the mean wind speeds in January and July:

	n	$\bar{y}$	s^2
January	32	23.4	26.42
July	35	16.2	23.875

SOLUTION

Let μ_1 denote the mean wind speed in January and μ_2 denote the mean wind speed in July. The problem is to construct a confidence interval for $\mu_1 - \mu_2$. Because we can assume that the distribution of wind speeds at this location is symmetric with tails that are not unusually long, we use $\bar{y}_1 - \bar{y}_2$ as the estimate of $\mu_1 - \mu_2$. From the summary data, we have

$$\bar{y}_1 - \bar{y}_2 = 23.4 - 16.2 = 7.2$$

and

$$SE(\bar{y}_1 - \bar{y}_2) = \sqrt{\frac{26.42}{32} + \frac{23.875}{35}} = 1.228$$

Because a 99% confidence interval is desired, t^* is the upper .005 ($1 - \alpha = .99$; $\alpha = .01$; $\alpha/2 = .005$) critical value found in the t-table with

$$df = \frac{(s_1^2/n_1 + s_2^2/n_2)^2}{\dfrac{(s_1^2/n_1)^2}{n_1 - 1} + \dfrac{(s_2^2/n_2)^2}{n_2 - 1}} = \frac{(26.42/32 + 23.875/35)^2}{\dfrac{(26.42/32)^2}{31} + \dfrac{(23.875/35)^2}{34}} \approx 63$$

From the t-table with 63 degrees of freedom, we find the upper .005 critical value to be $t^* = 2.66$. (We take the closest value, which corresponds to 60 degrees of freedom.) Therefore, the 99% confidence interval for $\mu_1 - \mu_2$ is

$$7.2 \pm (2.66)(1.228) \quad \text{or} \quad 7.2 \pm 3.266$$

Thus, we are reasonably sure that the difference between the mean wind speeds in the two months is somewhere in the interval (3.934, 10.466).

The length of the interval is rather large and may not provide much useful information. To decrease the length, we can reduce the confidence level from 99% to, say, 90%, but this may not be desirable. The other choice is to take larger samples. Given that the equipment is already in place to measure the wind speed, collecting more data would be a reasonable next step.

Hypothesis Test for $\mu_1 - \mu_2$ Based on $\bar{y}_1 - \bar{y}_2$

We next develop the test of hypothesis about the difference between two population means. The null and alternative hypotheses can take one of three forms:

Left-tailed—$H_0: \mu_1 - \mu_2 \geq 0$ versus $H_a: \mu_1 - \mu_2 < 0$

Right-tailed—$H_0: \mu_1 - \mu_2 \leq 0$ versus $H_a: \mu_1 - \mu_2 > 0$

Two-tailed—$H_0: \mu_1 - \mu_2 = 0$ versus $H_a: \mu_1 - \mu_2 \neq 0$

In the left-tailed test, for example, testing $H_0: \mu_1 - \mu_2 \geq 0$ is the same as testing $H_0: \mu_1 \geq \mu_2$. Therefore, the test of the difference between two population means is appropriate when we wish to determine whether the mean response under one condition is greater than (or less than in the case of a right-tailed test) the mean response under another condition. Also it is easy to see that the inference is for the difference between the two population means, $\mu_1 - \mu_2$. Given that the population distributions are symmetric and do not have long tails, the choice of a test statistic is the point estimator, $\bar{y}_1 - \bar{y}_2$.

In the test of $H_0: \mu_1 - \mu_2 = 0$, the problem is to determine whether $\bar{y}_1 - \bar{y}_2$ is significantly different from zero. As in single-parameter inference, significant departures of a test statistic from a given value are measured by first computing the "standardized" test statistic and then finding the associated p-value. In this case, the standardized test statistic is

> Recall that testing a null hypothesis, such as $H_0: \mu_1 - \mu_2 \geq 0$, is equivalent to testing $H_0: \mu_1 - \mu_2 = 0$ because the p-value is maximized when the null hypothesis is stated with an equal sign.

$$t = \frac{(\bar{y}_1 - \bar{y}_2) - (\mu_1 - \mu_2)}{\sqrt{s_1^2/n_1 + s_2^2/n_2}}$$

which has an approximate t distribution. Under the assumption that the null hypothesis is true (that is, $\mu_1 - \mu_2 = 0$), the test statistic becomes

$$t = \frac{\bar{y}_1 - \bar{y}_2}{\sqrt{s_1^2/n_1 + s_2^2/n_2}}$$

The p-value is the probability on the tail of the sampling distribution of the test statistic beyond the observed value of the test statistic. In testing $H_0: \mu_1 - \mu_2 \geq 0$

versus $H_a: \mu_1 - \mu_2 < 0$, we have p-value $= P(t < t_{obs})$, where t_{obs} is found by substituting the observed values of $n_1, n_2, \bar{y}_1, \bar{y}_2, s_1$, and s_2 into the formula for t.

The test is summarized in the box:

Two-Sample t-Test of $\mu_1 - \mu_2$ Based on $\bar{y}_1 - \bar{y}_2$

Application: Symmetric distributions with tails that are not excessively long. If the parent distributions deviate substantially from normality, then the sample sizes should be larger than 30.

Assumptions: The two samples are independent of each other.

Left-Tailed Test	Right-Tailed Test	Two-Tailed Test
$H_0: \mu_1 - \mu_2 \geq 0$	$H_0: \mu_1 - \mu_2 \leq 0$	$H_0: \mu_1 - \mu_2 = 0$
$H_a: \mu_1 - \mu_2 < 0$	$H_a: \mu_1 - \mu_2 > 0$	$H_a: \mu_1 - \mu_2 \neq 0$

Standardized test statistic:

$$t = \frac{\bar{y}_1 - \bar{y}_2}{\sqrt{s_1^2/n_1 + s_2^2/n_2}}$$

From the t-table with degrees of freedom given by

$$df = \frac{\left(s_1^2/n_1 + s_2^2/n_2\right)^2}{\dfrac{(s_1^2/n_1)^2}{n_1 - 1} + \dfrac{(s_2^2/n_2)^2}{n_2 - 1}}$$

the p-value is the tail probability most closely associated with the observed value of t, t_{obs}. If t_{obs} falls between two table values, then give the two associated probabilities as bounds for the p-value.

The p-value for the two-tailed test is calculated as in the single-tailed test and then doubled to account for both tails.

If a level of significance α is specified, then reject H_0 if p-value $< \alpha$.

EXAMPLE 9.7

Plastic grocery bags have almost replaced the standard brown paper bags at the supermarket. When the plastic bags were introduced, problems with tearing and ripping resulted in some very disturbed customers. One particular company was trying to increase the tensile strength of the bags and still hold down production costs by adjusting temperature and pressure in the production runs. These summary data are from two independent random samples and give the tensile strengths of plastic bags from two different production runs:

Sample 1: $n_1 = 32, \bar{y}_1 = 102.33, s_1 = 14.06$

Sample 2: $n_2 = 40, \bar{y}_2 = 118.19, s_2 = 24.44$

Assume the two population distributions that produced the data are symmetric with tails that are not excessively long. Determine whether there is a significant difference between the mean tensile strengths from the two production runs.

SOLUTION

Let μ_1 and μ_2 denote the mean tensile strengths from the two different production runs. The null and alternative hypotheses are:

$$H_0: \mu_1 - \mu_2 = 0 \quad \text{versus} \quad H_a: \mu_1 - \mu_2 \neq 0$$

The standardized test statistic is

$$t = \frac{\bar{y}_1 - \bar{y}_2}{\sqrt{s_1^2/n_1 + s_2^2/n_2}}$$

From the sample data we find

$$t_{\text{obs}} = \frac{102.33 - 118.19}{\sqrt{(14.06)^2/32 + (24.44)^2/40}} = -3.45$$

$$df = \frac{(s_1^2/n_1 + s_2^2/n_2)^2}{\dfrac{(s_1^2/n_1)^2}{n_1 - 1} + \dfrac{(s_2^2/n_2)^2}{n_2 - 1}} = \frac{\left[(14.06)^2/32 + (24.44)^2/40\right]^2}{\dfrac{\left[(14.06)^2/32\right]^2}{31} + \dfrac{\left[(24.44)^2/40\right]^2}{39}} \approx 64$$

Because the test is two-tailed,

$$p\text{-value} = 2P(t < -3.45)$$

From the t-table with 60 degrees of freedom (the closest value to 64), we see that 3.46 corresponds to a tail probability of .0005. Because this is so close to 3.45, we have

$$p\text{-value} \approx 2(.0005) = .001$$

This small p-value indicates that the difference between the mean tensile strengths of the plastic bags from the two different production runs is highly significant.

The highly significant difference in the mean tensile strengths of plastic bags from the two different production runs in the preceding example was not at all apparent from the summary data. A graphical illustration of the data, such as boxplots or dotplots, prior to any formal test may reveal differences that might indicate that a formal test is not required.

EXAMPLE 9.8

The raw data for the tensile strengths of plastic bags from the two production runs described in Example 9.7 are given in worksheet: **Tensile.mtw**. Complete side-by-side boxplots of the data and comment on any differences.

SOLUTION

Figure 9.1 shows side-by-side boxplots produced by Minitab. Both distributions appear symmetric without excessively long tails. The boxplots show that the median for run 1 is very near the first quartile of run 2 and the median for run 2 is greater than the third quartile of run 1. Notice also that the data in run 2 are considerably more variable than those in run 1. Thus, run 2, on average, produces a higher tensile strength than run 1, but it has considerably more variability. In all, the boxplots clearly show a difference in the two production runs. The test of significance in Example 9.7 simply confirms what we see in Figure 9.1.

FIGURE 9.1

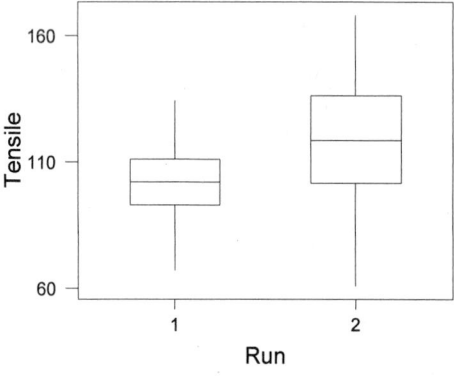

Computer Analysis

As demonstrated in Examples 9.6 and 9.7, the computations for the two-sample t-test are rather complicated, especially for calculating the degrees of freedom. Therefore, it is highly recommended that you use a computer to analyze data involving two or more samples. As you learn more about the analysis of statistical data, you will realize that the computer is almost indispensable. This is especially true in the later chapters of this book on regression analysis and the analysis of variance.

Figure 9.2 is the computer output of the two-sample t-test applied to the data in worksheet: **Tensile.mtw**. Notice that the summary statistics (after rounding) are as they were in Example 9.7. The procedure has computed a 95% confidence interval for the difference between the means and has confirmed the test statistic as $t = -3.45$. To find the p-value, the computer used 64 degrees of freedom where we used 60 (our table does not have 64 degrees of freedom). We got the same p-value because when the degrees of freedom become large, there is very little difference between the t critical values.

C O M P U T E R   T I P

To execute the two-sample *t*-test under the **Stat** menu, select **Basic Statistics** and click on **2-sample t**. When the 2-Sample t window appears, you must decide whether your data are in one column of the worksheet with subscripts describing the grouping in another column, or in two separate columns. If it is the former, click on **Samples in one column** and identify the column containing the data in the **Samples box** and the column containing the group's identifier in the **Subscripts box**. If your data are in two separate columns, click on **Samples in different columns**, identify the first and second columns, and click **OK**.

FIGURE 9.2   Two Sample T-Test and Confidence Interval

```
Twosample T for Tensile
Run    N      Mean      StDev    SE Mean
1     32     102.3      14.1       2.5
2     40     118.2      24.4       3.9

95% C.I. for mu 1 - mu 2: ( -25.0,  -6.7)
T-Test mu 1 = mu 2 (vs not =): T= -3.45  P=0.0010  DF=  64
```

Robustness of the Two-Sample *t*-Procedures

The development of the two-sample *t*-procedures relied on the assumption that the two parent populations are normally distributed. As was pointed out in the one-sample case, rarely will we ever see distributions that are exactly normally distributed. As it turns out, this is not a serious problem, however, because of the robustness of the *t*-procedures. In fact, if the sample sizes are reasonably close and the distributions are similar in shape, the two-sample *t*-procedures are more robust than the one-sample *t*-procedures. And as before, when we have large samples, the normality of the parent populations is not as important because of the Central Limit Theorem. Therefore, the *t*-procedures are valid over a large spectrum of underlying parent distributions. This does not mean all distributions, however. As pointed out previously, the statistical properties of $\bar{y}$ as an estimator of the population mean tend to deteriorate when there are outliers. The result is a loss of precision in estimating μ with the usual confidence interval based on $\bar{y}$. In Chapter 7, we pointed out that a more precise interval for symmetric, long-tailed distributions is obtained if the interval is based on $\bar{y}_T$. The same is true when we are estimating or testing the difference between population means. If the parent populations are symmetric, long-tailed distributions, then the confidence interval and hypothesis-testing problems should be based on $\bar{y}_{T1} - \bar{y}_{T2}$. This situation is considered in Section 9.6. If the distributions are skewed, we should abandon the mean as the parameter of concern and investigate the population median. The difference between population medians is considered at the end of Section 9.4.

A Conservative Approach for the Two-Sample t-Procedures

As previously suggested, a computer is recommended for the analysis of data for the two-sample t-procedures. If we are performing calculations without a computer, however, the formula for calculating degrees of freedom is somewhat complicated. In those cases it is recommended that we use the following conservative approach for the two-sample t-procedures:

> For a conservative approach to the two-sample t-procedures, the degrees of freedom are given by
>
> $$\text{df} = \text{smaller of } n_1 - 1 \text{ and } n_2 - 1$$

This conservative approach gives a confidence interval whose confidence level is slightly higher than what is reported and in hypothesis testing gives a p-value larger than its actual value. As the sample sizes increase, the confidence level of the interval becomes closer to its stated level and the p-value for a test becomes more accurate.

EXERCISES 9.3

9.29   Construct these confidence intervals for $\mu_1 - \mu_2$ based on $\bar{y}_1 - \bar{y}_2$ from the given summary data using the conservative approach:
   a. 90% interval

$$n_1 = 65, \ \bar{y}_1 = 252.8, \ s_1 = 48.65$$

$$n_2 = 62, \ \bar{y}_2 = 215.4, \ s_2 = 152.83$$

   b. 98% interval

$$n_1 = 33, \ \bar{y}_1 = 15.3, \ s_1 = 2.55$$

$$n_2 = 34, \ \bar{y}_2 = 16.8, \ s_2 = 7.47$$

   c. Are the intervals found in parts **a** and **b** equally valid? Explain.

9.30   These data are arranged in a back-to-back stem and leaf plot:

Worksheet: Exec9-30.mtw

```
              5 | 0   1
        7   9 | 5 | 5
            4 | 6 | 1
    5   7   8 | 6 | 5   8
        0   1 | 7 | 0   3   4
    6   7   8 | 7 | 5   7   8
        0   2 | 8 | 0   1
            8 | 8 |
              | 9 | 2
            9 | 9 | 7
```

a. If these data are independent random samples from two populations, do you think it is reasonable to base inferences about $\mu_1 - \mu_2$ on $\bar{y}_1 - \bar{y}_2$?
b. Calculate the standard error of $\bar{y}_1 - \bar{y}_2$.
c. Find a conservative 95% confidence interval for the difference between the population means.

9.31   With $\bar{y}_1 - \bar{y}_2$ as the test statistic and the following summary statistics, test the null hypothesis $H_0: \mu_1 - \mu_2 = 0$ versus $H_a: \mu_1 - \mu_2 < 0$:

$$n_1 = 53, \ \bar{y}_1 = 6.15, \ s_1 = 17.23$$

$$n_2 = 48, \ \bar{y}_2 = 8.65, \ s_2 = 8.16$$

a. Calculate the observed value of the standardized test statistic.
b. Is this a one-tailed or a two-tailed test? Calculate the p-value using the conservative approach.
c. Based on the p-value calculated in part **b**, should the null hypothesis be rejected?

9.32   The standardized norm on a science test for tenth-graders has a mean of 75 and a standard deviation of 20. In the Grover County school district, 123 tenth-grade students were randomly divided into two groups. One group of 61 students received instruction through a traditional lecture class, and the other group of 62 students received instruction in an experimental class. The average grade on the science test in the traditional group was 77.2 and the standard deviation was 19.6; the average in the experimental group was 78.6 and the standard deviation was 42.4. Is there statistical evidence of a difference between the two groups? Formulate null and alternative hypotheses to evaluate the difference between the two groups. Test the hypothesis and be sure to give the observed value of the test statistic and its p-value as well as a summary report. Is this a controlled experiment or an observational study?

9.33   A study was conducted to determine whether persons in suburban district I have a higher mean income than those in suburban district II. Random samples were taken in each district with these results (measured in $1,000s):

District I: $n_1 = 38, \bar{y}_1 = 28.65, s_1 = 9.38$

District II: $n_2 = 42, \bar{y}_2 = 25.94, s_2 = 5.47$

Determine with a test of hypothesis whether the mean annual income in district I exceeds the mean annual income in district II. Assume that the population distributions satisfy the conditions for the two-sample t-test.

9.34   The FDA tests the tobaccos in two different types of cigars for nicotine content and obtains these results (in milligrams):

Brand A: $n_1 = 23, \bar{y}_1 = 85.3, s_1 = 12.44$

Brand B: $n_2 = 25, \bar{y}_2 = 89.8, s_2 = 26.67$

Do these results indicate that there is a difference between the mean nicotine contents of the two types of cigars? Assume that the population distributions satisfy the conditions for the two-sample t-test.

9.35   The back-to-back stem and leaf plot at the top of page 570 gives the grades from two introductory statistics classes that meet at different times:

Worksheet: Statclas.mtw

9 A.M.		2 P.M.
0	6	0
8 7 7 6 6 6	6	
4 2	7	1 4
8 8 8 7 7	7	6 6 6 6 6 9 9
4 4 3 3 3 2 1	8	0 1 2 2 2 2 2 4
9 9 8 7 6 6 5 5 5	8	5 5 5 6 6 7 8 9 9
4 4 3 2 1	9	0 1 2 3
8	9	6

a. Do the data meet the requirements for inferences based on $\bar{y}_1 - \bar{y}_2$?
b. State the null and alternative hypotheses necessary for determining whether there is a significant difference between the mean grades from the two classes?
c. Use the descriptive statistics to calculate the observed value of the standardized test statistic that is used to test the hypotheses.

```
Descriptive Statistics

Variable       N       Mean     Median     TrMean     StDev     SEMean
9am           36       81.00     83.00      81.19       9.56      1.59
2pm           32       82.50     82.00      82.86       7.29      1.29


Variable      Min       Max         Q1         Q3
9am          60.00     98.00      74.75      87.75
2pm          60.00     96.00      76.75      87.75
```

d. Calculate the p-value and decide whether or not to reject the null hypothesis. State your conclusion.

9.36   This computer printout is for the test in Exercise 9.35.

```
Two Sample T-Test and Confidence Interval

Twosample T for 9am vs 2pm

            N       Mean       StDev     SE Mean
9am        36       81.00       9.56        1.6
2pm        32       82.50       7.29        1.3

95% C.I. for mu 9am - mu 2pm: ( -5.6,   2.6)
T-Test mu 9am = mu 2pm (vs not =): T= -0.73   P=0.47   DF=  64
```

a. Are the sample sizes large enough for the Central Limit Theorem to apply to these data?
b. Compare the test statistic given in the computer printout with the test statistic you found in Exercise 9.35. Are they the same? Compare the two t-statistics, their degrees of freedom, and the corresponding p-values.

9.37   In an experiment to study the effects of a particular drug on the number of errors in the maze-learning behavior of rats, these results were obtained:

	n	$\bar{y}$	s
Drug	12	18.67	11.21
Placebo	16	10.3	4.13

    a. State the null hypothesis that the drug has no effect on the errors.
    b. What assumptions are required to test the hypothesis in part **a** using the two-sample t-test?
    c. Assume that the assumptions in part **b** are met and proceed with the test.
    d. Calculate the p-value and state your conclusion from the test.

9.38   A store owner wants to know whether an advertising campaign has increased mean daily receipts. The daily receipts for the 2 weeks prior to the campaign were recorded as well as the receipts for a 2-week period after the campaign:

	n	$\bar{y}$	s
Before	12	$2,277	$375
After	12	$2,664	$938

    a. State the null and alternative hypotheses to determine whether the campaign increased mean daily receipts.
    b. What assumptions are required to test the hypothesis in part **a** using the two-sample t-test?
    c. Based on the values of the two standard deviations, do you think that the population variances are homogeneous or not? Is this a requirement for the two-sample t-test?
    d. Assume that the assumptions for the two-sample t-test in part **b** are met and proceed with the test. Be sure to calculate the p-value and state your conclusion from the test.

9.39   At a large industrial plant, employees were classified according to age and given a leadership exam. The data are the scores on the exam:

Worksheet: Leader.mtw

Under 35

25	13	9	46	25	30	17	20
17	20	37	25	26	23	20	17
18	26	11	36	30	12	32	54
24	20	16	8	21	37	31	26

Over 35

24	31	43	23	13	23	21	42	34
14	15	38	30	14	45	19	20	27
26	38	29	9	50	41	13	15	
16	68	32	7	9	28	30	51	

    a. State the null and alternative hypotheses to evaluate the claim that the mean score for the over-35 age group exceeds the mean score for the under-35 group.
    b. Based on the boxplots at the top of page 572, do you think the data meet the requirements for inferences based on $\bar{y}_1 - \bar{y}_2$?

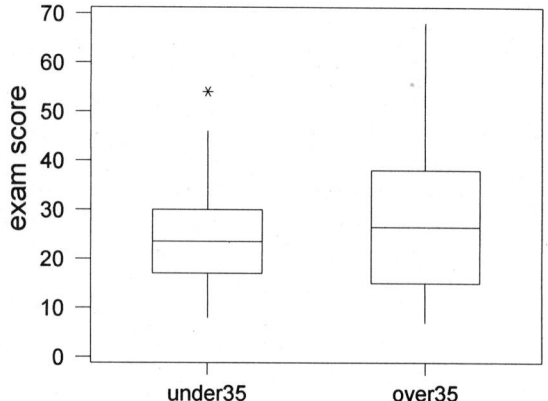

c. Use the descriptive statistics to complete the test of hypothesis given in part **a**. Summarize your results.

```
Descriptive Statistics

Variable      N     Mean    Median    TrMean    StDev    SEMean
under35       32    24.13    23.50     23.39     10.32     1.82
over35        34    27.59    26.50     26.77     14.04     2.41

Variable    Min      Max        Q1        Q3
under35     8.00    54.00     17.00     30.00
over35      7.00    68.00     15.00     38.00
```

9.40   Do the following boxplots indicate a difference between group A and group B? Give a verbal description of what you see in the boxplots. Is a test of significance required? Why or why not?

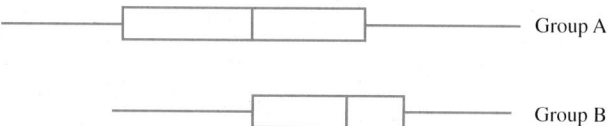

9.41   Do the following boxplots indicate a difference between group A and group B? Give a verbal description of what you see in the boxplots. Is a test of significance required? Why or why not?

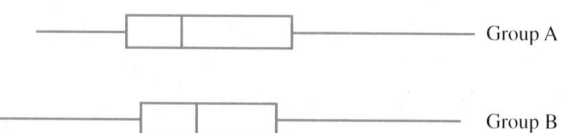

9.42   Exercise 9.10 of Section 9.1 described worksheet: **Smokyph.mtw**, which gives the water pH values and elevations for 75 water samples from the Great Smoky Mountains. The Descriptive Statistics command in Minitab confirms that the mean pH for the 42 low elevations is 7.2181 and the standard deviation is .5005, and the mean pH for the 33 high elevations is 7.0397 and the standard deviation is .3273. Construct a 90% confidence interval for the difference in the mean pH values at the low and high elevations. Does the interval contain both positive and negative values? What does this mean? If the confidence level is increased to 99%, what is the effect on the resulting confidence interval? Is it possible for zero to be inside a 99% confidence interval and not inside a 90% interval?

9.43   Exercise 9.11 of Section 9.1 described worksheet: **Indy500.mtw**, which gives the qualifying speeds and the numbers of previous starts for drivers in the 1995 Indy 500 race. In an effort to determine whether the qualifying speed is related to the number of previous starts, recall that the drivers were separated into two groups: those with fewer than four previous Indy 500 starts and those with five or more previous Indy 500 starts. In Exercise 9.11 of Section 9.1 you were to evaluate the distributions and to state the null and alternative hypotheses to determine whether there is a difference between the qualifying speeds for experienced and inexperienced drivers.

   a. Based on your results in Exercise 9.11, do you think that a two-sample t-test is appropriate for these data?

   b. Complete the test and determine whether the more experienced drivers have faster qualifying speeds.

9.44   Worksheet: **Fish.mtw** contains the lengths and numbers of fish caught with a small-mesh (35-mm diamond-mesh) and a large-mesh (87-mm diamond-mesh) codend (see Exercise 7.61 in Section 7.4). The data produced these statistics:

	n	$\bar{y}$	s
Small mesh	739	33.424	11.67
Large mesh	787	34.59	9.95

SOURCE: R. Millar, "Estimating the Size-Selectivity of Fishing Gear by Conditioning on the Total Catch," *Journal of the American Statistical Association* 87 (1992): 962–968.

   a. Formulate null and alternative hypotheses to examine the difference in the lengths of fish caught by the two different size codends.

   b. Based on the sample sizes, do you think that the Central Limit Theorem applies to these data?

   c. From the summary statistics, calculate the standardized test statistic and its p-value and make a decision.

   d. Based on your analysis, does the codend size make a difference in the size of fish caught?

9.45   Patients with small cell lung cancer received one of two types of treatments (arm A or arm B) as described in Exercise 6.30 of Section 6.2. Their survival times are recorded in worksheet: **Censored.mtw**. (SOURCE: Z. Ying, S. Jung, and L. Wei, "Survival Analysis with Median Regression Models," *Journal of the American Statistical Association* 90 (1995): 178–184.)

   a. Based on the side-by-side boxplots at the top of page 574, do you think the data meet the requirements for the two-sample t-procedures?

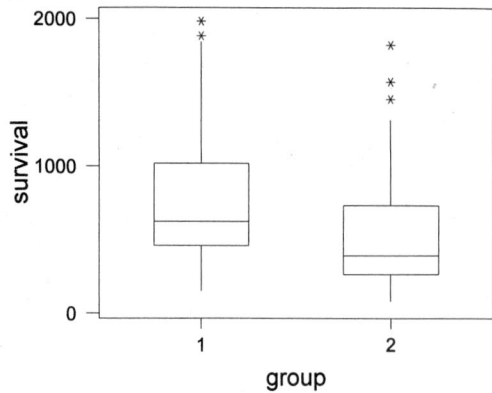

b. Notice from the descriptive statistics that there is a large difference between the respective means and medians of the two groups. What would cause such a large difference between a mean and a median?

Descriptive Statistics

Variable	N	Mean	Median	TrMean	StDev	SEMean
survivA	62	766.1	623.0	737.1	456.8	58.0
survivB	59	546.5	395.0	508.9	391.1	50.9

Variable	Min	Max	Q1	Q3
survivA	152.0	1980.0	461.8	1016.7
survivB	83.0	1820.0	265.0	731.0

c. Is it better to compare the population means or the population medians for these data? Explain your answer.
d. Without formally analyzing these data, do you think one treatment is better than the other in increasing survival time?

9.4 INFERENCE ABOUT THE DIFFERENCE BETWEEN TWO POPULATION CENTERS WHEN THE VARIANCES ARE EQUAL

In this section, we present inference procedures for comparing the centers of two population distributions when the population variances are homogeneous; that is, when σ_1^2 and σ_2^2 are equal. First we give the pooled t-procedures based on $\bar{y}_1 - \bar{y}_2$, and then we give a nonparametric test for cases where the test based on $\bar{y}_1 - \bar{y}_2$ is not appropriate.

Pooled
t-Procedures

In certain situations, such as some randomized experiments, it is safe to assume that the population variances are equal. In Example 9.6, we compared wind speed data in the months of January and July at a site proposed for a wind generator. From the summary data, we see that there is very little difference between the two sample variances: $s^2_{\text{January}} = 26.42$ and $s^2_{\text{July}} = 23.875$. In this problem, assuming that the population variances are equal is probably justified. In Example 9.7, however, $s^2_1 = (14.06)^2 = 197.6836$ and $s^2_2 = (24.44)^2 = 597.3136$. The larger sample variance is more than 3 times greater than the smaller sample variance. There may be some doubt whether the two population variances are equal. Some statisticians feel comfortable in assuming homogeneous population variances as long as one sample variance is not more than 4 times greater than the other, whereas others say not more than 3 times greater. We take a closer look at this issue later.

To develop the *pooled t-procedures*, let us assume that the population variances are equal. Then we can write the standard deviation of $\bar{y}_1 - \bar{y}_2$ as

> Because $\sigma^2_1 = \sigma^2_2$, the common value can be factored out of the two terms and then taken out from under the square root.

$$\sqrt{\frac{\sigma^2_1}{n_1} + \frac{\sigma^2_2}{n_2}} = \sigma \sqrt{\frac{1}{n_1} + \frac{1}{n_2}}$$

where σ is the common standard deviation. Because $\sigma^2_1 = \sigma^2_2 = \sigma^2$, we have that both s^2_1 and s^2_2 are unbiased estimators of σ^2. We can combine the two estimators using a weighted or pooled average to obtain a single unbiased estimator of σ^2:

$$s^2_p = \frac{(n_1 - 1)s^2_1 + (n_2 - 1)s^2_2}{n_1 + n_2 - 2}$$

Observe that s^2_1 is weighted by its degrees of freedom, $n_1 - 1$, and s^2_2 is weighted by its degrees of freedom, $n_2 - 1$. The *pooled variance*, s^2_p, then has

> Notice that if $n_1 = n_2$, then $s^2_p = (s^2_1 + s^2_2)/2$.

$$n_1 - 1 + n_2 - 1 = n_1 + n_2 - 2$$

degrees of freedom, which appears in the denominator. Using $s_p = \sqrt{s^2_p}$ as an estimate of σ, we have the standard error:

$$\text{SE}(\bar{y}_1 - \bar{y}_2) = s_p \sqrt{\frac{1}{n_1} + \frac{1}{n_2}}$$

If the parent distributions are normal, it follows that

$$t = \frac{(\bar{y}_1 - \bar{y}_2) - (\mu_1 - \mu_2)}{s_p \sqrt{1/n_1 + 1/n_2}}$$

has a Student's t distribution with $n_1 + n_2 - 2$ degrees of freedom. Consequently, the critical value, t^*, for the confidence interval and the p-value for the test of hypothesis will be found in the t-table (Table B.3).

In light of our discussion, we have the following pooled t inference procedures for $\mu_1 - \mu_2$:

Pooled t-Procedures Based on $\bar{y}_1 - \bar{y}_2$

Application: Symmetric distributions with tails that are not excessively long. If the parent distributions deviate substantially from normality, then the sample sizes should be larger than 30.

Assumptions: The two samples are independent of each other and the population variances are equal.

A $(1 - \alpha)100\%$ confidence interval for $\mu_1 - \mu_2$ is given by the limits

$$(\bar{y}_1 - \bar{y}_2) \pm t^* s_p \sqrt{\frac{1}{n_1} + \frac{1}{n_2}}$$

where

$$s_p = \sqrt{\frac{(n_1 - 1)s_1^2 + (n_2 - 1)s_2^2}{n_1 + n_2 - 2}}$$

and t^* is the upper $\alpha/2$ critical point of the t distribution with $n_1 + n_2 - 2$ degrees of freedom.

The test of hypothesis is called the **pooled t-test**:

Left-Tailed Test	Right-Tailed Test	Two-Tailed Test
H_0: $\mu_1 - \mu_2 \geq 0$	H_0: $\mu_1 - \mu_2 \leq 0$	H_0: $\mu_1 - \mu_2 = 0$
H_a: $\mu_1 - \mu_2 < 0$	H_a: $\mu_1 - \mu_2 > 0$	H_a: $\mu_1 - \mu_2 \neq 0$

Standardized test statistic:

$$t = \frac{\bar{y}_1 - \bar{y}_2}{s_p \sqrt{1/n_1 + 1/n_2}}$$

From the t-table with $n_1 + n_2 - 2$ degrees of freedom, the p-value is the tail probability most closely associated with the observed value of t, t_{obs}. If t_{obs} falls between two table values, then give the two associated probabilities as bounds for the p-value.

The p-value for the two-tailed test is calculated as in the single-tailed test and then doubled to account for both tails.

If a level of significance α is specified, then reject H_0 if p-value $< \alpha$.

EXAMPLE 9.9

Polychlorinated biphenyls (PCBs) are classified as health hazards. To study the concentration of PCBs in a river, a group of environmentalists measured the PCB levels (in parts per million) in fish at two locations on the river. Their results are given here:

	n	$\bar{y}$	s
Location A	8	25.2	3.8
Location B	8	23.1	4.2

Assuming that the variability of the concentration is the same in the two areas and that the distributions are close to normal, construct a 95% confidence interval to estimate the difference between the mean concentration levels at the two sites.

SOLUTION

Because $n_1 = n_2 = 8$, it is important that the parent populations be reasonably close to normally distributed. The degrees of freedom are $n_1 + n_2 - 2 = 14$, and for a 95% confidence interval we have $\alpha = .05$, so that $\alpha/2 = .025$. We then find from the t-table with 14 degrees of freedom that $t^* = 2.145$. The pooled variance is found as follows:

> Also, because $n_1 = n_2$, we have $s_p^2 = \dfrac{(3.8)^2 + (4.2)^2}{2}$

$$s_p^2 = \frac{(n_1 - 1)s_1^2 + (n_2 - 1)s_2^2}{n_1 + n_2 - 2} = \frac{7(3.8)^2 + 7(4.2)^2}{14} = 16.04$$

Therefore, the standard error is

$$\text{SE}(\bar{y}_1 - \bar{y}_2) = s_p\sqrt{\frac{1}{n_1} + \frac{1}{n_2}}$$

$$= \sqrt{16.04}\,\sqrt{\frac{1}{8} + \frac{1}{8}}$$

$$= 2.0025$$

The confidence interval becomes

$$(25.2 - 23.1) \pm 2.145(2.0025) \quad \text{or} \quad 2.1 \pm 4.295$$

which in turn gives the interval $(-2.195,\ 6.395)$.

Notice that the confidence interval runs from about -2.2 to $+6.4$, which includes both positive and negative values. This means that a test of hypothesis would not detect a significant difference (with $\alpha = .05$) between the two population means.

The hypothesis test for comparing two population means based on the pooled t-test is developed in a manner similar to the development of the confidence interval.

EXAMPLE 9.10

A random sample of 14 fourth-graders took a standardized achievement test immediately after an hour of recess. A second random sample of 16 fourth-graders took the same test after a 1-hour rest period. From the summary data on page 578, test the hypothesis of no difference between the mean scores for the two groups.

	n	$\bar{y}$	s
Recess group	14	56.5	6.2
Rest group	16	62.2	9.8

SOLUTION

Because the scores on most standardized tests are normally distributed, the test of hypothesis is based on the difference between the ordinary sample means. The null and alternative hypotheses are:

$$H_0: \mu_1 = \mu_2 \quad \text{versus} \quad H_a: \mu_1 \neq \mu_2$$

where μ_1 denotes the mean test score for the recess group and μ_2 denotes the mean test score for the rest group. The test statistic is

$$t = \frac{\bar{y}_1 - \bar{y}_2}{s_p\sqrt{1/n_1 + 1/n_2}}$$

The sampling distribution is t with degrees of freedom

$$n_1 + n_2 - 2 = 14 + 16 - 2 = 28$$

From the sample data we have for the observed value of t:

$$t_{obs} = \frac{56.5 - 62.2}{\sqrt{[13(6.2)^2 + 15(9.8)^2]/28}\sqrt{1/14 + 1/16}}$$

$$= \frac{-5.7}{(8.3245)(.366)}$$

$$= -1.87$$

From the t-table with 28 degrees of freedom, we see that

$$1.701 < 1.87 < 2.048$$

and therefore

> The p-value is doubled because this is a two-tailed test.

$$2(.025) < p\text{-value} < 2(.05) \quad \text{or} \quad .05 < p\text{-value} < .10$$

Thus, we can say that there is a mildly significant difference between the rest group and the recess group on the achievement test.

*Computer
Analysis*

C O M P U T E R   T I P

The pooled *t*-procedures are executed in the same manner as the two-sample *t*-test given in the preceding section. The only difference is that for the pooled *t*-procedures, you must click on the **Assume equal variance** box and then click on **OK**.

EXAMPLE 9.11

A math achievement test is given to a random sample of 25 high school students. The scores and gender (coded as 1 for girls and 2 for boys) are given here:

Worksheet: Achieve.mtw

Gender	1	2	2	1	2	1	1	1	1	2	2	1	2
Score	87	68	87	91	67	78	81	72	95	74	81	89	93

Gender	2	2	1	2	1	1	2	1	2	2	1	1
Score	60	78	93	74	83	74	92	75	81	62	85	95

Is there a significant difference between the scores for boys and girls? Store the data in the Minitab worksheet and conduct the *t*-test.

SOLUTION

Let μ_1 denote the mean score for girls and μ_2 denote the mean score for boys. Then the null and alternative hypotheses are

$$H_0: \mu_1 = \mu_2 \quad \text{versus} \quad H_a: \mu_1 \neq \mu_2$$

Rejection of the null hypothesis indicates a difference between the scores for boys and girls.

To conduct a pooled *t*-test, we should first check assumptions. The normal probability plots in Figure 9.3 show no unusual deviations from normality. The side-by-side boxplots

FIGURE 9.3

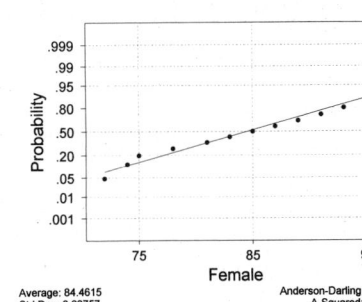

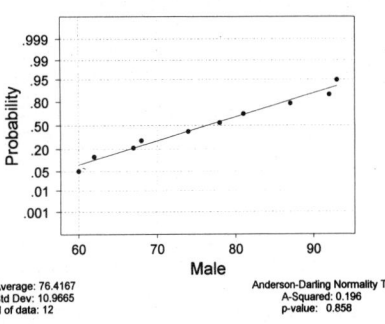

FIGURE 9.4

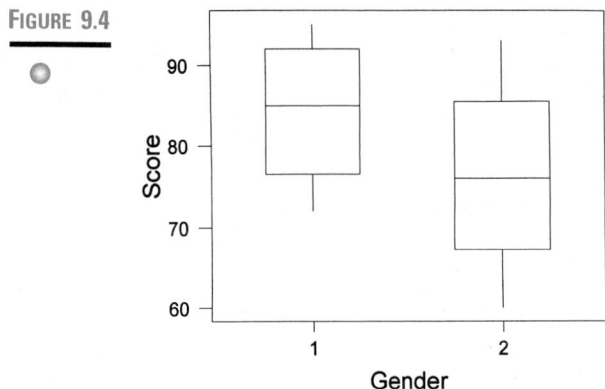

in Figure 9.4 show that the variances are reasonably homogeneous. The boxplots also show that girls generally scored higher than boys. To determine whether the difference is significant, we will conduct a pooled *t*-test with Minitab.

Figure 9.5 is the computer output. Notice that the two standard deviations are close in value. The ratio of the sample variances is $(11.0/8.04)^2 = 1.87$, which suggests that the population variances are homogeneous. Based on *p*-value $= .047$, there is evidence of a significant difference between the scores for boys and girls.

FIGURE 9.5

```
MTB > TwoT 95.0 'score' 'gender';
SUBC>    Alternative 0;
SUBC>    Pooled.

Two Sample T-Test and Confidence Interval

Twosample T for score
gender    N      Mean      StDev    SE Mean
1        13      84.46      8.04       2.2
2        12      76.4      11.0       3.2
95% C.I. for mu 1 - mu 2: ( 0.1,   16.0)
T-Test mu 1 = mu 2 (vs not =): T= 2.10   P=0.047   DF=  23
Both use Pooled StDev = 9.55
```

When to Use the Pooled t-Procedures

The pooled *t*-procedures assume that the populations are reasonably close to normally distributed and their variances are homogeneous. The normal probability plot is excellent for detecting deviations from normality. Although there is a formal test for the equality of population variances, we have chosen not to present it because it is so sensitive to moderate departures from normality (extremely nonrobust). Moser and Stevens (1992) suggest not to conduct any formal test for homogeneity and to use the pooled *t*-test only when the ratio of population variances is *known* to be near 1.0. In practice, the ratio of the population variances is rarely known; therefore, their suggestion may be too restrictive. Instead, we suggest that you informally compare the variability of the two samples by viewing side-by-side boxplots of the data and look at the ratio of the sample variances before applying either the pooled t- or the two-sample t-procedures. If the ratio is greater than 4, we should not assume homogeneous variances; if the ratio is less than 3, it is probably safe to assume homogeneous variances. If the ratio is between 3 and 4, as in Example 9.7, we should use the pooled t-procedure if the sample sizes are close; otherwise, use the two-sample t-procedure.

The Wilcoxon Rank Sum Test

If we are not sure whether the assumptions for the pooled t-procedures are met, there are certain situations when we may use the nonparametric *Wilcoxon rank sum test*. For the two-sample Wilcoxon rank sum test, no specific distributional assumptions are made except that if a difference exists between the distributions, then it is in their locations. This means that the two distributions should be similar in shape and have rather homogeneous variances.

For this test, the population median is the preferred measure of location to study. Inferences, then, are concerned with the difference between the population medians. We give a modified version of the test in which the ordinary pooled t-test is applied to the *rank-transformed* data (Conover and Iman, 1981). The resulting test gives a good approximation to the Wilcoxon test when both sample sizes are at least 10.

To rank-transform a data set, arrange the data in increasing order and assign ranks of 1, 2, 3, and so on. The test is then conducted on the ranks. If two or more observations have the same value (tied), then average ranks are given. For example, this data set

> Remember that the boxplot is an excellent tool for detecting a distribution's shape. Side-by-side boxplots are useful for comparing shapes.

| 8.6 | 11.2 | 9.3 | 7.4 | 11.2 | 10.4 | 9.1 | 12.1 |

has two entries of 11.2. They would normally be assigned ranks of 6 and 7, so each is assigned a rank of $(6 + 7)/2 = 6.5$. The completed transformed ranking (in the same order as the given data) is

| 2 | 6.5 | 4 | 1 | 6.5 | 5 | 3 | 8 |

To conduct the two-sample Wilcoxon rank sum test, the two samples are combined into one data set and then ranks are assigned. Then the pooled t-test is applied to the two sets of ranks.

The Two-Sample Wilcoxon Rank Sum Test

Application: Comparing locations of general populations

Assumptions: Population distributions are similar except for possibly different centers. Samples are independent; $n_1 \geq 10$ and $n_2 \geq 10$.

The test of hypothesis procedures are the same as the pooled t-test applied to the rank-transformed data.

The next example illustrates the test.

EXAMPLE 9.12

A study was undertaken to determine whether one's acquisition of a response is influenced by taking a particular drug. The dependent measurement is the number of trials required to master a given task. A group of 28 subjects are randomly assigned to the experimental and control groups. Those assigned to the experimental group are given the drug, and those assigned to the control group are given a placebo. After a given period of time, the numbers of trials to master the task are recorded as listed here:

Worksheet: Drug.mtw

Experimental group

17	15	5	14	18	3	16
13	15	16	17	8	19	

Control group

14	10	1	12	11	14	8	10
2	12	16	12	15	4	12	

Test the hypothesis to see whether the drug has an effect on the dependent measurement.

SOLUTION

Figure 9.6 is a back-to-back stem and leaf plot, and Figure 9.7 shows side-by-side boxplots. The boxplots indicate that both distributions are skewed left. Because the distributions are not symmetric (but have the same shape), we use the two-sample Wilcoxon rank sum test to test the equality of the population medians. From the back-to-back stem and leaf plot, it is easy to rank the combined samples.

FIGURE 9.6

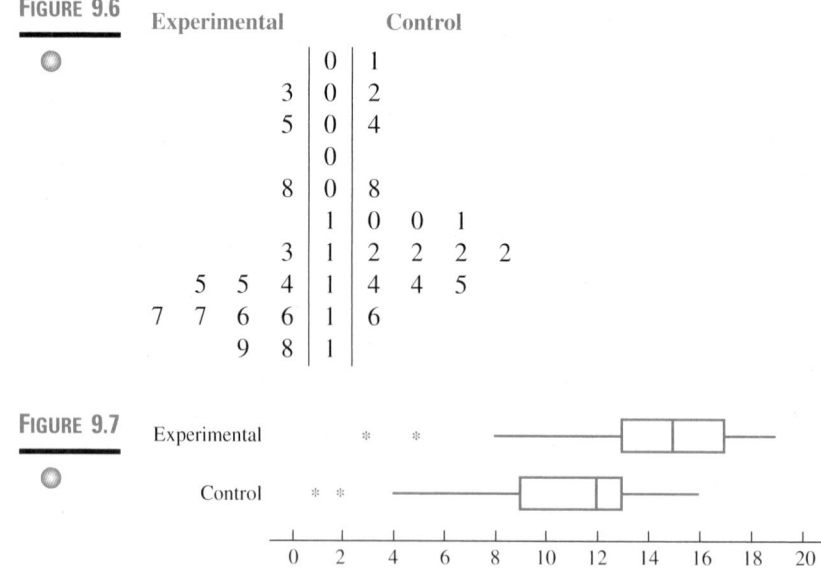

Experimental				Control			
			0	1			
		3	0	2			
		5	0	4			
			0				
		8	0	8			
			1	0	0	1	
		3	1	2	2	2	2
	5 5 4	1	4	4	5	
7 7 6 6	1	6			
	9 8	1			

FIGURE 9.7

Figure 9.8 is the same stem and leaf plot with the observations replaced by their ranks.

FIGURE 9.8

```
        Experimental                          Control
                          0 | 1
                    3     0 | 2
                    5     0 | 4
                          0 |
                  6.5     0 | 6.5
                          1 | 8.5   8.5   10
                 15       1 | 12.5  12.5  12.5  12.5
        20   20   17      1 | 17    17    20
 25.5  25.5  23   23      1 | 23
             28   27      1 |
```

The averages and standard deviations for the ranks in the two groups are given here:

Experimental: $\bar{r}_1 = 18.346$, $s_{r_1} = 8.594$ Control : $\bar{r}_2 = 11.167$, $s_{r_2} = 6.377$

$$s_p = \sqrt{\frac{12(8.594)^2 + 14(6.377)^2}{26}} = 7.482$$

Given that the drug may have a positive or a negative effect on the number of trials to master the task, we conduct a two-tailed test.

Let θ_1 = the median number of trials to complete the task with the drug and θ_2 = the median without the drug. The null and alternative hypotheses are:

$$H_0: \theta_1 - \theta_2 = 0 \quad \text{versus} \quad H_a: \theta_1 - \theta_2 \neq 0$$

The test statistic is the usual t-test applied to the rank-transformed data. Thus, we have

$$t_{obs} = \frac{18.346 - 11.167}{7.482\sqrt{1/13 + 1/15}} = 2.53$$

From the t-table with 26 degrees of freedom, we find that $2(.005) < p\text{-value} < 2(.01)$, which gives $.01 < p\text{-value} < .02$. Thus, we can declare a significant difference between the two groups; that is, there is sufficient evidence to say that the drug has an effect on the number of trials needed to master the task.

> The null hypothesis for the Wilcoxon rank sum test can be stated more generally as H_0: the population distributions are identical. In this example, the distributions differ at most in their medians; hence, the null hypothesis is stated as an equality of population medians.

Side-by-side boxplots are excellent diagnostic tools to reveal the presence of outliers that indicate a long-tailed distribution. To investigate the difference between population means when the parent distributions are symmetric with long tails, it has been suggested that the inference procedure be based on the difference between trimmed means. The large-sample test based on trimmed means is presented in Section 9.6. In the small-sample case, however, we use the Wilcoxon rank sum test when the t-test is in doubt and, in particular, when the parent distributions are symmetric with long tails. The test is illustrated in the next example.

EXAMPLE 9.13

These are the social adjustment scores for a rural group and a city group of children. Test to see whether there is a significant difference between the social adjustment scores for the two groups of children.

Worksheet: Rural.mtw

Rural

55 57 62 58 34 52 63 84
50 56 98 58 54 60 55 51

City

61 59 64 42 58 65 81 67 69
23 63 51 65 68 53 61 68

SOLUTION

The null and alternative hypotheses are:

$$H_0: \mu_1 = \mu_2 \quad \text{versus} \quad H_a: \mu_1 \neq \mu_2$$

where μ_1 denotes the mean social adjustment score for the rural group and μ_2 denotes the mean social adjustment score for the city group. Figure 9.9 is a back-to-back stem and leaf plot, and Figure 9.10 shows side-by-side boxplots for the two groups. Because of the long tails, we should not conduct a t-test. Noting that the two distributions are similar in shape, we compare the population means using the Wilcoxon rank sum test.

FIGURE 9.9

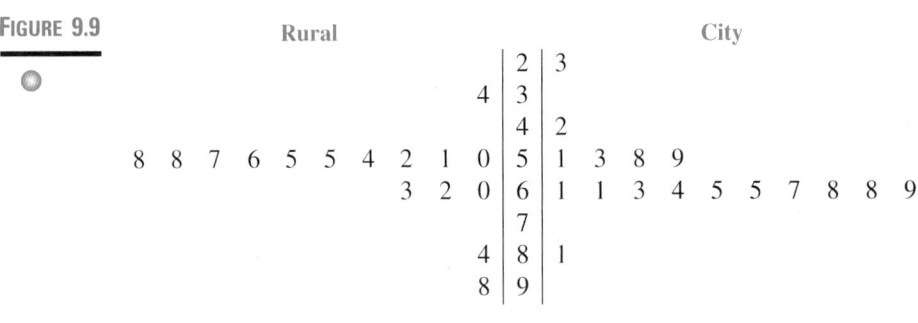

FIGURE 9.10

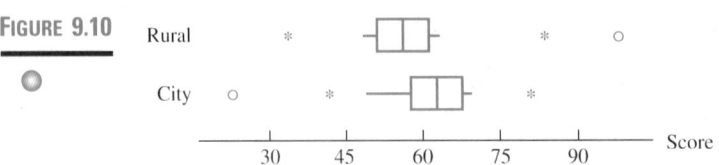

Replacing the data in the stem and leaf plot with ranks, we get Figure 9.11.

FIGURE 9.11

| | | Rural | | | | | | | | | 2 | 1 | | | City | | | | | | | | |

```
                                                          Rural                                    City

                                                                   2 | 1
                                                                2  3
                                                                   4 | 3
              15  15  13  12  10.5  10.5  9      7  5.5    4  5 | 5.5    8    15  17
                                        22.5  21  18  6 | 19.5  19.5  22.5  24  25.5  25.5  27  28.5  28.5  30
                                                                   7
                                                            32  8 | 31
                                                            33  9
```

From the ranks, we get these averages and standard deviations:

Rural: $n_1 = 16, \bar{r}_1 = 14.375, s_{r_1} = 9.099$

City: $n_2 = 17, \bar{r}_2 = 19.471, s_{r_2} = 9.783$

The test statistic is

$$t = \frac{\bar{r}_1 - \bar{r}_2}{s_p \sqrt{1/n_1 + 1/n_2}}$$

$$t_{obs} = \frac{14.375 - 19.471}{9.458 \sqrt{1/16 + 1/17}} = -1.55$$

From the t-table and $16 + 17 - 2 = 31$ degrees of freedom, we find that

$$2(.05) < p\text{-value} < 2(.10) \quad \text{or} \quad .10 < p\text{-value} < .20$$

which indicates no significant difference between the rural and city social adjustment scores.

For the Wilcoxon rank sum test, the two samples must be combined and ranked before we can proceed with a pooled t-test on the ranks. For computer analysis, this is best accomplished if the data are in one column and the groups specified in another column.

EXAMPLE 9.14

A new method of making concrete blocks has been proposed. To test whether or not the new method increases the compressive strength, ten sample blocks are made by each method. The compressive strengths in pounds per square inch are listed here:

Worksheet: Concrete.mtw

New Method

| 152 | 147 | 134 | 146 | 138 | 156 | 145 | 137 | 157 | 160 |

Old Method

| 132 | 146 | 151 | 127 | 137 | 125 | 138 | 141 | 127 | 137 |

Store the data in a column of the Minitab worksheet. In an adjacent column, store a group identifier such as a 1 for the new method and a 2 for the old method. Test the hypothesis that the new method is no better than the old using the two-sample Wilcoxon rank sum test.

SOLUTION

If we let μ_1 denote the mean compressive strength with the new method and μ_2 denote the mean compressive strength with the old method, then the null and alternative hypotheses are

$$H_0: \mu_1 \leq \mu_2 \quad \text{versus} \quad H_a: \mu_1 > \mu_2$$

Because this is a right-tailed test, we must choose the *greater than* alternative in the 2-sample t window.

Note that if the null hypothesis is rejected, then the alternative hypothesis is established, which says that the mean compressive strength is greater under the new method than under the old method. We now store the data in C1 (name it Strength) and the method coded as 1 or 2 in C2 of the Minitab worksheet.

C O M P U T E R   T I P

To rank the data in C1, choose the **Manip** menu and select **Rank**. When the Rank window appears, **Select** C1 for the **Rank data in** box and choose C3 for the **Store ranks in** box and select **OK**. Now we have the ranks in C3 and the codes for the methods in C2. To complete Wilcoxon's rank sum test, simply apply the two-sample *t*-test to the ranks in C3 with the subscripts in C2.

In the Session window the commands are:

MTB > **Rank 'Strength' c3.**
MTB > **TwoT 95.0 'Ranks' 'Method';**
SUBC> **Alternative 1;**
SUBC> **Pooled.**

Figure 9.12 is the computer output. Based on this *p*-value, there is highly significant evidence that the new method has increased the compressive strength.

FIGURE 9.12 Two Sample T-Test and Confidence Interval

```
Twosample T for Ranks
Method   N       Mean      StDev    SE Mean
1       10      13.60      5.16       1.6
2       10       7.40      5.05       1.6

95% C.I. for mu 1 - mu 2: ( 1.4,  11.0)
T-Test mu 1 = mu 2 (vs >): T= 2.72  P=0.0071  DF=  18
Both use Pooled StDev = 5.11
```

EXERCISES 9.4

9.46 These data are arranged in a back-to-back stem and leaf plot:

Worksheet: Exer9-46.mtw

			0	14	1		
				15			
			5	16			
				17	3		
		8	4	18	2	7	
8	5	3		19	0	8	
		2	0	20	1	3	7
		9	7	21	2	8	9
			3	22			

a. If these data are independent random samples from two populations, do you think that inferences should be about $\mu_1 - \mu_2$ or $\theta_1 - \theta_2$? Explain your reasoning.
b. Based on your answer to part **a**, conduct either a pooled t-test or the two-sample Wilcoxon rank sum test.

9.47 It is believed that the mean amount of coffee dispensed by vending machine A in the student lounge is less than that of machine B in the cafeteria. These summary statistics were obtained from samples of each machine:

	n	$\bar{y}$	s
Machine A	10	9.8	1.4
Machine B	12	10.1	.8

a. State the null and alternative hypotheses to determine whether the mean amount of coffee dispensed by vending machine A in the student lounge is less than that of machine B in the cafeteria.
b. To test the claim with the pooled t-test, we must make certain assumptions. What are two of those assumptions?
c. Assume that the assumptions in part **b** are met and proceed with the test. Is there statistical evidence to support the conjecture?

9.48 With $\bar{y}_1 - \bar{y}_2$ as the test statistic and the following summary statistics, test the null hypothesis $H_0: \mu_1 - \mu_2 \geq 0$ versus $H_a: \mu_1 - \mu_2 < 0$.

$$n_1 = 45, \bar{y}_1 = 115.3, s_1 = 15.92$$

$$n_2 = 45, \bar{y}_2 = 123.6, s_2 = 17.65$$

    a. Calculate the standard error of the test statistic.

    b. Calculate the observed value of the standardized test statistic.

    c. Is this a one-tailed or a two-tailed test? Calculate the p-value.

    d. Based on the p-value, should the null hypothesis be rejected?

9.49 An instructor of an introductory history course is interested in knowing, in general, whether the average grades on the final exam differ from the fall semester to the spring semester. From the summary statistics, construct a 90% confidence interval based on $\bar{y}_1 - \bar{y}_2$ for the difference between the two population means:

	n	$\bar{y}$	s
Fall	150	82.4	11.56
Spring	150	84.2	11.44

Does the interval contain both positive and negative values? What does this mean about the average grades for the fall and spring classes?

9.50 Strength tests on two types of wool fabric produced these results:

Worksheet: Wool.mtw

Type 1

| 138 | 127 | 148 | 134 | 125 | 136 | 152 | 110 | 137 | 160 |

Type 2

| 134 | 137 | 135 | 140 | 130 | 134 | 120 | 157 | 162 | 114 |

    a. State the null and alternative hypotheses to see whether there is a significant difference between the mean strengths of the two types of wool.

    b. If we want to conduct a pooled t-test with small samples, the data should come from normal populations with homogeneous variances. Based on these normal probability plots of the two samples, is the normality assumption met?

Normal Probability Plot	Normal Probability Plot

Average: 136.7
Std Dev: 14.3686
N of data: 10

Anderson-Darling Normality Tes
A-Squared: 0.203
p-value: 0.828

Average: 136.3
Std Dev: 14.5987
N of data: 10

Anderson-Darling Normality Test
A-Squared: 0.407
p-value: 0.281

    c. Based on the side-by-side boxplots of the two samples, do the variances of the two samples appear to be homogeneous?

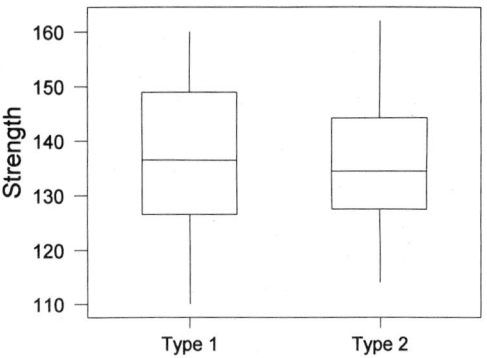

d. Given this computer printout, complete the pooled *t*-test and state your conclusion.

```
Two Sample T-Test and Confidence Interval

Twosample T for Type 1 vs Type 2

            N      Mean     StDev    SE Mean
Type 1     10     136.7     14.4       4.5
Type 2     10     136.3     14.6       4.6

95% C.I. for mu Type 1 - mu Type 2: ( -13.2, 14.0)
T-Test mu Type 1 = mu Type 2 (vs not =): T= 0.06   P=0.95   DF= 18
Both use Pooled StDev = 14.
```

9.51   Nationally, 25% of all college freshmen enroll in some type of remedial math course. To determine whether men and women differ in their pre-enrollment ability, a math placement exam (scored from 0 to 40) was given to a sample of 35 incoming freshmen women and a sample of 42 incoming freshmen men. These are their scores:

Worksheet: Remedial.mtw

Females

28	21	4	20	16	19	39
22	5	18	17	21	19	3
11	22	19	18	17	35	21
18	16	5	19	21	20	20
40	21	19	20	22	38	3

Males

18	22	16	14	2	16	18
20	22	19	11	35	18	22
4	21	20	15	19	17	38
18	16	4	19	18	16	17
33	16	19	18	21	20	17
17	13	14	19	15	38	40

a. Based on side-by-side boxplots of the two samples at the top of page 590, does there appear to be a significant difference between the mean abilities of men and women?

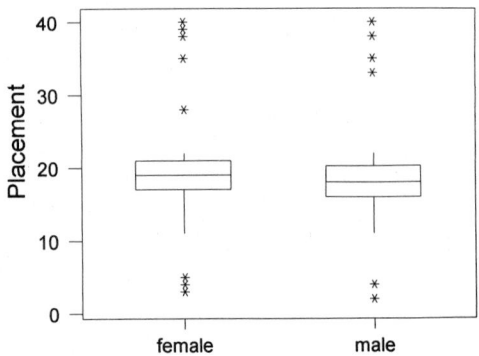

b. Use the descriptive statistics for the two samples to construct a 98% confidence interval based on $\bar{y}_1 - \bar{y}_2$ for the mean difference in ability for men and women.

```
Descriptive Statistics

Variable        N     Mean    Median    TrMean    StDev    SEMean
female         35    19.34     19.00     19.10     9.09      1.54
male           42    18.93     18.00     18.71     8.03      1.24


Variable      Min      Max        Q1        Q3
female       3.00    40.00     17.00     21.00
male         2.00    40.00     16.00     20.25
```

c. Does the interval from part **b** contain both positive and negative values? Does this suggest a significant difference between the abilities of men and women?

d. The confidence interval in part **b** is based on $\bar{y}_1 - \bar{y}_2$. From the appearance of the boxplots, can you think of an alternative procedure for finding a confidence interval for $\mu_1 - \mu_2$?

9.52   Measurements of viscosity for a certain substance were taken on two days:

Worksheet: Viscosit.mtw

First day

35.4   38.3   34.0   36.2   37.2   32.8   36.0   35.2   36.0   31.3   35.7

Second day

37.0   38.6   35.1   37.1   36.2   36.8   37.6   34.8   35.8   38.2   32.1

a. From the boxplots of the two samples, do you think the populations are sufficiently skewed that inferences should be made about the population medians instead of the population means?

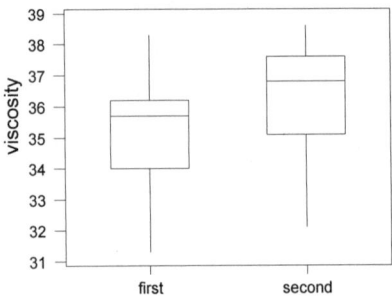

b. Investigate the normal probability plots of the two samples. Does it appear that the normality assumption for the pooled t-test has been violated?

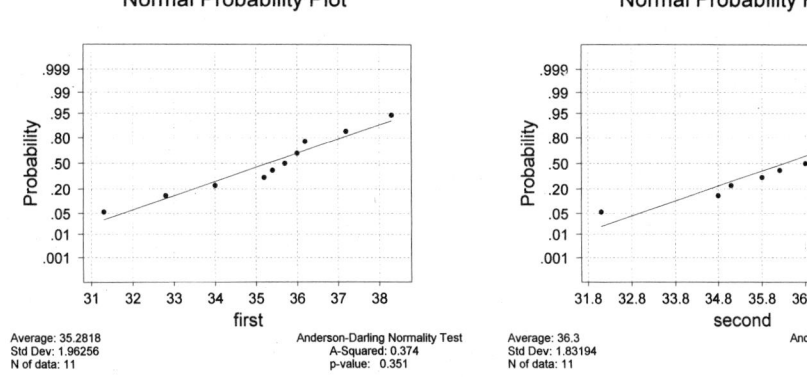

c. In terms of the population means, state the null and alternative hypotheses to determine whether the population has changed from one day to the next.

d. Using the computer printout, complete the pooled t-test and state your conclusion.

```
Two Sample T-Test and Confidence Interval

Twosample T for first vs second

                N       Mean      StDev      SE  Mean
first     11       35.28      1.96          0.59
second    11       36.30      1.83          0.55

95% C.I. for mu first - mu second: ( -2.71,   0.67)
T-Test mu first = mu second (vs not =): T= -1.26   P=0.22   DF= 20
Both use Pooled StDev = 1.90
```

9.53 A manufacturer thinks that the amount of carbon monoxide emitted by its smoke stacks is less than in the smoke from its competitor's stacks. The EPA released these readings:

Worksheet: Monoxide.mtw

Manufacturer

2.7   3.1   3.1   2.9   2.5   3.4   3.4   3.4   2.4

Competitor

3.7   3.0   3.5   3.8   2.8   3.5   3.4   3.6   2.7   3.7

a. State the null and alternative hypotheses to test the manufacturer's claim.

b. If we want to conduct a pooled t-test with small sample sizes, the data should come from a population that is approximately normally distributed with homogeneous variances. From the boxplots of the two samples (at the top of page 592), evaluate the normality assumption.

c. Do the variances of the two samples appear to be homogeneous?

d. The pooled t-test is robust except when the parent distributions are severely skewed or have unusually long tails. In light of your answers to parts **b** and **c**, do you think that the pooled t-test can be applied to these data?

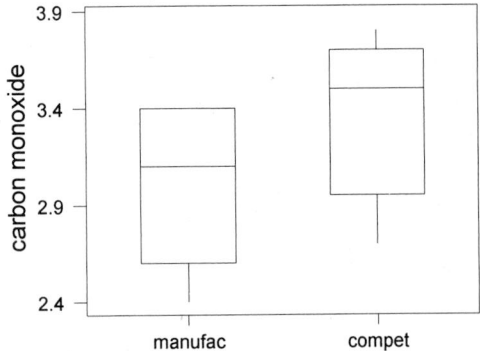

e. From the descriptive statistics, complete the pooled *t*-test and state your conclusion.

```
Descriptive Statistics

Variable          N        Mean     Median    TrMean      StDev     SEMean
manufac           9       2.989      3.100     2.989      0.389      0.130
compet           10       3.370      3.500     3.400      0.395      0.125


Variable        Min        Max         Q1         Q3
manufac       2.400      3.400      2.600      3.400
compet        2.700      3.800      2.950      3.700
```

9.54 Two independent random samples were selected from two populations with the following values:

Worksheet: Exer9-54.mtw

	Sample A							
6	7							
7	2							
7								
8	2	3						
8	6	7						
9	0	1	2	2	3	3	4	
9	5	5	6	8	8	9		
10								
10	5	7						
11								
11	8							
12								
12	5							

	Sample B							
6	4	9						
7	0							
7	5							
8	0	1	4					
8	5	6	7	7	8	9	9	
9	0	1	3	4				
9	5	8						
10	2							
10								
11	3							
11	8							
12								
12								

a. Do the samples appear to come from normal populations?

b. Do the variances appear to be homogeneous?

c. Would boxplots for the two samples help in answering the questions in parts **a** and **b**? Explain.

d. It is quite possible that the two samples came from long-tailed distributions. Is it advisable to test the hypothesis $\mu_1 = \mu_2$ using the pooled *t*-test?

e. Complete an analysis of the data by performing the two-sample Wilcoxon rank sum test.

9.55 A distributor of auto gears wishes to determine with which of two manufacturers of gears he should do business. He obtained, from the two manufacturers, these numbers of defective gears in the production of 100 gears per day for 20 consecutive days.

Worksheet: Autogear.mtw

Manufacturer A

16	25	15	26	21	22	17	26	23	20
28	32	43	18	16	36	21	16	29	26

Manufacturer B

28	24	28	42	17	31	26	33	26	24
32	36	38	26	25	30	21	36	18	34

a. State the null and alternative hypotheses to see whether there is a significant difference in the mean numbers of defective gears per day from the two manufacturers.

b. If we wish to conduct a pooled t-test with small sample sizes, the data should come from approximately normally distributed populations with homogeneous variances. Do the boxplots of the two samples indicate that the normality assumption is in question? Do the variances of the two samples appear to be homogeneous?

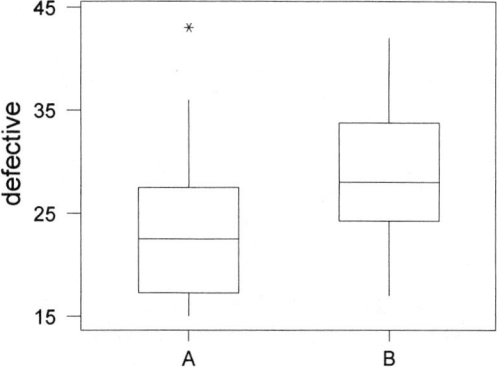

c. From the descriptive statistics, complete the pooled t-test and state your conclusion.

```
Descriptive Statistics

Variable        N      Mean    Median    TrMean    StDev    SEMean
A              20     23.80     22.50     23.22     7.32      1.64
B              20     28.75     28.00     28.67     6.59      1.47


Variable      Min       Max        Q1        Q3
A           15.00     43.00     17.25     27.50
B           17.00     42.00     24.25     33.75
```

9.56 Exercise 8.54 in Section 8.3 described the process of manufacturing integrated circuits (chips) used in computers. Recall that in one part of the process, a thin layer of silicon oxide is placed on the surface of a wafer. The thickness of the oxide layer is critical to the performance of the resulting chips. Two wafers are randomly selected from each of 30 lots. Four measurements of the thickness of the oxide layer on each wafer are then taken. Columns 1 and 2 of worksheet: **Chipavg.mtw** (page 594) contain the average of the four measurements for each of the two wafers selected from the 30 lots.

Worksheet: Chipavg.mtw

Wafer1

940.0	1,042.5	942.5	1,007.5	985.0	925.0	1,072.5	1,070.0
1,047.5	982.5	1,050.0	1,057.5	1,057.5	1,070.0	970.0	997.5
1,107.5	932.5	960.0	1,000.0	1,105.0	1,037.5	1,037.5	
1,015.0	1,002.5	972.5	1,070.0	1,005.0	842.5	1,057.5	

Wafer2

985.0	1,052.5	990.0	990.0	917.5	930.0	1,120.0	1,082.5
1,057.5	1,002.5	1,020.0	1,060.0	1,010.0	1,030.0	1,002.5	1,005.0
1,095.0	1,000.0	1,042.5	1,007.5	1,067.5	1,097.5	1,015.0	
1,102.5	1,057.5	1,097.5	1,007.5	930.0	887.5	955.0	

SOURCE: E. Yashchin, "Likelihood Ratio Methods for Monitoring Parameters of a Nested Random Effect Model," *Journal of the American Statistical Association* 90 (1995): 729–738.

a. From the data set, construct side-by-side boxplots of the two samples. Comment on the shapes of the two distributions. Are the two samples equally variable?

b. Construct normal probability plots of the two samples. Does normality seem reasonable in either distribution?

c. Based on your results in parts **a** and **b**, complete the pooled *t*-test, the two-sample *t*-test, or the two-sample Wilcoxon rank sum test to determine whether the mean thickness of the oxide layer is different for the two sets of wafers.

9.57   Patients with small cell lung cancer received one of two types of treatments (arm A or arm B) as described in Exercise 6.30 in Section 6.2. Their survival times are recorded in worksheet: **Censored.mtw**. [SOURCE: Ying et al. (1995)].

a. In Exercise 9.45 in Section 9.3, you determined that the two distributions are similarly skewed right. Does this mean that inferences should be about the population means or the population medians?

b. State null and alternative hypotheses for determining whether the median survival time is greater for arm A.

c. Test the hypotheses from part **b**. Be sure to give the *p*-value and state your conclusion.

9.58   In Exercise 6.12 in Section 6.1, 25 schizophrenic patients were classified as psychotic or nonpsychotic after being treated with an antipsychotic drug. Cerebrospinal fluid was taken from each patient and assayed for dopamine b-hydroxylase (DBH) activity. We wish to compare the listed DBH activity [units are nmol/(ml)(h)/(mg) of protein] for the two groups.

Worksheet: Dopamine.mtw

Nonpsych (group 1)

.0104	.0105	.0112	.0116	.0130
.0145	.0154	.0156	.0170	.0180
.0200	.0200	.0210	.0230	.0252

Psychotic (group 2)

.0150	.0204	.0208	.0222	.0226
.0245	.0270	.0275	.0306	.0320

SOURCE: D. E. Sternberg, D. P. Van Kammen, and W. E. Bunney, "Schizophrenia: Dopamine b-Hydroxylase Activity and Treatment Response," *Science* 216 (1982): 1423–1425.

For ease of calculation, the data values in the worksheet are multiplied by 10,000, so that .0104, for example, becomes 104. In Exercise 6.12 in Section 6.1, boxplots and normal probability plots of the two samples were constructed. Based on those results, it was concluded that both groups have approximately normally distributed parent populations with homogeneous variances. Does one group tend to have higher DBH activity? Conduct a formal test of significance to evaluate the antipsychotic drug. Be sure to identify the test you are using, give the *p*-value, and state your conclusion.

9.59   Exercise 2.57 in Section 2.3 described a special diet mixed with a drug compound designed to reduce low-density lipoproteins (LDL) cholesterol that was fed to a treatment group of quail. A placebo group of quail were fed the same special diet for the same period of time but without the drug compound.

Worksheet: Quail.mtw

Placebo

64	49	54	64	97	66	76	44	71	89
70	72	71	55	60	62	46	77	86	71

Treatment

40	31	50	48	152	44	74	38	81	64

SOURCE: J. McKean and T. Vidmar, "A Comparison of Two Rank-Based Methods for the Analysis of Linear Models," *The American Statistician* 48 (1994): 220–229.

In Exercises 7.41 and 7.42 of Section 7.3, you analyzed these data and determined that if we remove the one outlier in the treatment group (152), then both groups are reasonably close to normally distributed. Remove the outlier and conduct a formal test of significance to evaluate the drug compound. Be sure to identify the test you are using, give the *p*-value, and state your conclusion.

9.5   INFERENCE ABOUT THE DIFFERENCE BETWEEN TWO POPULATION CENTERS USING MATCHED SAMPLES

Suppose we wish to evaluate a new method of reading instruction for students with low ability. An important part of the evaluation is a comparison of reading achievement scores for students using the experimental method with those using the standard method. For the comparison, one possibility is to independently and randomly assign students to the two methods and compare their reading achievement scores at the end of the project. The difference in the mean achievement test scores, $\mu_1 - \mu_2$, can be investigated using the methods of Section 9.3 or 9.4.

   Although we are dealing with students who have low ability, they will still have varying levels of ability prior to exposure to the method of instruction. Given that they are independently and randomly assigned to the two groups, it is likely that the students assigned to the two groups will be comparable in reading skills; however, there is no guarantee. A difference in reading achievement at the end of the project might be due only to their different levels of skill prior to their instruction. The *matched pairs experiment* is a method of assigning subjects to groups that prevents this from happening and guarantees equivalent groups prior to the instruction.

Matched Pairs
t-Test

The matched pairs test is also called the paired difference test.

Suppose a reading skills test is given to all students *before* they are assigned to the groups. The students are then paired according to reading skill. One of each pair is assigned to the experimental group, and the other is assigned to the control group. A comparison of *matched pairs* of achievement test scores then gives a fair evaluation of the new method of instruction.

Analysis of the matched pairs experiment is straightforward. We reduce the matched pairs to a single sample by taking the difference between the two observations in each pair. These differences form a single sample that can be analyzed using the one-sample techniques of Chapter 8.

Suppose that the matched pairs of observations are

$$(x_1, y_1), (x_2, y_2), \ldots, (x_n, y_n)$$

where x_i is the score for the *i*th subject in group 1 and y_i is the score for the matching *i*th subject in group 2. Let $d_i = y_i - x_i$ for $i = 1, 2, \ldots, n$ be the difference in the scores. Then $d_1, d_2, \ldots, d_n$ form a single sample from which the mean, $\overline{d}$, and the standard deviation, s_d, can be calculated.

If we let $\mu_d = \mu_1 - \mu_2$ be the mean of the population of differences, then the null hypothesis can be stated as

$$H_0: \mu_d = 0$$

If the population of differences is normally distributed, then the test statistic

$$t = \frac{\overline{d}}{s_d / \sqrt{n}}$$

is distributed as Student's *t* with $n - 1$ degrees of freedom. The remainder of the test proceeds like the one-sample *t*-test described in Chapter 8.

The confidence interval for μ_d and the test of hypothesis are summarized in the box:

Inferences for $\mu_1 - \mu_2$ Based on Matched Samples

Application: Matched samples

Assumptions: If the sample size is small, then the difference scores should be approximately normally distributed.

Confidence interval: A $(1 - \alpha)100\%$ confidence interval for $\mu_1 - \mu_2$ based on matched samples is

$$\overline{d} \pm t^* \frac{s_d}{\sqrt{n}}$$

(continued)

> **Matched pairs *t*-test:**
>
> | Left-Tailed Test | Right-Tailed Test | Two-Tailed Test |
> |---|---|---|
> | $H_0: \mu_d \geq 0$ | $H_0: \mu_d \leq 0$ | $H_0: \mu_d = 0$ |
> | $H_a: \mu_d < 0$ | $H_a: \mu_d > 0$ | $H_a: \mu_d \neq 0$ |
>
> Standardized test statistic:
>
> $$t = \frac{\overline{d}}{s_d / \sqrt{n}}$$
>
> From the *t*-table with $n - 1$ degrees of freedom, find the *p*-value most closely associated with t_{obs}. If t_{obs} falls between two table values, then give the two associated probabilities as bounds for the *p*-value.
>
> The *p*-value for the two-tailed test is calculated as in the one-tailed test and then doubled to account for both tails.
>
> If a level of significance α is specified, then reject H_0 if *p*-value $< \alpha$.

EXAMPLE 9.15

A group of 24 low-ability students of the same age were tested for reading skills and then paired according to ability. One member of each pair was assigned to an experimental reading group and the other to a control group. The experimental group was taught using a new method of reading instruction, and the control group was taught using the standard method that normally is used to teach low-ability students. After the period of instruction, all students were given a reading achievement test with these results:

Worksheet: Lowabil.mtw

Pair	1	2	3	4	5	6	7	8	9	10	11	12
Experimental	82	65	63	71	48	74	61	65	73	92	57	66
Control	78	60	64	66	51	68	61	59	71	88	50	57

Perform a test of hypothesis to determine whether the experimental group performed significantly better than the control group.

SOLUTION

Because the data are matched, we investigate the difference scores:

Pair	1	2	3	4	5	6	7	8	9	10	11	12
Difference	4	5	-1	5	-3	6	0	6	2	4	7	9

The dot graph of the difference scores in Figure 9.13 exhibits a degree of skewness to the left. With so few observations, however, it is difficult to discern the shape of the underlying distribution. We will assume that it does not deviate appreciably from normality and conduct the matched pairs *t*-test.

FIGURE 9.13

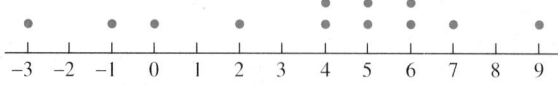

To determine whether the experimental group scored higher than the control group, a test of significance is performed with these hypotheses:

$$H_0: \mu_d \leq 0 \quad \text{versus} \quad H_a: \mu_d > 0$$

where $\mu_d = \mu_1 - \mu_2$, $\mu_1 =$ mean response for the experimental group, and $\mu_2 =$ mean response for the control group. From the differences, we easily find

> When calculating $\overline{d}$, be sure that the negative values are added in as negatives.

$$\overline{d} = 3.667 \qquad s_d = 3.525$$

and therefore,

$$t_{obs} = \frac{3.667}{3.525/\sqrt{12}} = 3.604$$

From the t-table with $n - 1 = 11$ degrees of freedom, we find that the t-score corresponding to .005 is 3.106. Because

$$t_{obs} = 3.604 > 3.106$$

we have p-value $< .005$. With such a small p-value, the evidence is highly significant that the new method of reading instruction for low ability students is better than the old one.

Before–After Experiment

Another application of the matched pairs design is the *before–after experiment*, in which the same subject serves in both groups. The subject is measured both before and after the treatment.

EXAMPLE 9.16

A group of women in a large city were given instructions on self-defense. Prior to the course, they were tested to determine their self-confidence. After the course they were given the same test. A high score on the test indicates a high degree of self-confidence. Do these self-confidence scores indicate that the course significantly increased their self-confidence?

Worksheet: Selfdefe.mtw

Woman	1	2	3	4	5	6	7	8	9
Before course	6	10	8	6	5	4	3	8	5
After course	7	12	7	5	8	6	5	8	6

SOLUTION

The data are matched, so difference scores are found by

$$\text{Difference} = (\text{after score}) - (\text{before score})$$

The null and alternative hypotheses are

$$H_0: \mu_d \leq 0 \quad \text{versus} \quad H_a: \mu_d > 0$$

where $\mu_d = \mu_1 - \mu_2$, μ_1 = mean self-confidence score after the course, and μ_2 = mean self-confidence score before the course. Rejection of the null hypothesis indicates that $\mu_1 > \mu_2$, which says that the self-defense course increased their self-confidence.
These differences are used to test the hypotheses:

Woman	1	2	3	4	5	6	7	8	9
Difference	1	2	-1	-1	3	2	2	0	1

from which we find

$$\overline{d} = 1.0 \qquad s_d = 1.414$$

The test statistic is

$$t_{obs} = \frac{1.0}{1.414/\sqrt{9}} = 2.122$$

From the t-table with 8 degrees of freedom, we find

$$.025 < p\text{-value} < .05$$

because $1.860 < 2.122 < 2.306$. Thus, there is evidence to reject H_0. Therefore, we can say that the self-defense course was effective in increasing self-confidence.

Computer Analysis

In Chapters 7 and 8, we introduced the one-sample t-procedures in Minitab. These commands also can be used to analyze the difference between two population means when the samples are matched. We have Minitab first compute the difference between the two observations in each pair and then perform the usual one-sample t-test or t confidence interval.

EXAMPLE 9.17

A psychologist who is interested in testing the relationship between stress and short-term memory administered a test to 12 subjects prior to their exposure to a stressful situation, and then retested them after the stress situation. From the following data can we conclude that the stress situation decreases one's performance on a test that measures short-term memory?

Worksheet: Stress.mtw

Subject	1	2	3	4	5	6	7	8	9	10	11	12
Pre-stress	13	15	9	13	15	17	13	16	11	13	9	12
Post-stress	10	14	7	15	11	14	13	14	9	14	9	10

Compute the difference in the scores and perform a one-tailed t-test.

SOLUTION

Let μ_1 denote the mean pre-stress test score and μ_2 the mean post-stress test score. Also let $\mu_d = \mu_2 - \mu_1$. Then the null and alternative hypotheses are:

$$H_0: \mu_d \geq 0 \quad \text{versus} \quad H_a: \mu_d < 0$$

Rejection of H_0 establishes the alternative, which says that the mean post-stress test score is significantly lower than the mean pre-stress test score.

C O M P U T E R   T I P

Having stored the pre-stress scores in C1 and the post-stress scores in C2 of the Minitab worksheet, we can compute the difference scores and store them in a third column in the following manner.

Under the **Calc** menu, select **Mathematical Expressions**. In the **Variable box** of the Mathematical Expressions window, identify the column where you wish to store the difference scores; for example, type **C3**. In the **Expression box**, type the mathematical expression **C2 - C1**, and then click on **OK**.

An alternative method is to use the **LET** command in the Session window. Simply type

MTB > **LET C3 = C2 − C1**

at the MTB prompt.

Having created the difference column in C3, we next apply the one-sample t- or z-test as before.

Figure 9.14 gives the output from the one-sample t-test with a left-tailed alternative. The p-value of .012 is significant, and thus we reject the null hypothesis. There is statistical evidence that the mean post-stress test score is significantly lower than the mean pre-stress test score.

FIGURE 9.14

```
MTB > Let c3 = c2 - c1
MTB > Name c3 'Differ'
MTB > TTest 0.0 'Differ';
SUBC>    Alternative -1.


T-Test of the Mean

Test of mu = 0.000 vs mu < 0.000

Variable   N    Mean  StDev  SE Mean   T    P-Value
Differ     12  -1.333 1.775   0.512  -2.60   0.012
```

Wilcoxon Signed Rank Test for the Matched Pairs Experiment

> A sufficient condition for the difference distribution to be symmetric is for the two distributions to be identical in shape and spread.

It is possible that the distribution of difference scores will deviate substantially from normality. In those cases, we have the *Wilcoxon signed rank test*. The null hypothesis for the Wilcoxon signed rank test states that the population distributions from which the matched pairs came are identical. If the two distributions differ only in location, then the test is a test of equality of means (as in the matched pairs *t*-test) or medians, depending on the measure of location that is used. Although this test does not require normality of the difference scores, it does assume that the distribution is symmetric.

As in the Wilcoxon rank sum test, we will give the Conover–Iman (1981) version of the Wilcoxon signed rank test, whereby the usual *t*-test is applied to the ranks. This rank-transform test provides a good approximation to the Wilcoxon signed rank test when the sample size is at least 10.

To conduct a test of hypothesis, the absolute values of the d_i's are ranked, and if the original d_i was negative, then the corresponding rank is given a negative sign. If a d_i is 0 (tied observations), then d_i is discarded from the analysis and the sample size is reduced by 1. The remainder of the test is applying the ordinary *t*-test to the signed ranks. The test statistic is

$$t = \frac{\bar{r}}{s_r / \sqrt{n}}$$

where $\bar{r}$ and s_r are the mean and standard deviation of the signed ranks, respectively. It is closely approximated by Student's *t*-distribution with $n - 1$ degrees of freedom when $n \geq 10$. The test is illustrated in the next example.

EXAMPLE 9.18

To evaluate a new fabric softener, a consumer organization purchased two of each type of ten different garments for children. The ten pieces of clothing were washed 15 times in the experimental softener. The like ten pieces were washed 15 times without a softener. After the washings, the two groups of clothing were measured for softness. (A high number indicates that the garment is judged more soft than one with a lower number.)

Worksheet: Fabric.mtw

Type garment	1	2	3	4	5	6	7	8	9	10
With softener	12	3	12	16	4	24	11	17	19	8
Without softener	8	4	15	14	6	21	10	15	22	7

Test to see whether the softener significantly increases the softness.

Solution

Clearly, the data are matched pairs, so we find the differences:

Garment	1	2	3	4	5	6	7	8	9	10
Difference	4	−1	−3	2	−2	3	1	2	−3	1

Figure 9.15

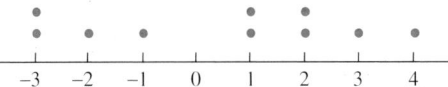

The dot graph of the differences in Figure 9.15 indicates that the normality assumption might not be met, yet symmetry seems plausible. From these observations, it seems reasonable to use the Wilcoxon signed rank test. The null and alternative hypotheses are

$$H_0: \mu_d \leq 0 \quad \text{versus} \quad H_a: \mu_d > 0$$

where $\mu_d = \mu_1 - \mu_2$, μ_1 = mean softness measurement with the softener, and μ_2 = mean softness measurement without the softener. Table 9.1 is useful in calculating the test statistic. Remember that tied observations are given average ranks. From the signed ranks, we find

$$\bar{r} = .9 \qquad s_r = 6.42$$

Table 9.1

d_i	$|d_i|$	Rank $|d_i|$	Signed ranks
4	4	10	10
−1	1	2	−2
−3	3	8	−8
2	2	5	5
−2	2	5	−5
3	3	8	8
1	1	2	2
2	2	5	5
−3	3	8	−8
1	1	2	2

Thus,

$$t_{obs} = \frac{.9}{6.42/\sqrt{10}} = .443$$

From the t-table with 9 degrees of freedom, we see that

$$p\text{-value} > .10$$

which indicates that there is insufficient evidence to claim that the softener is effective in making the clothing more soft.

C O M P U T E R   T I P

With the **Mathematical Expressions** menu or with the **Let** command in the Session window, Minitab can easily calculate the signed ranks from the original data.

As before, calculate the differences and store them in a column such as C3. Then in the **Variable box** of the Mathematical Expressions window, identify a column to store the signed ranks, such as **C4**. In the **Expression box**, type the following expression and then click on **OK**:

signs(C3)*rank(absolute(C3))

Or, if you prefer, in the Session window, simply type

Let C4 = signs(C3)*rank(absolute(C3))

Having stored the signed ranks in C4, you execute the ordinary one-sample t-test on column C4.

As pointed out in Chapter 1, the matched pairs experiment is a special case of the randomized block experiment. In the randomized block design, subjects are separated into blocks based on some extraneous variable that has a confounding effect on the dependent variable. After the blocking is complete, the subjects within each block are randomly assigned to the treatments. In the matched pairs experiment, each pair of subjects is a block, with one of the pair being randomly assigned to group 1 and the other to group 2.

EXERCISES 9.5

9.60   Explain why it would not be possible to conduct a before–after experiment on the data in Example 9.15.

9.61   As a test of learning ability, nine randomly selected eighth-graders were given a spelling test. After a 2-week course of instruction, they were given a similar test. Here are the test scores:

Worksheet: Spelling.mtw

Subject	1	2	3	4	5	6	7	8	9
Before	90	72	80	57	64	70	98	76	59
After	95	79	90	60	62	70	99	80	58

a. In what way are the two samples matched?

b. For the matched pairs t-test, the difference scores should be approximately normally distributed. Does this normal probability plot of the difference scores indicate that normality is a reasonable assumption in this case?

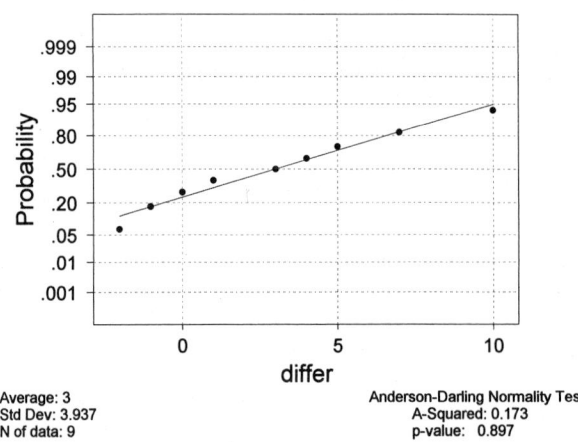

Normal Probability Plot

Average: 3
Std Dev: 3.937
N of data: 9

Anderson-Darling Normality Test
A-Squared: 0.173
p-value: 0.897

c. With so few scores it may be difficult to verify the normality assumption. Assume that it is met and complete the matched pairs t-test to determine whether there was improvement in the spelling scores.

9.62   To test the effectiveness of a drug to relieve asthma, a group of subjects was randomly given a drug and a placebo on two different occasions. After 1 hour an asthmatic relief index was obtained for each subject, with these results:

Worksheet: Asthmati.mtw

Subject	1	2	3	4	5	6	7	8	9
Drug	28	31	17	22	12	32	24	18	25
Placebo	32	33	19	26	17	30	26	19	25

a. To conduct a matched pairs *t*-test or confidence interval, the samples should be matched. In what way are these samples matched?
b. To conduct a matched pairs *t*-test, what is assumed about the distribution of the difference scores? Does the dotplot of the difference scores indicate that the assumption is met?

c. Assume that all assumptions are met. Do the data indicate that the drug significantly reduced the asthmatic relief index? Use the matched pairs *t*-test.

9.63 A study was designed to determine the effect of a certain movie on the moral attitude of young children. The scores are the ratings from 0 to 20 on a moral attitude scale recorded before and after the children viewed the film. A high score is associated with high morality.

Worksheet: Movie.mtw

Subject	1	2	3	4	5	6	7	8	9	10	11	12
Before	14	16	15	18	15	17	19	17	17	16	19	15
After	14	18	16	17	16	19	20	18	19	15	18	16

a. In what way are these samples matched?
b. To conduct a matched pairs *t*-test, what is assumed about the distribution of the difference scores? Based on the dotplot of the difference scores, can you tell whether the assumption is met?

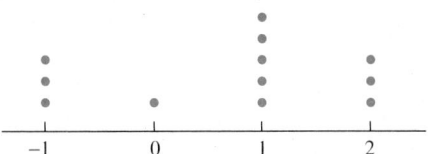

c. Assume that all assumptions are met and construct a 99% confidence interval for μ_d, the mean of the population of difference scores.
d. Based on the interval in part **c**, can you conclude that the movie had an effect on the moral attitude of the children?

9.64 The superintendent of school district A thinks that his students, on the whole, have better study habits than the students in school district B. Eleven students from each district were paired according to IQ and then their study habits were scored by an independent party. The results are given here:

Worksheet: Habits.mtw

A	105	109	115	112	124	107	121	112	104	101	114
B	115	103	110	125	99	121	119	106	100	97	105

a. To conduct a matched pairs t-test, the samples should be matched. In what way are these samples matched?
b. To conduct a matched pairs t-test, what is assumed about the distribution of the difference scores? Based on the dotplot and normal probability plot of the difference scores, do you think that the assumption is met?

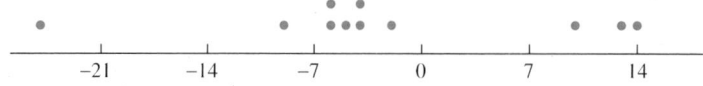

Normal Probability Plot

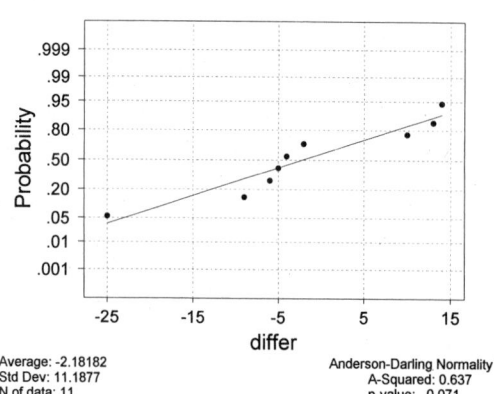

Average: -2.18182
Std Dev: 11.1877
N of data: 11

Anderson-Darling Normality Test
A-Squared: 0.637
p-value: 0.071

c. Listed here are the difference scores and the signed ranks. If you think that all the assumptions from part **b** are met, complete a matched pairs t-test of the hypothesis $H_0: \mu_d = 0$; otherwise, conduct the Wilcoxon signed rank test.

Difference	10	−6	−5	13	−25	14	−2	−6	−4	−4	−9
Signed rank	8.0	−5.5	−4.0	9.0	−11.0	10.0	−1.0	−5.5	−2.5	−2.5	−7.0

d. Based on your analysis, can you conclude that there is a difference in the study habits of students at the two schools?

9.65 In a study using identical twins, one twin was given a drug and then given an intelligence test while under the influence of the drug. The other twin was given the same intelligence test under drug-free conditions. Here are the test scores:

Worksheet: Twin.mtw

Twin A (no drug)	83	74	67	64	70	67	81	64	72
Twin B (drug)	78	74	63	66	68	63	77	65	70

a. In what way are the two samples matched?
b. For the matched pairs t-test, the difference scores should be approximately normally distributed. Construct a dotplot of the difference scores. Is normality a reasonable assumption in this case?
c. With so few scores it may be difficult to verify the normality assumption. Assume that it is met and complete the matched pairs t-test to determine whether the drug had an influence on the test scores.

9.66 A company wished to study the effectiveness of a coffee break on the productivity of its workers. The productivities of nine randomly selected workers were measured on a day with no coffee break and later on a day when the workers were given a 10-minute coffee break. The scores measuring productivity are listed here:

Worksheet: Coffee.mtw

Worker	1	2	3	4	5	6	7	8	9
Without coffee break	23	35	29	33	43	32	41	38	40
With coffee break	28	38	29	37	42	30	43	37	39

a. In what way are the samples matched?
b. Construct a dotplot of the difference scores to see whether the assumptions for the matched pairs *t*-test are met.
c. Use the Wilcoxon signed rank test to determine whether a coffee break increases productivity.

9.67 To evaluate a speed reading course, a group of 15 subjects was asked to read two comparable articles before and after the course. These are their scores on a reading comprehension test:

Worksheet: Speed.mtw

Subject	1	2	3	4	5	6	7	8	9	10	11	12	13	14	15
Before	57	80	64	72	90	59	76	98	70	57	94	77	46	71	89
After	60	90	62	79	95	58	80	99	75	64	63	80	79	73	91

a. Outliers may have an adverse effect on the matched pairs *t*-test. Examine the boxplot and normal probability plot of the difference scores to check the normality assumption. Are there any outliers?

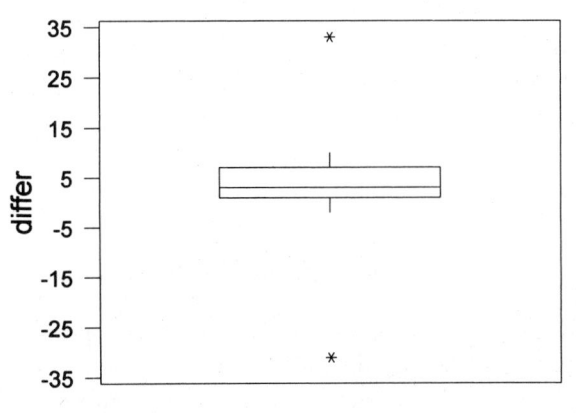
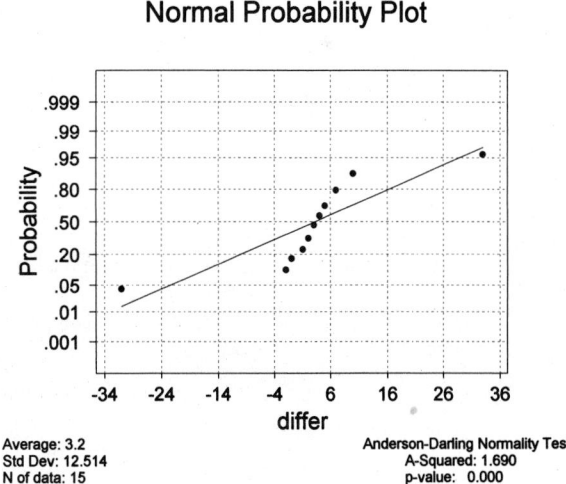

Average: 3.2
Std Dev: 12.514
N of data: 15

Anderson-Darling Normality Test
A-Squared: 1.690
p-value: 0.000

b. Based on your findings in part **a**, should we conduct the matched pairs *t*-test or the Wilcoxon signed rank test to determine whether the course was beneficial?
c. Examine the two computer printouts of the data (at the top of page 608). The first is the matched pairs *t*-test, and the second is the Wilcoxon signed rank test. What are the null and alternative hypotheses? What are the two *p*-values for the two tests? What conclusion can you draw from these two different procedures? Which is the proper analysis of the data?

```
T-Test of the Mean

Test of mu = 0.00 vs mu > 0.00

Variable      N    Mean   StDev   SE Mean      T   P-Value
differ       15    3.20   12.51      3.23   0.99      0.17


T-Test of the Mean

Test of mu = 0.00 vs mu > 0.00

Variable      N    Mean   StDev   SE Mean      T   P-Value
signrnks     15    5.40    7.55      1.95   2.77    0.0075
```

d. Based on your analysis, did the speed reading course increase comprehension?

9.68 Twelve sets of identical twins were taught music recognition by two techniques (each twin was taught with a different method). At the end of the course, their improvement scores were recorded.

Worksheet: Music.mtw

Twin set	1	2	3	4	5	6	7	8	9	10	11	12
Method 1	7	4	6	1	5	1	6	3	4	6	5	3
Method 2	2	4	3	2	3	4	2	4	4	3	5	2

a. Construct a dotplot of the difference scores. Is normality a reasonable assumption in this case?

b. Assume that the normality assumption is met. Complete the matched pairs t-test to determine whether there is a statistical difference between the improvement scores for the two methods.

9.69 A sample of ten students enrolled in a course in German were asked to copy a passage written in German. After experimental instruction in German, the same ten students were asked to copy the same passage. The numbers of errors are listed here:

Worksheet: German.mtw

Student	1	2	3	4	5	6	7	8	9	10
Errors before	10	6	8	7	7	12	4	0	7	10
Errors after	6	4	5	3	6	8	0	1	8	5

a. Construct a dotplot of the difference scores to see whether the assumptions for the matched pairs t-test are met.

b. Use the Wilcoxon signed rank test to test the hypothesis that the mean numbers of errors made before and after the instruction are not significantly different.

9.70 Two psychiatrists were asked to rate each of 20 prison inmates on their rehabilitative potential using a scale from 0 to 15. The larger the rating, the greater the potential for rehabilitation.

Worksheet: Rehab.mtw

Inmate	1	2	3	4	5	6	7	8	9	10
Psychiatrist 1	7	12	7	5	8	6	5	8	5	9
Psychiatrist 2	5	10	8	6	5	4	5	6	5	8

Inmate	11	12	13	14	15	16	17	18	19	20
Psychiatrist 1	6	7	3	5	7	9	5	11	4	10
Psychiatrist 2	9	7	5	4	8	4	8	9	3	12

a. Construct a dotplot, boxplot, and normal probability plot of the difference scores. Does it appear that the assumptions for the matched pairs *t*-test are met?
b. Proceed with the matched pairs *t*-test to determine whether or not there is a significant difference between the mean rating scores given by the two psychiatrists?

9.71   One of the most popular data sets to be analyzed by statisticians is from Charles Darwin's study of cross-fertilized and self-fertilized plants. In his experiment, Darwin planted a pair of seedlings, one produced by cross-fertilization and the other by self-fertilization, on opposite sides of a pot to determine which grew the fastest. The data are the heights of the plants after a fixed time for 15 pairings.

Worksheet: Darwin.mtw

Pot	Cross	Self
1	23.500	17.375
1	12.000	20.375
1	21.000	20.000
2	22.000	20.000
2	19.125	18.375
2	21.500	18.625
3	22.125	18.625
3	20.375	15.250
3	18.250	16.500
3	21.625	18.000
3	23.250	16.250
4	21.000	18.000
4	22.125	12.750
4	23.000	15.500
4	12.000	18.000

SOURCE: C. Darwin, *The Effect of Cross- and Self-fertilization in the Vegetable Kingdom*, 2nd ed. (London: John Murray, 1876).

a. Construct a stem and leaf plot of each sample. Are the data symmetric or skewed?
b. Are the samples matched?
c. Create a column of difference scores, and construct a stem and leaf plot of the resulting data. Do you think the data are more suited for analysis by the paired *t*-test or the Wilcoxon signed rank test? Explain.
d. Perform the Wilcoxon signed rank test. Is there evidence of a significant difference in the heights of the plants from the two methods of fertilization?

9.6   ROBUST INFERENCE ABOUT THE DIFFERENCE BETWEEN TWO POPULATION MEANS (OPTIONAL)

Confidence Interval for the Difference Between Population Means Based on $\bar{y}_{T1} - \bar{y}_{T2}$

Under the condition that the parent distributions are symmetric with long tails, inferences about the difference between means should be based on trimmed means. Such inferences are natural extensions of the one-sample inference problems and modifications of the inference procedures based on ordinary means given in Section 9.3. For large independent samples, the inference procedures based on $\bar{y}_1 - \bar{y}_2$ are modified by replacing the summary data $n_1, \bar{y}_1, s_1$, and $n_2, \bar{y}_2$, and s_2 by the trimmed summary data $k_1, \bar{y}_{T1}, s_{T1}$ and $k_2, \bar{y}_{T2}$, and s_{T2}, respectively (see Koopmans, 1981, p. 314). Because we assume large samples ($n_1 > 30, n_2 > 30$), the sampling distribution of the standardized statistic

$$z = \frac{(\bar{y}_{T1} - \bar{y}_{T2}) - (\mu_1 - \mu_2)}{\sqrt{s_{T1}^2/k_1 + s_{T2}^2/k_2}}$$

is approximately normally distributed. Thus, the p-value for hypothesis testing and the critical value for confidence intervals are found in the standard normal distribution table.

The confidence interval for $\mu_1 - \mu_2$ based on $\bar{y}_{T1} - \bar{y}_{T2}$ is summarized here:

Large-Sample Confidence Interval for $\mu_1 - \mu_2$ Based on $\bar{y}_{T1} - \bar{y}_{T2}$

Application: Symmetric distributions with long tails

Assumptions: The two samples are independent of each other: $n_1 > 30$ and $n_2 > 30$

Estimator: $\bar{y}_{T1} - \bar{y}_{T2}$

Standard error:

$$SE(\bar{y}_{T1} - \bar{y}_{T2}) = \sqrt{\frac{s_{T1}^2}{k_1} + \frac{s_{T2}^2}{k_2}}$$

A $(1 - \alpha)100\%$ confidence interval for $\mu_1 - \mu_2$ is given by the limits

$$(\bar{y}_{T1} - \bar{y}_{T2}) \pm z^* SE(\bar{y}_{T1} - \bar{y}_{T2})$$

EXAMPLE 9.19

To compare two methods of producing textiles, a textile mill counted the numbers of imperfections in the pieces of material produced by the two methods. Random samples of size 31 and 34 were taken from method A and method B, respectively. From the following data on the numbers of imperfections, construct a 95% confidence interval for the difference in means for the two methods:

Worksheet: Textile.mtw

Method A

7	9	8	10	7	9	8	6
2	11	14	8	9	12	4	6
9	10	8	11	10	9	11	8
9	8	12	10	7	9	11	

Method B

5	8	7	22	6	7	2	6	6
10	7	6	9	13	4	3	6	7
7	4	6	8	6	6	8	4	
11	9	7	6	17	7	4	6	

SOLUTION

Figure 9.16 is an ordered back-to-back stem and leaf plot, and Figure 9.17 shows side-by-side box plots for the two data sets. Due to the presence of outliers in the data, the confidence interval is based on 10% trimmed means.

FIGURE 9.16

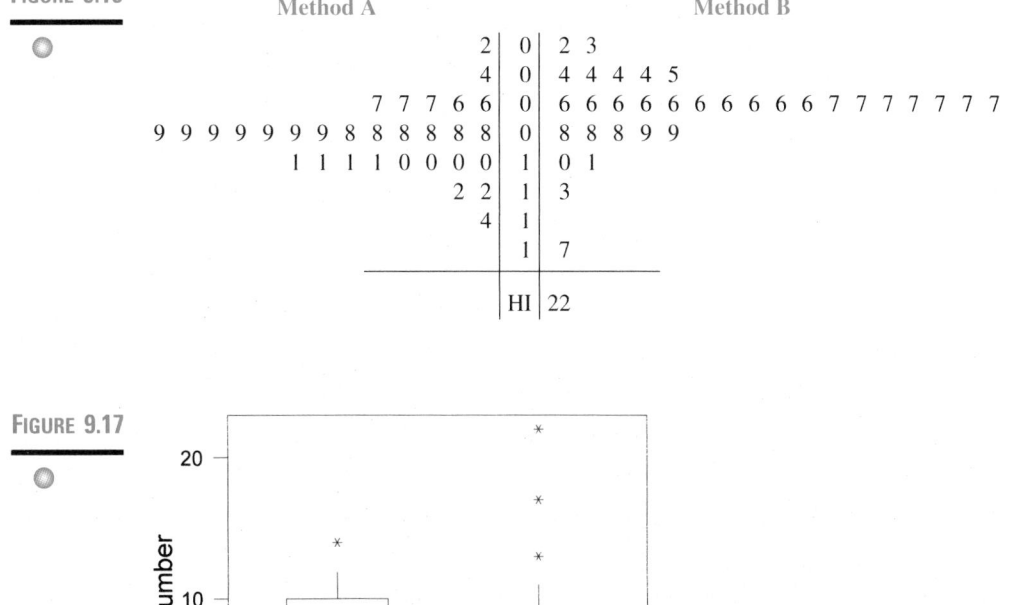

```
          Method A                                        Method B
                              2 | 0 | 2 3
                              4 | 0 | 4 4 4 4 5
                    7 7 7 6 6 | 0 | 6 6 6 6 6 6 6 6 6 6 7 7 7 7 7 7 7
    9 9 9 9 9 9 9 8 8 8 8 8 8 | 0 | 8 8 8 9 9
          1 1 1 1 0 0 0 0 | 1 | 0 1
                      2 2 | 1 | 3
                        4 | 1 |
                            1 | 7
                        ────────────
                           HI | 22
```

FIGURE 9.17

Note, however, that the boxplot for method B gives the appearance of being skewed right, which violates the condition that the population distributions are symmetric. We continue, however, because the trimmed mean is a robust estimator and yields satisfactory results even when the parent distribution is skewed. If both distributions were skewed in the same direction, we would compare the population medians using the Wilcoxon rank sum test given in Section 9.4.

From the raw data, we calculate these summary statistics:

	n	$\bar{y}$	s	k	$\bar{y}_T$	s_T
Method A	31	8.774	2.390	23	8.913	1.730
Method B	34	7.353	3.829	26	6.692	2.210

From the 10% trimmed means, we have

$$\bar{y}_{TA} - \bar{y}_{TB} = 8.913 - 6.692 = 2.221$$

and from the s_T's, we have

$$SE(\bar{y}_{TA} - \bar{y}_{TB}) = \sqrt{\frac{(1.730)^2}{23} + \frac{(2.210)^2}{26}} = .564$$

Because a 95% confidence interval is desired, the table value is $z^* = 1.96$. Therefore, the 95% confidence interval for $\mu_1 - \mu_2$ is

$$2.221 \pm (1.96)(.564) = 2.221 \pm 1.105$$

The interval is (1.116, 3.326).

For comparison purposes, we calculate the confidence interval based on $\bar{y}_1 - \bar{y}_2$:

$$\bar{y}_1 - \bar{y}_2 = 8.774 - 7.353 = 1.421$$

$$SE(\bar{y}_1 - \bar{y}_2) = \sqrt{\frac{(2.390)^2}{31} + \frac{(3.829)^2}{34}} = .7845$$

Therefore, the 95% confidence interval based on the ordinary means is

$$1.421 \pm (1.96)(.7845) \quad \text{or} \quad 1.421 \pm 1.538$$

which gives the interval $(-.117, 2.959)$. Figure 9.18 gives a comparison of the two confidence intervals.

Note in Figure 9.18 that the interval based on the difference between the ordinary means is longer (less precision) and includes zero. The fact that zero is inside the interval means that in a test of hypothesis of the equality of the two means, the null hypothesis is not rejected. On the other hand, the interval based on the difference in the trimmed means is more precise and is well above zero. A test based on the trimmed means indicates a significant difference between the two mean responses, which implies that method B is a better way to produce textiles (significantly fewer imperfections). However, this is not without a cost. The boxplot in Figure 9.17 points out that occasionally method B results in a large number of imperfections. Thus, in this example, the trimmed mean is useful in detecting a difference, and the graphical display points out certain peculiarities that might otherwise go undetected.

FIGURE 9.18

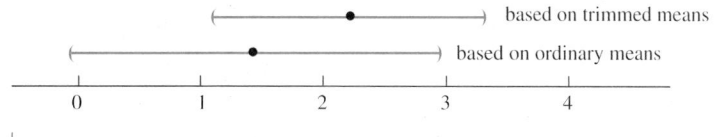

Hypothesis Test for the Difference Between Population Means Based on $\bar{y}_{T1} - \bar{y}_{T2}$

Just as in confidence interval estimation, the procedures based on $\bar{y}_1 - \bar{y}_2$ given in Section 9.3 are modified to construct the test for the difference between population means when the parent distributions are symmetric with long tails.

The test is summarized in the box:

Large-Sample Hypothesis Test of $\mu_1 - \mu_2$ Based on $\bar{y}_{T1} - \bar{y}_{T2}$

Application: Symmetric distributions with long tails

Assumptions: The two samples are independent of each other; $n_1 > 30$ and $n_2 > 30$.

Left-Tailed Test	Right-Tailed Test	Two-Tailed Test
$H_0: \mu_1 - \mu_2 \geq 0$	$H_0: \mu_1 - \mu_2 \leq 0$	$H_0: \mu_1 - \mu_2 = 0$
$H_a: \mu_1 - \mu_2 < 0$	$H_a: \mu_1 - \mu_2 > 0$	$H_a: \mu_1 - \mu_2 \neq 0$

Standardized test statistic:
$$z = \frac{\bar{y}_{T1} - \bar{y}_{T2}}{\sqrt{s_{T1}^2/k_1 + s_{T2}^2/k_2}}$$

From the z-table, we find

p-value $= P(z < z_{\text{obs}})$ p-value $= P(z > z_{\text{obs}})$ p-value $= 2P(z < z_{\text{obs}})$ if $z_{\text{obs}} < 0$
 or p-value $= 2P(z > z_{\text{obs}})$ if $z_{\text{obs}} > 0$

If a level of significance α is specified, then reject H_0 if p-value $< \alpha$.

EXAMPLE 9.20   Do physically active women have stronger bones than nonactive women as they grow older? Research by doctors at the University of North Carolina at Chapel Hill indicates that they do, which suggests that active women are less likely to have bone fractures as they grow older. A study of 300 women reported in the February 1985 issue of the *Journal of Orthopedic Research* indicates that older athletic women, aged 55 to 75, have arm and spine bone measurements in the same range as younger athletic women. Measurements of bone density (page 614) were taken from a random sample of 70 women, 35 of whom are physically active and the remaining are considered nonactive:

Worksheet: Bones.mtw

Active women

213	227	211	208	155	204	216	219	224
202	207	212	184	214	245	210	192	218
219	163	230	214	209	210	226	203	208
207	217	257	212	203	232	221	215	

Nonactive women

201	205	187	208	203	265	201	210	219
205	173	202	199	216	192	207	243	194
201	217	270	209	185	176	202	211	213
208	236	214	209	162	204	206	213	

Is there statistical evidence to support the hypothesis that active women have greater bone density?

Solution

Figure 9.19 is an ordered back-to-back stem and leaf plot, and Figure 9.20 shows side-by-side boxplots for the two data sets. Both distributions appear symmetric with long tails (an abundance of outliers), which suggests that we base the inference on the difference between the trimmed means. We will trim 10%.

Figure 9.19

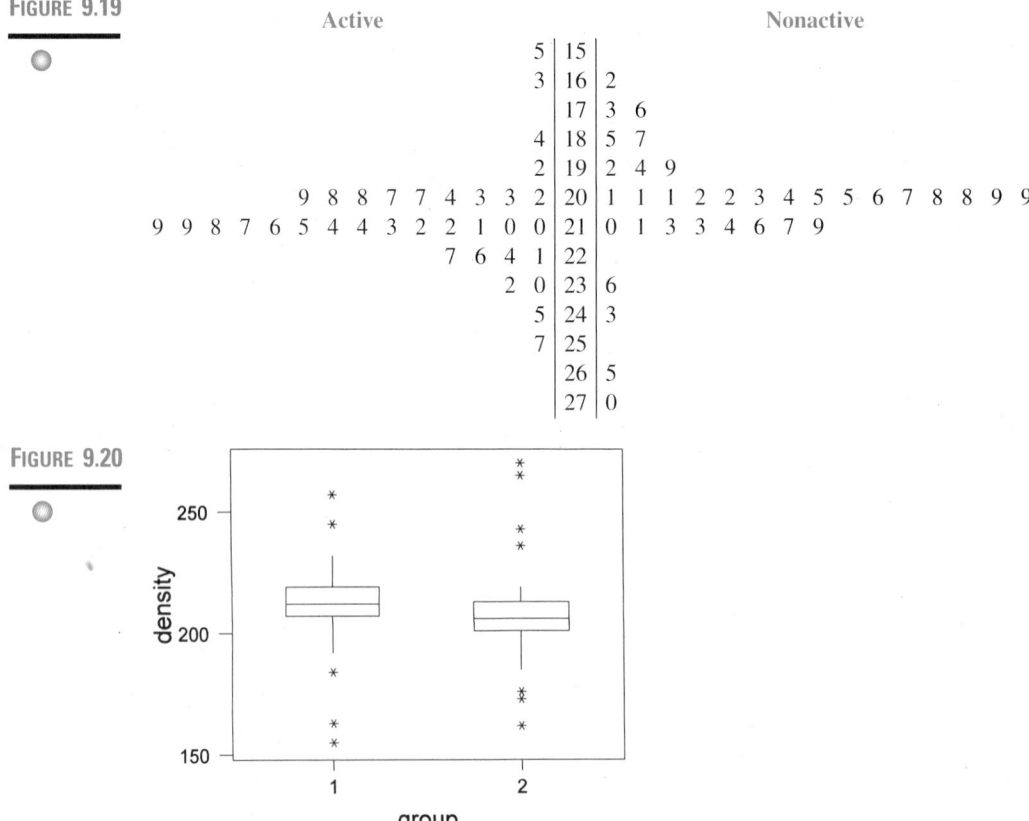

```
              Active                                      Nonactive
                                  5 | 15 |
                                  3 | 16 | 2
                                    | 17 | 3  6
                                  4 | 18 | 5  7
                                  2 | 19 | 2  4  9
          9  8  8  7  7  4  3  3  2 | 20 | 1  1  1  2  2  3  4  5  5  6  7  8  8  9  9
 9  9  8  7  6  5  4  4  3  2  2  1  0 | 21 | 0  1  3  3  4  6  7  9
                          7  6  4  1 | 22 |
                                2  0 | 23 | 6
                                  5 | 24 | 3
                                  7 | 25 |
                                    | 26 | 5
                                    | 27 | 0
```

Figure 9.20

The research hypothesis is that the mean bone density for active women is greater than the mean bone density for nonactive women. Thus, we have

$$H_0: \mu_1 - \mu_2 \leq 0 \quad \text{versus} \quad H_a: \mu_1 - \mu_2 > 0$$

The test statistic is

$$z = \frac{\bar{y}_{T1} - \bar{y}_{T2}}{\sqrt{s_{T1}^2/k_1 + s_{T2}^2/k_2}}$$

The data give these values:

	$\bar{y}_{T.10}$	s_T
Active	212.926	9.8198
Nonactive	205.778	11.7445

$$\bar{y}_{T1} - \bar{y}_{T2} = 212.926 - 205.778 = 7.148$$

and from the s_T's, we have

$$SE(\bar{y}_{T1} - \bar{y}_{T2}) = \sqrt{\frac{s_{T1}^2}{k_1} + \frac{s_{T2}^2}{k_2}}$$

$$= \sqrt{\frac{(9.8198)^2}{27} + \frac{(11.7445)^2}{27}} = 2.95$$

The test statistic is

$$z_{obs} = \frac{7.148}{2.95} = 2.42$$

Because the test is right-tailed, we have

$$p\text{-value} = P(z > 2.42) = 1 - .9922 = .0078$$

Therefore, the null hypothesis is rejected. We have highly significant evidence that active women have greater bone density than nonactive women.

E X E R C I S E S   9 . 6

9.72   Construct each confidence interval for $\mu_1 - \mu_2$ based on 10% trimmed means from the given summary data:

a. 90% interval

$$n_1 = 44, \bar{y}_{T1} = 36.7, s_{T1} = 5.24$$
$$n_2 = 48, \bar{y}_{T2} = 34.6, s_{T2} = 6.77$$

b. 98% interval

$$n_1 = 50, \bar{y}_{T1} = 177.5, s_{T1} = 25.83$$
$$n_2 = 50, \bar{y}_{T2} = 185.2, s_{T2} = 23.68$$

c. Which interval has greater validity?

9.73   Using 10% trimmed means, test each null hypothesis using the given summary data:
       a. $H_0: \mu_1 - \mu_2 \geq 0$ versus $H_a: \mu_1 - \mu_2 < 0$

$$n_1 = 35, \bar{y}_{T1} = 55.8, s_{T1} = 15.56$$
$$n_2 = 35, \bar{y}_{T2} = 62.4, s_{T2} = 15.38$$

b. $H_0: \mu_1 - \mu_2 = 0$ versus $H_a: \mu_1 - \mu_2 \neq 0$

$$n_1 = 120, \bar{y}_{T1} = 467, s_{T1} = 79$$
$$n_2 = 135, \bar{y}_{T2} = 482, s_{T2} = 65$$

9.74   Two independent random samples were selected from two populations:

Worksheet: Exer9-74.mtw

Sample A									Sample B									
2	9								2	4	9							
3	2	4							3	0								
3									3	5								
4	0	2	3						4	0	1	3	3	4	4			
4	5	6	7	8	9	9			4	5	6	6	7	7	8	8	9	9
5	0	0	1	2	2	3	3	4	4	5	0	1	3	3	4	4		
5	5	5	6	8	8	9			5	5	6	8						
6									6	2								
6	5	7							6									
7	4								7	3	4							
7	8								7	7								

a. Do the stem and leaf plots suggest that inferences about $\mu_1 - \mu_2$ should be based on ordinary means or trimmed means?
b. How would boxplots help in answering part **a**?
c. Construct a 95% confidence interval for $\mu_1 - \mu_2$ based on 20% trimmed means.

9.75   Use the data in Exercise 9.74 to test the hypothesis of no difference between the population means using the 20% trimmed means.

9.76   Can computers help children with learning disabilities become better writers? The results of studies by researchers at Claremont Graduate School in Claremont, California, suggest that they can. In a year-long classroom study, learning-disabled children who wrote on word processors improved dramatically compared with learning-disabled children who used pen and paper. These are the creative writing scores of learning-disabled children using computers and using pen and paper: (These data are similar to those from the Claremont study.)

Worksheet: Writing.mtw

			Computer								Pen and Paper				
80	75	86	66	92	113	70	91	75	78	73	90	85	70	76	100
100	83	88	64	84	88	60	93	65	93	77	61	55	79	74	83
85	102	94	94	87	87	89	85	76	74	105	83	82	78	53	79
117	92	96	89	82	85	93	90	77	101	85	80	75	60	50	72

a. Consider the side-by-side boxplots of the data, and comment on the shapes of the two distributions.

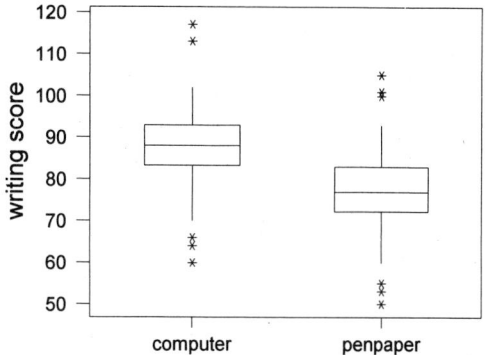

b. Would you base a test of hypothesis concerning the difference in population means on ordinary means or on trimmed means?

9.77 Do the data in Exercise 9.76 indicate that computers tend to improve the creative writing scores of learning-disabled children? Test the hypothesis that the creative writing scores of the computer group exceeded the scores of the noncomputer group. Use the test statistic suggested in part **b** of Exercise 9.76 and these summary measures based on 10% trimming:

$$n_1 = 32, \bar{y}_{T1} = 87.83, s_{T1} = 5.97$$

$$n_2 = 32, \bar{y}_{T2} = 76.96, s_{T2} = 11.3$$

9.78 In a study related to the one described in Exercise 9.76, the computer aptitude scores of 32 learning-disabled teens was compared with the scores of 32 "normal achievers":

Worksheet: Computer.mtw

		Learning Disabled								Normal Achievers					
70	73	67	74	77	66	50	74	72	78	67	71	87	69	73	74
65	69	72	79	97	72	55	70	88	76	77	72	71	93	61	73
74	75	86	78	69	51	73	94	70	74	96	75	75	65	69	70
76	71	70	66	79	68	73	76	60	66	71	67	56	77	79	53

a. Organize the data in back-to-back stem and leaf plots and side-by-side boxplots.
b. Comment on the shapes of the two distributions.
c. Would you base a test of hypothesis concerning the difference between population means on ordinary means or on trimmed means?

9.79   Using the data in Exercise 9.78, test the hypothesis that there is no difference between the computer aptitude scores of the learning-disabled and the normal achievers. Use the test statistic suggested in part **c** of Exercise 9.78.

9.80   Exercise 9.51 in Section 9.4 investigated the pre-enrollment abilities of freshmen women and men on a math placement exam. The scores are in worksheet: **Remedial.mtw**.

a. Based on the boxplots of the two samples, do you think that the confidence interval based on $\bar{y}_1 - \bar{y}_2$ found in Exercise 9.51 of Section 9.4 is appropriate, or would the robust procedures given in this section be more appropriate?

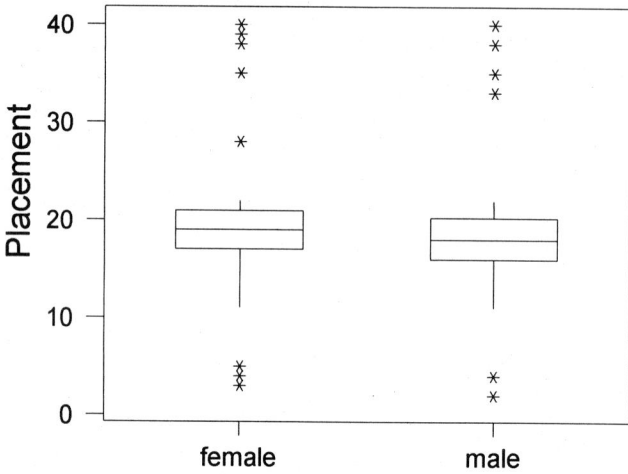

b. These summary measures are based on 10% trimming:

$$n_1 = 35, \bar{y}_{T1} = 18.889, s_{T1} = 5.833$$

$$n_2 = 42, \bar{y}_{T2} = 18.031, s_{T2} = 2.417$$

Construct a 98% confidence interval for the difference between the means using the 10% trimmed data.

c. This computer analysis of the original data is from Exercise 9.51 in Section 9.4. Compare the results found in part **b** with these results.

```
Two Sample T-Test and Confidence Interval

Twosample T for female vs male
            N        Mean      StDev    SE Mean
female     35       19.34       9.09        1.5
male       42       18.93       8.03        1.2

98% C.I. for mu female - mu male: ( -4.3,   5.1)
T-Test mu female = mu male (vs not =): T= 0.21   P=0.83   DF=  68
```

9.7   CHOOSING THE RIGHT PROCEDURE

Several procedures for comparing two population parameters are presented in this chapter. You have learned how to compare two Bernoulli proportions, two population means, and two population medians. For proportions, only the large-sample case is presented. The small-sample comparison case is not presented because most applications use large samples. A typical application might involve an analysis of survey results and rarely is a survey conducted with a small sample. This is not to say that small-sample analysis of proportions is unimportant, however. In some situations, such as clinical trials studies, the sample sizes are extremely small. These are special situations, however, and most likely you will cover the details of the comparison of proportions with small samples in a second course in statistics.

The remainder of this section is devoted to comparing the centers of two population distributions. Choosing the right procedure depends on certain characteristics of the two samples and on the shapes of the parent distributions. Are the samples independent? Samples are independent when two separate random samples are chosen from two separate populations. If the samples are not independent, they may be matched in such a way that each observation in one sample is naturally paired with a corresponding observation in the other sample. For example, if the same subject is measured both before and after a treatment, then obviously the data are matched because the two measurements are taken on the same subject.

Independent Samples

For independent samples, we presented four inference procedures. There is considerable overlap in the applications of these procedures, so deciding which to use in a particular situation can be confusing. Some procedures are very restrictive in their applications (the pooled t-test) and others apply to a wide variety of situations. To simplify the selection process, you should understand the basic conditions that apply to each procedure.

Pooled t-Test   As mentioned, the pooled t-test, presented in Section 9.4, is the most restrictive procedure of all. For small samples, it requires that (1) the parent distributions are normally distributed and (2) the population variances are equal. Because of the robustness of the t-procedure, however, these conditions can be relaxed somewhat and the test (or confidence interval) will still be valid. The conditions for which they are not valid are long tails and extreme skewness. In those cases you should avoid the t-test and use one of the other procedures.

Two-Sample t-Test   The two-sample t-test, presented in Section 9.3, is similar to the pooled t-test except that it does not require that the population variances be equal. It should be used to test the equality of population means when the parent distributions are reasonably close to normally distributed. It can be used in place of the pooled t-test as well. Recent statistical studies have shown that it is almost as powerful as the pooled t-test even when the population variances are equal. Therefore, it is recommended that we use the pooled t-test only in those cases where we are reasonably sure that the population variances are equal.

If the sample sizes are large, the Central Limit Theorem applies and says that the sampling distribution of $\bar{y}_1 - \bar{y}_2$ is approximately normally distributed. Because the

Central Limit Theorem applies regardless of the shape of the underlying parent distribution, there are hardly any restrictions on using the two-sample t-test when the sample sizes are large. On the other hand, recall that the t-test is used for making inferences about $\mu_1 - \mu_2$. If the parent distributions are skewed, it has been suggested that we make inferences about the difference between the population medians, $\theta_1 - \theta_2$, instead of the difference between the population means. If this is the case, you should use the Wilcoxon procedure in place of the t-test. Furthermore, you should use a trimmed mean to estimate the population mean when the parent distribution is symmetric with long tails. Thus, you should use the t-test when evidence suggests that the parent distributions are symmetric without excessively long tails.

Wilcoxon Test   The two-sample Wilcoxon rank sum test does not require any specific shape of the parent distributions, but it does require that the two populations be similarly shaped. For example, if population 1 is skewed left, then population 2 should be skewed left. In fact, the test requires that if a difference exists between the two populations, then it is reflected only in the centers of the two distributions. On the surface, it appears that the applications for the Wilcoxon test are very limited. In reality, however, in many cases where a comparison is in order, the two population distributions are similar in shape. For example, the distributions of the weights of men and women are similar in shape; it is just that men weigh more than women, which means that the distributions differ in location but not in shape. So, the applications for the Wilcoxon test are numerous, and in fact, many statisticians suggest that it should be used in all cases except where the parent distributions are known to be normally distributed. For small samples, this is good advice. For large samples, however, the Wilcoxon test should be used in those situations where the parent distributions are similarly skewed and leave the symmetric cases for the t-test and the z-test based on trimmed means.

z-Test Based on Trimmed Means   The z-test based on trimmed means is a large-sample, robust procedure. It is robust in the sense that it may be applied to a wide variety of applications. It is best, however, in those situations where the parent distributions are symmetric with long tails. If the parent distributions are skewed, you should use the Wilcoxon procedure; if the parent distributions are symmetric with normal or short tails, use the t-test.

Matched Samples

For matched samples, we presented two procedures: the matched pairs t-test and the Wilcoxon signed rank test.

Matched Pairs t-Test   As in all previous t-tests, the matched pairs t-test is derived assuming that the underlying distribution (in this case the distribution of difference scores) is normally distributed. Again, however, robustness of the t distribution allows us to relax this condition, especially when the sample size is increased, and apply it to those situations where the distribution of difference scores does not have excessively long tails. In fact, for large samples, the t distribution approaches the standard normal distribution, and thus the test can be thought of as a z-test for large samples and applied to a wide variety of cases. Though not presented here, the one-sample z-test based on trimmed means may be applied in those cases where the sample size is large and the distribution of difference scores is symmetric with long tails.

Wilcoxon Signed Rank Test   For small samples, when we have reservations about using the *t*-test, we can apply the Wilcoxon signed rank test. The only condition is that the difference scores be symmetrically distributed.

Summary

Clearly, all cases that you will confront in the practice of statistics have not been addressed in this chapter. Numerous other testing procedures can be applied in special cases. What you have been exposed to are the standard tests that you will see in routine comparison problems. As implied in the discussion, two or more tests may be applied to a particular problem. Your job is to apply the procedure you think best. An exploratory approach to data analysis is the only way you can critically appraise the situation and come up with a reasonable plan of attack.

To help in your decision, consider the following steps when trying to decide which procedure is best for a particular situation.

Choosing the Right Procedure

1. Determine whether you have independent or matched samples.
2. In the case of independent samples, analyze the samples to determine the shapes of the two parent distributions. In the case of matched samples, analyze the difference scores to determine the shape of the difference population.
3. Based on the shape(s) of the parent distribution(s), decide whether the inference should be for means or for medians. In other words, are the parent distributions symmetric or skewed? If they are symmetric, are they normally distributed, symmetric with short tails, or symmetric with long tails?
4. If the inference is for the population medians, then the nonparametric Wilcoxon procedures are to be used.
5. If the inference is for the population means, determine whether the standard *t*-procedures apply. That is, if your sample sizes are small, do the normality and homogeneous variances assumptions seem reasonable? If not, apply the Wilcoxon procedure. If the sample sizes are large, examine the length of the tails. For symmetric, long-tailed distributions, use the *z*-test based on trimmed means; otherwise, use the *t*-test based on ordinary means.

9.8   SUMMARY AND REVIEW

KEY CONCEPTS

✓   Three basic inference problems are addressed in this chapter: the comparison of two population proportions, the comparison of two population centers using independent samples, and the comparison of two population centers using matched samples.

✓ Only the large-sample inference procedures for the difference between two population proportions are given. The confidence interval is

$$(p_1 - p_2) \pm z^* \sqrt{\frac{p_1(1 - p_1)}{n_1} + \frac{p_2(1 - p_2)}{n_2}}$$

and the test statistic for testing the hypothesis that $\pi_1 = \pi_2$ is

$$z = \frac{p_1 - p_2}{\sqrt{p(1 - p)}\sqrt{1/n_1 + 1/n_2}}$$

where

$$p = \frac{x_1 + x_2}{n_1 + n_2}$$

is the pooled estimate of π.

✓ The confidence interval for $\mu_1 - \mu_2$ based on $\bar{y}_1 - \bar{y}_2$ is

$$(\bar{y}_1 - \bar{y}_2) \pm t^* \text{SE}(\bar{y}_1 - \bar{y}_2)$$

✓ The test statistic for testing $\mu_1 - \mu_2$ is

$$t = \frac{\bar{y}_1 - \bar{y}_2}{\text{SE}(\bar{y}_1 - \bar{y}_2)}$$

✓ If the population variances are equal, then

$$\text{SE}(\bar{y}_1 - \bar{y}_2) = s_p \sqrt{\frac{1}{n_1} + \frac{1}{n_2}}$$

and

$$t = \frac{(\bar{y}_1 - \bar{y}_2) - (\mu_1 - \mu_2)}{s_p \sqrt{1/n_1 + 1/n_2}}$$

has a Student's t distribution with $n_1 + n_2 - 2$ degrees of freedom.

✓ If the population variances are unequal, then

$$\text{SE}(\bar{y}_1 - \bar{y}_2) = \sqrt{\frac{s_1^2}{n_1} + \frac{s_2^2}{n_2}}$$

and

$$t = \frac{(\bar{y}_1 - \bar{y}_2) - (\mu_1 - \mu_2)}{\sqrt{s_1^2/n_1 + s_2^2/n_2}}$$

has an approximate t distribution with degrees of freedom given by

$$\text{df} = \frac{\left(s_1^2/n_1 + s_2^2/n_2\right)^2}{\dfrac{(s_1^2/n_1)^2}{n_1 - 1} + \dfrac{(s_2^2/n_2)^2}{n_2 - 1}}$$

✓ The *Wilcoxon rank sum test* is an alternative to the two-sample *t*-test when the assumption of normality is not met. For the Wilcoxon rank sum test, the usual *t*-test is applied to the rank-transformed data.

✓ When the samples are matched, the test of the difference between population means is called the *matched pairs t-test*. The test statistic is

$$t = \frac{\bar{d}}{s_d / \sqrt{n}}$$

which has a Student's *t* distribution with $n - 1$ degrees of freedom. Again, the assumption of normality must be met for this *t*-test unless the sample size is large. If the assumption is not met, then the *Wilcoxon signed rank test* is recommended. Here the usual *t*-test is applied to the signed ranks.

✓ Large-sample inference procedures for $\mu_1 - \mu_2$ based on $\bar{y}_{T1} - \bar{y}_{T2}$ are the same as those given earlier except that the ordinary sample means and standard deviations are replaced with the trimmed means and trimmed standard deviations.

LEARNING GOALS

Having completed this chapter, you should be able to:

1. Perform inferences on the difference between two population proportions. *Section 9.2*
2. Construct a confidence interval for $\mu_1 - \mu_2$ based on $\bar{y}_1 - \bar{y}_2$ when the population variances are unequal. *Section 9.3*
3. Test hypotheses about the difference between two population means based on $\bar{y}_1 - \bar{y}_2$ when the population variances are unequal. *Section 9.3*
4. Perform inferences about $\mu_1 - \mu_2$ based on $\bar{y}_1 - \bar{y}_2$ when the population variances are equal. *Section 9.4*
5. Perform a two-sample Wilcoxon rank sum test. *Section 9.4*
6. Perform inferences on the difference between two population centers using matched samples by conducting a matched pairs *t*-test or a Wilcoxon signed rank test. *Section 9.5*
7. Perform large-sample inferences about $\mu_1 - \mu_2$ based on $\bar{y}_{T1} - \bar{y}_{T2}$. *Section 9.6*
8. Choose the correct procedure for analyzing data. *Section 9.7*

QUESTIONS
FOR REVIEW

Use the following problems to test your skills:

9.81 The U.S. Department of Agriculture lists the average size of the farms (in acres) in each state. In a comparison of the farms in two regions of Iowa, these data were obtained:

Region 1: $n_1 = 65, \bar{y}_1 = 315, s_1 = 32.7$
Region 2: $n_2 = 60, \bar{y}_2 = 302, s_2 = 28.4$

a. Are the standard deviations close enough that we should pool them together to estimate the standard error of $\bar{y}_1 - \bar{y}_2$?
b. Construct a 90% confidence interval for $\mu_1 - \mu_2$ based on $\bar{y}_1 - \bar{y}_2$.
c. Does the interval contain both positive and negative values? What does this tell us about the sizes of the farms in the two regions of Iowa?

9.82 Life expectancy tables indicate that women live longer than men. To examine this theory, a statistics class, over a 1-month period, randomly selected from the local newspaper obituaries of ten men and ten women. An analysis of the data gave these values:

Women: $n_1 = 10, \bar{y}_1 = 79.7, s_1 = 11.32$
Men: $n_2 = 10, \bar{y}_2 = 76.4, s_2 = 9.64$

With $\bar{y}_1 - \bar{y}_2$ as the test statistic, test

$$H_0: \mu_1 - \mu_2 \geq 0 \quad \text{versus} \quad H_a: \mu_1 - \mu_2 < 0$$

where μ_1 = mean life expectancy of men and μ_2 = mean life expectancy of women.
a. In words, what does the null hypothesis say?
b. Should we pool the standard deviations together to estimate the standard error of $\bar{y}_1 - \bar{y}_2$?
c. What is the p-value of the test statistic?
d. State your conclusion.

9.83   Test the hypothesis that there is no significant difference between the population medians using the following data arranged in a back-to-back stem and leaf plot. Because both distributions appear to be skewed, use the Wilcoxon rank sum test. Why do we suggest a test of the equality of the population medians instead of the population means?

Worksheet: Exer9-83.mtw

```
            |  21 | 0   4   7
    3   6   9 |  22 | 5   6   9   9
        5   8 |  23 | 1   4   8
    2   6   9 |  24 | 5   8
0   3   5   8 |  25 | 0   7
    3   7   7 |  26 | 3
        0   6 |  27 | 0   1
            8 |  28 |
            7 |  29 | 2
            2 |  30 | 7
              |  31 |
            4 |  32 |
```

9.84   In a random sample of 200 young drivers, 54 were judged to be careless drivers. In a random sample of 200 adult drivers, 38 were judged to be careless. Find a 99% confidence interval for the difference between the percents of young and adult drivers judged to be careless.

9.85   In an experiment to evaluate a new variety of corn, 12 plots of land were divided in half, with the new variety planted on one half and a standard variety planted on the other half. Here are the yields obtained on the 12 plots of land:

Worksheet: Corn.mtw

Plot	1	2	3	4	5	6	7	8	9	10	11	12
New variety	110	103	95	94	87	119	102	93	87	98	105	117
Standard	102	86	88	75	89	102	105	88	83	89	100	110

a. To test the hypothesis that the new variety has a significantly higher yield per acre than the standard variety, what procedure do you recommend?
b. The boxplot shows the difference between the yields for the new variety and the standard variety. Does it appear that the assumptions for an inference based on the t distribution are violated?

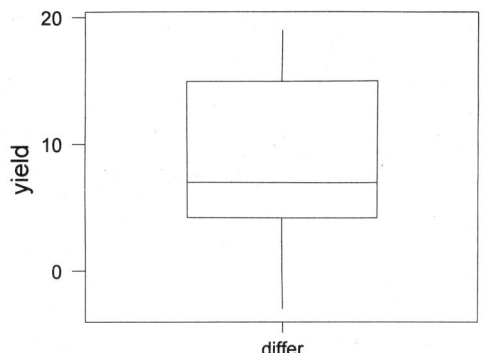

c. State the null and alternative hypotheses necessary to conduct the test of significance suggested in part **a**.

d. Calculate the test statistic and its *p*-value.

e. Write your conclusion.

9.86   Do more women choose diet soft drinks than men? In a random sample of 160 women, 94 preferred diet soft drinks to regular soft drinks. In a random sample of 135 men, 71 preferred diet to regular soft drinks. Do the data imply that significantly more women than men prefer diet soft drinks? Perform the test of significance.

9.87   Do students who attend private high schools spend more time on homework than students who attend public high schools? Listed here are the numbers of hours per week spent on homework for a random sample of 15 private high school students and a random sample of 15 public high school students:

Worksheet: Homework.mtw

Private

21.3	16.8	8.5	12.6	15.8	19.3	18.5	24.6
18.3	12.9	15.7	18.4	18.7	22.6	20.5	

Public

15.3	17.4	12.3	10.7	16.4	11.3	17.6	13.9
20.2	16.8	23.6	14.2	5.7	18.8	9.4	

a. Do the boxplots indicate that the assumptions for a *t*-interval or a *t*-test have been violated?

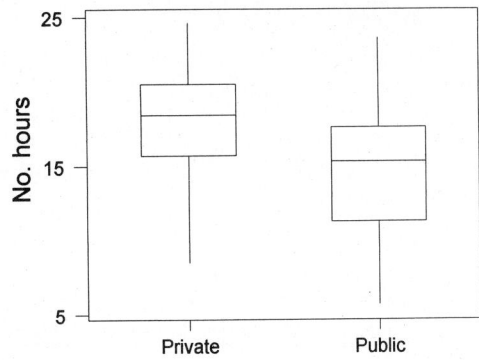

    b. Should an inference about the difference between the population means be based on independent or matched samples?

    c. Find a 98% confidence interval for the mean difference between the numbers of hours per week spent on homework for private and public high school students.

    d. Based on the confidence interval from part **c**, what can be said about the number of hours per week spent on homework?

9.88 Use the data in Exercise 9.87 to test the hypothesis that private high school students spend significantly more time on homework per week than public high school students. Be sure to state the hypotheses and give your conclusion. Compare these results to the conclusion given in part **d** of Exercise 9.87.

9.89 Construct a 95% confidence interval for $\mu_1 - \mu_2$ based on $\bar{y}_{T1.10} - \bar{y}_{T2.10}$ using these summary data:

$$n_1 = 42, \bar{y}_{T1.10} = 134.6, s_{T1} = 26.82$$
$$n_2 = 44, \bar{y}_{T2.10} = 127.8, s_{T2} = 23.47$$

Under what circumstances do you recommend a confidence interval for $\mu_1 - \mu_2$ based on $\bar{y}_{T1.10} - \bar{y}_{T2.10}$?

9.90 Sixteen children were selected from a first-grade class and paired according to IQ such that one member of each pair had attended kindergarten and the other had not. A reading test was given to all 16 children, with these scores:

Worksheet: Kinder.mtw

Pair	1	2	3	4	5	6	7	8
Kindergarten	83	74	67	64	70	67	81	64
No kindergarten	78	74	63	66	68	63	77	65

Is there evidence that kindergarten is beneficial for reading skills?

9.91 A 1984 study by the Centers for Disease Control in Atlanta indicates that giving aspirin to children who have the chicken pox or the flu can cause Reye's syndrome. Dr. Sidney M. Wolfe of the Public Citizen Health Research Group said it is the most convincing study to date to link aspirin and the life-threatening illness.

    The study traced 29 cases of Reye's syndrome and 143 control cases of children who did not develop the illness. Twenty-eight of the 29 children who contracted Reye's syndrome took aspirin, and 64 of the control group took aspirin. Construct a 95% confidence interval for the difference between the proportions of those who took aspirin in the Reye's syndrome group and those who took aspirin in the control group.

9.92 Can the computer help high school students prepare for the SAT exam? An educational consultant, George Hopmeier in Milton, Florida, studied 90 Florida high school students. Half of the students, without computer coaching, averaged 370 on the SAT math and had a standard deviation of 125. The other half, who used computer coaching, averaged 407 and had a standard deviation of 80. Is there evidence that computer coaching improved SAT math scores? To answer this question, you need to formulate null and alternative hypotheses and conduct a test of significance. Be sure to give the p-value and state your conclusion.

9.93 To test the effect of a physical fitness course on one's physical ability, the numbers of sit-ups that a person could do in 1 minute, both before and after the course, were recorded. Nine randomly selected participants could do the numbers of sit-ups listed here:

Worksheet: Fitness.mtw

Subject	1	2	3	4	5	6	7	8	9
Before	28	31	17	22	12	32	24	18	25
After	32	33	19	26	17	30	26	19	25

Test the hypothesis that there was an increase in the number of sit-ups.

9.94 Sixty-four individuals randomly selected from a metropolitan area were asked to indicate their preference for the candidates for a political office. A week later, one of the candidates visited the area. Following the visit, a second random sample of 64 individuals was asked to indicate their preference for the same list of candidates. Thirty-two of the 64 respondents favored the candidate before his visit, whereas 40 of the 64 favored him after his visit. Is there evidence to indicate that personal contact has a positive impact?

9.95 A study by the American Association for Counseling and Development found that in 1983, 59% of all 11th-graders were interested in business and tech jobs. Suppose the study was based on a random sample of 400 11th-graders and we wish to determine whether the percent has changed significantly since then. A random sample of 400 11th-graders this year found that 251 were interested in similar jobs. With a test of hypothesis, determine whether significantly more are interested now than in the past.

9.96 According to a survey by the Road Information Program, Ohio is the pothole capital of the United States with an estimated 6.8 million potholes. Potholes are caused by water seeping into cracks, freezing, and swelling the pavement. Suppose we wish to compare two types of material for making pavement. A measurement is devised to test the resistance to water. From the following data, determine whether there is a significant difference in the resistance to water for the two types of materials:

	n	$\bar{y}$	s
Material A	25	63.8	16.55
Material B	25	87.2	35.75

9.97 Carolyn S. Hartsough, a University of California, Berkeley, educational psychologist, found that out of 301 hyperactive children, 26% of the mothers has poor health during pregnancy. This compares with 16% of mothers of 191 normal children. Is the percentage for hyperactive children significantly greater than the percentage for normal children? Does this study show that poor health during pregnancy causes hyperactive children?

9.98 High levels of blood fat contribute to atherosclerosis (hardening of the arteries), which is an underlying cause of heart attacks and stroke. Dr. Robert Knopp of the University of Washington School of Medicine in Seattle analyzed 381 diabetics and found that women had higher levels of LDL, or "bad cholesterol," than men and lower levels of HDL, or "good cholesterol." From the following simulated data, determine whether the level of LDL in women is greater than that in men:

Women: $n_1 = 201, \bar{y}_1 = 52.35, s_1 = 6.55$
Men: $n_2 = 180, \bar{y}_2 = 47.28, s_2 = 17.29$

COMPUTER
EXERCISES

9.99 Fourteen employees who have not completed high school were given a reading test. Afterward, they were given formal vocabulary training and then retested on their reading skills. From the test scores given at the top of page 628, determine whether there is any difference between the scores before and after the vocabulary training.

Worksheet: Vocab.mtw

Employee	1	2	3	4	5	6	7	8	9	10	11	12	13	14
First test	84	55	43	64	72	65	72	52	49	80	38	93	77	60
Second test	86	52	50	72	70	67	80	50	62	81	56	90	78	64

Store the data in two columns of your worksheet and have the computer calculate the differences and conduct a *t*-test. What is your justification for applying a *t*-test to the difference scores?

9.100   Here are the results of quality control tests on two manufacturing processes:

Worksheet: Quality.mtw

Process I	1.5	2.5	3.4	2.3	3.2	2.8	1.9	
Process II	2.5	3.0	2.7	4.0	3.5	2.0	1.8	3.7

Use a computer to determine whether the mean results of the two processes are equivalent.

9.101   A fourth-grade teacher thinks that his students are on the whole better spellers than his colleague's students. Ten students were randomly selected from each teacher's class and given a standardized spelling test. Use the following results and a computer to determine whether the fourth-grade teacher is justified in his claim.

Worksheet: Spellers.mtw

Fourth-grade class

105	109	115	112	124	107	121	112	104	119

Colleague's class

115	103	110	125	99	121	119	106	100	123

9.102   A developer of housing projects would like to reduce his costs on kitchen cabinets. He obtained these cost estimates from two suppliers in 20 prospective homes:

Worksheet: Cabinets.mtw

Home	1	2	3	4	5	6	7	8	9	10
Supplier A	380	560	425	389	568	651	595	455	540	520
Supplier B	325	470	420	375	574	595	570	475	560	500

Home	11	12	13	14	15	16	17	18	19	20
Supplier A	375	468	492	510	379	750	520	480	394	624
Supplier B	362	465	445	490	350	780	512	465	382	614

Do the data indicate a significant difference between the estimates from the two suppliers?

9.103   Dr. Marvin Moser of Yale University School of Medicine says that automated blood pressure machines in airports and shopping malls are generally unreliable. He suggests having your blood pressure checked regularly by an expert or using an at-home device. Suppose 15 adult men check their blood pressure on an automated machine, and then they have an expert check it. From the following scores, determine whether the mean difference in diastolic blood pressure is significant:

Worksheet: Blood.mtw

Subject	1	2	3	4	5	6	7	8	9	10	11	12	13	14	15
Machine	68	82	94	106	92	80	76	74	110	93	86	65	74	84	100
Expert	72	84	89	100	97	88	84	70	103	84	86	63	69	87	93

Note that this test may not evaluate accuracy. It will tell you whether the machine tends to give high readings or tends to give low readings. However, the machine could be unreliable with some readings unusually high and some unusually low. The differences may cancel each other out, in which case the test statistic would be insignificant. Can you think of a better way to check the accuracy of the machines?

9.104 Worksheet: **Commute.mtw** contains commuting times to work in urban areas for 1980 and 1990. (The data are based on a 1994 study by the Federal Highway Administration.) In Computer Exercise 1.100 of Section 1.5 you were asked to construct side-by-side dotplots of the two samples. Here are those dotplots:

Character Dotplot

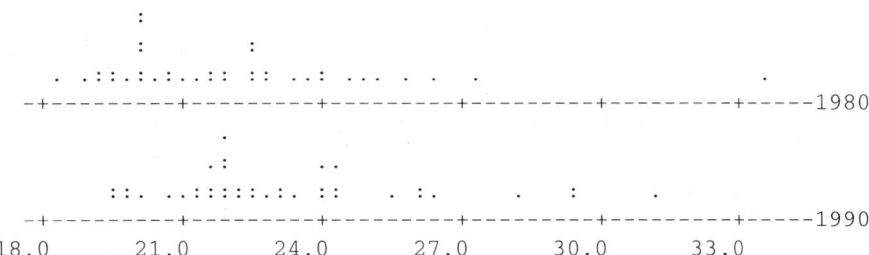

a. Are the dotplots sufficiently skewed that we should compare the median commuting times instead of the mean commuting times?
b. Stack the two samples into one column of the worksheet and store the codes in an adjacent column. Rank the stacked column and then apply the pooled t-test to the ranks.
c. What is the name for the procedure of applying the ordinary pooled t-test to the ranked data?
d. Interpret the results of the analysis. Is there evidence that the commuting time has changed from 1980 to 1990?

9.105 Here are the mean arterial blood pressures of 11 subjects before and after they received oxytocin. Does oxytocin affect arterial blood pressure?

Worksheet: Oxytocin.mtw

Subject	1	2	3	4	5	6	7	8	9	10	11
Before	95	173	94	97	81	100	97	104	72	101	83
After	55	90	36	59	46	46	49	92	23	55	49

9.106 Microparticles, small quantities of micron-sized bits of dust, have been injected into the atmosphere by volcanic eruptions, dust storms, and micrometeorite infall. The particles are extremely small and would remain suspended indefinitely, except that they are scrubbed from the atmosphere by snow because they are nuclei for tiny ice crystals. It has been hypothesized that the concentration of microparticles in snow should be uniform over the earth. Samples of snow were obtained from the permanent snowfields of the Greenland ice cap and from Antarctica. The snow was melted and the volume concentrations (in parts per billion) were as listed here:

Worksheet: Snow.mtw

Antarc

3.7	2.0	1.3	3.9	.2	1.4	4.2	4.9
.6	1.4	4.4	3.2	1.7	2.1	4.2	3.5

(*continued*)

, Greenld

3.7	7.8	1.9	2.0	1.1	1.3	1.9	3.7	3.4
1.6	2.4	1.3	2.6	3.7	2.2	1.8	1.2	.8

SOURCE: J. Davis, *Statistics and Data Analysis in Geology* (New York: Wiley, 1986).

a. Construct side-by-side boxplots of the two samples. Do you notice any outliers? Do you think the conditions for a *t*-test are satisfied?

b. Remove any outliers by placing a * in the cell that contains the outlier. Repeat part **a**. Do you now think that the conditions for a *t*-test are satisfied?

c. If we conduct a two-sample *t*-test, should we pool the standard deviations together or conduct an unpooled *t*-test?

d. Based on your answer to part **c**, complete the test. Be sure to state your hypotheses and give a conclusion.

CAUTIONS IN INFERENCE

When we hear the expression *confidence interval*, we think of the formula that is used to construct the interval, and we think of the numerical realization of the interval when we substitute in the values from the sample. For example,

$$\bar{y} \pm t^* \frac{s}{\sqrt{n}}$$

is the formula for a confidence interval for μ, the mean of a normally distributed population. In the formula, $\bar{y}$ and s are random variables, which means that their values vary from sample to sample. Because they are random variables, we can speak of the *probability* that the interval will or will not have the population mean somewhere inside it. After we substitute in numerical values and get an interval, such as (479.5, 523.8), we can no longer attach a probability statement. It either happened that μ is inside the interval or did not happen; probability is no longer relevant. Knowing, however, that the numerical interval was obtained from a formula that has a probability of, say, 95% of covering the mean, we have a certain degree of confidence that the numerical interval indeed contains the unknown value of μ. This is why we call the resulting interval a 95% *confidence* interval.

The *p*-value in a test of significance is a measure of the statistical evidence against the null hypothesis. The farther the test statistic is from the hypothesized value of the parameter, the smaller will be the *p*-value. Thus, small *p*-values suggest strong evidence against the null hypothesis. Some statisticians insist on setting a level of significance α prior to testing and rejecting the null hypothesis only when *p*-value $< \alpha$. This is perfectly acceptable, but you should avoid the practice of indiscriminately rejecting H_0 when the *p*-value crosses the α barrier and declaring the results significant. For example, when $\alpha = .05$ (a very popular value), what is the difference between *p*-value $= .049$ and *p*-value $= .051$? Are the results significant only in the first case and not in the second? It may be that adopting the *criterion for rejecting* presented at the end of Section 8.1 is a more reasonable approach to hypothesis testing.

Using the inference procedures developed in this book requires that you be aware of certain cautions:

- In theory, when the sample size is small, the confidence interval based on $\bar{y}$ and the *t*-test require that the sample be randomly selected from a normally distributed population. In practice, however, populations are never perfectly normally distributed. Because of the robustness of the *t*-procedures, this is not a serious problem except when the parent distribution is either skewed or long-tailed. Therefore, it is important that you use the descriptive tools to diagnose the shape of the underlying parent distribution.
- Even when the sample size is large, you should be aware of skewed and long-tailed distributions. The sample mean and standard deviation are not resistant measures. Outliers have an adverse effect on inferences based on $\bar{y}$. For this reason, we recommend inferences for the population median when the parent distribution is skewed and robust procedures based on $\bar{y}_T$ when the parent distribution has long tails.

- The probability statements that pertain to confidence intervals and tests of significance require that the data result from a simple random sample from the population. Certainly, no probability statement can be made about a convenience or self-selected sample. Even in the cases of stratified and cluster samples, new formulas are developed for inferences for means and proportions. You will find those formulas in more advanced statistics books.

HYPOTHESIS TESTING VERSUS CONFIDENCE INTERVALS

Both confidence interval estimation and hypothesis testing deal with inferences about unknown population parameters. Deciding which to use in a particular application depends on the intent of the investigation. Do you need to gather information about the parameter, or do you ultimately have to make a decision concerning the parameter? Gathering information about the parameter involves confidence intervals. Determining the truth of a particular conjecture about the parameter involves hypothesis testing.

When we calculate a p-value associated with a test statistic, we are assessing the strength of evidence against the null hypothesis. If, in addition, a level of significance, α, is given, then we reject the null hypothesis in favor of the alternative when p-value $< \alpha$. If p-value $\geq \alpha$, we fail to reject the null hypothesis. In this context, we are using hypothesis testing as a decision making tool. Prior to collecting any data, we specify the level at which we wish to test the null hypothesis—namely, the level of significance. If that level is attained by the sample data, then we reject the hypothesis. It is clear that in the process of testing a hypothesis, we are making a decision about a specified value of the population parameter. On the other hand, there is no specified value of the parameter when we are finding a confidence interval. We are attempting to narrow down the possibilities for the parameter by finding an interval that should contain the parameter with a reasonable degree of certainty. We are simply estimating the value of the parameter. The next examples illustrate the difference between hypothesis testing and confidence interval estimation.

- A politician wants to know what percent of the voters in her district are in favor of her running for a second term. The politician will make a decision on the basis of the results of a poll, but the question at hand is: What percent are in favor? Thus, a confidence interval is appropriate.
- The FDA is testing a new dietary supplement to see whether it dissolves cholesterol deposits in arteries. They will decide that either the supplement is effective or it is not. A test of hypothesis is appropriate.
- Government agencies that investigate such quantities as unemployment rates, inflation, and gross national product need estimates from confidence intervals.
- A claim is made that more women smoke than men. A test of hypothesis would shed light on this question.
- A textile company wishes to compare a new manufacturing process with the old one. Is the new technique an improvement over the old one? Clearly, hypothesis testing is appropriate.
- A consumer agency analyzing the rising cost of medical care would investigate the mean fees for various operations with confidence intervals.
- Does vitamin C help prevent colds? An experiment could be designed where vitamin C is compared with a placebo and analyzed using hypothesis testing.
- A congressional committee investigating fatal traffic accidents caused by drinking would most likely be interested in confidence intervals.

- In the study of a virus, 60 mice were injected with the virus but not treated. A like group of 60 mice were injected with the virus and treated with an experimental drug. A comparison study of the effectiveness of the experimental drug could be conducted with hypothesis testing.
- A business report of the inventory value of a warehouse of household carpet would most likely consist of confidence interval estimates of the mean inventory value of the carpet.

In general, confidence intervals give information and hypothesis testing helps make decisions.

EQUIVALENCE
OF HYPOTHESIS
TESTING AND
CONFIDENCE
INTERVALS

Although confidence intervals and hypothesis testing are different inference procedures, they are closely related. They are different ways of expressing the same information contained in a sample. For example, if a 99% confidence interval for the proportion of fatal accidents caused by drinking has the limits .45 to .55, then a test of the hypothesis that the population proportion is .5 would certainly not be rejected. On the other hand, the hypothesis that it is .6 would be rejected. In fact, any claim that the proportion is a specific value between .45 and .55 would not be rejected and outside those limits would be rejected at the 1% level of significance.

The same applies to testing population means. For instance, we fail to reject the null hypothesis $H_0: \mu = \mu_0$ against the two-sided alternative $H_a: \mu \neq \mu_0$ when

$$-1.96 \leq \frac{\bar{y} - \mu_0}{\sigma/\sqrt{n}} \leq +1.96 \qquad \text{(level of significance} = .05)$$

This is equivalent to saying, fail to reject H_0 if

$$\bar{y} - 1.96\frac{\sigma}{\sqrt{n}} \leq \mu_0 \leq \bar{y} + 1.96\frac{\sigma}{\sqrt{n}}$$

Notice that the endpoints of the interval are the same endpoints for a 95% confidence interval for μ. That is to say, we fail to reject the above null hypothesis if and only if μ_0 is inside the confidence interval. We reject H_0 if and only if μ_0 is not inside the interval. In other words, the confidence interval consists of all the values of μ_0 for which the null hypothesis $H_0: \mu = \mu_0$ is not rejected. Because it covers both areas of inference, confidence interval inference is a more comprehensive inference procedure than is hypothesis testing.

STATISTICAL
SIGNIFICANCE
VERSUS
PRACTICAL
SIGNIFICANCE

In all of hypothesis testing, we reject the null hypothesis when the test statistic falls in the rejection region. We then say that the results of the study are statistically significant. This means that the sample information produced a value for the test statistic that is not consistent with the hypothesized value of the population parameter. This does not necessarily mean, however, that the observed difference between the test statistic and the hypothesized parameter is significant in any practical sense.

Consider, for example, the conjecture at the beginning of Section 8.1 that the mean per capita income in a certain rural county is $15,000 versus the alternative that it is more than $15,000. Suppose a random sample of 1,000 persons in the county had a mean income of $15,100 and the standard deviation was $800. The resulting test statistic is $t = 3.95$, which has p-value $\approx .00003$. This is highly significant evidence to reject the null hypothesis. Thus, on the basis of this sample, we declare that the mean per capita income in this county is statistically significantly greater than $15,000. But from a practical point of view, is the difference of $100 really significant? We would have to say, in this case, that the statistical significance has no practical significance. You should be cautioned that with very large samples, small differences become statistically significant.

UNIT EXERCISES

1. A telephone survey of 507 students at a university with an enrollment of 10,188 was conducted on April 15, 1992, by the student development office. The survey examined how student employment influences academic performance. The sample was stratified by classification.
   a. Do the formulas for confidence intervals developed in Chapter 7 apply to these data? Explain.
   b. It was stated in the report that "Enough responses were obtained to be 95% confident that the sample accurately reflects the entire student population." What do you think is meant by this? Is it stated correctly?

2. A USA Today/CNN/Gallup national telephone poll of 1,022 adults was conducted on September 6–7, 1994, on the issue of allowing Haitian and Cuban refugees into the United States.
   a. The poll stated that 79% believed that we should not allow Cuban refugees into the United States. Is it true that 79% of all adults in the United States think that we should not allow Cuban refugees into the United States?
   b. The margin of error was stated to be ± 3 percentage points. What information is needed to derive this figure?
   c. From the provided information, can you construct a confidence interval for the percentage of all adults who think that we should not allow Cuban refugees into the United States? If so, what is the interval? What is its confidence level?

3. Suppose one wishes to estimate the mean monthly expenditures of women students on campus. The standard deviation of the monthly expenditures is $27.
   a. How large a sample is needed so that the margin of error in a 99% confidence interval estimate is no greater than $10?
   b. Suppose a random sample of 45 women was selected and the average monthly expenditure was found to be $128. Find a 99% confidence interval based on $\bar{y}$ for the mean monthly expenditure of all women students.
   c. Based on your answer to part **b**, would the hypothesis $H_0: \mu = \$135$ be rejected with $\alpha = .01$?

4. The average violent crime rate (number per 100,000 population) in 25 randomly selected areas of the South was 486 and the standard deviation was 94. Assuming the distribution is normal, find a 99% confidence interval for the mean violent crime rate in the South.

5. The registrar at a university would like to estimate the percent of students registered for the spring semester who plan to attend summer school.
   a. If the registrar wishes to be 98% confident of obtaining a sample percent within 3% of the actual percent that plan to attend summer school, what sample size is needed?
   b. The registrar selected a random sample of 500 students and found that 94 plan to attend summer school. Find a 98% confidence interval for the actual percent of students who plan to attend summer school.
   c. Based on the confidence interval obtained in part **b**, would the hypothesis $H_0: \pi = .15$ be rejected with $\alpha = .02$?

6. A laboratory tested 15 batteries manufactured by a company and found these lifetimes (in hours):

   Worksheet: Battery.mtw

19	18	26	17	22	16	25	20
17	18	17	19	20	18	19	

   Construct a 95% confidence interval for the median lifetime of all batteries produced by the company. Is normality of the parent distribution required for the confidence interval?

7. The water quality in Charlotte, North Carolina, was recently compared with the water quality in several other large U.S. cities. Overall, the Charlotte water quality was ranked low in purity because of the comparatively high amounts of trihalomethanes, a chemical linked to cancer. The amount of trihalomethanes in the water was found to range from 20 to 80 parts per million (ppm). A sample of 100 water specimens was randomly selected across Charlotte, and the amount of trihalomethanes (in ppm) was measured for each water specimen. The random sample yielded an average of 51 ppm and a standard deviation of 16 ppm. Based on these results, construct a 99% confidence interval for the mean amount of trihalomethanes for the city of Charlotte.

8. A class of 50 eighth-graders has completed a standardized reading test on which their scores had a mean of 107.5 and a standard deviation of 10.5. The national mean score on the test is 100. Set up the null and alternative hypotheses to test whether this class is "superior in reading ability."

9. A university administrator believes that their applicants for admission are significantly above the national norm of 450 on SAT math. He randomly pulls 100 records from their applicant pool and finds their average SAT Math is 463 and the standard deviation is 42.6. Assuming SAT scores are somewhat normal in shape, is this enough evidence to verify his claim?

10. In a survey of 100 randomly selected families in California, it was found that the mean medical expense for the year was $1,640 and the standard deviation was $260. An insurance company claims that the mean medical expense for all California families is at least $1,700 per year. Test the claim using the data from the 100 families.

11. A supermarket is trying to decide whether to accept or reject a shipment of tomatoes. It is impossible to check all the tomatoes for size, but the store wants an average weight of 8 ounces (neither too large nor too small). A random sample of 400 tomatoes yields an average weight of 7.85 ounces and a standard deviation of 1.15 ounces. Should the supermarket reject the shipment?

12. The U.S. Postal Service claims that at least 80% of the letters mailed in New York City destined for Los Angeles are delivered within 2 working days. To verify this claim, suppose that 100 letters were mailed from New York to various destinations in the Los Angeles area, and that 76 were delivered within 2 working days. Is there evidence to dispute the U.S. Postal Service's claim?

13. It is commonly believed that the mean life span of animals held in captivity is greater in an open environment than in a caged one. From the data, determine whether there is statistical evidence to support the conjecture:

	n	$\bar{y}$	s
Open environment	35	7.6	1.4
Caged environment	62	5.9	3.7

14. A survey of 70 randomly selected households in city A showed that the average monthly cost of electricity was $84.38 and the standard deviation was $24.82. A survey of 80 randomly selected households in city B found that the average monthly cost of electricity was $95.46 and the standard deviation was $32.75. Find a 99% confidence interval for the mean difference in the costs of electricity for homes in the two areas.

15. The pooled t-test is based on certain assumptions. Describe those assumptions and how you might check to see whether they are met in a specific problem.

16. Two testing instruments are designed to evaluate one's aptitude for a certain job. Both instruments (A and B) are scored on the same scale and administered to two groups of subjects. These are the summary statistics:

	n	$\bar{y}$	s^2
A	16	352	9,606
B	21	420	8,120

Which testing instrument would you choose? Why?

17. Do cigarette smokers care if their smoking bothers nonsmokers? Suppose we want to estimate the percent of smokers who would light up a cigarette without asking permission.
    a. How large a sample should we select to estimate the percent of smokers who would light up without permission to within 4% of its true value with 98% confidence?
    b. Ignoring the sample size found in part **a**, researchers selected a simple random sample of 900 smokers and asked, "Would you light a cigarette indoors without asking if anyone minds?" From the sample, 500 of the smokers responded yes. Based on these results, find a 98% confidence interval for the true percent of smokers who would light up without asking permission. Interpret your results.

18. A random sample of size 200 produced a sample proportion of .65. A 99% confidence interval for the population proportion is $.65 \pm .087$. What do we call the .087?

19. To investigate the self-concept of a group of college administrators, a psychologist gave a standardized exam to 16 administrators and found that their average self-concept score was 22 and the standard deviation was 3. If the national norm for the self-concept score is 20, can we say that this group of college administrators has an unusually high self-concept? What assumption did you make about the underlying parent distribution?

20. A sample of 80 observations yielded $\bar{y}_{T.20} = 2.6$ and $s_T = .84$. With $\bar{y}_T$ as the test statistic, use this information to test the hypotheses:

$$H_0: \mu \leq 2.5 \quad \text{versus} \quad H_a: \mu > 2.5$$

Be sure to calculate the p-value and interpret the results.

21. A researcher wants to compare the ages at which male and female children begin to walk. He knows that the walking ages for both genders range from 8 to 24 months. To estimate the mean difference in age at which they begin to walk to within $\frac{1}{2}$ month with 95% confidence, how many children will he need?

COMPUTER EXERCISES

22. A manufacturer of solar-powered calculators claims that the mean life expectancy of their calculator is at least 48 months. A random sample of 19 such calculators produced the following life times (in months):

Worksheet: Calculat.mtw

| 38.4 | 42.8 | 48.9 | 47.1 | 43.2 | 50.8 | 56.3 | 48.1 | 49.4 | 53.2 |
| 44.8 | 42.5 | 50.1 | 49.7 | 48.2 | 52.4 | 49.7 | 55.4 | 50.3 | |

Assess the shape of the distribution by constructing a boxplot and a normal probability plot. Point out any unusual behavior in the data that would indicate problems with interpreting a confidence interval for the population mean. Find a 95% confidence interval for the mean life of the calculators and evaluate whether or not the manufacturer's claim is valid.

23. Construct boxplots for the two samples listed here. Assess whether there is a difference between the variances of the two underlying populations. Can we assume that the population variances are equal?

    Worksheet: Variance.mtw

Sample 1	21	32	16	19	27	22	28	24
Sample 2	33	28	41	48	37	26	36	

    If the boxplots are not conclusive, calculate descriptive statistics for each sample and determine whether one standard deviation is more than twice the other. If not, pool the two standard deviations together and test the equality of the two population means. Is there a significant difference between the population means? Do the boxplots indicate a difference between the means? Was the test of significance really required in this exercise?

24. The table lists the pretest and posttest scores on the MLA listening test in Spanish for ten high school Spanish teachers who attended an intensive summer institute:

    Worksheet: Spanish.mtw

Subject	1	2	3	4	5	6	7	8	9	10
Pretest	30	28	31	26	20	30	34	15	28	20
Posttest	29	30	32	30	16	25	31	18	33	25

    We hope to show that attending the institute improves listening skills. State the null and alternative hypotheses accordingly. Ascertain whether or not the data satisfy the conditions for a matched pairs t-test by constructing a boxplot and a normal probability plot. If the conditions are satisfied, carry out the test of significance; otherwise, use the Wilcoxon signed rank test.

25. These are the percents of high school dropouts for each state:

    Worksheet: Dropouts.mtw

Ala	12.6	Neb	7.0
Alaska	10.9	Nev	15.2
Ariz	14.4	NH	9.4
Ark	11.4	NJ	9.6
Calif	14.2	NM	11.7
Colo	9.8	NY	9.9
Conn	9.0	NC	12.5
Del	10.4	ND	4.6
Fla	14.3	Ohio	8.9
Ga	14.1	Okla	10.4
Hawaii	7.5	Ore	11.8
Idaho	10.4	Pa	9.1
Ill	10.6	RI	11.1
Ind	11.4	SC	11.7
Iowa	6.6	SD	7.7
Kan	8.7	Tenn	13.4
Ky	13.3	Texas	12.9
La	12.5	Utah	8.7
Maine	8.3	Vt	8.0
Md	10.9	Va	10.0
Mass	8.5	Wash	10.6
Mich	10.0	WVa	10.9
Miss	11.8	Wis	7.1
Mo	11.4	Wyo	6.9
Mont	8.1		

Assess the shape of the distribution with a boxplot and a normal probability plot. Then, construct a 99% confidence interval for the center of the data.

26. To estimate the income made by a typical tobacco farmer, a random sample of 50 tobacco farmers was selected. These incomes were recorded for the 50 farmers:

Worksheet: Tobacco.mtw

$6,280	9,690	7,858	8,820	6,500	7,468	8,719	6,790	8,650	9,400
7,843	12,170	9,760	9,280	14,897	5,438	9,980	7,654	10,190	7,823
9,840	5,790	6,874	10,690	9,450	11,657	6,470	19,357	6,794	7,865
8,747	9,347	8,785	7,589	12,768	8,658	24,860	9,793	6,680	8,749
9,845	7,895	14,678	8,980	5,897	9,879	8,370	8,530	10,250	8,450

Comment on the distributional shape of the incomes of tobacco farmers and decide what parameter best describes a "typical" income. Find a 99% confidence interval for that parameter.

27. A grocery store chain is trying to decide how many checkout aisles there should be in each store. From the following checkout times of 40 randomly selected customers, test the hypothesis that the mean checkout time is at least 7 minutes per customer (data are recorded in minutes). Be sure to examine the data for outliers and skewness prior to testing the hypothesis. The analysis of the data should suggest the proper testing procedure.

Worksheet: Checkout.mtw

5.6	3.8	7.9	1.3	4.2	8.8	5.9	9.4	15.6	4.8
1.2	6.5	3.5	7.2	5.3	7.6	6.6	9.2	8.4	14.1
6.3	9.1	6.3	2.5	6.3	7.1	6.7	7.8	5.8	6.5
4.6	6.8	7.9	5.6	6.4	6.2	6.9	5.5	7.3	5.1

28. Checking the processor speed is very important when shopping for a personal computer. Two competing computer companies would like to compare the processor speeds of their computers. The processor scores listed here are derived from several tests using medium-sized instruction mixes. Each score indicates how well the computer's CPU executes common computer applications. These are the results of 15 tests on each brand:

Worksheet: Processo.mtw

Brand A

15,883	16,767	15,799	15,843	15,582	16,235	15,966	16,092
15,870	15,678	15,541	16,128	16,099	15,743	15,692	

Brand B

16,870	15,931	15,818	16,343	17,214	16,435	15,936	16,386
15,846	15,399	16,848	16,950	15,892	16,432	16,021	

Use the data to test the equivalence of the processor speeds of the two brands of computers. Be sure to first examine the data for outliers and skewness with side-by-side boxplots.

29. A study was conducted to compare the effectiveness of two brands of fly spray. The two brands were sprayed on either end of a board treated with honey in nine different environmental conditions. (Different environmental conditions were used because the fly population differs in each environment.) The dependent measure is the number of flies that land on the board in a 30-minute period.

Worksheet: Flies.mtw

| Environmental | Number of Flies | |
Condition	Brand A	Brand B
1	23	36
2	17	22
3	28	25
4	48	60
5	10	16
6	36	34
7	15	28
8	22	22
9	94	104

Test the hypothesis of no difference in the effectiveness of the two brands of fly spray.

DISCUSSION EXERCISES

30. A construction engineer would like to determine whether a new type of cement has a better bonding quality than the mix he currently uses. Should he use hypothesis testing or confidence interval estimation? Explain your answer.

31. A health department official would like to determine the severity of the recent flu epidemic. Should she use hypothesis testing or confidence interval estimation? Explain your answer.

32. A team of Environmental Protection Agency scientists sampled the water in 225 lakes in the Adirondack Mountains in New York to assess the extent of acid rain pollution in the lakes. Should they use hypothesis testing or confidence interval estimation? Explain your answer.

33. Traveling employees spend an average of $167.21 a day on lodging, rental car, and food, according to a report by Corporate Travel/RIT. The most expensive city for travel is New York City with an average daily cost of $320.56. Two popular cities for travel are Orlando, Florida, and Washington, D.C. Describe how you would determine the average daily costs for travel in these two cities. Consider only lodging, rental car, and food. Describe how you would determine the average cost of each. Then describe how you would statistically compare the costs of travel in the two cities.

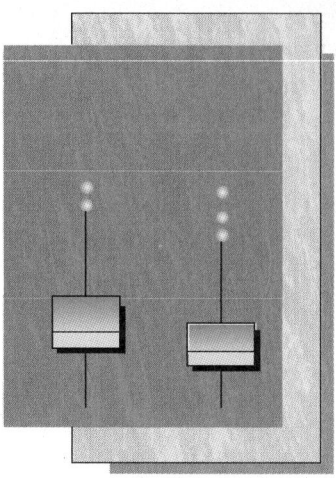

10

ANALYSIS OF
CATEGORICAL DATA

In most of the problems we have considered up to this point, the data have been the result of some measurement. Statistical inference procedures for variables such as IQ, reaction time, and earnings per share were presented in Chapters 7–9. The results of surveys and some experiments, however, can only be classified into categories and not quantified. A public poll, for example, records the opinions of respondents on a particular issue; the data are the frequency counts for the categories: for, against, and no opinion. An experiment that studies childhood leukemia records the cancer history of the parents; the data are frequency counts for categorical variables that indicate the parents' history of cancer. An admissions officer at a university records the gender of the applicants and their intended major. Each is a categorical variable, and the data are frequency counts for the categories of the variables. The objective of this chapter is to present methods for analyzing data of this type.

CONTENTS

STATISTICAL INSIGHT

HOW DOES RACE RELATE TO TYPE OF DRUG OFFENSE FOR FEDERAL PRISONERS?

Federal inmates in prison for a drug offense are usually sentenced for possessing or trafficking in heroin, crack, cocaine, or marijuana. A 1994 report, *Comparing Federal and State Prison Inmates*, by the U.S. Department of Justice classified 1991 federal inmates by type of drug offense and race. The results in Table 10.1 suggest that blacks, for example, are more likely to be convicted for dealing in crack than are either whites or Hispanics. On the other hand, very few blacks are convicted for dealing in marijuana as compared with whites and Hispanics. Whites are more likely to be convicted for dealing in cocaine

and marijuana. Also there seems to be a very high conviction rate for Hispanics dealing in cocaine.

A first question to ask from data of this type is whether or not the two classifying variables are independent. The data in this study seem to suggest that the variables "type of drug offense" and "race" are dependent. In this chapter, we develop the chi-square distribution, which allows us to statistically evaluate the relationship between categorical variables (see computer Exercise 10.67 in Section 10.4). In the next chapter, we will investigate relationships between numerical variables.

TABLE 10.1   Numbers of federal inmates convicted for dealing in illegal drugs in 1991

		Type of Drug Offense			
		Heroin	Crack	Cocaine	Marijuana
	White	407	106	4,525	2,825
Race	Black	1,156	2,513	4,439	442
	Hispanic	1,314	348	7,297	2,675

SOURCE: C. Wolf Harlow, *Comparing Federal and State Prison Inmates*, NCJ-145864, U.S. Department of Justice, Bureau of Justice Statistics, 1994.

10.1   THE CHI-SQUARE GOODNESS-OF-FIT TEST

Thus far, we have applied inference procedures to numerical variables. Now we consider the analysis of categorical variables. As described in Chapter 2, a categorical variable is organized into categories, where the data are the frequency counts of the various categories. A Time Warner survey of 425 randomly selected 8- to 12-year-old schoolchildren in the November 1992 issue of *Sports Illustrated for Kids* found that 130 aspired to be a teacher, 105 a doctor, 80 a lawyer, 60 a police officer, and 50 a firefighter. In the following table, the categories are the occupations, and the data are the numbers of children who chose the categories:

Occupation	Teacher	Doctor	Lawyer	Police	Firefighter	Total
Number of children	130	105	80	60	50	425

In another example, the Red Cross classifies a group of potential donors according to blood type. The categories are the different blood types, and the data are the numbers of donors who have the various blood types. Categories can also be defined by ranges of values of a numerical variable, such as income level being classified as low, medium, or high.

Categorical data, often called qualitative data, are statistically analyzed by means of the *chi-square* (χ^2) *statistic* introduced by Karl Pearson in 1900. A single variable, like occupation given above, is analyzed with the chi-square *goodness-of-fit test*. Pearson developed the chi-square goodness-of-fit test while studying the randomness associated with the game of roulette in Monte Carlo.

The goodness-of-fit test consists of determining whether the frequency counts in the categories of the variable agree with a specified distribution. For example, are the occupations chosen equally by children? If they are chosen equally, then each occupation has a probability of $\frac{1}{5}$ of being chosen. Out of 425 children, we would expect 85 ($\frac{1}{5}$ of 425) to choose each category. Do the results of the Time Warner survey lead us to reject the conjecture that the occupations are chosen equally? To answer this question, we need to conduct a statistical test of hypothesis. First, however, we describe the *multinomial experiment*, which is an extension of the binomial experiment.

The Multinomial Experiment

The Multinomial Experiment

1. The experiment consists of n identical, independent trials.
2. The outcome of each trial falls into one of k categories.
3. The probabilities associated with the k outcomes, denoted by $\pi_1, \pi_2, \ldots, \pi_k$, remain the same from trial to trial. Since there are only k possible outcomes, we have

$$\pi_1 + \pi_2 + \cdots + \pi_k = 1$$

4. The experimenter records the values $o_1, o_2, \ldots, o_k$, where o_j ($j = 1, 2, \ldots, k$) is equal to the observed number of trials in which the outcome is in category j. Note that

$$n = o_1 + o_2 + \cdots + o_k$$

Let us determine whether our example fits the description of a multinomial experiment.

1. The n identical trials are the 425 randomly chosen children, and because the choice of any one child does not depend on the choice of any other child, the trials are independent.
2. Each child chooses one of the $k = 5$ occupations. (In the actual survey, the children were able to choose other occupations, but for simplicity we are assuming that each child must choose one of the five occupations.)
3. The probabilities with which the children choose the five occupations are denoted as π_1, π_2, π_3, π_4, and π_5, and they remain the same from one child to the next.
4. The observed counts are

$$o_1 = 130 \qquad o_2 = 105 \qquad o_3 = 80 \qquad o_4 = 60 \qquad o_5 = 50$$

which sums to $n = 425$.

The four conditions are satisfied; thus, the example describes a multinomial experiment.

As stated, we determine whether the occupations are chosen equally with a test of hypothesis. If they are chosen equally, then the null hypothesis

$$H_0: \pi_1 = \pi_2 = \pi_3 = \pi_4 = \pi_5 = \frac{1}{5}$$

should be true. The alternative hypothesis is stated as

$$H_a: \text{At least one } \pi_i \neq \frac{1}{5}$$

To develop a test statistic to test the hypothesis, we compare the observed values, o_i, with the values we expect if the hypothesis is true. If H_0 is true, then out of 425 children, we *expect* each occupation to be chosen $(\frac{1}{5})(425) = 85$ times.

Expected Number of Outcomes in a Multinomial Experiment

Out of n trials of a multinomial experiment, the expected number of outcomes to fall in category j is $e_j = n\pi_j$. Remember that the expected cell count is calculated assuming H_0 is true.

We realize, however, that the choices could be evenly distributed and not all $o_j = 85$ exactly. There could be some discrepancies in a random sample of only 425 children.

The χ^2 Test Statistic

The test statistic should consider the differences between the observed o_j's and those expected under the null hypothesis. The chi-square (χ^2) statistic, proposed by Karl Pearson, measures the amount of disagreement between the observed data and the expected data.

Pearson χ^2 Test Statistic

$$\chi^2 = \sum \frac{(o_j - e_j)^2}{e_j}$$

where the sum is over all categories, with o_j being the observed frequency count and e_j the expected frequency count in category j.

> Remember that k is the number of categories. There are five occupations to choose from; therefore, $k = 5$.

If there are large differences between o_j and e_j, then χ^2 will be large, which in turn suggests that the null hypothesis should be rejected. But how large does it have to get for us to have significant evidence to reject the null hypothesis? This can be answered only by investigating the sampling distribution of the χ^2 statistic.

When n is large and H_0 is true, the sampling distribution of χ^2 is known to be approximately *chi-square with $k - 1$ degrees of freedom*. Figure 10.1 illustrates a typical chi-square density curve. The curve begins at 0 and is skewed right. As the degrees of freedom increase, the distribution stretches out along the horizontal axis.

FIGURE 10.1   Chi-square distribution

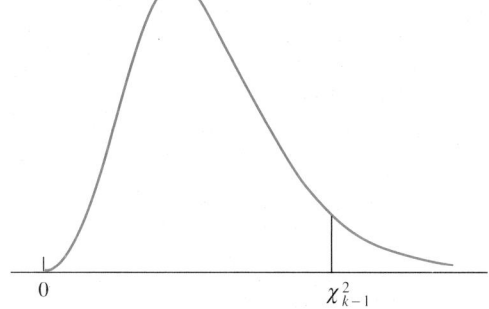

0 χ^2_{k-1}

Table B.4 in Appendix B gives the upper-tail probabilities of the χ^2 distribution for various degrees of freedom. For example, when the degrees of freedom are 16, then 10% of the area lies above 23.5, 5% lies above 26.3, and 1% lies above 32.0. The table also gives the lower-tail probabilities. For 16 degrees of freedom, we see that 95% of the area lies above 7.96; consequently, 5% lies below 7.96.

Goodness-of-Fit Test

For the χ^2 goodness-of-fit test, we are interested in only large values of χ^2 and the upper-tail probabilities. Returning to our example, we have

$$\chi^2_{obs} = \frac{(130 - 85)^2}{85} + \frac{(105 - 85)^2}{85} + \frac{(80 - 85)^2}{85} + \frac{(60 - 85)^2}{85} + \frac{(50 - 85)^2}{85}$$

$$= \frac{1}{85}[(45)^2 + (20)^2 + (-5)^2 + (-25)^2 + (-35)^2]$$

$$= 50.59$$

From the χ^2 table with $k - 1 = 5 - 1 = 4$ degrees of freedom, we see that .5% of the area lies above 14.9. Thus, less than .5% of the area lies above 50.59; that is, p-value $< .005$, and therefore there is sufficient evidence to reject H_0. Based on these sample results, there is highly significant evidence that the occupations are not chosen equally by schoolchildren between 8 and 12 years old.

In this example, we are investigating statistically how well the observed data fit the conjectured probabilities in the null hypothesis. This is why the test is called the χ^2 goodness-of-fit test. The general form of the test is given in the box:

Chi-Square Goodness-of-Fit Test

Application: Multinomial experiments

Assumptions:

1. The experiment satisfies the properties of a multinomial experiment.
2. No expected cell count, e_j, is less than 1, and no more than 20% of the e_j's are less than 5. (This is so that the χ^2 approximation will be good.)

$$H_0: \pi_1 = p_1, \pi_2 = p_2, \ldots, \pi_k = p_k$$

where $p_1, p_2, \ldots, p_k$ are the hypothesized values of the multinomial probabilities.

$$H_a: \text{At least one of the multinomial probabilities}$$
$$\text{does not equal the hypothesized value}$$

Test statistic:

$$\chi^2 = \sum \frac{(o_j - e_j)^2}{e_j}$$

where $e_j = np_j$.

The test is a right-tailed test, where the p-value is found in the χ^2 table with $k - 1$ degrees of freedom. Usually the exact value cannot be found, but bounds for it can be found from the closest values to the observed value of the χ^2 statistic.

EXAMPLE 10.1

A local grocery store stocks four brands of cola. Suppose that, nationally, brand A commands 40% of the market, brand B has 35%, brand C has 20%, and brand D has 5%. Of 2,000 colas sold during 1 week in the store, 615 were brand A, 804 were brand B, 383 were brand C, and 198 were brand D. Do the data collected at the local grocery store fit the national percentages?

SOLUTION

Let

$$\pi_1 = \text{proportion of people who buy brand A at the grocery store}$$
$$\pi_2 = \text{proportion of people who buy brand B at the grocery store}$$
$$\pi_3 = \text{proportion of people who buy brand C at the grocery store}$$
$$\pi_4 = \text{proportion of people who buy brand D at the grocery store}$$

The null hypothesis, which says that the local percentages are the same as the national percentages, is

$$H_0: \pi_1 = .40, \ \pi_2 = .35, \ \pi_3 = .20, \ \pi_4 = .05$$

and the alternative hypothesis is

$$H_a: \text{At least one of the proportions differs from its hypothesized value}$$

The test statistic is

$$\chi^2 = \sum \frac{(o_j - e_j)^2}{e_j}$$

where

$$e_1 = (2{,}000)(.40) = 800$$
$$e_2 = (2{,}000)(.35) = 700$$
$$e_3 = (2{,}000)(.20) = 400$$
$$e_4 = (2{,}000)(.05) = 100$$

Because all of these values are greater than 5 and the conditions of a multinomial experiment are met, we conduct a χ^2 goodness-of-fit test.

The observed value of χ^2 is

$$\chi^2_{obs} = \frac{(615 - 800)^2}{800} + \frac{(804 - 700)^2}{700} + \frac{(383 - 400)^2}{400} + \frac{(198 - 100)^2}{100}$$
$$= 154.995$$

From the χ^2 table with 3 degrees of freedom, we see, as depicted in Figure 10.2, that p-value $< .005$ because any χ^2 value greater than 12.84 has a tail probability less than .005. This is highly significant evidence that H_0 should be rejected. Clearly, the shoppers at the local grocery store do not choose their brands of cola the same as the rest of the nation. The manager of the grocery store is well advised not to follow the national percentages when stocking cola.

FIGURE 10.2

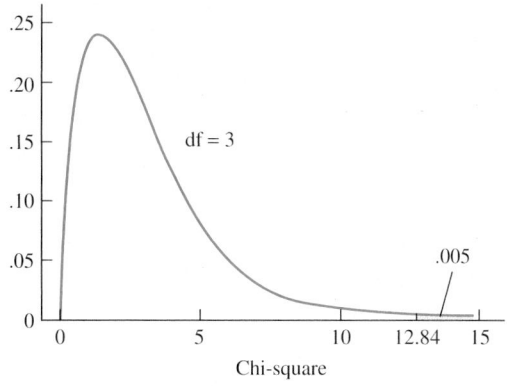

The sampling distribution of χ^2 is only approximately chi-square with $k - 1$ degrees of freedom. The approximation becomes better as the value of n gets larger. The value of n must be large enough to guarantee that no e_j ($e_j = np_j$) is less than 1 and no more than 20% of the e_j's are less than 5 (Cochran, 1954). If these conditions are not met, you should combine cells so that the assumption is satisfied.

EXAMPLE 10.2

Suppose a high school teacher gives a standardized English grammar test to her class of 25 students, with these scores as results:

Worksheet: Grammar.mtw

56	58	40	77	87	75	61	70	73
71	66	69	67	68	60	72	73	
61	64	66	84	72	52	65	67	

> Percentiles divide data into 100 parts, deciles divide it into 10 parts, and quartiles divide it into 4 parts.

She wishes to compare her class to the national standard. The test manual reports these decile scores:

Decile	1	2	3	4	5	6	7	8	9
Score	45.0	56.8	62.5	66.1	68.7	71.3	74.0	78.5	84.2

(That is, 10% scored below 45.0, 10% from 45.0 to 56.8, etc.) Does her class differ significantly from the national standard?

SOLUTION

First we set up the categories as shown in Table 10.2 on page 648. Because each category is defined by the decile scores, 10% of the data should be within each category. Therefore, the expected number in each category is

$$e_j = 25(.10) = 2.5$$

which violates the basic assumption that no more than 20% of the expected cell counts should be less than 5.

TABLE 10.2

Category	Number of scores
below 45.0	1
45.0–56.8	2
56.8–62.5	4
62.5–66.1	4
66.1–68.7	3
68.7–71.3	3
71.3–74.0	4
74.0–78.5	2
78.5–84.2	1
above 84.2	1

This example illustrates how categories can be defined by ranges of values of a numerical variable so that they can be analyzed with the chi-square statistic.

A solution to the problem is to combine the ten categories into five categories, where each category covers 20% of the data. Then the expected number in each category is

$$e_j = 25(.20) = 5$$

Table 10.3 gives the observed and expected numbers of scores in each of the combined categories. The χ^2 statistic can be calculated as follows:

$$\chi^2_{obs} = \frac{(3-5)^2}{5} + \frac{(8-5)^2}{5} + \frac{(6-5)^2}{5} + \frac{(6-5)^2}{5} + \frac{(2-5)^2}{5} = 4.8$$

TABLE 10.3

Category	Observed number of scores	Expected number of scores
below 56.8	3	5
56.8–66.1	8	5
66.1–71.3	6	5
71.3–78.5	6	5
above 78.5	2	5

With 4 degrees of freedom, p-value $> .10$ $(4.8 < 7.78)$, and therefore there is insufficient evidence to reject

$$H_0: \pi_1 = \pi_2 = \pi_3 = \pi_4 = \pi_5 = .20$$

Her class does not differ significantly from the national standard.

EXERCISES 10.1

10.1 For a χ^2 distribution with 15 degrees of freedom, determine the upper critical value for each tail probability:

a. .05   b. .01   c. .95

10.2 Find the p-value associated with these observed values of the chi-square statistic with the specified degrees of freedom:

a. $\chi^2_{obs} = 8.4$, df $= 4$ b. $\chi^2_{obs} = 3.7$, df $= 7$
c. $\chi^2_{obs} = 16.9$, df $= 3$ d. $\chi^2_{obs} = 13.5$, df $= 5$

10.3 For a χ^2 distribution with 12 degrees of freedom, find the two values that locate the middle 95% of the distribution.

10.4 Compute the value of the chi-square statistic for each table of observed and expected values. Does there appear to be a significant difference between the observed and expected frequencies?

a.

j	1	2	3
o_j	15	24	6
e_j	15	20	10

b.

j	1	2	3	4
o_j	23	36	51	70
e_j	36	45	63	36

10.5 A multinomial experiment with $k = 5$ and $n = 500$ yielded these results:

Category	1	2	3	4	5
o_j	92	97	106	85	120

Is there evidence to reject the hypothesis that the categories are equally likely?

10.6 A multinomial experiment with four possible outcomes and 100 trials produced these data:

Category	1	2	3	4
o_j	44	29	21	6

Is there evidence to reject this hypothesis:

$$\pi_1 = 40\%, \quad \pi_2 = 30\%, \quad \pi_3 = 20\%, \quad \pi_4 = 10\%$$

10.7 A local official claims that of all voters in the district, 40% are Democrats, 45% are Republican, 7% are conservative, 5% are liberal, and the remaining 3% are classified as Other. These party preferences were found from a sample of 1,200 voters:

Party	Democrat	Republican	Conservative	Liberal	Other
Number of voters	504	523	72	70	31

Are the sample results consistent with the claim made by the official?

10.8 In 1882, R. Wolf tossed a die 20,000 times and recorded the number of times each side faced up, as listed at the top of page 650. Was the die fair?

Die	1	2	3	4	5	6
Frequency	3,407	3,631	3,176	2,916	3,448	3,422

SOURCE: R. Wolf, *Vierteljahresschrift Naturforschl Ges. Zurich* 207 (1882): 242.

10.9 A large jar has red, black, blue, and white marbles in it. One hundred marbles are drawn from the jar with these results:

Color	Red	Black	Blue	White
Frequency	28	19	22	31

Is there statistical evidence that the proportions of marbles of the four colors are the same?

10.10 The numbers of books borrowed from a public library for a particular week are given here:

Day	Monday	Tuesday	Wednesday	Thursday	Friday
Books	125	105	120	114	136

Determine whether the number of books borrowed depends on the day of the week.

10.11 Five different strains of flies were tested for their resistance to a particular chemical agent. A large resort area was sprayed with the chemical. Afterward, 1,000 dead flies were randomly selected and classified according to their strain, as follows:

Strain	1	2	3	4	5	Total
Number killed	265	178	301	115	141	1,000

Assuming that each strain is equally prevalent in the resort area, is there evidence that some strains are more resistant to the chemical agent than others?

10.12 According to Information Resources (*USA Today*, April 18, 1994), these are the market shares for the different cereal companies:

Company	Kellogg	General Mills	Post	Quaker Oats	Other
Share	35%	29%	12%	7%	17%

A survey of 740 shoppers selected randomly from grocery stores in California asked for their favorite brands of cereal. From the following data, determine whether the cereal companies have the same market share in California as nationwide.

Company	Kellogg	General Mills	Post	Quaker Oats	Other
Number of preferences	220	163	109	95	153

10.13 Should companies be self-insured and provide their own group health plans for their employees? A survey of 461,208 employees of private companies by Medstat Systems at the request of the *Wall Street Journal* (June 17, 1994) found that 31% of the employees had no medical claims, 50% had claims less than $1,000, 14% had claims between $1,000 and $5,000, and 5% had claims that exceeded $5,000. A small company obtained the health records of its 250 employees and found that 48 had no claims, 88 had claims less than $1,000, 55 had claims between $1,000 and $5,000, and the rest had claims in excess of $5,000. Do these data suggest that the claims from the employees of this company follow the same distribution as that found by the *Wall Street Journal*?

10.14 Are coal mining disasters more likely to happen in certain months? These data were obtained on 191 coal-mining disasters that occurred from March 15, 1851, to March 22, 1962, inclusive:

Month	Number of disasters
January	14
February	20
March	20
April	13
May	14
June	10
July	18
August	15
September	11
October	16
November	16
December	24

SOURCE: R. Jarrett, "A Note on the Intervals Between Coal Mining Disasters," *Biometrika* 66 (1979): 191–193.

Can you conclude from the data that accidents are equally likely in all 12 months of the year? State your hypothesis, give the test statistic and its *p*-value, and provide your conclusion.

10.15 The following results were obtained in a 1962 national survey of 452 college students on the issue of student academic integrity:

Copied from another student during a test without his/her knowledge	20%
Copied from another student during a test with his/her knowledge	12%
Used unpermitted crib notes during a test	16%
Used unfair methods to learn what was on a test before it was given	35%
Helped someone else cheat on a test	22%

The same questions were used in a 1993 survey of 1,782 students, with these results:

Copied from another student during a test without his/her knowledge	45%
Copied from another student during a test with his/her knowledge	32%
Used unpermitted crib notes during a test	27%
Used unfair methods to learn what was on a test before it was given	29%
Helped someone else cheat on a test	38%

SOURCE: Appalachian State University, *Academic Integrity, Student Life and Learning*, Report Vol. 4, No. 1, October 25, 1994.

Can the chi-square goodness-of-fit test be used to compare the results of the 1993 survey with the results found in 1962? Explain why or why not.

10.16 Here are the percents of total deaths in 1990 from various causes:

Cause	Percent
Heart disease	33.5
Cancer	23.4
Stroke	6.7
Accidents	4.3
Lung disease	4.1
Other	28.0

SOURCE: U.S. Department of Health and Human Services, *Monthly Vital Statistics Report*, Vol. 39, No. 13, August 28, 1991.

In 1993, there were 2,268,000 deaths distributed as follows:

Cause	Number
Heart disease	739,860
Cancer	530,870
Stroke	149,740
Accidents	88,630
Lung disease	101,090
Other	657,810

SOURCE: *World Almanac and Book of Facts*, 1995.

With a test of significance, determine whether the 1993 figures are consistent with the 1990 percents.

10.2   THE CHI-SQUARE TEST OF INDEPENDENCE

When frequency count data are classified according to two or more variables or populations, they are called *cross-tabulated data* and displayed in a contingency table (introduced in Chapter 3). For example, subjects might be classified in a contingency table according to their religious preferences and their attitudes about abortion. College students might be classified according to class rank and study habits (good, average, bad).

In this section, we investigate the chi-square test of independence, where subjects in a single population are classified according to two different variables. In Section 10.3, we investigate the chi-square test of homogeneity, where subjects from several different populations are classified according to a single variable. In both cases, the chi-square statistic measures the discrepancy between actual counts and expected counts.

Test of Independence

For the test of independence, the aim is to determine whether there is any dependency between the classifying variables. Suppose that 200 randomly selected people are asked about their views on gun control and their preferred political party. Are their views on gun control affected by their political party preference? Table 10.4 is a contingency table of the distribution of the frequency counts from the sample of 200 people.

TABLE 10.4

		Opinion on gun control			
		Favor	Oppose	No opinion	Total
Political Party	Democrat	44	48	18	110
	Republican	32	48	10	90
	Total	76	96	28	200

Note that we have a single sample that is classified two ways.

The test is constructed in such a way that the classifying variables are assumed to be independent until the data prove otherwise. The null hypothesis is

H_0: Political party and opinion on gun control are independent

and the alternative hypothesis is

H_a: Political party and opinion on gun control are dependent

To utilize the χ^2 test statistic, we must demonstrate that the conditions of a multinomial experiment are satisfied.

We denote the probabilities of falling into the six cells (categories) as depicted in Table 10.5. The cell probability π_{D1} denotes the probability that a randomly selected subject is a Democrat and favors gun control; that is,

A cell is the intersection of a specific row and a specific column.

$$\pi_{D1} = P(\text{Democrat and favor})$$

The other five cell probabilities are defined similarly.

TABLE 10.5

		Opinion on gun control			
		Favor	Oppose	No opinion	Total
Political Party	Democrat	π_{D1}	π_{D2}	π_{D3}	π_D
	Republican	π_{R1}	π_{R2}	π_{R3}	π_R
	Total	π_1	π_2	π_3	1

As for satisfying the conditions of a multinomial experiment, we have that the 200 people constitute the n independent trials. Each subject falls into one of the six cells. The probabilities of falling into the cells are $\pi_{D1}, \pi_{D2}, \pi_{D3}, \pi_{R1}, \pi_{R2},$ and π_{R3}, and they remain the same for all subjects. Thus, the conditions of a multinomial experiment are satisfied.

The row and column probabilities are called *marginal probabilities*. The row marginal probability π_D is the probability that a Democrat is selected regardless of his or her opinion on gun control. We have that

$$\pi_D = P(\text{Democrat}) = \pi_{D1} + \pi_{D2} + \pi_{D3}$$

Similarly,

$$\pi_R = P(\text{Republican}) = \pi_{R1} + \pi_{R2} + \pi_{R3}$$
$$\pi_1 = P(\text{favor}) = \pi_{D1} + \pi_{R1}$$
$$\pi_2 = P(\text{oppose}) = \pi_{D2} + \pi_{R2}$$
$$\pi_3 = P(\text{no opinion}) = \pi_{D3} + \pi_{R3}$$

The existing data suggest that

$$\pi_D = P(\text{Democrat}) = \frac{110}{200}$$

$$\pi_R = P(\text{Republican}) = \frac{90}{200}$$

Also, not accounting for political party, we have

$$\pi_1 = P(\text{favor}) = \frac{76}{200}$$

$$\pi_2 = P(\text{oppose}) = \frac{96}{200}$$

$$\pi_3 = P(\text{no opinion}) = \frac{28}{200}$$

Recall from Chapter 4 that if two events are independent, then the probability of the joint occurrence of the two is the product of their probabilities. Thus, if political party and opinion are independent, then the probability of being Democrat and favoring gun control is

$$\pi_{D1} = P(\text{Democrat and favor})$$

$$= P(\text{Democrat})P(\text{Favor})$$

$$= \left(\frac{100}{200}\right)\left(\frac{76}{200}\right)$$

Therefore, out of 200 subjects, we would expect

$$200\left(\frac{110}{200}\right)\left(\frac{76}{200}\right) = \frac{(110)(76)}{200}$$

to fall in the first cell; that is,

$$E[\text{Democrat and favor}] = \frac{(110)(76)}{200} = 41.8$$

In a similar fashion,

$$E[\text{Democrat and oppose}] = \frac{(110)(96)}{200} = 52.8$$

$$E[\text{Democrat and no opinion}] = \frac{(110)(28)}{200} = 15.4$$

Note that the expected cell count, under the condition that the two classifying variables are independent, is

$$e_j = \frac{RC}{n}$$

where R is the row total, C is the column total, and n is the grand total. Thus,

$$E[\text{Republican and favor}] = \frac{(90)(76)}{200} = 34.2$$

$$E[\text{Republican and oppose}] = \frac{(90)(96)}{200} = 43.2$$

$$E[\text{Republican and no opinion}] = \frac{(90)(28)}{200} = 12.6$$

Table 10.6 gives the observed cell counts with the expected cell counts in parentheses.

TABLE 10.6

		Opinion on gun control			
		Favor	Oppose	No opinion	Total
Political Party	Democrat	44 (41.8)	48 (52.8)	18 (15.4)	110
	Republican	32 (34.2)	48 (43.2)	10 (12.6)	90
	Total	76	96	28	200

Because the expected cell counts were calculated under the assumption that H_0 is true (political party and opinion are independent), a large discrepancy between the observed and expected cell counts should lead to the rejection of H_0 and thus establish that the two classifications are dependent. To compare the differences between observed and expected counts, we calculate the χ^2 statistic as in the previous section:

$$\chi^2_{\text{obs}} = \frac{(44 - 41.8)^2}{41.8} + \frac{(48 - 52.8)^2}{52.8} + \frac{(18 - 15.4)^2}{15.4} + \frac{(32 - 34.2)^2}{34.2}$$
$$+ \frac{(48 - 43.2)^2}{43.2} + \frac{(10 - 12.6)^2}{12.6}$$
$$= 2.2025$$

To determine whether the observed value of χ^2 is statistically large, we must determine its p-value from the χ^2 table. But first we must determine the number of degrees of freedom for the test statistic. Note that in Table 10.6 the expected cell counts in the first row sum to 110. Therefore, we had to calculate only two of the three expected cell counts in the first row. Also the two expected cell counts in the first column sum to 76, so only one had to be calculated. Without much difficulty we see that of the six expected cell counts, only two had to be calculated, and the remaining four were obtained by subtraction. We say that, of the six pieces of data, only two are free to vary and, therefore, the degrees of freedom are 2. In general, the degrees of freedom are

$$\text{df} = (\text{number of rows} - 1)(\text{number of columns} - 1)$$

From the χ^2 table with 2 degrees of freedom, we see that the observed χ^2 statistic of 2.2025 falls below 4.61, which is the value associated with a tail probability of .10. Thus, we have

$$p\text{-value} > .10$$

which says that there is insufficient evidence to reject H_0. We have failed to show a dependence between political party and opinion on gun control.

We now summarize the test.

Chi-Square Test of Independence

Application: Test the independence of two classifying variables

Assumptions:

1. The experiment satisfies the properties of a multinomial experiment.

2. No expected cell count is less than 1, and no more than 20% of the cell counts are less than 5.

H_0: The two classifications are independent

H_a: The two classifications are dependent

Test statistic:

$$\chi^2 = \sum \frac{(o_j - e_j)^2}{e_j}$$

where o_j represents the observed cell frequencies and e_j represents the expected cell frequencies given by

$$e_j = \frac{RC}{n}$$

where R = row total, C = column total, and n = grand total or the total number of subjects.

The test is a right-tailed test, where the p-value is found in the χ^2 table with $(r - 1)(c - 1)$ degrees of freedom (r denotes the number of rows and c denotes the number of columns).

EXAMPLE 10.3

A random sample of 400 undergraduate college students were classified according to class and study habits. From the data in Table 10.7, test to see whether the two classifications are independent.

TABLE 10.7

		Study Habits			
		Good	Average	Bad	Total
	Freshman	20	42	58	120
	Sophomore	25	48	32	105
Class	Junior	31	28	35	94
	Senior	24	27	30	81
	Total	100	145	155	400

SOLUTION

The null hypothesis is

$$H_0: \text{Class rank and study habits are independent}$$

and the alternative hypothesis is

$$H_a: \text{Class rank and study habits are dependent}$$

To find the value of the test statistic, we must first find the expected cell counts with the formula

$$e_j = \frac{RC}{n}$$

For the first cell in the first row and first column, we have

$$e_1 = \frac{(120)(100)}{400} = 30$$

The next cell in the first row has expected count

$$e_2 = \frac{(120)(145)}{400} = 43.5$$

The expected count in the third cell in the first row is

$$e_3 = \frac{(120)(155)}{400} = 46.5$$

Continuing, we get the expected cell counts shown in Table 10.8 at the top of page 658. Note that all the expected cell frequencies exceed 5, so the conditions for the χ^2 approximation are satisfied.

Substituting the observed and expected values into the formula for the χ^2 statistic, we find

$$\chi^2_{obs} = \frac{(20 - 30)^2}{30} + \frac{(42 - 43.5)^2}{43.5} + \cdots + \frac{(30 - 31.39)^2}{31.39}$$

$$= 15.221$$

TABLE 10.8

		Study Habits			
		Good	Average	Bad	Total
	Freshman	30	43.5	46.5	120
	Sophomore	26.25	38.06	40.69	105
Class	Junior	23.5	34.08	36.42	94
	Senior	20.25	29.36	31.39	81
	Total	100	145	155	400

The degrees of freedom for the χ^2 statistic are

$$df = (r - 1)(c - 1) = (4 - 1)(3 - 1) = 6$$

From the χ^2 table with 6 degrees of freedom, we see that

$$14.45 < 15.221 < 16.81$$

and thus $.01 < p$-value $< .025$. There is statistical evidence to reject H_0. The data suggest that study habits of students are related to their class rank.

COMPUTER TIP

The analysis of the data in Example 10.3 can easily be done with the Chisquare command in Minitab.

First, key the observed counts as they appear in Example 10.3 in C1, C2, C3 of the worksheet. From the **Stat** menu, select **Tables** and then select **Chisquare Test**. In the Chisquare Test window, **select** the three columns and click on **OK**. Figure 10.3 contains the Minitab output. Note that there is a slight disagreement between the results found in Example 10.3 and the results found by Minitab. These differences are due to round-off error. Also note that Minitab computes the actual p-value of .019.

FIGURE 10.3   Expected counts are printed below observed counts

```
                  C1        C2        C3      Total
        1         20        42        58       120
                30.00     43.50     46.50

        2         25        48        32       105
                26.25     38.06     40.69

        3         31        28        35        94
                23.50     34.08     36.42

        4         24        27        30        81
                20.25     29.36     31.39

Total   .        100       145       155       400

ChiSq =     3.333 +   0.052 +   2.844 +
            0.060 +   2.595 +   1.855 +
            2.394 +   1.083 +   0.056 +
            0.694 +   0.190 +   0.061 = 15.216
df = 6, p = 0.019
```

In the preceding example, the frequencies for the contingency table were already given. In the next example, we show how Minitab can construct the table of frequencies from the data.

EXAMPLE 10.4

A survey was conducted in a voting district to record (among several other variables) the political party and gender of the respondents. The political party and gender were coded as follows and stored in Minitab worksheet: **Politic.mtw**:

Party	Gender
1—Democrat	1—female
2—Republican	2—male
3—Other	

Organize the data in a contingency table according to the variables political party and gender. Perform a chi-square test of independence on the two variables.

SOLUTION

The null and alternative hypotheses are

H_0: Political party and gender are independent

H_a: Political party and gender are dependent

COMPUTER TIP

In Menu mode, from the Stat menu select **Tables** followed by **Cross Tabulation**. In the Cross Tabulation window, **Select** the two columns, C1 and C2. (If more than two columns are selected, a separate chi-square test is done for each two-way table.) Check **Chisquare analysis**, click on **Above and expected count**, and click **OK**.

In the Session window, the form of the command is

MTB > Table C1 C2;
SUBC > ChiSquare 2.

(The 2 says to print observed and expected counts. If it is omitted, only observed counts are printed.)

Figure 10.4 gives the Minitab output.

FIGURE 10.4

```
MTB > Table 'Party' 'Gender';
SUBC>   ChiSquare 2.

Tabulated Statistics


    ROWS: Party     COLUMNS: Gender

                1          2        ALL

    1          49         55        104
            49.50      54.50     104.00


    2          64         73        137
            65.21      71.79     137.00


    3           6          3          9
             4.28       4.72       9.00


    ALL       119        131        250
            119.00     131.00     250.00

    CHI-SQUARE =      1.365   WITH D.F. =     2

        CELL CONTENTS --
                        COUNT
                        EXP FREQ
```

From the chi-square table, we see that

$$p\text{-value} > .10.$$

There is insufficient evidence to reject H_0. The two classifications are assumed to be independent.

Application

Hot Hand in Basketball?

The hot hand theory states that a basketball player (with the hot hand) has a greater chance of hitting a second free throw after a previous hit than after a previous miss. Tversky and Gilovich (1989) concluded from their analysis that "a player's chances of hitting are largely independent of the outcome of his or her previous shot." Larkey, Smith, and Kadane (1989) challenged Tversky and Gilovich's conclusion. Based on televised NBA games during the 1987–1988 season, they concluded that at least one player, Vinnie Johnson of the Detroit Pistons, was a streak shooter. Hooke (1989) discounted both studies and claimed that the real situation is much more complex than either of the hypothesized models; it may be an individual thing. Some players are "slaves to their recent past" and others "can ignore it altogether."

These data were collected on Larry Bird and Rick Robey of the 1980–1981 and 1981–1982 Boston Celtics:

Bird

		Second shot	
		Hit	Miss
First	Hit	251	34
shot	Miss	48	5

Robey

		Second shot	
		Hit	Miss
First	Hit	54	37
shot	Miss	49	31

Figures 10.5 and 10.6 on page 662 show chi-square tests of independence of the first and second free-throw shots for Bird and Robey. In neither case is there a significant p-value (.602 for Bird and .799 for Robey). We conclude from these data that making or missing the second shot is independent of the outcome of the first shot. If their data are combined, however, as in Figure 10.7, we find a significant p-value $= .047$, which says that the second shot does indeed depend on the outcome of the first shot.

FIGURE 10.5   Larry Bird data

Expected counts are printed below observed counts

```
          bird ht   bird ms    Total
    1        251        34       285
           252.12     32.88

    2         48         5        53
            46.88      6.12

Total        299        39       338

ChiSq =   0.005 +   0.038 +   0.027 +   0.203 = 0.273
df = 1,  p = 0.602
```

FIGURE 10.6   Rick Robey data

Expected counts are printed below observed counts

```
         robey ht robey ms    Total
    1         54        37        91
            54.81     36.19

    2         49        31        80
            48.19     31.81

Total        103        68       171

ChiSq =   0.012 +   0.018 +   0.014 +   0.021 = 0.065
df = 1,  p = 0.799
```

FIGURE 10.7   Combined data

Expected counts are printed below observed counts

```
          comb ht   comb ms    Total
    1        305        71       376
           296.96     79.04

    2         97        36       133
           105.04     27.96

Total        402       107       509

ChiSq =   0.218 +   0.818 +   0.616 +   2.313 = 3.964
df = 1,  p = 0.047
```

Simpson's Paradox

What does this mean? For one thing, it reiterates Simpson's paradox: Two contingency tables with insignificant results, when collapsed to one table, become significant. This may also simply show that there are data to support either theory; that is, looking at the problem one way we can justify the hot hand theory, and looking at it another way we can contradict the hot hand theory. The jury is still out on this issue.

Whether you believe in the hot hand theory or not, it is difficult to ignore the final seconds of the first game of the 1995 NBA championship between the Houston Rockets and the Orlando Magic. Orlando was up by 3 points when Nick Anderson (Magic) was fouled. Just one hit out of his two free throws would have put the game out of reach for Houston. He missed both and was fouled again retrieving the ball. He missed the next two tries. By Anderson missing all four free throws (cold hand?), Houston was able to make a last-second 3-point shot to put the game into overtime. Houston went on to win the game and eventually swept the series in four straight games.

EXERCISES 10.2

10.17   Based on the computer output, test the independence of the two variables displayed in this contingency table:

		Variable A		
		Level 1	Level 2	Level 3
Variable B	Level 1	65	39	16
	Level 2	133	156	61

```
Chi-Square Test

Expected counts are printed below observed counts

         level1   level2   level3     Total
    1        65       39       16       120
          50.55    49.79    19.66

    2       133      156       61       350
         147.45   145.21    57.34

Total       198      195       77       470

ChiSq =   4.129 +   2.337 +   0.681 +
          1.415 +   0.801 +   0.234 = 9.597
df = 2, p = 0.008
```

10.18 Using the accompanying computer printout, test the independence of the two variables displayed in this contingency table.

		Variable A		
		Level 1	Level 2	Level 3
	Level 1	572	418	451
Variable B	Level 2	352	379	315
	Level 3	256	278	205

```
Chi-Square Test

Expected counts are printed below observed counts

       level1    level2    level3     Total
   1      572       418       451      1441
        527.09    480.18    433.73

   2      352       379       315      1046
        382.60    348.56    314.84

   3      256       278       205       739
        270.31    246.26    222.43

Total    1180      1075       971      3226

ChiSq =  3.827 +  8.053 +  0.688 +
         2.448 +  2.659 +  0.000 +
         0.758 +  4.092 +  1.366 = 23.890
df = 4, p = 0.000
```

10.19 In Table 3.2 of Section 3.1 (reproduced here), automobile dealers were classified according to the type of dealership and the service rendered to customers on their 15,000-mile checkup.

Worksheet: Dealers.mtw

Dealership	Replace parts before they are needed	Perform only services recommended by manufacturer	Total
Honda Accord	19 (9.8)	2 (11.2)	21
Saturn	4 (8.9)	15 (10.1)	19
Ford Taurus	8 (9.8)	13 (11.2)	21
Dodge Caravan	11 (9.8)	10 (11.2)	21
Mazda Miata	12 (9.8)	9 (11.2)	21
Toyota Lexus	3 (8.9)	16 (10.1)	19
Total	57	65	122

The numbers in parentheses are the expected numbers if the two classifications are independent. Use the chi-square test to determine whether the discrepancies between the actual counts and the expected counts indicate that there is a dependency between the two classifications.

10.20 A group of college students were classified as either left- or right-handed and as either mathematically inclined or not. From the data in the table, test the hypothesis that the two classifications are independent.

	Predominant hand	
	Left	Right
Mathematically inclined	12	93
Not mathematically gifted	7	108

10.21   Suppose a number of patients were treated for cancer with these results:

		Tumor regression?	
		Yes	No
Toxic	Yes	15	5
reaction?	No	4	22

Determine whether there is a relationship between the presence of a toxic reaction and tumor regression. Are the expected cell counts greater than 5? Why is this important?

10.22   An experimental psychologist wishes to study the effects that three different drugs have on one's ability to learn a list of nonsense syllables. Sixty subjects were categorized as to the type of drug they had been receiving for the past 3 months and their ability to memorize the list of nonsense syllables.

		Drug		
		A	B	C
Ability to	Low	12	6	6
memorize the list	Medium	8	5	11
of syllables	High	2	7	3

Determine whether the type of drug is related to the subject's ability to memorize the list of nonsense syllables. Are all of the required assumptions for the χ^2 test met?

10.23   Numerous seasonal allergy medicines are advertised to relieve sneezing, runny nose, watery eyes, and congestion. In most cases, however, there are associated side effects. In double-blind, controlled clinical trials, Marion Merrell Dow Inc., the maker of Seldane-D, reported the following results in June 1991:

Worksheet: Allergy.mtw

		Medication		
		Seldane-D	Pseudoephedrine	Placebo
	Insomnia	97	77	12
Adverse Event	Headache	65	49	43
	Drowsiness	27	14	22

SOURCE: Marion Merrell Dow Inc, Kansas City, MO 64114.

a. Is there statistical dependence between the medication and the adverse events?
b. Discard the insomnia data and determine, as in part **a**, whether there is statistical dependence between the medication and the adverse events headache and drowsiness.
c. In light of the results found in part **b**, what is causing the dependence found in part **a**?
d. Marion Merrell Dow reported that there was no significant difference in drowsiness between those who took Seldane-D and those who took a placebo. Do the data support this claim?

10.24   In 1984, history was made when Geraldine Ferraro was nominated as the first woman vice-presidential candidate. For the first time ever, the gender of the candidate was an issue. On August

7 and 9, 1984, *Time* magazine conducted a telephone survey of 1,000 voters. In each contingency table, determine whether the gender of the respondent is independent of his or her choice.

a. If the election were tomorrow, for whom would you vote?

Worksheet: Ferraro1.mtw

	Reagan/Bush	Mondale/Ferraro	Undecided
Men	245	140	115
Women	205	160	135

b. Who would be a better vice president?

Worksheet: Ferraro2.mtw

	Bush	Ferraro	Undecided
Men	245	155	100
Women	185	235	80

10.25 Does the desire to participate in class projects relate to a child's academic achievement? To study this issue, a sample of 80 third-grade students were asked whether they wished to participate in the science project program. The students were then classified according to their academic standing. From these data, is there evidence of a relationship between the desire to participate in the science project program and academic standing? (Expected cell counts are in parentheses.)

		Desire to participate in Science Project?	
		Yes	No
Academic Standing	Below Average	17 (17.23)	9 (8.77)
	Average	14 (17.89)	13 (9.11)
	Above Average	22 (17.89)	5 (9.11)
	Total	53	27

10.26 A political pollster would like to determine whether the voters' feelings on a local referendum are related to their views on freedom of the press. A sample of 237 voters were asked how they plan to vote on the referendum, and they were asked to choose the one of the following that most closely represents their view on freedom of the press:

A. The press is at liberty to report anything it sees fit.
B. The press should not report anything that would jeopardize anyone's life.
C. The press should not report anything that would jeopardize anyone's life or reputation.

The contingency table lists the number of voters in each category:

Worksheet: Referend.mtw

		Opinion on referendum			
		For	Against	Undecided	Total
Response to question	A	24	29	7	60
	B	68	39	12	119
	C	47	8	3	58

Do the data show a relationship between opinion on the referendum and view on freedom of the press?

10.27 USA Today/CNN/Gallup national telephone polls found the following opinions on an embargo against Cuba:

	Dec. 1993	June 1994
Favor	660	700
Oppose	260	240

The headline stated "Embargo against Cuba gains favor."

a. Conduct a chi-square test of independence on the data.

b. Is the change of opinion from December 1993 to June 1994 large enough to justify the claim made in the headline?

10.28   Are snoring and heart disease related? The accompanying data are the results of a study reported in the *British Medical Journal*. Subjects were classified according to the amount they snored (as reported by their spouses) and whether they had a history of heart disease.

Worksheet: Snore.mtw

		Nonsnorer	Occasional snorer	Snores nearly every night	Snores every night	Total
Heart disease?	Yes	24	35	21	30	110
	No	1,355	603	192	224	2,374

SOURCE: P. Norton and E. Dunn, "Snoring As a Risk Factor for Disease," *British Medical Journal* 291 (1985): 630–632.

Determine whether snoring and heart disease are statistically related.

10.29   Prior to the enactment of seat belt legislation in the province of Alberta, Canada, data were collected on 86,769 automobile accident reports to determine the effectiveness of seat belts in preventing injury. The table gives the injury level of the driver and whether or not he or she was wearing a seatbelt:

Worksheet: Seatbelt.mtw

		Injury level			
		None	Minimal	Minor	Major/Fatal
Seatbelt?	Yes	12,813	647	359	42
	No	65,963	4,000	2,642	303

SOURCE: J. Jobson, *Applied Multivariate Data Analysis* (New York: Springer-Verlag, 1992), p. 18.

Is there statistical evidence that seat belts help prevent injury? Be sure to formulate null and alternative hypotheses, give the *p*-value, and state your conclusion.

10.30   A study of the absenteeism of bus drivers in the transit system for the city of Edmonton, Alberta, produced these data:

Worksheet: Bus.mtw

	Shift				
	A.M.	Noon	P.M.	Swing	Split
Absent	454	208	491	160	1,599
Present	5,806	2,112	3,989	3,790	10,754

SOURCE: J. Jobson, *Applied Multivariate Data Analysis* (New York: Springer-Verlag, 1992), p. 67.

Based on the data, is the attendance of bus drivers dependent on the shift? Be sure to state the null and alternative hypotheses. Give the *p*-value of the test statistic and write your conclusion.

10.3 THE CHI-SQUARE TEST OF HOMOGENEITY

In the test of independence, we attempt to determine whether two characteristics (variables) associated with the subjects in a *single* population are independent. For example, subjects randomly selected from a population may be classified according to their views on gun control and their preferred political party. We then determine whether their views on gun control are independent of their preferred political party. The chi-square test of homogeneity, presented in this section, attempts to determine whether *several* populations are similar or *homogeneous* with respect to some variable. With the variable as one classification and the populations as the other, the data form a two-way contingency table. The assumptions and statements of the null and alternative hypotheses are different from those for the test of independence, but the details of the analysis are the same. We illustrate with an example.

EXAMPLE 10.5

Suppose that over a 2-year period, 120 patients with heart disease were treated with one of two drugs (A or B). After a period of time, each patient's condition was rated as no change, improved, or greatly improved. Table 10.9, a contingency table, gives the distribution of frequency counts. Determine whether the patients' conditions are similar with respect to the two drugs.

TABLE 10.9

		Patient's Condition			
		No change	Improved	Greatly Improved	Total
Drug	A	15	22	33	70
	B	20	18	12	50
	Total	35	40	45	120

SOLUTION

The null and alternative hypotheses are:

H_0: The proportions of patients falling into the three categories are the same for drug A and drug B

H_a: The proportions of patients falling into the three categories are not the same for drug A and drug B

If we denote the probabilities of falling into the three categories (cells) for drug A as π_{A1}, π_{A2}, and π_{A3} and for drug B as π_{B1}, π_{B2}, and π_{B3}, then the null hypothesis can be stated as:

$$H_0: \pi_{A1} = \pi_{B1}, \quad \pi_{A2} = \pi_{B2}, \quad \pi_{A3} = \pi_{B3}$$

From the two samples, one of size 70 from the population of patients who received drug A and the other of size 50 from the population of patients who received drug B, we wish to determine whether the two populations are homogeneous with respect to the cell probabilities.

Because the sample sizes are determined before the data are collected, the row (or column) marginal totals in the test of homogeneity are fixed quantities with values that are the sizes of the samples taken from the populations. In the test of independence, however, the grand total is the only fixed quantity with a value that is the size of the sample taken from the single population. Thus, a feature that distinguishes the test of homogeneity from the test of independence is whether or not the marginal totals are fixed or random quantities.

Under the assumptions that the marginal totals are fixed and the null hypothesis is true, the expected cell counts for the test of homogeneity are calculated just as they were in the test for independence; namely, the expected cell counts are

$$e_j = \frac{RC}{n}$$

where R is the row total, C is the column total, and n is the grand total. Thus, in the example, we have

$$E[\text{A and no change}] = \frac{(70)(35)}{120} = 20.42$$

$$E[\text{A and improved}] = \frac{(70)(40)}{120} = 23.33$$

$$E[\text{A and greatly improved}] = \frac{(70)(45)}{120} = 26.25$$

$$E[\text{B and no change}] = \frac{(50)(35)}{120} = 14.58$$

$$E[\text{B and improved}] = \frac{(50)(40)}{120} = 16.67$$

$$E[\text{B and greatly improved}] = \frac{(50)(45)}{120} = 18.75$$

Table 10.10 gives the observed cell counts with the expected cell counts in parentheses.

TABLE 10.10

		Patient's Condition			
		No Change	Improved	Greatly Improved	Total
Drug	A	15 (20.42)	22 (23.33)	33 (26.25)	70
	B	20 (14.58)	18 (16.67)	12 (18.75)	50
	Total	35	40	45	120

The test statistic is computed exactly as in the preceding section; that is,

$$\chi^2 = \sum \frac{(o_j - e_j)^2}{e_j}$$

From the data we have

$$\chi^2_{obs} = \frac{(15 - 20.42)^2}{20.42} + \frac{(22 - 23.33)^2}{23.33} + \frac{(33 - 26.25)^2}{26.25} + \frac{(20 - 14.58)^2}{14.58}$$
$$+ \frac{(18 - 16.67)^2}{16.67} + \frac{(12 - 18.75)^2}{18.75}$$
$$= 7.801$$

Because there are two rows and three columns, the degrees of freedom are

$$df = (2 - 1)(3 - 1) = 2$$

From the χ^2 table, we see that the observed χ^2 statistic of 7.801 falls between 7.38 and 9.21, the values associated with tail probabilities of .025 and .01, respectively. Thus, we have

$$.01 < p\text{-value} < .025$$

which indicates that there is evidence to reject H_0. We conclude that the patients' conditions depend on the drug they received.

Here is a summary of the test procedure:

Chi-Square Test of Homogeneity

Application: Contingency table with fixed marginal totals

Assumptions:

1. A random sample is selected from each of the row category populations. The sample sizes (row marginal totals) are fixed prior to sampling.
2. No expected cell count is less than 1, and no more than 20% of the cell counts are less than 5.

H_0: The populations are homogeneous with respect to the variable of classification

H_a: The populations are not homogeneous

(continued)

Test statistic:

$$\chi^2 = \sum \frac{(o_j - e_j)^2}{e_j}$$

where o_j represents the observed cell frequencies and e_j represents the expected cell frequencies given by

$$e_j = \frac{RC}{n}$$

where R = row total, C = column total, n = grand total or the total number of subjects.

The test is a right-tailed test, where the p-value is found in the χ^2 table with $(r - 1)(c - 1)$ degrees of freedom (r denotes the number of rows and c denotes the number of columns).

E X E R C I S E S   1 0 . 3

10.31 From the contingency table data and the computer analysis, test the hypothesis that the proportions that fall into the three categories are the same for the three populations.

		Category			
		1	2	3	Total
	1	35	16	29	80
Population	2	29	21	30	80
	3	25	27	28	80

Chi-Square Test

Expected counts are printed below observed counts

	Cat1	Cat2	Cat3	Total
1	35	16	29	80
	29.67	21.33	29.00	
2	29	21	30	80
	29.67	21.33	29.00	
3	25	27	28	80
	29.67	21.33	29.00	
Total	89	64	87	240

(*continued*)

```
ChiSq =   0.959 +   1.333 +   0.000 +
          0.015 +   0.005 +   0.034 +
          0.734 +   1.505 +   0.034 = 4.621
df = 4, p = 0.329
```

Be sure to state the hypotheses and give the *p*-value. Does it appear that the row marginal totals are fixed? Do the expected cell frequencies exceed 5? Is there evidence to reject the null hypothesis?

10.32　From the data and the computer analysis, test the hypothesis that the proportions that fall into the four categories are the same for the three populations:

		Category				
		1	2	3	4	Total
	1	16	38	5	41	100
Population	2	24	41	12	23	100
	3	19	36	15	30	100

Chi-Square Test

Expected counts are printed below observed counts

	Cat1	Cat2	Cat3	Cat4	Total
1	16	38	5	41	100
	19.67	38.33	10.67	31.33	
2	24	41	12	23	100
	19.67	38.33	10.67	31.33	
3	19	36	15	30	100
	19.67	38.33	10.67	31.33	
Total	59	115	32	94	300

```
ChiSq =   0.684 +   0.003 +   3.010 +   2.982 +
          0.955 +   0.186 +   0.167 +   2.216 +
          0.023 +   0.142 +   1.760 +   0.057 = 12.184
df = 6, p = 0.059
```

Be sure to state the hypotheses and give the *p*-value. What assumptions must be met for the chi-square test of homogeneity? Have any been violated? Is there evidence to reject the null hypothesis?

10.33　A pollster sampled 200 voters, 100 from District 1 and 100 from District 2, to determine their opinion on an upcoming referendum. The results of the survey are given in the contingency table, with the expected counts in parentheses.

	Opinion on Referendum			
	Favor	Against	Undecided	Total
District 1	72 (66)	21 (27.5)	7 (6.5)	100
District 2	60 (66)	34 (27.5)	6 (6.5)	100

Is there evidence that the two districts will vote differently in the referendum?

a. Identify the variable of classification.
b. Identify the populations.
c. Formulate your hypotheses.
d. Calculate the test statistic and its p-value and state your conclusion.
e. Are any assumptions for the chi-square test in question?

10.34   A study was conducted to compare two treatments for smokers who wish to stop smoking. Three hundred smokers were divided between the two methods with these results:

	Stop smoking	Smoke less	Smoke the same
Treatment A	44	38	68
Treatment B	33	42	75

Do the data suggest that the two treatments are equally effective in helping smokers stop smoking?
a. Identify the variable of classification.
b. Identify the populations.
c. Formulate your hypotheses.
d. Calculate the test statistic and its p-value and state your conclusion.
e. Are any assumptions for the chi-square test in question?

10.35   The collegiate record for the single-season average rushing yards per game is held by Marcus Allen, who in 1981 with Southern Cal rushed for an average of 212.9 yards per game. Generally when a record is broken, the back who carries the ball receives the credit in the press release; many believe, however, that the linemen should receive the credit. Samples of 50 football players and 70 members of the press were asked who should receive the most credit: the back or the linemen. From these data, determine whether their views are similar.

Who should receive credit?

	Linemen	Back	Both	Total
Players	22	14	14	50
Press	6	48	16	70

a. Identify the variable of classification.
b. Identify the populations.
c. Conduct the chi-square test. Be sure to state the hypotheses, give the p-value, and present your conclusion.

10.36   The justice system in the 75 largest counties in the United States disposed of 540 spouse-murder cases in 1988. Can we conclude from the following data that wife defendants were less likely to be convicted and to receive severe sentences than husband defendants?

Worksheet: Spouse.mtw

		Defendant	
		Husband	Wife
	Not prosecuted	35	35
Result	Pleaded guilty	146	87
	Convicted at trial	130	69
	Acquitted at trial	7	31

SOURCE: Bureau of Justice Statistics, *Spouse Murder Defendants in Large Urban Counties*, Executive Summary, NCJ-156831, September 1995.

a. Identify the variable of classification.
b. Identify the populations.

c. Conduct the chi-square test. Be sure to state the hypotheses, give the p-value, and present your conclusion.

10.37   It is reported that a Vietnam veteran is much more likely to commit suicide than a nonveteran. From the list of those eligible for the draft in 1970, a sample of 100 Vietnam veterans and a sample of 100 nonveterans were selected. The 200 were asked whether they had ever contemplated suicide. The results are recorded in the contingency table.

		Vietnam veteran?	
		Yes	No
Contemplated	Yes	32	11
suicide?	No	68	89

Determine with a test of significance whether there is statistical evidence to suggest that the proportion of veterans who have contemplated suicide is different from the proportion of nonveterans. What is the variable of classification and what are the populations?

10.38   An experiment is designed to study the side effects of two drugs used as treatments for a certain ailment. A group of 90 subjects are assigned to two drug groups. After the specified drug is given, the side effects are classified as follows:

	Side effects			
	Major	Minor	None	Total
Drug A	13	15	17	45
Drug B	8	21	16	45

Are the side effects distributed the same for the two drugs? State your hypotheses accordingly. Are the marginal totals fixed? Complete the test, give the p-value, and state your conclusion.

10.39   Do male and female college students differ in their favorite sport? Random samples of 100 college women and 100 college men were asked to name their favorite sport. The results are recorded in the contingency table:

Worksheet: Sports.mtw

	Favorite Sport			
	Football	Basketball	Baseball	Tennis
Male	33	38	24	5
Female	38	21	15	26

Are the favorite sports distributed the same for men and women?
a. Identify the variable of classification.
b. Identify the populations.
c. Formulate your hypotheses.
d. Calculate the test statistic and its p-value and state your conclusion.
e. Are any assumptions for the chi-square test in question?

10.40   Are foreign cars safer than domestic cars? The Insurance Institute for Highway Safety (IIHS) and the Highway Loss Data Institute (HLDI) collect vehicle loss data from major insurers and the federal government to produce fatality ratings for most makes of automobiles. The fatality ratings are based on actual occupant deaths per registered vehicle. These data are based on foreign and domestic vehicles (pickup trucks included) for the 1988–1992 model years.

Worksheet: Vehicle.mtw

Fatality rating

	Much better than average	Above average	Average	Below average	Much worse than average
Foreign	11	0	10	4	12
Domestic	30	9	38	7	30

SOURCE: Insurance Institute for Highway Safety and the Highway Loss Data Institute, 1995.

Based on these data, are the fatality ratings similar for foreign and domestic vehicles?
   a. Identify the variable of classification.
   b. Identify the populations.
   c. Formulate your hypotheses.
   d. Calculate the test statistic and its p-value and state your conclusion.
   e. Are any assumptions for the chi-square test in question?

10.41   Determine whether the insurance injury ratings for Chevrolet vehicles improved from 1990 to 1993 by analyzing these data:

Worksheet: Chevy.mtw

Chevrolet Injury Frequency

	Much better than average	Above average	Average	Below average	Much worse than average
1988–1990	16	5	5	3	4
1991–1993	12	2	12	2	6

SOURCE: Insurance Institute for Highway Safety and the Highway Loss Data Institute, 1995.

   a. Identify the variable of classification.
   b. Identify the populations. Are the row marginal totals fixed? Explain.
   c. Are any assumptions for the chi-square test in question?
   d. Adjust the data so that the assumptions from part **c** are met.
   e. Formulate your hypotheses.
   f. Calculate the test statistic and its p-value and state your conclusion.

10.4   SUMMARY AND REVIEW

KEY CONCEPTS

✓   When data consist of frequency counts and satisfy the properties of a multinomial experiment, the statistic

$$\chi^2 = \sum \frac{(o_j - e_j)^2}{e_j}$$

is used to test hypotheses about the data. We discussed three applications of this test: the χ^2 goodness-of-fit test, the χ^2 test of independence, and the χ^2 test of homogeneity. To use the χ^2 statistic in any of the three cases, the expected cell counts must be found. For the goodness-of-fit test, the expected counts are

$$e_j = n\pi_j$$

where n is the number of subjects and π_j is the hypothesized probability of an observation falling in category j.

✓ For the test of independence and the test of homogeneity, the expected cell counts are found by

$$e_j = \frac{RC}{n}$$

where R = row total, C = column total, and n = total number of subjects.

✓ The test statistic is approximately distributed as χ^2 with the degrees of freedom being $k - 1$ in the goodness-of-fit test and $(r - 1)(c - 1)$ in the test of independence and the test of homogeneity. In all cases, the approximation is valid if no expected cell count is less than 1 and no more than 20% of the cell counts are less than 5.

LEARNING GOALS   Having completed this chapter, you should be able to:

1. Organize data corresponding to a single categorical variable into a frequency table. *Section 10.1*
2. Conduct a χ^2 goodness-of-fit test. *Section 10.1*
3. Recognize the characteristics of a multinomial experiment. *Sections 10.1 and 10.2*
4. Organize data corresponding to two categorical variables into a contingency table. *Section 10.2*
5. Conduct a χ^2 test of independence. *Section 10.2*
6. Calculate the p-value associated with the χ^2 distribution. *Sections 10.1–10.3*
7. Conduct a χ^2 test of homogeneity. *Section 10.3*

QUESTIONS
FOR REVIEW   Use the following problems to test your skills:

10.42   A multinomial experiment with three categories resulted in this table:

Category	1	2	3
Count	48	102	150

Test the hypothesis that $\pi_1 = \frac{1}{6}$, $\pi_2 = \frac{2}{6}$, $\pi_3 = \frac{3}{6}$.

10.43   From the computer printout, determine whether or not the row and column classifications in the contingency table are independent. Is this a chi-square test of independence or a test of homogeneity?

		Column			
		1	2	3	4
Row	1	15	26	18	31
	2	19	21	25	32

Chi-Square Test

Expected counts are printed below observed counts

	C1	C2	C3	C4	Total
1	15	26	18	31	90
	16.36	22.62	20.70	30.32	

(continued)

```
  2        19        21        25        32        97
          17.64     24.38     22.30     32.68

Total      34        47        43        63        187

ChiSq =  0.114 +  0.505 +  0.351 +  0.015 +
         0.105 +  0.469 +  0.326 +  0.014 = 1.899
df = 3, p = 0.594
```

10.44   From these data, test the hypothesis that the proportions that fall into the three categories are the same for the two populations. (Expected cell counts are in parentheses.) Is this a test of independence or a test of homogeneity?

	Category			
	1	2	3	Total
Population 1	22 (25.5)	33 (30.5)	45 (44)	100
Population 2	29 (25.5)	28 (30.5)	43 (44)	100

10.45   Teenage suicide is a serious problem in the United States. The family economic status of a sample of 38 teenagers who committed suicide was distributed as follows:

Class	Upper	Upper middle	Lower middle	Lower
Number of suicides	14	10	8	6

Is there statistical evidence that the proportions in the four classes differ?

10.46   People who are said to have a Type A personality are outgoing and always on the go. A Type B personality is the opposite. Suppose 50 Type A and 50 Type B persons were classified according to their risk of a heart attack with these results:

	Risk of a heart attack			
	Low	Average	High	Total
Type A	9	27	14	50
Type B	12	31	7	50

Determine whether the risk of a heart attack is distributed the same for Type A and Type B personalities. Is this a test of independence or a test of homogeneity?

10.47   In a local referendum to legalize the sale of beer, voters were surveyed about their opinions on the referendum and their views on the moral issue of selling alcoholic beverages. From the contingency table, determine whether their moral values are related to their opinion on the referendum:

Worksheet: Drink.mtw

		Opinion on referendum		
		For	Against	Undecided
	OK	95	83	21
Drinking is:	Tolerated	73	71	18
	Immoral	12	46	8

10.48   Do the demographic characteristics of developmental students differ for 2-year colleges and 4-year colleges? The results given here are from a study commissioned by the Exxon Education Foundation and conducted by the National Center for Developmental Education in 1992 [*Research in Developmental Education*, Vol. 11, No. 2, 1994]. Based on the computer printout, is there statistical evidence that the distribution of race is the same for the two types of colleges? Be sure to state the hypothesis, give the *p*-value, and write a conclusion.

Worksheet: Develop.mtw

		College	
		Two-year	Four-year
	African American	545	986
	American Indian	24	66
Race	Asian	71	66
	Latino	142	230
	White	1,587	1,939

```
Chi-Square Test

Expected counts are printed below observed counts

         two-year   four-yr    Total
    1        545        986      1531
          641.26     889.74

    2         24         66        90
           37.70      52.30

    3         71         66       137
           57.38      79.62

    4        142        230       372
          155.81     216.19

    5       1587       1939      3526
         1476.86    2049.14

Total       2369       3287      5656

ChiSq = 14.448 + 10.413 +
         4.976 +  3.586 +
         3.232 +  2.329 +
         1.224 +  0.882 +
         8.215 +  5.920 = 55.227
df = 4, p = 0.000
```

10.49   As of December 31, 1993, there were 2,710 prisoners in state and federal prisons under the sentence of death. Here is the distribution of the prisoners according to regions of the country:

Region	Number under sentence of death	Percent of U.S. resident population aged 18 and under
Northeast	181	20%
Midwest	421	24
South	1,500	35
West	608	21

Are the prisoners on death row distributed similarly to the general population as far as region of the country is concerned? State your null and alternative hypotheses to examine the conjecture. Give the value of the test statistic and its p-value and state your conclusion.

10.50   Is the abortion rate higher in the Northeast and West than it is in the Midwest and South? Worksheet: **Abortion.mtw** contains the 1992 abortion rates for each state and the District of Columbia (*USA Today*, June 16, 1994). In the table, the rate is coded as low if it is less than or equal to 20 abortions per 1,000 women and high if it is greater than 20 per 1,000 women. From the data and the chi-square test, is there evidence that the abortion rate is dependent on the region of the country?

```
               Rate

Region   Low      High      ALL

  NE       3         8        11
         5.82      5.18     11.00

  MW       9         3        12
         6.35      5.65     12.00

   S       9         6        15
         7.94      7.06     15.00

   W       6         7        13
         6.88      6.12     13.00

 ALL      27        24        51
         27.00     24.00     51.00

ChiSq =  1.369 +   1.540 +
         1.103 +   1.241 +
         0.141 +   0.159 +
         0.113 +   0.127 = 5.793
df = 3, p = 0.123
```

10.51   In a study related to the one in Exercise 10.50, the number of facilities that performed abortions in 1988 was compared to the number in 1992. From the data and the accompanying computer printout on page 680, is there evidence that the facilities performing abortions have changed from 1988 to 1992? State the null and alternative hypotheses, give the p-value and write your conclusion.

	1988	1992
Public hospital	230	180
Private hospital	810	675
Clinic	885	889
Physician's office	657	636

SOURCE: *USA Today*, June 16, 1994.

```
Chi-Square Test

Expected counts are printed below observed counts


            C1        C2      Total
    1       230       180       410
          213.35    196.65


    2       810       675      1485
          772.73    712.27


    3       885       889      1774
          923.11    850.89


    4       657       636      1293
          672.82    620.18


Total      2582      2380      4962


ChiSq =   1.300 +   1.410 +
          1.798 +   1.951 +
          1.573 +   1.707 +
          0.372 +   0.403 = 10.514
df = 3, p = 0.015
```

10.52   A multinomial experiment with four categories resulted in this table:

Category	1	2	3	4	Total
Frequency	29	74	32	65	200

Test the hypothesis that $\pi_1 = \pi_3 = .20$ and $\pi_2 = \pi_4 = .30$.

10.53   Mutual funds continually revise their holdings, hoping to improve on their portfolio. At one point, a certain mutual fund classified its holdings like this:

Stock type	Income	Growth	High Risk
Number	48	39	43

Is there statistical evidence that its holdings are equally divided among the three types of stocks?

10.54   A sample of 120 diamonds from a diamond mine were classified as low grade, medium grade, and high grade:

Grade	Low	Medium	High	Total
Number of diamonds	52	46	22	120

It is suspected that the percents of low- and medium-grade diamonds are the same and each is twice that of high grade. Is there evidence to support this conjecture?

10.55 A company sales manager wishes to know whether all of his salespeople are contributing the same effort to the company. He records the number of sales made by each of his six salespeople over a given time. Are the sales equally distributed among the salespeople?

Salesperson	A	B	C	D	E	F	Total
Number of sales	34	21	44	52	37	42	230

10.56 The contingency table shows the results of an experiment designed to study the effects of vaccinating laboratory animals against a particular disease:

		Got the disease?	
		Yes	No
Vaccinated?	Yes	12	38
	No	21	29

Test the independence of disease and vaccination.

10.57 Executives of small, medium, and large corporations were polled on the issue of the economy. From the data, determine whether the size of the corporation is related to the outlook on the economy.

		Economic Outlook		
		Optimistic	Cautious	Pessimistic
Size of Corporation	Small	38	22	14
	Medium	25	20	11
	Large	20	8	12

10.58 In a 14-year study, 200 female college graduates were divided into two groups according to their career. Those in male-dominated fields such as law, government, business, science, and medicine were classified in the professional group. Those who never worked or who had careers in teaching, nursing, secretarial, and social work were classified in the traditional group. From the data, determine whether the marital status is independent of career.

	Married?	
	Yes	No
Professional	65	35
Traditional	85	15

10.59 It is believed that in many species of mammals and birds, a greater proportion of young males than females die when resources are scarce. For example, due to its greater size, the domestic ram requires 15% more food than the ewe. To study this further, a group of scientists watched three populations of red deer in three areas where the food supply was considered low, medium, and high, to record the number of deer that die before their first birthday. From the table at the top of page 682, which gives the numbers of red deer that died before their first birthday, determine whether there is any association between the gender of the dead deer and the food supply.

		Males	Females
	Low	23	17
Food supply	Medium	12	10
	High	7	8

COMPUTER EXERCISES

10.60   The average cost of a funeral ranges from around $2,000 in the West to as much as $3,000 to $4,000 in the central states. One hundred people in each region who had buried a loved one were surveyed and asked to complete a questionnaire regarding the cost of the funeral. From the contingency table, determine whether the perceived cost of a funeral is the same across the regions of the country:

Worksheet: Funeral.mtw

		Cost		
		Less than expected	About as expected	More than expected
	West	15	60	25
Region	Central	20	38	42
	South	34	44	22
	East	12	40	48

10.61   An article in the *Wall Street Journal* (March 8, 1994) reported that black women are outdoing black men in corporate America. For example, in 1982, there were 1.2 black female professionals for every black male professional, compared with 1.8 in 1994. Is the difference because black women are better educated than black men? Examine these data to compare the educational levels of black women to those of black men.

Worksheet: Blackedu.mtw
Counts based on Bureau of Census data.

	High school dropout	High school graduate	Some college	Bachelor's degree	Graduate degree
Female	486	659	691	208	96
Male	496	530	435	134	65

a. State the null and alternative hypotheses to determine whether there is a difference in the education levels of black women and men.
b. Is this a test of independence or a test of homogeneity?
c. What is the test statistic? What is its observed value?
d. What is the *p*-value? Should the null hypothesis be rejected?
e. Write your conclusion.

10.62   A summary of the birthday selections in the 1970 draft lottery is given here. The data are the numbers of selections in each month that were for the first half of potential draftees—that is, those with draft numbers less than or equal to 183. If the draft were truly random, then each month should have been selected with a probability proportional to the number of days in the month. Test the hypothesis that the months were selected with probabilities proportional to the number of days in the months.

Worksheet: Draft.mtw

Month	Number of selections in first half	Month	Number of selections in first half
January	13	July	14
February	12	August	18
March	9	September	19
April	11	October	13
May	14	November	21
June	14	December	25

10.63   Compared with boys, girls begin to decline in academic achievement at about age 13 and are below boys in math ability by age 17 (based on interviews of 285 educators and students in ten cities by the National Coalition of Advocates for Students). Many believe that the decline in math ability of girls is due to their own lack of confidence. Samples of 200 13-year-old girls and 200 13-year-old boys were asked how they perceived their math ability, with these results:

Worksheet: Ability.mtw

	Hopeless	Below average	Average	Above average	Superior
Girls	56	61	54	21	8
Boys	35	43	61	42	19

Is there evidence that 13-year-old girls have less self-confidence than boys as far as math ability is concerned?

10.64   The number of hazardous waste sites and the number located in communities with above-average minority populations are given in worksheet: **Toxic.mtw**. From the following summarized data, determine whether the number of hazardous waste sites located in communities with a large number of minorities is dependent on the region of the country.

		Minority population	
		Above average	Below average
	Northeast	29	50
Region	Midwest	133	43
	South	63	99
	West	85	28

10.65   In 1984, many states began to implement mandatory seat belt laws in an effort to cut down on traffic fatalities. Here are the results of a 1984 Gallup poll that asked the question: "Thinking about the last time you got into a car, did you use a seat belt, or not?"

	May 18–21, 1984			
	Yes, did	No, did not	No opinion	Number of interviews
National	25%	74%	1%	(1,516)
Sex				
Male	25	74	1	755
Female	25	74	1	761
Age				
Total under 30	28	71	1	329
18–24 years	25	75	*	173
25–29 years	33	66	1	156
30–49 years	26	74	*	569
Total 50 and older	22	77	1	609
50–64 years	23	76	1	331
65 and older	20	78	2	278
Region				
East	24	75	1	382
Midwest	26	74	*	395
South	18	81	1	416
West	35	64	1	323

(continued)

	Yes, did	No, did not	No opinion	Number of interviews
Race				
Whites	26	73	1	1,320
Non-whites	19	80	1	196
Blacks	18	82	*	165
Hispanics	31	69	*	65
Education				
College graduates	39	60	1	310
College incomplete	31	69	*	359
High school graduates	21	79	*	529
Less than H.S. graduates	16	83	1	316
Politics				
Republicans	32	67	1	410
Democrats	23	77	*	654
Independents	16	83	1	430
Occupation				
Prof. and business	32	67	1	433
Clerical and sales	23	76	1	115
Manual workers	21	78	1	545
Non-labor force	20	79	1	289
Income				
$40,000 and over	35	64	1	226
30,000–39,999	30	70	*	182
20,000–29,999	24	76	*	299
10,000–19,999	23	76	1	427
Under $10,000	17	83	*	306
Religion				
Protestants	21	78	1	876
Catholics	28	72	*	420
Labor Union				
Union family	26	73	1	333
Non-union family	25	74	1	1,183
Urbanization				
Center cities	28	72	*	440
Suburbs	29	70	1	514
Rural areas	17	82	1	562

* Less than 1%
SOURCE: Gallup Report No. 226, July 1984.

a. From the data, is there statistical evidence that the frequency of wearing seat belts is the same in all regions of the country?

b. Is one age group more likely to wear seat belts than others?

c. Is the response to the question independent of the income of the respondents?

10.66   When you take your car for its annual inspection, what type of inspection station should you choose? In Mecklenburg County (Charlotte), North Carolina, 183 inspection stations inspected at least 500 vehicles in 1992. The table indicates the type of inspection station and the percent of vehicles that passed inspection with perfect scores. Is there statistical evidence that the percent of vehicles that pass with perfect scores depends on the type of inspection station?

Worksheet: Inspect.mtw

		Percent that passed		
		< 70%	70%–84%	≥ 85%
	Auto inspection station	1	8	4
	Auto repair dealer	16	13	7
Type of	Tire dealer	16	11	6
station	Gas station	19	21	6
	Car care center	4	4	4
	New car dealer	1	5	28

SOURCE: *The Charlotte Observer*, December 13, 1992.

a. State your hypotheses.
b. Give the value of the test statistic and its *p*-value and state your conclusion.
c. Are any of the assumptions for the chi-square test violated?
d. Why might you remove the new car dealers from the analysis?
e. Remove the new car dealers and recalculate the chi-square statistic. What happened?

10.67 STATISTICAL INSIGHT REVISITED The Statistical Insight in the beginning of the chapter was concerned with whether the type of drug offense for federal prisoners is related to their race. The data, obtained from a 1994 report by the Department of Justice, are in worksheet: **Inmate.mtw**.

a. State your hypotheses to evaluate the relationship between race and type of drug offense.
b. What is the value of the test statistic? Is it unusually large?
c. Can we reject the null hypothesis?
d. What cells have the greatest difference between the actual and expected frequency counts?
e. Can we say, for example, that blacks are more likely to be convicted for a crack drug offense than either whites or Hispanics? What other similar observations can you make from the data?

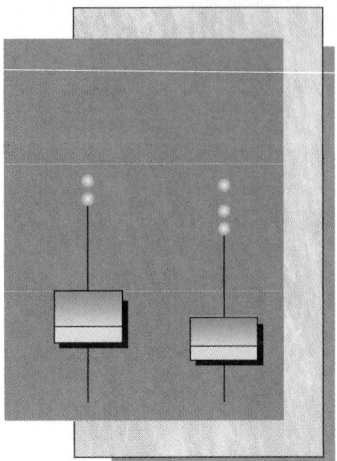

11

REGRESSION ANALYSIS

In Chapter 3, we introduced bivariate data and the study of association. We now revisit the association problem with the idea of using a statistical model to relate two or more variables in a single population. For example, the grade point averages of students in general are most likely related to their intelligence as measured by their IQ scores, study habits, and motivation. The statistical model is the formula that shows how the variables are related. A wide variety of statistical models relate a dependent variable to different independent variables; we restrict our attention, however, to models that are linear in nature.

In this chapter, we view the model as the mechanism that produces data that relate the variables. Using the inference tools developed previously, we statistically evaluate the data with the goal of finding the linear model that best fits the data. With the exception of Section 11.4, this chapter is devoted to the simple linear model that relates the dependent variable to a single independent variable.

CONTENTS

Statistical Insight

Scots heart-attack rate near top of global league table

By Bryan Christie
HEALTH CORRESPONDENT

WOMEN in Glasgow have the world's highest rate of heart attacks and are almost nine times more likely to suffer cardiac failure than their counterparts in Spain.

A report published today also shows that Glaswegian women have higher rates than some southern European men, although heart disease has traditionally been seen as a male problem.

A comparison of heart attack rates in 21 countries has confirmed huge international differences, with Scotland at, or close to, the top of the global league table.

Rates among men find those in Glasgow occupying third place behind two provinces in Finland and just in front of Belfast.

The study, reported today in *Circulation*, the journal of the American Heart Association, is based on the largest heart disease research project ever undertaken. The World Health Organisation's Monica project is monitoring patterns of heart disease over a ten-year period in more than 20 million people living across four continents.

Huge international differences have already been found in death rates from heart disease, but the reliability of such statistics has been challenged. Instead, the current study examined the number of people who survived a heart attack between 1985–87 and found that the same differences existed.

Women in Glasgow were well out in front with 256 heart attacks for every 100,000 women, ahead of Belfast (197), Australia (188), Finland (165), France and China (37), and Spain (30). Men in Spain and parts of southern France have lower rates than Scottish women.

High smoking rates among women in Glasgow are thought to lie behind this problem. Other familiar heart disease related problems of high blood pressure and high cholesterol are also common in the city.

Prof. Hugh Tunstall-Pedoe, of Dundee University, who led the group which prepared the report, said the lower rates in Mediterranean countries could be due to a diet which relied heavily on fruit and vegetables.

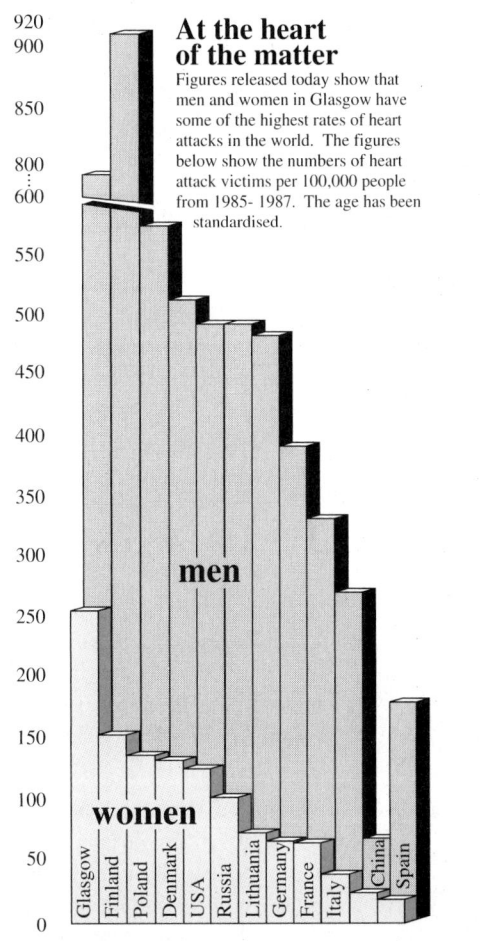

At the heart of the matter

Figures released today show that men and women in Glasgow have some of the highest rates of heart attacks in the world. The figures below show the numbers of heart attack victims per 100,000 people from 1985- 1987. The age has been standardised.

However, he added that the whole question of the precise causes of heart disease remains controversial and the aim of the Monica study is to follow changes in behaviour over time to see if there is a resultant decrease in the amount of heart disease. "We want to find out what the engine is that is driving these changes," he said.

Reductions have already been recorded in rates among men in Finland, but no comparable improvement has been detected in Scotland. Measuring such changes will provide an important indicator to help scientists understand more about heart disease—one of the major killers in the western world.

Prof. Tunstall-Pedoe, of the university's cardiovascular epidemiology unit, said the current findings arose out of a remarkable international collaboration which began in the 1980s and united East and West, in spite of difficulties posed by the Cold War.

"The objective of the project is to look next at the reasons why these rates are changing dramati-

cally in different populations. We expect to finalise these results in approximately four years from now but this will depend on the continuation of national and international support during difficult financial times."

The Scottish part of the project has been funded until now by the Scottish Office but indications have been given that future funding may be difficult to obtain. Professor Tunstall-Pedoe said negotiations on this matter are continuing.

SOURCE: *The Scotsman*, July 12, 1994.

What factors contribute to the elevated rate of heart attacks among the women of Glasgow, Scotland? The article points to the high rate of smoking by Glaswegian women and possibly elevated blood pressure and cholesterol levels.

In this chapter, you will learn how to investigate relationships between variables such as

heart attack rate, smoking rate, blood pressure, and cholesterol level. You will learn how to formulate models relating the variables and how to statistically evaluate them. And you will compute measures of the strength of the relationship and use the model to make predictions.

11.1   THE LINEAR REGRESSION MODEL

In Chapter 3, the scatterplot helped us visually examine the extent to which two numerical variables are related. You learned that although the points of the scatterplot may not lie in a perfectly straight line, often a straight line adequately describes the relationship between the variables. The method of least squares was used to find the equation of the *regression* line that best fits the data.

The Population Regression Line

We now extend the study of regression analysis by examining *statistical models*. In our study of univariate statistics, we viewed the parent population as the population that produces the univariate data in the sample. We now ask the question: Is there a *population regression line* that produces the sample data in a scatterplot? If so, then we refer to the model that relates the response variable y to the predictor variable x as the *linear regression model*.

Linear Regression Model

Given the response variable y and the predictor variable x, the linear regression model that relates the two is

$$y = \beta_0 + \beta_1 x + \varepsilon$$

where β_0 and β_1 are the regression coefficients and ε is the error term.

The Error Term   *Error* is used here in the sense that ε encompasses all of the variation in the response variable y that is *not* modeled by the linear expression $\beta_0 + \beta_1 x$. Without ε, the model would be

$$y = \beta_0 + \beta_1 x$$

and the data, (x_i, y_i), exhibited in a scatterplot would fall in a perfect straight line. The error is then the quantity that causes the points to fall away from the line. The smaller the error, the closer the data are modeled with the straight line. To make inferences about the regression model, we must make certain assumptions about the random error term.

Assumptions About the Error Term in the Linear Regression Model $y = \beta_0 + \beta_1 x + \varepsilon$

The error term, ε, is a random variable that satisfies these conditions:

1. It has mean value 0 ($\mu_\varepsilon = 0$).
2. It has standard deviation σ, which does not depend on x.
3. It has a normal distribution.
4. Any two observed values of ε are independent of each other.

Implications for y   Because y is a linear function of ε, it is also a random variable. The assumptions about ε have implications for the random variable y that are illustrated in Figure 11.1.

FIGURE 11.1

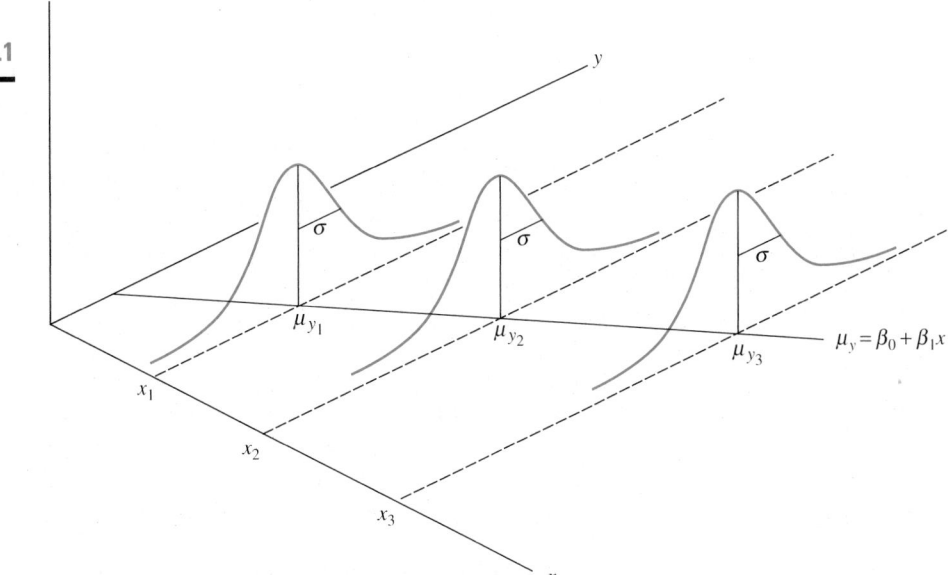

The first assumption implies that the mean value of y is

$$\mu_y = \beta_0 + \beta_1 x \qquad \text{(the population regression line)}$$

This says that the means of the distribution curves drawn in Figure 11.1 are connected with this population regression line.

The second assumption is that y has standard deviation σ and is the same for all values of x. This is illustrated in Figure 11.1, where all of the distribution curves have the same standard deviation, σ, and hence the same spread.

The third assumption states that y is normally distributed, so the distribution curves in Figure 11.1 are drawn as normal curves.

The fourth assumption implies that the observed values, $e_1, e_2, \ldots, e_n$, found when we collect the data should distribute themselves randomly about 0. The value e_i is the difference between the actual observed y_i and the predicted y_i obtained from the estimated regression line. Recall from Chapter 3 that these values are called the *residuals*. By examining the residuals, we can check the above assumptions about the error term. For example, we could construct a normal probability plot of the residuals to check assumption 3. More is said about this in Section 11.5.

EXAMPLE 11.1

In 1991, the states were given the chance to have their eighth-grade math proficiency scores compared on a state-by-state basis. Here are the scores for a group of states that chose to participate:

Worksheet: Mathpro.mtw

State	Conn.	Del.	D.C.	Ga.	Hawaii	Ind.	Md.	NH
Score	23.1	16.1	2.8	14.7	12.5	17.5	17.1	22.5

State	NJ	NY	NC	Ore.	Pa.	RI	Va.
Score	22.8	16.2	9.2	22.6	19.1	15.3	18.9

SOURCE: National Assessment of Educational Progress.

(Thirty-five states actually participated in the study. These are the ones that participated and also where a majority of the students took the SAT exam.) Four years later, when the eighth-graders were seniors in high school, these are the SAT math scores for the same states:

Worksheet: Mathpro.mtw

State	Conn.	Del.	D.C.	Ga.	Hawaii	Ind.	Md.	NH
Score	472	464	443	446	480	466	479	486

State	NJ	NY	NC	Ore.	Pa.	RI	Va.
Score	475	472	455	491	462	462	469

SOURCE: The College Board.

Suppose that the linear regression model

$$y = \beta_0 + \beta_1 x + \varepsilon$$

with $\beta_0 = 440$, $\beta_1 = 1.7$, and $\sigma = 10$ describes the relationship between the proficiency scores and the SAT math scores. Explain how the eighth-grade proficiency scores relate to the SAT math scores 4 years later by giving the equation for the population regression line and graphing it. For a given value of x, describe the possibilities for y.

SOLUTION

The population regression line is

$$\mu_y = 440 + 1.7x$$

where x represents the proficiency score and μ_y is the mean SAT math score. Figure 11.2 is a graph of the equation. For any x, y is a normally distributed random variable with mean

$$\mu_y = \beta_0 + \beta_1 x = 440 + 1.7x$$

and standard deviation $\sigma_y = 10$. We see that the mean depends on the value of x; in fact, the means are connected by the population regression line.

FIGURE 11.2

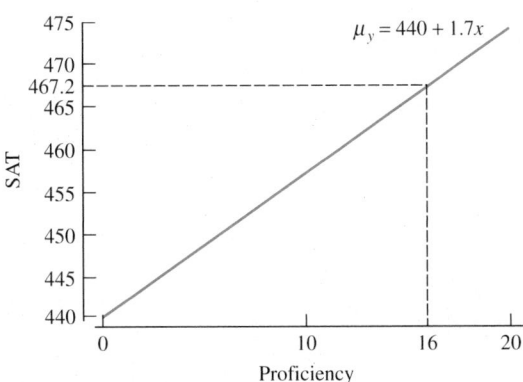

When a particular state has a proficiency score of, say 16, the mean SAT math score is

$$\mu_y = 440 + 1.7(16) = 467.2$$

See Figure 11.2. Furthermore, using the empirical rule when $x = 16$, we are reasonably sure (95% confident) that y will be between 447.2 and 487.2, which is found from $467.2 \pm 2(10)$. Notice that Delaware and New York had proficiency scores close to 16

(16.1 and 16.2, respectively). Their 1995 SAT math scores were 464 and 472; both are inside the given interval.

If the linear model is the correct model in the example, and if indeed $\beta_0 = 440$, $\beta_1 = 1.7$, and $\sigma = 10$, then we know what values to expect for y for a given value of x. In practice, however, we do not know for sure that the linear model is the correct model relating x and y, and certainly, we do not know the numerical values of β_0, β_1, and σ. We must use sample data to verify that the linear model is correct and to estimate β_0, β_1, and σ.

Estimating the Values in the Population Regression Line

Having collected sample data, $(x_1, y_1), (x_2, y_2), \ldots, (x_n, y_n)$, we graph them in a scatterplot to see whether the linear regression model seems plausible. If a straight line appears to serve our purposes, we use the sample data to estimate the regression parameters.

Estimating β_0 and β_1

The least squares method, presented in Chapter 3, gives estimates of β_0 and β_1.

The population regression line is estimated by the least squares regression line given by

$$\hat{y} = b_0 + b_1 x$$

where

$$b_1 = \frac{s_{xy}}{s_x^2} = \frac{\sum(x_i - \bar{x})(y_i - \bar{y})}{\sum(x_i - \bar{x})^2}$$

and

$$b_0 = \bar{y} - b_1 \bar{x}$$

EXAMPLE **11.2**

A study at the Stanford University School of Medicine suggests that coffee drinking may be linked to heart disease. In particular, the report says that coffee drinking is strongly related to elevated levels of apolipoprotein B, a cholesterol-associated protein linked to heart disease. From the following data on 15 adult men over age 35 who drink from one to five cups of coffee per day, create a scatterplot and calculate the least squares estimates of β_0 and β_1 for the model

$$y = \beta_0 + \beta_1 x + \varepsilon$$

Worksheet: Apolipop.mtw

Subject	Cups of coffee per day (x)	Level of apolipoprotein B (y)
1	1	23
2	1	19
3	1	13
4	2	21
5	2	18
6	2	25
7	3	26
8	3	32
9	3	28
10	4	35
11	4	27
12	4	33
13	5	33
14	5	37
15	5	38

SOLUTION

The scatterplot in Figure 11.3 indicates that the linear model is a reasonable model to relate the level of apolipoprotein B to coffee drinking. From the data, we find the summary statistics:

$$\bar{x} = 3, \quad \bar{y} = 27.2, \quad \sum(x_i - \bar{x})^2 = 30, \quad \sum(y_i - \bar{y})^2 = 780.4,$$

$$\sum(x_i - \bar{x})(y_i - \bar{y}) = 137$$

FIGURE 11.3

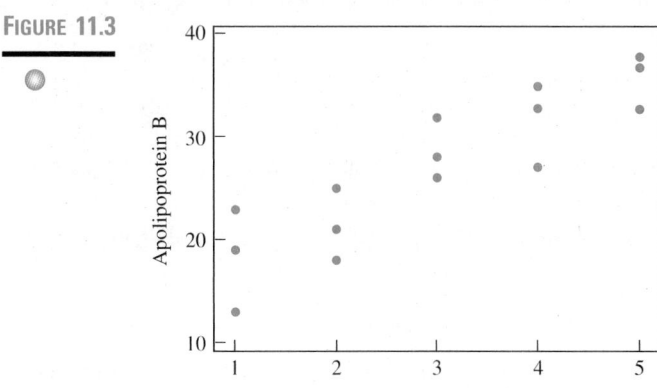

The value of $\Sigma(y_i - \bar{y})^2$ is not needed to calculate the least squares estimates. It is included so that the correlation can be calculated if needed.

Therefore, the least squares estimates are

$$b_1 = \frac{137}{30} = 4.567 \qquad b_0 = 27.2 - (4.567)(3) = 13.5$$

The least squares regression line is

$$\hat{y} = 13.5 + 4.567x$$

The slope of the regression line, 4.567, gives the average increase in the level of apolipoprotein B for each additional cup of coffee per day. This does not mean that if you normally drink two cups of coffee per day and then one day you drink an additional cup, your apolipoprotein B level will increase 4.567 units. For those who normally drink two cups per day, the average apolipoprotein B level is estimated to be

$$\hat{y} = 13.5 + 4.567(2) = 22.634$$

and for those who normally drink three cups per day, the average is estimated to be

$$\hat{y} = 13.5 + 4.567(3) = 27.201$$

The difference is the slope, 4.567. If the linear trend continues, then for each additional cup of coffee, we expect that the average apolipoprotein B level will increase by 4.567 units.

Estimating σ^2

As previously stated, a residual is the difference between the actual y_i and the predicted y_i found by the equation

$$\hat{y}_i = b_0 + b_1 x_i$$

That is, the residual associated with the data point (x_i, y_i) is

$$e_i = y_i - \hat{y}_i$$

The magnitude of a residual is directly related to the magnitude of σ^2. In fact, the residuals are realizations of the random variable ε, which has a normal distribution with mean 0 and variance σ^2. The larger σ^2 is, the larger the residuals will be.

The sum of squares due to error (also called the residual sum of squares) is given by

$$\text{SSE} = \sum e_i^2 = \sum (y_i - \hat{y}_i)^2$$

When divided by its degrees of freedom, $n - 2$, it provides us with an unbiased estimator of σ^2.

Mean Square Error

The mean square error given by

$$MSE = \frac{SSE}{n - 2}$$

is an unbiased estimator of σ^2. The square root of MSE is the estimated standard deviation:

$$s = \sqrt{MSE}$$

Why $n - 2$ Degrees of Freedom?   Sums of squares such as SSE have degrees of freedom associated with them. Recall that for a random sample $x_1, x_2, \ldots, x_n$, the sum of squares, $\sum(x_i - \bar{x})^2$, was divided by $n - 1$ to obtain an unbiased estimator of the population variance. We used $n - 1$ instead of n as the divisor because one degree of freedom is lost when the sample mean, $\bar{x}$, is used to estimate the population mean μ. Substituting for $\hat{y}_i$, we have

$$SSE = \sum(y_i - b_0 - b_1 x_i)^2$$

and we see that the two quantities, β_0 and β_1, have been estimated by b_0 and b_1, resulting in a loss of two degrees of freedom. Thus, the degrees of freedom for SSE are $n - 2$. Dividing SSE by $n - 2$ then gives an unbiased estimator of σ^2.

This formula for SSE involves calculating all of the predicted y's and then subtracting them from the actual y's to obtain the residuals, which are in turn squared and summed. If you are using a calculator, this is very tedious and time-consuming. An equivalent equation, which is easier to use, is

$$SSE = \sum(y_i - \bar{y})^2 - b_1 \sum(x_i - \bar{x})(y_i - \bar{y})$$

Note that the same quantities are used in calculating the regression equation and the correlation coefficient.

EXAMPLE 11.3

Calculate SSE and the estimated standard deviation from the summary information in Example 11.2.

SOLUTION

Using the previously calculated summary information, we have

$$SSE = 780.4 - (4.567)(137) = 154.721$$

Since there were 15 subjects, we have $n - 2 = 13$. The estimate of σ is then

$$s = \sqrt{\frac{\text{SSE}}{n-2}} = \sqrt{\frac{154.721}{13}} = 3.45$$

Finding the regression coefficients b_0 and b_1 and the estimated standard deviation s in Examples 11.2 and 11.3 involved computing these summary statistics:

$$\bar{x}, \quad \bar{y}, \quad \text{SS}_x = \sum(x_i - \bar{x})^2, \quad \text{SS}_y = \sum(y_i - \bar{y})^2, \quad \text{SP}_{xy} = \sum(x_i - \bar{x})(y_i - \bar{y})$$

from the sample data. It is instructive to go through the calculations at least one time as in those examples. In most cases, however, we depend on the computer to execute the calculations and thus leave more time for us to interpret the results.

Figure 11.4 is a partial regression printout associated with the data in worksheet: **Apolipop.mtw**. The regression coefficients appear in two places. They are given in the regression equation, and then they are summarized in the second column of the table. The first column of the table gives the predictors, which are the constant associated with b_0 and the dependent variable coffee associated with b_1. In the column Coef are the numerical values of b_0 and b_1. Below the table, the computed value of s is printed out. Notice that these values agree with the quantities found in Examples 11.2 and 11.3. The remaining values in the printout are discussed in the next section.

FIGURE 11.4   Regression Analysis

```
The regression equation is
apolipB = 13.5 + 4.57 coffee

Predictor        Coef        Stdev      t-ratio          p
Constant       13.500       2.089         6.46      0.000
coffee         4.5667       0.6300        7.25      0.000

s = 3.450        R-sq = 80.2%       R-sq(adj) = 78.6%
```

From our discussion on the linear regression model, we extract six main tasks associated with regression analysis.

Main Tasks in Regression Analysis

1. Construct a scatterplot of the bivariate data.
2. Propose a statistical model that relates the response and predictor variables.
3. Estimate the parameters in the model.

(continued)

4. Test the parameters in the model.

5. Check the assumptions of the model.

6. Use the model for making predictions and for generally describing the relationship between the two variables.

In the next section, we focus on task 4, testing the parameters in the linear regression model.

EXERCISES 11.1

11.1   Given $y = 17 - 4x + \varepsilon$, find the residual for each data point:

a. $(4, 3)$ b. $(-1, 20)$ c. $(3, 5)$

11.2   Given $y = 92.88 + 12.93x + \varepsilon$, find the residual for each data point:

a. $(2, 100)$ b. $(-1, 72)$ c. $(10, 195)$

11.3   For a given geographic region, a proposed model relating the cost of a new home, y, to the square feet of heated floor space, x, is $y = \beta_0 + \beta_1 x + \varepsilon$, with $\beta_0 = 65,280$, $\beta_1 = 24.3$, and $\sigma = 2,300$.

a. What is the mean cost for 1,500 square feet?

b. Using the empirical rule, when $x = 1,500$, give an interval that contains the cost approximately 95% of the time. Out of 100 randomly selected homes in this region, how many do you think will have prices outside this interval?

c. How much do you expect the cost to increase if the area is increased from 1,500 to 1,600 square feet?

d. A new home with 1,500 square feet was purchased for $97,500. Is this cost consistent with the interval found in part **b**?

11.4   The linear model relating state SAT math scores, y, to eighth-grade proficiency scores, x, proposed in Example 11.1 was for states that use the SAT as their dominant college admissions test. For those states where the SAT is not dominant, consider the model $y = \beta_0 + \beta_1 x + \varepsilon$, with $\beta_0 = 482$, $\beta_1 = 2.48$, and $\sigma = 25$.

a. What is the mean SAT math score when the proficiency score is 15?

b. Using the empirical rule, when $x = 15$, give an interval that contains the SAT math score 95% of the time.

c. How much do you expect the SAT score to increase if the proficiency score increases from 15 to 16 points?

d. Michigan and Ohio had proficiency scores very close to 15. Their SAT math scores were 537 and 510, respectively. Are these scores consistent with the interval found in part **b**?

11.5   As soon as a new personal computer hits the market, its price begins to drop. For example, when computers with Intel's Pentium processor were introduced, the average price exceeded $5,000. Six months later, the average price had dropped to around $3,000. For a new model to be released, a proposed model relating its cost (y) to the time on the market (x, in months) for the first 6 months is $y = \beta_0 + \beta_1 x + \varepsilon$, with $\beta_0 = 4,800$, $\beta_1 = -375$, and $\sigma = 45$.

a. What is the mean cost 3 months after the computer is introduced? What is the initial cost?

b. After a computer is on the market 3 months, use the empirical rule to give an interval that contains the cost 95% of the time. Out of 100 randomly selected computer stores that sell this particular computer, how many do you think have prices outside this interval?

c. How much do you expect the cost to decrease each month?

d. An individual purchased a computer after it had been on the market 3 months for $3,750. Is this cost consistent with the interval found in part **b**?

e. Is it reasonable to use this model to predict the cost after the computer has been on the market for 1 year? Explain your answer.

11.6 To study how bone density changes as children age, researchers at Citrus Hill Plus Calcium compiled data for five prominent research journals. They found these percents of peak bone density for different aged children:

Worksheet: Citrus.mtw

Age (x)	2	4	6	8	10	12	14	16	18
Percent of peak bone density (y)	43	49	51	56	63	71	82	91	95

These are the summary statistics:

$$\bar{x} = 10, \quad \bar{y} = 66.78, \quad \sum(x_i - \bar{x})^2 = 240, \quad \sum(y_i - \bar{y})^2 = 2{,}893.56,$$
$$\sum(x_i - \bar{x})(y_i - \bar{y}) = 822$$

Graph the data in a scatterplot and obtain the least squares estimates of β_0 and β_1 for the model $y = \beta_0 + \beta_1 x + \varepsilon$. Calculate SSE and the estimated standard deviation. Is it reasonable to extrapolate the percent of peak bone density to age 20?

11.7 In the National Football League draft held each spring, professional teams pick new players from the college ranks. A lot of effort goes into rating the players to determine the best prospects at each position. Probably the most widely used rating tool is the speed at which the player can run 40 yards. Here are ratings (y) provided by National Scouting and USA Today research (*USA Today*, April 20, 1994) and the times (x, in seconds) for the 40-yard dash for the top offensive linemen in the 1994 draft:

Worksheet: NFLdraft.mtw

Rating	6.5	6.1	6.0	5.7	5.5	5.2	5.0	7.6	7.2	7.0
Time	4.94	5.27	5.27	5.14	5.09	5.23	5.30	5.15	5.20	5.20

Rating	6.5	6.2	6.1	6.0	5.9	5.5	5.3	7.1	7.0	6.4
Time	5.18	5.23	5.35	5.29	5.20	5.25	5.18	5.06	5.36	5.06

Rating	6.4	6.3	6.2	6.0	5.9	5.7	5.5	5.4	5.2
Time	5.26	5.36	5.29	5.03	5.25	5.29	5.56	5.29	5.26

The summary statistics are:

$$\bar{x} = 5.226, \quad \bar{y} = 6.083, \quad \sum(x_i - \bar{x})^2 = .406, \quad \sum(y_i - \bar{y})^2 = 12.101,$$
$$\sum(x_i - \bar{x})(y_i - \bar{y}) = -.595$$

Do you think that the time for the 40-yard dash is important in determining the rating of a player? To answer this question, graph the data in a scatterplot and calculate the correlation coefficient. Obtain the least squares estimates of β_0 and β_1 for the model $y = \beta_0 + \beta_1 x + \varepsilon$ and calculate SSE and the estimated standard deviation. In the next section, we will statistically evaluate the model.

11.8 The Children's Defense Fund, a private advocacy group, claims that more than 25% of all children who live in cities with populations over 100,000 were impoverished in 1989 (*USA Today*, Aug. 12, 1992). Listed here are the 20 cities with populations over 100,000 that were identified by the Children's Defense Fund as having the highest rates of poverty:

Worksheet: Poverty.mtw

City	Poverty percent	Crime rate
Detroit	46.6	13
Laredo, TX	46.4	4
New Orleans	46.3	9
Flint, MI	44.6	12
Miami	44.1	17
Hartford, CT	43.8	12
Gary, IN	43.0	8
Cleveland	43.0	7
Atlanta	42.9	18
Dayton, OH	40.9	6
St. Louis	39.7	18
Buffalo	38.8	9
Rochester, NY	38.4	5
Milwaukee	37.8	5
Newark, NJ	37.6	19
Cincinnati	37.4	7
Fresno, CA	36.9	8
Shreveport	36.7	6
Waco, TX	36.6	8
Macon, GA	36.1	4

SOURCE: Children's Defense Fund and Bureau of Justice Statistics.

In addition to the percent of children who live in poverty, the crime rate (per 1,000) is given for each city. Does it appear that there is a linear relationship between the poverty rate and the crime rate? What tools do you have to examine the relationship between these two variables? Perform some preliminary analysis of the data to determine whether it is worthwhile to pursue the linear relationship.

11.9 A Swedish research study claims to have shown a relationship between the risk of heart disease and the waist-to-hip ratio in men and women. The study suggests that as the waist-to-hip ratio increases, so does the risk of heart disease. The 13-year study of about 1,500 Swedish men with a high waist-to-hip ratio showed their risk of coronary heart disease, stroke and death was 4 times that for men of more normal proportions. A similar study of 1,500 women showed that those with a high waist-to-hip ratio had a risk of heart attack 8 times greater than slimmer women. From the provided information, do you think that this describes an association problem or a comparison problem? Describe the variables that you think are of interest in the heart disease study and explain how they would be used in an association problem (in a comparison problem). What is the basic difference between an association problem and a comparison problem?

11.10 The ability to predict rainfall in an agricultural state such as Kansas would be invaluable for farmers. In an effort to predict the average monthly (May through September) rainfall (in inches) for an area in west central Kansas, researchers measured the rainfall in four (eight in the original study) surrounding counties. The data are for 35 consecutive years ending in 1970.

Worksheet: Rainks.mtw

rain	x1	x2	x3	x4
2.97	1.85	2.36	2.17	2.77
1.88	1.72	1.93	1.25	1.50
2.19	1.94	3.35	2.30	2.64

(*continued*)

rain	x1	x2	x3	x4
1.68	.78	1.23	.97	1.66
2.40	2.51	2.04	3.19	2.17
3.40	3.55	4.46	3.82	4.06
2.05	2.08	1.67	1.94	2.22
2.11	1.40	1.83	1.94	2.55
2.66	2.36	1.88	2.78	3.18
1.49	2.25	2.34	1.82	2.12
2.45	2.77	3.21	2.57	3.18
2.45	1.70	2.33	2.18	1.95
2.41	2.69	3.22	2.62	3.28
4.10	3.71	3.81	3.30	4.13
3.82	2.98	3.50	3.40	4.28
4.78	4.66	4.45	4.46	4.79
1.25	.95	1.16	.90	1.25
1.88	1.45	1.76	1.33	1.98
1.49	1.36	1.95	2.50	1.50
2.02	2.27	2.28	3.21	2.24
1.36	1.05	1.35	.86	1.39
3.38	2.73	2.92	2.40	4.30
3.16	3.94	2.70	4.30	2.99
2.23	1.94	2.14	2.27	3.52
1.76	1.88	1.71	2.25	2.65
3.33	2.84	2.84	2.62	5.44
2.30	2.16	3.07	2.91	3.46
2.59	3.96	2.46	3.02	4.14
2.28	1.90	2.11	1.84	2.75
3.41	3.94	3.89	3.70	3.86
2.56	1.49	2.71	1.44	2.49
2.60	2.92	2.63	3.32	2.78
2.19	2.21	2.40	2.27	3.76
2.48	3.57	2.62	3.18	2.44
2.23	2.63	2.13	2.12	2.55

SOURCE: R. Picard and K. Berk, "Data Splitting," *The American Statistician* 44, No. 2 (1990): 140–147.

a. Here is a correlation matrix of all pairwise correlations among the five variables. The intersection of a row and a column gives the correlation between the two identified variables. For example, the correlation between rain and x3 is .748, and the correlation between x1 and x3 is .883. Examine the correlations and determine which variable is the best predictor variable for rainfall.

```
Correlations (Pearson)

              rain        x1        x2        x3
    x1       0.793
    x2       0.818     0.769
    x3       0.748     0.883     0.746
    x4       0.818     0.720     0.734     0.638
```

b. Examine the regression computer printout. What is the response variable? What is the predictor variable? Is this the predictor variable that you chose in part **a**? What is the regression equation? What is the estimated standard deviation?

```
Regression Analysis

The regression equation is
rain = 0.909 + 0.660 x1

Predictor          Coef        Stdev      t-ratio         p
Constant         0.9091       0.2275         4.00     0.000
x1               0.65988      0.08816        7.48     0.000

s = 0.4884        R-sq = 62.9%      R-sq(adj) = 61.8%
```

c. This is a scatterplot that accompanies the regression output in part **b**. Graph the regression line on the scatterplot. Does it appear that the line fits the data?

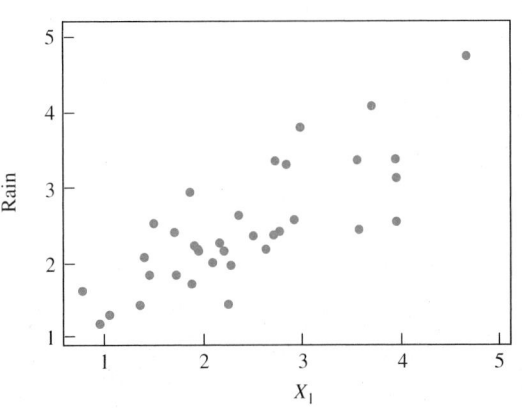

11.11 The data given here are from analyses of the magnesium concentrations in stream samples that were collected along a river. Sampling locations were identified on an aerial photograph, and later the distances between samples were measured.

Worksheet: Magnesiu.mtw

distance	0	1,820	2,542	2,889	3,460	4,586	6,020
magnesium	6.44	8.61	5.24	5.73	3.81	4.05	2.95

distance	6,841	7,232	10,903	11,098	11,922	12,530	14,065
magnesium	2.57	3.37	3.84	2.86	1.22	1.09	2.36

distance	14,937	16,244	17,632	19,002	20,860	22,471
magnesium	2.24	2.05	2.23	.42	.87	1.26

SOURCE: J. Davis, *Statistics and Data Analysis in Geology*, 2nd. ed. (New York: Wiley, 1986), p. 146.

a. From the scatterplot does it appear that a straight line would fit the data? Are there any unusual observations in the scatterplot?

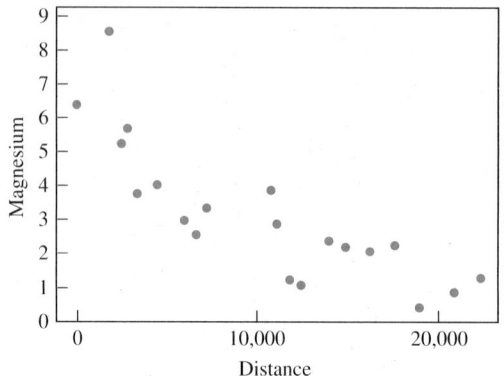

b. A regression computer printout is presented here. What is the regression equation? Graph it on the scatterplot. Does it fit the data reasonably well? What is the estimated standard deviation?

```
Regression Analysis

The regression equation is
magnesiu = 5.77 -0.000252 distance

Predictor        Coef        Stdev      t-ratio          p
Constant       5.7678       0.4952        11.65      0.000
distance -0.00025184  0.00004027        -6.25      0.000

s = 1.195       R-sq = 68.5%      R-sq(adj) = 66.7%
```

11.2 INFERENCE ABOUT THE REGRESSION COEFFICIENTS

The parameter β_0 in the linear regression model

$$y = \beta_0 + \beta_1 x + \varepsilon$$

is the *intercept* or *constant term* that gives the expected value of y when x is zero. In many applications, however, β_0 is not of interest because zero is out of the range of reasonable values for x. In those cases, we simply view β_0, along with its estimate b_0, as an initializing quantity that gives the mean value of y when x is not present.

On the other hand, β_1 is extremely important. It is the slope of the regression line (often referred to as the *regression coefficient*) and gives the increase in y when x is increased by one unit. This was revealed in Example 11.2, where it was pointed out that men who drink three cups of coffee a day have an estimated apolipoprotein B level that is 4.567 units higher than those who normally drink two cups a day. Remember, however, that 4.567 is the value of b_1, an estimate of β_1. We have yet to show that it is a good estimator; in fact, we have not shown, in this example, that the linear model is our final choice to relate y to x.

Hypothesis Test for β_1

To statistically evaluate β_1, and the linear model, we begin by testing the null hypothesis

$$H_0: \beta_1 = 0$$

If, indeed, $\beta_1 = 0$, then we can say that y is *not* linearly related to x because the model would be

$$y = \beta_0 + \varepsilon$$

which says simply that the values of y are randomly distributed about β_0. The value of x would be irrelevant as a predictor of y.

To proceed with the inference about β_1, we use b_1 as the test statistic. As in all previous tests of hypotheses, we need to know the sampling distribution of the test statistic. To derive the sampling distribution of b_1, we must invoke the assumptions given in Section 11.1 about the error term in the linear regression model.

Assumptions Needed to Make Inferences About the Regression Model

1. y is related to x by the linear regression model

$$y = \beta_0 + \beta_1 x + \varepsilon$$

2. As a random variable, y is normally distributed with a mean of $\mu_y = \beta_0 + \beta_1 x$ and a standard deviation of σ.
3. Realizations of the random variable y are independent.

Recall from Section 11.1 that b_0, b_1, and s^2 ($=$ MSE) are unbiased estimators of β_0, β_1, and σ^2, respectively. Using the preceding assumptions (in particular, the normality assumption), we find that the standardized test statistic

$$t = \frac{b_1 - \beta_1}{\text{SE}(b_1)}$$

has a t distribution with $n - 2$ degrees of freedom.

The test statistic, as presented, allows us to test any reasonable value for β_1. Because we generally are testing $\beta_1 = 0$, however, the test statistic simplifies to

$$t = \frac{b_1}{\text{SE}(b_1)}$$

For computational purposes, it can be shown that the standard deviation of b_1 is

$$\frac{\sigma}{\sqrt{\Sigma(x_i - \overline{x})^2}}$$

Using s to estimate σ, we have

$$\text{SE}(b_1) = \frac{s}{\sqrt{\Sigma(x_i - \overline{x})^2}}$$

Because s has $n - 2$ degrees of freedom, the t distribution has $n - 2$ degrees of freedom. The test of hypothesis about β_1 is summarized here:

Test of Hypothesis About β_1

Assumptions: The assumptions listed earlier

Left-Tailed Test	Right-Tailed Test	Two-Tailed Test
H_0: $\beta_1 \geq 0$	H_0: $\beta_1 \leq 0$	H_0: $\beta_1 = 0$
H_a: $\beta_1 < 0$	H_a: $\beta_1 > 0$	H_a: $\beta_1 \neq 0$

Standardized test statistic:

$$t = \frac{b_1}{\text{SE}(b_1)}$$

where $\text{SE}(b_1) = s/\sqrt{\Sigma(x_i - \overline{x})^2}$.

From the t-table with $n - 2$ degrees of freedom, find the p-value most closely associated with t_{obs}. If t_{obs} falls between two table values, then give the two associated probabilities as bounds for the p-value.

The p-value for the two-tailed test is calculated as above and then doubled to account for both tails.

If a level of significance α is specified, reject H_0 if p-value $< \alpha$.

We illustrate the test in the next example.

EXAMPLE 11.4

A study to relate the number of years of experience of New York City street-patrolpersons with the average number of tickets they give per week found these data:

Worksheet: Patrol.mtw

Patrolperson	1	2	3	4	5	6	7	8	9	10
years of experience	3	8	2	15	5	20	1	10	7	12
tickets per week	42	30	54	12	32	8	75	28	20	15

Find the least squares estimates of the regression coefficients in the model $y = \beta_0 + \beta_1 x + \varepsilon$, and test the hypothesis that $\beta_1 = 0$.

SOLUTION

Figure 11.5 is a partial computer printout from a regression routine. We see the regression equation followed by a detailed analysis for the two regression coefficients. We find

$$b_0 = 55.938 \qquad b_1 = -2.9322$$

FIGURE 11.5

```
The regression equation is
tickets = 55.9 - 2.93 years

Predictor        Coef        Stdev        t-ratio            p
Constant       55.938        6.215           9.00        0.000
years         -2.9322       0.6150          -4.77        0.000

s = 11.21         R-sq = 74.0%        R-sq(adj) = 70.7%
```

Under the column Stdev are the standard errors of b_0 and b_1; in particular,

$$\text{SE}(b_1) = .6150$$

To test the hypothesis

$$H_0\text{: } \beta_1 = 0 \quad \text{versus} \quad H_a\text{: } \beta_1 \neq 0$$

we need the observed value of the test statistic

$$t = \frac{b_1}{\text{SE}(b_1)}$$

We can calculate it as follows:

$$t_{obs} = \frac{-2.9322}{.6150} = -4.77$$

however, we see it computed in the t-ratio column of the computer printout.

The t-ratio has 8 degrees of freedom, and therefore p-value $< 2(.001) = .002$. In fact, the computer calculates the p-value as .000, which clearly is less than .002. We reject the null hypothesis and conclude that there is a significant linear relationship between the number of years of experience and the number of tickets given per week. The regression equation $\hat{y} = 55.94 - 2.93x$ can now be used to predict the number of tickets based on the number of years of experience of a patrolperson. For example, someone with 5 years of experience is expected to give each week

$$\hat{y} = 55.94 - 2.93(5) = 41.29 \text{ tickets}$$

We could leave this problem at this point and conclude that we have found the model that best describes the relationship between the two variables under consideration. A scatterplot of the data, however, reveals that there is more to the problem. If you investigate a scatterplot of the data, you will notice a curvilinear pattern to the data. In Section 11.5, we reexamine the scatterplot and a *residual plot* in an effort to improve the model.

Pearson's correlation, ρ, measures the linear relationship between two variables. If the null hypothesis $H_0: \beta_1 = 0$ is not rejected, then there is no statistically significant linear relationship between x and y and hence $\rho = 0$. In fact, testing $H_0: \beta_1 = 0$ is equivalent to testing $H_0: \rho = 0$.

Even if $\rho = 0$, it is possible that y could depend on x in a nonlinear way, such as

$$y = \beta_0 + \beta_2 x^2 + \varepsilon$$

Using x^2 and y as the data, we can analyze the significance of β_2 in the same manner as β_1 above.

Confidence Interval for β_1

In addition to the test of hypothesis, a confidence interval for the regression coefficient can be constructed. Following the construction of the t interval for the population mean, μ, we have that the form of the confidence interval for β_1 is

$$b_1 \pm (t \text{ critical value})\text{SE}(b_1)$$

As in the test of hypothesis, the t critical value is found in the t-table with $n - 2$ degrees of freedom.

A $(1 - \alpha)100\%$ **confidence interval for β_1** is given by the limits

$$b_1 \pm t^* \text{SE}(b_1)$$

where t^* is the upper critical value associated with an upper $\alpha/2$ probability found in the t-table with $n - 2$ degrees of freedom.

EXAMPLE 11.5

In Example 11.2, the simple linear model $y = \beta_0 + \beta_1 x + \varepsilon$ was proposed to describe the relationship between a man's level of apolipoprotein B, a cholesterol-associated protein, and the number of cups of coffee that he normally drinks per day. From the computer printout in Figure 11.6, find a 98% confidence interval for β_1.

FIGURE 11.6

```
Regression Analysis

The regression equation is
apolipB = 13.5 + 4.57 coffee

Predictor          Coef         Stdev       t-ratio          p
Constant         13.500         2.089          6.46      0.000
coffee           4.5667        0.6300          7.25      0.000

s = 3.450        R-sq = 80.2%        R-sq(adj) = 78.6%
```

SOLUTION

From the printout we find that $b_1 = 4.5667$ (which agrees with Example 11.2), and the standard error of b_1 is found under the Stdev column to be .6300. For a 98% confidence interval and 13 ($n = 15$) degrees of freedom, we find from the t-table that $t^* = 2.65$. Therefore, the confidence interval for β_1 is

$$4.5667 \pm (2.65)(.63) \quad \text{or} \quad 4.5667 \pm 1.6695$$

The interval is (2.8972, 6.2362), and we are 98% confident that β_1 is somewhere inside this interval. Notice that the interval does not include 0; this indicates that a test of H_0: $\beta_1 = 0$ would be rejected at a level of significance of 2% (the complement of the 98% confidence). In fact, the p-value for the hypothesis is given to be .000 in the printout. Obviously, the p-value is not 0 but is so small that it is less than .0005 and, when rounded to three decimal places, is .000.

Computer Analysis

C O M P U T E R   T I P

To perform a regression analysis using Minitab, first enter the bivariate data, (x_i, y_i), in the Minitab worksheet. For example, store the x values in C1 and the y values in C2. From the **Stat** menu, click on **Regression**. From the Regression window, **Select** the **Response** variable (y) and the **Predictor** variable (x). Also select any of the options, such as Residuals, Fits, Coefficients, and so on, that you need. (Later we will discuss these options; at this point, you need not select any of them.) Finally click on **OK**.

EXAMPLE 11.6

Perform a computer analysis of the data provided in Example 11.1.

SOLUTION

Recall that the problem was to relate state SAT math scores to eighth-grade proficiency scores taken 4 years earlier. The states, SAT math scores, and proficiency scores have been stored in C1, C2, and C3 of Minitab worksheet: **Mathpro.mtw**. Figure 11.7 gives the Minitab output from the Regression command.

FIGURE 11.7

```
Regression Analysis

The regression equation is
Sat-M1 = 439 + 1.73 Profic1

15 cases used 4 cases contain missing values

Predictor        Coef        Stdev      t-ratio         p
Constant      439.270        8.371        52.47     0.000
Profic1        1.7291       0.4774         3.62     0.003

s = 9.919        R-sq = 50.2%        R-sq(adj) = 46.4%

Analysis of Variance

SOURCE          DF            SS           MS           F         p
Regression       1        1290.6       1290.6       13.12     0.003
Error           13        1279.1         98.4
Total           14        2569.7

Unusual Observations
Obs.  Profic1    Sat-M1      Fit   Stdev.Fit    Residual   St.Resid
  3       2.8    443.00   444.11        7.11       -1.11     -0.16 X
  5      12.5    480.00   460.88        3.25       19.12      2.04R

R denotes an obs. with a large st. resid.
X denotes an obs. whose X value gives it large influence.
```

Regression Analysis

From the output in Figure 11.7 we first see the least squares regression line

```
Sat-M1 = 439 + 1.73 Profic1
```

The intercept is $b_0 = 439$, and the slope is $b_1 = 1.73$. In the data set, there were 19 states but only 15 contained usable information. Next a table is printed that gives more accurate values for b_0 and b_1, with their standard deviations (standard errors), t statistics, and the p-value. The t-ratio is obtained by dividing the coefficient by its standard error. As explained previously, this is the t-ratio for testing the significance of the regression coefficient. For example, to test

$$H_0: \beta_1 = 0 \quad \text{versus} \quad H_a: \beta_1 \neq 0$$

the test statistic is $t_{obs} = 3.62$, which is significant with p-value $= .003$.

Although we seldom have need to test

$$H_0: \beta_0 = 0 \quad \text{versus} \quad H_a: \beta_0 \neq 0$$

the computer output has the necessary information. The test statistic is $b_0 = 439.270$ and t-ratio $= 52.47$, which is highly significant with p-value $< .001$. This means that the regression line does not go through the origin.

Next, we see $s = 9.919$. This is an estimate of σ that is used in the preceding t-tests for β_0 and β_1. The coefficient of determination, R-$sq = 50.2\%$ was introduced in Chapter 3. It measures the percent of variability in the dependent variable that is explained by the regression equation. In this case, we can say that 50.2% of the variability in SAT scores is explained by proficiency scores. The *adjusted R-sq* $= 46.4\%$ is an adjustment to the value of R^2 based on the number of variables in the equation. Here there is only one independent variable in the equation; hence, there is very little difference between *R-sq* and *R-sq(adj)*. It is in multiple regression studies that the value of *R-sq(adj)* becomes relevant.

Analysis of Variance

Next, in Figure 11.7 an analysis of variance (ANOVA) table is printed. An ANOVA table is a summary table of the variability in the data. It is very useful in tabulating the results for hypothesis testing in more complicated statistical analysis problems. The main idea of an ANOVA table is to partition the total variability of the dependent variable into different sources that are relevant to the study. In a regression analysis, the variability in the dependent variable y is measured by SSTotal $= \Sigma(y_i - \bar{y})^2$. The ANOVA table partitions this *total variability* into two components: SSReg and SSE. In the first column of the ANOVA table, the different sources of variability are given, the second column gives the associated degrees of freedom, and in the third column are the numerical values of the sum of squares.

We have already seen SSE; it is a measure of the variability in y that is explained by the random error term, ε, in the linear regression model. SSReg $=$ SSTotal $-$ SSE is the amount of variability in y that is explained by the regression component $\beta_1 x$ in the linear regression model.

As explained previously, every sum of squares has associated degrees of freedom. SSTotal $= \Sigma(y_i - \bar{y})^2$ has $n - 1$ degrees of freedom, and we recall that SSE has $n - 2$ degrees of freedom. Because SSReg $=$ SSTotal $-$ SSE, the degrees of freedom for SSReg are $(n - 1) - (n - 2) = 1$, which represents the one independent variable x. Later when we discuss multiple regression with several independent variables, we will see that SSReg has degrees of freedom equal to the number of independent variables in the model.

The ratio of a sum of squares to its degrees of freedom is called a *mean square* (average sum of squares). As before, MSE $=$ SSE$/(n - 2)$. In a similar fashion, MSReg $=$ SSReg$/$df.

F Distribution

It was pointed out previously that MSE is an *unbiased* estimator of σ^2. If the null hypothesis, $H_0: \beta_1 = 0$, is true, then MSReg is also an *unbiased* estimator of σ^2. That is, when H_0 is true, both mean squares estimate the same thing, and consequently their ratio should be close to 1. In the example, however, MSReg = 1,290.6 and MSE = 98.4 [notice that $s^2 = (9.919)^2 = 98.4$], which gives a ratio of 13.12, considerably greater than 1. To test $H_0: \beta_1 = 0$, an alternative to the *t*-ratio is to investigate the magnitude of the ratio

$$F = \frac{\text{MSReg}}{\text{MSE}}$$

As illustrated above, values of F close to 1 support the null hypothesis, and significantly large values of F indicate that $\beta_1 \neq 0$ and hence the null hypothesis should be rejected.

The next question is, How large does F have to be for us to reject H_0? For example, is 13.12 large enough to reject H_0? The answer depends on the sampling distribution of F.

If the null hypothesis is true and if the assumptions associated with the linear model are satisfied, then the sampling distribution is called the *F distribution* after the renowned English statistician, Sir Ronald Fisher (1890–1962). The F distribution is very much like the chi-square distribution. It assumes only nonnegative values and is skewed right. Because the F statistic is composed of the ratio of two mean squares, which have their own associated degrees of freedom, we say that the F distribution has degrees of freedom associated with its numerator and its denominator. Like other sampling distributions, the shape depends on the degrees of freedom. In particular, the amount of skewness depends on the degrees of freedom in the numerator and the denominator. Figure 11.8 shows the shape of a typical F distribution with the critical value associated with the upper-tail probability of α given as F^*. The critical values for the F distribution are found in Table B.5 in Appendix B in very much the same way that critical values for the chi-square distribution are found.

FIGURE 11.8

Density curve for an F distribution with critical value F^*

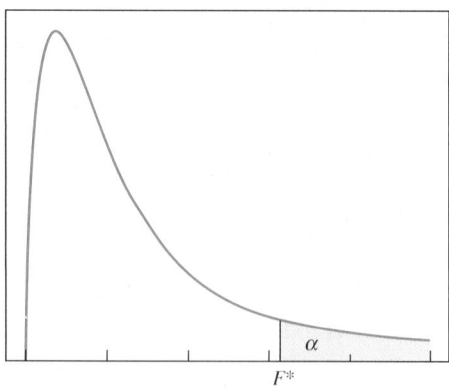

EXAMPLE **11.7**   Suppose the degrees of freedom associated with the numerator are 1 and the degrees of freedom associated with the denominator are 13.

(a) Find the F critical values associated with upper-tail probabilities of .05, .025, and .01.

(b) Find the approximate p-values associated with an F of 2.5, an F of 5.6, and an F of 13.12.

SOLUTION

(a) From Table B.5 with degrees of freedom (1, 13), we find

$$F^*_{.05} = 4.67$$

$$F^*_{.025} = 6.41$$

$$F^*_{.01} = 9.07$$

(b) Also, from Table B.5, we have $F^*_{.10} = 3.14$. Because 2.5 is less than 3.14, we have that its p-value is greater than .10. Because 5.6 is between 4.67 and 6.41, we have that its p-value is between .025 and .05. Because 13.12 is greater than 9.07, we have that its p-value is less than .01. In fact, from the printout in Figure 11.7, we see that its p-value = .003.

Equivalence of F and t^2

Notice in the printout in Figure 11.7 that the p-value for the F statistic is the same (p-value = .003) as the p-value for the t statistic. This is because in regression analysis with only one independent variable, the t-test and the F-test yield the same results. Specifically, if a t statistic has k degrees of freedom, then t^2 is an F statistic with $(1, k)$ degrees of freedom. Consequently, we can test $H_0: \beta_1 = 0$ with a t-test with $n - 2$ degrees of freedom or with an F-test with $(1, n - 2)$ degrees of freedom. Notice in the printout that

$$t^2 = (3.62)^2 = 13.10 \approx 13.12 = F$$

The discrepancy is because 3.62 has been rounded off. A more exact value is $t = 1.7291/.4774 = 3.6219$.

Unusual Observations

To complete the analysis of the computer output in Figure 11.7, we describe the section on Unusual Observations. Two unusual observations are identified in the printout. One is identified as having a large standardized residual, and the other is identified because its x value gives it great influence. An observation with a large standardized residual naturally lies far from the fitted regression line and thus is classified as a regression outlier. As always, you should investigate outliers to determine their cause. If it can be determined that the outlier does not belong to the data set, then it should be removed and the regression equation recomputed. Notice in the regression plot in Figure 11.9

that the identified outlier lies well above the regression line with a standardized residual of 2.04. There appears to be another outlier that lies well below the regression line with a standardized residual of -1.96. It is not identified by the computer as being an unusual observation, however, because its standardized residual is less than 2.0 in absolute value. It probably should also be classified as an unusual observation because it cancels the effect of the other outlier. If either observation is removed from the data set, the regression equation will be altered substantially. If both are removed, then the regression equation will not be affected very much. An important lesson is that you should visually examine the scatterplot for unusual behavior in addition to examining the computer printout.

FIGURE 11.9

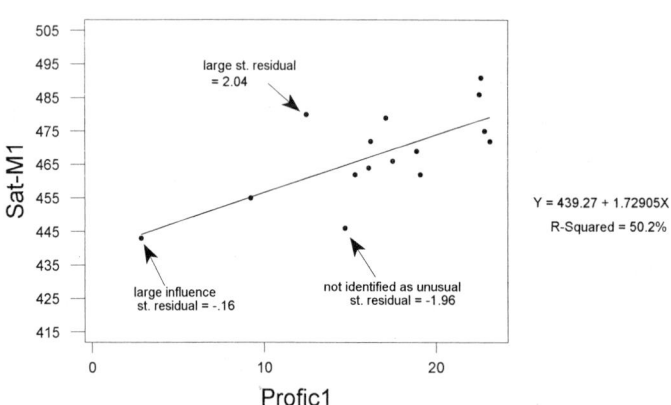

The other unusual observation identified by the computer is categorized as an *influential observation*. In regression analysis, an influential observation is an observation whose removal might have a substantial effect on the regression equation. Notice in this case that the observation is not a regression outlier in the sense that it has a large standardized residual $(= -.16)$ but rather is an outlier because its x value is far removed from the remaining x values. Least squares regression is not resistant to outliers and thus may be influenced by observations with large x values. Because of the influence of a large x value, it will pull the regression line toward it and thus will have a small residual. Having a small residual, it will not be detected with the usual residual analysis. It is for this reason that we use a computer with a regression diagnostic procedure in conjunction with a residual analysis.

You will learn how to conduct a residual analysis in Section 11.5 when we check the model specifications. First, however, we investigate inferences about the regression line in Section 11.3 and study the topic of multiple linear regression in Section 11.4.

EXERCISES 11.2

11.12   Examine the table to determine whether the model $y = \beta_0 + \beta_1 x + \varepsilon$ fits the data reasonably well.

x	6	9	11	15	11
y	5	10	8	14	12

a. Recall that $\sum(x_i - \bar{x})^2 = (n-1)s_x^2$. From the descriptive statistics for x and y, find $\sum(x_i - \bar{x})^2$.

```
Descriptive Statistics

Variable        N      Mean    Median   TrMean   StDev    SEMean
x               5     10.40     11.00    10.40    3.29      1.47
y               5      9.80     10.00     9.80    3.49      1.56

Variable      Min      Max        Q1       Q3
x            6.00    15.00      7.50    13.00
y            5.00    14.00      6.50    13.00
```

b. From the regression printout, find b_1, $SE(b_1)$, and the estimated standard deviation of the regression line.

```
Regression Analysis

The regression equation is
y = 0.07 + 0.935 x

Predictor          Coef        Stdev      t-ratio         p
Constant          0.074        3.151         0.02     0.983
x                0.9352       0.2916         3.21     0.049

s = 1.916        R-sq = 77.4%           R-sq(adj) = 69.9%
```

c. From your answer to part **a** and the estimated standard error found in part **b**, verify that $SE(b_1) = s / \sqrt{\sum(x_i - \bar{x})^2}$.

d. Find a 95% confidence interval for β_1.

e. Examine the computer printout and conduct a t-test to test the hypothesis that $\beta_1 = 0$. What is the p-value?

11.13   Use the following data and the computer printout at the top of page 714 to test the significance of the slope of the regression line: $\mu_y = \beta_0 + \beta_1 x$.

x	11	6	11	15	14	15	8	16
y	8	5	10	12	12	9	5	11

```
Regression Analysis

The regression equation is
y = 0.78 + 0.685 x

Predictor         Coef        Stdev       t-ratio           p
Constant         0.783        1.903         0.41       0.695
x                0.6848       0.1526        4.49       0.004

s = 1.464        R-sq = 77.0%        R-sq(adj) = 73.2%
```

a. What is the estimated standard deviation of the regression line?
b. What is the regression equation?
c. Verify that the t-ratio to test the hypothesis $\beta_1 = 0$ is $b_1/\text{SE}(b_1)$.
d. Complete the t-test to test the hypothesis $\beta_1 = 0$. Be sure to give the p-value.
e. Based on the computer printout, is there evidence to reject the hypothesis $H_0: \beta_0 = 0$? Does this mean that the regression line should go through the origin $(0, 0)$?

11.14   From the computer printout in Exercise 11.13, find a 95% confidence interval for β_1.

11.15   A retailer of satellite dishes would like to know the impact of advertising on her sales. For 6 months she records the number of ads run in the newspaper and the number of sales:

Worksheet: Adsales.mtw

Month	March	April	May	June	July	August
Number of ads	0	3	5	8	10	9
Number of sales	5	7	12	15	17	15

```
Regression Analysis

The regression equation is
sales = 4.67 + 1.23 ads

Predictor         Coef        Stdev       t-ratio           p
Constant         4.6748       0.8018        5.83       0.004
ads              1.2272       0.1176       10.44       0.000

s = 1.017        R-sq = 96.5%        R-sq(adj) = 95.6%

Analysis of Variance

SOURCE        DF           SS           MS          F         p
Regression     1        112.70       112.70     108.93     0.000
Error          4          4.14         1.03
Total          5        116.83
```

a. Examine the accompanying scatterplot. Does there appear to be a linear trend in the data?

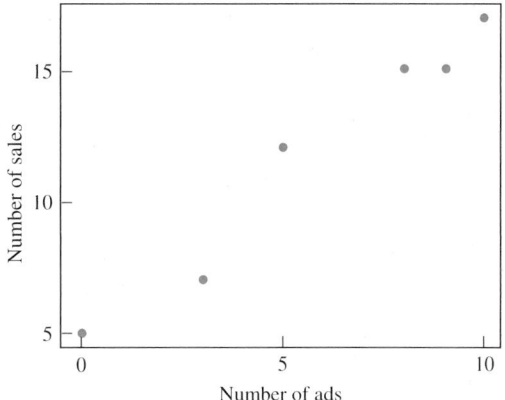

b. Obtain the least squares estimates of β_0 and β_1 for the linear model $y = \beta_0 + \beta_1 x + \varepsilon$.
c. Test the hypothesis $\beta_1 = 0$. Do the results of this test indicate that the linear model is reasonable? What is the value of the coefficient of determination?
d. Predict the number of sales if she runs six ads in June.
e. The collected data are for March through August. What impact does this have on using the model to predict the number of sales in January?

11.16   In an attempt to cut down on traffic through the city, a free shuttle service from the suburban areas to downtown was provided. At the beginning very few people took advantage of the shuttle service, and by the end of 1 year, a decision had to be made whether or not to add more shuttles. Here are the numbers of people who rode on the shuttle and the numbers of automobiles in the downtown area on 15 different occasions throughout the year:

Worksheet: Shuttle.mtw

Number using shuttle	160	180	240	280	440
Number of autos downtown	2,460	2,730	2,560	2,600	2,290

Number using shuttle	370	490	620	850	840
Number of autos downtown	2,370	2,410	2,040	1,820	1,950

Number using shuttle	970	1,230	1,140	1,290	1,350
Number of autos downtown	1,870	1,460	1,330	1,390	1,250

a. Which variable is the dependent variable: the number of people who rode the shuttle or the number of automobiles?
b. Construct a scatterplot of the data. Does there appear to be a linear trend?
c. Obtain the least squares estimates of β_0 and β_1 for the model $y = \beta_0 + \beta_1 x + \varepsilon$.
d. Test the hypothesis $\beta_1 = 0$. Do the results of this test indicate that a linear trend is significant?
e. For each additional person who rides the shuttle, how many automobiles are removed from the downtown area?

      f. Predict the number of automobiles downtown if 1,000 people ride the shuttle.

      g. Would this model be useful in predicting the number of automobiles downtown if 3,000 people ride the shuttle? Explain.

11.17   A cigarette manufacturer recorded the nicotine contents (in milligrams) and sales figures (in $100,000) for eight major brands of cigarettes:

Worksheet: Nicotine.mtw

Nicotine	.86	1.38	1.67	.25	.59	1.17	1.00	.66
Sales	24	65	83	34	59	62	85	38

      a. Which variable is the dependent variable: nicotine content or sales?

      b. Construct a scatterplot of the data. Does there appear to be a linear relationship between the two variables?

      c. Obtain the least squares estimates of β_0 and β_1 for the model $y = \beta_0 + \beta_1 x + \varepsilon$.

      d. Test the hypothesis $\beta_1 = 0$. Do the results of this test indicate that sales increase when the nicotine content is increased?

      e. How much do sales increase for each additional unit of nicotine?

11.18   An owner of a chicken farm would like to test a new feed supplement that is supposed to increase egg production. He randomly selects 12 groups of 100 chickens each and feeds them various levels of the feed supplement. Listed are the amounts of feed supplement and the number of eggs per day.

Worksheet: Eggs.mtw

Group	1	2	3	4	5	6	7	8	9	10	11	12
Feed supplement	10	10	10	15	15	15	20	20	20	25	25	25
Number of eggs	78	84	81	85	79	95	98	96	89	84	93	87

      a. Construct a scatterplot of the data. Does there appear to be an upward trend that would predict the number of eggs?

      b. From the scatterplot in part **a**, does it appear that a linear trend would continue upward if the feed supplement is increased to 30 units?

      c. Obtain the least squares estimates of β_0 and β_1 for the model $y = \beta_0 + \beta_1 x + \varepsilon$.

      d. Test the hypothesis $\beta_1 = 0$. Do the results of this test indicate that a linear trend is significant?

      e. How many additional eggs are expected to be produced for each additional unit of feed supplement?

      f. Predict the number of eggs if the feed supplement is set at 20 units.

      g. Would this model be useful in predicting the number of eggs if the feed supplement is increased to 40 units? Explain.

11.19   In Exercise 11.7 of Section 11.1, a linear model was proposed to relate the ratings of top offensive linemen for the NFL draft to their times in the 40-yard dash. Here is the Minitab output from the Regression routine.

Worksheet: NFLdraft.mtw

```
Regression Analysis

The regression equation is
Rating = 13.7 - 1.47 forty

Predictor        Coef       Stdev     t-ratio         p
Constant        13.744      5.291        2.60     0.015
forty           -1.466      1.012       -1.45     0.159

s = 0.6449      R-sq = 7.2%        R-sq(adj) = 3.8%

Analysis of Variance

SOURCE         DF          SS          MS         F         p
Regression      1       0.8724      0.8724      2.10     0.159
Error          27      11.2290      0.4159
Total          28      12.1014

Unusual Observations
Obs.    forty    Rating     Fit   Stdev.Fit   Residual   St.Resid
  1      4.94     6.500    6.502     0.313      -0.002      -0.00 X
  8      5.15     7.600    6.194     0.142       1.406       2.24R
 27      5.56     5.500    5.593     0.359      -0.093      -0.17 X

R denotes an obs. with a large st. resid.
X denotes an obs. whose X value gives it large influence.
```

a. Find the regression equation. Does it agree with your answer from Exercise 11.7?
b. Locate the t statistic for testing H_0: $\beta_1 = 0$. What is its p-value? Should you reject or accept the null hypothesis? What does this mean about the proposed linear model?
c. If the rating is a valid measure of the potential of a future offensive lineman in the NFL, should scouts be concerned with his time in the 40-yard dash?

11.20 This Minitab printout is for the Regression routine for the data in Exercise 11.8 of Section 11.1. The problem was to determine whether the poverty level of children in large cities can be predicted with a linear model based on the crime rate of the city.

Worksheet: Poverty.mtw

```
Regression Analysis

The regression equation is
Poverty = 39.1 + 0.181 Crime

Predictor        Coef       Stdev     t-ratio         p
Constant        39.116      1.858       21.05     0.000
Crime           0.1810     0.1710        1.06     0.304

s = 3.667       R-sq = 5.9%        R-sq(adj) = 0.6%
```

(continued)

```
Analysis of Variance

SOURCE        DF            SS            MS          F          p
Regression    1          15.06         15.06        1.12      0.304
Error         18        242.02         13.45
Total         19        257.07
```

a. In the case of the simple linear model, the square root of R^2 is the correlation coefficient. What is the correlation between the poverty rate and the crime rate? What percent of the variability in the poverty rate is explained by the linear model? Is it reasonable?

b. Locate the t statistic for testing H_0: $\beta_1 = 0$. What is its p-value? Should you reject or accept the null hypothesis? What does this mean about the proposed linear model?

c. Should we use this regression equation to predict children's poverty levels from the crime rates?

11.21   How significant is the earned run average (ERA) for pitchers in major league baseball in determining the number of wins for the team? To investigate, consider the numbers of wins (out of 162 games) and the earned run averages for the National League teams at the end of the 1990 season:

Worksheet: Wins.mtw

Team	Wins	ERA
Atlanta	65	4.58
Chicago	77	4.34
Cincinnati	91	3.39
Houston	75	3.61
Los Angeles	86	3.72
Montreal	85	3.37
New York	91	3.43
Philadelphia	77	4.07
Pittsburgh	95	3.40
St. Louis	70	3.87
San Diego	75	3.68
San Francisco	85	4.08

a. Construct a scatterplot of the data. Does it appear that a linear model would describe the relationship between the number of wins and the earned run average?

b. From the Minitab printout of the Regression routine, find the regression equation. What is the change in the number of wins for a unit change in the earned run average?

```
Regression Analysis

The regression equation is
wins = 144 - 16.5 era

Predictor        Coef        Stdev       t-ratio         p
Constant        143.50       19.40          7.40       0.000
era             -16.469       5.086        -3.24       0.009

s = 6.752        R-sq = 51.2%        R-sq(adj) = 46.3%
```

(continued)

```
Analysis of Variance

SOURCE          DF              SS              MS              F           p
Regression      1           478.10          478.10          10.49       0.009
Error          10           455.90           45.59
Total          11           934.00
```

c. Locate the t statistic for testing H_0: $\beta_1 = 0$. What is its p-value? Should we reject or accept the null hypothesis? What does this mean about the proposed linear model?

d. Should we use this regression equation to predict the number of wins based on the earned run average of the pitchers?

11.22 The data relating SAT math scores and eighth-grade proficiency scores presented in Example 11.1 and revisited in Example 11.6 are graphed in the scatterplot. The regression line found in Example 11.6 has been superimposed over the scatterplot. How well does the line fit the data? Do there appear to be any violations of the assumptions necessary for making inferences about the regression model?

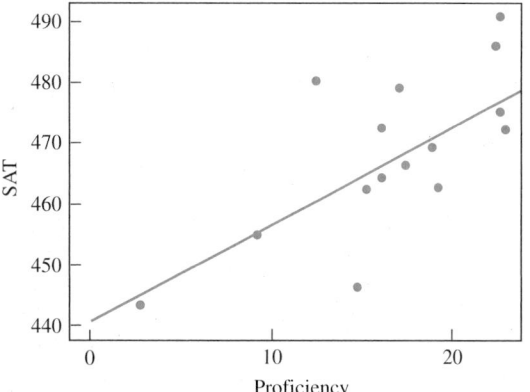

11.23 Example 11.6 gave the computer output relating state SAT math scores and eighth-grade proficiency scores. The analysis was based on data from states where a majority of students take the SAT exam. In Minitab worksheet: **Mathpro.mtw**, the same data are given for states where a majority do not take the SAT exam. The data are listed under the variables Sat-M2 and Profic2. Conduct an analysis of these data similar to the analysis given in Example 11.6.

a. Compare your output with that given in Example 11.6. What is R^2 for your data? Is the F statistic significant?

b. Of the two data sets, which has a linear model that better fits the data?

c. Create side-by-side boxplots of the SAT math scores for the two groups. Is one more variable than the other? Are the medians similar?

11.24 In the early 1960s, tremendous amounts of waste water from the Rocky Mountain Arsenal (a manufacturing plant for producing various military compounds) were forced, under high pressure, into a deep injection well near Denver, Colorado. Unfortunately, the well, drilled into basement rocks, penetrated the shear zone of a major fault along the Rocky Mountains. There is evidence that the high pressure injection of waste fluids lubricated and mobilized the fault and triggered earthquakes. The data are the month-by-month volumes of injected water (in million gallons) and the numbers of earthquakes detected in Denver each month between March 1962 and October 1965.

Worksheet: Wastewat.mtw

Index	1	2	3	4	5	6	7	8	9	10	11	12
Water	4.2	7.2	8.4	8.0	5.2	6.0	5.0	5.6	4.0	3.6	6.0	7.6
Earthquakes	*	2	12	35	23	29	24	8	6	20	25	22

Index	13	14	15	16	17	18	19	20	21	22	23	24
Water	7.8	6.4	3.6	4.0	3.4	2.4	3.9	.0	.0	.0	.0	.0
Earthquakes	21	42	21	8	6	10	11	12	4	2	5	2

Index	25	26	27	28	29	30	31	32	33	34	35	36
Water	.0	.0	.0	.0	.0	.0	.6	1.8	2.4	2.0	2.0	1.7
Earthquakes	9	9	2	4	4	5	2	14	2	7	1	30

Index	37	38	39	40	41	42	43	44
Water	1.6	3.6	4.0	6.4	8.9	5.4	6.4	3.8
Earthquakes	9	19	11	38	62	48	87	5

SOURCE: J. C. Davis, *Statistics and Data Analysis in Geology*, 2nd ed. (New York: Wiley, 1986), p. 228; and G. E. Bardwell, "Some Statistical Features of the Relationship Between Rocky Mountain Arsenal Waste Disposal and Frequency of Earthquakes," *Geological Society of America, Engineering Geology Case Histories*, No. 8 (1970): 33–37.

a. Bardwell presented the two accompanying times series graphs as evidence of a relationship between the volume of injected water and the number of earthquakes. Examine the peaks and valleys of the two graphs. How many prominent peaks do you see? Do they appear in approximately the same place on the two graphs? Near the middle of the data, no waste water was injected into the well. Do you see this in the first graph? What do you see in the second graph during this same period? Do you think that Bardwell demonstrated a valid relationship between the two variables?

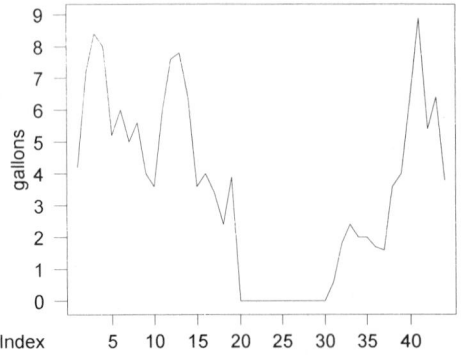 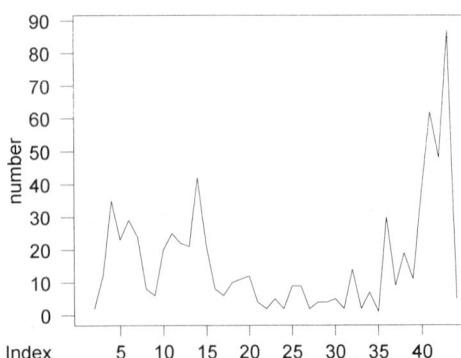

b. Does the scatterplot of the two variables demonstrate a relationship between the two variables?

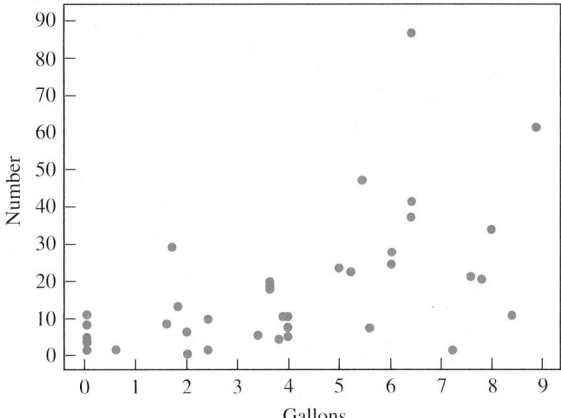

c. Examine the computer printout. What is the correlation between the two variables? What is the regression equation? Is the straight line a good fit for the data? Are there any unusual observations? Where are they on the scatterplot?

```
Correlations (Pearson)

Correlation of gallons and number = 0.602

Regression Analysis

The regression equation is
number = 3.59 + 3.79 gallons

43 cases used 1 cases contain missing values

Predictor        Coef        Stdev     t-ratio          p
Constant        3.590        3.483        1.03      0.309
gallons        3.7905       0.7843        4.83      0.000

s = 14.33      R-sq = 36.3%      R-sq(adj) = 34.7%

Analysis of Variance

SOURCE         DF           SS          MS          F          p
Regression      1       4800.2      4800.2      23.36      0.000
Error          41       8424.9       205.5
Total          42      13225.1
```

(continued)

```
Unusual Observations
Obs.  gallons   number      Fit  Stdev.Fit  Residual  St.Resid
  2      7.20     2.00    30.88       3.66    -28.88     -2.08R
 43      6.40    87.00    27.85       3.18     59.15      4.23R
```

R denotes an obs. with a large st. resid.

11.25   This regression printout is related to the problem, presented in Examples 3.11 and 3.20 of Chapter 3, of associating the value of a company's brand name with its revenue.

```
Worksheet: Name.mtw

Regression Analysis

The regression equation is
value = - 0.889 + 2.02 revenue

Predictor        Coef       Stdev     t-ratio         p
Constant       -0.8889      0.4174      -2.13     0.039
revenue         2.0244      0.1158      17.49     0.000


s = 2.096       R-sq = 88.4%      R-sq(adj) = 88.1%

Analysis of Variance

SOURCE          DF          SS          MS          F         p
Regression       1      1344.1      1344.1     305.82     0.000
Error           40       175.8         4.4
Total           41      1519.9

Unusual Observations
Obs.   revenue    value      Fit  Stdev.Fit  Residual  St.Resid
  1      15.4   31.200   30.287      1.553     0.913      0.65 X
  2       8.4   24.400   16.116      0.779     8.284      4.26R
 11       6.0    3.700   11.257      0.539    -7.557     -3.73R
```

R denotes an obs. with a large st. resid.
X denotes an obs. whose X value gives it large influence.

a. Is there sufficient evidence to reject the hypothesis H_0: $\beta_1 = 0$? What is the p-value?

b. What evidence in the printout suggests that the straight line provides a strong relationship between revenue and value?

c. The computer has identified three unusual observations. Explain why they are called unusual. Identify them on this scatterplot.

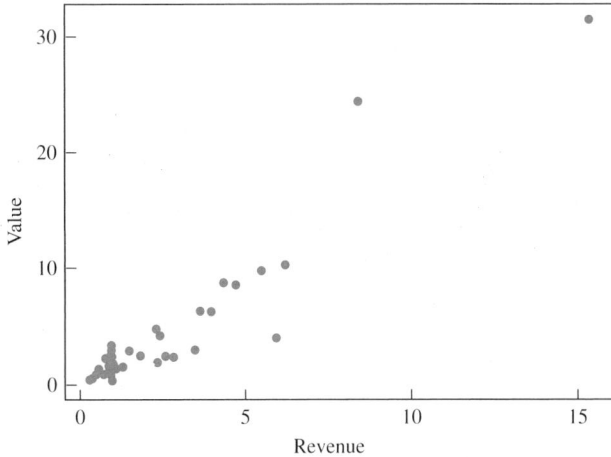

11.3 ESTIMATION AND PREDICTION

If the linear model adequately describes the relationship between x and y, then for a particular value of x, the estimated regression line, $\hat{y} = b_0 + b_1 x$, gives an estimate of the corresponding value of y. We can think of the estimate in two ways. First, it is an estimate of the mean value of y, $\mu_y = \beta_0 + \beta_1 x$, for the given value of x. That is, if we were able to take all possible values of y for that particular value of x and average them, $\hat{y}$ would be an estimate of that average. Second, the value of $\hat{y}$ predicts a future value of y for the given value of x. In each case, we should ask: How precise is the estimate (prediction)? In particular, can we give a *confidence interval estimate* of μ_y, and can we give a *prediction interval* for a future value of y? We first look at the confidence interval.

Confidence Interval Estimate of μ_y

For the linear regression model

$$y = \beta_0 + \beta_1 x + \varepsilon$$

the expected y for a given x—say, x' —is

$$\mu_y = \beta_0 + \beta_1 x'$$

This expected y is further estimated by

$$\hat{y} = b_0 + b_1 x'$$

Using $b_0 + b_1 x'$ as an estimate, we can make inferences about μ_y. To find a confidence interval for μ_y, we need to know the standard error of our estimate $\hat{y}$. It is a rather complicated formula:

$$SE(\hat{y}) = s\sqrt{\frac{1}{n} + \frac{(x' - \bar{x})^2}{SS_x}}$$

With the basic assumptions of independent, normally distributed errors with common variance, we have these formulas:

For a given x', a $(1 - \alpha)100\%$ **confidence interval for the expected response** μ_y is given by the limits

$$b_0 + b_1 x' \pm t^* SE(\hat{y})$$

where

$$SE(\hat{y}) = s\sqrt{\frac{1}{n} + \frac{(x' - \bar{x})^2}{SS_x}}$$

is the standard error of the estimate of μ_y and t^* is the upper $\alpha/2$ critical value found in the t-table with $n - 2$ degrees of freedom.

EXAMPLE 11.8

A manufacturer of thermal pane windows wished to analyze the heat loss (in BTUs) through the windows at different outside temperatures. In a controlled environment, the heat loss through three windows was measured at four different outside temperatures (in degrees centigrade). From these data, obtain the estimated regression line:

Worksheet: Thermal.mtw

Temperature	-10	-10	-10	0	0	0	10	10	10	20	20	20
Heat loss	58	62	54	39	41	36	22	26	20	10	13	15

Graph the data in a scatterplot and find a 95% confidence interval for the expected heat loss when the outside temperature is 10° C.

SOLUTION

A scatterplot of the data suggests that a straight line provides a reasonable approximation of the relationship between the two variables. Substituting in the formulas for b_0 and b_1, we find

$$b_0 = 40.6 \qquad b_1 = -1.52$$

so that

$$\hat{y} = b_0 + b_1 x = 40.6 - 1.52x$$

Figure 11.10 is the scatterplot of the data with the predicting line also graphed.

FIGURE 11.10

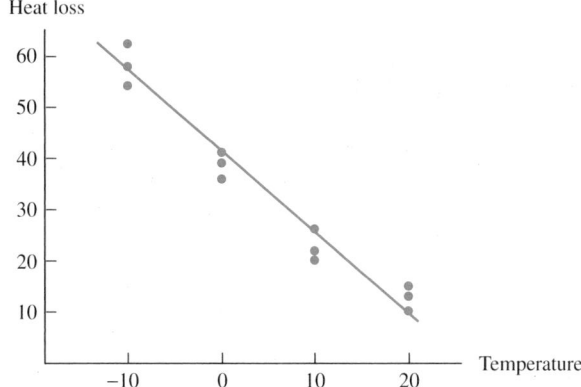

When the outside temperature is 10° C, we would expect the heat loss to be

$$\hat{y} = 40.6 - 1.52(10) = 25.4 \text{ BTUs}$$

For a 95% confidence interval estimate of the heat loss, we add and subtract the margin of error given by

$$t^* \text{SE}(\hat{y}) = t^* s \sqrt{\frac{1}{n} + \frac{(x' - \bar{x})^2}{\text{SS}_x}}$$

From the t-table with $n - 2 = 10$ degrees of freedom, we have

$$t^* = 2.228$$

Also,

$$\bar{x} = 5$$

$$\text{SS}_x = 1{,}500$$

$$s = \sqrt{\frac{\text{SSE}}{n-2}} = \sqrt{\frac{142.4}{10}} = 3.774$$

Substituting in all known values, we have that the 95% confidence interval for the expected heat loss when the outside temperature is 10° C is

$$25.4 \pm 2.228(3.774) \sqrt{\frac{1}{12} + \frac{(10 - 5)^2}{1{,}500}} = 25.4 \pm 2.659$$

which gives the interval (22.74, 28.06). When the outside temperature is 10°C, we are confident that the average heat loss is somewhere between 22.74 and 28.06 BTUs.

EXAMPLE 11.9

Most of the details for this solution are left out. When you are asked to find confidence intervals for several values of x, it is best to use a computer.

Using the data in Example 11.8, construct 95% confidence intervals for the expected heat loss when the outside temperature is -10, 0, 10, 20, 30, and 40 degrees. Graph the results together with the least squares regression line.

SOLUTION

The solution is given in Table 11.1. Figure 11.11 is the graph of the regression line and the confidence intervals for the various values of x'. Notice that the standard error is smaller when the value of x' is closer to $\bar{x}\,(= 5)$. This in turn results in a narrower confidence interval. In general, we can say that predictions are more accurate in the vicinity of $\bar{x}$. The farther x' is from $\bar{x}$, the greater the possibility for error when predicting y. Extreme

TABLE 11.1

x'	$\hat{y}$	Standard Error	95% C.I.
-10	55.8	1.82	(51.74, 59.86)
0	40.6	1.19	(37.94, 43.26)
10	25.4	1.19	(22.74, 28.06)
20	10.2	1.82	(6.14, 14.26)
30	-5.0	2.67	$(-10.95, .95)$
40	-20.2	3.58	$(-28.18, -12.22)$

FIGURE 11.11

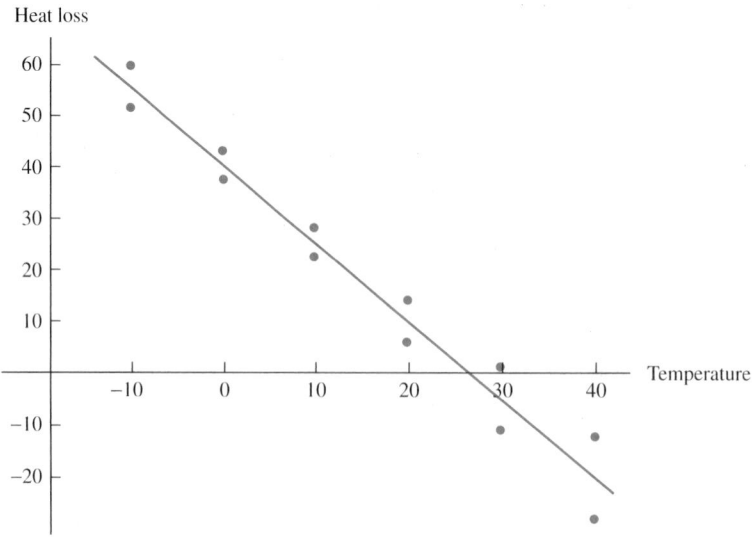

care must be taken when predicting for values of x' beyond the original data (called extrapolation) because the linear relationship may not continue beyond the range of available data.

Prediction Interval for a New y

The difference between the confidence interval for μ_y and the prediction interval for a future value of y is that there is more variability in predicting a single score than in estimating the average of many scores. That is to say, individual scores vary more than averages. The greater variability in a prediction interval is reflected in the standard error of the estimate. The standard error for predicting an individual y for a given x' is

$$s\sqrt{1 + \frac{1}{n} + \frac{(x' - \bar{x})^2}{SS_x}}$$

Notice that the only difference between this and the standard error of the estimate of μ_y is an extra 1 under the square root sign. It makes the resulting prediction interval wider.

For a given x', a $(1 - \alpha)100\%$ **prediction interval for an individual y** is given by the limits

$$b_0 + b_1 x' \pm t^* s \sqrt{1 + \frac{1}{n} + \frac{(x' - \bar{x})^2}{SS_x}}$$

EXAMPLE 11.10

Use the data in Example 11.8 to predict the heat loss when the outside temperature is $10°$ C. Construct a 95% prediction interval for y.

SOLUTION

As before, the estimated heat loss is

$$\hat{y} = 40.6 - 1.52(10) = 25.4 \text{ BTUs}$$

To get a 95% prediction interval of the heat loss, we add and subtract the margin of error to obtain

$$25.4 \pm 2.228(3.774)\sqrt{1 + \frac{1}{12} + \frac{(10 - 5)^2}{1,500}} = 25.4 \pm 8.82$$

which gives the interval (16.58, 34.22). Notice that the margin of error is more than 3 times what it was for the confidence interval estimate (8.82 versus 2.659). The result is a much wider prediction interval.

You should have noticed that the computations for the confidence intervals and predicting intervals are rather tedious. In practice, we leave these calculations up to the computer.

C O M P U T E R   T I P

To construct confidence intervals and prediction intervals with Minitab all that is required is that the values for which predictions are desired be stored in a separate column of the worksheet. In the Regression window, after specifying the response and predictor variables, click on **Options**. In the Regression Options window, insert the column containing the values for predictions in the **Prediction intervals for new observations** block and click on **OK**. Click **OK** again in the Regression window and the resulting output will appear in the Session window.

To graph the confidence and prediction bands, click on **Regression** under the Stat menu and, instead of clicking on Regression again, click on **Fitted Line Plot**. In the Fitted Line Plot window, from the list of variables, **Select** the **Response** and **Predictor** variables. Click on **Display confidence bands** and click on **Display prediction bands** and then click **OK**.

The heat loss data from Example 11.8 have been stored in Minitab worksheet: **Thermal.mtw** under columns named "loss" and "temp". The x' values from Example 11.9 have been stored in C3 of the worksheet. The output generated from the Regression command with the prediction interval option is presented in Figure 11.12.

FIGURE 11.12   Regression Analysis

```
The regression equation is
loss = 40.6 - 1.52 temp

Predictor        Coef        Stdev      t-ratio          p
Constant       40.600        1.193        34.02      0.000
temp          -1.52000      0.09743       -15.60      0.000

s = 3.774       R-sq = 96.1%      R-sq(adj) = 95.7%

Analysis of Variance

SOURCE         DF          SS           MS          F          p
Regression      1        3465.6       3465.6      243.37      0.000
Error          10         142.4         14.2
Total          11        3608.0
```

(*continued*)

FIGURE 11.12
(continued)

```
     Fit    Stdev.Fit       95.0% C.I.             95.0% P.I.
    55.80        1.82  (   51.74,   59.86)  (   46.46,   65.14)
    40.60        1.19  (   37.94,   43.26)  (   31.78,   49.42)
    25.40        1.19  (   22.74,   28.06)  (   16.58,   34.22)
    10.20        1.82  (    6.14,   14.26)  (    0.86,   19.54)
    -5.00        2.67  (  -10.95,    0.95)  (  -15.30,    5.30)
   -20.20        3.58  (  -28.18,  -12.22)  (  -31.79,   -8.61) XX
X  denotes a row with X values away from the center
XX denotes a row with very extreme X values
```

Notice that the complete regression analysis is printed along with 95% confidence intervals and 95% prediction intervals for the data that have been previously stored. The confidence and prediction bands are graphed in Figure 11.13. It is easy to see that the prediction band is much wider than the confidence band.

FIGURE 11.13

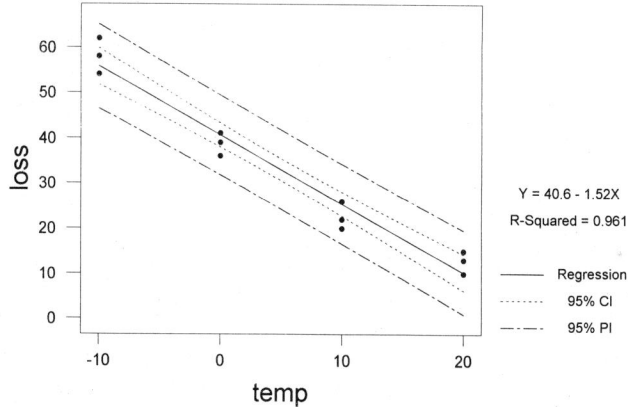

Figure 11.13 — Regression Plot. Y = 40.6 - 1.52X, R-Squared = 0.961.

Application

The incubation period of acquired immune deficiency syndrome (AIDS) is the time from infection of the individual with human immunodeficiency virus (HIV) to the clinical manifestation of full-blown AIDS. The distribution of the incubation period and its dependence on variables such as gender and age are of great interest to epidemiologists worldwide. Worksheet: **Aids.mtw** contains the incubation times (duration in months) and ages at time of infection for 295 patients who were thought to have been infected with HIV by a blood transfusion. These data are made available by the Centers for Disease Control (CDC) in Atlanta, Georgia [Kalbfleisch and Lawless (1989)].

The scatterplot of the durations of the incubation times and the ages of the 295 patients in Figure 11.14 (page 730) reveals no distinct pattern to suggest that duration is dependent on the age of the patient. Examining the scatterplot more closely, however, and this time focusing only on the duration for children under 10 years old, we do see a pattern. It appears that for children under the age of 10, there is a very definite linear

FIGURE 11.14

Pearson correlation of
duration and age = 0.159

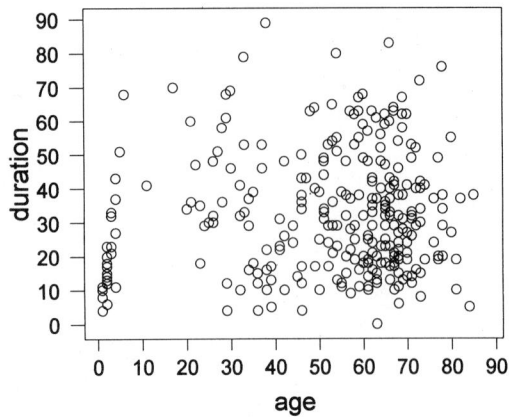

relationship between the duration of the incubation period and the age of the patient. Current theories suggest that the immunocompetence is low for very young patients and for elderly patients; therefore, we divide the data set consisting of two variables, duration and age, into three groups corresponding to these age categories:

Children: ages 1 to 9

Adults: ages 10 to 59

Elderly: ages 60 and above

Examining the three scatterplots in Figure 11.15, we see that a straight line definitely fits the data corresponding to children but not the other two age groups. In Figure 11.16 we see that the linear regression model for the children group is highly significant with $R^2 = 78.8\%$ and an F statistic of 126.74, giving p-value = .000[+].

FIGURE 11.15

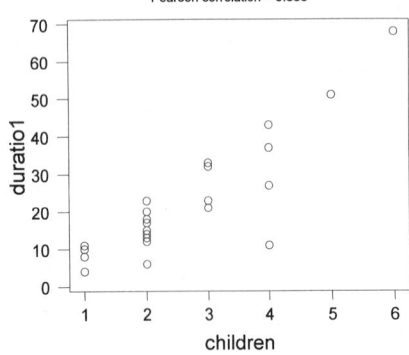

FIGURE **11.15**
(continued)

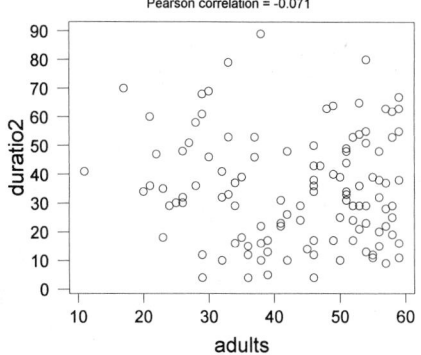

Pearson correlation = -0.071

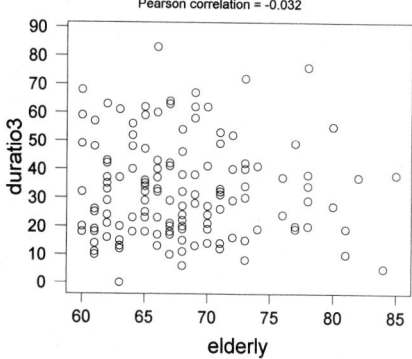

Pearson correlation = -0.032

FIGURE **11.16**   Regression Analysis

The regression equation is
duratio1 = - 2.80 + 9.62 children

Predictor	Coef	Stdev	t-ratio	p
Constant	-2.795	2.301	-1.21	0.233
children	9.6191	0.8544	11.26	0.000

s = 6.635   R-sq = 78.8%   R-sq(adj) = 78.2%

Analysis of Variance

SOURCE	DF	SS	MS	F	p
Regression	1	5579.9	5579.9	126.74	0.000
Error	34	1496.9	44.0		
Total	35	7076.7			

Unusual Observations

Obs.	children	duratio1	Fit	Stdev.Fit	Residual	St.Resid
17	6.00	68.00	54.92	3.30	13.08	2.27RX
28	4.00	11.00	35.68	1.78	-24.68	-3.86R

R denotes an obs. with a large st. resid.
X denotes an obs. whose X value gives it large influence.

FIGURE **11.17**

Regression Plot

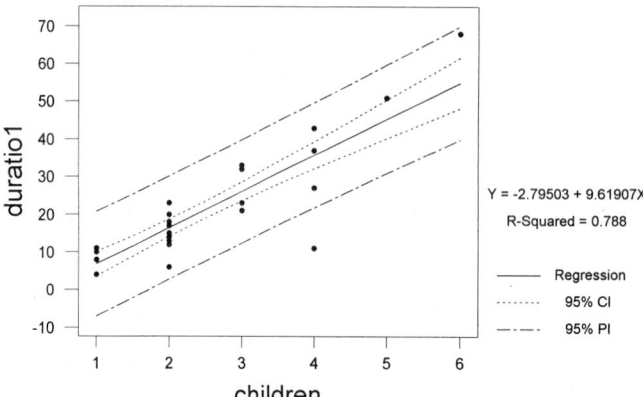

It appears from the prediction bands in Figure 11.17 that it is possible to use the regression equation to predict the duration of the incubation period for children older than 6. Judging from the overall scatterplot in Figure 11.14, do you think this is wise? In fact, judging from the second and third scatterplots in Figure 11.15, do you think that we can predict the duration of the incubation period for adults or elderly based on their age?

EXERCISES 11.3

11.26   A least squares equation relating x and y is $\hat{y} = 29.02 - .54x$. Additionally, we have

$$n = 15, \quad \sum x_i = 195, \quad SS_x = 265, \quad s = 7.58$$

  a. Verify that SSE = 746.9332.
  b. Construct 95% confidence intervals for the expected y for $x' = 10, 12, 14,$ and 15.
  c. Graph the regression line and the confidence limits about the line for the four values of x' similar to Figure 11.10.

11.27   A least squares equation relating x and y is $\hat{y} = 18.06 + .67x$. Additionally, we have

$$n = 24, \quad \sum x_i = 480, \quad SS_x = 1,280, \quad s = 9.25$$

  a. Verify that SSE = 1,882.375.
  b. Construct 95% prediction intervals for a new y for $x' = 10, 15, 20, 25,$ and 30.
  c. Graph the regression line and the prediction limits about the line for the five values of x' similar to Figure 11.10.

11.28   From each set of summary data, construct the least squares equation relating x and y. Also, construct 95% confidence intervals for the expected y or prediction intervals for a new y for the specified values of x'.

a. $n = 8, \sum x_i = 27, s_x = 1.685, b_0 = -6.85, b_1 = 8.92, s = 2.85$, confidence intervals for $x' = 2, 4, 6$

b. $n = 5, \sum x_i = 0, s_x = 2.24, b_0 = 6.00, b_1 = -1.80, s = .6325$, prediction intervals for $x' = -2, 0, +2$

11.29 From the following data we have constructed a scatterplot and found the least squares estimates of β_0 and β_1 for the model $y = \beta_0 + \beta_1 x + \varepsilon$.

Worksheet: Ex11–29.mtw

x	10	10	10	20	20	20	30	30	30	40	40	40
y	16	20	14	22	26	23	28	30	25	31	35	29

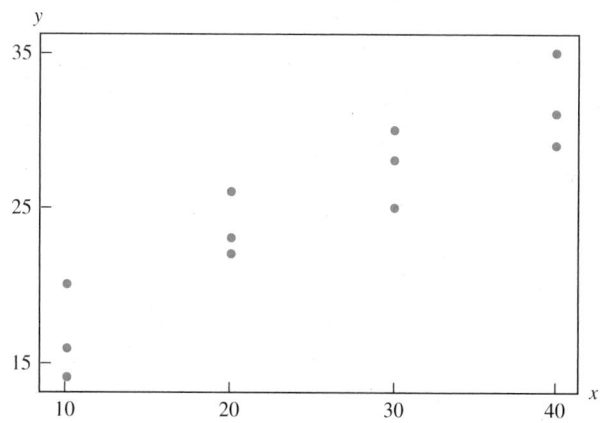

```
The regression equation is
y = 12.7 + 0.490 x

Predictor        Coef        Stdev      t-ratio         p
Constant       12.667       1.827         6.93     0.000
x             0.49000     0.06672         7.34     0.000

s = 2.584      R-sq = 84.4%      R-sq(adj) = 82.8%
```

a. Does the regression equation fit the data in the scatterplot?

b. Find 90% confidence intervals for the expected y for $x' = 10, 20, 25, 30$, and 40.

c. For what value of x' is the confidence interval the narrowest? Why?

11.30 In Exercise 3.45a of Section 3.4, the least squares estimates of β_0 and β_1 in the model $y = \beta_0 + \beta_1 x + \varepsilon$ obtained from the following data were found to be $b_0 = -1.67$ and $b_1 = 3$. Additionally it has been determined that $s = 1.354$.

Worksheet: Ex3–13.mtw

x	2	1	3	2	3	1
y	4	0	6	6	8	2

a. Graph the regression line and 95% confidence intervals for the expected y when $x' = 0, 1,$ 2, 3, and 4.

b. For what value of x' is the confidence interval the narrowest?

11.31 In Exercise 3.45b of Section 3.4, you were asked to fit these data to the model $y = \beta_0 + \beta_1 x + \varepsilon$:

Worksheet: Ex3-13.mtw

x	1	3	4	6	8	9	11	14
y	1	2	4	4	5	7	8	9

The computer printout gives the regression equation.

```
The regression equation is
y = 0.545 + 0.636 x

Predictor        Coef       Stdev      t-ratio          p
Constant       0.5455      0.4588         1.19      0.279
x              0.63636     0.05669       11.22      0.000

s = 0.6513      R-sq = 95.5%      R-sq(adj) = 94.7%
```

a. How well does the regression line fit the data?

b. Find 95% prediction intervals for a new y for $x' = 2, 4, 6, 8,$ and 10.

11.32 These data match a child's age with the number of gymnastic activities he or she was able to complete successfully:

Worksheet: Gym.mtw

Age	2	3	4	4	5	6	7	7
Activities	5	5	6	3	10	9	11	13

a. Graph the data in a scatterplot. Does it appear that a straight line fits the data?

b. Based on the regression printout, does it appear that the straight line fits the data?

```
The regression equation is
number = - 0.03 + 1.64 age

Predictor        Coef       Stdev      t-ratio          p
Constant       -0.032      2.014        -0.02      0.988
age            1.6383      0.3988         4.11      0.006

s = 1.933       R-sq = 73.8%      R-sq(adj) = 69.4%
```

c. Should the regression line go through the origin? Explain. Is there evidence in the computer printout to support this conjecture?

d. To find 95% confidence intervals for the expected y, these values of x' have been stored in column 3 of the worksheet: 2, 4, 6, 8, 10. What general comments can you make about the difference between the confidence intervals and the prediction intervals?

```
      Fit   Stdev.Fit       95.0% C.I.           95.0% P.I.
     3.245     1.292   (  0.082,   6.408)  ( -2.447,   8.936)
     6.521     0.746   (  4.695,   8.347)  (  1.449,  11.593)
     9.798     0.846   (  7.727,  11.869)  (  4.633,  14.963)
    13.074     1.465   (  9.488,  16.661)  (  7.137,  19.012)
    16.351     2.202   ( 10.960,  21.742)  (  9.178,  23.524) XX
 X  denotes a row with X values away from the center
 XX denotes a row with very extreme X values
```

e. Which interval is the narrowest? What value of x' gives the narrowest interval?

11.33 An automobile dealer records the numbers of cars sold and the prime interest rates for six 2-month periods:

Worksheet: Prime.mtw

Prime rate	15	14	13	12	11	10
Sales	116	132	148	136	155	184

a. Graph the data in a scatterplot. Does it appear that a straight line fits the data?
b. Based on the regression printout, does it appear that the straight line fits the data?

```
The regression equation is
sales = 287 - 11.3 rate

Predictor        Coef       Stdev      t-ratio          p
Constant       286.95       32.67         8.78      0.001
rate          -11.343        2.589       -4.38      0.012

s = 10.83       R-sq = 82.8%       R-sq(adj) = 78.4%
```

c. Find 98% confidence intervals for the expected y for $x' = 10, 12, 14$, and 15.
d. At what value of x' is the confidence interval for the expected y narrowest?
e. What would happen to the width of the confidence intervals for the expected y if x' changes from 15 to 20? What does this say about predicting sales when the prime rate is 20%?

11.34 In Exercise 11.15 of Section 11.2, it was determined that the number of sales of satellite dishes by a retailer (worksheet: **Adsales.mtw**) could be accurately predicted with the regression line

$$\text{sales} = 4.67 + 1.23 \text{ ads}$$

Use this additional information:

$$s = 1.017, \quad SS_x = 74.83, \quad \bar{x} = 5.83, \quad n = 6$$

a. Construct a 98% confidence interval for the expected sales when six ads are placed.
b. If the retailer places six ads in July, what is a 98% prediction interval for the number of sales?

11.35 Exercise 11.16 of Section 11.2 demonstrated that the regression equation

$$\text{Autos} = 2,837 - 1.15 \text{ shuttle}$$

accurately fits the data (worksheet: **Shuttle.mtw**). Here is additional information:

$$s = 109.5, \quad s_x = 426, \quad \bar{x} = 697, \quad n = 15$$

Recall also that $SS_x = (n-1)s_x^2$.

    a. Construct a 99% confidence interval for the expected number of autos when 1,000 people ride the shuttle.

    b. If 1,000 people ride the shuttle next Friday, what is a 99% prediction interval for the number of cars?

11.36   In Exercise 11.17 of Section 11.2, there was a moderate linear relationship between sales figures and the nicotine content of a manufacturer's cigarettes (worksheet: **Nicotine.mtw**). You are given the equation

$$\text{Sales} = 24.9 + 33.1\,\text{nicotine}$$

and the additional information

$$SSE = 1,922.6, \quad \bar{x} = .947, \quad s_x = .458, \quad n = 8$$

Recall that $SS_x = (n-1)s_x^2$.

    a. Construct a 99% confidence interval for the expected sales when the nicotine content is 1.0.

    b. If the nicotine content is 1.0, what is a 99% prediction interval for the sales?

11.37   Refer to the computer printout in Exercise 11.21 of Section 11.2 (worksheet: **Wins.mtw**). You may want to use your calculator or the computer to find $\bar{x}$ and SS_x.

    a. What is the regression equation that relates wins to the earned run averages for professional baseball teams? Does it accurately describe the relationship?

    b. Construct a 99% confidence interval for the expected number of wins when the earned run average is 3.8.

    c. If the earned run average is 3.8, what is a 99% prediction interval for the number of wins?

11.38   The following confidence intervals and prediction intervals were obtained with the regression printout in Exercise 11.25 of Section 11.2. The problem was to relate the value of a company's brand name to its revenue (worksheet: **Name.mtw**). The three intervals are for revenues of $5, $10, and $20 billion.

```
    Fit   Stdev.Fit       95.0% C.I.              95.0% P.I.
   9.233      0.452   (   8.320,   10.146)  (   4.898,   13.568)
  19.355      0.951   (  17.433,   21.277)  (  14.702,   24.008) X
  39.599      2.077   (  35.401,   43.797)  (  33.634,   45.564) XX
X   denotes a row with X values away from the center
XX denotes a row with very extreme X values
```

    a. What is the predicted value of a company's brand name when its revenue is $5 billion? What is a 95% interval estimate of the mean brand name value?

    b. A new company has a projected a revenue of $5 billion. What is the predicted value of its brand name? What is a 95% interval estimate of its brand name value?

    c. Why are the limits of the intervals in parts **a** and **b** different?

    d. Shown here is a regression plot with 95% confidence intervals and prediction intervals. Explain what happens to the intervals as the value of revenue increases from 0 to 15.

Regression Plot

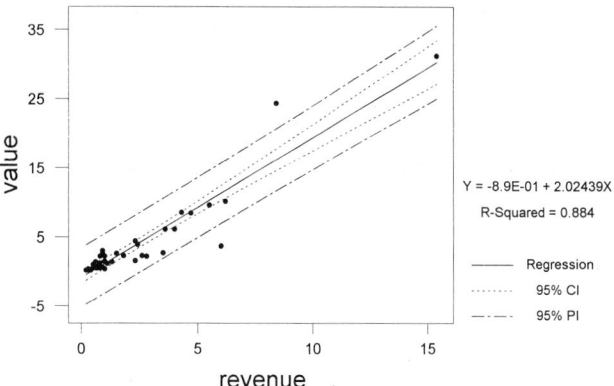

$Y = -8.9E-01 + 2.02439X$

R-Squared = 0.884

——— Regression

·········· 95% CI

—·—· 95% PI

11.4 MULTIPLE LINEAR REGRESSION

In many regression problems, the dependent variable y may be related to more than one independent variable. For example, one's blood pressure is dependent on age, weight, and physical condition. To obtain a meaningful prediction model, all variables that might affect the dependent variable should be measured. Then the independent variables $x_1, x_2, x_3, \ldots, x_k$ that have a significant effect on the dependent variable y can be incorporated into a regression model of the form

$$y = \beta_0 + \beta_1 x_1 + \beta_2 x_2 + \cdots + \beta_k x_k + \varepsilon$$

The equation is called a *multiple linear regression model*. The assumptions made about the error term, ε, in Section 11.1 also apply to the multiple regression model.

Data are collected on variables $y, x_1, x_2, \ldots, x_k$ and used to find the least squares estimates $b_0, b_1, b_2, \ldots, b_k$ of the parameters $\beta_0, \beta_1, \beta_2, \ldots, \beta_k$, respectively. The prediction equation is

$$\hat{y} = b_0 + b_1 x_1 + b_2 x_2 + \cdots + b_k x_k$$

Multiple Regression with a Computer

There are confidence interval estimates and hypothesis-testing procedures for the multiple regression coefficients, but the formulas are more difficult to work with than those in the simple case of a single independent variable. For this reason, multiple regression analysis is almost always performed with the aid of one of the standard statistical packages such as Minitab, SPSS, or SAS. We illustrate multiple regression in the next example.

EXAMPLE **11.11**

It is generally believed that blood pressure is related to one's age and weight. To study the possible effects of age and weight on systolic blood pressure, these data were recorded from a sample of 15 adult men:

Worksheet: Systolic.mtw

Age (x_1)	Weight (x_2)	Blood Pressure (y)
48	175	143
50	159	131
35	191	135
41	174	131
33	165	121
25	157	115
51	182	133
53	164	126
40	157	120
31	168	128
35	138	120
46	191	160
28	155	124
32	171	143
44	164	128

Perform a multiple regression analysis, using age and weight as independent variables and blood pressure as the dependent variable.

SOLUTION

To perform the analysis with Minitab, the data for x_1, x_2, and y must be stored in C1, C2, and C3 of the worksheet. From the Stat menu, select Regression. In the Regression window, select the response variable, C3, and select the two predictor variables, C1 and C2. Also click on Residuals, Standard Resid, and Fits. These values will be stored in C4, C5, and C6, respectively, for further analysis.

The output shown in Figure 11.18 appears in the Session window. We see that the fitted regression equation is

$$\text{Pressure} = 27.5 + .239 \text{ age} + .559 \text{ weight}$$

FIGURE 11.18

```
MTB > Name c4 = 'SRES1' c5 = 'FITS1' c6 = 'RESI1'
MTB > Regress 'Pressure' 2 'Age' 'Weight';
SUBC>    SResiduals 'SRES1';
SUBC>    Fits 'FITS1';
SUBC>    Constant;
SUBC>    Residuals 'RESI1'.


Regression Analysis

The regression equation is
Pressure = 27.5 + 0.239 Age + 0.559 Weight
```

(continued)

FIGURE 11.18
(continued)

```
Predictor        Coef        Stdev      t-ratio        p
Constant        27.45        25.39         1.08     0.301
Age            0.2392       0.2502         0.96     0.358
Weight         0.5594       0.1586         3.53     0.004

s = 7.946         R-sq = 58.7%      R-sq(adj) = 51.8%

Analysis of Variance

SOURCE          DF          SS          MS          F          p
Regression       2      1077.97      538.99       8.54      0.005
Error           12       757.76       63.15
Total           14      1835.73

SOURCE          DF      SEQ SS
Age              1      292.49
Weight           1      785.48

Unusual Observations
Obs. Age    Pressure    Fit    Stdev.Fit   Residual    St.Resid
12   46.0   160.00    145.30    4.14        14.70       2.17R
R denotes an obs. with a large st. resid.
```

The estimated coefficients and their standard errors are

$$b_0 = 27.45 \qquad SE(b_0) = 25.39$$
$$b_1 = .2392 \qquad SE(b_1) = .2502$$
$$b_2 = .5594 \qquad SE(b_2) = .1586$$

The value of $s = 7.946$ is an estimate of σ, the standard deviation associated with ε. The number of degrees of freedom associated with s is given by

$$df = n - (\text{number of independent variables}) - 1$$
$$= n - 3 = 12$$

Assuming that the ε's are independent and normally distributed about 0 with a standard deviation of σ, we can perform interval estimation and hypothesis testing about the regression coefficients β_0, β_1, and β_2. In particular, a 95% confidence interval for β_1 is given by

$$b_1 \pm t^* SE(b_1)$$

where t^* is determined with $n - 3$ degrees of freedom. With 12 degrees of freedom (and 95% confidence), $t^* = 2.179$. From the data in the printout we get

$$.2392 \pm (2.179)(.2502) = .2392 \pm .5452$$

which gives the interval $(-.3060, .7844)$. Note that the interval contains the value of 0. This means that the coefficient is insignificant.

The same results can be found by testing the hypothesis $\beta_1 = 0$. From the printout we see that the t-ratio for age is .96 and the p-value is .358, which indicates that the hypothesis $\beta_1 = 0$ should not be rejected. To illustrate the t-test, we take a closer look at the test for β_2. To test $H_0: \beta_2 = 0$ versus $H_a: \beta_2 \neq 0$, we use the t-ratio

$$t = \frac{b_2}{\text{SE}(b_2)}$$

as the test statistic. Its significance is evaluated by comparing it to the tail probabilities associated with a t distribution with $n - 3$ degrees of freedom. From the printout, we see that the observed value of the test statistic is $t_{\text{obs}} = 3.53$ and p-value $= .004$. Therefore, there is significant evidence to reject H_0 and conclude that the blood pressure of adult men is related to their weight. Because the observed value of t is positive, the blood pressure increases as weight increases.

In multiple regression, R^2 is called the *multiple coefficient of determination*. It is analogous to r^2, the coefficient of determination, in the simple linear regression case. That is, it is interpreted as being the percent of variability in the dependent variable that is explained by the regression model. In this example,

$$R^2 = 58.7\%$$

so that 58.7% of the variability in blood pressure is explained by the age and weight of the adult man. The remaining 41.3% is explained by other factors.

The analysis of variance table in Figure 11.18 gives the total variability, SSTotal, associated with the dependent variable and shows how it is divided between the regression model and the residual (error). That is, we have the total variability of y being

$$\text{SSTotal} = 1,835.73$$

of which

$$\text{SSReg} = 1,077.97$$

is due to the regression model and

$$\text{SSE} = 757.76$$

is due to the residual or error. Thus, the percent explained by the regression model is

$$R^2 = \frac{\text{SSReg}}{\text{SSTotal}}$$
$$= \frac{1,077.97}{1,835.73} = .5872$$

There are $n - 1 = 14$ degrees of freedom, 2 of which belong to regression and the remaining 12 belong to residual. It is interesting to note that $s = 7.946$ can be found by

$$s = \sqrt{\text{MSE}}$$
$$= \sqrt{63.15} = 7.9467$$

(the slight difference is due to round-off error), which is the estimate of the standard deviation σ.

Polynomial Regression

In some regression problems, a straight line does not adequately describe the relationship between y and an independent variable x. For instance, we may need a quadratic expression in the model. This is easily done with the multiple regression model

$$y = \beta_0 + \beta_1 x_1 + \beta_2 x_2 + \varepsilon$$

We simply associate the variable x with x_1 and its square, x^2, with x_2. So the *second-degree polynomial model*

$$y = \beta_0 + \beta_1 x + \beta_2 x^2 + \varepsilon$$

is a special case of the multiple regression model. To analyze the data with Minitab, all that is required is to read the data for y and x into the worksheet, perform the squared transformation on x and store the results in a third column, and then perform a regression analysis on y with the two predictors x and x^2.

EXAMPLE **11.12**

In Example 3.10, a very high linear correlation (.978) was found between the dose of growth stimulant and weight gain in laboratory animals. The scatterplot also showed the linear relationship. To further study the experiment, additional data were collected. The weight gains for doses of 7, 8, 9, and 10 were measured:

Worksheet: Growth.mtw

Dose	0	1	2	3	4	5	6	7	8	9	10
Weight gain	1.0	1.2	2.0	2.4	3.4	4.9	5.1	4.7	3.5	2.5	1.8

With the additional data, the correlation between dose and weight gain is only .410. Construct a scatterplot of the data and suggest an alternative method of analyzing the data.

SOLUTION

The scatterplot in Figure 11.19 (page 742) reveals the reason for the drastic drop in the correlation with the additional data. Clearly, the data do not exhibit a linear relationship

FIGURE 11.19

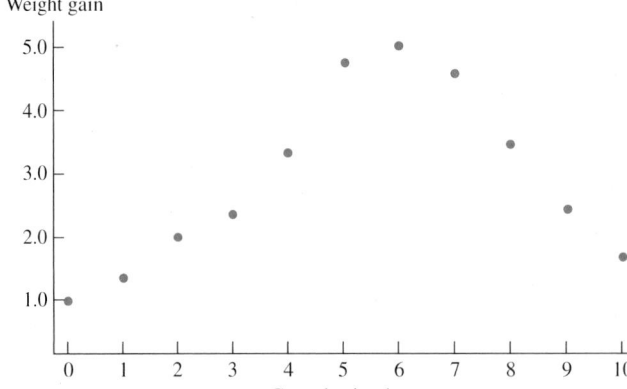

as they did when doses of only 0 through 6 were measured. We see that when the dose increases from 6 to 10, the weight gain begins to fall off. The scatterplot suggests that a quadratic term (second-degree) should be considered in the model. The Minitab solution, shown in Figure 11.20, gives the least squares estimates of β_0, β_1, and β_2 in the second-degree polynomial model

$$y = \beta_0 + \beta_1 x + \beta_2 x^2 + \varepsilon$$

C O M P U T E R   T I P

To create an x^2 column from x stored in C1 in the worksheet, first choose the **Calc** menu and then select **Mathematical Expressions**. In the Mathematical Expressions window, identify **C3** as the **new variable** and in the **Expression box**, type **C1 * C1** and then select **OK**.

FIGURE 11.20   Regression Analysis

```
The regression equation is
Gain = 0.133 + 1.46 Dosage - 0.128 C3

Predictor        Coef        Stdev       t-ratio          p
Constant       0.1329       0.5388          0.25      0.811
Dosage         1.4569       0.2507          5.81      0.000
C3            -0.12751      0.02415         -5.28      0.000

s = 0.7073       R-sq = 81.5%       R-sq(adj) = 76.8%
```

(continued)

FIGURE 11.20
(continued)

Analysis of Variance

SOURCE	DF	SS	MS	F	p
Regression	2	17.5855	8.7928	17.58	0.001
Error	8	4.0018	0.5002		
Total	10	21.5873			

SOURCE	DF	SEQ SS
Dosage	1	3.6364
C3	1	13.9491

Note that for this model the coefficient of determination is $R^2 = 81.5\%$, which means that 81.5% of the variability in weight gain is explained by x and x^2. The coefficient of determination for the simple linear model

$$y = \beta_0 + \beta_1 x + \varepsilon$$

is only

$$r^2 = (.410)^2 = .168 = 16.8\%$$

We see that the addition of x^2 to the model gives a reasonable explanation of the weight gain.

If necessary, an x^3 term can be added to the model. Polynomials with any power of x can be fitted in the same way. Using more than variables x, x^2, and x^3, however, runs the risk of fitting a curve to random fluctuations in the scatterplot.

Testing the Utility of the Model

In simple linear regression, the F-test is equivalent to the t-test because of the relationship $t^2 = F$ described at the end of Section 11.2. For multiple regression, however, we have a separate t-test for each of the regression coefficients. The F-test found in the ANOVA table is for testing that *all* regression coefficients are simultaneously equal to zero. That is, the statistic

$$F = \frac{\text{MSReg}}{\text{MSE}}$$

is for testing

$$H_0: \quad \beta_1 = \beta_2 = \cdots = \beta_k = 0$$

(notice that β_0 is not included in this hypothesis) against the alternative

$$H_a: \quad \beta_j \neq 0 \text{ for at least one } j = 1, 2, \ldots, k$$

Basically the null hypothesis says that none of the independent variables is a legitimate predictor of the dependent variable. Thus, we can evaluate the overall *utility* of the regression model by performing the above F-test. Large values of the F statistic (values significantly greater than 1) indicate that more of the variability in the dependent variable is explained by the regression part of the model than is explained by the error term. Therefore, we reject the null hypothesis when the F statistic exceeds the critical value in the F-table.

When H_0 is rejected, there is statistical evidence that at least one of the predictors is linearly related to y. Individual t-tests can evaluate the significance of each individual coefficient. You should be cautioned, however, that the individual t-test used to test each of the β_j's is an evaluation of the independent variable x_j assuming that all other independent variables have been incorporated into the model. That is, it is the contribution of x_j after the other variables have been entered into the model. Although x_j may not be useful in predicting y when other variables are in the model, it may be useful when it is the only variable in the model. Likewise, a certain variable, x_j, may be significant when used in conjunction with other variables but may not be a significant contributor when it is the only predictor.

EXAMPLE 11.13   Determine whether dose alone is a significant factor in determining the weight gain of laboratory animals studied in Example 11.12.

SOLUTION

The Minitab printout in Figure 11.21 shows that with only dose in the model, R^2 is only 16.8%. Also, the t-ratio of 1.35 [or $F = (1.35)^2 = 1.82$] with a p-value of .210 is insignificant. Therefore, dose alone is not a good predictor of weight gain. Figure 11.22 shows that dose-squared (C3) alone does not predict weight gain at all ($R^2 = 3.2\%$, $F = .3$ with p-value $= .598$). Referring back to Figure 11.20, however, we see that dose in combination with dose-squared (C3) is highly significant. The F-test of the overall model utility is $F = 17.58$ with p-value $= .001$, and both individual t-tests are significant.

FIGURE 11.21   Regression Analysis

The regression equation is
Gain = 2.05 + 0.182 Dosage

Predictor	Coef	Stdev	t-ratio	p
Constant	2.0455	0.7966	2.57	0.030
Dosage	0.1818	0.1347	1.35	0.210

s = 1.412   R-sq = 16.8%   R-sq(adj) = 7.6%

Analysis of Variance

SOURCE	DF	SS	MS	F	p
Regression	1	3.636	3.636	1.82	0.210
Error	9	17.951	1.995		
Total	10	21.587			

FIGURE 11.22   Regression Analysis

```
The regression equation is
Gain = 2.69 + 0.0076 C3

Predictor        Coef        Stdev      t-ratio         p
Constant       2.6871       0.6715        4.00      0.003
C3             0.00764      0.01399       0.55      0.598

s = 1.524        R-sq = 3.2%      R-sq(adj) = 0.0%

Analysis of Variance

SOURCE         DF           SS          MS           F         p
Regression      1        0.692       0.692        0.30     0.598
Error           9       20.895       2.322
Total          10       21.587
```

EXERCISES 11.4

11.39   A sample of $n = 25$ data points was used to find the least squares estimates of β_0, β_1, β_2, and β_3 in the multiple regression model

$$y = \beta_0 + \beta_1 x_1 + \beta_2 x_2 + \beta_3 x_3 + \varepsilon$$

From the data it was found that

$$b_0 = 5.6, \quad b_1 = -3.1, \quad b_2 = 4.9, \quad b_3 = -1.1$$

with

$$SE(b_1) = 2.67, \quad SE(b_2) = 1.34, \quad SE(b_3) = .48$$

a. Use the least squares equation to estimate y when $x_1 = 5$, $x_2 = .35$, and $x_3 = 1.4$.
b. Find a 95% confidence interval for β_1.
c. Test the null hypothesis $\beta_3 = 0$ against the alternative hypothesis $\beta_3 < 0$.

11.40   A sample of $n = 20$ data points was used to find the least squares estimates of β_0, β_1, and β_2 in the second-degree polynomial model

$$y = \beta_0 + \beta_1 x + \beta_2 x^2 + \varepsilon$$

The least squares estimates were

$$b_0 = 6.8, \quad b_1 = 2.5, \quad b_2 = 3.8$$

with

$$SE(b_1) = .94, \quad SE(b_2) = 2.65$$

    a. Determine whether the second-degree term belongs in the model.

    b. Determine whether the first-degree term belongs in the model.

    c. If the second-degree term is dropped from the model, does this mean that the final regression equation is $\hat{y} = 6.8 + 2.5x$?

11.41 A study was conducted to evaluate the effects of several factors on teachers' attitudes toward extracurricular activities. Two variables considered important are the years of teaching experience and the size of the school. Give the form of the proposed linear regression model that relates teachers' attitudes to years of experience and size of school.

11.42 In an effort to study the effects of different doses of a drug on the pulse rate of human subjects, four dose levels were given to 16 subjects, with four subjects at each dose level. It is believed that a second-degree polynomial model is needed. Give the form of the proposed linear regression model that relates pulse rate to drug dose.

11.43 In a poverty area, a study was conducted to investigate the effects of educational level and years living in the region on family income. Give the form of the proposed linear regression model that relates family income to educational level and years residing in the region.

11.44 Here is a computer printout associated with the problem described in Exercise 11.41:

Worksheet: Attitude.mtw

Regression Analysis

The regression equation is
attitude = 132 - 0.232 years - 0.810 size

Predict	Coef	Stdev	t-ratio	p
Constant	132.032	5.620	23.49	0.000
years	-0.2322	0.3468	-0.67	0.516
size	-0.81005	0.07494	-10.81	0.000

s = 4.906   R-sq = 90.7%   R-sq(adj) = 89.1%

Analysis of Variance

SOURCE	DF	SS	MS	F	p
Regression	2	2814.1	1407.0	58.46	0.000
Error	12	288.8	24.1		
Total	14	3102.9			

SOURCE	DF	SEQ SS
years	1	1.5
size	1	2812.6

Unusual Observations

Obs.	years	attitude	Fit	Stdev.Fit	Residual	St.Resid
7	2.0	40.00	48.13	3.62	-8.13	-2.46R

R denotes an obs. with a large st. resid.

    a. How many independent variables are there?

    b. What is the regression equation?

    c. What is the value of R^2?

    d. Give an estimate of σ.

    e. Construct a 98% confidence interval for β_2.

    f. Test the hypothesis $\beta_1 = 0$.

    g. Comment on the results.

11.45   This computer printout is associated with the problem described in Exercise 11.42. The simple linear regression model is used with only the dose level.

```
Regression Analysis

The regression equation is
pulse = 58.3 + 5.82 dose

Predictor        Coef       Stdev     t-ratio          p
Constant       58.250       2.936       19.84      0.000
dose            5.825       1.072        5.43      0.000

s = 4.795        R-sq = 67.8%     R-sq(adj) = 65.5%

Analysis of Variance

SOURCE         DF          SS          MS          F          p
Regression      1      678.61      678.61      29.52      0.000
Error          14      321.83       22.99
Total          15     1000.44
```

    a. What is the regression equation?

    b. What is the value of R^2?

    c. Test the hypothesis $\beta_1 = 0$.

    d. Evaluate the proposed model.

11.46   This is a computer printout associated with the problem described in Exercise 11.42, and it is a continuation of Exercise 11.45. The square of the dose level (dosesq) has been added to the linear regression model.

```
Regression Analysis

The regression equation is
pulse = 41.7 + 22.4 dose - 3.31 dosesq

Predictor        Coef       Stdev     t-ratio          p
Constant       41.688       4.669        8.93      0.000
dose           22.388       4.259        5.26      0.000
dosesq        -3.3125       0.8386      -3.95      0.002

s = 3.354        R-sq = 85.4%     R-sq(adj) = 83.1%
```

(continued)

```
Analysis of Variance

SOURCE          DF          SS          MS          F          p
Regression      2           854.17      427.09      37.96      0.000
Error           13          146.26      11.25
Total           15          1000.44


SOURCE          DF          SEQ SS
dose            1           678.61
dosesq          1           175.56
```

    a. What is the regression equation?
    b. What is the value of R^2?
    c. Test the hypothesis $\beta_1 = 0$.
    d. Construct a 90% confidence interval for β_2.
    e. Evaluate the proposed model. How does it compare with the simple linear model proposed in Exercise 11.42?

11.47 In Exercise 11.21 of Section 11.2, a simple linear model was proposed for predicting the number of wins for a professional baseball team based on the earned run averages of the pitchers. From the printout, we saw that $R^2 = 51.2\%$ and $t = -3.24$ was very significant. Following is an analysis of this problem with six additional variables entered into the model (batavg = batting average, rbi = runs batted in, stole = number of stolen bases, strkout = number struck out, caught = number caught stealing base, errors = number of errors, era = earned run average). The table lists the final 1990 National League baseball team statistics from a total of 162 games.

Worksheet: Wins.mtw

team	wins	batavg	rbi	stole	strkout	caught	errors	era
Atlanta	65	.250	636	92	1,010	55	158	4.58
Chicago	77	.263	649	151	869	50	124	4.34
Cincinnati	91	.265	644	166	913	66	102	3.39
Houston	75	.242	536	179	997	83	131	3.61
Los Angeles	86	.262	669	141	952	65	130	3.72
Montreal	85	.250	607	235	1,024	99	110	3.37
New York	91	.256	734	110	851	33	132	3.43
Philadelphia	77	.255	619	108	915	35	117	4.07
Pittsburgh	95	.259	693	137	914	52	134	3.40
St. Louis	70	.256	554	221	844	74	130	3.87
San Diego	75	.257	628	138	902	59	141	3.68
San Francisco	85	.262	681	109	973	56	107	4.08

```
Regression Analysis

The regression equation is
wins = - 39 + 342 batavg + 0.0717 rbi + 0.082 stole + 0.0647
strkout - 0.270 caught - 0.092 errors - 15.1 era
```

(continued)

```
Predictor        Coef       Stdev     t-ratio         p
Constant        -39.4       137.3       -0.29      0.788
batavg          341.6       391.2        0.87      0.432
rbi           0.07169     0.04394        1.63      0.178
stole          0.0815      0.1177        0.69      0.527
strkout       0.06472     0.06741        0.96      0.391
caught        -0.2697      0.3454       -0.78      0.479
errors        -0.0917      0.1225       -0.75      0.496
era           -15.114       4.394       -3.44      0.026

s = 3.511      R-sq = 94.7%     R-sq(adj) = 85.5%
```

Analysis of Variance

```
SOURCE        DF           SS          MS         F        p
Regression     7       884.70      126.39     10.25    0.020
Error          4        49.30       12.33
Total         11       934.00

SOURCE        DF        SEQ SS
batavg         1       203.26
rbi            1       198.64
stole          1       211.69
strkout        1         9.95
caught         1         9.54
errors         1       105.79
era            1       145.82
```

When six more variables are added, R^2 is increased from 51.2% to 94.7%, but none of the *t*-ratios for the added variables is significant! What would cause this to happen? Notice also that the overall *F* statistic (utility test) is less significant, with *p*-value = .02 versus *p*-value = .009 for the simple model. Enter only the two independent variables rbi and era into the model and see what happens. Would you judge this model better than the simple model that contains only era?

11.48   In Exercise 11.7 of Section 11.1, a linear model was proposed to relate the ratings of top offensive linemen for the NFL draft to their times to run the 40-yard dash. In Exercise 11.19 of Section 11.2, we learned that the ratings were not linearly related to the times ($R^2 = 7.2\%$, $F = 2.10$ with *p*-value = .159). We now add another variable: the weight of the player.

Worksheet: NFLdraft.mtw

Rating	6.5	6.1	6.0	5.7	5.5	5.2	5.0	7.6	7.2	7.0
Time	4.94	5.27	5.27	5.14	5.09	5.23	5.30	5.15	5.20	5.20
Weight	285	285	285	277	280	274	310	303	315	325

Rating	6.5	6.2	6.1	6.0	5.9	5.5	5.3	7.1	7.0	6.4
Time	5.18	5.23	5.35	5.29	5.20	5.25	5.18	5.06	5.36	5.06
Weight	325	290	321	305	313	305	301	289	317	315

Rating	6.4	6.3	6.2	6.0	5.9	5.7	5.5	5.4	5.2
Time	5.26	5.36	5.29	5.03	5.25	5.29	5.56	5.29	5.26
Weight	302	311	289	274	292	286	359	301	295

Complete a regression analysis on the ratings with both variables, 40-yard dash time and weight, in the model.

a. Write the equation for the linear model.

b. What is the value of R^2? Has it increased much?

c. Test the utility of the model by performing an F-test. Is it significant?

d. Test the significance of each of the variables with individual t-tests. Are both significant?

e. Explain why the 40-yard dash time is significant in the model that contains both variables and not significant when it is the only variable in the model.

11.49 The tensile strength of Kraft paper (in pounds per square inch) was measured for different percentages of hardwood in the batch of pulp that was used to produce the paper (Joglekar, et al., 1989).

Worksheet: Hardwood.mtw

Strength	6.3	11.1	20.0	24.0	26.1	30.0	33.8	34.0	38.1	39.9
Percent hardwood	1.0	1.5	2.0	3.0	4.0	4.5	5.0	5.5	6.0	6.5
Strength	42.0	46.1	53.1	52.0	52.5	48.0	42.8	27.8	21.9	
Percent hardwood	7.0	8.0	9.0	10.0	11.0	12.0	13.0	14.0	15.0	

a. Construct a scatterplot of tensile strength versus hardwood content.

b. Does it appear that a straight line fits the data?

c. Fit a simple linear regression model to the data. What is R^2? Is the F statistic significant?

d. Does your answer to part **c** agree with your answer to part **b**?

e. Fit a second-degree polynomial to the data. How much improvement is there in R^2? Is the F statistic significant? Is this a better model than the simple model in part **c**?

11.50 Exercise 11.10 in Section 11.1 attempted to find a model that relates the rainfall in an area in west central Kansas to the rainfall in a neighboring county. Now use the data obtained from four surrounding counties.

Worksheet: Rainks.mtw

Regression Analysis

The regression equation is
rain = 0.276 + 0.138 x1 + 0.303 x2 + 0.099 x3 + 0.301 x4

Predictor	Coef	Stdev	t-ratio	p
Constant	0.2764	0.2209	1.25	0.221
x1	0.1376	0.1636	0.84	0.407
x2	0.3027	0.1373	2.20	0.035
x3	0.0991	0.1563	0.63	0.531
x4	0.30088	0.09979	3.02	0.005

s = 0.3782   R-sq = 79.8%   R-sq(adj) = 77.1%

Analysis of Variance

SOURCE	DF	SS	MS	F	p
Regression	4	16.9400	4.2350	29.61	0.000
Error	30	4.2913	0.1430		
Total	34	21.2313			

(continued)

```
SOURCE          DF      SEQ SS
x1               1     13.3609
x2               1      2.2603
x3               1      0.0185
x4               1      1.3004
```

a. What is R^2 with all four variables in the model? Recall that R^2 was 62.9% with the one variable x1. By adding three more variables, have we improved the model?

b. Examine the *t*-ratios. Which ones are significant? Does this mean that we can reduce the number of variables in the model?

c. This is the correlation matrix that relates all five variables. Based on this and the *t*-ratios, what variables do you recommend for the model?

```
Correlations (Pearson)
          rain       x1       x2       x3
x1       0.793
x2       0.818    0.769
x3       0.748    0.883    0.746
x4       0.818    0.720    0.734    0.638
```

d. This is the regression printout with variables x2 and x4 in the model. Compare this model to the model that contains all four independent variables. What variables should be included in the model?

Regression Analysis

The regression equation is
rain = 0.307 + 0.449 x2 + 0.361 x4

```
Predictor      Coef       Stdev      t-ratio         p
Constant     0.3075      0.2210         1.39     0.174
x2           0.4495      0.1179         3.81     0.001
x4           0.36116     0.09535        3.79     0.001
```

s = 0.3889   R-sq = 77.2%   R-sq(adj) = 75.8%

Analysis of Variance

```
SOURCE          DF          SS          MS         F         p
Regression       2     16.3923      8.1962     54.20     0.000
Error           32      4.8390      0.1512
Total           34     21.2313
```

```
SOURCE          DF      SEQ SS
x2               1     14.2229
x4               1      2.1694
```

```
Unusual Observations
Obs.     x2       rain        Fit    Stdev.Fit    Residual    St.Resid
 16     4.45     4.7800     4.0376       0.1680      0.7424        2.12R
 26     2.84     3.3300     3.5487       0.2251     -0.2187       -0.69 X
```

11.51 In 1973, the National Football League arrived at an formula for rating quarterbacks. The formula is based on the completion percentage, the average yards gained, the percent of touchdowns, and the percent of interceptions. With the formula, quarterbacks in the NFL as far back as 1945 have been rated. Up to and including the 1995 season, only 14 NFL quarterbacks have achieved a rating in excess of 100. Here is the statistical information on all qualifying quarterbacks at the end of the 1995 regular season.

Worksheet: Quarbck.mtw

Name	Team	Atts	Comp	Comp%	yards	AvgGn	TDs	TD%	INT	INT%	Rating
Harbaugh, J.	Ind	314	200	63.7	2,575	8.20	17	5.4	5	1.6	100.7
Favre, B.	GB	570	359	63.0	4,413	7.74	38	6.7	13	2.3	99.5
Aikman, T.	Dal	432	280	64.8	3,304	7.65	16	3.7	7	1.6	93.6
Kramer, E.	Chi	522	315	60.3	3,838	7.35	29	5.6	10	1.9	93.5
Young, S.	SF	447	299	66.9	3,200	7.16	20	4.5	11	2.5	92.3
Mitchell, S.	Det	583	346	59.3	4,338	7.44	32	5.5	12	2.1	92.3
Moon, W.	Min	606	377	62.2	4,228	6.98	33	5.4	14	2.3	91.5
Marino, D.	Mia	447	286	64.0	3,378	7.56	22	4.9	13	2.9	91.2
George, J.	Atl	557	336	60.3	4,143	7.44	24	4.3	11	2.0	89.5
Testaverde, V.	Cle	392	241	61.5	2,883	7.35	17	4.3	10	2.6	87.8
Chandler, C.	Hou	356	225	63.2	2,460	6.91	17	4.8	10	2.8	87.8
ODonnell, N.	Pit	416	246	59.1	2,970	7.14	17	4.1	7	1.7	87.7
Everett, J.	NO	567	345	60.8	3,970	7.00	26	4.6	14	2.5	87.0
Elway, J.	Den	542	316	58.3	3,970	7.32	26	4.8	14	2.6	86.4
Brunell, M.	Jax	346	201	58.1	2,168	6.27	15	4.3	7	2.0	82.6
Hostetler, J.	Oak	286	172	60.1	1,998	6.99	12	4.2	9	3.1	82.2
Blake, J.	Cin	567	326	57.5	3,822	6.74	28	4.9	17	3.0	82.1
Kelly, J.	Buf	458	255	55.7	3,130	6.83	22	4.8	13	2.8	81.1
Humphries, S.	SD	478	282	59.0	3,381	7.07	17	3.6	14	2.9	80.4
Bono, S.	KC	520	293	56.3	3,121	6.00	21	4.0	10	1.9	79.5
Miller, C.	Ram	405	232	57.3	2,623	6.48	18	4.4	15	3.7	76.2
Brown, D.	NYG	456	254	55.7	2,814	6.17	11	2.4	10	2.2	73.1
Krieg, D.	Arz	521	304	58.3	3,554	6.82	16	3.1	21	4.0	72.6
Esiason, B.	NYJ	389	221	56.8	2,275	5.85	16	4.1	15	3.9	71.4
Frerotte, G.	Was	396	199	50.3	2,751	6.95	13	3.3	13	3.3	70.2
Peete, R.	Phi	375	215	57.3	2,326	6.20	8	2.1	14	3.7	67.3
Mirer, R.	Sea	391	209	53.5	2,564	6.56	13	3.3	20	5.1	63.7
Bledsoe, D.	NE	636	323	50.8	3,507	5.51	13	2.0	16	2.5	63.7
Collins, K.	Pan	432	214	49.5	2,717	6.29	14	3.2	19	4.4	62.0
Dilfer, T.	TB	416	224	53.8	2,762	6.64	4	1.0	18	4.3	59.8

SOURCE: Indra's Net, Inc. (http://net.indra.com).

a. Perform a multiple regression analysis using Rating as the dependent variable and Comp%, AvgGn, TD%, and INT% as the independent variables.

b. What is the value of R^2 with all four variables in the model? Is this to be expected? What can be said about the relationship between the regression equation and the formula used by the NFL?

c. Examine each of the t-ratios. Should any of the variables be dropped from the model?

d. Perform an analysis with only Comp% and INT% in the model. What is R^2 now? How much variability in the Rating is explained by these two variables? How much is explained by all four independent variables? Should all four variables be included in the model?

e. In Superbowl XXX, the Dallas Cowboys defeated the Pittsburgh Steelers by a score of 27 to 17. Use the multiple regression model with all four independent variables and the following data to find the Ratings for the two Superbowl quarterbacks. Are the results what you expected?

Name	Team	Atts	Comp	Comp%	yards	AvgGn	TDs	TD%	INT	INT%
Aikman, T.	Dal	23	15	65.2	209	9.09	1	4.3	0	.0
ODonnell, N.	Pit	49	28	57.1	239	4.88	1	2.0	3	6.1

11.5 CHECKING MODEL ADEQUACY

In the preceding sections, you learned how to find estimates and make inferences about regression coefficients, $\beta_0, \beta_1, \beta_2, \ldots, \beta_k$, in the linear regression model

$$y = \beta_0 + \beta_1 x_1 + \beta_2 x_2 + \cdots + \beta_k x_k + \varepsilon$$

You should be cautioned, however, that the conclusions drawn from the inference procedures can be seriously misleading if the assumptions made about the model are invalid. We present the assumptions again as a review.

Assumptions About the Error Term in the Linear Regression Model $y = \beta_0 + \beta_1 x_1 + \beta_2 x_2 + \cdots + \beta_k x_k + \varepsilon$

The random error term, ε, is a random variable that satisfies these conditions:

1. It has mean value 0 ($\mu_\varepsilon = 0$).
2. It has standard deviation σ, which does not depend on x.
3. It has a normal distribution.
4. Any two observed values of ε are independent of each other.

A regression study is not complete without an investigation of these assumptions.

In the simple linear regression case (one independent variable), you can use a scatterplot to visually check the linearity assumption by observing whether the data exhibit a straight-line behavior. Additionally, Pearson's correlation gives a measure of the degree of linearity. In the multiple regression case (and the simple linear regression case), the coefficient of determination, R^2, measures the amount of variability in y explained by the independent variables. The remaining variability that cannot be explained by the model is said to be explained by error and is measured by the residuals. We now present ways to evaluate the residuals.

Evaluating the Residuals

The residual associated with data point (x_i, y_i) is defined as

$$e_i = y_i - \hat{y}_i$$

where $\hat{y}_i$ is the predicted y_i. If the assumptions about the random error are met, then the e_i's should appear to have come from a normal population with mean 0 and standard deviation σ. It may be convenient to investigate the *standardized residuals*, which are found from e_i/s, where $s = \sqrt{\text{MSE}}$ is the estimate of σ. Under the normality assumption, the standardized residuals should appear to have come from a *standard normal distribution*. As such, most of the residuals should be between -2 and $+2$. In most computer programs for linear regression, the standardized residuals are printed out, so it is easy to see whether they lie between -2 and $+2$.

COMPUTER   CAUTION

When you request standardized residuals with the Minitab regression routine, you actually get *Studentized residuals*. Studentized residuals are very similar to standardized residuals except that the divisor is a modification of s that is adjusted for each individual residual. Residuals associated with observations that have unusually large x values are adjusted the most. For our purposes, however, we do not distinguish between standardized and Studentized residuals.

Graphically, the shape of the distribution of the residuals can be checked by plotting them in much the same way that raw data were plotted in previous chapters. By using the histogram, stem and leaf plot, boxplot, or a simple dot diagram, we can observe whether the data have a normal curve appearance.

Remember, the midsummary statistics can be used to check for symmetry (if the residuals are not symmetric, then clearly they are not normally distributed), and the normal probability plot can be used to check for normality. Also, the boxplot can be used to check for outliers and long tails. As might be expected, outliers tend to have an adverse effect on the least squares regression. Not only do they affect the precision of the inference procedure, but they also affect the actual equation of the straight line. Because the least squares procedure minimizes the sum of squares of the distances of all points to the line, an outlier tends to pull the line toward it in an unusual fashion. In this situation, a fitting technique such as a *resistant line* is needed. As you saw in Section 3.5, the resistant line tends to reduce, if not eliminate, the effect of outliers.

EXAMPLE 11.14

The data are the research and development (R&D) expenditures and the corresponding sales for a large company. (All numbers are in million dollars.)

Worksheet: RandD.mtw

R&D	1.2	2.4	3.1	4.0	4.9	6.3	1.8	2.7	3.3	3.7	4.2	5.4
Sales	55	48	32	21	10	41	43	44	21	32	12	8

Determine whether a linear relationship between the two variables seems plausible by evaluating the regression assumptions.

SOLUTION

The least squares equation that relates R&D expenditures, x, with sales, y, is

$$\hat{y} = 55.008 - 6.816x$$

The correlation is $r = -.634$, which gives $r^2 = .402$. Thus, approximately 40% of the variability in sales is explained by the R&D expenditures.

Using the least squares equation, we find the predicted y_i's and then the residuals in Table 11.2. The residuals seem reasonable except for the one large residual (28.934) corresponding to $(6.3, 41)$. From the scatterplot with the equation graphed in Figure 11.23, we see that $(6.3, 41)$ is an outlier, which explains the large residual.

TABLE 11.2

x_i	y_i	$\hat{y}_i$	e_i
1.2	55	46.829	8.171
2.4	48	38.649	9.351
3.1	32	33.878	−1.878
4.0	21	27.743	−6.743
4.9	10	21.609	−11.609
6.3	41	12.066	28.934
1.8	43	42.739	.261
2.7	44	36.604	7.396
3.3	21	32.515	−11.515
3.7	32	29.788	2.212
4.2	12	26.380	−14.380
5.4	8	18.201	−10.201

FIGURE 11.23

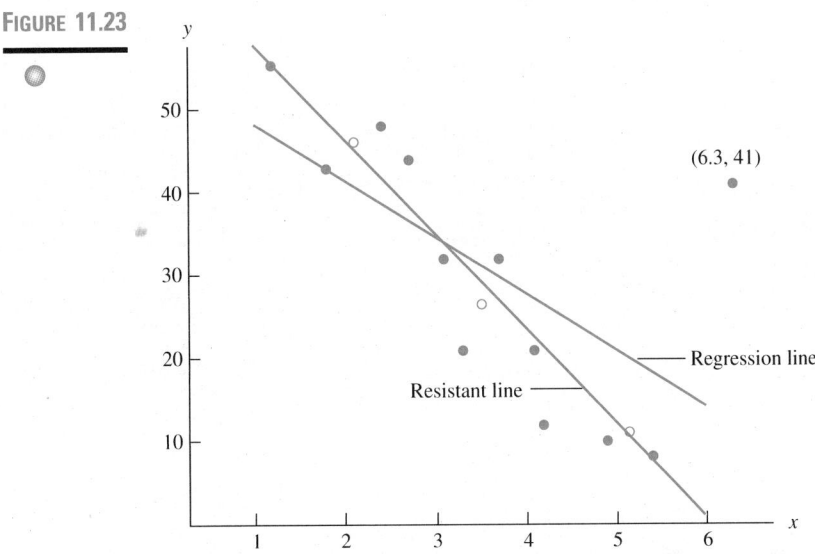

From the scatterplot, we can easily fit the resistant line to the medians of the three groups (identified by 0) and see that it is a more reasonable fit to the data than the least squares line. (The equation for the resistant line is $y = 68.952 - 11.475x$.)

Residual Plots

A plot of residuals against the predicted values, $\hat{y}_i$, amounts to an examination of y after the linear dependence on the independent variables is removed. Figure 11.24 gives three patterns of residual plots that might come up. In Figure 11.24(a), the points form a horizontal band about 0 as would be expected when there are no abnormalities. In Figure 11.24(b), the width of the band increases as the predicted y increases. This indicates

FIGURE 11.24   Possible residual plots

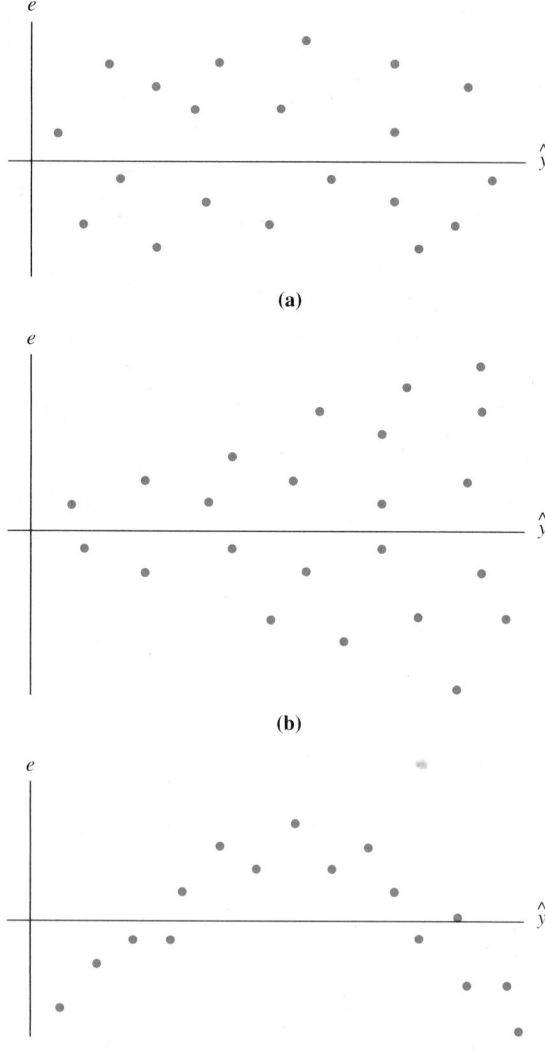

that the variance is not constant. A transformation of data might be in order. In Figure 11.24(c), the residuals exhibit a curvilinear pattern, which suggests that the errors are not independent. Often a nonlinear term should be considered in such a model.

Figure 11.25 presents the standardized residuals plotted against the predicted y_i's for the regression equation found in Example 11.14. Notice that the residual corresponding to (6.3, 41) shows up clearly as an outlier. Furthermore, the remaining residuals have a systematic pattern that needs to be rectified.

FIGURE 11.25

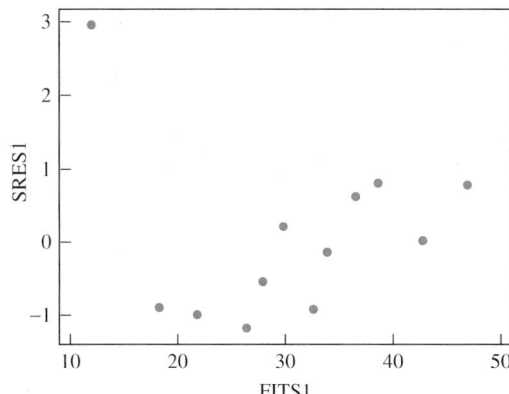

Removal of the outlier, (6.3, 41), results in a significant increase in r^2 (from 40.3% to 88.4%) and the much more reasonable residual plot shown in Figure 11.26. Notice that the residuals appear to be randomly distributed about 0, as they should be. Whether the outlier should be removed from the data set is a decision that the experimenter will have to make. Without the outlier, we have a reasonable model relating sales to R&D expenditures. If we choose to retain the outlier, then the model needs improvement. Perhaps additional independent variables should be considered.

FIGURE 11.26

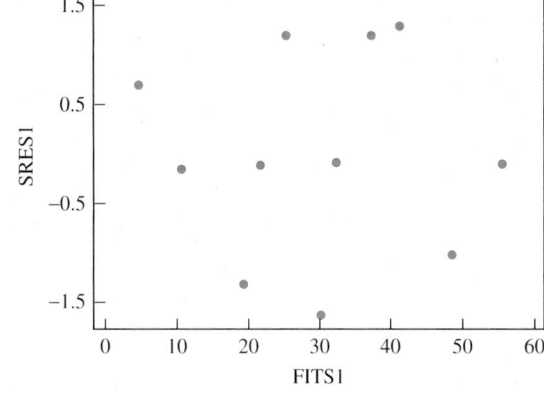

With a single independent variable in the model, the residual plot against $\hat{y}$ gives the same results as a residual plot against the independent variable. In the case of the multiple regression analysis, however, additional information can be obtained by plotting the residuals against each of the independent variables.

EXAMPLE 11.15

In an effort to determine the effective duration of a tranquilizer for animals, the concentrations of the substance in blood samples taken at various times after the injection are measured. From the data, determine the least squares equation that relates the concentration to the elapsed time after the injection, and conduct a residual analysis.

Elapsed time (hours)	1	2	3	6	12	18
Concentration (mg/ml)	1.8	1.4	1.2	.9	.5	.1

SOLUTION

From the sample data, we find the least squares equation:

$$\hat{y} = 1.6052 - .08884x$$

The correlation is $r = -.9625$, which suggests a strong linear relationship. Figure 11.27 is a scatterplot with the graphed equation. Using the equation, we find the predicted values of y and the residuals as listed in Table 11.3. Figure 11.28 is a graph of the residuals plotted against the independent variable. The residual plot shows a definite curvilinear pattern, which suggests that a higher-ordered x term is needed in the model.

FIGURE 11.27

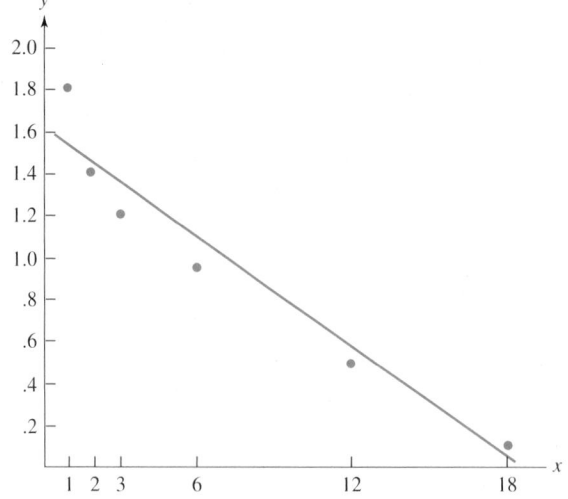

TABLE 11.3

x_i	y_i	$\hat{y}_i$	e_i
1	1.8	1.5164	.2836
2	1.4	1.4275	−.0275
3	1.2	1.3387	−.1387
6	.9	1.0722	−.1722
12	.5	.5391	−.0391
18	.1	.0061	.0939

FIGURE 11.28

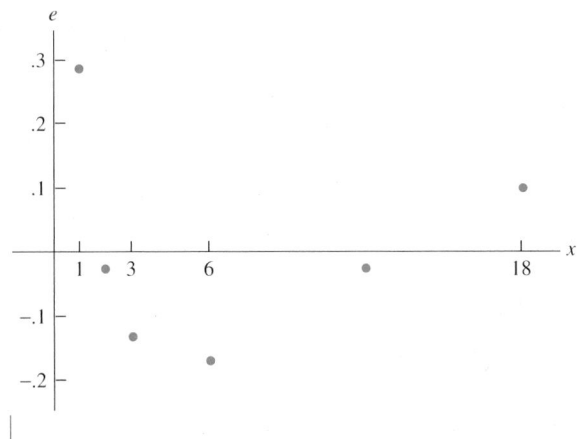

Residual Plot Against Time

Often data for a regression study are collected sequentially over a period of time. When the observations are correlated with time, the residuals are dependent. A plot with time on the horizontal axis and the residuals on the vertical axis will often show violations of the assumption of independence of the residuals. Figure 11.29 shows a definite relationship between the residuals and time. It appears that time should be incorporated into the model.

FIGURE 11.29 Residual plot against time

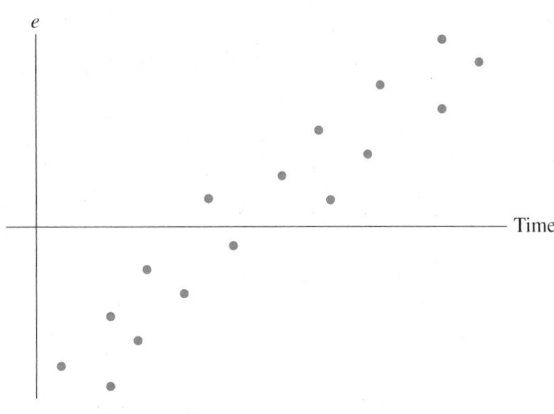

A discussion of time series models can be found in *Time Series Analysis, Forecasting and Control* (San Francisco: Holden-Day, 1970) by G. E. P. Box and G. M. Jenkins or in *Statistics for Experimenters* (New York: Wiley, 1978) by G. E. P. Box, W. G. Hunter, and J. S. Hunter.

C O M P U T E R   T I P

A residual analysis with Minitab is straightforward. After entering the data in the worksheet, perform a regression analysis as before. Make sure that in the Regression window you have checked **Standard resids**. and **Fits**. By checking these two boxes, you store the standardized residuals and predicted values (fits) in the next two columns of the worksheet. Click on **Regression** under the Stat menu again, but this time instead of clicking on Regression a second time, click on **Residual Plots**. In the Residual Plots window, from the list of variables, **Select** the **Residuals** and **Fits** variables, and then click **OK**. Prior to clicking **OK**, you may want to give a title to the residual plots in the **Title** box.

EXAMPLE 11.16

The effects of age and weight on systolic blood pressure were examined in Example 11.11 (see Figure 11.18). Perform a residual analysis of these data.

SOLUTION

Notice in Figure 11.18 that SRES1 and FITS1 have been stored, respectively, in C4 and C5 of the worksheet. These are the columns that should be selected to produce the residual plot analysis. The completed residual plots appear in Figure 11.30. The

FIGURE 11.30

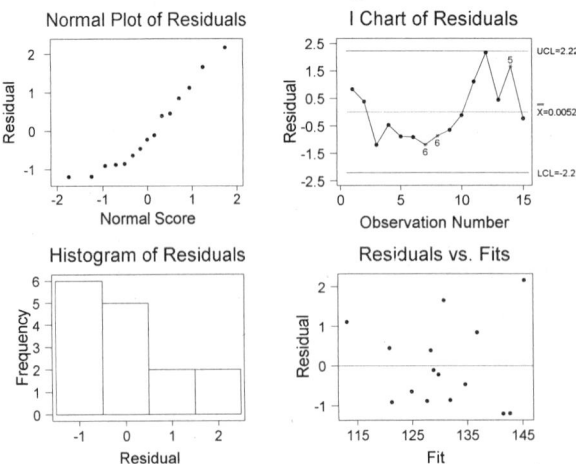

normal probability plot and the histogram indicate that there may be a problem with the normality assumption. The I Chart of Residuals and the Residuals vs. Fits plots appear to have nonrandom patterns, which again are violations of assumptions. Although the analysis presented in Example 11.11 justified the linear model, it appears from the residual analysis that more study needs to go into this problem. More advanced texts address issues such as variable selection and modification to yield the "best" regression equation.

Transformations   You have seen how the polynomial regression model can improve upon the simple linear model. In some cases, a simple transformation of one or both variables dramatically improves a model.

EXAMPLE 11.17   A simple linear regression model was proposed in Example 11.4 to relate the number of tickets given by New York City patrolpersons to their years of experience. Figure 11.31 gives a scatterplot, the regression printout, and a residual plot from the data. Evaluate the printout and the residual plots and make possible modifications to the model.

FIGURE 11.31

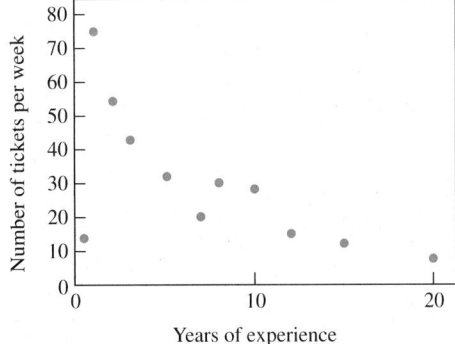

```
Regression Analysis

The regression equation is
tickets = 55.9 - 2.93 years

Predictor         Coef        Stdev       t-ratio          p
Constant        55.938        6.215          9.00      0.000
years          -2.9322       0.6150         -4.77      0.000

s = 11.21        R-sq = 74.0%      R-sq(adj) = 70.7%
```

(continued)

Analysis of Variance

SOURCE	DF	SS	MS	F	p
Regression	1	2855.4	2855.4	22.73	0.000
Error	8	1005.0	125.6		
Total	9	3860.4			

Unusual Observations

Obs.	years	tickets	Fit	Stdev.Fit	Residual	St.Resid
7	1.0	75.00	53.01	5.72	21.99	2.28R

R denotes an obs. with a large st. resid.

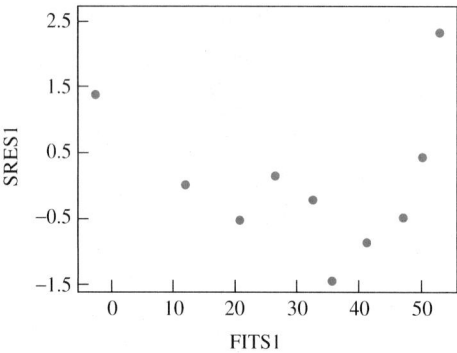

SOLUTION

It appears that we have a good model; $R^2 = 74\%$ and $F = 22.73$ is highly significant. The scatterplot and, to a greater extent, the residual plot, however, show a curvilinear trend in the data. The scatterplot has the appearance of the function $y = e^{-x}$. By taking the natural logarithm of both sides, we have $\ln(y) = -x$, which is a linear model. To execute the transformation, take the logarithm of each of the y observations. This is easily accomplished in Minitab by letting

```
C2 = loge(C1)
```

where C1 contains the y data and C2 then contains the $\ln(y)$ data.

Figure 11.32 gives a scatterplot of the transformed data, the regression printout for the linear model

$$\ln(y) = b_0 + b_1 x$$

and a residual plot. The scatterplot of years versus ln(tickets) shows a more linear pattern; R^2 has increased to 91.2%, and $F = 83.00$ is extremely significant. Additionally, the residual plot shows a rather random pattern, indicating that the transformed model is the proper model.

FIGURE 11.32

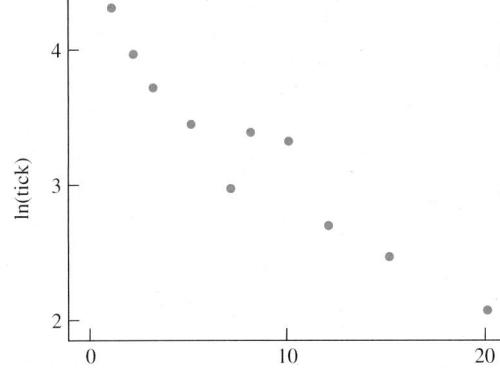

```
MTB > Regress 'lntick' 1 'years';
SUBC>    Constant.

Regression Analysis


The regression equation is
lntick = 4.15 - 0.109 years

Predictor        Coef        Stdev        t-ratio          p
Constant       4.1545       0.1207          34.42      0.000
years         -0.10884      0.01195         -9.11      0.000

s = 0.2177       R-sq = 91.2%       R-sq(adj) = 90.1%

Analysis of Variance

SOURCE          DF          SS           MS           F          p
Regression       1       3.9343       3.9343       83.00      0.000
Error            8       0.3792       0.0474
Total            9       4.3135
```

(continued)

FIGURE 11.32
(continued)

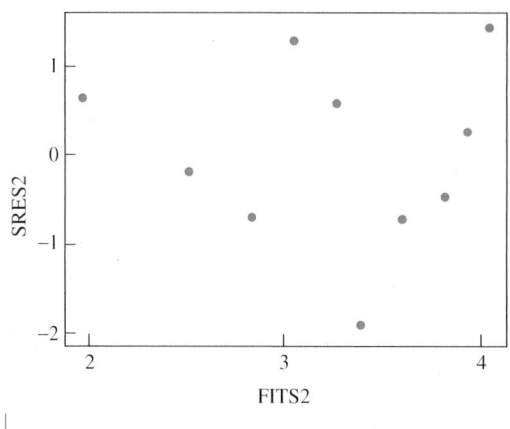

EXERCISES 11.5

11.52   The stem and leaf plot shows the residuals associated with a simple linear regression equation. Do the residuals appear to violate the assumptions stated for the linear regression model?

−3	22
−2	05, 18
−1	22, 35, 51, 74, 81
−0	04, 19, 31, 56, 64, 71, 82, 94
0	02, 15, 24, 48, 61, 75, 93
1	12, 34, 47, 63, 74, 92
2	34
3	21, 47

11.53   Shown here is a residual plot against the independent variable in a simple linear regression problem. What pattern do you see in the plot? Does it appear that the assumptions for the linear regression model have been violated?

11.54 From the residual plot against time, what observations can you make about the model?

11.55 The residual plot is for these data:

x	1	1	1	2	2	2	3	3	3
y	15	5	3	10	8	3	6	5	4

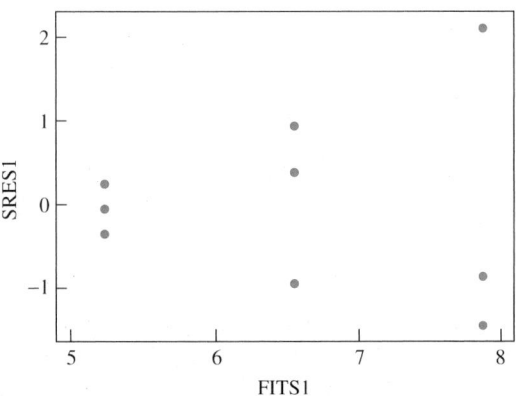

Does it appear that any assumptions for the linear regression model have been violated?

11.56 Examine the regression printout for these data:

x	1.2	2.3	3.2	3.9	4.6	6.5	9.8	15.5
y	1.3	2.6	4.8	5.9	7.1	12.1	16.3	24.5

```
Regression Analysis

The regression equation is
y = - 0.423 + 1.66 x
```

(*continued*)

```
Predictor           Coef        Stdev      t-ratio            p
Constant         -0.4230       0.5160        -0.82        0.444
x                 1.65924      0.07023       23.63        0.000

s = 0.8765        R-sq = 98.9%       R-sq(adj) = 98.8%
```

Analysis of Variance

```
SOURCE              DF          SS           MS           F          p
Regression           1       428.81       428.81       558.20     0.000
Error                6         4.61         0.77
Total                7       433.42
```

```
Unusual Observations
Obs.      x         y       Fit    Stdev.Fit    Residual    St.Resid
  6      6.5    12.100    10.362      0.313        1.738        2.12R
```

R denotes an obs. with a large st. resid.

a. Based on R^2 and the t-ratio (or F statistic), does it appear that the straight-line regression equation fits these data?
b. Do you anticipate any significant violation of the assumptions for the linear regression model?
c. Examine the residual plot. Do you detect any violation of assumptions?

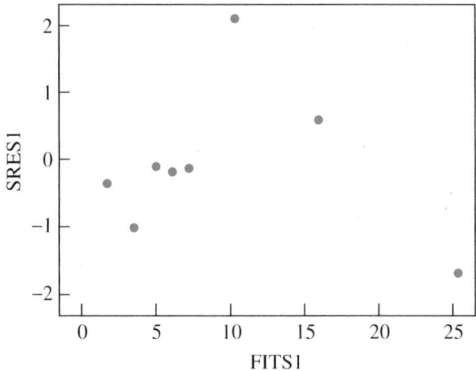

11.57   Conduct a residual analysis for the least squares solutions given in Exercise 11.32 of Section 11.3. The data, which are repeated here, match a child's age with the number of gymnastic activities he or she is able to complete successfully.

Worksheet: Gym.mtw

Age	2	3	4	4	5	6	7	7
Activities	5	5	6	3	10	9	11	13

11.58   Conduct a residual analysis for the least squares solutions given in Exercise 11.33 of Section 11.3. The data, which are repeated here, match an automobile dealer's number of cars sold with the prime interest rate for six 2-month periods.

Worksheet: Prime.mtw

Prime rate	15	14	13	12	11	10
Sales	116	132	148	136	155	184

11.59 A school psychologist believes that there is a linear relationship between the verbal test scores for eighth-graders and the number of library books they check out:

Worksheet: Verbal.mtw

Books	12	15	3	7	10	5	22	9	13	7	25	17	14	19	20
Scores	77	85	48	59	75	41	94	72	80	70	98	85	83	96	89

a. Based on the scatterplot, do you think the school psychologist is correct?

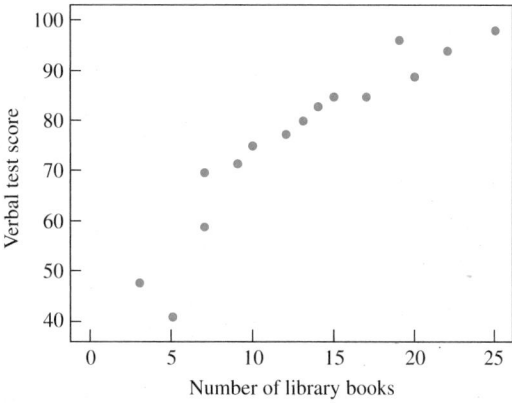

b. Examine the regression printout. Does it appear that a straight line fits the data? What is R^2? Is the regression coefficient significantly different from 0? What is the p-value?

```
Regression Analysis

The regression equation is
verbal = 45.3 + 2.38 number

Predictor      Coef       Stdev      t-ratio         p
Constant      45.346      4.022        11.28     0.000
number        2.3828      0.2751        8.66     0.000

s = 6.695       R-sq = 85.2%       R-sq(adj) = 84.1%

Analysis of Variance

SOURCE        DF          SS            MS         F        p
Regression     1        3363.6        3363.6     75.03    0.000
Error         13         582.8          44.8
Total         14        3946.4
```

(continued)

```
Unusual Observations
Obs. number   verbal      Fit  Stdev.Fit   Residual  St.Resid
     6      5.0   41.00   57.26      2.84     -16.26    -2.68R
R denotes an obs. with a large st. resid.
```

c. Examine this residual plot. Is there a random pattern to the residual or should we consider revising the model? What suggestions can you give?

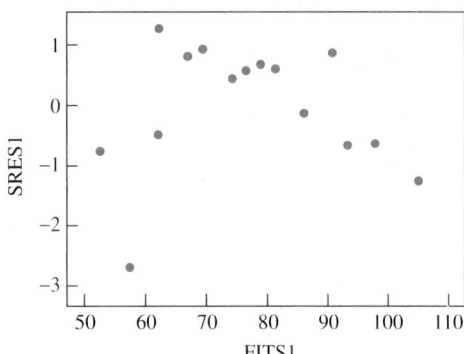

11.60   A residual analysis of Exercise 11.45 in Section 11.4 is shown here. Recall that the problem was to examine the effects of different doses of a drug on the pulse rates of human subjects. The model considered here is the simple linear regression model with dose level as the only independent variable.

a. Does it appear that the normality assumption has been violated?

b. Is there a random pattern to the Residuals vs. Fits plot?

c. What adjustments to the model should we consider?

Residual Model Diagnostics

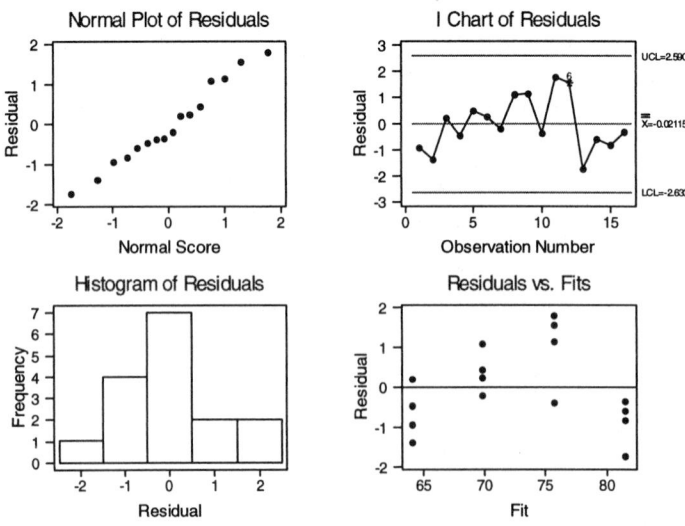

11.61 Here is a residual analysis of Exercise 11.46 in Section 11.4. We are reexamining the effects of drug dose on pulse rate (see Exercise 11.60). The model considered now is the second-degree polynomial model that contains both drug dose and the square of drug dose.

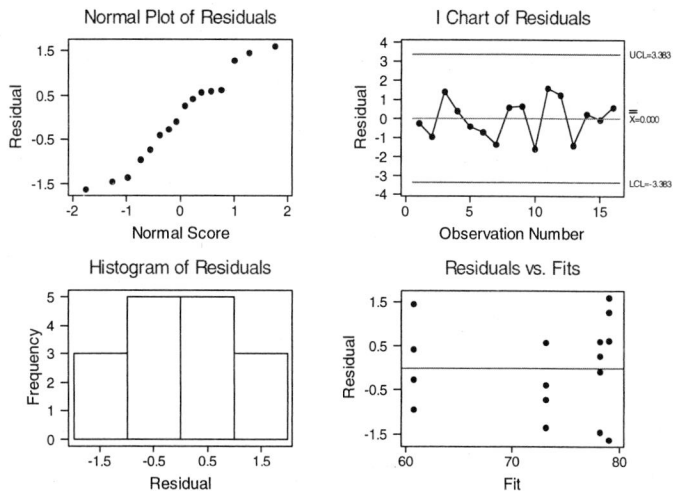

Residual Model Diagnostics

a. Does it appear that the normality assumption has been violated?
b. Is there a random pattern to the Residuals vs. Fits plot?
c. Does this appear to be an improvement on the simple linear regression model that contained only drug dose?

11.62 In Exercise 11.49 of Section 11.4, the tensile strength of Kraft paper was related to the percent of hardwood in the pulp that was used to produce the paper. A simple linear regression model and a second-degree polynomial model were considered. Reexamine the problem by constructing residual plots for both models (worksheet: **Hardwood.mtw**).

a. Is it clear that the simple linear regression model is inappropriate? What reasons can you give for its inadequacy?
b. Do the residuals in the polynomial model appear to have a random pattern? Is there reason to think that an alternative model should be considered?

11.63 Joglekar and colleagues (1989) considered the following windmill data that record the direct current (in volts) produced by given wind velocities (in miles per hour):

Worksheet: Windmill.mtw

Velocity	2.45	2.70	2.90	3.05	3.40	3.60	3.95	4.10	4.60
Output	.123	.500	.653	.558	1.057	1.137	1.144	1.194	1.562

Velocity	5.00	5.45	5.80	6.00	6.20	6.35	7.00	7.40	7.85
Output	1.582	1.501	1.737	1.822	1.866	1.930	1.800	2.088	2.179

(*continued*)

Velocity	8.15	8.80	9.10	9.55	9.70	10.00	10.20
Output	2.166	2.112	2.303	2.294	2.386	2.236	2.310

SOURCE: Joglekar et al., "Lack of Fit Testing When Replicates Are Not Available," *The American Statistician* 43, No. 3 (1989): 135–143.

```
Regression Analysis

The regression equation is
output = 0.131 + 0.241 velocity

Predictor        Coef       Stdev     t-ratio         p
Constant       0.1309      0.1260        1.04     0.310
velocity      0.24115     0.01905       12.66     0.000

s = 0.2361      R-sq = 87.4%       R-sq(adj) = 86.9%

Analysis of Variance

SOURCE        DF          SS           MS         F         p
Regression     1       8.9296       8.9296    160.26     0.000
Error         23       1.2816       0.0557
Total         24      10.2112

Unusual Observations
Obs. velocity    output     Fit   Stdev.Fit   Residual  St.Resid
   1       2.5    0.1230   0.7217     0.0845    -0.5987     -2.72R

R denotes an obs. with a large st. resid.
```

The simple linear model seems adequate, with $R^2 = 87.4\%$ and $F = 160.26$, which is highly significant. The residual plot, however, suggests that there is still a distinct pattern in the data. Perform a reciprocal transformation on the wind velocity data and recompute the regression analysis.

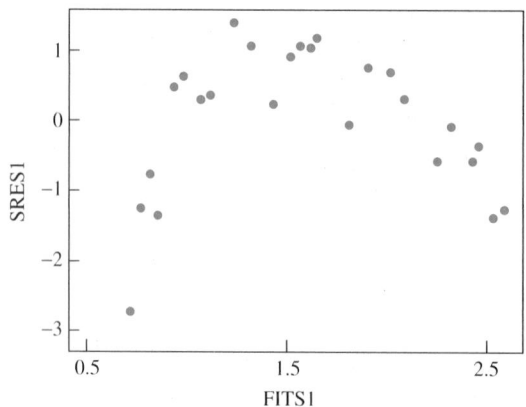

a. What is the new R^2? Was there improvement?

b. What is the new F statistic? Has it improved?

c. Create a new residual plot. Are the residuals now random?

11.64   *Consumer Reports* (October 1994) rated these toaster ovens:

Worksheet: Toaster.mtw

Toaster Brand	Rating	Cost
Black&D T660G	85	85
Toastmaster 336V	77	50
DeLonghi XU20L	75	130
Proctor-Silex 03030	75	48
Black&D SO2500G	75	92
Toastmaster 342	72	60
Munsey M88	70	56
Sears Kenmore 48216	70	70
Proctor-Silex 03010	70	41
Panasonic NT855U	68	70
DeLonghi XU14	65	69
Black&D TRO510	63	55
Black&D TRO400	61	50
Hamilton Beach 336	59	40
Toastmaster 319V	77	39
Black&D TRO200	60	40
Proctor-Silex 03008	59	35

a. Is there a significant linear relationship between the rating given by *Consumer Reports* and the cost of the toaster?

b. What is R^2? Is its value acceptable or should the model be revised in an effort to increase R^2?

c. Conduct a residual analysis of the model. Does the pattern of residuals seem random? Would a transformation on either x or y help?

11.65   In Example 11.14, we observed a single outlier that produced an unusually large residual. The resistant line provided a better fit to the data than did the regression line. Another alternative is to evaluate the regression line both with and without the outlier. Retrieve the data (worksheet: **RandD.mtw**) into the computer.

a. Perform a regression analysis with all the data. What is the regression equation? Is the regression coefficient significantly different from zero? What is the p-value?

b. What is the value of R^2? What percent of the variability in sales is explained by R&D expenditures?

c. What p-value is associated with the F statistic? Why is it the same as the p-value associated with the t statistic?

d. Are there any unusual observations? What are they and why are they identified as unusual?

e. Remove the one outlier and repeat parts **a–d**.

f. With the outlier removed, is the p-value for the t statistic and the F statistic more significant? Did R^2 improve? Are there any unusual observations?

g. The equation for the resistant line was $y = 71.82 - 12.656x$. Compare both regression lines with the resistant line.

11.66   Of interest to geologists in eastern Louisiana is the moisture content of marine muds that accumulate in small inlets on the Gulf Coast. The following measurements of the moisture contents of core samples were obtained by comparing the weight of a sample immediately after its removal from the core barrel with its weight after forced drying. The moisture content is expressed as grams of water per 100 grams of dried sediment. We wish to relate the moisture content to the depth of the core sample.

Worksheet: Moisture.mtw

Depth	0	5	10	15	20	25	30	35
Moisture	124	78	54	35	30	21	22	18
Depth	0	5	10	15	20	25	30	35
Moisture	137	84	50	32	28	24	23	20

SOURCE: J. C. Davis, *Statistics and Data Analysis in Geology*, 2nd ed. (New York: Wiley, 1986), pp. 177, 185.

a. Here is the correlation between depth and moisture content:

```
Correlations (Pearson)

Correlation of depth and moisture = -0.870
```

Does this correlation indicate that a straight line will fit the data?

b. Examine the scatterplot. Does it appear that a straight line fits the data?

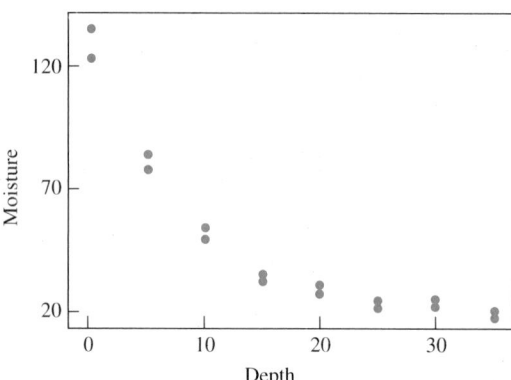

c. Examine the regression printout. What is the regression equation? Do the value of R^2 and the F statistic indicate that the equation fits the data?

```
Regression Analysis

The regression equation is
moisture = 97.3 - 2.78 depth

Predictor        Coef       Stdev      t-ratio          p
Constant       97.333       8.778        11.09      0.000
depth         -2.7762      0.4197        -6.62      0.000
```

(continued)

```
s = 19.23        R-sq = 75.8%        R-sq(adj) = 74.0%
```

Analysis of Variance

SOURCE	DF	SS	MS	F	p
Regression	1	16185	16185	43.76	0.000
Error	14	5178	370		
Total	15	21363			

Unusual Observations

Obs.	depth	moisture	Fit	Stdev.Fit	Residual	St.Resid
9	0.0	137.00	97.33	8.78	39.67	2.32R

R denotes an obs. with a large st. resid.

d. Examine this residual plot associated with the regression equation in part **c**. What does the residual plot tell about the simple linear regression equation? Should the equation be revised? In what way?

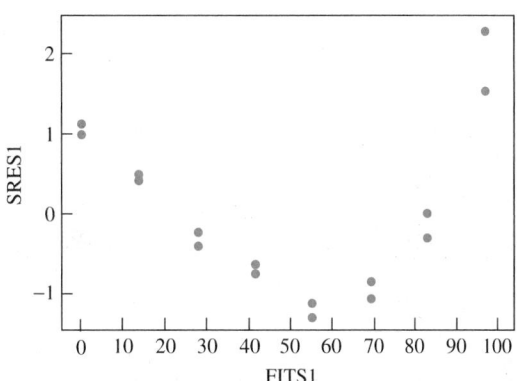

e. Modify the regression equation in part **c** to a logarithm model as in Example 11.17. Perform the regression analysis and conduct a residual analysis. Check R^2, t-ratios, and the F statistic, and explain any improvements to the model. What does the residual analysis tell you about the logarithm model?

f. Modify the regression equation in part **c** to a second-degree polynomial model. Perform the regression analysis and conduct a residual analysis. Check R^2, t-ratios, and the F statistic, and explain any improvements to the model. What does the residual analysis tell about the second-degree polynomial model?

g. Which model, the logarithm or the second-degree polynomial model, best fits the data?

11.6   SUMMARY AND REVIEW

✓   Regression is the study of the relationship between a *dependent variable y* and the *independent variables* $x_1, x_2, \ldots, x_k$. The regression equation is of the form

$$y = \beta_0 + \beta_1 x_1 + \beta_2 x_2 + \cdots + \beta_k x_k + \varepsilon$$

✓   If $k = 1$, the equation is called a *simple linear regression equation*. From the data (x_1, y_1), $(x_2, y_2), \ldots, (x_n, y_n)$, the *least squares estimates* of β_0 and β_1 are found.

✓   From the least squares equation, the *predicted value of y* is found:

$$\hat{y}_i = b_0 + b_1 x_i$$

where b_0 and b_1 are the least squares estimates.

✓   The *residual* associated with the data point (x_i, y_i) is $e_i = y_i - \hat{y}_i$. All the residuals are squared and summed to get SSE, the *sum of squares due to error*.

✓   The *coefficient of determination* is the square of the correlation coefficient and gives the percent of the variability in the dependent variable that is explained by the independent variable.

✓   A test of hypothesis about the regression coefficient, β_1, can be performed. The test statistic for testing $H_0: \beta_1 = 0$ is given by

$$t = \frac{b_1}{s/\sqrt{\text{SS}_x}}$$

and is distributed as a t distribution with $n - 2$ degrees of freedom when the basic assumptions are satisfied.

✓   Confidence intervals for β_1 can be constructed from the formula

$$b_1 \pm t^* \frac{s}{\sqrt{\text{SS}_x}}$$

✓   Confidence intervals for the expected response, μ_y, are found with the formula

$$b_0 + b_1 x' \pm t^* s \sqrt{\frac{1}{n} + \frac{(x' - \bar{x})^2}{\text{SS}_x}}$$

✓   A prediction interval for a new y is found with the formula

$$b_0 + b_1 x' \pm t^* s \sqrt{1 + \frac{1}{n} + \frac{(x' - \bar{x})^2}{\text{SS}_x}}$$

✓   *Multiple regression* is the study of the relationship between the dependent variable and several independent variables. The computations are best handled with a computer.

✓   An *analysis of the residuals* makes it possible to check out the assumptions that are made in a regression analysis. It also can point out possible deficiencies in the regression model and suggest alternatives.

LEARNING GOALS   Having completed this chapter, you should be able to

1. Identify the independent and dependent variables in a regression problem. *Section 11.1*
2. Construct a scatterplot and comment on the relationship between the variables. *Section 11.1*
3. Formulate the least squares equation that relates the independent and dependent variables. *Section 11.1*
4. Calculate the predicted values of y associated with the x values. *Section 11.1*
5. Calculate the sum of squares due to error. *Section 11.1*
6. Test hypotheses about the regression coefficient. *Section 11.2*
7. Compute a confidence interval for the regression coefficient. *Section 11.2*
8. Compute a confidence interval for the expected response μ_y. *Section 11.3*
9. Compute a prediction interval for a new value of y. *Section 11.3*
10. Analyze a multiple regression problem with a computer program. *Section 11.4*
11. Perform an analysis of residuals. *Section 11.5*

QUESTIONS
FOR REVIEW   Use the following problems to test your skills:

11.67   Identify the independent and dependent variables in each case:
   a. A study of the relationship between robbery rates and population density
   b. A study of the relationship between attitude scores and academic achievement scores
   c. A study of the relationship between the growth rate of rainbow trout and the number of fish per cubic yard of water
   d. A study of the relationship between expenditures per student and teachers' salaries

11.68   Graph the linear equations:
   a. $y = 3.1 + 4.7x$ b. $y = -7.3 + 5.5x$ c. $y = -8.3 - 2.1x$

11.69   Identify the slope and y-intercept in each equation:
   a. $2x + 3y = 6$ b. $-3.1y + 7.4x = 12$ c. $-5.6x - 4.1y = 10$

11.70   Graph the equations in Exercise 11.69.

11.71   Construct a scatterplot for each data set and comment on the relationship between x and y:

a. x	100	110	120	130	140	150				
y	7.8	6.1	5.4	5.2	4.7	3.5				
b. x	2.3	3.4	1.6	6.4	4.2	3.1	5.6	4.9		
y	.65	.82	.47	1.23	.92	.74	1.08	1.01		
c. x	41	52	37	26	45	32	49	55	22	30
y	1.2	2.8	.6	-1.7	2.9	-1.1	3.1	2.7	-2.4	-1.3

11.72   Find the least squares estimates of β_0 and β_1 in the linear model $y = \beta_0 + \beta_1 x_1 + \varepsilon$ for the data in Exercise 11.71.

11.73   For the data in Exercise 11.71 and the solutions found in Exercise 11.72, find the predicted y for the given values of x.
   a. $x = 100$ and $x = 160$ b. $x = 1.5$ and $x = 5.0$ c. $x = 30$ and $x = 60$

11.74 Using the solution to Exercise 11.72, find the residuals associated with the given data points:
a. $(120, 5.4)$ and $(150, 3.5)$ b. $(1.6, .47)$ and $(3.1, .74)$ c. $(52, 2.8)$ and $(32, -1.1)$

11.75 Graph the equation $y = -7.8 + 6.3x$. Find the residuals for $(5, 18)$ and $(5, 20)$.

11.76 Identify the independent variable and the dependent variable in a study that matches the time necessary for a subject to react with the number of alternatives he or she is given to react to.

11.77 Given the accompanying data, construct a scatterplot and comment on the general relationship between x and y.

x	2.5	3.4	4.7	5.2	6.8	7.6
y	10.3	14.2	17.5	22.6	24.8	29.0

11.78 Given the least squares equation $\hat{y} = 3.5 - 6.8x$, find the predicted y when $x = 3.5$. What is the residual for $(3.5, -20)$?

11.79 A proposed model that relates the failure time and the operating temperature of a piece of electronic equipment is $y = \beta_0 + \beta_1 x + \varepsilon$, with $\beta_0 = 200$, $\beta_1 = -.35$, and $\sigma = 1.5$.
a. What is the expected mean failure time when the operating temperature is 45?
b. When the operating temperature is 45, use the empirical rule to give an interval that would contain the mean failure time with 95% probability.
c. If the operating temperature changes by one unit, how much do you expect the failure time to change?

11.80 A manufacturer of a new insulation medium recorded the heat loss (in BTUs) through the insulation as the outside temperature (in degrees centigrade) dropped:

Worksheet: Insulate.mtw

Outside temp	-10	-10	0	0	10	10	20	20	30	30
Heat loss	96	91	84	82	68	75	49	51	28	24

a. Construct a scatterplot and find the least squares estimates of β_0 and β_1 in the model $y = \beta_0 + \beta_1 x_1 + \varepsilon$.
b. For each degree increase of outside temperature, how much does the heat loss change?
c. Calculate SSE and the estimated standard deviation.

11.81 In Exercise 11.80, find the predicted heat loss when the outside temperature is 15° C. Give a prediction interval for y.

11.82 Find a 95% confidence interval for β_1 in Exercise 11.80. Is 0 inside the interval? If it is, what does that mean about the relationship between the two variables?

11.83 Examine the scatterplot of these data:

Worksheet: IqGpa.mtw

IQ (x)	115	132	125	120	119	132	105	114	106	139	127	118
GPA(y)	2.2	3.3	3.0	2.6	2.9	3.5	2.2	2.7	3.7	1.8	3.7	2.4

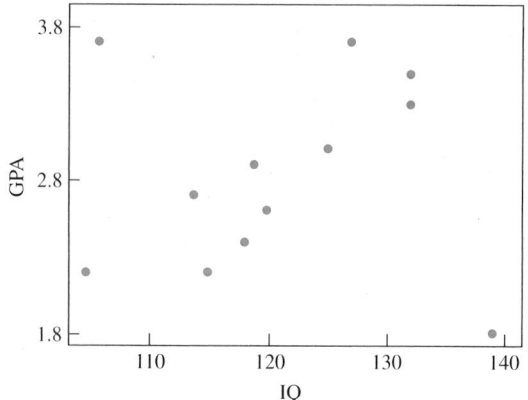

Would you recommend a resistant line or a regression line to fit the data? Explain your answer.

COMPUTER **11.84**
EXERCISES

11.84 To investigate the relationship between the numbers of books read during the term by third-graders and their final spelling scores, an educator collected the following data on 17 randomly selected students:

Worksheet: Books.mtw

Books	27	11	32	5	17	0	8	15	24
Score	85	81	98	61	92	36	59	84	90

Books	6	4	23	41	7	2	19	13
Score	70	72	95	99	78	58	87	80

Conduct a simple linear regression analysis of the data. Examine a scatterplot. Construct a residual plot based on the simple linear regression equation.
   a. Does the scatterplot show a straight line pattern?
   b. Is the regression coefficient significantly different from 0? What is the p-value? Did you do a t-test or an F-test? What is the difference between the two?
   c. Does the residual plot show a random pattern, or do you think the model needs to be revised? If so, what revision do you recommend?

11.85 Use the solution to Exercise 11.84 to find the predicted spelling scores for students who have read 10, 15, 20, 25, and 30 books. Obtain 95% confidence intervals and prediction intervals. Graph the confidence band and the prediction band on the scatterplot.

11.86 Fit a second-degree polynomial model to the data in Exercise 11.84. How much did R^2 improve? Interpret the t-ratios and the F statistic. Is this model an improvement on the simple linear regression model?

11.87 How much confidence do we have in the press to report the facts accurately? One study suggests that one's confidence level is related to the amount of education one has. Listed are the educational levels of 20 randomly selected persons and their degrees of confidence in the press (a score from 0 to 100: the higher the score, the more confidence):

Worksheet: Press.mtw

Education (years), x	12	12	14	8	10	12	11	12	16	14
Confidence, y	28	36	22	58	41	32	30	62	14	21
Education	8	12	15	12	9	12	10	14	12	16
Confidence	42	48	25	31	40	42	57	28	16	18

a. Construct a scatterplot and find the least squares estimates of β_0 and β_1 in the model $y = \beta_0 + \beta_1 x_1 + \varepsilon$.
b. Calculate SSE and the estimated standard deviation.
c. Test the hypothesis $\beta_1 = 0$.
d. Calculate R^2. How much of the variability in confidence is explained by education?
e. Based upon your responses, do you think the model in part **a** adequately describes the relationship between the two variables?

11.88 From Exercise 11.87, find the predicted degree of confidence for someone who has had 12 years of education. Repeat for someone who has had 16 years of education.

11.89 Find the residuals associated with the least squares equation from Exercise 11.87. Plot the residuals and comment on the results.

11.90 The owner of a department store decided to investigate the relationship between the amount lost due to shoplifting and the number of customers in the store. Unable to get a count of the number of shoppers, he examined the sales receipts:

Worksheet: Shoplift.mtw

Sales receipts ($1,000)	8.4	7.1	9.3	12.3	10.8	8.1	6.5	7.8
Shoplifting loss ($100)	16.2	12.3	19.8	18.4	14.6	15.8	11.4	13.1

a. Construct a scatterplot and find the least squares estimates of β_0 and β_1 in the model $y = \beta_0 + \beta_1 x_1 + \varepsilon$. Evaluate the model.
b. Find the predicted loss due to shoplifting if the sales receipts are $10,000.
c. Find the residuals associated with the least squares equation found in part **a**. Plot the residuals and comment on the results.
d. Test the hypothesis $\beta_1 = 0$.

11.91 An owner of a large retail store thinks that the monthly gross sales figures of her employees are related to their length of employment. Listed here are July gross sales figures and the number of months of employment for ten employees:

Worksheet: Retail.mtw

Months	10	22	8	16	31	2	13	36	18	6
Sales	3,860	4,230	2,650	5,170	4,970	4,780	3,120	4,690	4,920	2,150

Complete a simple linear regression analysis of the data.
a. What are the least squares estimates of β_0 and β_1 in the simple linear regression model?
b. What is the coefficient of determination?
c. Test the hypothesis $\beta_1 = 0$. Is there statistical evidence to reject the hypothesis? What is the p-value?
d. Perform a residual analysis. Does the residual plot indicate any violations of assumptions?

11.92   In "An Evaluation of Teacher Stress and Job Satisfaction" (*Education*, 105 No. 2), G. W. Sutton et al. reports a negative correlation between stress level and job satisfaction, suggesting a tendency for teachers to report a higher level of job satisfaction when their stress levels are low. (WSPT is the Wilson Stress Profile for Teachers.)

Worksheet: Jobsat.mtw

WSPT	90	78	85	65	94	82	96	79	80
Job satisfaction	3.6	5.3	4.7	8.9	3.2	4.0	3.8	6.2	6.5

a. Construct a scatterplot and find the least squares estimates of β_0 and β_1 in the model $y = \beta_0 + \beta_1 x_1 + \varepsilon$.
b. Test the hypothesis $\beta_1 = 0$.
c. What is the value R^2? How much of the variability in stress level is explained by job satisfaction? Based on R^2, do you think the proposed model adequately describes the relationship between the two variables?

11.93   A report recently stated that the prices of oranges were expected to rise sharply due to low projected harvests. Here are the average prices California growers charged for a 75-pound box of navel oranges and the size of the harvest (in millions of boxes) for six consecutive years.

Worksheet: Orange.mtw

Harvest	72	69	58	70	65	54
Price	$5.40	6.10	9.30	6.50	7.20	13.40

a. Construct a scatterplot and find the least squares estimates of β_0 and β_1 in the model $y = \beta_0 + \beta_1 x_1 + \varepsilon$.
b. Test the hypothesis $\beta_1 = 0$.
c. Calculate R^2. How much of the variability in price is explained by the size of the harvest?
d. Find the residuals associated with the least squares equation. Plot the residuals and comment on the results. Do you recommend revising the model?
e. Using the model found in part **a**, find the predicted price for a box of oranges if the projected harvest is 60 million boxes; 30 million boxes. Which prediction do you think is more reliable? Why?

11.94   The *IAAF/ATFS Track and Field Statistics Handbook* for the 1984 Los Angeles Olympics gives the national records for women in several races (worksheet: **Track.mtw**) (Dawkins, 1989).
a. From the data set, create a correlation matrix of all races. Are there any strong correlations?
b. Using the 100-meter, 200-meter, and 400-meter races, produce all possible scatterplots of pairs of columns. Does it appear that a straight line would adequately fit any two variables?
c. Would you normally expect a straight line relationship between the times for the 100-meter and 200-meter records? Explain.

11.95   The data are the numbers of adults on probation, in jail or prison, or on parole from 1980 through 1993.

Worksheet: Prison.mtw

year	total	probation	jail	prison	parole
1980	1,840,400	1,118,097	182,288	319,598	220,438
1981	2,006,600	1,225,934	195,085	360,029	225,539
1982	2,192,600	1,357,264	207,853	402,914	224,604
1983	2,475,100	1,582,947	221,815	423,898	246,440
1984	2,689,200	1,740,948	233,018	448,264	266,992
1985	3,011,500	1,968,712	254,986	487,583	300,203

(*continued*)

year	total	probation	jail	prison	parole
1986	3,239,400	2,114,621	272,735	526,436	325,638
1987	3,459,600	2,247,158	294,092	562,814	355,505
1988	3,714,100	2,356,483	341,893	607,766	407,977
1989	4,055,600	2,522,125	393,303	683,367	456,803
1990	4,348,000	2,670,234	403,019	743,382	531,407
1991	4,536,200	2,729,322	424,129	792,535	590,198
1992	4,763,200	2,811,611	441,781	851,205	658,601
1993	4,879,600	2,843,445	455,500	909,185	671,470

SOURCE: Bureau of Justice Statistics, *Correctional Populations in the United States*, NCJ-153849, April 1995.

a. Produce a correlation matrix of all variables. Are any of the variables correlated with the variable year?

b. Produce a scatterplot of the total number of prisoners versus year. Does it appear that a straight line will fit the data? Regress the total number of prisoners on year. What percent of the variability in the number of prisoners is explained by the year? Is there any reason to add additional variables to the model?

c. Create a residual plot. Does the pattern of residuals seem random?

d. Regress the number on parole on year. What percent of the variability is explained by year?

e. Produce a residual plot. Does the pattern of residuals seem random? Is there any reason to modify the model?

11.96   In Exercises 11.7, 11.19, and 11.48, a linear model was proposed to relate the ratings of top offensive linemen for the NFL draft to their times to run the 40-yard dash. We learned that the rating was not linearly related to the race time ($R^2 = 7.2\%$, $F = 2.10$ with p-value $= .159$) and found that if we added the additional variable, weight of the player, then the model improved significantly. Worksheet: **NFLdraf2.mtw** contains similar data on the top defensive linemen in the 1994 draft. Analyze these data in the same way; that is, determine whether a simple linear regression equation adequately relates the ratings to the 40-yard dash times.

a. What is R^2 with only one variable in the model?

b. Is the regression coefficient significantly different from 0?

c. If the model needs improvement, then try adding the weight of the player. What is R^2 for the multiple regression model with 40-yard dash time and weight in the model?

d. Are the regression coefficients significantly different from 0? What about the F-test? What is the p-value for each test?

11.97   The Galápagos Islands, a territory of the Republic of Ecuador, has much information related to the development and survival of different species. Hamilton et al. (1963) and Johnson and Raven (1973) studied the dependence of the number of plant species on the area of each of the Galápagos Islands. The variables in the table are (in order): island, observed number of species, observed native species, area of island, elevation of island, distance from nearest island, distance from Santa Cruz, and area of the adjacent island. We will not attempt to use all the variables in our analysis, but the data are listed so that you may explore further if you so desire.

Worksheet: Galapago.mtw

Island	species	native	area, km^2	elevat, m	dist1, km	distSC, km	areaadj, km^2
Baltra	58	23	25.09	*	.6	.6	1.84
Bartolome	31	21	1.24	109	.6	26.3	572.33
Caldwell	3	3	.21	114	2.8	58.7	.78

(continued)

Island	species	native	area, km²	elevat, m	dist1, km	distSC, km	areaadj, km²
Champion	25	9	.10	46	1.9	47.4	.18
Coamano	2	1	.05	*	1.9	1.9	903.82
Daphne Major	18	11	.34	119	8.0	8.0	1.84
Daphne Minor	24	*	.08	93	6.0	12.0	.34
Darwin	10	7	2.33	168	34.1	290.2	2.85
Eden	8	4	.03	*	.4	.4	17.95
Enderby	2	2	.18	112	2.6	50.2	.10
Espanola	97	26	58.27	198	1.1	88.3	.57
Fernandina	93	35	634.49	1,494	4.3	95.3	4,669.32
Gardner	58	17	.57	49	1.1	93.1	58.27
Gardner2	5	4	.78	227	4.6	62.2	.21
Genovesa	40	19	17.35	76	47.4	92.2	129.49
Isabela	347	89	4,669.32	1,707	.7	28.1	634.49
Narchena	51	23	129.49	343	29.1	85.9	59.56
Onslow	2	2	.01	25	3.3	45.9	.10
Pinta	104	37	59.56	777	29.1	119.6	129.49
Pinzon	108	33	17.95	458	10.7	10.7	.03
Las Plazas	12	9	.23	*	.5	.6	25.09
Rabida	70	30	4.89	367	4.4	24.4	572.33
San Cristobal	280	65	551.62	716	45.2	66.6	.57
San Salvador	237	81	572.33	906	.2	19.8	4.89
Santa Cruz	444	95	903.82	864	.6	.0	.52
Santa Fe	62	28	24.08	259	16.5	16.5	.52
Santa Maria	285	73	170.92	640	2.6	49.2	.10
Seymour	44	16	1.84	*	.6	9.6	25.09
Tortuga	16	8	1.24	186	6.8	50.9	17.95
Wolf	21	12	2.85	253	34.1	254.7	2.33

SOURCE: D. F. Andrews and A. M. Herzberg, *Data: A Collection of Problems from Many Fields for the Student and Research Worker*, *Springer Series in Statistics* (New York: Springer-Verlag, 1985).

a. Construct a scatterplot of the observed number of species and the area of the island.

b. Perform a simple linear regression analysis to see whether the observed number of species is linearly related to the area of the island. What is R^2? What is the F statistic? What is its p-value? Should we try another model?

c. The scatterplot should suggest that we need to apply some sort of transformation to the data. List two transformations that would apply to this problem.

d. Create a new variable that is the natural log of area. Complete the regression analysis using this variable as the predictor. What are your results? Is this an improvement over the previous model?

e. Create a new variable that is the square root of area. Complete the regression analysis using this variable as the predictor. What are your results? Is this an improvement over the model in part **d**?

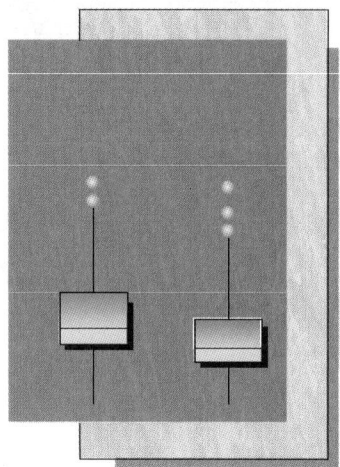

12

ANALYSIS OF VARIANCE

This chapter extends the methods of Chapters 8 and 9 to the comparison of more than two population distributions. It is one of the most widely used statistical procedures available.

The approach is somewhat different from the one-sample and two-sample problems, however, because we introduce the idea of comparing between-sample and within-sample variability. For example, subjects treated for some ailment with different types of drugs have different recovery rates, but is the *variability* in their recovery rates due to the different drugs or is it simply because they are different people? Analysis of variance is a procedure that attempts to determine how much of the variability is due to the treatment and how much is due to all other factors.

This chapter on analysis of variance is an elementary introduction to the vast field of experimental design. A second course in statistical methods begins where this chapter ends.

CONTENTS

STATISTICAL INSIGHT

THE SUPERCAR OLYMPICS

What country makes the fastest car? *Car and Driver* (July 1995) conducted tests of five supercars from five different countries: the United States (Dodge Viper RT/10), Germany (Porsche 911 Turbo), Great Britain (Lotus Esprit S4S), Italy (Ferrari F355), and Japan (Acura NSX-T).

Among the tests conducted by *Car and Driver* is a comparison of the top speeds achieved by the cars using as much distance as necessary and without exceeding the engine's redline. From six runs, three in each direction to cancel any grade or wind factors at the test facility, can we determine whether there is a difference in the top speeds attained by these five supercars? The table gives the speeds in miles per hour.

Worksheet: Supercar.mtw

Acura (1)	Ferrari (2)	Lotus (3)	Porsche (4)	Viper (5)
159.7	179.6	167.4	173.5	172.3
161.5	173.9	163.0	182.4	168.9
163.7	180.2	160.3	171.3	169.5
166.0	183.9	164.9	175.7	174.6
157.7	176.7	160.5	179.1	161.1
161.7	178.4	158.3	175.0	164.2

Our natural reaction is to look at the average of the six runs for each of the five cars. This is a good approach, but remember that these are only sample averages and they can vary about their respective population means. The problem is to determine whether the observed difference in the sample means is statistically significant and thus leads to the conclusion that the corresponding population means are different. In this chapter, we investigate the equality of several population means with a procedure called *analysis of variance*. See Exercise 12.75 in Section 12.5 for a continuation of the supercar example.

12.1   INTRODUCTION

In Chapter 9, we presented the methods for comparing two population means. There are situations, however, as in the Statistical Insight example, where we wish to compare more than two means. A college administrator wishes to compare the mean grade point averages of freshmen, sophomores, juniors, and seniors. A building contractor wants to compare the effectiveness of five different types of insulation. A physician wishes to compare the effectiveness of three different drugs that she is using to treat patients. An agricultural experiment station compares the yields of crops treated with four different types of fertilizer. There is an unending list of experiments in which we wish to compare more than two means.

Having learned the two-sample *t*-test in Chapter 9, it is quite natural to suggest that we simply compare all possible pairs of means when we wish to evaluate the equality of several population means. Certainly with a computer and the different samples stored in the columns of a worksheet, it is very easy to execute a *t*-test on any two columns of data. There is a basic flaw in this procedure, however. When we conduct many *t*-tests, it is difficult to determine the overall risk of a Type I error. Moreover, the more tests that you conduct, the greater is the risk of making a Type I error on at least one of the tests. When we compare five population means with a 5% level of significance, for example, the overall risk of at least one Type I error can exceed 40%. *Analysis of variance* provides us with the tools to compare the means of several populations with a single test where the overall risk of a Type I error is controlled.

As always, any numerical analysis of data should be preceded by exploratory graphs. Graphs such as dotplots and boxplots help us formulate good questions and interpret the information contained in the samples. For example, consider Figure 12.1, side-by-side boxplots of the data comparing the top speeds of the supercars. It appears that the Ferrari and the Porsche (cars 2 and 4) are superior to the other three cars. But is the Ferrari better than the Porsche? Is there a significant difference between the Acura and the Lotus (cars 1 and 3)? What about the Dodge Viper (car 5)?

FIGURE 12.1

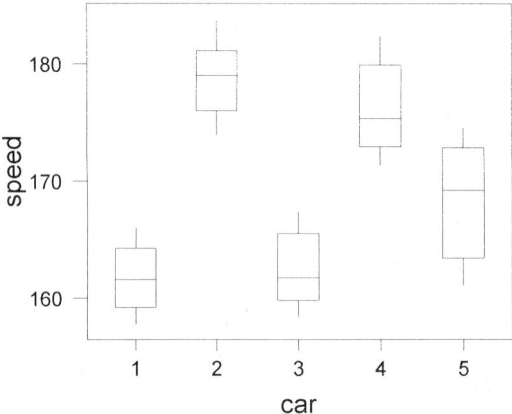

The side-by-side boxplots in Figure 12.1 show differences, but are the observed differences statistically significant? The role of analysis of variance is to perform a numerical test of significance that will test the equality of all the means. That is, the null hypothesis states that there is no difference in the top speeds of all five cars. If this null hypothesis is rejected, then we will attempt to determine where the differences actually exist. For example, if there is a difference in the means, then is there a significant difference between the Ferrari and the Porsche or between the Acura and the Lotus? You are asked to complete an analysis of these data in Exercise 12.75 in Section 12.5.

As in previous inference procedures, the numerical analysis performed depends on the characteristics of the underlying parent distributions. When it can be assumed that the parent distributions are normally distributed with homogeneous variances, the standard *F* procedure (presented in Section 12.2) is performed. Multiple comparison procedures that

Recall that homogeneous variances mean that the variances of the different populations are close in value.

are used to further analyze differences are given in Section 12.3. If the assumptions for the F procedure cannot be made about the parent distributions, we have the nonparametric Kruskal–Wallis test presented in Section 12.4. This procedure is recommended when the distributions are similar in shape but possibly have different centers.

12.2   COMPARISON OF SEVERAL MEANS: THE ONE-WAY ANALYSIS OF VARIANCE

One-Way ANOVA Design

The statistical design of comparing several population means using independent random samples is called *one-way analysis of variance* (ANOVA). The design, which describes the procedure for selecting sample data, is given in Table 12.1. In the array, $y_{i,j}$ denotes the response measurement obtained from experimental unit (subject) i in sample j, where $i = 1, 2, \ldots, n_j$ and $j = 1, 2, \ldots, k$. Note that there can be different sample sizes for the various samples. The grand sample size is $N = n_1 + n_2 + n_3 + \cdots + n_k$. The grand mean is found by dividing the grand total by the grand sample size, that is, $\overline{y}_G = T/N$.

TABLE 12.1

○

The grand mean is the average of the individual averages *only* when all sample sizes are equal.

	Sample 1	Sample 2	$\cdots$	Sample k	
	$y_{1,1}$	$y_{1,2}$		$y_{1,k}$	
	$y_{2,1}$	$y_{2,2}$		$y_{2,k}$	
	$y_{3,1}$	$y_{3,2}$		$y_{3,k}$	
	$\vdots$	$\vdots$		$\vdots$	
	$y_{n1,1}$	$y_{n2,2}$	$\cdots$	$y_{nk,k}$	
Total	T_1	T_2		T_k	Grand total $= T$
Average	$\overline{y}_1$	$\overline{y}_2$		$\overline{y}_k$	Grand mean $= \overline{y}_G$
Variance	s_1^2	s_2^2		s_k^2	

We will illustrate one-way ANOVA with a simple example. Suppose we wish to compare the average gas mileages of standard four-wheel drive pickup trucks manufactured by Chevrolet, Dodge, and Ford. An experiment is designed in which five vehicles of each type are randomly and independently selected from the population of four-wheel drive trucks. (Choosing the samples randomly and independently satisfies the conditions of a one-way ANOVA design.) Each vehicle is driven in a stationary position for the equivalent of 500 miles. The gasoline consumed is measured and the miles per gallon computed. The results appear in Table 12.2 on page 786.

Are there any differences in the gas mileages of the three types of pickups? The sample means are certainly different (14.9, 14.4, and 14.5). We must remember, however, that they are sample means, which could be different estimates of a common population mean. It may be that the population mean miles per gallon for each of the three types of vehicles is 14.6, and we have three different sample means that estimate it. We would

TABLE 12.2

Worksheet: Trucks.mtw

	Chevy	Dodge	Ford	
	15.2	14.8	15.1	
	15.4	14.4	14.3	
	14.8	14.3	14.6	
	14.4	14.1	13.9	
	14.7	14.4	14.6	
Total	74.5	72.0	72.5	Grand total = 219.0
Average	14.9	14.4	14.5	Grand mean = 14.6
Variance	.16	.065	.195	

Because the sample sizes are equal, the grand mean is the average of the three averages. This is not the case if the sample sizes are unequal.

say that $\mu_1 = \mu_2 = \mu_3 = 14.6$, where μ_1, μ_2, and μ_3 represent the overall miles per gallon for each brand of trucks. On the other hand, there may indeed be a difference in the population means, and the sample means are reflecting those differences.

To get another view of the problem consider the side-by-side boxplots of the data in Figure 12.2. Notice that the entire boxplot for the Dodge trucks is contained within the middle 50% of the data (box portion of the boxplot) for the Ford trucks. This suggests that there is no difference between the miles per gallon for Dodge and Ford trucks. But what about the Chevrolet trucks? The median miles per gallon is at or above most of the data for Dodge trucks and approximately 75% of the data for Ford trucks. Does this mean that there is a significant difference between the mean miles per gallon for the different makes of trucks?

FIGURE 12.2

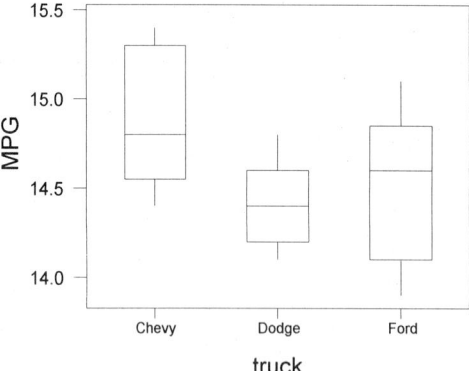

Basically, the question is this: Are the *observed* differences in the *sample* means different enough to conclude that the *population* means are different? The question can be answered by testing the null hypothesis

$$H_0: \mu_1 = \mu_2 = \mu_3$$

versus the alternative hypothesis

$$H_a: \text{At least two } \mu_i\text{'s differ}$$

As in all tests of hypotheses, an appropriate test statistic is needed to give a unique numerical measure of how much the sample means differ. To better understand the formulation of the test statistic, imagine for the moment that the miles per gallon for the vehicles are as in Table 12.3. Notice that the averages for the three samples are the same as in Table 12.2, but now there is no variability of the measurements within any of the samples. This is clearly demonstrated with the dotplots in Figure 12.3. There is a clear distinction between the samples and therefore a clear distinction between the population means. The data suggest that Chevrolets, in general, get 14.9 miles per gallon, Dodges get 14.4 miles per gallon, and Fords get 14.5 miles per gallon. These data lead to the rejection of H_0.

TABLE 12.3

	Chevy	Dodge	Ford	
	14.9	14.4	14.5	
	14.9	14.4	14.5	
	14.9	14.4	14.5	
	14.9	14.4	14.5	
	14.9	14.4	14.5	
Total	74.5	72.0	72.5	Grand total = 219.0
Average	14.9	14.4	14.5	Grand mean = 14.6
Variance	.0	.0	.0	

FIGURE 12.3   Character dotplot

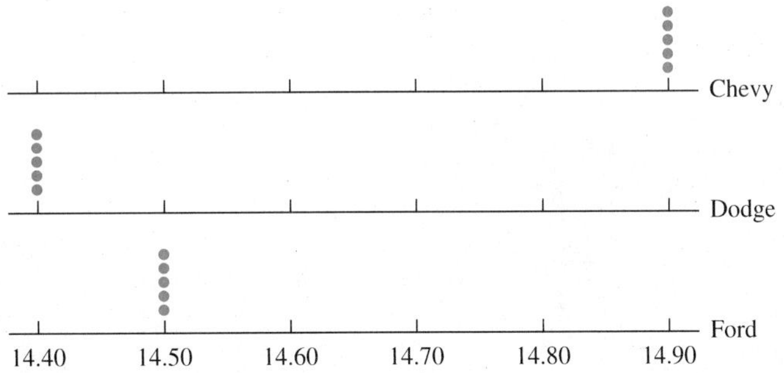

On the other hand, suppose now that the data are as in Table 12.4 (page 788). Again the averages are 14.9, 14.4, and 14.5, yet there is no clear distinction between the samples. From Figure 12.4, we see that any measurement in sample 2, for example, could easily have been from sample 1 or 3. It is as if all the data came from a population with a lot of variability and a single mean of 14.6 rather than three samples from distinct populations. There is insufficient information to reject H_0.

TABLE 12.4

	Chevy	Dodge	Ford	
	16.6	13.2	13.5	
	16.8	14.7	16.9	
	13.2	15.7	15.4	
	15.3	12.3	12.8	
	12.6	16.1	13.9	
Total	74.5	72.0	72.5	Grand total = 219.0
Average	14.9	14.4	14.5	Grand mean = 14.6
Variance	3.71	2.63	2.71	

FIGURE 12.4   Character dotplot

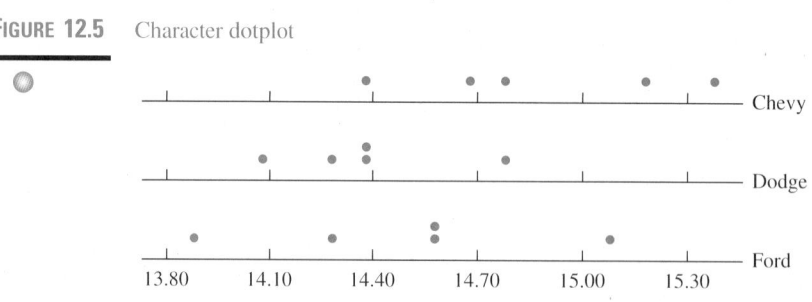

The data sets of Tables 12.3 and 12.4 illustrate *variability between the samples* and *variability within the samples*. In Table 12.3, the variation between the samples is large in comparison with the variation within the samples (in fact, as previously observed, there is no variation within the samples). Yet in Table 12.4, the variation between the samples is slight in comparison with the variation within the samples.

From the original data in Table 12.2, it is not clear (see Figure 12.5 or the boxplots in Figure 12.2) whether the variation between samples is statistically greater than the variation within the samples. We must investigate with a test of significance. The test statistic used to test the equality of the population means should compare the variation between the samples with the variation within the samples. First, however, we need to learn how to measure the variations between and within samples.

FIGURE 12.5   Character dotplot

Within-Sample Variability

If there is only one sample, then the sample variance, s^2, is a measure of the variability within that sample. In the case of multiple samples, a reasonable combined measure of variation within the samples is a pooling of all the individual sample variances. Indeed, if $s_1^2, s_2^2, \ldots, s_k^2$ represent the sample variances from k samples and $n_1, n_2, \ldots, n_k$ are the associated sample sizes, then

$$s_p^2 = \frac{(n_1 - 1)s_1^2 + (n_2 - 1)s_2^2 + \cdots + (n_k - 1)s_k^2}{n_1 + n_2 + \cdots + n_k - k}$$

is a pooled measure of the variability within the k samples. Notice that it is an extension of the pooled variance used in the pooled t-test in Chapter 9. The numerator of s_p^2 is called the *sum of squares for error* (SSE) because it is a combined measure of errors within each sample. The denominator is the degrees of freedom associated with s_p^2.

Recall that s^2 is obtained from n measurements and has $n - 1$ degrees of freedom. Because s_p^2 is obtained from k different s^2's, its degrees of freedom are

$$(n_1 - 1) + (n_2 - 1) + \cdots + (n_k - 1) = n_1 + n_2 + \cdots + n_k - k$$

Dividing the degrees of freedom into the sum of squares gives the variance, s_p^2, which is also called the *mean square error* (MSE).

The sum of squares for error is

$$\text{SSE} = (n_1 - 1)s_1^2 + (n_2 - 1)s_2^2 + \cdots + (n_k - 1)s_k^2$$
$$= \sum(y_{i,1} - \bar{y}_1)^2 + \sum(y_{i,2} - \bar{y}_2)^2 + \cdots + \sum(y_{i,k} - \bar{y}_k)^2$$

and has $n_1 + n_2 + \cdots + n_k - k$ degrees of freedom. Dividing the degrees of freedom into the sum of squares for error yields the **mean square for error**:

$$\text{MSE} = \frac{\text{SSE}}{\text{df}} = \frac{(n_1 - 1)s_1^2 + (n_2 - 1)s_2^2 + \cdots + (n_k - 1)s_k^2}{n_1 + n_2 + \cdots + n_k - k} = s_p^2$$

EXAMPLE 12.1

Calculate the sum of squares for error for the data in Table 12.2. Also give the degrees of freedom associated with the sum of squares for error and calculate the mean square for error.

SOLUTION

From Table 12.2, we find that

$$s_1^2 = .16, \quad s_2^2 = .065, \quad s_3^2 = .195$$

Each sample size is 5, so that $n_j - 1 = 4$ for $j = 1, 2, 3$. Therefore,

$$\text{SSE} = 4(.16) + 4(.065) + 4(.195) = 4(.42) = 1.68$$

The degrees of freedom are

$$n_1 + n_2 + \cdots + n_k - k = 5 + 5 + 5 - 3 = 12$$

Dividing SSE by 12 gives

$$\text{MSE} = \frac{\text{SSE}}{\text{df}} = \frac{1.68}{12} = .14$$

Thus, a pooled measure of the variability within the three samples is .14.

Between-Sample Variability

To measure the variability between the samples, we simply need to calculate the variation across the sample means. If $\bar{y}_1, \bar{y}_2, \ldots, \bar{y}_k$ are the sample means of the k samples and $\bar{y}_G$ is the overall sample mean, then $(\bar{y}_1 - \bar{y}_G), (\bar{y}_2 - \bar{y}_G), \ldots, (\bar{y}_k - \bar{y}_G)$ are the k deviations of the sample means from their grand mean. Summing the squared deviations and dividing by their degrees of freedom, $k - 1$, we obtain the between-sample variation:

$$s_b^2 = \frac{\sum n_j(\bar{y}_j - \bar{y}_G)^2}{k - 1}$$

Recall that each $\bar{y}_j$ came from a sample of size n_j; hence, each squared deviation is weighted by its corresponding sample size to get s_b^2.

The numerator

$$\sum n_j(\bar{y}_j - \bar{y}_G)^2$$

is called the between-sample sum of squares. It is more commonly known as the *sum of squares for treatments* (SST) because the various samples arise from the different treatments in the experiment. The denominator, $k - 1$, is the associated degrees of freedom.

The **sum of squares for treatments** is

$$\text{SST} = \sum n_j(\bar{y}_j - \bar{y}_G)^2$$

and has $k - 1$ degrees of freedom. Dividing the degrees of freedom into the sum of squares yields the **mean square for treatments**:

$$\text{MST} = \frac{\text{SST}}{\text{df}} = \frac{\sum n_j(\bar{y}_j - \bar{y}_G)^2}{k - 1} = s_b^2$$

EXAMPLE 12.2   Calculate the sum of squares for treatments for the data in Table 12.2. Also give the degrees of freedom associated with the sum of squares for treatments, and calculate the mean square for treatments.

SOLUTION

From the data, we have

$$\bar{y}_1 = 14.9, \quad \bar{y}_2 = 14.4, \quad \bar{y}_3 = 14.5, \quad \bar{y}_G = 14.6$$

Each $n_j = 5$, so

$$\text{SST} = 5(14.9 - 14.6)^2 + 5(14.4 - 14.6)^2 + 5(14.5 - 14.6)^2$$

$$= 5[.09 + .04 + .01] = .7$$

The degrees of freedom are $k - 1 = 3 - 1 = 2$. Dividing into SST, we have

$$\text{MST} = \frac{\text{SST}}{\text{df}} = \frac{.7}{2} = .35$$

which is a measure of the variability between the three samples. If the treatments in an experiment have different effects on the data, then MST will be large relative to MSE. Notice here that MST is more than twice as large as MSE.

The F-Test

In the one-way design, we compare several population means by testing

$$H_0: \mu_1 = \mu_2 = \cdots = \mu_k \quad \text{versus} \quad H_a: \text{At least two } \mu_j\text{'s differ}$$

As previously noted, the test statistic used to test H_0 should compare the relative magnitude of the between-sample variability with the within-sample variability. This is done by calculating the ratio

$$F = \frac{\text{MST}}{\text{MSE}}$$

If the k populations have homogeneous variances, that is, if

$$\sigma_1^2 = \sigma_2^2 = \cdots = \sigma_k^2$$

then MSE, being a pooled estimate of all the individual sample variances, is an unbiased estimator of the common variance. Additionally, if the null hypothesis is true, then MST is also an unbiased estimator of the common variance, resulting in a value of F very close to 1. When H_0 is not true, however, MST tends to overestimate the common variance because it includes between-sample variability. The result is a large F value. Thus, a significantly large F value indicates that the variability between the samples (MST) is *significantly* larger than the variability within the samples (MSE), which in turn indicates that the null hypothesis H_0 should be rejected.

The sampling distribution of F is not easily obtainable when we are sampling from arbitrary populations. If the null hypothesis is true, however, and if the samples are *independent* and from *normal populations* with *equal variances*, then the sampling distribution is an F distribution, which was introduced in Chapter 11.

Assumptions for the F Distribution

1. The samples are randomly and independently selected from their respective populations.
2. The sampled populations are normally distributed.
3. The variances of the sampled populations are equal; that is,

$$\sigma_1^2 = \sigma_2^2 = \cdots = \sigma_k^2$$

When these assumptions are satisfied and when the null hypothesis is true, the test statistic

$$F = \frac{\text{MST}}{\text{MSE}}$$

has an F distribution with $k - 1$ and $n_1 + n_2 + \cdots + n_k - k$ degrees of freedom in the numerator and denominator, respectively. Because the test is a comparison of two sources of variation, the procedure is called analysis of variance.

To interpret large F values and assess the statistical evidence against H_0, we need to determine the tail probability (p-value) of the F distribution. We refer to the degrees of freedom associated with the numerator (the degrees of freedom for the mean square for treatments) as df_n and the degrees of freedom associated with the denominator (the degrees of freedom for the mean square for error) as df_d. The upper-tail critical F values, F^*, for various values of df_n, df_d, and α are given in Table B.5 of Appendix B.

Analysis of Variance for the One-Way ANOVA

Assumptions: As stated earlier

$$H_0: \mu_1 = \mu_2 = \cdots = \mu_k \quad \text{versus} \quad H_a: \text{At least two } \mu\text{'s differ}$$

Test statistic:

$$F = \frac{\text{MST}}{\text{MSE}}$$

The test is a right-tailed test where the p-value is found in the F-table with $k - 1$ and $n_1 + n_2 + \cdots + n_k - k$ degrees of freedom. Unless you are using a computer, the exact p-value cannot be found with the tables, but bounds for it can be found by using the closest value to the observed value of the F statistic.

If a level of significance α is specified, then reject H_0 if p-value $< \alpha$.

Equivalence of F and t^2   If $k = 2$, notice that the assumptions for the F distribution are the same as the assumptions given for the pooled t-test in Section 9.4. In this case, it can be shown algebraically that the t statistic and the F statistic are related by the equation $t^2 = F$. In other words, the two-tailed pooled t-test is a special case of the F-test when $k = 2$.

Checking the basic assumptions for the F-test involves checking the normality of each of the k populations and checking that the variances of the k populations are homogeneous. The normality assumption is best checked with normal probability plots of each of the k samples. A simple way to check the equal variances assumption is to compare the smallest sample variance with the largest sample variance. As long as the largest s^2 is not more than 4 times greater than the smallest s^2, we can assume that the variances are homogeneous.

Checking Assumptions

1. Examine dotplots and boxplots for unusual behavior.
2. Check the normality assumption by constructing normal probability plots for each of the k samples.
3. Check the homogeneity of variances assumption by comparing the largest sample variance with the smallest sample variance. If the largest s^2 is no more than 4 times greater than the smallest s^2, assume that the assumption is satisfied. Equivalently, the largest standard deviation, s, should be no more than twice as large as the smallest standard deviation.

EXAMPLE 12.3

Returning to the earlier discussion of the gas mileage of pickup trucks, we wish to test the equality of the mean miles per gallon for the three major brands of trucks. Use the data in Table 12.2 to complete the test.

SOLUTION

Here, μ_1, μ_2, and μ_3 denote the mean miles per gallon for Chevrolet, Dodge, and Ford trucks, respectively. The null hypothesis says that they are the same; that is,

$$H_0: \mu_1 = \mu_2 = \mu_3 \quad \text{versus} \quad H_a: \text{At least two } \mu\text{'s differ}$$

Before proceeding with the F-test, we should check the assumptions. The normal probability plots of the three samples in Figure 12.6 (page 794) support the normality assumption. The smallest s^2 is .065, and the largest, $s^2 = .195$, is not more than 4 times as large. It is reasonable to assume that the variances are homogeneous.

In Example 12.1, we found that SSE = 1.68 with 12 degrees of freedom, so MSE = 1.68/12 = .14. In Example 12.2, we found that SST = .7 with 2 degrees of freedom, so MST = .7/2 = .35. Therefore, we have

$$F = \frac{\text{MST}}{\text{MSE}} = \frac{.35}{.14} = 2.5$$

FIGURE 12.6

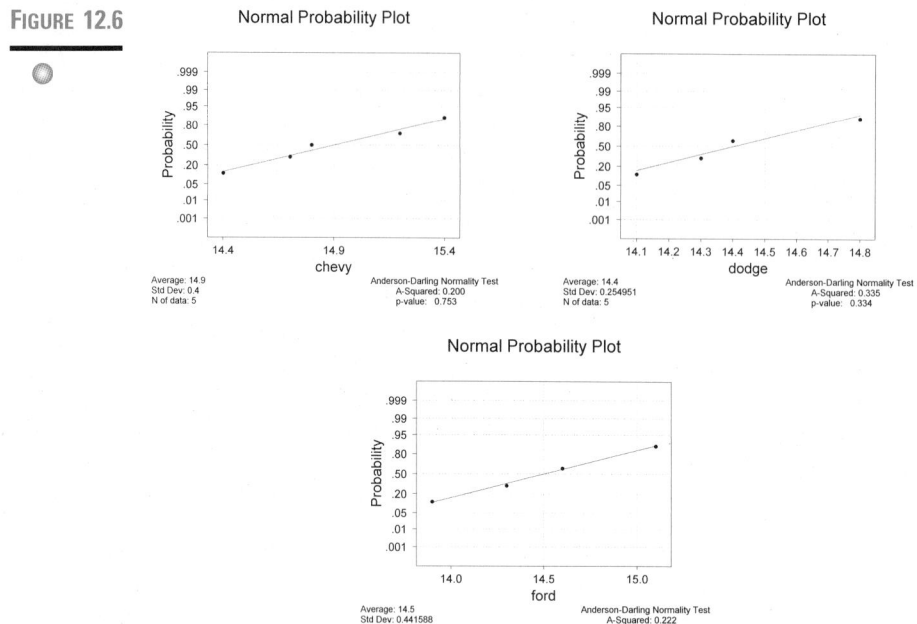

From Table B.5, we find that the 10% critical value for an F with 2, 12 degrees of freedom is 2.81. Because $2.5 < 2.81$, we have p-value $> .10$; hence, we fail to reject H_0. Based on the sample information, there is insufficient evidence to say that the trucks differ in gas mileage.

Total Variability

As we have noted, SST and SSE measure the variability between samples and within samples, respectively. Combining the two, we get a measure of the *total variability* in the data. If we ignore the distinction between the different samples and treat the data as one large sample with a single mean $\overline{y}_G$, then the total variability is simply the sum of the squared deviations of all the observations from the grand mean.

The **total sum of squares** (SSTotal) in a data array consisting of k samples is the total of the squared deviations between each observation and the overall mean. If $y_{i,j}$ denotes the ith observation in the jth sample and $\overline{y}_G$ denotes the overall mean, then

$$\text{SSTotal} = \sum\sum(y_{i,j} - \overline{y}_G)^2$$

(The double summation denotes a sum over subscripts i, j—that is, over rows and columns of the data array.)

The degrees of freedom associated with SSTotal is the number of observations less 1:

$$df = n_1 + n_2 + \cdots + n_k - 1$$

An important relationship of SSTotal, SST, and SSE is

$$\text{SSTotal} = \text{SST} + \text{SSE}$$

EXAMPLE 12.4

The data in Table 12.2 have been combined into a single sample. Calculate the total sum of squares and show that it is the sum of the treatment sum of squares and the error sum of squares.

15.2	15.4	14.8	14.4	14.7	14.8	14.4	14.3
14.1	14.4	15.1	14.3	14.6	13.9	14.6	

SOLUTION

Figure 12.7 gives the descriptive statistics for the combined data. Notice that the overall mean is $\bar{y}_G = 14.6$. Therefore, the total sum of squares is the sum of the deviations of all the scores from 14.6; that is,

$$\text{SSTotal} = (15.2 - 14.6)^2 + (15.4 - 14.6)^2 + \cdots + (14.6 - 14.6)^2$$
$$= 2.38$$

FIGURE 12.7 Descriptive Statistics

```
Variable          N        Mean    Median    TrMean    StDev    SEMean
Combined         15      14.600    14.600    14.592    0.412     0.106

Variable        Min        Max        Q1        Q3
Combined     13.900     15.400    14.300    14.800
```

Alternatively, we have that the standard deviation of all the data is $s = .412$ and it has $n - 1 = 14$ degrees of freedom. Therefore,

$$\text{SSTotal} = (n - 1)s^2 = (14)(.412)^2 = 2.38$$

Recall also that SSE = 1.68 and SST = .7, so we get

$$\text{SSE} + \text{SST} = 1.68 + .7 = 2.38 = \text{SSTotal}$$

ANOVA Table

The results of an analysis of variance are summarized in an ANOVA table. The form of the table for the one-way design is presented in Table 12.5.

TABLE 12.5

SV	SS	df	MS	F	*p*-value
Treatment	SST	$k-1$	MST	MST/MSE	
Error	SSE	$N-k$	MSE		
Total	SSTotal	$N-1$			

> Remember that $N = n_1 + n_2 + n_3 + \cdots + n_k$. Also notice that the degrees of freedom for treatment and error sum to the total degrees of freedom.

The first column lists the possible *sources of variation* (SV) in the data. The second column gives the numerical value of the source (i.e., the *sum of squares*). The third column gives the *degrees of freedom* associated with each sum of squares. The fourth column gives the *mean square* (or s^2), which is obtained by dividing the sum of squares by the degrees of freedom. The fifth and sixth columns give the *F*-ratio and its *p*-value, respectively.

EXAMPLE 12.5

Complete an ANOVA table for the truck data.

SOLUTION

Organizing the results of the previous examples, we get Table 12.6. Because *p*-value > .10, there is insufficient evidence to say that the trucks differ in gas mileage.

TABLE 12.6

SV	SS	df	MS	F	*p*-value
Treatment	.7	2	.35	2.5	> .10
Error	1.68	12	.14		
Total	2.38	14			

Confidence Intervals

After conducting an analysis of variance, we can find a confidence interval for any one of the treatment means. The procedure is the same as for a one-sample confidence interval based on $\bar{y}$ given in Chapter 7, except that the estimate of σ^2 is obtained from the pooled variance MSE given in the ANOVA table.

A $(1 - \alpha)100\%$ confidence interval for the mean of treatment j is given by the limits

$$\overline{y}_j \pm t^* \sqrt{\frac{\text{MSE}}{n_j}}$$

where t^* is the upper $\alpha/2$ critical value found in the t-table with $N - k$ degrees of freedom.

EXAMPLE 12.6

An experiment was conducted to study the reaction effects of four drugs on a nervous disorder. Twenty-eight subjects with the nervous disorder were independently and randomly assigned to the four drug groups, seven to each group. Unfortunately, two subjects in group 1 and one in group 4 were unable to complete the experiment. The reaction times for an experimental task were recorded for the remaining 25 subjects after they were administered their drug.

Worksheet: Nervous.mtw

	Drug Type		
1	2	3	4
3	5	6	2
5	7	5	4
4	3	7	3
6	4	9	4
4	5	6	2
	3	7	5
	6	8	

From the side-by-side boxplot diagram in Figure 12.8, it appears that the mean reaction time for subjects given drug 3 is greater than the means for the other drugs.

FIGURE 12.8   Side-by-side boxplots of reaction times to four drugs

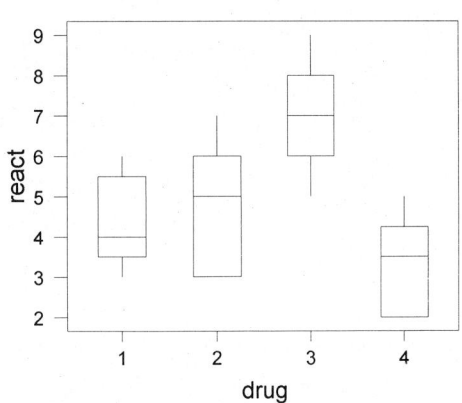

To investigate further, complete an ANOVA table and test the hypothesis of equality of mean reaction times for the four drugs. Then find individual 95% confidence intervals for the mean reaction time for each of the drugs.

SOLUTION

If we let μ_j denote the mean reaction time to drug j for $j = 1, 2, 3$, and 4, then the null hypothesis is

$$H_0: \mu_1 = \mu_2 = \mu_3 = \mu_4$$

and the alternative hypothesis is

$$H_a: \text{At least two } \mu\text{'s differ}$$

From the data, we get Figure 12.9. A p-value $= .001$ means that we should reject H_0 in favor of the alternative. Thus, there is highly significant evidence that the mean reaction effects of the four drugs are different. From the appearance of the boxplots, we speculate that the reaction effect of drug 3 is greater than those of the other drugs. Can we also conclude that the reaction effect of drug 4 is significantly less than those of the other drugs?

FIGURE 12.9

Descriptive Statistics

Variable	drug	N	Mean	Median	TrMean	StDev	SEMean
react	1	5	4.400	4.000	4.400	1.140	0.510
	2	7	4.714	5.000	4.714	1.496	0.565
	3	7	6.857	7.000	6.857	1.345	0.508
	4	6	3.333	3.500	3.333	1.211	0.494

Variable	drug	Min	Max	Q1	Q3
react	1	3.000	6.000	3.500	5.500
	2	3.000	7.000	3.000	6.000
	3	5.000	9.000	6.000	8.000
	4	2.000	5.000	2.000	4.250

One-Way Analysis of Variance

Analysis of Variance on react

Source	DF	SS	MS	F	p
drug	3	43.02	14.34	8.18	0.001
Error	21	36.82	1.75		
Total	24	79.84			

Individual confidence intervals may help answer this question. From the ANOVA table, we find MSE $= 1.75$ and df $= N - k = 21$. For a 95% confidence interval, we find from the t-table with 21 degrees of freedom, $t^* = 2.08$. The confidence intervals for the mean reaction time for the four drugs are:

$$4.400 \pm 2.08\sqrt{\frac{1.75}{5}} \qquad (3.169, 5.631)$$

$$4.714 \pm 2.08\sqrt{\frac{1.75}{7}} \qquad (3.674, 5.754)$$

$$6.857 \pm 2.08\sqrt{\frac{1.75}{7}} \qquad (5.817, 7.897)$$

$$3.333 \pm 2.08\sqrt{\frac{1.75}{6}} \qquad (2.210, 4.456)$$

Figure 12.10 shows the graphs of the four intervals on a common number line. Clearly, the mean reaction time for drug 3 is greater than that of any of the other drugs. The graphs also suggest that the mean reaction time for drug 4 is less than that of drugs 1 and 2. In the next section, we take a closer look at this problem and evaluate what differences are statistically significant.

FIGURE 12.10

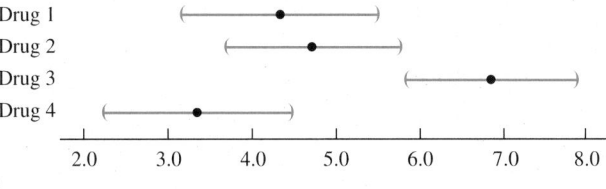

C O M P U T E R   T I P

To perform a one-way analysis of variance with Minitab, we have two different forms for the input data. Either all the response data are in one column of the worksheet with a group code value in a separate column, or the response data for the different groups are in separate columns. To execute the ANOVA, from the **Stat** menu select **ANOVA**. If the response data are all in one column, select **Oneway**; otherwise, select **Oneway (Unstacked)**. When the Oneway Analysis of Variance window appears, **Select** the **Response** variable column and then insert the column containing the codes in the **Factor** box. In the unstacked version, simply **Select** each column in the **Response** box. Finally, click **OK**.

EXAMPLE 12.7

In almost any magazine one can find advertisements for several different brands of cigarettes, each claiming to have the lowest tar and nicotine. To evaluate cigarettes, the Federal Trade Commission (FTC) uses "smoking machines" that measure the tar, nicotine, and carbon monoxide in each cigarette. Carbon monoxide has been linked to heart disease, tar has been linked to cancer, and nicotine is addictive.

Suppose that the amount of tar (measured in milligrams) is recorded for 25 cigarettes randomly selected from each of four brands:

Worksheet: Cigar.mtw

Brand A	Brand B	Brand C	Brand D
.41	.43	.52	.43
.48	.49	.48	.55
.44	.52	.67	.71
.37	.65	.49	.65
.31	.63	.38	.47
.40	.55	.57	.63
.53	.38	.70	.40
.32	.41	.45	.55
.35	.58	.47	.56
.52	.63	.55	.32
.40	.53	.49	.54
.51	.57	.56	.58
.53	.68	.51	.42
.29	.44	.60	.39
.39	.35	.41	.58
.48	.53	.65	.46
.58	.52	.57	.48
.46	.43	.39	.52
.59	.57	.54	.38
.50	.48	.61	.47
.47	.55	.53	.53
.51	.41	.68	.61
.33	.44	.50	.59
.56	.53	.58	.63
.61	.44	.49	.44

Perform an analysis of variance on the cigarette data to determine whether one brand is "lowest" in tar.

SOLUTION

First we look at boxplots of the data. The side-by-side boxplots in Figure 12.11 show that all samples are reasonably symmetric and have no outliers. The variances appear to be homogeneous. Normality does not seem out of the question. To further assess normality, we construct normal probability plots of each sample in Figure 12.12. In no case should we reject normality.

FIGURE **12.11**

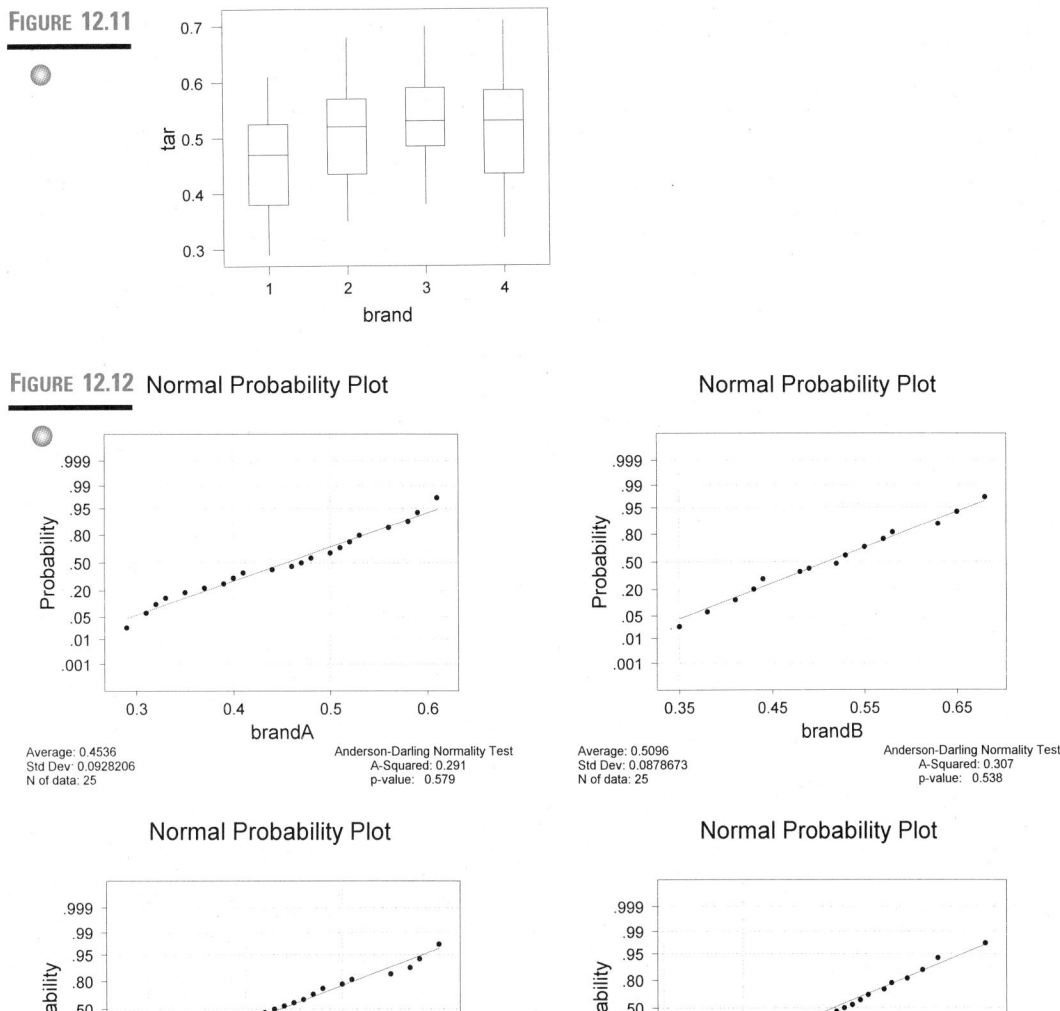

FIGURE **12.12**   Normal Probability Plot

Normal Probability Plot

Normal Probability Plot

Normal Probability Plot

Finally, we complete the one-way analysis of variance in Figure 12.13 (page 802). Notice that the standard deviations support the homogeneous variances assumption. The F statistic is 3.72 with p-value $= .014$. We should reject the hypothesis of equal mean tar contents for the four brands of cigarettes. The individual confidence intervals suggest that brand 1 has a mean tar content lower than the other brands. What can be said about the comparison of brands 2, 3, and 4? We look closer at this question with the multiple comparisons test in the next section.

FIGURE 12.13   One-Way Analysis of Variance

```
Analysis of Variance on tar
Source      DF        SS        MS        F        p
brand        3    0.09260   0.03087    3.72    0.014
Error       96    0.79670   0.00830
Total       99    0.88930
                                  Individual 95% CIs For Mean
                                  Based on Pooled StDev
Level   N      Mean      StDev  -------+---------+---------+---------
   1   25   0.45360    0.09282  (-------*------)
   2   25   0.50960    0.08787                (------*------)
   3   25   0.53560    0.08617                      (------*------)
   4   25   0.51560    0.09713                 (------*------)
                                  -------+---------+---------+---------
Pooled StDev =  0.09110               0.450     0.500     0.550
```

EXERCISES 12.2

12.1 Examine the stem and leaf plots and descriptive statistics:

Worksheet: ABC.mtw

Group A	Group B	Group C

```
0 |              0 | 8          0 |
1 | 0 2 4        1 | 2          1 |
2 | 3 5 6 7 7    2 | 2 4 4 6    2 | 2 4 6
3 | 1 3 6 7 8    3 | 0 3 5 5 8 9  3 | 5 8 9 9
4 | 0 2 3        4 | 1 4 5      4 | 0 3 5 6
5 |              5 | 0 2        5 | 3 5 7
6 | 4 5                         6 | 0 2
```

Descriptive Statistics

Variable	N	Mean	Median	TrMean	StDev	SEMean
GroupA	18	32.94	32.00	32.37	15.12	3.56
GroupB	17	32.82	35.00	33.20	12.39	3.00
GroupC	16	42.75	41.50	42.86	12.44	3.11

Variable	Min	Max	Q1	Q3
GroupA	10.00	65.00	24.50	40.50
GroupB	8.00	52.00	24.00	42.50
GroupC	22.00	62.00	35.75	54.50

a. Judging from the stem and leaf plots, do you think the variances of the three groups are homogeneous?

b. Is it plausible that the samples are from normally distributed populations?

c. Judging from the standard deviations of the three groups, do you think the variances are homogeneous?

d. From the standard deviations, calculate the mean square for error.

e. Can the grand mean be calculated by averaging the means of the three groups given in the descriptive statistics? Explain.

12.2   Here are side-by-side boxplots of the data given in Exercise 12.1.

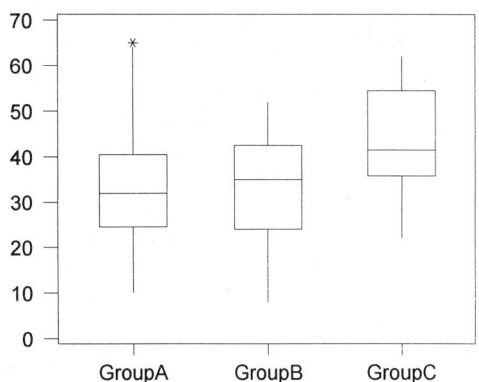

a. Does it appear that the variances are homogeneous?

b. Based on the boxplots, do you think there is statistical evidence to reject the hypothesis that the population means are equal?

12.3   This analysis of variance table is for the data in Exercise 12.1.

```
One-Way Analysis of Variance

Analysis of Variance
Source      DF       SS         MS           F         p
Factor       2      1069       534         2.96      0.061
Error       48      8662       180
Total       50      9731
                                 Individual 95% CIs For Mean
                                 Based on Pooled StDev
  Level    N     Mean    StDev   ---+---------+---------+---------+-
GroupA    18    32.94    15.12   (--------*--------)
GroupB    17    32.82    12.39   (--------*--------)
GroupC    16    42.75    12.44                  (---------*---------)
                                 ---+---------+---------+---------+-
Pooled StDev =   13.43          28.0      35.0      42.0      49.0
```

a. What is the sum of squares for error?

b. What is the mean square for error?

c. What is the sum of squares for treatments?

d. What is the mean square for treatments?

e. The pooled standard deviation is 13.43. What is another expression for the square of this figure?

f. State the null and alternative hypotheses for which this analysis is intended.

g. What is the F-ratio? What is its p-value? Should the null hypothesis be rejected?

12.4   Summary statistics associated with data calculated from three groups in a one-way ANOVA are given at the top of page 804.

Worksheet: Groups.mtw

```
Descriptive Statistics

    Variable      N      Mean    Median    TrMean     StDev    SEMean
    GroupA       23     66.35     67.00     66.29      7.67      1.60
    GroupB       26     75.69     74.50     75.58     10.08      1.98
    GroupC       25     71.64     73.00     71.52     10.06      2.01


    Variable     Min       Max        Q1        Q3
    GroupA     53.00     81.00     62.00     71.00
    GroupB     54.00    100.00     67.75     83.25
    GroupC     56.00     90.00     62.00     78.00
```

   a. Can we conclude that the homogeneous variances assumption is met?

   b. Give the degrees of freedom associated with the pooled variance, s_p^2.

   c. Compute the mean square for error.

12.5 This analysis of variance table is for the data in Exercise 12.4.

```
One-Way Analysis of Variance

Analysis of Variance
Source       DF         SS        MS         F         p
Factor        2     1067.5     533.7      6.05     0.004
Error        71     6260.5      88.2
Total        73     7328.0
                                Individual 95% CIs For Mean
                                Based on Pooled StDev
  Level    N     Mean     StDev  ------+---------+---------+---------+
  GroupA  23   66.348     7.667  (-------*-------)
  GroupB  26   75.692    10.079                        (------*-------)
  GroupC  25   71.640    10.058                 (------*-------)
                                 ------+---------+---------+---------+
Pooled StDev =     9.390         65.0      70.0      75.0      80.0
```

   a. State the null hypothesis for which this analysis is intended.

   b. What is the F-ratio? What is its p-value? Should the null hypothesis be rejected?

12.6 The side-by-side boxplots are for the groups in Exercises 12.4 and 12.5. In Exercise 12.5, the null hypothesis was rejected. Do these boxplots illustrate what means are different? Explain.

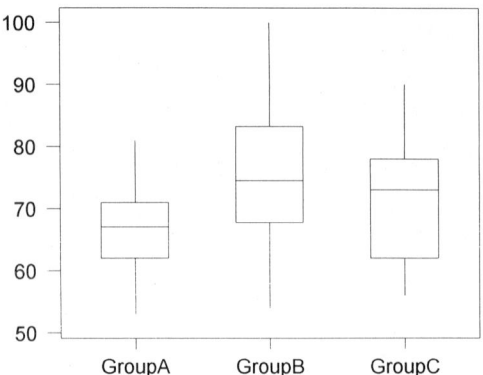

12.7   In Example 9.11 of Section 9.4, a math achievement test was given to a random sample of male and female high school students. A two-tailed, pooled *t*-test was applied to the data (worksheet: **Achieve.mtw**) with these results.

```
Two Sample T-Test and Confidence Interval

Twosample T for Score
Gender   N      Mean      StDev     SE Mean
1        13     84.46     8.04        2.2
2        12     76.4      11.0        3.2

95% C.I. for mu 1 - mu 2: ( 0.1,  16.0)
T-Test mu 1 = mu 2 (vs not =): T= 2.10  P=0.047  DF=  23
Both use Pooled StDev = 9.55
```

Here is a one-way analysis of variance applied to the same data:

```
One-Way Analysis of Variance

Analysis of Variance on Score
Source   DF       SS        MS          F         p
Gender   1      403.9     403.9       4.43     0.047
Error    23    2098.1      91.2
Total    24    2502.0
                                  Individual 95% CIs For Mean
                                  Based on Pooled StDev
Level    N     Mean     StDev    ---+---------+---------+---------+--
    1   13   84.462     8.038                   (--------*--------)
    2   12   76.417    10.967    (--------*---------)
                                 ---+---------+---------+---------+--
Pooled StDev = 9.551            72.0      78.0      84.0      90.0
```

a. Compare the *T* statistic and its *p*-value with the *F* statistic and its *p*-value. Is there a relationship between *T* and *F*? What about the *p*-values?

b. Compare the pooled standard deviations for the two test procedures. Are they the same? Find the mean square for error (MSE) and compare it with the pooled standard deviation. What, if any, is the relationship between the two?

c. What general observations can you make about the pooled *t*-test and the *F*-test when there are only two groups to compare?

12.8   Can music steady the scalpel? Psychologist Karen Allen, at the State University of New York–Buffalo studied 50 male surgeons under simulated stress. She randomly divided them into three groups: listening to music they chose, listening to her choice of Pachelbel's Canon in D, and no music. She reported in the *Journal of the American Medical Association* that the surgeons stayed relaxed with their own music, their blood pressure rose with her music, and the surgeons jumped to near hypertension with no music.

a. Is this a one-way analysis of variance design?

b. What is the response variable?

c. Verbally describe the null hypothesis.

d. What assumptions are necessary to test the hypothesis?

e. What is the test statistic?

f. What is the distribution of the test statistic under the assumption that the null hypothesis is true?

g. Is there a control group?

12.9   Exercise 8.54 in Section 8.3 described the process of manufacturing integrated circuits (chips) used in computers. In one part of the process, a thin layer of silicon oxide is placed on the surface of a wafer. The thickness of the oxide layer is critical to the performance of the resulting chips. Two wafers are randomly selected from each of 30 lots. Four measurements of the thickness of the oxide layer on each wafer are then taken (Yashchin, 1995). In worksheet: **Chips.mtw**, columns C1–C4 contain the four measurements on the first wafer, and columns C5–C8 contain the four measurements on the second of the two wafers selected from the 30 lots.

   a. From the side-by-side boxplots, is there evidence of a difference in the measurements of the thicknesses of the two wafers?

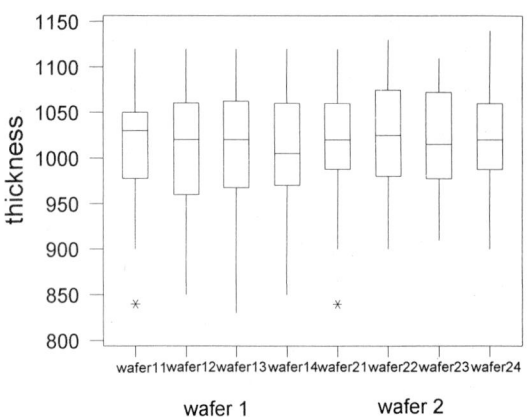

   b. Review the assumptions for the F distribution. Is it appropriate to compare the means of the observations obtained from the eight measurements using a one-way analysis of variance?

12.10   A psychologist is studying the effect of drug and electroshock therapy on a subject's ability to solve simple tasks. The number of tasks completed in a 10-minute period is recorded for the subjects who have been randomly assigned to four treatment groups: drug with electroshock, drug without electroshock, no drug with electroshock, and no drug and no electroshock.

Worksheet: Shkdrug.mtw

Drug w/Shock	Drug wo/Shock	No Drug w/Shock	No Drug/No Shock
3	2	1	5
1	3	2	3
2	4	2	2
1	0	0	3
4	2	1	4
3	4	2	6
2	2	3	5
4	4	1	4
2	5	0	6
1	6	3	3
3	2	1	4
4	3	3	5
5	5	2	6
3	6	0	4
4	2	1	2
3	3	2	4

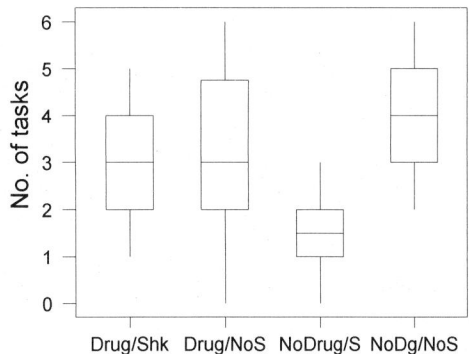

a. State the null and alternative hypotheses to determine whether the drug and electroshock therapies have an effect on the ability to solve simple tasks.
b. From the appearance of the boxplots, do you think the null hypothesis should be rejected?
c. Judging from the boxplots, do you think the variances are homogeneous?
d. Should we apply the ordinary F-test to these data or should we look for an alternative test?

12.11 This analysis of variance is for the data in Exercise 12.10. Judging from the standard deviations, do you think the variances are homogeneous? Should the null hypothesis be rejected? Interpret the decision.

```
One-Way Analysis of Variance

Analysis of Variance
Source      DF       SS       MS           F        p
Factor       3    58.12    19.38        11.01    0.000
Error       60   105.63     1.76
Total       63   163.75
                                Individual 95% CIs For Mean
                                Based on Pooled StDev
   Level    N    Mean   StDev ----+---------+---------+---------+--
Drug/Shk   16   2.812   1.223                 (----*-----)
Drug/NoS   16   3.312   1.662                    (-----*----)
NoDrug/S   16   1.500   1.033   (----*-----)
NoDg/NoS   16   4.125   1.310                           (----*-----)
                                ----+---------+---------+---------+--
Pooled StDev =   1.327          1.2       2.4       3.6       4.8
```

12.12 The following error scores were obtained for four groups of experimental animals running a maze under different experimental conditions:

Worksheet: Maze.mtw

Condition A	Condition B	Condition C	Condition D
16	20	9	15
12	18	11	14
15	22	14	18
13	17	15	20
15	21	8	16
14	19	10	17
15	18	11	17
14	18	10	16

a. State the null and alternative hypotheses for an ANOVA test.
b. Based on the normal probability plots, do you think the normality assumption for an analysis of variance is met?

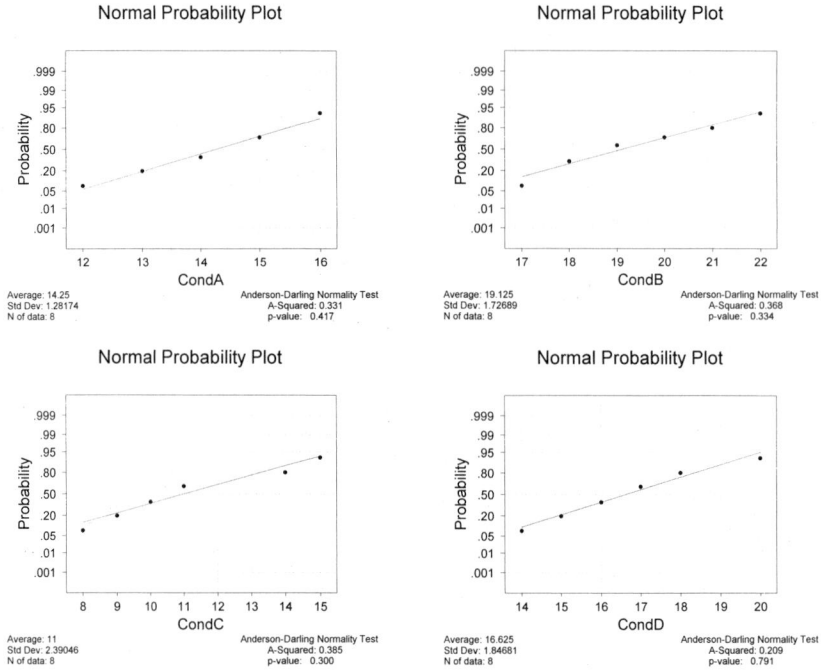

c. Based on the boxplots, does it appear that the variances are somewhat homogeneous?

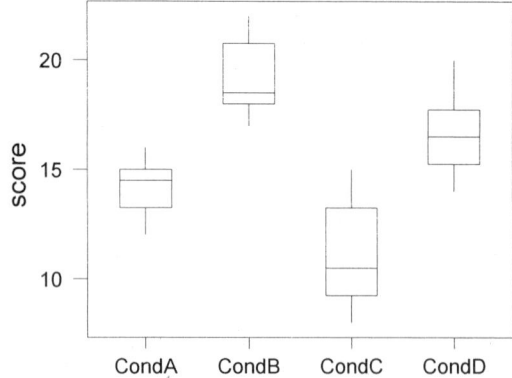

d. Do you think the null hypothesis should be rejected? Explain your answer.
e. From the ANOVA table at the top of the next page, test the hypothesis that there is no difference in the mean error scores for the four experimental conditions. Be sure to give the *p*-value and state your conclusion.

```
One-Way Analysis of Variance

Analysis of Variance
Source      DF       SS        MS        F         p
Factor       3    287.75     95.92     27.90     0.000
Error       28     96.25      3.44
Total       31    384.00
                                  Individual 95% CIs For Mean
                                  Based on Pooled StDev
Level    N      Mean   StDev    --------+---------+---------+--------
CondA    8    14.250   1.282                (----*---)
CondB    8    19.125   1.727                                (----*---)
CondC    8    11.000   2.390    (----*---)
CondD    8    16.625   1.847                        (---*----)
                                  --------+---------+---------+--------
Pooled StDev =   1.854               12.0      15.0      18.0
```

12.13   Use the information in the ANOVA table in Exercise 12.12 to construct individual 95% confidence intervals for each of the conditions.

12.14   According to IRS statements, the 1995 salaries (in $1,000) of the members of the boards of directors of three different universities are listed here:

Worksheet: Board.mtw

University A	University B	University C
70	30	100
120	90	900
85	80	300
200	250	90
60	70	1,200
310	55	260
90	180	60

a. Based on the side-by-side boxplots, do you think the assumptions for the F-test are satisfied?

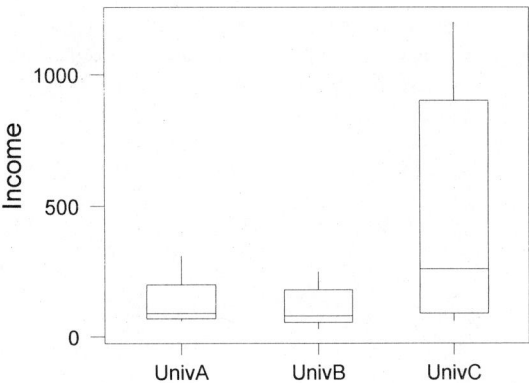

b. Based on the standard deviations given in the descriptive statistics at the top of page 810, do you think the variances are homogeneous?

```
Descriptive Statistics

Variable        N       Mean    Median    TrMean    StDev    SEMean
UnivA           7      133.6      90.0     133.6     90.8      34.3
UnivB           7      107.9      80.0     107.9     78.3      29.6
UnivC           7        416       260       416      451       170


Variable      Min       Max        Q1        Q3
UnivA        60.0     310.0      70.0     200.0
UnivB        30.0     250.0      55.0     180.0
UnivC          60      1200        90       900
```

   c. Do you recommend an F-test in this situation?

   d. Look at the means for the three universities. Is it necessary to conduct a formal test of the hypothesis that the mean salaries for the three groups are the same?

12.15   Three randomly selected groups of chickens are fed three different rations. These are the weight gains during a specified period of time:

Worksheet: Chicken.mtw

Ration 1	Ration 2	Ration 3
4	3	6
4	4	7
7	5	7
3	4	7
2	6	6
5	4	8
4	5	5
5	6	6
2	7	7
3	6	6
6	5	7
4	5	5
5	5	6

   a. State the null and alternative hypotheses for an ANOVA test.

   b. Examine the dotplots. Is there evidence that the normality assumption is violated? Are the variances homogeneous? In all, does it appear that the assumptions for the F-test are satisfied?

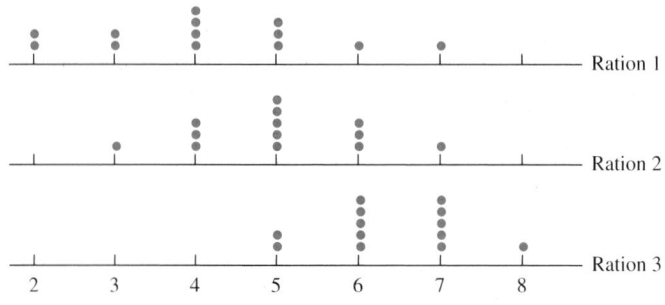

c. From the ANOVA table, test the hypothesis (stated in part **a**) that there is no difference in the average weight gains for the three rations. Be sure to state your conclusion.

```
One-Way Analysis of Variance

Analysis of Variance
Source      DF      SS      MS          F         p
Factor       2    32.97   16.49      12.17     0.000
Error       36    48.77    1.35
Total       38    81.74
                                Individual 95% CIs For Mean
                                Based on Pooled StDev
 Level    N    Mean   StDev  ------+---------+---------+---------+
Ration1  13   4.154   1.463  (------*-----)
Ration2  13   5.000   1.080            (------*------)
Ration3  13   6.385   0.870                        (------*-----)
                                ------+---------+---------+---------+
Pooled StDev =   1.164           4.0       5.0       6.0       7.0
```

12.16   Suppose that four groups of 11 students each used a different method of programmed learning to study statistics. A standard test was administered to the four groups and graded on a 15-point scale. Given these results, determine whether there was a significant difference in the results of the four methods.

Worksheet: Program.mtw

Method I	Method II	Method III	Method IV
3	5	7	4
5	7	5	6
6	7	6	6
8	7	8	7
4	6	7	6
3	6	6	5
5	8	9	5
6	4	8	5
4	6	7	6
6	7	7	5
3	5	8	4

a. State the null and alternative hypotheses for an ANOVA test.
b. Construct side-by-side boxplots. Does it appear that there is a difference in the four methods?
c. Construct normal probability plots for each method. Does the normality assumption seem to be satisfied?
d. In all, do you think the assumptions for the F-test are satisfied?
e. Complete an ANOVA table.
f. Test the hypothesis that there is no difference in the four methods.

12.17   A laminectomy is the surgical removal of a ruptured disk or all or part of the bony arch of a segment of the spine. The table at the top of page 812 gives the median costs of laminectomies at hospitals across the state of North Carolina in 1992. Each hospital has been classified as a rural, regional, or metropolitan hospital.

Worksheet: Laminect.mtw

Rural (1)	Regional (2)	Metropol (3)
5,223	7,455	5,807
5,782	4,142	6,943
8,892	6,278	4,925
7,405	9,158	4,556
8,460	7,733	2,696
4,353	4,288	6,177
9,296	7,054	5,477
5,227	8,308	5,656
6,535	11,391	5,621
10,238	3,339	4,324
4,346	6,801	5,493
4,973	3,835	6,306
11,191	7,224	5,073
5,225	5,445	8,435
4,027	5,841	5,450
6,538		

SOURCE: North Carolina Medical Database Commission, Department of Insurance, *Consumer's Guide to Hospitalization Charges in North Carolina Hospitals*, August 1994.

a. From the side-by-side boxplots, does it appear that the assumptions for an *F*-test are satisfied?

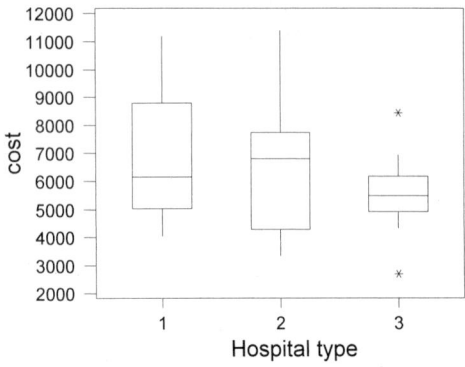

b. Construct normal probability plots for each of the three samples. Is normality rejected in any case?

c. Calculate descriptive statistics for each sample. Is any variance more than 4 times greater than any other variance? Is the homogeneous variances assumption satisfied?

d. State the null hypothesis that there is no difference in the costs at the three types of hospitals.

e. Test the hypothesis from part **d**. Be sure to state your conclusion.

12.18   *El Niño* (Spanish for "the Christ Child") refers to unusually warm ocean currents in the Pacific that appear around Christmas time and last for several months. Such effects as monsoon rains in the central Pacific and severe droughts and disastrous forest fires in Indonesia and Australia have been linked to El Niño. One hypothesis is that the warm phase of El Niño tends to suppress hurricanes, whereas a cold phase encourages hurricanes. The following information is the numbers of storms and resulting hurricanes from 1950 through 1995. Each year is classified as a warm, cold, or neutral El Niño year.

Worksheet: Hurrican.mtw

Year	Storms	Hurricanes	El Niño	Year	Storms	Hurricanes	El Niño
1950	13	11	cold	1973	7	4	cold
1951	10	8	warm	1974	7	4	cold
1952	7	6	neutral	1975	8	6	cold
1953	14	6	warm	1976	8	6	warm
1954	11	8	cold	1977	6	5	warm
1955	12	9	cold	1978	11	5	neutral
1956	8	4	neutral	1979	8	5	neutral
1957	8	3	warm	1980	11	9	neutral
1958	10	7	neutral	1981	11	7	neutral
1959	11	7	neutral	1982	5	2	warm
1960	7	4	neutral	1983	4	3	warm
1961	11	8	cold	1984	12	5	neutral
1962	5	3	neutral	1985	11	7	cold
1963	9	7	neutral	1986	6	4	warm
1964	12	6	cold	1987	7	3	warm
1965	6	4	warm	1988	12	5	cold
1966	11	7	neutral	1989	11	7	cold
1967	8	6	cold	1990	14	8	neutral
1968	7	4	neutral	1991	8	4	warm
1969	17	12	neutral	1992	6	3	warm
1970	10	5	cold	1993	8	4	warm
1971	13	6	cold	1994	7	3	warm
1972	4	3	warm	1995	17	10	cold

SOURCE: National Hurricane Center

a. Construct side-by-side boxplots of the numbers of storms associated with cold, warm, and neutral El Niño years. Does it appear that the cold phase encourages storms or that the warm phase suppresses storms?

b. Do the conditions for a one-way analysis of variance seem to be met? That is, is there any unusual behavior in any of the three samples to suggest that the assumptions for the F-test are not satisfied?

c. Perform an ANOVA on the number of storms. What is your conclusion?

d. What is the correlation between the number of storms and the number of hurricanes?

e. Perform an ANOVA on the number of hurricanes. How do these results differ from the analysis of the number of storms in part **c**? Based on the correlation found in part **d**, are these results consistent with the results found in part **c**?

12.3   TUKEY'S MULTIPLE COMPARISON PROCEDURE

The analysis of variance test presented in the preceding section tests the equality of the population means. If the null hypothesis

$$H_0: \mu_1 = \mu_2 = \cdots = \mu_k \quad \text{versus} \quad H_a: \text{At least two } \mu\text{'s differ}$$

is not rejected, then we conclude that the population means are essentially the same. If H_0 is rejected, however, the question still remains as to which means are different. For example, if the hypothesis of equality of the gas mileages of the three makes of pickup trucks had been rejected, then we would like to know which truck got the best mileage and which got the worst. When differences do exist we can further analyze them with a *multiple comparison test*. Remember that we conduct a multiple comparison test only *after* the null hypothesis is rejected in an analysis of variance test.

There are several multiple comparison tests, but the one we consider is Tukey's procedure, which utilizes the *Studentized range* (Neter et al., 1996):

$$q = \frac{\bar{y}_{max} - \bar{y}_{min}}{\sqrt{MSE/n}}$$

The multiple comparison test is needed only after an analysis of variance has led to rejection of the equality of the population means.

where $\bar{y}_{max}$ is the largest mean among an ordered group of means and $\bar{y}_{min}$ is the smallest in the group.

We will not study the statistical properties of the Studentized range, but we illustrate its use in Tukey's multiple comparison test.

The upper-tail critical value, $q_\alpha(k, v)$, of the Studentized range is needed. We find it in Table B.6 in Appendix B by locating k, the number of means, along the top of the table and the value of v, the degrees of freedom associated with MSE, in the left-hand column. The upper-tail critical values of the Studentized range are given for $\alpha = .05$ and .01. For example, if $k = 6$ and $v = 20$, then $q_{.05}(6, 20) = 4.45$ and $q_{.01}(6, 20) = 5.51$.

The test is summarized in the box.

Tukey's Multiple Comparison Procedure

For a specified value of α, calculate

$$W = q_\alpha(k, v)\sqrt{\frac{MSE}{n}}$$

where

$$n = \text{number of observations in each sample}$$
$$MSE = \text{mean square error from the ANOVA table}$$
$$k = \text{number of different population means}$$
$$v = \text{degrees of freedom associated with MSE}$$
$$q_\alpha(k, v) = \text{critical value of the Studentized range found in Table B.6}$$
$$\text{in Appendix B}$$

To conduct Tukey's procedure, complete these steps:

1. Rank the sample means from highest to lowest and arrange the population means in the same order.

(continued)

---

2. Compute the difference between the largest and smallest sample means: $\bar{y}_{\text{largest}} - \bar{y}_{\text{smallest}}$. If the difference exceeds W, then the corresponding population means are declared significantly different. Proceed to compute the difference between the largest and the next smallest sample mean: $\bar{y}_{\text{largest}} - \bar{y}_{\text{2nd smallest}}$. As above, if the difference exceeds W, then declare the corresponding population means different. Continue to make comparisons with the largest sample mean, $\bar{y}_{\text{largest}} - \bar{y}_{\text{3rd smallest}}$ and so on, until a difference fails to exceed W. Once a difference between two sample means is less than W, the corresponding population means, and all means between, are declared nonsignificant.

3. Make comparisons with the next largest sample mean, $\bar{y}_{\text{2nd largest}} - \bar{y}_{\text{smallest}}$ and so on, using the same procedures as in step 2. Continue until all possible comparisons are made.

4. Summarize the results by drawing a line under the population means that are declared nonsignificant.

EXAMPLE 12.8

An ecologist wishes to investigate the levels of mercury pollution in five major lakes. He catches ten lake trout from each lake and measures the concentration of mercury present in each fish. Table 12.7 is a summary of the measurements (in parts per million). Perform an analysis of variance on the means. If a significant difference exists, analyze the means with Tukey's multiple comparison procedure.

TABLE 12.7

	Lake				
	1	2	3	4	5
n	10	10	10	10	10
$\bar{y}$	4.1	3.7	2.4	4.6	3.4
s	.82	1.06	.68	1.44	.93

SOLUTION

The null hypothesis is

$$H_0: \mu_1 = \mu_2 = \mu_3 = \mu_4 = \mu_5$$

where μ_j is the mean concentration of mercury present in the fish from lake j and $j = 1$, 2, 3, 4, 5.

The ANOVA table is given in Table 12.8 (page 816). The p-value indicates that the null hypothesis should be rejected, which means that the mean concentrations of mercury in the various lakes are different. We now use Tukey's multiple comparison procedure to examine the differences.

TABLE 12.8

SV	SS	df	MS	F	p-value
Treatment	27.32	4	6.83	6.57	$< .01$
Error	46.772	45	1.0394		
Total	74.092	49			

We will compute W using $\alpha = .01$. We have $n = 10$, $k = 5$, $v = 45$, and MSE $= 1.0394$. From Table B.6 in Appendix B, we do not find $q_{.01}(5, 45)$ because the value for 45 degrees of freedom is not listed. We use the closest value of q, which is for 40 degrees of freedom. Thus, we have

$$q_{.01}(5, 45) \approx 4.93$$

Note that as we go from 40 to 60 degrees of freedom the value of q does not change much. Thus, very little is lost by using the value of q for 40 degrees of freedom. From the previous information, we have

$$W = q_{.01}(5, 45)\sqrt{\frac{\text{MSE}}{n}}$$

$$= 4.93\sqrt{\frac{1.0394}{10}} = 1.589$$

Ranking the sample means from highest to lowest, we have:

Population mean	μ_4	μ_1	μ_2	μ_5	μ_3
Sample mean	4.6	4.1	3.7	3.4	2.4

Because $\bar{y}_4$ is the largest, comparisons are conducted with it first. The analysis is presented in Table 12.9.

TABLE 12.9

Comparison	Difference	W	Conclusion
$\bar{y}_4 - \bar{y}_3$	2.2	1.589	significant
$\bar{y}_4 - \bar{y}_5$	1.2	1.589	not significant
$\bar{y}_1 - \bar{y}_3$	1.7	1.589	significant
$\bar{y}_1 - \bar{y}_5$	.7	1.589	not significant
$\bar{y}_2 - \bar{y}_3$	1.3	1.589	not significant
$\bar{y}_5 - \bar{y}_3$	1.0	1.589	not significant

We now arrange the population means in the ranked order and draw a line under the means that we judged not significantly different:

$$\mu_4 \qquad \mu_1 \qquad \mu_2 \qquad \mu_5 \qquad \mu_3$$

The line from μ_4 to μ_5 signifies that population means μ_4, μ_1, μ_2, and μ_5 do not differ yet are significantly larger than μ_3. The line from μ_1 to μ_5 is a subset of the line from μ_4 to μ_5 and therefore is not needed. Also the line from μ_5 to μ_3 is a subset of the line from μ_2 to μ_3 and is not needed. The final summary is

$$\mu_4 \qquad \mu_1 \qquad \mu_2 \qquad \mu_5 \qquad \mu_3$$

which says that μ_4 and μ_1 are significantly larger than μ_3.

Tukey's procedure can be modified for unequal sample sizes by comparing $\bar{y}_i - \bar{y}_j$ to W_{ij}, where

$$W_{ij} = q_\alpha(k, v)\sqrt{\frac{\text{MSE}}{2}}\sqrt{\frac{1}{n_i} + \frac{1}{n_j}}$$

Note that $W_{ij} = W$ when $n_i = n_j$. The remainder of the procedure is as before.

C O M P U T E R   T I P

To perform Tukey's multiple comparison with Minitab, we proceed as before with the one-way analysis of variance. That is, from the **Stat** menu, select **ANOVA** followed by **Oneway**. As before, select your **Response** and **Factor** columns, but in addition click on **Comparisons**. In the Multiple Comparisons window, click on **Tukey's family error rate**. The default error rate is 5%. It can be changed, however, by replacing the 5 with the desired value. Notice that in addition to Tukey's comparison, there are four other multiple comparisons. They are described in more advanced statistics books. Having selected Tukey's procedure, click on **OK**. When it returns to the Oneway window, click on **OK** again.

EXAMPLE 12.9

In Example 12.7, the tar contents of four major brands of cigarettes were compared with an analysis of variance procedure. The results indicated that a significant difference exists between the different brands. Analyze the means further with Tukey's multiple comparison procedure.

SOLUTION

Retrieve worksheet: **Cigar.mtw**. Following the computer tip, perform the usual one-way analysis of variance. This time click on Comparisons and check Tukey's procedure. Immediately after the ANOVA table and the individual confidence intervals in the Session window, we get the array shown in Figure 12.14. The output, consisting of an array of confidence intervals, does not exactly match what was described previously for a Tukey comparison. The results, however, are the same; it is a matter of interpreting the output. First, notice that the Family error rate is the 5% specified in the Multiple Comparisons window. The individual error rate of .0104 is the probability that a given confidence interval will not contain the true difference between two group means. The family error rate is the probability that among all the confidence intervals, at least one will not contain the true difference in group means.

FIGURE 12.14

Tukey's pairwise comparisons

```
        Family error rate = 0.0500
     Individual error rate = 0.0104

  Critical value = 3.70

  Intervals for (column level mean) - (row level mean)
               1           2           3

      2 -0.12341
          0.01141

      3 -0.14941     -0.09341
         -0.01459      0.04141

      4 -0.12941     -0.07341     -0.04741
          0.00541      0.06141      0.08741
```

Next, the critical value, $q_{.05}(4, 96) = 3.70$, is printed. The six groups of numbers in the array are the confidence intervals for all possible pairwise comparisons of the means. For example, the group

```
-0.12341
 0.01141
```

at the intersection of column 1 and row 2 is the confidence interval $(-.12341, .01141)$ for $\mu_1 - \mu_2$. The fact that 0 is inside the interval means that there is no significant difference between μ_1 and μ_2. In fact, the only interval that does not include 0 is the interval $(-.14941, -.01459)$ for $\mu_1 - \mu_3$. Consequently, there is a significant difference between μ_1 and μ_3. We conclude that the mean tar contents of brands 1 and 3 are different.

If we arrange the means in order of magnitude based on the sample means, Tukey's procedure, outlined previously, yields these summary bars:

μ_3 μ_4 μ_2 μ_1

We get the same results; μ_1 and μ_3 are significantly different, which suggests that the mean tar content of cigarette brand 1 is significantly less than that of brand 3.

EXERCISES 12.3

12.19 Find $q_\alpha(k, v)$ for the given values of α, k, and v:
    a. $\alpha = .01, k = 7, v = 15$
    b. $\alpha = .05, k = 3, v = 14$
    c. $\alpha = .01, k = 10, v = 30$

12.20 Here are the summary data from an experiment involving a treatment with four levels:

	Treatment						
	A1	A2	A3	A4	Source	SS	df
n	8	8	8	8	Treatment	131.375	3
$\sum y_i$	28	22	62	50	Error	41.0	28

    a. Complete the ANOVA table.
    b. Is there a significant difference in the treatment means?
    c. Analyze the means using Tukey's multiple comparison procedure with $\alpha = .01$.

12.21 Twenty subjects were randomly assigned to four reducing diets (five per diet) for a period of 6 weeks. The weight losses were recorded and an analysis of variance performed:

Diet	A	B	C	D	Source	SS	df
Weight lost	26	42	65	22	Diet	228.55	3
					Error	84.1	16

    a. Complete the ANOVA table.
    b. Is there a significant difference in the treatment means?
    c. Analyze the means using Tukey's multiple comparison procedure with $\alpha = .05$.

12.22 These reaction times (in tenths of a second) were recorded (top of page 820) for a group of subjects after each had been given a drug for pain.

Drug A	Drug B	Drug C
4	9	8
7	11	6
6	12	7
3	8	6
4	10	5
3	11	7

```
Descriptive Statistics

Variable         N        Mean     Median    TrMean     StDev     SEMean
DrugA            6       4.500      4.000     4.500      1.643     0.671
DrugB            6      10.167     10.500    10.167      1.472     0.601
DrugC            6       6.500      6.500     6.500      1.049     0.428

Variable       Min         Max          Q1         Q3
DrugA        3.000       7.000       3.000      6.250
DrugB        8.000      12.000       8.750     11.250
DrugC        5.000       8.000       5.750      7.250
```

a. Based on the descriptive statistics and the boxplots, would you say there is a significant difference in the reaction times for the three drugs?

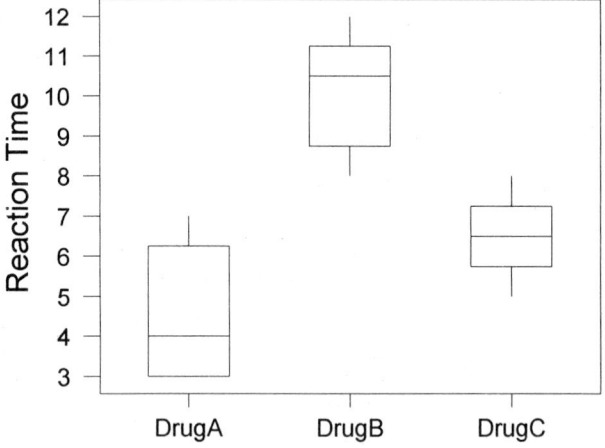

b. From the individual standard deviations, calculate MSE.

c. Analyze the means using Tukey's multiple comparison procedure with $\alpha = .01$.

12.23   In Example 12.6, an experiment was conducted to study the reaction effects of four drugs on a nervous disorder. The analysis of variance indicates that a significant difference exists in the mean reaction effects. Using an α of .05, perform Tukey's multiple comparison procedure to determine exactly what means differ.

```
One-Way Analysis of Variance

Analysis of Variance on react
Source   DF      SS     MS      F       p
drug      3    43.02  14.34   8.18    0.001
Error    21    36.82   1.75
Total    24    79.84
                            Individual 95% CIs For Mean
                            Based on Pooled StDev
Level    N    Mean   StDev -------+---------+---------+---------
   1     5   4.400   1.140            (------*-------)
   2     7   4.714   1.496           (-----*------)
   3     7   6.857   1.345                           (------*-----)
   4     6   3.333   1.211   (------*------)
                            -------+---------+---------+---------
Pooled StDev = 1.324              3.2       4.8       6.4
```

12.24   Review Example 12.5 to determine whether a multiple comparison test is necessary for analyzing the miles per gallon for the different makes of trucks. Explain your answer.

12.25   A psychologist determined that drug and electroshock therapies have a significant effect on one's ability to solve simple tasks (see Exercises 12.10 and 12.11 in Section 12.2). Use the results from Exercise 12.11 to pinpoint exactly what groups differ with Tukey's multiple comparison procedure ($\alpha = .05$).

12.26   In Exercise 12.12 in Section 12.2, the error scores for four groups of experimental animals running a maze were analyzed with analysis of variance. From the results of that exercise, determine whether a multiple comparison test is necessary. If it is, complete the test and indicate what differences exist in the four means.

12.27   In Exercise 12.16 in Section 12.2, grades on a standard test designed to evaluate different methods of programmed learning for four groups of statistics students were examined with an analysis of variance test. From the results of that exercise, determine whether a multiple comparison test is necessary. If it is, perform the test and indicate what differences exist in the four methods.

12.28   An analysis of variance test on an experiment involving five treatment levels yielded this ANOVA table:

SV	SS	df	MS	F
Treatment	3,740	4		
Error	6,650	35		

Complete the ANOVA table. From the following summary information, analyze the means using Tukey's multiple comparison test:

Sample	1	2	3	4	5
Size	8	8	8	8	8
Total	964	737	851	593	925

12.29   An experiment was designed to compare three different cleansing agents. Forty-five subjects with similar skin conditions were randomly assigned to three groups of 15 subjects each. A patch of skin on each individual was exposed to a contaminant and then cleaned with one of the cleansing agents. After 8 hours, the residual contaminants were measured:

Worksheet: Clean.mtw

Cleansing Agent

A	B	C
2	6	5
4	7	6
3	9	5
3	8	4
2	6	7
4	6	5
5	8	6
3	6	5
2	7	4
4	8	6
3	5	7
2	6	6
5	7	7
4	8	6
3	8	7

Here are the results of an analysis of variance test and Tukey's procedure to compare the mean residual contaminants left by the three cleansing agents:

```
One-Way Analysis of Variance

Analysis of Variance on clean
Source    DF        SS       MS        F        p
agent      2    108.13    54.07    47.44    0.000
Error     42     47.87     1.14
Total     44    156.00
                                Individual 95% CIs For Mean
                                Based on Pooled StDev
Level    N      Mean    StDev  --+---------+---------+---------+---
    1   15     3.267    1.033  (---*--)
    2   15     7.000    1.134                            (---*--)
    3   15     5.733    1.033                    (--*---)
                                --+---------+---------+---------+---
Pooled StDev = 1.068            3.0       4.5       6.0       7.5

Tukey's pairwise comparisons

    Family error rate = 0.0500
Individual error rate = 0.0193

Critical value = 3.44
```

(continued)

```
Intervals for (column level mean) - (row level mean)

                    1              2

          2      -4.682
                 -2.785


          3      -3.415         0.318
                 -1.518         2.215
```

Is there a significant difference in the mean residual contaminants left by the three cleansing agents? Summarize the results of Tukey's multiple comparison procedure.

12.30   Multiple comparison procedures, other than Tukey's procedure, are well known. Fisher's LSD (least significant difference) is one of the oldest comparison procedures. It is based on the ordinary t distribution, whereas Tukey's procedure is based on the Studentized range distribution. Fisher's LSD is conducted in the same way as Tukey's procedure. The only modification is to replace the critical value of the Studentized range, $q_\alpha(k, v)$, in the equation for W (or W_{ij} in the case of unequal sample sizes) with $\sqrt{2}\,t_v^*$, where t_v^* is the $\alpha/2$ critical value found in the t-table with v degrees of freedom. That is, compare the differences between pairs of sample means to

$$\text{LSD} = \sqrt{2}\,t_v^* \sqrt{\frac{\text{MSE}}{n}}$$

or if the sample sizes are unequal, compare the differences to

$$\text{LSD}_{ij} = \sqrt{2}\,t_v^* \sqrt{\frac{\text{MSE}}{2}} \sqrt{\frac{1}{n_i} + \frac{1}{n_j}}$$

$$= t_v^* \sqrt{\text{MSE}} \sqrt{\frac{1}{n_i} + \frac{1}{n_j}}$$

Apply Fisher's LSD to Example 12.8 and see whether the results are different from those of Tukey's procedure.

12.4   THE KRUSKAL–WALLIS TEST

The one-way analysis of variance presented in Section 12.2 is appropriate to use when the parent distributions are normally distributed and have equal variances. An alternative to the ordinary F-test is the nonparametric Kruskal–Wallis test. The only assumption made about the parent distributions is that they are similar in shape. Then the Kruskal–Wallis test can be used to test equality of location. For example, if the parent distributions are similarly skewed, then the Kruskal–Wallis test can be used to test the equality of their population medians.

     The procedure for the Kruskal–Wallis test that is presented here is the ordinary F-test applied to the rank summary data, n_j, $\bar{y}_{Rj}$, and s_{Rj}. This is an approximation of the original Kruskal–Wallis test, and therefore it is suggested that there be *5 or more observations in each sample.* For samples smaller than 5, exact tables for the Kruskal–Wallis test can be found in Conover (1980). It is also required that the samples be random and independent and can be ranked. As in the two-sample Wilcoxon test of Chapter 9, ranks are assigned to the combined samples.

EXAMPLE 12.10

A study was conducted to compare the hostility levels of high school students in rural, suburban, and urban areas. A psychological test, the Hostility Level Test (HLT), was used to measure the degree of hostility. Fifteen students were randomly selected from each type of school and given the HLT. The data are summarized in the side-by-side ordered stem and leaf plots:

Worksheet: Hostile.mtw

```
           Rural            Suburban              Urban

    1 | 6
    2 | 1                2
    3 | 3                7                3
    4 |
    5 | 1  3            2                3  4
    6 | 3  4  4  6  7  8  8   3  5  7
    7 | 2  5  7         0  2  3  4  6  8  9   2  4  6
    8 |                 2  3            0  2  3  3  4  6  7  8
    9 |                                 2
```

Is there evidence to indicate that the hostility levels of students differ in the three environments?

SOLUTION

The midsummary statistics in Table 12.10, together with the shapes of the stem and leaf plots, suggest that all the distributions are skewed left. Thus, to compare the hostility levels of students from the different environments, we suggest that a test of equality of the population *medians* be conducted using the Kruskal–Wallis test.

> The normality assumption has been violated; hence, the ordinary *F*-test should not be used. The distributions are similarly skewed, so the Kruskal–Wallis test is suggested.

TABLE 12.10

	Rural	Suburban	Urban
Median	64	72	82
Mid Q	60	70.5	79
Mid E	50.25	62.5	70.5
Midrange	46.5	52.5	62.5

Let θ_1, θ_2, and θ_3 denote the median hostility levels for students in the three environments. The null hypothesis is

$$H_0: \theta_1 = \theta_2 = \theta_3$$

and the alternative hypothesis is

$$H_a: \text{At least two } \theta_i\text{'s differ}$$

As stated earlier, the form of the Kruskal–Wallis test presented in this text is the ordinary F-test applied to the ranks. From the side-by-side ordered stem and leaf plots, the ranks can easily be assigned. Remember that the ranks are assigned to the combined samples, and tied values are given the average of the ranks that normally would have been assigned. In Table 12.11, we denote the ranks for each data value in parentheses. From the ranks of the data, we get the summary results shown in Table 12.12 on page 826.

TABLE 12.11

○

	Rural
1	6(1)
2	1(2)
3	3(4.5)
4	
5	1(7)　　3(9.5)
6	3(12.5)　4(14.5)　4(14.5)　6(17)　　7(18.5)　8(20.5)　8(20.5)
7	2(24)　　5(29)　　7(32)
8	
9	

	Suburban
1	
2	2(3)
3	7(6)
4	
5	2(8)
6	3(12.5)　5(16)　　7(18.5)
7	0(22)　　2(24)　　3(26)　　4(27.5)　6(30.5)　8(33)　　9(34)
8	2(36.5)　3(39)
9	

	Urban
1	
2	
3	3(4.5)
4	
5	3(9.5)　　4(11)
6	
7	2(24)　　4(27.5)　6(30.5)
8	0(35)　　2(36.5)　3(39)　　3(39)　　4(41)　　6(42)　　7(43)　　8(44)
9	2(45)

TABLE 12.12		**Rural**	**Suburban**	**Urban**	
	n	15	15	15	
	R_j	227.0	336.5	471.5	$T = 1{,}035$
	$\bar{y}_R$	15.133	22.433	31.433	
R_j is analogous to T_j in that it denotes the total of all the ranks in group j.	s_R	9.330	11.450	13.436	
	s_R^2	87.052	131.102	180.531	

Table 12.13 is the complete analysis of variance on the ranks. The p-value is less than 1%, and therefore the null hypothesis is rejected. There is a significant difference in the median hostility scores for the three school environments.

TABLE 12.13	SV	SS	df	MS	F	p-value
	Treatment	1,999.90	2	999.950	7.524	< .01
	Error	5,581.60	42	132.895		

Unlike the ordinary F-test, the Kruskal–Wallis test can be applied to nonnormal distributions. Although no specific distributional shape is required, it is assumed that the distributions are similar in shape. For example, in Example 12.10, all three distributions were skewed left. When the distributions are similar in shape, the Kruskal–Wallis test is sensitive to location and can be used to test equality of location, as was illustrated in the example. Violations of the assumptions for the ordinary F-test result in a loss of power (the ability to detect significant differences) of the test. In the case of a long-tailed distribution, the loss may be to the point that the test fails to detect actual differences in populations. The Kruskal–Wallis test, which is based on ranks, is resistant to outliers and thus may detect differences that might otherwise go undetected.

EXAMPLE 12.11   Three engineering universities wished to compare the salaries of their graduates 10 years after graduation. Seventeen graduates were randomly and independently selected from each of the three universities. Their salaries (in $1,000) are recorded in the side-by-side ordered stem and leaf plots:

Worksheet: Engineer.mtw

	University A	University B	University C
3	5	6	5
4	2 6	9	0
5	0 1 1 4 4 5 6 8 8	3 5 7 7	6
6	0 7 9	0 2 2 4 6 8	0 2 4 4 7 7 9
7		0 1 3 5	0 2 2 8
8	9		0 4
9		1	3
10	4		

Is there statistical evidence to indicate that the salaries of the graduates of the three engineering schools differ?

SOLUTION

First we run an ordinary F-test of the equality of population means. Let μ_1, μ_2, and μ_3 denote the mean salaries ten years after graduation at universities A, B, and C, respectively. The null hypothesis is

$$H_0: \mu_1 = \mu_2 = \mu_3$$

Having retrieved the data from worksheet: **Engineer.mtw**, we get the complete analysis of variance printout in Figure 12.15. A p-value greater than 10% (.292) indicates that we should not reject the null hypothesis. Therefore the ordinary F-test shows no significant difference in mean salaries of the graduates of the three universities.

FIGURE 12.15 One-Way Analysis of Variance

```
Analysis of Variance on salary
Source    DF      SS      MS        F        p
school     2     528     264     1.26    0.292
Error     48   10039     209
Total     50   10567
                          Individual 95% CIs For Mean
                          Based on Pooled StDev
Level    N    Mean    StDev  ----+---------+---------+---------+---
    1   17   58.76    16.60  (-----------*-----------)
    2   17   62.88    12.11        (-----------*-----------)
    3   17   66.65    14.32             (-----------*-----------)
                              ----+---------+---------+---------+---
Pooled StDev = 14.46          54.0      60.0      66.0      72.0
```

The side-by-side boxplots of the data in Figure 12.16 (page 828), however, show numerous outliers, which indicates that the assumptions for the ordinary F-test have been violated. The distributions appear to be long-tailed and similar in shape. We should therefore conduct the Kruskal–Wallis test.

FIGURE 12.16

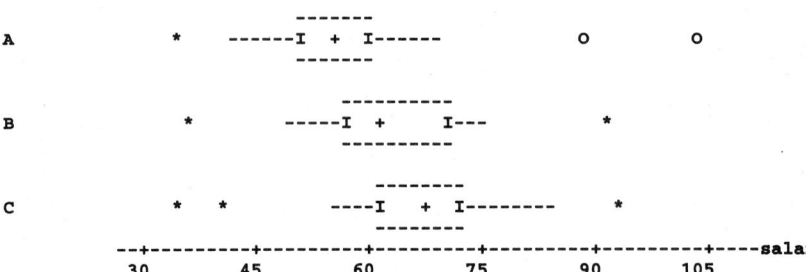

```
Character Boxplot

University

             --------
A       *    ------I  +  I------          O          O
             --------

               ----------
B         *    -----I  +     I---                *
               ----------

             --------
C      *   *    ----I  +  I--------          *
             --------
      --+---------+---------+---------+---------+---------+----salary
        30        45        60        75        90       105
```

C O M P U T E R   T I P

There is a Minitab command for the Kruskal–Wallis test; however, the test statistic, though equivalent, is not the same as the one used in this text. If you recall, we simply apply the ordinary F-test to the rank-transformed data. Thus, to perform a Kruskal–Wallis test, we store the data in one column of the worksheet with the levels in a second column, rank the data, and then apply the one-way ANOVA. The data column is ranked by the following procedure: From the **Manip** menu, select **Rank**. In the Rank window, select the column to be ranked in the **Rank data in** box, select the column to **Store ranks in**, and click **OK**.

The completed analysis of variance table for the Kruskal–Wallis test in Figure 12.17 is obtained by applying the one-way analysis of variance procedure to the ranks of the data. The p-value is .043, which indicates that H_0 should be rejected. There is statistical evidence that the mean salaries of the three university graduates differ. Because of the long-tailed distributions, the ordinary F-test failed to detect a difference in the means of the three distributions (p-value $= .292$), whereas the Kruskal–Wallis test, the more powerful test in this example, was able to detect the differences.

FIGURE 12.17   One-Way Analysis of Variance

```
Analysis of Variance on ranks
Source    DF        SS        MS         F          p
school     2      1354       677      3.36      0.043
Error     48      9683       202
Total     50     11037
```

(continued)

FIGURE **12.17**
(continued)

```
                                     Individual 95% CIs For Mean
                                     Based on Pooled StDev
Level    N     Mean   StDev  -----+---------+---------+---------+-
    1    17    19.29  14.73  (--------*--------)
    2    17    26.88  13.61              (--------*-------)
    3    17    31.82  14.25                     (--------*-------)
                             -----+---------+---------+---------+-
Pooled StDev =       14.20        16.0      24.0      32.0      40.0
```

As in any analysis of variance, if the null hypothesis is rejected, a multiple comparison test should be run to further analyze the differences. Tukey's multiple comparison test proceeds just as before, except that it is applied to the ranks.

C O M P U T E R   T I P

Minitab can perform the standard Kruskal–Wallis test that is found in nonparametric textbooks. Under the **Stat** menu, select **Nonparametrics** and then choose **Kruskal-Wallis**. In the Kruskal–Wallis window, select the **Response** and **Factor** columns and click **OK**.

Figure 12.18 is the output of the Kruskal–Wallis test for Example 12.11. Notice that p-value $= .047$ is almost identical to p-value $= .043$ that we obtained using the rank transformed Kruskal–Wallis test.

FIGURE **12.18**   Kruskal-Wallis Test

```
LEVEL        NOBS     MEDIAN   AVE. RANK    Z VALUE
    1         17      14.50      19.3        -2.28
    2         17      26.00      26.9         0.30
    3         17      33.00      31.8         1.98
OVERALL       51                 26.0

H = 6.13   d.f. = 2   p = 0.047
H = 6.13   d.f. = 2   p = 0.047 (adjusted for ties)
```

It may appear from our discussion that the Kruskal–Wallis is superior to the ordinary F-test and should be used in most applications. This is far from the truth. The ordinary F-test is a very robust test that can be used in a wide variety of situations. Certainly, if the specified assumptions for the F-test are met, then it should be used because, in that case, it is the most powerful test. It is only when the data grossly violate assumptions—that is, highly skewed or unusually long tails—that you should resort to the Kruskal–Wallis test.

EXERCISES 12.4

12.31   What is the Kruskal–Wallis test? Describe how the rank-transformation form of the test is conducted.

12.32   Under what conditions do you choose the Kruskal–Wallis test over the ordinary F-test?

12.33   These summary data were obtained by collectively assigning ranks to three independent samples. Complete an ANOVA table on the ranks for the Kruskal–Wallis test. Include the p-value.

n	10	10	10	
R_j	124.5	147.0	193.5	$T = 465$
$\bar{y}_R$	12.45	14.70	19.35	$\bar{y}_G = 15.5$
s_R	8.67	8.43	8.72	

12.34   The following ranks were collectively assigned to three independent samples. Using the descriptive statistics, complete an ANOVA table on the ranks for the Kruskal–Wallis test. Include the p-value.

Sample 1	Sample 2	Sample 3
1	2	5
4	3	9
6	7	12
10	8	18
11	13	20
14	15	22
16	17	23
19	21	24

```
Descriptive Statistics of Ranks

Variable      N      Mean    Median    TrMean     StDev    SEMean
Sample1       8     10.12     10.50     10.12      6.17      2.18
Sample2       8     10.75     10.50     10.75      6.82      2.41
Sample3       8     16.63     19.00     16.63      7.09      2.51

Variable     Min      Max        Q1        Q3
Sample1     1.00    19.00      4.50     15.50
Sample2     2.00    21.00      4.00     16.50
Sample3     5.00    24.00      9.75     22.75
```

12.35   Independent random samples were selected from three populations:

Worksheet: Ex12-35.mtw

```
3 | 5                 3 | 5  6  9          3 | 5  6  7  9
4 | 2  3  4           4 | 0  1  3  4        4 | 0  1  2  2  4
4 | 5  6  6  7  8      4 | 6  7  8  9        4 | 5  6  9
5 | 0  1  4           5 | 1  3              5 | 2
5 | 6                 5 |                    5 | 7
6 | 4                 6 | 3                   6 | 4
6 | 8                 6 | 5                   6 |
```

a. From the side-by-side boxplots, comment on the shapes of the distributions. Are they similar?

```
Character Boxplot


                          --------------
  1   ---------------I    +         I---------------------        *
                          -------------


                       -----------------
  2   ---------I            +       I---------------------    *
                       -----------------


                     ---------------
  3   --------I    +          I----------------                *
                     -------------
      --+---------+---------+---------+---------+---------+--Sample
      36.0      42.0      48.0      54.0      60.0      66.0
```

b. To test the equality of the centers of the distributions, should an ordinary F-test or the Kruskal–Wallis test be conducted?

12.36 Using the data in Exercise 12.35, test the hypothesis that the population medians are equal with the test you suggested in part **b** of Exercise 12.35.

12.37 Independent random samples were selected from three populations:

Worksheet: Ex12-37.mtw

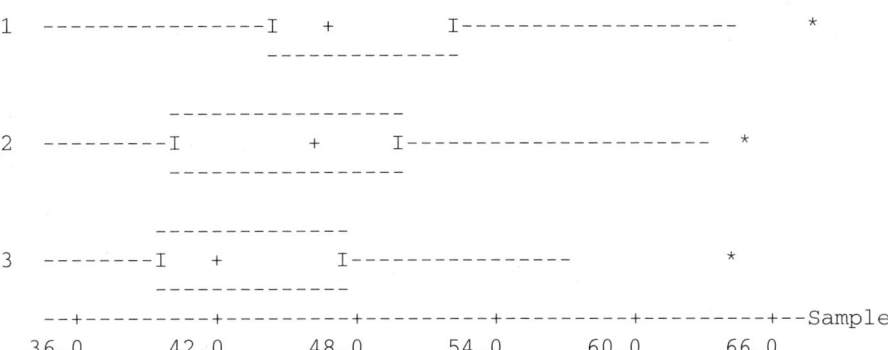

```
14 |                    14 |                   14 |
15 | 3                  15 | 7                 15 | 3
16 | 4                  16 | 5  7              16 | 5  7
17 | 1  3  6  8         17 | 0  1  2  5  7  9  17 | 1  5  6  9
18 | 0  1  1  2  3  5  9 18 | 0  2  7  9        18 | 2  3  4  5  7  9
19 | 2  4  6            19 | 4  6              19 | 2  6
20 | 0                  20 |                   20 | 0
```

a. Examine the side-by-side boxplots of the data. Does any boxplot show severe skewness? Does any boxplot show extreme outliers? Do these boxplots suggest that the population variances are homogeneous? Are they similar in shape but possibly centered in different places? What about the normality assumption?

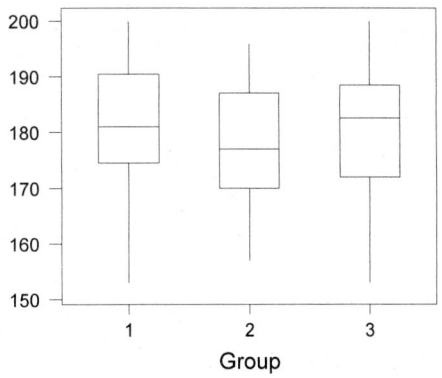

b. Based on your responses to part **a**, are the assumptions for the ordinary *F*-test satisfied? Are the assumptions for the Kruskal–Wallis test satisfied? Which one is the recommended procedure to test the equality of centers of the three distributions?

12.38   Here are two analysis of variance procedures applied to the data in Exercise 12.37. Which one is the ordinary *F*-test and which one is the Kruskal–Wallis test? Based on your answer to Exercise 12.37, which one is the proper way to analyze these data? Does either test reject the hypothesis that the populations have the same center?

```
One-Way Analysis of Variance

Analysis of Variance on Samples
Source    DF      SS       MS        F         p
Group      2     115       58      0.42     0.658
Error     45    6150      137
Total     47    6265
                              Individual 95% CIs For Mean
                              Based on Pooled StDev
Level    N     Mean    StDev --------+---------+---------+--------
   1    17   181.06    11.81                (----------*-----------)
   2    15   177.40    10.98  (-----------*-----------)
   3    16   180.25    12.19         (-----------*----------)
                              --------+---------+---------+--------
Pooled StDev = 11.69            175.0     180.0     185.0

One-Way Analysis of Variance

Analysis of Variance on Ranks
Source    DF      SS       MS        F         p
Group      2     246      123      0.62     0.543
Error     45    8951      199
Total     47    9197
                              Individual 95% CIs For Mean
                              Based on Pooled StDev
Level    N     Mean    StDev -------+---------+---------+--------
   1    17    26.35    13.91               (-----------*----------)
   2    15    21.17    13.84 (-----------*------------)
   3    16    25.66    14.54         (----------*-----------)
                              -------+---------+---------+--------
Pooled StDev = 14.10            18.0      24.0      30.0
```

12.39   The carbon monoxide level was measured (in parts per million) at three industrial sites at randomly selected times. Is there a significant difference in the carbon monoxide levels at the three sites?

Worksheet: Carbon.mtw

Site A	.106	.127	.132	.105	.117	.109	.107	.109
Site B	.121	.119	.115	.120	.117	.134	.118	.142
Site C	.119	.110	.106	.118	.115	.121	.109	.134

a. Here are side-by-side boxplots and normal probability plots of the three samples. Is normality a reasonable assumption in each case? How do you classify the shapes of the distributions? Are the three distributions similar in shape?

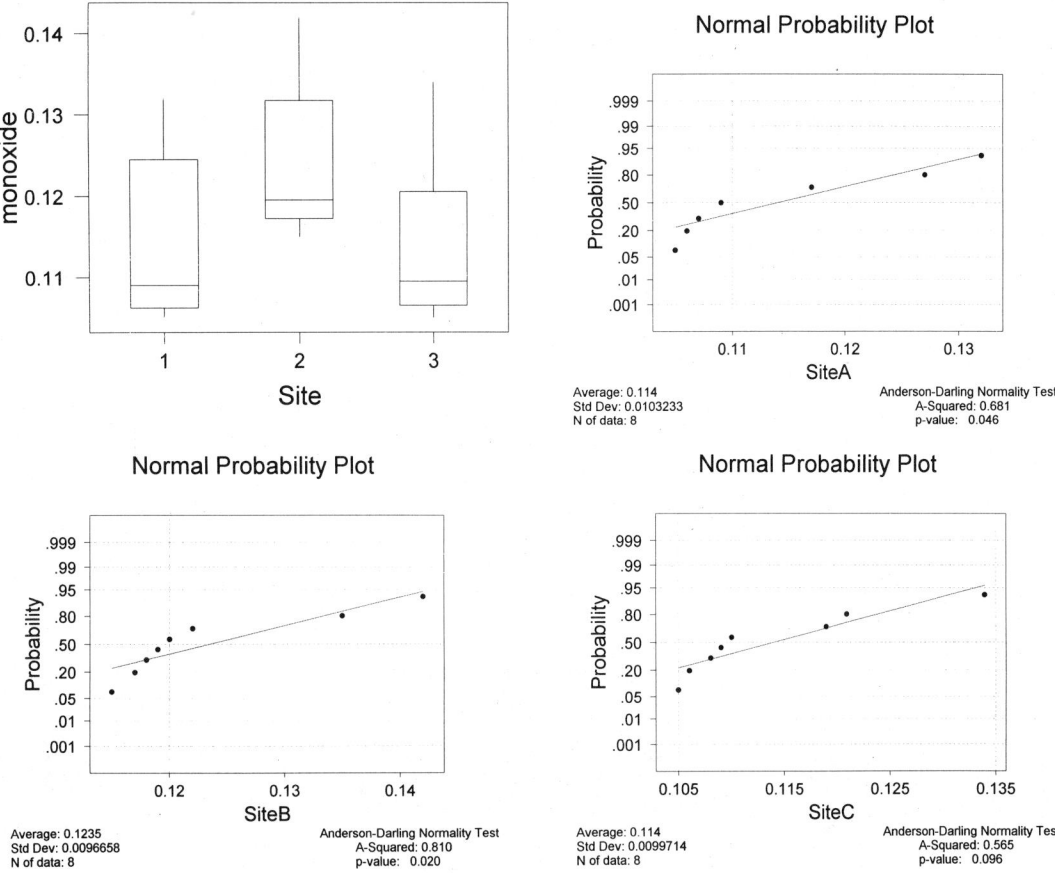

b. Should the ordinary F-test or the Kruskal–Wallis test be used to compare the centers of the three distributions? Explain your reasoning.

12.40 Here are two analysis of variance procedures applied to the data in Exercise 12.39. Which one is the ordinary F-test and which one is the Kruskal–Wallis test? Based on your answer to Exercise 12.39, which one is the proper way to analyze these data? Does either test reject the hypothesis that the populations have the same center? For the Kruskal–Wallis test, does the null hypothesis state that the population means are equal or that the population medians are equal? Write your conclusion.

```
One-Way Analysis of Variance

Analysis of Variance on monoxide
Source    DF        SS         MS        F         p
Site       2  0.0004813  0.0002407      2.41    0.114
Error     21  0.0020960  0.0000998
Total     23  0.0025773
```

(*continued*)

```
                              Individual 95% CIs For Mean
                              Based on Pooled StDev
   Level   N     Mean    StDev --------+---------+---------+------
      1    8   0.11400  0.01032 (----------*----------)
      2    8   0.12350  0.00967                 (---------*----------)
      3    8   0.11400  0.00997 (----------*---------)
                              --------+---------+---------+------
   Pooled StDev =  0.00999           0.1120    0.1190    0.1260
```

One-Way Analysis of Variance

Analysis of Variance on Ranks

```
Source   DF     SS      MS        F        p
Site      2   244.6   122.3     2.85    0.080
Error    21   901.4    42.9
Total    23  1146.0
                              Individual 95% CIs For Mean
                              Based on Pooled StDev
   Level   N    Mean    StDev ----------+---------+---------+-----
      1    8   9.938   7.317  (---------*---------)
      2    8  17.000   4.728               (--------*---------)
      3    8  10.563   7.272  (---------*---------)
                              ----------+---------+---------+-----
   Pooled StDev =   6.552             10.0      15.0      20.0
```

12.41 To get extra spending money, students are often eager to sell their used textbooks back to the bookstore at the end of the semester. In one university community, three independent bookstores will buy used books. In an effort to compare the amounts that the bookstores will pay for used books, several students were randomly selected as they exited a bookstore and asked how much they were paid for their used books. The recorded amounts are given in side-by-side stem and leaf plots:

Worksheet: Bookstor.mtw

```
       Bookstore A              Bookstore B              Bookstore C
   1 |                                           0 |
   1 |                     6 |                    7 |
   2 | 3                   4 |                      |
   2 | 7                                          6 | 8
   3 | 1                   4 |                    0 | 0 1 2 3 3 4
   3 | 5 6 6 8 9           5 | 6 9                5 | 6 7 7 9
   4 | 0 1 1 2 3 4 4       0 | 1 1 3 4            0 | 1 4
   4 | 5 6 6 7 8           5 | 5 6 7 8 9 9        7 |
   5 | 0 1 4               1 | 3                  2 |
   5 | 6                                          9 |
   6 | 4                   3 |                    4 |
   6 | 8                   5 |                    6 |
```

The stem and leaf plots suggest that the distributions are long-tailed. To compare the centers of the distributions, is the ordinary F-test or the Kruskal–Wallis test recommended?

12.42 Compare the two analysis of variance tests that have been applied to the data in Exercise 12.41. Which one is the ordinary F-test and which one is the Kruskal–Wallis test? Using the test you recommended in Exercise 12.41, can we conclude that the mean amount paid for used books is different for the three bookstores? Explain why one test shows a significant difference and the other one does not.

```
One-Way Analysis of Variance

Analysis of Variance on Dollars
Source  DF    SS    MS      F      p
Store    2   573   287   2.19  0.120
Error   69  9023   131
Total   71  9596
                        Individual 95% CIs For Mean
                        Based on Pooled StDev
Level   N   Mean   StDev -----+---------+---------+---------+-
    1  26  43.65   10.09                 (--------*--------)
    2  22  43.36   10.90                 (--------*--------)
    3  24  37.54   13.16  (--------*--------)
                          -----+---------+---------+---------+-
Pooled StDev = 11.44      35.0      40.0      45.0      50.0
```

```
One-Way Analysis of Variance

Analysis of Variance on Ranks
Source     DF      SS      MS      F      p
Store       2    2908    1454   3.56  0.034
Error      69   28153     408
Total      71   31061
                           Individual 95% CIs For Mean
                           Based on Pooled StDev
Level   N   Mean   StDev -+---------+---------+---------+-----
    1  26  40.67   19.11                 (-------*-------)
    2  22  41.36   19.46                 (-------*--------)
    3  24  27.52   21.94 (--------*-------)
                         -+---------+---------+---------+-----
Pooled StDev =   20.20   20        30        40        50
```

12.43   A home heating contractor sells three types of oil heaters. To compare the heating units, these efficiency ratings were obtained on samples of each type of heater:

Worksheet: Heating.mtw

Type A	Type B	Type C
75	73	60
71	83	63
74	70	74
86	66	56
77	54	61
84	71	73
76	74	71
75	76	62
57	92	91
96	75	64

a. Construct side-by-side boxplots and normal probability plots of the three samples. Is normality a reasonable assumption in each case? Are there many outliers?

b. Should the ordinary F-test or the Kruskal–Wallis test be used to compare the three distributions?

c. Complete the test recommended in part **b** to determine whether there is statistical evidence of a difference in the mean efficiency ratings of the three types of heating units. Be sure to write a conclusion and justify your choice of statistical test.

12.44   The owners of a soon-to-be-built motor lodge wished to evaluate three prospective locations for the business. On randomly selected occasions, they recorded the numbers of vehicles that passed prospective sites in 1-hour periods of time:

Worksheet: Lodge.mtw

Site A	Site B	Site C
162	165	165
154	193	160
174	178	155
148	184	168
150	157	140
148	165	151
185	204	163
157	195	175
164	183	182
172	189	150
159	179	139
193	160	164
160	198	181
159	185	176
173	215	177

a. Construct normal probability plots of the three samples. Is normality a reasonable assumption in each case?

b. Construct boxplots of the three samples. Does the variability in the three distributions appear the same?

c. Should the ordinary F-test or the Kruskal–Wallis test be used to compare the three distributions?

d. Complete the test recommended in part **c** to determine whether there is a significant difference in the amount of traffic that passes the three points.

12.45   Due to bad debts and risky investments, the U.S. savings and loans industry lost nearly $48 billion from 1987 to 1990. Many of the nation's largest savings and loans (S&Ls) went broke, thus creating a financial crisis for the U.S. government. The Congressional Budget Office estimated it would cost $215 billion to clean up the mess. Recovery was slow and fragile. Some banks were able to return to profitability on their own; others were taken over by federal regulators. In 1991, S&Ls operating under government control cut losses to $2 billion, and those in private hands earned a profit of $1.8 billion. By June 1992, 326 of the nation's 2,188 S&Ls were still listed as being in financial difficulty. A key measure of the health of a savings and loan company is its problem asset ratio. Problem assets include bad loans and repossessed property. The ratio measures those assets against tangible capital plus reserves set aside for loan losses. S&Ls with a problem asset ratio greater than 100% are listed as troubled. A lower number is better, and the national average is 89.91%. Following are the problem asset ratios for those S&Ls in California, New York, and Texas that were listed as being financially troubled in 1992:

Worksheet: Saving.mtw

California	New York	Texas
103.45	124.23	34,033.33
188.29	176.08	417.50
107.55	168.98	416.08
139.72	110.83	564.74
101.15	271.56	127.70
104.98	107.16	162.94
118.80	118.19	534.48
168.02	232.74	335.10
1,246.44	163.15	381.59
140.11	120.65	261.91
547.77	187.36	145.67
264.63	207.53	132.29
533.59	278.45	938.13
653.93	474.43	139.62
122.11	123.60	234.63
109.20		367.93
764.61		219.20
112.57		115.70
277.93		141.00
572.66		293.78
378.88		105.41
		100.26
		428.08
		171.92
		812.77
		127.74
		150.91
		131.59
		113.73

a. Construct side-by-side boxplots of the numbers in the three columns. Describe the boxplots. Can you classify the shapes of the three distributions?

b. Do you recommend the ordinary F-test to compare the means of the three distributions? Explain why or why not.

c. Notice that the first value in the Texas column is an extreme outlier. Should the analysis be conducted with or without that observation?

d. Remove the outlier described in part **c** and then construct side-by-side boxplots for the three columns. Can you now classify the shapes for the three distributions?

e. Do you recommend the ordinary F-test to compare the means of the three distributions, or should we compare the medians with the Kruskal–Wallis test?

f. Proceed with the test you recommended in part **e**. First analyze with the extreme Texas outlier and then without it. Does it matter whether the outlier is included or not?

g. From the analysis in part **f**, determine whether there is a statistically significant difference in the problem asset ratios for the three states. Be sure to state your conclusion in nonstatistical terms.

12.5 SUMMARY AND REVIEW

KEY CONCEPTS

✓ *Analysis of variance* allows one to statistically analyze several population means at one time. The *one-way ANOVA design* is an experiment in which independent random samples are obtained from several populations. The *total variability* in the data for a one-way design is partitioned into two parts: *within-sample* variability and *between-sample* variability. The within-sample variability is measured by the *sum of squares for error* and has $n_1 + n_2 + \cdots + n_k - k$ degrees of freedom, where n_j is the sample size for the sample from the *j*th population. The between-sample variability is measured by the *sum of squares for treatments* and has $k - 1$ degrees of freedom. Dividing the degrees of freedom into the sum of squares yields a *mean square*. Dividing the mean square for treatments by the mean square for error produces the *F statistic*, which is used to test the hypothesis of equality of the population means.

✓ To conduct the *F-test*, certain assumptions about the populations should be satisfied: They should be normally distributed and have homogeneous variances. If the assumptions are not met, then the *Kruskal–Wallis test* should be used.

✓ If an analysis of variance test indicates a significant difference in the population means, then *Tukey's multiple comparison procedure* is used to further analyze the means.

✓ If you are using a calculator, these computational formulas for the sums of squares are useful (refer to the notation in Table 12.1):

$$\text{Total sum of squares} = \text{TSS} = \sum\sum y_{i,j}^2 - \frac{T^2}{N}$$

$$\text{Sum of squares for treatments} = \text{SST} = \sum\left(\frac{T_j^2}{n_j}\right) - \frac{T^2}{N}$$

$$\text{Sum of squares for error} = \text{SSE} = \text{TSS} - \text{SST}$$

The value $C = T^2/N$ appears frequently and is called the correction term.

LEARNING GOALS

Having completed this chapter, you should be able to:

1. Identify a one-way analysis of variance experimental design. *Section 12.2*
2. Work with the data array for the one-way design. *Section 12.2*
3. Identify the assumptions that are necessary for an *F*-test. *Section 12.2*
4. Complete an ANOVA table from given data. *Section 12.2*
5. Perform an *F*-test on data. *Section 12.2*
6. Conduct Tukey's multiple comparison procedure. *Section 12.3*
7. Determine when the Kruskal–Wallis test should be conducted. *Section 12.4*
8. Perform a Kruskal–Wallis test. *Section 12.4*

QUESTIONS
FOR REVIEW

Use the following problems to test your skills:

12.46 Describe the conditions under which you would use each test:
 a. The ordinary *F*-test
 b. The Kruskal–Wallis test

12.47 Suppose an exploratory analysis of data from four populations suggested that the populations were symmetric with tails that are not excessively long. What mode of analysis would you use: the ordinary *F*-test or the Kruskal–Wallis test?

12.48 Suppose an exploratory analysis of data from three populations suggested that the populations were similarly skewed. What mode of analysis would you use: the ordinary F-test or the Kruskal–Wallis test?

12.49 Find the F values associated with upper-tail probabilities of .05, .025, and .01 in each case:
   a. $df_n = 3$ and $df_d = 24$
   b. $df_n = 5$ and $df_d = 35$
   c. $df_n = 7$ and $df_d = 42$

12.50 Given that $df_n = 4$ and $df_d = 28$, find the approximate p-values associated with each F:
   a. 2.5   b. 3.4   c. 5.6

12.51 Find $q_\alpha(k, v)$ from Table B.6 for the given values of $\alpha, k,$ and v:
   a. $\alpha = .01, k = 5, v = 20$
   b. $\alpha = .05, k = 3, v = 15$
   c. $\alpha = .01, k = 8, v = 40$

12.52 From the summary data, calculate SSE, SST, and SSTotal. Also give the associated degrees of freedom and compute the mean squares.

	Group A	Group B	Group C
n	14	16	18
$\bar{y}$	141.8	136.5	125.3
s	11.35	14.68	13.82

12.53 From the summary data in Exercise 12.52, construct an analysis of variance table. Compute the p-value associated with the F-value.

12.54 From these summary data, complete an ANOVA table. Include the p-value.

n	13	13	13	13
$\bar{y}$	16.5	18.3	14.7	19.5
s	6.72	4.94	4.31	5.82

12.55 This partially completed ANOVA table is for a one-way completely randomized design:

Source	SS	df	MS	F	p-value
Treatment	428	4			
Total	1,460	32			

   a. How many treatment levels are there?
   b. Complete the ANOVA table.
   c. Is there a significant difference in the treatment levels?

12.56 From the descriptive statistics (page 840) for the following data sets, calculate SSE, SST, and SSTotal. Also give the associated degrees of freedom and compute the mean squares.

Worksheet: Rev12-56.mtw

Group A	Group B	Group C
10	10 | 5	10 | 6
11 | 0	11 | 4	11
12 | 1 4 7	12 | 2 3 4 7	12 | 3 4
13 | 0 3 5 5 8	13 | 1 2 3 5 8 8	13 | 3 5 7 8
14 | 1 3 6	14 | 1 5	14 | 1 5 8
15 | 0 2	15 | 0	15 | 0 3 5
16 | 4	16	16 | 0

```
Descriptive Statistics

Variable         N      Mean    Median    TrMean    StDev    SEMean
GroupA          15    136.60    135.00    136.54    13.58     3.51
GroupB          15    130.53    132.00    131.00    11.80     3.05
GroupC          14    139.14    139.50    140.17    14.58     3.90

Variable       Min       Max        Q1        Q3
GroupA      110.00    164.00    127.00    146.00
GroupB      105.00    150.00    123.00    138.00
GroupC      106.00    160.00    130.75    150.75
```

12.57   Using the results of Exercise 12.56, construct an analysis of variance table. Compute the F statistic and its p-value.

12.58   These summary data resulted from an experiment involving a treatment with four levels:

Treatment					Source	SS	df
	A1	A2	A3	A4			
n	12	12	12	12	Treatment	756.5	3
Σy_i	135	152	147	184	Error	261.3	44

Analyze the means using Tukey's multiple comparison procedure with $\alpha = .01$.

12.59   Examine the following data obtained from an experiment involving three treatments:

Worksheet: Rev12-59.mtw

Treatment 1	Treatment 2	Treatment 3
21	41	35
24	44	37
31	38	33
42	37	46
38	42	42
31	48	38
36	39	37
34	32	30

a. Based on the boxplots, would you say that the distributions are symmetric? Would you rule out normality for any one of the distributions?

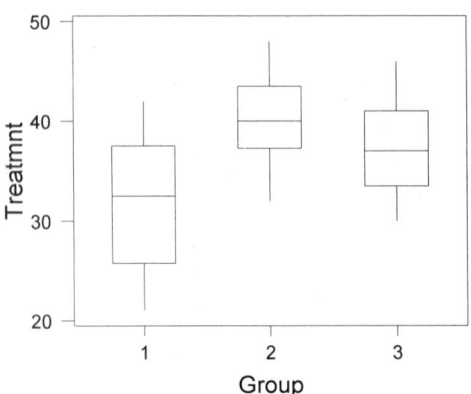

b. For all three samples, use the descriptive statistics to calculate the midrange and the midQ and compare them with the mean, median, and trimmed mean. Are they about the same? If so, would you say that the distributions are symmetric?

Descriptive Statistics

Variable	N	Mean	Median	TrMean	StDev	SEMean
treat1	8	32.12	32.50	32.12	7.00	2.47
treat2	8	40.12	40.00	40.12	4.82	1.71
treat3	8	37.25	37.00	37.25	5.01	1.77

Variable	Min	Max	Q1	Q3
treat1	21.00	42.00	25.75	37.50
treat2	32.00	48.00	37.25	43.50
treat3	30.00	46.00	33.50	41.00

c. Based on the boxplots, would you say that the variances of the three distributions are homogeneous? Examine the standard deviations. Do they rule out homogeneous variances?

d. Do you think that the assumptions for the ordinary F-test are satisfied? What about the assumptions for the Kruskal–Wallis test?

e. Based on the boxplots, do you think there is significance evidence to reject the hypothesis that the medians of the three treatments are the same?

12.60   Independent random samples were selected from three populations:

Worksheet: Rev12-60.mtw

```
 5 |                         5 | 8                5 | 7
 6 | 2                       6 | 3                6 | 0
 6 | 8                       6 | 7                6 |
 7 | 0                       7 |                  7 | 0 3 4
 7 | 6 8                     7 | 6 8              7 | 5 5 6 7 8 8 8 9
 8 | 0 1 2 2 3 3 4 4         8 | 0 1 3 3 4        8 | 0 1 3 4
 8 | 5 6 7 9                 8 | 5 6 7 8 9 9      8 |
 9 |                         9 | 1 4              9 | 3
 9 | 8                       9 |                  9 | 8
10 | 2                      10 | 2               10 | 0
10 | 7                      10 | 7               10 |
```

a. Use the boxplots and normal probability plots to comment on the shapes of the distributions.

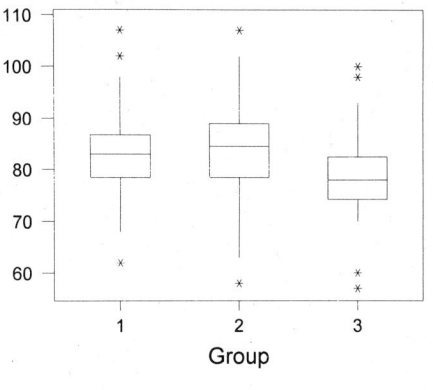

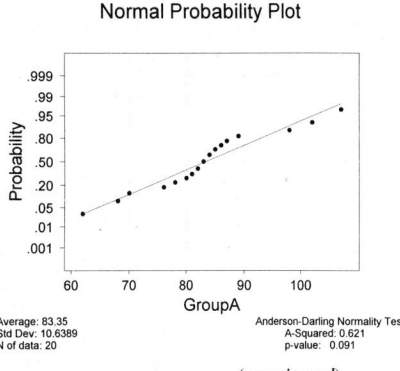

Normal Probability Plot

Average: 83.35
Std Dev: 10.6389
N of data: 20

Anderson-Darling Normality Test
A-Squared: 0.621
p-value: 0.091

(*continued*)

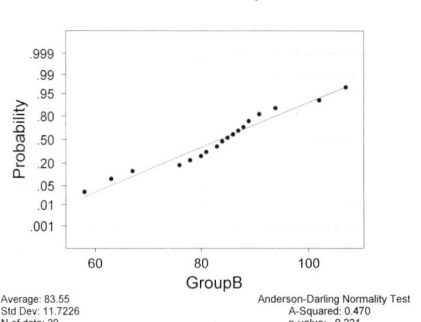

Normal Probability Plot — GroupB
Average: 83.55
Std Dev: 11.7226
N of data: 20
Anderson-Darling Normality Test
A-Squared: 0.470
p-value: 0.221

Normal Probability Plot — GroupC
Average: 78.45
Std Dev: 10.4452
N of data: 20
Anderson-Darling Normality Test
A-Squared: 0.738
p-value: 0.045

b. To test the equality of the population means, should an ordinary F-test or the Kruskal–Wallis test be conducted? Explain your reasoning.

12.61   Radiocarbon dating is a method of determining the age of archaeological sites. The data are the "ages" recorded as years before 1983 (the data were obtained in 1983)—that is, B.C. + 1983—for samples taken from one site at the archaeological excavation of the Danebury iron-age hill fort. Each of the 65 observations is associated with a pottery shard or fragment that has been classified into one of four phases, referred to as Ceramic Phases 1–4. The phases are thought to be abutting, nonoverlapping periods of stylistically consistent production. Determine whether the mean "ages" for the four phases differ.

Worksheet: Archaeo.mtw

Phase 1	Phase 2	Phase 3	Phase 4	
2,530	2,290	2,230	2,140	2,170
2,420	2,330	2,060	2,030	2,370
2,160	2,340	2,210	2,100	2,120
2,770	2,270	2,120	2,110	2,150
2,370	2,140	2,380	2,060	1,980
2,440	2,300	2,220	1,990	2,260
2,330	2,120	2,210	2,170	2,090
2,300	2,580	2,090	2,040	2,120
2,460	2,180	2,210	2,160	2,000
2,210		2,470	2,200	2,130
2,450		2,520	2,300	1,900
		2,330	2,060	
		2,090		
		2,110		
		2,250		
		2,280		
		2,300		

SOURCE: B. Cunliffe, "Danebury: An Iron-Age Hill Fort in Hampshire," Research Report 2, Council for British Archaeology, London, 1984; and J. Naylor and A. Smith, "An Archaeological Inference Problem," *Journal of the American Statistical Association* 83 (1988): 588–595.

a. Examine the boxplots. Does it appear that the median ages for the four phases differ? Does anything in the boxplots suggest that the assumptions for the ordinary F-test are not satisfied?

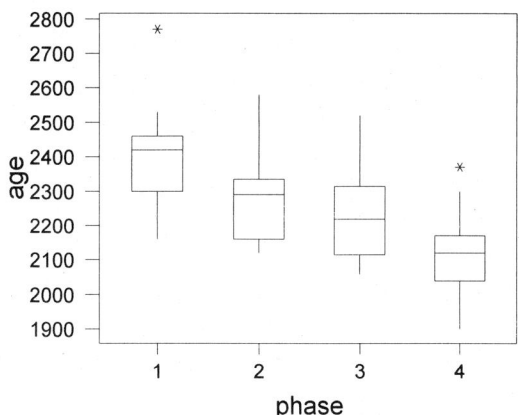

b. Examine the analysis of variance table. Is there statistical evidence to reject the null hypothesis that the mean ages are the same? What is the p-value? Based on the confidence intervals, do you think that the periods associated with the four phases are nonoverlapping?

```
One-Way Analysis of Variance

Analysis of Variance
Source    DF       SS       MS        F        p
Factor     3   661866   220622    13.00    0.000
Error     56   950028    16965
Total     59  1611894
                             Individual 95% CIs For Mean
                             Based on Pooled StDev
 Level    N    Mean   StDev  ---------+---------+---------+-------
phase1   11  2403.6   164.7                        (-----*------)
phase2    9  2283.3   137.7              (------*-------)
phase3   17  2240.0   131.4           (-----*----)
phase4   23  2115.2   106.8   (---*----)
                             ---------+---------+---------+-------
Pooled StDev = 130.2            2160      2280      2400
```

c. Examine the printout from Tukey's multiple comparison procedure. Which pairs of means are not statistically different? What does this say about the periods being nonoverlapping?

```
Tukey's pairwise comparisons

    Family error rate =  0.0500
Individual error rate = 0.0106
```

(*continued*)

```
Critical value = 3.74

Intervals for (column level mean) - (row level mean)
                 1           2           3

         2      -35
                275

         3       30         -99
                297         185

         4      162          33          15
                415         304         235
```

12.62   These summary data were obtained by collectively assigning ranks to three independent samples. Complete an ANOVA table on the ranks for the Kruskal–Wallis test. Include the p-value.

n	18	18	18	
R_j	372.5	568.5	544.0	$T = 1,485$
$\bar{y}_R$	20.694	31.583	30.222	
s_R	5.336	6.712	8.344	

12.63   Exercise 12.45 of Section 12.4 examined the problem asset ratios for savings and loan companies in California, New York, and Texas that were listed as being financially troubled in 1992 (worksheet: **Saving.mtw**). An ordinary F-test is applied to the data:

```
One-Way Analysis of Variance

Analysis of Variance
Source    DF          SS          MS         F        p
Factor     2    22690724    11345362      0.64    0.532
Error     62   1.103E+09    17784536
Total     64   1.125E+09
                                    Individual 95% CIs For Mean
                                    Based on Pooled StDev
  Level    N     Mean    StDev   ----+---------+---------+---------+-
  calif   21      322      300        (-----------*-----------)
  newyork 15      191       97    (--------------*--------------)
  texas   29     1452     6270                  (----------*---------)
                                    ----+---------+---------+---------+-
  Pooled StDev = 4217             -1500         0      1500      3000
```

Notice that the sample means for the three samples are 322, 191, and 1,452, and yet the F-test is insignificant ($F = .64$, p-value $= .532$). With such a large difference between the sample means (the mean for Texas more than 4 times larger than the other two), why is the F-test insignificant? Have we violated any assumptions? Explain.

12.64   A psychologist designs an experiment to study the effect of shock treatment on the amount of time one takes to complete a difficult task. Subjects were randomly assigned to three groups. Group 1 received no shock, group 2 received a medium shock, and group 3 received a severe shock. The dependent measure is the number of attempts to complete the task.

Worksheet: Shock.mtw

Group 1	Group 2	Group 3
6	11	14
3	9	15
4	7	12
8	14	16
6	10	15
3	9	18
7	13	16
9	11	13
7	10	15

a. State the null and alternative hypotheses to evaluate the effect of shock.
b. What assumptions are necessary for the analysis to be valid?
c. Based on the boxplots, does it appear that the assumptions are satisfied? Do you think shock had a significant effect?

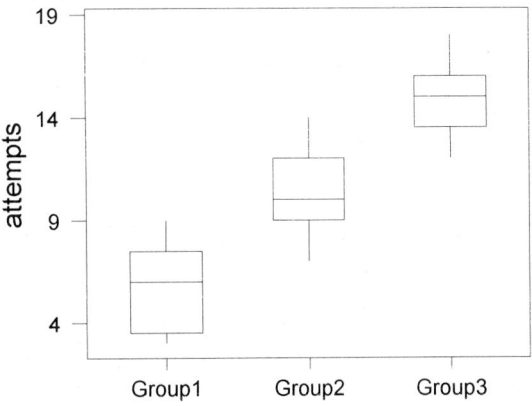

d. Using the accompanying information, complete the ANOVA table and test the hypotheses stated in part **a**.

```
One-Way Analysis of Variance

Analysis of Variance
Source    DF        SS        MS       F       p
Factor     2     364.52
Error     24      98.00
Total     26     462.52
                                Individual 95% CIs For Mean
                                Based on Pooled StDev
  Level    N     Mean     StDev   --------+---------+---------+-------
  Group1   9    5.889     2.147   (---*---)
  Group2   9   10.444     2.128               (---*---)
  Group3   9   14.889     1.764                           (---*---)
                                  --------+---------+---------+-------
Pooled StDev = 2.021                  7.0      10.5      14.0
```

12.65 To assess the impact of the level of impurities in a particular ingredient on the solubility of an aspirin tablet, a scientist wishes to test the null hypothesis that the mean dissolving time is the same regardless of the impurity level. In test batches, these dissolving times (in seconds) were obtained:

Worksheet: Asprin.mtw

Level of Impurity

1%	5%	10%
2.0	1.9	2.3
1.8	2.3	2.3
1.7	2.2	2.2
1.9	1.9	2.1
2.1	2.2	2.6

```
Descriptive Statistics

Variable         N      Mean    Median    TrMean    StDev     SEMean
1%               5    1.9000    1.9000    1.9000    0.1581    0.0707
5%               5    2.1000    2.2000    2.1000    0.1871    0.0837
10%              5    2.3000    2.3000    2.3000    0.1871    0.0837


Variable        Min       Max        Q1        Q3
1%           1.7000    2.1000    1.7500    2.0500
5%           1.9000    2.3000    1.9000    2.2500
10%          2.1000    2.6000    2.1500    2.4500
```

a. State the null and alternative hypotheses for an ANOVA test.
b. Does it appear that the assumptions for the F-test are satisfied?
c. Complete an ANOVA table.
d. Test the hypothesis that there is no difference in the dissolving times for the different impurity levels.

12.66 Independent random samples were selected from three populations:

Worksheet: Rev12-66.mtw

```
0 |                 0 |                 0 | 0  1
0 |                 0 | 5  6  7          0 | 5  6  8
1 | 1  2  4  4      1 | 0  2  4  4       1 | 0  1  2  3
1 | 5  6  7  8  9   1 | 5  6  6  9       1 | 5  6  8
2 | 0  2  3         2 | 0  3            2 | 4
2 | 5              2 | 7               2 | 5
3 | 0              3 | 2               3 | 1
3 | 5              3 |                 3 |
```

a. Comment on the shapes of the distributions.
b. To test the equality of the centers of the distributions, should an ordinary F-test or the Kruskal–Wallis test be conducted?
c. Test the hypothesis that the population medians are equal using the test you suggested in part **b**.

COMPUTER
EXERCISES **12.67** To compare the efficiency of the pit crews of major NASCAR teams, the durations of pit stops were measured for three of the top crews. These are the times (in seconds) for 12 randomly selected pit stops:

Worksheet: NASCAR.mtw

Team A	Team B	Team C
25	25	30
22	30	35
18	24	32
30	26	26
24	22	37
15	15	43
40	32	36
23	46	40
10	20	35
20	28	25
45	35	55
25	25	33

a. Complete side-by-side boxplots and comment on the shapes of the distributions. Are the tails of the distributions unusually long? What about symmetry?

b. Construct normal probability plots of the three samples. Do you detect any significant departures from normality?

c. To test the equality of the centers of the distributions, should an ordinary F-test or the Kruskal–Wallis test be conducted?

12.68   Using the test that you recommended in part **c** of Exercise 12.67, perform an analysis to determine whether a significant difference exists in the times it takes for a pit stop for the three crews.

12.69   The delay times (in minutes) were recorded for 20 flights selected randomly from four major air carriers:

Worksheet: Delay.mtw

Carrier A	Carrier B	Carrier C	Carrier D
20	15	20	25
14	17	27	17
12	10	22	10
20	36	35	5
17	18	26	22
30	20	24	35
19	5	15	19
7	16	17	24
22	20	10	3
18	13	25	20
10	42	45	15
15	15	20	40
13	8	16	16
5	17	12	10
19	10	5	9
25	4	21	19
45	19	32	45
14	25	23	15
40	12	10	5
10	5	15	10

a. Construct normal probability plots and boxplots of the four samples.

b. Based on the results of part **a**, should the analysis of the data be an ordinary F-test or the Kruskal–Wallis test?

12.70   Using the test that you recommended in part **b** of Exercise 12.69, perform an analysis to determine whether a significant difference exists in the delay times for the four air carriers.

12.71   Perform Tukey's multiple comparison procedure on the means you found in Exercise 12.70 or is one really needed?

12.72   A study was conducted to evaluate three treatments for arthritis. Arthritis sufferers were randomly assigned to three groups. After their treatment, the times until they experienced relief were measured:

Worksheet: Arthriti.mtw

Treatment A	Treatment B	Treatment C
40	73	50
35	32	75
47	47	34
52	52	47
31	34	87
61	60	45
92	77	38
46	42	25
50	20	86
49	81	39
93	75	42
84	35	30
72	25	75
43	40	36
80	90	90
85	33	32
30	30	89

a. Construct normal probability plots and boxplots of the three samples. Comment on the shapes of the distributions.

b. Based on the results of part **a**, should the analysis of the data be an ordinary F-test or the Kruskal–Wallis test?

12.73   Using the test that you recommended in part **b** of Exercise 12.72, perform an analysis to determine whether a significant difference exists in the relief times for the three treatments for arthritis.

12.74   These are the median costs of appendectomies at hospitals across North Carolina in 1992. Each hospital has been classified as a rural, regional, or metropolitan hospital.

Worksheet: appendec.mtw

Rural(1)

3,821	3,981	3,931	5,582	4,591	3,840	4,053	5,104	4,673	3,935	5,442
5,159	2,861	5,012	4,891	4,887	3,597	4,210	4,175	3,628	4,519	5,693
8,000	6,439	3,987	3,997	2,671	4,683	4,896				

Regional (2)

5,498	6,046	4,775	4,844	4,026	4,347	6,389	2,659	4,072	3,441	3,677
5,299	3,822	5,119	4,071	4,336						

(*continued*)

Metropol (3)

4,257   6,163   6,266   2,478   2,251   5,143   4,532   4,212   3,556   3,362   4,508
3,366   5,506   4,950

SOURCE: North Carolina Medical Database Commission, Department of Insurance, *Consumer's Guide to Hospitalization Charges in North Carolina Hospitals*, August 1994.

a. From the side-by-side boxplots, does it appear that the assumptions for an *F*-test are satisfied?

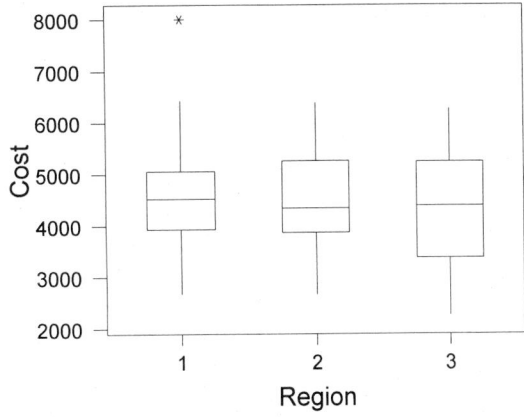

b. Construct normal probability plots for each of the three samples. Is normality rejected in any case?
c. Calculate the sample variance for each sample. Is any variance more than 4 times larger than any other variance? Is the homogeneous variances assumption satisfied?
d. State the null hypothesis that there is no difference in the costs at the three types of hospitals.
e. Test the hypothesis you stated in part **d**. Be sure to state your conclusion.

12.75   STATISTICAL INSIGHT REVISITED   Who is the winner of the Supercar Olympics? The top speeds of five supercars were compared by *Car and Driver* magazine. The data on six runs each are given in worksheet: **Supercar.mtw**. Retrieve the data.

a. Recreate the side-by-side boxplots shown in Figure 12.1. Based on these boxplots, do you see any reason not to conduct an analysis of variance with the ordinary *F*-test? Do the distributions appear reasonably symmetric? Are their variances reasonably homogeneous? Do you think the normality assumption is satisfied?
b. State the null and alternative hypotheses to compare the population means. Complete a one-way analysis of variance. Is there statistical evidence to reject the null hypothesis? What is the *p*-value?
c. Complete Tukey's multiple comparison procedure. Determine which means are not statistically different and determine which means are statistically different. Complete the summary analysis of the means. Are the results what you expected from looking at the boxplots?

CAUTIONS IN
REGRESSION

Multicollinearity problems in multiple regression occur when two or more of the predictor variables are highly correlated. Not only is it pointless to have two highly correlated predictor variables in the model (they tell you the same thing about the response variable), but it also creates serious problems in finding the least squares estimates of the regression parameters. In fact, if two perfectly related variables are used as predictors in a multiple regression model, then it is impossible to find unique least squares estimates of their corresponding regression parameters. If you try to use two highly correlated variables in your model, Minitab, for example, will automatically throw out one of the variables and give you a warning.

Detecting simple multicollinearity (two variables are directly related) is not that difficult. In addition to the correlation coefficient, a scatterplot of two variables will reveal how closely related they are. It is good practice to create a correlation matrix of all possible pairs of variables in a multiple regression problem and check for correlations close to 1. The scatterplots and correlation coefficients, however, will not detect more complicated relationships among several predictor variables. Most computer programs, Minitab included, will calculate various indicators of potential difficulties in regression problems. The *variance inflation factor* (VIF) is such an indicator that detects serious multicollinearity problems. It is beyond the scope of this textbook to discuss the variance inflation factor and other indicators, but if you intend to conduct a multiple regression analysis, you should learn more about the finer details of this widely used statistical tool.

When the simple linear regression model is used to fit sample data in which the x values are close to one another, small changes in the observed y's can cause the estimates, b_0 and b_1, to lose precision in estimating β_0 and β_1. This becomes apparent when we look at the standard error of the estimator b_1. Notice in the formula

$$SE(b_1) = \frac{s}{\sqrt{SS_x}}$$

that SS_x appears in the denominator. Recall that SS_x measures the variability of the x data; thus, if the x values are close to one another, then SS_x will be small, resulting in $SE(b_1)$ being large. The consequence is that the estimate b_1 can vary widely as an estimator of β_1 and thus loses precision. In designing a regression study, you should collect data so that a wide variety of x values are measured.

A final caution for multiple regression analysis pertains to the relationship between the coefficient of determination, R^2, and the inclusion of predictor variables in the model. Desirable traits of a multiple regression analysis are a large R^2 and a small value for $s \, (= \sqrt{MSE})$. As a new variable is added to the model, the value of R^2 automatically increases. This is good because it means that more of the variability in the dependent variable, y, is explained by the predictor variables. On the other hand, it means that the value of R^2 can become arbitrarily close to 1; in fact, as the number of variables approaches the number of observations, the value of R^2 approaches 1. Therefore, you should be careful in adding variables to the model; add a variable only if by its inclusion the value of R^2 increases an appreciable amount and the value of s decreases. A strategy used by some is to include those variables that produce the largest *adjusted* R^2, which will either increase or decrease when a predictor is added to the model. The value of R^2

is adjusted to compensate for the number of variables in the model. The adjusted R^2 will decrease when the gain obtained by adding a variable is nullified by the loss in degrees of freedom associated with SSE. With a computer, involved statistical procedures, such as stepwise regression analysis, are available to select the variables for inclusion in a regression model. You will learn more about this as you continue your studies in statistics.

CAUTIONS IN EXPERIMENTS

Analysis of variance is the procedure used to compare treatments in an experiment. If the treatments assigned to the subjects are not controlled by the experimenter, then the study is called *observational*. Drawing conclusions from observational studies can be risky. An observed difference might be due to some extraneous factor that is uncontrollable. Rejecting H_0 and concluding that group 1, for instance, outperformed group 2 do not necessarily mean that the treatment *caused* the difference. If the evidence mounts up from several different studies, such as in the smoking–lung cancer issue, then you might feel more comfortable in reaching the conclusion that the treatment caused the difference.

Even the well-designed randomized, controlled experiment can lead to significant differences and yet not establish causality. In theory it should because you are able to control the treatment. Certainly, if you start with equivalent groups and treat all groups the same except for the treatment, then observed differences should be caused by the treatment. In practice, however, it is very difficult to control all outside forces on the subjects so that the only difference between the groups is the treatment.

On January 27, 1988, the *New England Journal of Medicine* reported that "aspirin cuts the risk of heart attack" (Greenhouse and Greenhouse, 1988). This conclusion was based on a nationwide study of 22,071 U.S. physicians in which approximately half (11,037) were assigned at random to receive one Bufferin brand aspirin tablet every other day. The other half (11,034) received a placebo tablet. The experiment was double-blind in that neither the participant nor the experimenter knew who received the treatment and who received the placebo. Five years later the study was halted and declared a resounding success because it was determined that those in the "aspirin" group had one-third as many fatal heart attacks as those in the "placebo" group. The excitement over this new discovery was short-lived, however, when 3 days later a British study of heart attacks concluded, from a similar study conducted in England, that there was no significant difference between their "aspirin" and "control" groups. The 6-year British study involved 5,139 doctors, where two-thirds (3,429) received a higher dose of aspirin (500 mg versus 325 mg) than the physicians in the U.S. study. Furthermore, the British study was not double-blind and had a control group instead of a placebo group. The 1,710 physicians in the control group were told "to avoid aspirin and products containing aspirin unless some specific indication for aspirin was thought to have developed." It is clear that the two studies are not equivalent, but the findings of the British study seem to discount the randomized, controlled U.S. study.

CHECKING INDEPENDENCE

The assumption of independence among the observations in a sample is extremely important in statistical inference. In confidence interval estimation and hypothesis testing, the formula for the standard error of the estimator is derived based on the assumption that the observations are independent. If they are not independent, then the procedures are of no value. Random sampling and random allocation of subjects to groups are keys to assuring that the observations are independent. Time-ordered graphs of the observations will reveal patterns that may suggest dependence among the observations.

CHECKING NORMALITY

Throughout statistical inference, the assumption of normality continually comes up. It is the basis of the t-test, the chi-square test, and the F-test. Assessing normality then must be important. We introduced just a few of the ways that have been developed to determine whether a sample has come from a normally distributed population. First, you should look at a dotplot, stem and leaf plot, or histogram of your data for obvious indications of nonnormality. Look for skewness, outliers, and gaps in the data. The boxplot shows skewness and points out the outliers. When you have access to a computer, one of the best methods for detecting normality is the normal probability plot. Not only will it check your data for normality, but by studying the pattern of the plot, you can detect skewed, short-tailed, and long-tailed distributions.

CHECKING HOMOGENEITY

Homogeneity (or equality) of variances is another basic assumption that appears in the t-test of equality of means, the analysis of variance, and regression analysis. In many ways, it is more important than the normality assumption. For instance, theoretical studies by Box (1954) showed that the actual significance level for a stated 5% level of significance F-test could become as high as 14% when the sample sizes were unequal and the largest group variance was 3 times the smallest group variance. In studies of the effect of nonnormality, Ito (1980) obtained actual significance levels ranging from 3% to 6% for F-tests conducted at the 5% level of significance. In both cases, the results improved when the group sizes were equal.

There are formal tests for equality of variances, but they are not presented in this book simply because they are extremely sensitive to departures from normality. This lack of robustness makes the tests of little use. In place of formal tests, as in checking normality, it is suggested that you explore the data graphically to verify the assumption of homogeneity. Side-by-side boxplots and dotplots visually show differences in variability. In regression analysis, residual plots are designed to detect heterogeneity of variances. For a numerical evaluation, recall that the largest of the sample variances should be no more than 4 times greater than the smallest sample variance.

ROBUSTNESS OF t AND F

As pointed out previously, the validity of confidence interval estimates and tests of hypothesis for linear regression analysis and analysis of variance depends on the stated assumptions being met. Violating any one of those assumptions can lead to erroneous conclusions. You have learned that a statistical procedure is *robust* if the conclusions are unaffected by violations of the basic assumptions. The more robust the procedure, the more the assumptions can be violated without serious consequences. The t procedures and the F-test for analysis of variance and regression are remarkably robust against nonnormality and are reasonably robust to moderate deviations from the homogeneous variances assumption. They are extremely robust when the sample sizes are equal or near equal. What this means is that even when the underlying populations are not normal, you usually can proceed with the t-test or F-test without worry. Minor departures from normality and homogeneous variances generally will not cause serious problems in obtaining confidence interval estimates and in tests of significance. Exceptions to this are, of course, heavily skewed and unusually long-tailed distributions. In these two cases, you should avoid using the t-test and F-test and use one of the more efficient nonparametric or robust procedures instead.

TRANSFORMA-TIONS TO MEET ASSUMPTIONS

Another option for analyzing nonnormal populations is to perform some sort of transformation on the data so that the transformed data more readily meet the stated assumptions. In the case of skewed distributions, the *logarithm* transformation tends to pull in the skewed tail, which results in a more symmetric and possibly normal distribution. The standard t or F procedure is then applied to the transformed data. This technique is sometimes less than desirable because it may be difficult to interpret the results in terms of the transformed data. On the positive side, however, often one transformation, such as a logarithm or square root, will correct the nonnormality of the data as well as correct for differences in variability.

STATISTICAL SIGNIFICANCE VERSUS PRACTICAL SIGNIFICANCE

We close with one final caution illustrated with an example. A problem that has persisted over the years is the identification of effective predictors of success in college mathematics courses. J. D. House proposed using noncognitive predictors, such as academic self-concept, achievement expectancies, and motivation, to develop a model to predict student achievement in college mathematics. His sample consisted of 958 students (488 men and 470 women) who began as new freshmen at a large university during the same fall semester. This table is taken from Table 2 of his article. It is a summary of multiple regression analysis of mathematics grades; * means p-value $< .01$.

Step	Variable	R^2	F
1	Self-rating of mathematical ability	.077	80.28*
2	ACT composite score	.103	27.05*
3	Self-rating of overall academic ability	.112	9.80*
4	Self-confidence in intellectual ability	.119	7.05*
5	Self-rating of drive to achieve	.122	3.62
6	Expect to make at least a B average	.122	.26
7	Number of years of high school math	.122	.02
8	Expect to graduate with honors	.122	.00

SOURCE: J. Daniel House, "Noncognitive Predictors of Achievement in Introductory College Mathematics," *Journal of College Student Development* 36, No. 2 (1995): 171–181.

House used ordinary multiple regression analysis to evaluate the relative contribution of each variable to grades in mathematics. Notice that with one variable, self-rating of mathematical ability, in the model, $R^2 = .077$ and $F = 80.28$. With 958 observations and one variable in the model, the degrees of freedom for the F statistic are 1 in the numerator and 956 in the denominator; an F of 80.28 is extremely significant. But is it practically significant? Only 7.7% of the variability in grades is explained by the variable! When the second variable, ACT composite score, is added, the F is still significant but falls to 27.05 and R^2 increases to only 10.3%. (Recall that R^2 will always increase when a variable is added to the model.) With the first four variables in the model, R^2 is only 11.9% and the F statistic falls to 7.05. With 953 degrees of freedom in the denominator, $F = 7.05$ is still highly significant, but when we explain less than 12% of the variability, we must wonder whether the model has any practical use in predicting success in mathematics. When we add three variables to the first variable, the gain in R^2 is only 4.2% ($.119 - .077$). Is the addition of these three variables worthwhile? The point of this discussion is that with 900+ degrees of freedom, almost anything will be statistically significant. To judge its practical significance, however, we must rely on our common sense and not so much on our computer.

1. Construct a scatterplot of these data:

   Worksheet: Unit4-1.mtw

x	1	2	2	3	3	4	5	5	6	6	7	8	8	9
y	10	9	8	8	7	6	6	5	3	4	2	1	2	5

   a. Does it appear that a straight line will fit the data?
   b. Are there any influential observations?
   c. Sketch a line on the scatterplot that you think is closest to all the points.
   d. What effect will influential observations have on the line?

2. Physicians recommend one of four types of medication for a certain ailment. To determine whether one medication is preferred over another, a random sample of 60 physicians was polled on what medications they recommend to their patients. Here is a summary of the number of recommendations:

	A	B	C	D	Total
Medication Recommendations	16	12	21	11	60

   a. What statistical test is appropriate for analyzing these data?
   b. State the null hypothesis that the medications are recommended equally.
   c. Test the hypothesis and determine whether there is statistical evidence that the medications are not recommended equally.

3. The reading abilities of fourth-grade students in two different school districts were measured. The number of students falling in the three categories are recorded:

	Reading Ability		
	Low	Average	High
District A	21	77	38
District B	34	64	30

   a. To determine whether there is evidence that the school districts differ in reading ability, should we use the chi-square test of independence or test of homogeneity?
   b. State the null and alternative hypotheses to test the claim of a difference.
   c. Complete the test and state your conclusion.

4. Consider these data:

x	1	2	3	4	5
y	8	6	5	3	1

   a. Find the least squares estimates of β_0 and β_1 in the linear model $y = \beta_0 + \beta_1 x_1 + \varepsilon$.
   b. Find the residuals.
   c. Find the sum of squares due to error.

5. A national standardized test has 350 as the first quartile, 425 as the median, and 560 as the third quartile. A sample of 20 subjects had these scores:

Worksheet: Standard.mtw

| 300 | 480 | 402 | 523 | 628 | 324 | 561 | 349 | 476 | 538 |
| 647 | 572 | 621 | 385 | 491 | 587 | 401 | 346 | 565 | 421 |

a. From the national standard, what percent should score below 350? Between 350 and 425? Above 560?

b. What test should be conducted to determine whether the sample differs significantly from the national standard?

c. State the hypothesis and conduct the test.

d. Does the sample differ significantly from the national standard?

6. This contingency table shows the relationship between scores above and below the median on an examination and ratings of job performance for 100 employees of a company.

		Rating		
		Below	Average	Above
Score	Above Median	11	23	36
	Below Median	14	7	9

Test the hypothesis that job performance is independent of examination results.

7. Fifty people were asked to choose their favorite California wine from four different brands. Their preferences are recorded in the table. Is there statistical evidence that one wine is favored over the others?

Wine	A	B	C	D	total
Number choosing	18	9	11	12	50

8. A research study was designed to compare three methods of therapy for mental patients. From the data, determine whether the rating is independent of the method of therapy.

Worksheet: Therapy.mtw

		Rating		
		Improved greatly	Improved some	Not improved
Therapy	A	5	16	24
	B	9	21	18
	C	15	24	8

9. A psychologist who is evaluating a new testing instrument speculates that the mental ages of a certain group of subjects should be equally distributed in these categories: less than 85, 86 to 95, 96 to 110, 111 to 125, and above 125. A random sample of 35 subjects had the following mental ages:

Worksheet: Instrum.mtw

99	93	87	102	108	93	113	89	115
96	128	105	131	81	109	94	120	136
84	99	126	123	107	95	129	117	84
125	114	136	89	101	140	124	100	

Do the data support the psychologist's theory?

10. Which mail-order catalogs are the best? In an effort to evaluate catalog companies, *Consumer Reports* consulted more than 88,000 readers about their mail-order shopping. The accompanying ratings of 45 clothing catalog companies are based on 140,000 purchases of clothing. The companies were rated on overall satisfaction, the value of the item compared to the price, whether it was delivered in a timely fashion, and whether or not items were in stock. Value, delivery, and stock are rated from 1 (better) to 5 (worse).

Worksheet: Mailord.mtw

Company	Satisfac	Value	Deliv	Stock
L.L. Bean	90	2	1	3
Patagonia	89	4	1	3
Lands' End	88	1	1	2
REI	87	2	1	1
Cabela's	86	1	2	2
Road Runner	85	1	1	2
Talbots	85	3	1	2
Shepler's	85	2	3	2
Eddie Bauer	84	2	1	3
J. Peterman	83	4	1	2
Norm Thompson	83	3	1	1
The Tog Shop	83	2	4	1
Brooks Brothers	82	3	3	1
Orvis	82	3	3	1
Gander Mountain	82	2	2	3
Appleseed's	82	3	2	2
Austad's	81	2	2	3
Jos. A. Bank	81	3	3	3
Nieman Marcus	81	4	2	3
Huntington Cloth	80	3	2	1
J.C. Penney	80	2	1	3
J. Crew	80	4	1	3
Bachrach	80	4	3	3
L'Eggs Showcase	79	1	5	3
Clifford & Willis	78	3	3	4
Cheyenne Outfit	77	3	4	3
Sears	75	2	1	3
King-Size	75	5	3	4
Spiegel	74	4	2	4
Lerner Direct	74	3	3	5
Bloomingdales	73	4	5	5
Brownstone Studio	72	4	5	5
Willow Ridge	72	2	4	3
Victoria's Secret	72	5	4	3
Bedford Fair	72	3	4	3
Old Pueblo Traders	71	3	5	4
Blair Corp.	71	3	5	2
James River Trader	71	3	3	4
Roaman's	70	3	4	5
Lane Bryant	70	4	4	5
Haband	70	2	5	1

Company	Satisfac	Value	Deliv	Stock
International Male	68	5	5	5
Chadwick's of Boston	68	3	5	5
Avon Fashions	67	3	5	5
Fingerhut	59	5	5	2

SOURCE: *Consumer Reports*, October 1994.

a. To determine whether "satisfaction" is different for different levels of "value," examine the one-way analysis of variance. Does "satisfaction" depend on the different levels of "value"? In what way? At what "value" level is "satisfaction" the lowest? The highest? Do you think that "satisfaction" is different for "value" levels 1 through 4? Is this what you expected?

```
One-Way Analysis of Variance

Analysis of Variance on satisfac
Source    DF       SS       MS        F        p
value      4     673.8    168.5     4.68    0.003
Error     40    1439.4     36.0
Total     44    2113.2
                                Individual 95% CIs For Mean
                                Based on Pooled StDev
Level    N      Mean    StDev   --+---------+---------+---------+--
  1      4    84.500    3.873                      (-------*------)
  2     11    80.818    6.210                   (----*----)
  3     17    76.118    5.915              (---*---)
  4      9    78.000    6.164               (----*-----)
  5      4    68.500    6.952   (-------*------)
                                --+---------+---------+---------+--
Pooled StDev =    5.999         64.0      72.0      80.0      88.0
```

b. To determine whether "satisfaction" is different for different levels of "delivery," examine the one-way analysis of variance. Does "satisfaction" depend on the different levels of "delivery"? In what way? At what "delivery" level is "satisfaction" the lowest? The highest? Is this what you expected?

```
One-Way Analysis of Variance

Analysis of Variance on satisfac
Source     DF       SS       MS        F        p
delivery    4     1302.4    325.6    16.06    0.000
Error      40      810.8     20.3
Total      44     2113.2
                                Individual 95% CIs For Mean
                                Based on Pooled StDev
Level    N      Mean    StDev   ---------+---------+---------+-------
  1     12    84.083    4.295                        (---*----)
  2      7    80.857    3.579                    (-----*----)
  3      9    78.667    4.528                (----*----)
  4      7    73.714    4.716        (-----*-----)
  5     10    69.800    5.095   (---*----)
                                ---------+---------+---------+-------
Pooled StDev =    4.502                  72.0      78.0      84.0
```

c. The multiple regression analysis (page 858) has "satisfaction" as the dependent variable and "value," "delivery," and "stock" as independent variables. Is the overall model significant?

What is the *p*-value? What percent of the variability in "satisfaction" is explained by the independent variables? Should any of the variables be removed from the model?

```
The regression equation is
satisfac = 93.5 - 1.46 value - 2.91 delivery - 1.01 stock

Predictor        Coef       Stdev     t-ratio        p
Constant       93.485       1.929       48.47    0.000
value         -1.4610      0.5878       -2.49    0.017
delivery      -2.9067      0.4262       -6.82    0.000
stock         -1.0132      0.5153       -1.97    0.056

s = 3.882      R-sq = 70.8%      R-sq(adj) = 68.6%

Analysis of Variance

SOURCE         DF          SS          MS        F        p
Regression      3     1495.26      498.42    33.07    0.000
Error          41      617.98       15.07
Total          44     2113.24

SOURCE         DF      SEQ SS
value           1      490.55
delivery        1      946.45
stock           1       58.27

Unusual Observations
Obs.    value    satisfac      Fit    Stdev.Fit    Residual    St.Resid
 27      2.00      75.000    84.617      1.089       -9.617      -2.58R
 45      5.00      59.000    69.621      1.761      -10.621      -3.07R

R denotes an obs. with a large st. resid.
```

d. Write a short paragraph explaining how the different variables relate to the customers' satisfaction.

11. When owners register a purebred Arabian horse with the Arabian Horse Registry of America, they must specify one of four major colors: gray, chestnut, bay, or black. By selective breeding, owners can breed for specific colors. The data are the numbers of horses registered under the four colors for 10-year periods starting in 1950. Examine the computer printout and determine whether there is statistical evidence that the color preference has changed over the years. Be sure to state the null and alternative hypotheses, give the test statistic and its *p*-value, and write a conclusion.

Worksheet: Arabian.mtw

Year	Gray	Chestnut	Bay	Black
1950	243	337	167	2
1960	790	806	469	14
1970	3,837	2,638	2,222	71
1980	9,380	5,366	6,084	325
1990	5,803	4,089	6,051	643

SOURCE: Arabian Horse Registry of America, Westminster, Colorado.

```
Chi-Square Test

Expected counts are printed below observed counts

           C1        C2        C3        C4     Total
   1       243       337       167         2       749
        304.43    200.94    227.61     16.02

   2       790       806       469        14      2079
        845.01    557.75    631.79     44.46

   3      3837      2638      2222        71      8768
       3563.75   2352.26   2664.50    187.49

   4      9380      5366      6084       325     21155
       8598.44   5675.41   6428.78    452.37

   5      5803      4089      6051       643     16586
       6741.37   4449.65   5040.31    354.67

Total    20053     13236     14993      1055     49337

ChiSq = 12.396 + 92.129 + 16.141 + 12.266 +
         3.581 +110.496 + 41.944 + 20.865 +
        20.951 + 34.711 + 73.488 + 72.378 +
        71.040 + 16.868 + 18.491 + 35.862 +
       130.618 + 29.231 +202.664 +234.404 = 1250.525
df = 12, p = 0.000
```

12. Biologist Gregor Mendel (1822–1884) is considered to be the father of modern-day genetics. One of his famous experiments involved crossing round yellow pea plants with wrinkled green pea plants. He speculated that he would obtain plants bearing peas in one of four categories. Mendel's theory claimed that the expected frequencies of the categories would be in the proportion 9:3:3:1. These are the probabilities associated with the proportions and the actual numbers observed by Mendel:

Type	Probability	Observed frequency
Round yellow	9/16	315
Round green	3/16	108
Wrinkled yellow	3/16	101
Wrinkled green	1/16	32

a. Test Mendel's theory. Be sure to state the null and alternative hypotheses, give the test statistic and its p-value, and write a conclusion.

b. R. A. Fisher and other well-known statisticians thought that Mendel's observed data were "too good to be true." What do you suppose they meant by this statement?

COMPUTER EXERCISES

13. A sample of 550 voters were classified by political party and opinion on the administration's foreign policy, with the results shown at the top of page 860.

Worksheet: Voters.mtw

| | | Opinion | |
		For	Against
Party	Republican	167	103
	Democrat	96	145
	Other	15	24

Test the hypothesis of independence of political party and opinion on the foreign policy by reading the data into Minitab and conducting a chi-square analysis. Interpret the output.

14. In a pre-election poll to study the influence that age has on voter preference for two presidential candidates, these results were obtained:

Worksheet: Election.mtw

		For Candidate A	For Candidate B	Undecided
Age	20–29	67	117	16
	30–49	109	74	17
	over 49	118	64	18

Test the independence of age and choice of candidate.

15. A school psychologist believes that there is a linear relationship between verbal test scores for eighth-graders and the number of library books they check out. These data were collected from 15 students:

Worksheet: Library.mtw

Books	12	15	3	7	10	5	22	9	13	7	25	17	14	19	20
Scores	77	85	48	59	75	41	94	72	80	70	98	85	83	96	89

a. Graph the data in a scatterplot. Does it appear that a straight line will fit the data?
b. Construct the least squares estimates of β_0 and β_1 for $y = \beta_0 + \beta_1 x + \varepsilon$.
c. Find 90% confidence intervals for the expected y and prediction intervals for a new y when $x' = 10, 15, 20,$ and 25.

16. Fifty residents were asked whether they agree or disagree with the current school board policy on busing. The table gives the gender and opinion of each respondent (M—male, F—female; f—for, a—against):

Worksheet: Busing.mtw

Gender	M	F	F	M	F	M	M	M	F	F	F	M	F	M	F	M	M
Opinion	f	a	f	f	f	a	f	a	f	f	a	f	f	f	a	a	f

Gender	F	M	F	F	M	F	F	F	M	M	F	M	F	F	M	F	M
Opinion	f	f	a	f	f	a	f	f	a	f	f	f	a	a	f	a	f

Gender	M	M	F	F	M	F	F	M	M	M	F	F	M	F	F	M
Opinion	f	a	a	f	f	a	f	f	a	f	f	a	f	a	a	f

Retrieve the worksheet and construct a two-way table with gender and opinion. Conduct a chi-square test of independence. Interpret the output.

17. Consider these data:

Worksheet: Unit4-17.mtw

x_1	20	25	30	35	40	45	50	55	60
x_2	12	19	10	6	9	15	10	6	3
y	1.6	2.1	2.7	3.3	4.0	4.7	4.5	4.9	5.1

a. Construct scatterplots for y versus x_1 and y versus x_2.
b. Use the Regression command to find the least squares estimates of β_0, β_1, and β_2 in the model $y = \beta_0 + \beta_1 x_1 + \beta_2 x_2 + \varepsilon$.
c. What is the value of R^2? What does this tell you about the proposed model?
d. State the hypothesis for testing the utility of the model. Perform the test. Is it significant?
e. Investigate each of the t-tests on the individual regression coefficients. What conclusions can you make?

18. Concern over the academic performance of student athletes prompted the following study. The SAT scores (combined verbal and math) of athletes at randomly selected major public universities, minor public universities, and private universities were recorded:

Worksheet: Academic.mtw

Major	Minor	Private
960	1040	980
870	790	1060
1130	820	920
940	670	1240
730	840	1130
640	960	970
820	780	740
650	850	990
920	940	950
840	990	880
940	1160	1270
1050	890	900
750	750	1050
840	930	840
1280	870	960
930	1300	1370
750	950	1100
880	800	750
790	840	960
1030	920	850

a. Construct normal probability plots and boxplots of the data in the three samples. Comment on the shapes of the distributions.
b. Based on the results of part **a**, should the analysis of the data be an ordinary F-test or the Kruskal–Wallis test?

19. Using the test that you recommended in part **b** of Exercise 18, perform an analysis to determine whether a significant difference exists between the SAT scores for student athletes at the different types of school.

20. A survey was given to 556 adults 18 years of age and older to determine whether premarital sex is permissible. From the following results, determine whether their opinion is independent of their age.

Worksheet: Premarit.mtw

		View Toward Premarital Sex		
		OK	Not OK	No Opinion
	18–21	63	41	2
	22–29	58	54	4
Age	30–39	46	82	4
	40–54	24	79	6
	above 55	11	77	5

a. Retrieve the worksheet and conduct a chi-square analysis.
b. Interpret the output.

21. A study was conducted to compare the amount of time it takes to deliver a package for three major overnight delivery companies. For each company, a sample of 20 packages were shipped from the same point to the same destination. Following is the amount of time (in hours) to deliver the packages.

Worksheet: Overnigh.mtw

Company A	Company B	Company C
26	26	24
24	27	27
34	30	25
25	28	22
23	23	30
20	24	26
25	27	24
26	29	23
30	32	27
22	25	31
27	26	28
23	37	25
25	26	33
22	22	23
28	27	25
21	26	38
24	25	26
31	24	23
24	28	22
25	30	24

a. Construct normal probability plots and boxplots of the three samples. Comment on the shapes of the distributions.
b. Based on the results of part **a**, should the analysis of the data be an ordinary F-test or the Kruskal–Wallis test?

22. Using the test that you recommended in part **b** of Exercise 21, perform an analysis to determine whether a significant difference exists between the amount of time it takes to deliver packages for the three companies.

23. The following table classifies a group of people according to income bracket and time elapsed since they last consulted with a physician. Is there evidence of an association between the income bracket and the duration of time since the last visit with a physician?

Worksheet: Bracket.mtw

	Last consulted with physician		
Income bracket	Less than 6 mos.	6 mos. to 1 yr.	More than 1 yr.
less than $10,000	192	35	41
$10,000 to $19,999	124	43	65
$20,000 to $29,999	135	51	64
$30,000 to $39,999	174	62	108
$40,000 or more	121	65	52

24. A study was conducted to investigate the relationship of aptitude test scores and years of education on productivity in a factory. The results for eight employees follow.

Worksheet: Product.mtw

Education level	8	12	13	11	8	14	12	10
Aptitude score	9	17	20	19	20	23	18	15
Productivity	23	35	29	33	40	32	28	26

Use the Regression command to:
a. Find the least squares estimates of β_0, β_1, and β_2.
b. Find the coefficient of determination.
c. Test the hypothesis that $\beta_1 = 0$.
d. Test the hypothesis that $\beta_2 = 0$.
e. Conduct a residual analysis.
f. Considering your answers to parts **c**, **d**, and **e**, should the model be modified? Explain.

25. An experiment was conducted to study the effect of four stimuli on reaction time. The reaction times (in seconds) were as follows:

Worksheet: Stimuli.mtw

Stimuli			
A	B	C	D
.52	.56	.85	.51
.87	.72	.98	.57
.73	.65	.76	.64
.65	.57	.65	.68
.68	.54	1.09	.59
.77	.68	.89	.94
.52	.83	.79	.68
.88	.63	.94	.66
.72	.76	.80	.73
.94	.74	1.14	.65
.79	.81	.93	.60
.74	.87	.74	.88

a. Construct normal probability plots of the four samples. Is normality a reasonable assumption in each case?
b. Construct boxplots of the four samples. Does the variability in the four distributions appear the same?

c. Should the ordinary F-test or Kruskal–Wallis test be used to compare the four distributions?

d. Complete the test recommended in part **c** to determine whether there is a significant difference in the reaction times due to the four stimuli.

26. For the linear model

$$y = 250 + 3.5x + \varepsilon$$

and the fixed values $x = 2$ and $\sigma = 5$, simulate 500 different values for y. Complete a histogram, boxplot, and normal probability plot for the generated values.

DISCUSSION EXERCISES

1. On September 22, 1993, President Clinton presented his health-care plan to Congress. The next day a poll of 500 adult Americans taken for Time/CNN by Yankelovich Partners Inc. showed that 57% approved, 31% opposed, and 12% were "not sure." Five weeks later a second poll of 500 adult Americans revealed that 43% approved, 36% opposed, and 21% were "not sure."

To statistically analyze these data, should we conduct a chi-square test of independence or a chi-square test of homogeneity? That is, in using the chi-square statistic, are we investigating the independence of two variables in a single population or are we attempting to determine whether two population distributions are homogeneous with respect to a variable? Write a paragraph explaining your reasoning. Complete the test that you think is appropriate and give a summary of the results.

2. It took from the beginning of time until 1950 for the Earth to acquire its first 2.5 billion people. The population of the Earth doubled to 5 billion in the next 40 years. At the current rate of population growth, the United Nations projects that the population by the year 2150 will be over 125 times the present population size (*Discover*, November 1992, pp. 114–119).

Following are the population sizes from the beginning of time to current time and United Nations projections through 2150.

Worksheet: Worldpop.mtw

Year	Billions of people
1	.3
1650	.5
1850	1.131
1950	2.516
1975	4.079
1990	5.311
2000	6.463
2025	10.978
2050	21.161
2075	46.261
2100	109.405
2125	271.138
2150	694.213

Graph these data in a bar chart. Does it appear that the United Nations used a linear model to come up with the projected population sizes? Do you think it is reasonable to use current population growth trends to project into the future for 150 years? The area of the surface of the Earth is 196,938,800 square miles, oceans included. How many people would there be for each square mile by the year 2150? Do you think the Earth could sustain that many people? Is extrapolation an issue here? Write a paragraph addressing these issues.

3. In the summer of 1975, an area of Florida was used in a cloud seeding experiment. The goal was to determine whether seeding clouds with silver iodide would increase rainfall. In the experiment, 24 days were judged suitable for seeding. Twelve of the days were randomly

selected for seeding and the remaining days were used as controls. The response variable was the amount of rainfall that fell on the target area for a 6-hour period.

Worksheet: Cloud.mtw

Control	Seeded
12.85	5.52
6.11	6.29
3.61	2.45
.47	5.06
4.56	2.76
6.35	4.05
5.74	4.84
4.45	11.86
3.66	4.22
1.16	5.45
.82	2.02
.28	1.09

SOURCE: Woodley, W., Simpson, J., Biondini, R., and Berkeley, J. (1977). "Rainfall Results Florida Area Cumulus Experiment." *Science*, 195, pp. 735–742. The data also appear in Hand et al., *A Handbook of Small Data Sets*, Chapman and Hall, London.

a. From the following printout of the two-sample t-test, is there statistical evidence that the cloud seeding increased rainfall?

```
Two Sample T-Test and Confidence Interval

Twosample T for C1
C2   N      Mean    StDev   SE Mean
0    12     4.17    3.52    1.0
1    12     4.63    2.78    0.80


95% C.I. for mu 0 - mu 1: ( -3.1,  2.22)
T-Test mu 0 = mu 1 (vs not =): T= -0.36  P=0.72  DF=  22
Both use Pooled StDev = 3.17
```

b. From the following printout of the one-way analysis of variance, is there statistical evidence that the cloud seeding increased rainfall?

```
One-Way Analysis of Variance

Analysis of Variance on C1
Source    DF      SS      MS       F       p
C2         1     1.3     1.3     0.13    0.724
Error     22   221.1    10.0
Total     23   222.3
                               Individual 95% CIs For Mean
                               Based on Pooled StDev
Level     N     Mean    StDev   --+---------+---------+---------+----
   0      12    4.172   3.519   (---------------*---------------)
   1      12    4.634   2.777      (--------------*-------------)
                               --+---------+---------+---------+----
Pooled StDev =  3.170           2.4       3.6       4.8       6.0
```

c. In comparing the results, is there a difference between the two printouts?

d. In Chapter 11 we showed that simple linear regression is a special case of multiple regression when there is only one independent variable. In fact, we showed that the relationship between the t statistic and the F statistic is $F = t^2$. Is this relationship also true in analysis of variance when there are only two groups? Explain the relationship based on the accompanying two printouts.

REFERENCES

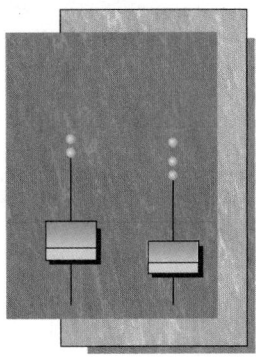

Andrews, D. F., and A. M. Herzberg. *Data—A Collection of Problems from Many Fields for the Student and Research Worker.* New York: Springer-Verlag, 1985.

Bellout, G. "Is Vitamin C Really Good for Colds?" *Consumer Reports,* February 1976, pp. 66–70.

Box, G. E. P. "Some Theorems on Quadratic Forms Applied in the Study of Analysis of Variance Problems, I. Effect of Inequality of Variance in the One-Way Classification." *Annals of Mathematical Statistics,* 25 (1954), pp. 290–302.

Box, G. E. P., G. Hunter, and J. S. Hunter. *Statistics for Experimentation.* New York: Wiley, 1978.

Box, G. E. P., and G. M. Jenkins. *Time Series Analysis, Forecasting and Control.* San Francisco: Holden-Day, 1970.

Chambers, J. M., W. S. Cleveland, B. Kleiner, and P. A. Tukey. *Graphical Methods for Data Analysis.* Boston: Duxbury Press, and Belmont, Calif.: Wadsworth, 1983.

Cochran, W. G. "Some Methods for Strengthening the Common χ^2 Tests," *Biometrics,* 10 (1954), pp. 417–451.

Conover, W. J. *Practical Nonparametric Statistics,* 2d ed. New York: Wiley, 1980.

Conover, W. J. and R. L. Iman. "Rank Transformation as a Bridge Between Parametric and Nonparametric Statistics," *The American Statistician,* 35, no. 3 (1981).

Corley, W. C., director. *Crime in North Carolina—1979—Uniform Crime Report.* Raleigh: State of North Carolina Department of Justice, 1979.

Cunliffe, B. *Danebury: An Iron-Age Hill Fort in Hampshire.* Research Report 2. London: Council for British Archaeology, 1984.

Daniel, C., and F. Wood. *Fitting Equations to Data.,* 2d ed. New York: Wiley, 1980.

Davis, J. *Statistics and Data Analysis in Geology,* 2d ed. New York: Wiley, 1986.

Dawkins, B. "Multivariate Analysis of National Track Records," *The American Statistician,* 43, no. 2 (1989), pp. 110–115.

Famighetti, R., ed. *The World Almanac & Book of Facts—1995.* New Jersey: Funk & Wagnalls Corp., 1994.

Fienberg, S. E. "Randomization and Social Affairs: The 1970 Draft Lottery," *Science,* 171 (January 22, 1971), pp. 255–261.

Freedman, D., R. Pisani, R. Purves, and A. Adhikari. *Statistics,* 2d ed. New York: W. W. Norton, 1991.

Freeh, L. J., director. *Crime in the United States—Uniform Crime Reports for the United States.* Washington, D.C.: Federal Bureau of Investigation, U.S. Department of Justice, 1995.

Gnanadesikan, R., ed. *Statistical Data Analysis—Proceedings of Symposia in Applied Mathematics—Volume 28.* Providence, R.I.: American Mathematical Society, 1983.

Greenhouse, J. B., and S. M. Greenhouse. "An Aspirin a Day . . .?" *Chance,* 1, no. 4 (Fall 1988), pp. 24–31.

Gross, A. M. "Confidence Interval Robustness with Long-Tailed Symmetric Distributions," *Journal of the American Statistical Association,* 71, no. 354 (1976).

Haack, D. G. *Statistical Literacy—A Guide to Interpretation.* Boston: Duxbury Press, 1979, p. 35.

Hamilton, T. H., I. Rubinoff, R. H. Barth, Jr., and G. L. Bush. "Species Abundance: Natural Regulation of Insular Variation," *Science,* 142 (1963), pp. 1575–1577.

Hand, D. J., et al., eds. *A Handbook of Small Data Sets.* London: Chapman and Hall, 1994.

Hartwig, F., and B. E. Dearing. *Exploratory Data Analysis.* Beverly Hills, Calif.: Sage Publications, 1979.

Hoaglin, D. C., F. Mosteller, and J. W. Tukey. *Understanding Robust and Exploratory Data Analysis.* New York: Wiley, 1983.

Hooke, R. "Basketball, Baseball, and the Null Hypothesis," *Chance,* 2, no. 4 (1989), pp. 35–37.

Hora, S. C., and W. J. Conover. "The *F* Statistic in the Two-Way Layout with Rank-Score Transformed Data," *Journal of the American Statistical Association,* 79, no. 387 (1984).

Huber, P. J. "Robust Statistics: A Review," *The Annals of Mathematical Statistics,* 43, no. 4 (1972).

Iman, R. L., S. C. Hora, and W. J. Conover, "Comparison of Asymptotically Distribution-Free Procedures for the Analysis of Complete Blocks," *Journal of the American Statistical Association,* 79, no. 387 (1984).

Ito, P. K. "Robustness of ANOVA and MANOVA Test Procedures," *Handbook of Statistics,* vol. 1, P. R. Krishnaiah, ed. Amsterdam: North-Holland, 1980, pp. 199–236.

Joglekar, G., J. H. Schuenemeyer, and V. LaRiccia. "Lack-of-Fit Testing When Replicates Are Not Available," *The American Statistician,* 43, no. 3 (1989), pp. 135–143.

Johnson, M. P., and P. H. Raven. "Species Number and Endemism: The Galápagos Archipelago Revisited," *Science,* 179, pp. 893–895.

Kalbfleisch, J., and J. Lawless. "An Analysis of the Data on Transfusion-Related AIDS," *Journal of the American Statistical Association,* 84 (1989), pp. 360–372.

Kaufman, P., A. Herlihy, J. Elwood, J. Mitch, W. Overton, M. Sale, J. Messer, K. Cougan, D. Pech, K. Reckhow, A. Kinney, S. Christie, D. Brown, C. Hagley, and H. Jager. *Chemical Characteristics of Streams in the Mid-Atlantic and Southeastern U.S., Vol. I: Population Descriptions and Physico-Chemical Relationships,* EPA/600/3088/021a. Washington, D.C.: U.S. Environmental Protection Agency, 1988.

Kempthorne, O. "Teaching of Statistics: Content Versus Form," *The American Statistician,* 31, no. 1 (1980).

Koopmans, L. H. *An Introduction to Contemporary Statistics.* Boston: Duxbury Press, 1981.

Larkey, P. D., R. A. Smith, and J. B. Kadane. "It's OK to Believe in the 'Hot Hand,'" *Chance,* 2, no. 4 (1989), pp. 22–30.

Maguire, K., and A. L. Pastore, eds. *Sourcebook of Criminal Justice Statistics—1994.* Albany, N.Y.: Criminal Justice Research Center, 1995.

McEntire, A., and A. N. Kitchens, "A New Focus for Educational Improvement Through Cognitive and Other Structuring of Subconscious Personal Axioms," *Education,* 105, no. 2 (1984).

Meier, P. "The Biggest Public Health Experiment Ever: The 1954 Field Trial of the Salk Polio Vaccine." *Statistics: A Guide to the Unknown,* 2d ed., J. Tanur, et al., eds. San Francisco: Holden-Day, 1978.

Miller, J. C., and J. N. Miller. *Statistics for Analytical Chemistry,* 2d ed. New York: Halsted Press, 1988.

Moore, D. S. *Statistics—Concepts and Controversies,* 3d ed. San Francisco: W. H. Freeman, 1991.

Moser, B. K., and G. R. Stevens. "Homogeneity of Variances in the Two-Sample Means Test," *The American Statistician,* 46 (1992), pp. 19–21.

Mosteller, F. "The Teaching of Statistics: Classroom and Platform Performance," *The American Statistician,* 34, no. 1 (1980).

Mosteller, F., and J. W. Tukey. *Data Analysis and Regression, A Second Course in Statistics.* Reading, Mass.: Addison-Wesley, 1977.

Naylor, J. and A. Smith. "An Archaeological Inference Problem," *Journal of the American Statistical Association,* 83 (1988), pp. 588–595.

Neter, J., M. H. Kutner, C. J. Nachtsheim, and W. Wasserman. *Applied Linear Statistical Models,* 4th ed. Chicago: Irwin, 1996.

Noether, G. E. "The Role of Nonparametrics in Introductory Statistics Courses," *The American Statistician,* 34, no. 1 (1980).

Roll, C. W., Jr., and A. H. Cantril. *Polls: Their Use and Misuse in Politics.* New York: Basic Books, 1972.

Rousseeuw, P., and A. Leroy. *Robust Regression and Outlier Detection,* New York: Wiley, 1987.

Ryan, B. F., and B. L. Joiner. *Minitab Handbook Third Edition.* Boston: Duxbury Press, 1994.

Sacks, J., and D. Ylvisaker, "A Note on Huber's Robust Estimation of a Location Parameter," *The Annals of Mathematical Statistics,* 43, no. 4 (1972).

Schmoyer, R. "Permutation Tests for Correlation in Regression Errors," *Journal of the American Statistical Association,* 89 (1994), pp. 1507–1516.

Soofi, E., N. Ebrahimi, and M. Habibullah. "Information Distinguishability with Application to Analysis of Failure Data," *Journal of the American Statistical Association,* 90 (1995), pp. 657–668.

Staudte, R., and S. Sheather. *Robust Estimation and Testing.* New York: Wiley, 1990.

Stigler, S. M. "Do Robust Estimators Work with 'Real' Data?" *The Annals of Statistics,* 5, no. 6 (1977).

Sutton, G. W., T. J. Huberty, and R. Price. "An Evaluation of Teacher Stress and Job Satisfaction," *Education,* 105, no. 2 (1984).

Tanur, J., et al., eds. *Statistics: A Guide to the Unknown,* 3d ed. San Francisco: Holden-Day, 1989.

Tong, H. *Threshold Models in Nonlinear Time Series Analysis.* New York: Springer-Verlag, 1983.

Tukey, J. W. *Exploratory Data Analysis.* Reading, Mass.: Addison-Wesley, 1977.

Tukey, J. W., and D. H. McLaughlin. "Less Vulnerable Confidence and Significance Procedures for Location Based on a Single Sample: Trimming/Winsorization 3," *Sankhyā Series A,* 25 (1963), pp. 331–352.

Tversky, A., and T. Gilovich. "The Cold Facts about the 'Hot Hand' in Basketball," *Chance,* 2, no. 1 (1989), pp. 16–21.

Velleman, P. F., and D. C. Hoaglin. *Applications, Basics, and Computing of Exploratory Data Analysis.* North Scituate, Mass.: Duxbury Press, 1981.

vos Savant, M. "Ask Marilyn," *Parade,* September 9, 1990, p. 15.

vos Savant, M. "Ask Marilyn," *Parade,* December 2, 1990, p. 25.

vos Savant, M. "Ask Marilyn," *Parade,* February 17, 1991, p. 12.

vos Savant, M. "Ask Marilyn," *Parade,* July 7, 1991, p. 28.

Wade, N. "IQ and Heredity: Suspicion of Fraud Beclouds Classic Experiment," *Science,* 194 (1976), pp. 916–919.

Wheeler, M. *Lies, Damn Lies, and Statistics.* New York: Liveright, 1976.

Wilner, D. M., et al. *The Housing Environment and Family Life.* Baltimore: Johns Hopkins Press, 1962.

Yashchin, E. "Likelihood Ratio Methods for Monitoring Parameters of a Nested Random Effect Model," *Journal of the American Statistical Association,* 90 (1995), pp. 729–738.

Ying, Z., S. Jung, and L. Wei. "Survival Analysis with Median Regression Models," *Journal of the American Statistical Association,* 90 (1995), pp. 178–184.

Yuen, K., and W. J. Dixon. "The Approximate Behavior and Performance of the Two-Sample Trimmed *t.*" *Biometrika,* 60, no. 2 (1973), p. 369.

APPENDIX

A

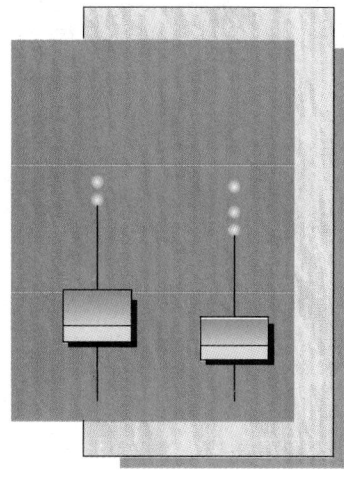

DATA SETS

ANSCOMBE.MTW	Exercise 3.42 Exercise 3.60 Exercise 3.69	Four historical data sets by Anscombe to illustrate that data can be very similar yet very different
ANXIETY.MTW	Exercise 3.90	Math test scores versus anxiety scores before the test
APOLIPOP.MTW	Example 11.2	Level of apolipoprotein B and number of cups of coffee consumed per day for 15 adult males
APPEND.MTW	Exercise 2.137	Median costs of an appendectomy at 20 hospitals in North Carolina
APPENDEC.MTW	Exercise 12.74	Median costs of appendectomies at three different types of North Carolina hospitals
APTITUDE.MTW	Exercise 3.15 Exercise 3.36 Exercise 3.50 Exercise 3.64	Aptitude test scores versus productivity in a factory
ARABIAN.MTW	Exercise U4.11	Colors of registered Arabian horses from 1950 through 1990
ARCHAEO.MTW	Example 2.23 Example 7.9 Exercise 7.109 Exercise 12.61	Radiocarbon ages of observations taken from an archaeological site
ARTHRITI.MTW	Exercise 12.72	Time of relief for three treatments of arthritis
ARTIFICI.MTW	Exercise 2.123	Durations of operation for 15 artificial heart transplants
ASPRIN.MTW	Exercise 12.65	Dissolving time versus level of impurities in aspirin tablets
ASTHMATI.MTW	Exercise 9.62	Asthmatic relief index on 9 subjects given a drug and a placebo
ATTITUDE.MTW	Exercise 11.44	Years of experience, size of school, and attitude toward extracurricular activities for 15 teachers
ATTORNEY.MTW	Example 3.6	Number of convictions reported by U.S. attorney's offices
AUTOGEAR.MTW	Exercise 9.55	Number of defective auto gears produced by two manufacturers
BASEBALL.MTW	Exercise 1.103	Baseball salaries for four different major league teams
BATTERY.MTW	Exercise U3.6	Lifetimes of 15 batteries
BIGTEN.MTW	Exercise 1.104 Exercise 3.89	Graduation rates for student athletes and nonathletes in the Big Ten Conference
BIOLOGY.MTW	Exercise 2.48	Test scores on first exam in biology class
BIRTH.MTW	Example 2.11	Live birth rates in 1990 for all states
BLACKEDU.MTW	Exercise 10.61	Education level of blacks by gender
BLKLUNG.MTW	Example 7.15	Medical expenses of a sample of 16 coal miners recovering from black lung
BLOOD.MTW	Exercise 9.103	Blood pressure of 15 adult males taken by machine and by an expert
BOARD.MTW	Exercise 12.14	Incomes of board members from three different universities
BONES.MTW	Example 9.20	Bone density measurements of 35 physically active and 35 non-active women
BOOKS.MTW	Exercise 11.84	Number of books read and final spelling scores for 17 third graders
BOOKSTOR.MTW	Exercise 12.41	Prices paid for used books at three different bookstores
BOOTSTRP.MTW	Exercise 5.36	Worksheet to illustrate a sampling distribution by the bootstrap method
BRACKET.MTW	Exercise U4.23	Last time consulted with physician versus income bracket
BRAIN.MTW	Example 3.7 Exercise 3.25 Example 3.23 Exercise 3.70	Brain weight versus body weight of 28 animals

FAITHFUL.MTW	Exercise 6.13 Exercise 7.89	Waiting times between successive eruptions of the Old Faithful geyser
FAMILY.MTW	Exercise 3.81	Size of family versus cost per person per week for groceries
FERRARO1.MTW	Exercise 10.24	Choice of presidental ticket in 1984 by gender
FERRARO2.MTW	Exercise 10.24	Choice of vice presidental candidate in 1984 by gender
FERTILIT.MTW	Exercise 2.150 Exercise U2.45	Fertility rates of all 50 states and D.C.
FIRSTCHI.MTW	Exercise 2.88 Exercise 6.7	Ages of women at the birth of their first child
FISH.MTW	Exercise 7.61 Exercise 7.107 Exercise 9.44	Length and number of fish caught with small and large mesh codend
FITNESS.MTW	Exercise 9.93	Number of sit-ups before and after a physical fitness course
FLIES.MTW	Exercise U3.29	Number of flies landing on boards treated with two different brands of fly spray
FLUID.MTW	Exercise 7.51	Breakdown times of an insulating fluid under various levels of voltage stress
FOOD.MTW	Exercise U1.22 Exercise 6.62	Annual food expenditures for 40 single households in Ohio
FOOTBALL.MTW	Exercise 2.84 Exercise 2.107	Worth of teams in the National Football League
FRAMINGH.MTW	Exercise 2.55 Exercise 2.74 Exercise 2.94 Exercise 4.128 Exercise 7.36	Cholesterol values of 62 subjects in the Framingham Heart Study
FRESHMAN.MTW	Exercise 8.65	Ages of a random sample of 30 college freshmen
FUEL.MTW	Exercise 6.1	Fuel-efficiency ratings for the 10 highest-rated cars and trucks
FUNERAL.MTW	Exercise 10.60	Cost of funeral by region of country
GALAPAGO.MTW	Exercise 11.97	Number of plant species on the Galápagos Islands versus several variables related to area and elevation
GALAXIE.MTW	Example 6.4	Velocities of 82 galaxies in the Corona Borealis region
GANNETT.MTW	Exercise 3.92	Ranking of states by Gannett News service versus SAT scores
GASOLINE.MTW	Exercise 2.44	Price of regular unleaded gasoline obtained from 25 service stations
GERMAN.MTW	Exercise 9.69	Number of errors in copying a German passage before and after an experimental course in German
GOLF.MTW	Exercise 6.39	Distances a golf ball can be driven by 20 professional golfers
GOVERNOR.MTW	Exercise 6.67	Annual salaries for state governors in 1994
GPA.MTW	Example 3.17	High school GPA versus college GPA
GRADES.MTW	Exercise 2.138	Test grades in a beginning statistics class
GRADUATE.MTW	Exercise 1.38 Exercise 1.102	Graduation rates for student athletes in the Southeastern Conference
GRAMMAR.MTW	Example 10.2	Grammar test scores for 25 students
GREENRIV.MTW	Exercise 8.69	Varve thickness from a sequence through an Eocene lake deposit in the Rocky Mountains
GRNRIV2.MTW	Exercise 8.57	Thickness of a varved section of the Green River oil shale deposit near a major lake in the Rocky Mountains

GROCERIE.MTW	Exercise 2.147	Per resident sales of groceries for the 50 states
GROUPS.MTW	Exercises 12.4–12.6	An illustration of analysis of variance
GROWTH.MTW	Example 11.12	Growth stimulant and weight gain in 11 lab animals
GYM.MTW	Exercise 3.31 Exercise 3.83 Exercise 11.32 Exercise 11.45	Children's age versus number of completed gymnastic activities
HABITS.MTW	Exercise 9.64	Study habits of students in two matched school districts
HALFWAY.MTW	Example 2.10	Yearly per bed rental costs in 35 halfway houses
HAPTOGLO.MTW	Example 8.9	Haptoglobin concentration in blood serum of 8 healthy adults
HARDWARE.MTW	Exercise 7.68	Daily receipts for a small hardware store for 31 working days
HARDWOOD.MTW	Example 3.22 Exercise 11.49 Exercise 11.62	Tensile strength of kraft paper for different percentages of hardwood in the batches of pulp
HEALTH.MTW	Exercise 7.66	Claim amounts for a random sample of 30 submitted medical claims
HEAT.MTW	Exercise 2.36	Primary heating sources of homes on Indian reservations versus all households
HEATING.MTW	Exercise 12.43	Fuel efficiency ratings for three types of oil heaters
HISTORY.MTW	Exercise 3.35	Ranking of professor and final grades by students in a beginning history class
HOMES.MTW	Chapter 3 SI Exercise 7.110 Exercise 7.111	Median prices of single-family homes in 65 metropolitan statistical areas
HOMEWORK.MTW	Exercise 9.87	Number of hours per week spent on homework for private and public high school students
HONDA.MTW	Chapter 8 SI Exercise 8.125	Miles per gallon for a Honda Civic on 35 different occasions
HOSTILE.MTW	Example 12.10	Hostility levels of high school students from rural, suburban, and urban areas
HOUSING.MTW	Exercise 6.68 Exercise 7.60	Median home prices for 1984 and 1993 in 37 markets across the U.S.
HURRICAN.MTW	Exercise 12.18	Number of storms, hurricanes, and El Niño effects from 1950 through 1995
ICEBERG.MTW	Exercise 3.59	Number of icebergs sighted each month south of Newfoundland and south of the Grand Banks in 1920
INCOME.MTW	Exercise 2.26	Percent change in personal income from 1st to 2nd quarter in 1994
INDIAN.MTW	Exercise 3.87	Educational attainment versus per capita income and poverty rate for American Indians living on reservations
INDIAPOL.MTW	Exercise 2.154	Average miles per hour for the winners of the Indianapolis 500 race
INDY500.MTW	Exercise 9.11 Exercise 9.43	Qualifying miles per hour and number of previous starts for drivers in 79th Indianapolis 500 race
INFLATIO.MTW	Exercise 3.22 Exercise 3.39	Private pay increase of salaried employees versus inflation rate
INLETOIL.MTW	Exercise 7.84 Exercise 8.51	Inlet oil temperature through a valve
INMATE.MTW	Chapter 10 SI Exercise 10.67	Type of drug offense by race

INSPECT.MTW	Exercise 10.66	Percent of vehicles passing inspection by type inspection station
INSTRUM.MTW	Exercise U4.9	Mental ages as reflected by a new testing instrument
INSULATE.MTW	Exercise 11.80	Heat loss through a new insulating medium
IQ.MTW	Exercise 6.37	IQ scores on a random sample of 100 citizens in a particular community
IQGPA	Exercise 11.83	GPA versus IQ for 12 individuals
IRISES.MTW	Example 6.3	R. A. Fisher's famous data on sepal length of a species of *iris setosa*
JAIL.MTW	Exercise 2.146 Exercise 3.91	Number of inmates in jails across the U.S. in 1983 and 1993
JDPOWER.MTW	Exercise 3.24 Exercise 3.41	Number of problems reported per 100 cars in 1994 versus 1995
JOB.MTW	Chapter 2 SI Exercise 2.155	Starting salaries for 28 different majors
JOBSAT.MTW	Exercise 11.92	Job satisfaction and stress level for 9 school teachers
KENNEDY.MTW	Exercise 7.34	Number of visitors to the John F. Kennedy Library for 12 randomly selected months
KIDSMOKE.MTW	Exercise 5.97	Smoking habits of boys and girls ages 12 to 18
KILOWATT.MTW	Example 6.14	Rates per kilowatt-hour for each of the 50 states and D.C.
KINDER.MTW	Exercise 9.90	Reading scores for first grade children who attended kindergarten versus those who did not
LAMINECT.MTW	Exercise 12.17	Median costs of laminectomies at hospitals across North Carolina in 1992
LEAD.MTW	Example 2.13	Lead levels in children's blood whose parents worked in a battery factory
LEADER.MTW	Exercise 9.39	Leadership exam scores by age for employees in an industrial plant
LETHAL.MTW	Example 8.12	Survival time of mice injected with an experimental lethal drug
LIBRARY.MTW	Exercise U4.15	Verbal test scores and number of library books checked out for 15 eighth graders; illustrates predicting intervals
LIFE.MTW	Exercise 2.19	Life expectancy of men and women in U.S.
LIFESPAN.MTW	Exercise 3.18 Exercise 3.55	Life span of electronic components used in a spacecraft versus heat
LODGE.MTW	Exercise 12.44	Measured traffic at three prospective locations for a motor lodge
LOTTERY.MTW	Exercise 2.149 Exercise 6.53	Commissions paid lottery agents in 18 states
LOWABIL.MTW	Example 9.15	Reading skills of 24 matched low-ability students
LOWTEMP.MTW	Exercise U1.18	Low temperatures in major cities in the east
MAGNESIU.MTW	Exercise 11.11	Magnesium concentration and distances between samples
MAILORD.MTW	Exercise U4.10	Ratings of 45 clothing catalog companies versus value, delivery, and whether items are in stock
MALPRACT.MTW	Exercise 7.50	Amounts awarded in 17 malpractice cases
MANAGER.MTW	Exercise 7.57	Advertised salaries offered general managers of major corporations in 1995
MARKED.MTW	Exercise 8.124	Percent of marked cars in 65 police departments in Florida
MARKET.MTW	Example 8.13	Starting salaries of 65 master's level marketing majors
MATH.MTW	Exercise 2.81	Standardized math test scores for 30 students
MATHCOMP.MTW	Exercise 6.41	Standardized math competency for a group of entering freshmen at a small community college

PARENTED.MTW	Exercise 2.25	Education backgrounds of parents of entering freshmen at a state university
PATROL.MTW	Example 11.4 Example 11.17	Years of experience and number of tickets given by patrolpersons in New York City
PEARSON.MTW	Exercise 3.30	Karl Pearson's data on heights of brothers and sisters
PGMOVIE.MTW	Exercise 7.70	Ages of 9 randomly selected people attending a PG-rated movie
PHONE.MTW	Exercise 8.60	Length of long-distance phone calls for a small business firm
POISON.MTW	Exercise 2.133	Number of poisonings reported to 16 poison control centers
POLITIC.MTW	Example 10.4	Political party and gender in a voting district
POLLUTIO.MTW	Exercise 7.35	Air pollution index for 15 randomly selected days for a major western city
POROSITY.MTW	Exercise 6.55 Exercise 7.78	Porosity measurements on 20 samples of Tensleep Sandstone, Pennsylvanian from Bighorn Basin in Wyoming
POVERTY.MTW	Exercise 11.8 Exercise 11.20	Percent poverty and crime rate for selected cities
PRECINCT.MTW	Exercise 3.16 Exercise 3.52 Exercise 3.65	Robbery rates versus percent low income in 8 precincts
PREJUDIC.MTW	Example 6.16 Example 7.11	Racial prejudice measured on a sample of 25 high school students
PREMARIT.MTW	Exercise U4.20	Age versus view toward premarital sex
PRESIDEN.MTW	Exercise 2.146	Ages at inauguration and death of U.S. presidents
PRESIDEN.XLS	Exercise 2.146	Ages at inauguration and death of U.S. presidents—Excel worksheet
PRESS.MTW	Exercise 11.87	Degree of confidence in the press versus education level for 20 randomly selected persons
PRIME.MTW	Exercise 11.33 Exercise 11.58	Prime interest rate versus number of cars sold by automobile dealer
PRISON.MTW	Exercise 11.95	Number of adults on probation, in jail or prison, or on parole from 1980 through 1993
PROCESSO.MTW	Exercise U3.28	Processor speeds of competing brands of computers
PRODUCT.MTW	Exercise U4.24	Effect of aptitude and education on productivity in a factory
PROGNOST.MTW	Exercise 8.64	Klopfer's prognostic rating scale for subjects receiving behavior modification therapy
PROGRAM.MTW	Exercise 12.16	Effects of four different methods of programmed learning for statistics students
PSAT.MTW	Exercise 3.19	PSAT scores versus SAT scores
PSYCH.MTW	Exercise 2.41	Correct responses for 24 students in a psychology experiment
PSYCHOL.MTW	Exercise 8.76	Psychology test scores for 36 beginning psychology students
PUERTO.MTW	Exercise 6.42 Exercise 7.40	Weekly incomes of a random sample of 50 Puerto Rican families in Miami
QUAIL.MTW	Exercise 2.57 Exercise 2.76 Exercise 2.106 Exercise 7.41 Exercise 9.59	Plasma LDL levels in two groups of quail
QUALITY.MTW	Exercise 9.100	Quality control test scores on two manufacturing processes

SCALES.MTW	Example 2.26	Readings obtained from a 100-pound weight placed on four brands of bathroom scales
SCHIZOP2.MTW	Exercise 8.123	Exam scores for 17 patients to assess the learning ability of schizophrenics after taking a specified dose of a tranquilizer
SCHIZOPH.MTW	Example 8.10	Standardized exam scores for 13 patients to investigate the learning ability of schizophrenics after a specified dose of a tranquilizer
SEATBELT.MTW	Exercise 10.29	Injury level versus seatbelt usage
SELFDEFE.MTW	Example 9.16	Self-confidence scores for 9 women before and after instructions on self-defense
SENIOR.MTW	Exercise 2.102 Exercise 4.126	Reaction times of 30 senior citizens applying for drivers license renewals
SENTENCE.MTW	Exercise 2.151	Sentences of 41 prisoners convicted of a homicide offense
SHKDRUG.MTW	Exercise 12.10	Effects of a drug and electroshock therapy on the ability to solve simple tasks
SHOCK.MTW	Exercise 12.64	Effect of experimental shock on time to complete difficult task
SHOPLIFT.MTW	Exercise 11.90	Sales receipts versus shoplifting losses for a department store
SHORT.MTW	Exercise 8.84 Exercise 8.86	James Short's measurements of the parallax of the sun
SHUTTLE.MTW	Exercise 11.16 Exercise 11.35	Number of people riding shuttle versus number of automobiles in the downtown area
SIMPSON.MTW	Example 6.5	Grade point averages of men and women participating in various sports—an illustration of Simpson's paradox
SITUP.MTW	Exercise 2.46	Maximum number of sit-ups by participants in an exercise class
SKIN.MTW	Exercise 6.31	Survival times of closely and poorly matched skin grafts on burn patients
SLC.MTW	Exercise 6.72	Sodium-lithium countertransport activity on 190 individuals from six large English kindreds
SMOKYPH.MTW	Exercise 8.42 Exercise 8.71 Exercise 9.10 Exercise 9.42	Water pH levels of 75 water samples taken in the Great Smoky Mountains
SNORE.MTW	Exercise 10.28	Snoring versus heart disease
SNOW.MTW	Exercise 9.106	Concentration of microparticles in snowfields of Greenland and Antarctica
SOCCER.MTW	Exercise 2.45	Weights of 25 soccer players
SOCIAL.MTW	Exercise 8.67	Median income level for 25 social workers from North Carolina
SOPHOMOR.MTW	Exercise U1.26	Grade point averages, SAT scores, and final grade in college algebra for 20 sophomores
SOUTH.MTW	Exercise 2.108	Murder rates for 30 cities in the South
SPANISH.MTW	Exercise U3.24	Pre- and post-test scores on the MLA listening test in Spanish for 10 high school Spanish teachers
SPEED.MTW	Exercise 9.67	Speed reading scores before and after a course on speed reading
SPELLERS.MTW	Exercise 9.101	Standardized spelling test scores for two fourth grade classes
SPELLING.MTW	Exercise 9.61	Spelling scores for 9 eighth graders before and after a 2-week course of instruction
SPORT.MTW	Exercise U1.11	Favorite sport by a sample of 1,100 people

TORT.MTW	Exercise 6.10	The number of torts, average number of months to process a tort, and county population from the court files of the nation's largest counties
TOXIC.MTW	Exercise 2.14 Exercise 2.54 Exercise 2.73 Exercise 2.92 Exercise 3.93 Exercise 7.86 Exercise 7.87 Exercise 10.64	Hazardous waste sites near minority communities
TOXIC.XLS	Exercise 2.14	Hazardous waste sites near minority communities—Excel worksheet
TRACK.MTW	Exercise 3.94 Exercise 6.70	National Olympic records for women in several races
TRACK.XLS	Exercise 3.94 Exercise 11.94	National Olympic records for women in several races—Excel worksheet
TRACK15.MTW	Exercise 2.21	Olympic winning times for the men's 1500-meter run
TRAVEL.MTW	Exercise 7.72	Travel times for 32 commuters
TREES.MTW	Exercise 2.52 Exercise 2.70	Number of trees in 20 grids
TRUCKS.MTW	Table 12.2 Example 12.1–12.5	Miles per gallon for standard 4-wheel drive trucks manufactured by Chevrolet, Dodge, and Ford
TV.MTW	Chapter 3 SI Exercise 3.98	Percent of students that watch more than 6 hours of TV per day versus national math test scores
TWIN.MTW	Exercise 9.65	Intelligence test scores for identical twins in which one twin is given a drug
UNIT4-1.MTW	Exercise U4.1	Illustration of a scatterplot
UNIT4-17.MTW	Exercise U4.17	Illustrates multiple regression
VACATION.MTW	Exercise 8.40 Exercise 8.122	Number of days of paid holidays and vacation leave for sample of 35 textile workers
VACCINE.MTW	Exercise 2.129	Reported serious reactions due to vaccines in 11 southern states
VARIANCE.MTW	Exercise U3.23	Checking homogeneous variances
VEHICLE.MTW	Exercise 10.40	Fatality ratings for foreign and domestic vehicles
VERBAL.MTW	Exercise 11.59	Verbal test scores and number of library books checked out for 15 eighth graders
VICTORIA.MTW	Exercise 3.95	Number of sunspots versus mean annual level of Lake Victoria Nyanza from 1902 to 1921
VISCOSIT.MTW	Exercise 9.52	Viscosity measurements of a substance on two different days
VISUAL.MTW	Exercise 6.4	Visual acuity of a group of subjects tested under a specified dose of a drug
VOCAB.MTW	Exercise 9.99	Reading scores before and after vocabulary training for 14 employees who did not complete high school
VOTERS.MTW	Exercise U4.13	Political party versus opinion on foreign policy
WASTEWAT.MTW	Exercise 11.24	Volume of injected waste water from Rocky Mountain Arsenal and number of earthquakes near Denver
WHEAT.MTW	Exercise 3.21	Price of a bushel of wheat versus the national weekly earnings of production workers

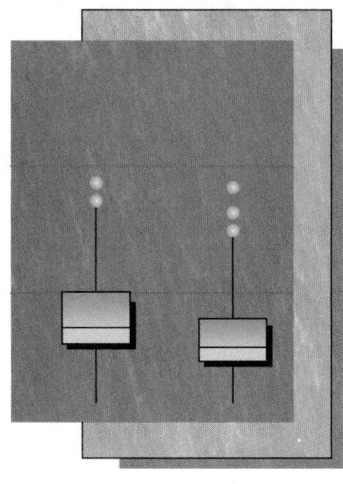

TABLE B.1   Binomial Probabilities

n	k	.01	.05	.10	.20	.30	.40	.50	.60	.70	.80	.90	.95	.99
								π						
2	0	980	902	810	640	490	360	250	160	090	040	010	002	0+
	1	020	095	180	320	420	480	500	480	420	320	180	095	020
	2	0+	002	010	040	090	160	250	360	490	640	810	902	980
3	0	970	857	729	512	343	216	125	064	027	008	001	0+	0+
	1	029	135	243	384	441	432	375	288	189	096	027	007	0+
	2	0+	007	027	096	189	288	375	432	441	384	243	135	029
	3	0+	0+	001	008	027	064	125	216	343	512	729	857	970
4	0	961	815	656	410	240	130	062	026	008	002	0+	0+	0+
	1	039	171	292	410	412	346	250	154	076	026	004	0+	0+
	2	001	014	049	154	265	346	375	346	265	154	049	014	001
	3	0+	0+	004	026	076	154	250	346	412	410	292	171	039
	4	0+	0+	0+	002	008	026	062	130	240	410	656	815	961
5	0	951	774	590	328	168	078	031	010	002	0+	0+	0+	0+
	1	048	204	328	410	360	259	156	077	028	006	0+	0+	0+
	2	001	021	073	205	309	346	312	230	132	051	008	001	0+
	3	0+	001	008	051	132	230	312	346	309	205	073	021	001
	4	0+	0+	0+	006	028	077	156	259	360	410	328	204	048
	5	0+	0+	0+	0+	002	010	031	078	168	328	590	774	951
6	0	941	735	531	262	118	047	016	004	001	0+	0+	0+	0+
	1	057	232	354	393	303	187	094	037	010	002	0+	0+	0+
	2	001	031	098	246	324	311	234	138	060	015	001	0+	0+
	3	0+	002	015	082	185	276	312	276	185	082	015	002	0+
	4	0+	0+	001	015	060	138	234	311	324	246	098	031	001
	5	0+	0+	0+	002	010	037	094	187	303	393	354	232	057
	6	0+	0+	0+	0+	001	004	016	047	118	262	531	735	941
7	0	932	698	478	210	082	028	008	002	0+	0+	0+	0+	0+
	1	066	257	372	367	247	131	055	017	004	0+	0+	0+	0+
	2	002	041	124	275	318	261	164	077	025	004	0+	0+	0+
	3	0+	004	023	115	227	290	273	194	097	029	003	0+	0+
	4	0+	0+	003	029	097	194	273	290	227	115	023	004	0+
	5	0+	0+	0+	004	025	077	164	261	318	275	124	041	002
	6	0+	0+	0+	0+	004	017	055	131	247	367	372	257	066
	7	0+	0+	0+	0+	0+	002	008	028	082	210	478	698	932
8	0	923	663	430	168	058	017	004	001	0+	0+	0+	0+	0+
	1	075	279	383	336	198	090	031	008	001	0+	0+	0+	0+
	2	003	051	149	294	296	209	109	041	010	001	0+	0+	0+
	3	0+	005	033	147	254	279	219	124	047	009	0+	0+	0+
	4	0+	0+	005	046	136	232	273	232	136	046	005	0+	0+
	5	0+	0+	0+	009	047	124	219	279	254	147	033	005	0+
	6	0+	0+	0+	001	010	041	109	209	296	294	149	051	003
	7	0+	0+	0+	0+	001	008	031	090	198	336	383	279	075
	8	0+	0+	0+	0+	0+	001	004	017	058	168	430	663	923
9	0	914	630	387	134	040	010	002	0+	0+	0+	0+	0+	0+
	1	083	299	387	302	156	060	018	004	0+	0+	0+	0+	0+
	2	003	063	172	302	267	161	070	021	004	0+	0+	0+	0+
	3	0+	008	045	176	267	251	164	074	021	003	0+	0+	0+
	4	0+	001	007	066	172	251	246	167	074	017	001	0+	0+

TABLE B.1 Binomial Probabilities (continued)

n	k	.01	.05	.10	.20	.30	.40	.50	.60	.70	.80	.90	.95	.99
								π						
	5	0+	0+	001	017	074	167	246	251	172	066	007	001	0+
	6	0+	0+	0+	003	021	074	164	251	267	176	045	008	0+
	7	0+	0+	0+	0+	004	021	070	161	267	302	172	063	003
	8	0+	0+	0+	0+	0+	004	018	060	156	302	387	299	083
	9	0+	0+	0+	0+	0+	0+	002	010	040	134	387	630	914
10	0	904	599	349	107	028	006	001	0+	0+	0+	0+	0+	0+
	1	091	315	387	268	121	040	010	002	0+	0+	0+	0+	0+
	2	004	075	194	302	233	121	044	011	001	0+	0+	0+	0+
	3	0+	010	057	201	267	215	117	042	009	001	0+	0+	0+
	4	0+	001	011	088	200	251	205	111	037	006	0+	0+	0+
	5	0+	0+	001	026	103	201	246	201	103	026	001	0+	0+
	6	0+	0+	0+	006	037	111	205	251	200	088	011	001	0+
	7	0+	0+	0÷	001	009	042	117	215	267	201	057	010	0+
	8	0+	0+	0+	0+	001	011	044	121	233	302	194	075	004
	9	0+	0+	0+	0+	0+	002	010	040	121	268	387	315	091
	10	0+	0+	0+	0+	0+	0+	001	006	028	107	349	599	904
11	0	895	569	314	086	020	004	0+	0+	0+	0+	0+	0+	0+
	1	099	329	384	236	093	027	005	001	0+	0+	0+	0+	0+
	2	005	087	213	295	200	089	027	005	001	0+	0+	0+	0+
	3	0+	014	071	221	257	177	081	023	004	0+	0+	0+	0+
	4	0+	001	016	111	220	236	161	070	017	002	0+	0+	0+
	5	0+	0+	002	039	132	221	226	147	057	010	0+	0+	0+
	6	0+	0+	0+	010	057	147	226	221	132	039	002	0+	0+
	7	0+	0+	0+	002	017	070	161	236	220	111	016	001	0+
	8	0+	0+	0+	0+	004	023	081	177	257	221	071	014	0+
	9	0+	0+	0+	0+	001	005	027	089	200	295	213	087	005
	10	0+	0+	0+	0+	0+	001	005	027	093	236	384	329	099
	11	0+	0+	0+	0+	0+	0+	0+	004	020	086	314	569	895
12	0	886	540	282	069	014	002	0+	0+	0+	0+	0+	0+	0+
	1	107	341	377	206	071	017	003	0+	0+	0+	0+	0+	0+
	2	006	099	230	283	168	064	016	002	0+	0+	0+	0+	0+
	3	0+	017	085	236	240	142	054	012	001	0+	0+	0+	0+
	4	0+	002	021	133	231	213	121	042	008	001	0+	0+	0+
	5	0+	0+	004	053	158	227	193	101	029	003	0+	0+	0+
	6	0+	0+	0+	016	079	177	226	177	079	016	0+	0+	0+
	7	0+	0+	0+	003	029	101	193	227	158	053	004	0+	0+
	8	0+	0+	0+	001	008	042	121	213	231	133	021	002	0+
	9	0+	0+	0+	0+	001	012	054	142	240	236	085	017	0+
	10	0+	0+	0+	0+	0+	002	016	064	168	283	230	099	006
	11	0+	0+	0+	0+	0+	0+	003	017	071	206	377	341	107
	12	0+	0+	0+	0+	0+	0+	0+	002	014	069	282	540	886
13	0	878	513	254	055	010	001	0+	0+	0+	0+	0+	0+	0+
	1	115	351	367	179	054	011	002	0+	0+	0+	0+	0+	0+
	2	007	111	245	268	139	045	010	001	0+	0+	0+	0+	0+
	3	0+	021	100	246	218	111	035	006	001	0+	0+	0+	0+
	4	0+	003	028	154	234	184	087	024	003	0+	0+	0+	0+

TABLE B.1   Binomial Probabilities (continued)

n	k	.01	.05	.10	.20	.30	.40	.50	.60	.70	.80	.90	.95	.99
	5	0+	0+	006	069	180	221	157	066	014	001	0+	0+	0+
	6	0+	0+	001	023	103	197	209	131	044	006	0+	0+	0+
	7	0+	0+	0+	006	044	131	209	197	103	023	001	0+	0+
	8	0+	0+	0+	001	014	066	157	221	180	069	006	0+	0+
	9	0+	0+	0+	0+	003	024	087	184	234	154	028	003	0+
	10	0+	0+	0+	0+	001	006	035	111	218	246	100	021	0+
	11	0+	0+	0+	0+	0+	001	010	045	139	268	245	111	007
	12	0+	0+	0+	0+	0+	0+	002	011	054	179	367	351	115
	13	0+	0+	0+	0+	0+	0+	0+	001	010	055	254	513	878
14	0	869	488	229	044	007	001	0+	0+	0+	0+	0+	0+	0+
	1	123	359	356	154	041	007	001	0+	0+	0+	0+	0+	0+
	2	008	123	257	250	113	032	006	001	0+	0+	0+	0+	0+
	3	0+	026	114	250	194	085	022	003	0+	0+	0+	0+	0+
	4	0+	004	035	172	229	155	061	014	001	0+	0+	0+	0+
	5	0+	0+	008	086	196	207	122	041	007	0+	0+	0+	0+
	6	0+	0+	001	032	126	207	183	092	023	002	0+	0+	0+
	7	0+	0+	0+	009	062	157	209	157	062	009	0+	0+	0+
	8	0+	0+	0+	002	023	092	183	207	126	032	001	0+	0+
	9	0+	0+	0+	0+	007	041	122	207	196	086	008	0+	0+
	10	0+	0+	0+	0+	001	014	061	155	229	172	035	004	0+
	11	0+	0+	0+	0+	0+	003	022	085	194	250	114	026	0+
	12	0+	0+	0+	0+	0+	001	006	032	113	250	257	123	008
	13	0+	0+	0+	0+	0+	0+	001	007	041	154	356	359	123
	14	0+	0+	0+	0+	0+	0+	0+	001	007	044	229	488	869
15	0	860	463	206	035	005	0+	0+	0+	0+	0+	0+	0+	0+
	1	130	366	343	132	031	005	0+	0+	0+	0+	0+	0+	0+
	2	009	135	267	231	092	022	003	0+	0+	0+	0+	0+	0+
	3	0+	031	129	250	170	063	014	002	0+	0+	0+	0+	0+
	4	0+	005	043	188	219	127	042	007	001	0+	0+	0+	0+
	5	0+	001	010	103	206	186	092	024	003	0+	0+	0+	0+
	6	0+	0+	002	043	147	207	153	061	012	001	0+	0+	0+
	7	0+	0+	0+	014	081	177	196	118	035	003	0+	0+	0+
	8	0+	0+	0+	003	035	118	196	177	081	014	0+	0+	0+
	9	0+	0+	0+	001	012	061	153	207	147	043	002	0+	0+
	10	0+	0+	0+	0+	003	024	092	186	206	103	010	001	0+
	11	0+	0+	0+	0+	001	007	042	127	219	188	043	005	0+
	12	0+	0+	0+	0+	0+	002	014	063	170	250	129	031	0+
	13	0+	0+	0+	0+	0+	0+	003	022	092	231	267	135	009
	14	0+	0+	0+	0+	0+	0+	0+	005	031	132	343	366	130
	15	0+	0+	0+	0+	0+	0+	0+	0+	005	035	206	463	860
16	0	851	440	185	028	003	0+	0+	0+	0+	0+	0+	0+	0+
	1	138	371	329	113	023	003	0+	0+	0+	0+	0+	0+	0+
	2	010	146	274	211	073	015	002	0+	0+	0+	0+	0+	0+
	3	0+	036	142	246	146	047	008	001	0+	0+	0+	0+	0+
	4	0+	006	051	200	204	101	028	004	0+	0+	0+	0+	0+
	5	0+	001	014	120	210	162	067	014	001	0+	0+	0+	0+
	6	0+	0+	003	055	165	198	122	039	006	0+	0+	0+	0+
	7	0+	0+	0+	020	101	189	175	084	018	001	0+	0+	0+
	8	0+	0+	0+	006	049	142	196	142	049	006	0+	0+	0+

TABLE B.1 Binomial Probabilities (continued)

n	k	.01	.05	.10	.20	.30	.40	.50	.60	.70	.80	.90	.95	.99
	9	0+	0+	0+	001	018	084	175	189	101	020	0+	0+	0+
	10	0+	0+	0+	0+	006	039	122	198	165	055	003	0+	0+
	11	0+	0+	0+	0+	001	014	067	162	210	120	014	001	0+
	12	0+	0+	0+	0+	0+	004	028	101	204	200	051	006	0+
	13	0+	0+	0+	0+	0+	001	008	047	146	246	142	036	0+
	14	0+	0+	0+	0+	0+	0+	002	015	073	211	274	146	010
	15	0+	0+	0+	0+	0+	0+	0+	003	023	113	329	371	138
	16	0+	0+	0+	0+	0+	0+	0+	0+	003	028	185	440	851
17	0	843	418	167	022	002	0+	0+	0+	0+	0+	0+	0+	0+
	1	145	374	315	096	017	002	0+	0+	0+	0+	0+	0+	0+
	2	012	158	280	191	058	010	001	0+	0+	0+	0+	0+	0+
	3	001	042	156	239	124	034	005	0+	0+	0+	0+	0+	0+
	4	0+	008	060	209	187	080	018	002	0+	0+	0+	0+	0+
	5	0+	001	018	136	208	138	047	008	001	0+	0+	0+	0+
	6	0+	0+	004	068	178	184	094	024	003	0+	0+	0+	0+
	7	0+	0+	001	027	120	193	148	057	010	0+	0+	0+	0+
	8	0+	0+	0+	008	064	161	186	107	028	002	0+	0+	0+
	9	0+	0+	0+	002	028	107	186	161	064	008	0+	0+	0+
	10	0+	0+	0+	0+	010	057	148	193	120	027	001	0+	0+
	11	0+	0+	0+	0+	003	024	094	184	178	068	004	0+	0+
	12	0+	0+	0+	0+	001	008	047	138	208	136	018	001	0+
	13	0+	0+	0+	0+	0+	002	018	080	187	209	060	008	0+
	14	0+	0+	0+	0+	0+	0+	005	034	124	239	156	042	001
	15	0+	0+	0+	0+	0+	0+	001	010	058	191	280	158	012
	16	0+	0+	0+	0+	0+	0+	0+	002	017	096	315	374	145
	17	0+	0+	0+	0+	0+	0+	0+	0+	002	022	167	418	843
18	0	835	397	150	018	002	0+	0+	0+	0+	0+	0+	0+	0+
	1	152	376	300	081	013	001	0+	0+	0+	0+	0+	0+	0+
	2	013	168	284	172	046	007	001	0+	0+	0+	0+	0+	0+
	3	001	047	168	230	105	025	003	0+	0+	0+	0+	0+	0+
	4	0+	009	070	215	168	061	012	001	0+	0+	0+	0+	0+
	5	0+	001	002	151	202	115	033	004	0+	0+	0+	0+	0+
	6	0+	0+	005	082	187	166	071	014	001	0+	0+	0+	0+
	7	0+	0+	001	035	138	189	121	037	005	0+	0+	0+	0+
	8	0+	0+	0+	012	081	173	167	077	015	001	0+	0+	0+
	9	0+	0+	0+	003	038	128	186	128	038	003	0+	0+	0+
	10	0+	0+	0+	001	015	077	167	173	081	012	0+	0+	0+
	11	0+	0+	0+	0+	005	037	121	189	138	035	001	0+	0+
	12	0+	0+	0+	0+	001	014	071	166	187	082	005	0+	0+
	13	0+	0+	0+	0+	0+	004	033	115	202	151	022	001	0+
	14	0+	0+	0+	0+	0+	001	012	061	168	215	070	009	0+
	15	0+	0+	0+	0+	0+	0+	003	025	105	230	168	097	001
	16	0+	0+	0+	0+	0+	0+	001	007	046	172	284	168	013
	17	0+	0+	0+	0+	0+	0+	0+	001	013	081	300	376	152
	18	0+	0+	0+	0+	0+	0+	0+	0+	002	018	150	397	835

TABLE B.1   Binomial Probabilities (continued)

n	k	.01	.05	.10	.20	.30	.40	.50	.60	.70	.80	.90	.95	.99
19	0	826	377	135	014	001	0+	0+	0+	0+	0+	0+	0+	0+
	1	159	377	285	069	009	001	0+	0+	0+	0+	0+	0+	0+
	2	014	179	285	154	036	005	0+	0+	0+	0+	0+	0+	0+
	3	001	053	180	218	087	018	002	0+	0+	0+	0+	0+	0+
	4	0+	011	080	218	149	047	007	0+	0+	0+	0+	0+	0+
	5	0+	002	027	164	192	093	022	002	0+	0+	0+	0+	0+
	6	0+	0+	007	096	192	145	052	008	0+	0+	0+	0+	0+
	7	0+	0+	001	044	152	180	096	024	002	0+	0+	0+	0+
	8	0+	0+	0+	017	098	180	144	053	008	0+	0+	0+	0+
	9	0+	0+	0+	005	051	146	176	098	022	001	0+	0+	0+
	10	0+	0+	0+	001	022	098	176	146	051	005	0+	0+	0+
	11	0+	0+	0+	0+	008	053	144	180	098	017	0+	0+	0+
	12	0+	0+	0+	0+	002	024	096	180	152	044	001	0+	0+
	13	0+	0+	0+	0+	0+	008	052	145	192	096	007	0+	0+
	14	0+	0+	0+	0+	0+	002	022	093	192	164	027	002	0+
	15	0+	0+	0+	0+	0+	0+	007	047	149	218	080	011	0+
	16	0+	0+	0+	0+	0+	0+	002	018	087	218	180	053	001
	17	0+	0+	0+	0+	0+	0+	0+	005	036	154	285	179	014
	18	0+	0+	0+	0+	0+	0+	0+	001	009	069	285	377	159
	19	0+	0+	0+	0+	0+	0+	0+	0+	001	014	135	377	826
20	0	818	358	122	012	001	0+	0+	0+	0+	0+	0+	0+	0+
	1	165	377	270	058	007	0+	0+	0+	0+	0+	0+	0+	0+
	2	016	189	285	137	028	003	0+	0+	0+	0+	0+	0+	0+
	3	001	060	190	205	072	012	001	0+	0+	0+	0+	0+	0+
	4	0+	013	090	218	130	035	005	0+	0+	0+	0+	0+	0+
	5	0+	002	032	175	179	075	015	001	0+	0+	0+	0+	0+
	6	0+	0+	009	109	192	124	037	005	0+	0+	0+	0+	0+
	7	0+	0+	002	054	164	166	074	015	001	0+	0+	0+	0+
	8	0+	0+	0+	022	114	180	120	036	004	0+	0+	0+	0+
	9	0+	0+	0+	007	065	160	160	071	012	0+	0+	0+	0+
	10	0+	0+	0+	002	031	117	176	117	031	002	0+	0+	0+
	11	0+	0+	0+	0+	012	071	160	160	065	007	0+	0+	0+
	12	0+	0+	0+	0+	004	036	120	180	114	022	0+	0+	0+
	13	0+	0+	0+	0+	001	015	074	166	164	054	002	0+	0+
	14	0+	0+	0+	0+	0+	005	037	124	192	109	009	0+	0+
	15	0+	0+	0+	0+	0+	001	015	075	179	175	032	002	0+
	16	0+	0+	0+	0+	0+	0+	005	035	130	218	090	013	0+
	17	0+	0+	0+	0+	0+	0+	001	012	072	205	190	060	001
	18	0+	0+	0+	0+	0+	0+	0+	003	028	137	285	189	016
	19	0+	0+	0+	0+	0+	0+	0+	0+	007	058	270	377	165
	20	0+	0+	0+	0+	0+	0+	0+	0+	001	012	122	358	818

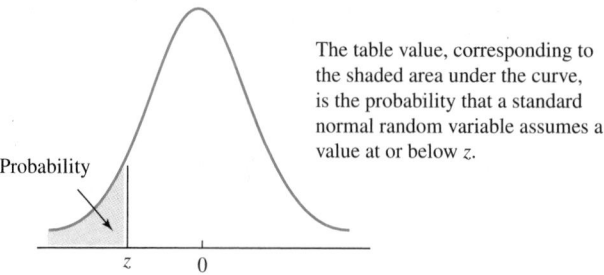

The table value, corresponding to the shaded area under the curve, is the probability that a standard normal random variable assumes a value at or below z.

Second decimal place in z

z	.00	.01	.02	.03	.04	.05	.06	.07	.08	.09
−5.0	.0000003									
−4.5	.000003									
−4.0	.00003									
−3.5	.0002									
−3.4	.0003	.0003	.0003	.0003	.0003	.0003	.0003	.0003	.0003	.0002
−3.3	.0005	.0005	.0005	.0004	.0004	.0004	.0004	.0004	.0004	.0003
−3.2	.0007	.0007	.0006	.0006	.0006	.0006	.0006	.0005	.0005	.0005
−3.1	.0010	.0009	.0009	.0009	.0008	.0008	.0008	.0008	.0007	.0007
−3.0	.0013	.0013	.0013	.0012	.0012	.0011	.0011	.0011	.0010	.0010
−2.9	.0019	.0018	.0018	.0017	.0016	.0016	.0015	.0015	.0014	.0014
−2.8	.0026	.0025	.0024	.0023	.0023	.0022	.0021	.0021	.0020	.0019
−2.7	.0035	.0034	.0033	.0032	.0031	.0030	.0029	.0028	.0027	.0026
−2.6	.0047	.0045	.0044	.0043	.0041	.0040	.0039	.0038	.0037	.0036
−2.5	.0062	.0060	.0059	.0057	.0055	.0054	.0052	.0051	.0049	.0048
−2.4	.0082	.0080	.0078	.0075	.0073	.0071	.0069	.0068	.0066	.0064
−2.3	.0107	.0104	.0102	.0099	.0096	.0094	.0091	.0089	.0087	.0084
−2.2	.0139	.0136	.0132	.0129	.0125	.0122	.0119	.0116	.0113	.0110
−2.1	.0179	.0174	.0170	.0166	.0162	.0158	.0154	.0150	.0146	.0143
−2.0	.0228	.0222	.0217	.0212	.0207	.0202	.0197	.0192	.0188	.0183
−1.9	.0287	.0281	.0274	.0268	.0262	.0256	.0250	.0244	.0239	.0233
−1.8	.0359	.0351	.0344	.0336	.0329	.0322	.0314	.0307	.0301	.0294
−1.7	.0446	.0436	.0427	.0418	.0409	.0401	.0392	.0384	.0375	.0367
−1.6	.0548	.0537	.0526	.0516	.0505	.0495	.0485	.0475	.0465	.0455
−1.5	.0668	.0655	.0643	.0630	.0618	.0606	.0594	.0582	.0571	.0559
−1.4	.0808	.0793	.0778	.0764	.0749	.0735	.0721	.0708	.0694	.0681
−1.3	.0968	.0951	.0934	.0918	.0901	.0885	.0869	.0853	.0838	.0823
−1.2	.1151	.1131	.1112	.1093	.1075	.1056	.1038	.1020	.1003	.0985
−1.1	.1357	.1335	.1314	.1292	.1271	.1251	.1230	.1210	.1190	.1170
−1.0	.1587	.1562	.1539	.1515	.1492	.1469	.1446	.1423	.1401	.1379
−0.9	.1841	.1814	.1788	.1762	.1736	.1711	.1685	.1660	.1635	.1611
−0.8	.2119	.2090	.2061	.2033	.2005	.1977	.1949	.1922	.1894	.1867
−0.7	.2420	.2389	.2358	.2327	.2296	.2266	.2236	.2206	.2177	.2148
−0.6	.2743	.2709	.2676	.2643	.2611	.2578	.2546	.2514	.2483	.2451
−0.5	.3085	.3050	.3015	.2981	.2946	.2912	.2877	.2843	.2810	.2776
−0.4	.3446	.3409	.3372	.3336	.3300	.3264	.3228	.3192	.3156	.3121
−0.3	.3821	.3783	.3745	.3707	.3669	.3632	.3594	.3557	.3520	.3483
−0.2	.4207	.4168	.4129	.4090	.4052	.4013	.3974	.3936	.3897	.3859
−0.1	.4602	.4562	.4522	.4483	.4443	.4404	.4364	.4325	.4286	.4247
−0.0	.5000	.4960	.4920	.4880	.4840	.4801	.4761	.4721	.4681	.4641

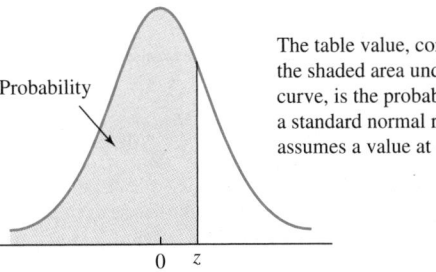

Probability

The table value, corresponding to the shaded area under the curve, is the probability that a standard normal random variable assumes a value at or below z.

Second decimal place in z

z	.00	.01	.02	.03	.04	.05	.06	.07	.08	.09
0.0	.5000	.5040	.5080	.5120	.5160	.5199	.5239	.5279	.5319	.5359
0.1	.5398	.5438	.5478	.5517	.5557	.5596	.5636	.5675	.5714	.5753
0.2	.5793	.5832	.5871	.5910	.5948	.5987	.6026	.6064	.6103	.6141
0.3	.6179	.6217	.6255	.6293	.6331	.6368	.6406	.6443	.6480	.6517
0.4	.6554	.6591	.6628	.6664	.6700	.6736	.6772	.6808	.6844	.6879
0.5	.6915	.6950	.6985	.7019	.7054	.7088	.7123	.7157	.7190	.7224
0.6	.7257	.7291	.7324	.7357	.7389	.7422	.7454	.7486	.7517	.7549
0.7	.7580	.7611	.7642	.7673	.7704	.7734	.7764	.7794	.7823	.7852
0.8	.7881	.7910	.7939	.7967	.7995	.8023	.8051	.8078	.8106	.8133
0.9	.8159	.8186	.8212	.8238	.8264	.8289	.8315	.8340	.8365	.8389
1.0	.8413	.8438	.8461	.8485	.8508	.8531	.8554	.8577	.8599	.8621
1.1	.8643	.8665	.8686	.8708	.8729	.8749	.8770	.8790	.8810	.8830
1.2	.8849	.8869	.8888	.8907	.8925	.8944	.8962	.8980	.8997	.9015
1.3	.9032	.9049	.9066	.9082	.9099	.9115	.9131	.9147	.9162	.9177
1.4	.9192	.9207	.9222	.9236	.9251	.9265	.9279	.9292	.9306	.9319
1.5	.9332	.9345	.9357	.9370	.9382	.9394	.9406	.9418	.9429	.9441
1.6	.9452	.9463	.9474	.9484	.9495	.9505	.9515	.9525	.9535	.9545
1.7	.9554	.9564	.9573	.9582	.9591	.9599	.9608	.9616	.9625	.9633
1.8	.9641	.9649	.9656	.9664	.9671	.9678	.9686	.9693	.9699	.9706
1.9	.9713	.9719	.9726	.9732	.9738	.9744	.9750	.9756	.9761	.9767
2.0	.9772	.9778	.9783	.9788	.9793	.9798	.9803	.9808	.9812	.9817
2.1	.9821	.9826	.9830	.9834	.9838	.9842	.9846	.9850	.9854	.9857
2.2	.9861	.9864	.9868	.9871	.9875	.9878	.9881	.9884	.9887	.9890
2.3	.9893	.9896	.9898	.9901	.9904	.9906	.9909	.9911	.9913	.9916
2.4	.9918	.9920	.9922	.9925	.9927	.9929	.9931	.9932	.9934	.9936
2.5	.9938	.9940	.9941	.9943	.9945	.9946	.9948	.9949	.9951	.9952
2.6	.9953	.9955	.9956	.9957	.9959	.9960	.9961	.9962	.9963	.9964
2.7	.9965	.9966	.9967	.9968	.9969	.9970	.9971	.9972	.9973	.9974
2.8	.9974	.9975	.9976	.9977	.9977	.9978	.9979	.9979	.9980	.9981
2.9	.9981	.9982	.9982	.9983	.9984	.9984	.9985	.9985	.9986	.9986
3.0	.9987	.9987	.9987	.9988	.9988	.9989	.9989	.9989	.9990	.9990
3.1	.9990	.9991	.9991	.9991	.9992	.9992	.9992	.9992	.9993	.9993
3.2	.9993	.9993	.9994	.9994	.9994	.9994	.9994	.9995	.9995	.9995
3.3	.9995	.9995	.9995	.9996	.9996	.9996	.9996	.9996	.9996	.9997
3.4	.9997	.9997	.9997	.9997	.9997	.9997	.9997	.9997	.9997	.9998
3.5	.9998									
4.0	.99997									
4.5	.999997									
5.0	.9999997									

TABLE B.3   Critical Values of Student's t Distribution

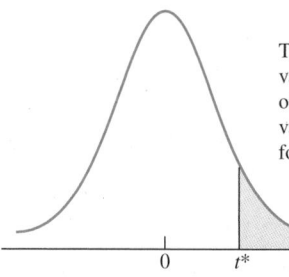

The table values are the critical values for Student's t for an area of α in the right-hand tail. Critical values for the left-hand tail are found by symmetry.

Degrees of freedom	Amount of α in one-tail							
	.2	.1	.05	.025	.01	.005	.0025	.001
1	1.376	3.078	6.314	12.706	31.821	63.657	127.3	318.3
2	1.061	1.886	2.920	4.303	6.965	9.925	14.09	22.33
3	.978	1.638	2.353	3.182	4.541	5.841	7.453	10.21
4	.941	1.533	2.132	2.776	3.747	4.604	5.598	7.173
5	.920	1.476	2.015	2.571	3.365	4.032	4.773	5.893
6	.906	1.440	1.943	2.447	3.143	3.707	4.317	5.208
7	.896	1.415	1.895	2.365	2.998	3.499	4.029	4.785
8	.889	1.397	1.860	2.306	2.896	3.355	3.833	4.501
9	.883	1.383	1.833	2.262	2.821	3.250	3.690	4.297
10	.879	1.372	1.812	2.228	2.764	3.169	3.581	4.144
11	.876	1.363	1.796	2.201	2.718	3.106	3.497	4.025
12	.873	1.356	1.782	2.179	2.681	3.055	3.428	3.930
13	.870	1.350	1.771	2.160	2.650	3.012	3.372	3.852
14	.868	1.345	1.761	2.145	2.624	2.977	3.326	3.787
15	.866	1.341	1.753	2.131	2.602	2.947	3.286	3.733
16	.865	1.337	1.746	2.120	2.583	2.921	3.252	3.686
17	.863	1.333	1.740	2.110	2.567	2.898	3.222	3.646
18	.862	1.330	1.734	2.101	2.552	2.878	3.197	3.611
19	.861	1.328	1.729	2.093	2.539	2.861	3.174	3.579
20	.860	1.325	1.725	2.086	2.528	2.845	3.153	3.552
21	.859	1.323	1.721	2.080	2.518	2.831	3.135	3.527
22	.858	1.321	1.717	2.074	2.508	2.819	3.119	3.505
23	.858	1.319	1.714	2.069	2.500	2.807	3.104	3.485
24	.857	1.318	1.711	2.064	2.492	2.797	3.091	3.467
25	.856	1.316	1.708	2.060	2.485	2.787	3.078	3.450
26	.856	1.315	1.706	2.056	2.479	2.779	3.067	3.435
27	.855	1.314	1.703	2.052	2.473	2.771	3.057	3.421
28	.855	1.313	1.701	2.048	2.467	2.763	3.047	3.408
29	.854	1.311	1.699	2.045	2.462	2.756	3.038	3.396
30	.854	1.310	1.697	2.042	2.457	2.750	3.030	3.385
40	.851	1.303	1.684	2.021	2.423	2.704	2.971	3.307
50	.849	1.299	1.676	2.009	2.403	2.678	2.937	3.261
60	.848	1.296	1.671	2.000	2.390	2.660	2.915	3.232
80	.846	1.292	1.664	1.990	2.374	2.639	2.887	3.195
120	.845	1.289	1.658	1.980	2.358	2.617	2.860	3.160
∞	.842	1.282	1.645	1.960	2.326	2.576	2.807	3.090

TABLE B.4 Critical Values of the Chi-Square Distribution

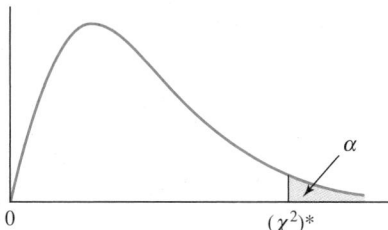

The table values are the critical values for the chi-square distribution for which the area to the right under the curve is equal to α.

0 $(\chi^2)*$

Amount of α in right-hand tail

df	.995	.990	.975	.950	.900	.200	.100	.050	.025	.010	.005	.0025	.001
1	0.0000393	0.000157	0.000982	0.00393	0.0158	1.64	2.71	3.84	5.02	6.63	7.88	9.14	10.8
2	0.0100	0.0201	0.0506	0.103	0.211	3.22	4.61	5.99	7.38	9.21	10.6	12.0	13.8
3	0.0717	0.115	0.216	0.352	0.584	4.64	6.25	7.81	9.35	11.3	12.8	14.3	16.3
4	0.207	0.297	0.484	0.711	1.0636	5.99	7.78	9.49	11.1	13.3	14.9	16.4	18.5
5	0.412	0.554	0.831	1.15	1.61	7.29	9.24	11.1	12.8	15.1	16.7	18.4	20.5
6	0.676	0.872	1.24	1.64	2.20	8.56	10.6	12.6	14.5	16.8	18.5	20.3	22.5
7	0.989	1.24	1.69	2.17	2.83	9.80	12.0	14.1	16.0	18.5	20.3	22.0	24.3
8	1.34	1.65	2.18	2.73	3.49	11.0	13.4	15.5	17.5	20.1	22.0	23.8	26.1
9	1.73	2.09	2.70	3.33	4.17	12.2	14.7	16.9	19.0	21.7	23.6	25.5	27.9
10	2.16	2.56	3.25	3.94	4.87	13.4	16.0	18.3	20.5	23.2	25.2	27.1	29.6
11	2.60	3.05	3.82	4.58	5.58	14.6	17.3	19.7	21.9	24.7	26.8	28.7	31.3
12	3.07	3.57	4.40	5.23	6.30	15.8	18.5	21.0	23.3	26.2	28.3	30.3	32.9
13	3.57	4.11	5.01	5.90	7.04	17.0	19.8	22.4	24.7	27.7	29.8	31.9	34.5
14	4.07	4.66	5.63	6.57	7.79	18.2	21.1	23.7	26.1	29.1	31.3	33.4	36.1
15	4.60	5.23	6.26	7.26	8.55	19.3	22.3	25.0	27.5	30.6	32.8	35.0	37.7
16	5.14	5.81	6.91	7.96	9.31	20.5	23.5	26.3	28.8	32.0	34.3	36.5	39.3
17	5.70	6.41	7.56	8.67	10.1	21.6	24.8	27.6	30.2	33.4	35.7	38.0	40.8
18	6.26	7.01	8.23	9.39	10.9	22.8	26.0	28.9	31.5	34.8	37.2	39.4	42.3
19	6.84	7.63	8.91	10.1	11.7	23.9	27.2	30.1	32.9	36.2	38.6	40.9	43.8
20	7.43	8.26	9.59	10.9	12.4	25.0	28.4	31.4	34.2	37.6	40.0	42.3	45.3
21	8.03	8.90	10.3	11.6	13.2	26.2	29.6	32.7	35.5	38.9	41.4	43.8	46.8
22	8.64	9.54	11.0	12.3	14.0	27.3	30.8	33.9	36.8	40.3	42.8	45.2	48.3
23	9.26	10.2	11.7	13.1	14.8	28.4	32.0	35.2	38.1	41.6	44.2	46.6	49.7
24	9.89	10.9	12.4	13.8	15.7	29.6	33.2	36.4	39.4	43.0	45.6	48.0	51.2
25	10.5	11.5	13.1	14.6	16.5	30.7	34.4	37.7	40.6	44.3	46.9	49.4	52.6
26	11.2	12.2	13.8	15.4	17.3	31.8	35.6	38.9	41.9	45.6	48.3	50.8	54.1
27	11.8	12.9	14.6	16.2	18.1	32.9	36.7	40.1	43.2	47.0	49.6	52.2	55.5
28	12.5	13.6	15.3	16.9	18.9	34.0	37.9	41.3	44.5	48.3	51.0	53.6	56.9
29	13.1	14.3	16.0	17.7	19.8	35.1	39.1	42.6	45.7	49.6	52.3	55.0	58.3
30	13.8	15.0	16.8	18.5	20.6	36.3	40.3	43.8	47.0	50.9	53.7	56.3	59.7
40	20.7	22.2	24.4	26.5	29.1	47.3	51.8	55.8	59.3	63.7	66.8	69.7	73.4
50	28.0	29.7	32.4	34.8	37.7	58.2	63.2	67.5	71.4	76.2	79.5	82.7	86.7
60	35.5	37.5	40.5	43.2	46.5	69.0	74.4	79.1	83.3	88.4	92.0	95.3	99.6
70	43.3	45.4	48.8	51.7	55.3	79.7	85.5	90.5	95.0	100.0	104.2	107.8	112.3
80	51.2	53.5	57.2	60.4	64.3	90.4	96.6	101.9	106.6	112.3	116.3	120.1	124.8
90	59.2	61.8	65.6	69.1	73.3	101.1	107.6	113.1	118.1	124.1	128.3	132.2	137.2
100	67.3	70.1	74.2	77.9	82.4	111.7	118.5	124.3	129.6	135.8	140.2	144.3	149.4

TABLE B.5   Critical Values of the F Distribution ($\alpha = .10$)

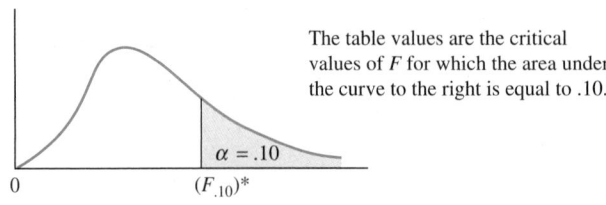

The table values are the critical values of F for which the area under the curve to the right is equal to .10.

$\alpha = .10$

0 $(F_{.10})^*$

Degrees of freedom for numerator

		1	2	3	4	5	6	7	8	9
	1	39.86	49.50	53.59	55.83	57.24	58.20	58.91	59.44	59.86
	2	8.53	9.00	9.16	9.24	9.29	9.33	9.35	9.37	9.38
	3	5.54	5.46	5.39	5.34	5.31	5.28	5.27	5.25	5.24
	4	4.54	4.32	4.19	4.11	4.05	4.01	3.98	3.95	3.94
	5	4.06	3.78	3.62	3.52	3.45	3.40	3.37	3.34	3.32
	6	3.78	3.46	3.29	3.18	3.11	3.05	3.01	2.98	2.96
	7	3.59	3.26	3.07	2.96	2.88	2.83	2.78	2.75	2.72
	8	3.46	3.11	2.92	2.81	2.73	2.67	2.62	2.59	2.56
	9	3.36	3.01	2.81	2.69	2.61	2.55	2.51	2.47	2.44
	10	3.29	2.92	2.73	2.61	2.52	2.46	2.41	2.38	2.35
	11	3.23	2.86	2.66	2.54	2.45	2.39	2.34	2.30	2.27
	12	3.18	2.81	2.61	2.48	2.39	2.33	2.28	2.24	2.21
	13	3.14	2.76	2.56	2.43	2.35	2.28	2.23	2.20	2.16
	14	3.10	2.73	2.52	2.39	2.31	2.24	2.19	2.15	2.12
	15	3.07	2.70	2.49	2.36	2.27	2.21	2.16	2.12	2.09
	16	3.05	2.67	2.46	2.33	2.24	2.18	2.13	2.09	2.06
	17	3.03	2.64	2.44	2.31	2.22	2.15	2.10	2.06	2.03
	18	3.01	2.62	2.42	2.29	2.20	2.13	2.08	2.04	2.00
	19	2.99	2.61	2.40	2.27	2.18	2.11	2.06	2.02	1.98
	20	2.97	2.59	2.38	2.25	2.16	2.09	2.04	2.00	1.96
	21	2.96	2.57	2.36	2.23	2.14	2.08	2.02	1.98	1.95
	22	2.95	2.56	2.35	2.22	2.13	2.06	2.01	1.97	1.93
	23	2.94	2.55	2.34	2.21	2.11	2.05	1.99	1.95	1.92
	24	2.93	2.54	2.33	2.19	2.10	2.04	1.98	1.94	1.91
	25	2.92	2.53	2.32	2.18	2.09	2.02	1.97	1.93	1.89
	26	2.91	2.52	2.31	2.17	2.08	2.01	1.96	1.92	1.88
	27	2.90	2.51	2.30	2.17	2.07	2.00	1.95	1.91	1.87
	28	2.89	2.50	2.29	2.16	2.06	2.00	1.94	1.90	1.87
	29	2.89	2.50	2.28	2.15	2.06	1.99	1.93	1.89	1.86
	30	2.88	2.49	2.28	2.14	2.05	1.98	1.93	1.88	1.85
	40	2.84	2.44	2.23	2.09	2.00	1.93	1.87	1.83	1.79
	60	2.79	2.39	2.18	2.04	1.95	1.87	1.82	1.77	1.74
	120	2.75	2.35	2.13	1.99	1.90	1.82	1.77	1.72	1.68
	∞	2.71	2.30	2.08	1.94	1.85	1.77	1.72	1.67	1.63

Degrees of freedom for denominator

TABLE B.5   Critical Values of the F Distribution ($\alpha = .10$) (continued)

Degrees of freedom for numerator

		10	12	15	20	24	30	40	60	120	∞
	1	60.19	60.71	61.22	61.74	62.00	62.26	62.53	62.79	63.06	63.33
	2	9.39	9.41	9.42	9.44	9.45	9.46	9.47	9.47	9.48	9.49
	3	5.23	5.22	5.20	5.18	5.18	5.17	5.16	5.15	5.14	5.13
	4	3.92	3.90	3.87	3.84	3.83	3.82	3.80	3.79	3.78	3.76
	5	3.30	3.27	3.24	3.21	3.19	3.17	3.16	3.14	3.12	3.10
	6	2.94	2.90	2.87	2.84	2.82	2.80	2.78	2.76	2.74	2.72
	7	2.70	2.67	2.63	2.59	2.58	2.56	2.54	2.51	2.49	2.47
	8	2.54	2.50	2.46	2.42	2.40	2.38	2.36	2.34	2.32	2.29
	9	2.42	2.38	2.34	2.30	2.28	2.25	2.23	2.21	2.18	2.16
	10	2.32	2.28	2.24	2.20	2.18	2.16	2.13	2.11	2.08	2.06
	11	2.25	2.21	2.17	2.12	2.10	2.08	2.05	2.03	2.00	1.97
	12	2.19	2.15	2.10	2.06	2.04	2.01	1.99	1.96	1.93	1.90
	13	2.14	2.10	2.05	2.01	1.98	1.96	1.93	1.90	1.88	1.85
	14	2.10	2.05	2.01	1.96	1.94	1.91	1.89	1.86	1.83	1.80
	15	2.06	2.02	1.97	1.92	1.90	1.87	1.85	1.82	1.79	1.76
	16	2.03	1.99	1.94	1.89	1.87	1.84	1.81	1.78	1.75	1.72
	17	2.00	1.96	1.91	1.86	1.84	1.81	1.78	1.75	1.72	1.69
	18	1.98	1.93	1.89	1.84	1.81	1.78	1.75	1.72	1.69	1.66
	19	1.96	1.91	1.86	1.81	1.79	1.76	1.73	1.70	1.67	1.63
	20	1.94	1.89	1.84	1.79	1.77	1.74	1.71	1.68	1.64	1.61
	21	1.92	1.87	1.83	1.78	1.75	1.72	1.69	1.66	1.62	1.59
	22	1.90	1.86	1.81	1.76	1.73	1.70	1.67	1.64	1.60	1.57
	23	1.89	1.84	1.80	1.74	1.72	1.69	1.66	1.62	1.59	1.55
	24	1.88	1.83	1.78	1.73	1.70	1.67	1.64	1.61	1.57	1.53
	25	1.87	1.82	1.77	1.72	1.69	1.66	1.63	1.59	1.56	1.52
	26	1.86	1.81	1.76	1.71	1.68	1.65	1.61	1.58	1.54	1.50
	27	1.85	1.80	1.75	1.70	1.67	1.64	1.60	1.57	1.53	1.49
	28	1.84	1.79	1.74	1.69	1.66	1.63	1.59	1.56	1.52	1.48
	29	1.83	1.78	1.73	1.68	1.65	1.62	1.58	1.55	1.51	1.47
	30	1.82	1.77	1.72	1.67	1.64	1.61	1.57	1.54	1.50	1.46
	40	1.76	1.71	1.66	1.61	1.57	1.54	1.51	1.47	1.42	1.38
	60	1.71	1.66	1.60	1.54	1.51	1.48	1.44	1.40	1.35	1.29
	120	1.65	1.60	1.55	1.48	1.45	1.41	1.37	1.32	1.26	1.19
	∞	1.60	1.55	1.49	1.42	1.38	1.34	1.30	1.24	1.17	1.00

Degrees of freedom for denominator (row labels)

TABLE B.5 Critical Values of the F Distribution ($\alpha = .05$)

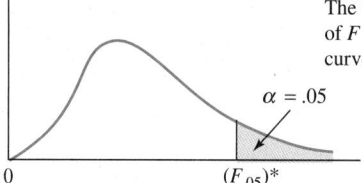

The table values are critical values of F for which the area under the curve to the right is equal to .05.

$\alpha = .05$

0 $(F_{.05})^*$

Degrees of freedom for numerator

	1	2	3	4	5	6	7	8	9
1	161.4	199.5	215.7	224.6	230.2	234.0	236.8	238.9	240.5
2	18.51	19.00	19.16	19.25	19.30	19.33	19.35	19.37	19.38
3	10.13	9.55	9.28	9.12	9.01	8.94	8.89	8.85	8.81
4	7.71	6.94	6.59	6.39	6.26	6.16	6.09	6.04	6.00
5	6.61	5.79	5.41	5.19	5.05	4.95	4.88	4.82	4.77
6	5.99	5.14	4.76	4.53	4.39	4.28	4.21	4.15	4.10
7	5.59	4.74	4.35	4.12	3.97	3.87	3.79	3.73	3.68
8	5.32	4.46	4.07	3.84	3.69	3.58	3.50	3.44	3.39
9	5.12	4.26	3.86	3.63	3.48	3.37	3.29	3.23	3.18
10	4.96	4.10	3.71	3.48	3.33	3.22	3.14	3.07	3.02
11	4.84	3.98	3.59	3.36	3.20	3.09	3.01	2.95	2.90
12	4.75	3.89	3.49	3.26	3.11	3.00	2.91	2.85	2.80
13	4.67	3.81	3.41	3.18	3.03	2.92	2.83	2.77	2.71
14	4.60	3.74	3.34	3.11	2.96	2.85	2.76	2.70	2.65
15	4.54	3.68	3.29	3.06	2.90	2.79	2.71	2.64	2.59
16	4.49	3.63	3.24	3.01	2.85	2.74	2.66	2.59	2.54
17	4.45	3.59	3.20	2.96	2.81	2.70	2.61	2.55	2.49
18	4.41	3.55	3.16	2.93	2.77	2.66	2.58	2.51	2.46
19	4.38	3.52	3.13	2.90	2.74	2.63	2.54	2.48	2.42
20	4.35	3.49	3.10	2.87	2.71	2.60	2.51	2.45	2.39
21	4.32	3.47	3.07	2.84	2.68	2.57	2.49	2.42	2.37
22	4.30	3.44	3.05	2.82	2.66	2.55	2.46	2.40	2.34
23	4.28	3.42	3.03	2.80	2.64	2.53	2.44	2.37	2.32
24	4.26	3.40	3.01	2.78	2.62	2.51	2.42	2.36	2.30
25	4.24	3.39	2.99	2.76	2.60	2.49	2.40	2.34	2.28
30	4.17	3.32	2.92	2.69	2.53	2.42	2.33	2.27	2.21
40	4.08	3.23	2.84	2.61	2.45	2.34	2.25	2.18	2.12
60	4.00	3.15	2.76	2.53	2.37	2.25	2.17	2.10	2.04
120	3.92	3.07	2.68	2.45	2.29	2.17	2.09	2.02	1.96
∞	3.84	3.00	2.60	2.37	2.21	2.10	2.01	1.94	1.88

Degrees of freedom for denominator

TABLE B.5 Critical Values of the F Distribution ($\alpha = .05$) (continued)

				Degrees of freedom for numerator						
	10	**12**	**15**	**20**	**24**	**30**	**40**	**60**	**120**	**∞**
1	241.9	243.9	245.9	248.0	249.1	250.1	251.1	252.2	253.3	254.3
2	19.40	19.41	19.43	19.45	19.45	19.46	19.47	19.48	19.49	19.50
3	8.79	8.74	8.70	8.66	8.64	8.62	8.59	8.57	8.55	8.53
4	5.96	5.91	5.86	5.80	5.77	5.75	5.72	5.69	5.66	5.63
5	4.74	4.68	4.62	4.56	4.53	4.50	4.46	4.43	4.40	4.36
6	4.06	4.00	3.94	3.87	3.84	3.81	3.77	3.74	3.70	3.67
7	3.64	3.57	3.51	3.44	3.41	3.38	3.34	3.30	3.27	3.23
8	3.35	3.28	3.22	3.15	3.12	3.08	3.04	3.01	2.97	2.93
9	3.14	3.07	3.01	2.94	2.90	2.86	2.83	2.79	2.75	2.71
10	2.98	2.91	2.85	2.77	2.74	2.70	2.66	2.62	2.58	2.54
11	2.85	2.79	2.72	2.65	2.61	2.57	2.53	2.49	2.45	2.40
12	2.75	2.69	2.62	2.54	2.51	2.47	2.43	2.38	2.34	2.30
13	2.67	2.60	2.53	2.46	2.42	2.38	2.34	2.30	2.25	2.21
14	2.60	2.53	2.46	2.39	2.35	2.31	2.27	2.22	2.18	2.13
15	2.54	2.48	2.40	2.33	2.29	2.25	2.20	2.16	2.11	2.07
16	2.49	2.42	2.35	2.28	2.24	2.19	2.15	2.11	2.06	2.01
17	2.45	2.38	2.31	2.23	2.19	2.15	2.10	2.06	2.01	1.96
18	2.41	2.34	2.27	2.19	2.15	2.11	2.06	2.02	1.97	1.92
19	2.38	2.31	2.23	2.16	2.11	2.07	2.03	1.98	1.93	1.88
20	2.35	2.28	2.20	2.12	2.08	2.04	1.99	1.95	1.90	1.84
21	2.32	2.25	2.18	2.10	2.05	2.01	1.96	1.92	1.87	1.81
22	2.30	2.23	2.15	2.07	2.03	1.98	1.94	1.89	1.84	1.78
23	2.27	2.20	2.13	2.05	2.01	1.96	1.91	1.86	1.81	1.76
24	2.25	2.18	2.11	2.03	1.98	1.94	1.89	1.84	1.79	1.73
25	2.24	2.16	2.09	2.01	1.96	1.92	1.87	1.82	1.77	1.71
30	2.16	2.09	2.01	1.93	1.89	1.84	1.79	1.74	1.68	1.62
40	2.08	2.00	1.92	1.84	1.79	1.74	1.69	1.64	1.58	1.51
60	1.99	1.92	1.84	1.75	1.70	1.65	1.59	1.53	1.47	1.39
120	1.91	1.83	1.75	1.66	1.61	1.55	1.50	1.43	1.35	1.25
∞	1.83	1.75	1.67	1.57	1.52	1.46	1.39	1.32	1.22	1.00

Degrees of freedom for denominator

TABLE B.5   Critical Values of the F Distribution ($\alpha = .025$)

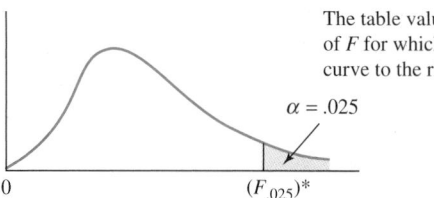

The table values are critical values of F for which the area under the curve to the right is equal to .025.

$\alpha = .025$

0 $(F_{.025})^*$

		Degrees of freedom for numerator							
	1	**2**	**3**	**4**	**5**	**6**	**7**	**8**	**9**
1	648	800	864	900	922	937	948	957	963
2	38.51	39.00	39.17	39.25	39.30	39.33	39.36	39.37	39.39
3	17.44	16.04	15.44	15.10	14.88	14.73	14.62	14.54	14.47
4	12.22	10.65	9.98	9.60	9.36	9.20	9.07	8.98	8.90
5	10.01	8.43	7.76	7.39	7.15	6.98	6.85	6.76	6.68
6	8.81	7.26	6.60	6.23	5.99	5.82	5.70	5.60	5.52
7	8.07	6.54	5.89	5.52	5.29	5.12	4.99	4.90	4.82
8	7.57	6.06	5.42	5.05	4.82	4.65	4.53	4.43	4.36
9	7.21	5.71	5.08	4.72	4.48	4.32	4.20	4.10	4.03
10	6.94	5.46	4.83	4.47	4.24	4.07	3.95	3.85	3.78
11	6.72	5.26	4.63	4.28	4.04	3.88	3.76	3.66	3.59
12	6.55	5.10	4.47	4.12	3.89	3.73	3.61	3.51	3.44
13	6.41	4.97	4.35	4.00	3.77	3.60	3.48	3.39	3.31
14	6.30	4.86	4.24	3.89	3.66	3.50	3.38	3.29	3.21
15	6.20	4.77	4.15	3.80	3.58	3.41	3.29	3.20	3.12
16	6.12	4.69	4.08	3.73	3.50	3.34	3.22	3.12	3.05
17	6.04	4.62	4.01	3.66	3.44	3.28	3.16	3.06	2.98
18	5.98	4.56	3.95	3.61	3.38	3.22	3.10	3.01	2.93
19	5.92	4.51	3.90	3.56	3.33	3.17	3.05	2.96	2.88
20	5.87	4.46	3.86	3.51	3.29	3.13	3.01	2.91	2.84
21	5.83	4.42	3.82	3.48	3.25	3.09	2.97	2.87	2.80
22	5.79	4.38	3.78	3.44	3.22	3.05	2.93	2.84	2.76
23	5.75	4.35	3.75	3.41	3.18	3.02	2.90	2.81	2.73
24	5.72	4.32	3.72	3.38	3.15	2.99	2.87	2.78	2.70
25	5.69	4.29	3.69	3.35	3.13	2.97	2.85	2.75	2.68
30	5.57	4.18	3.59	3.25	3.03	2.87	2.75	2.65	2.57
40	5.42	4.05	3.46	3.13	2.90	2.74	2.62	2.53	2.45
60	5.29	3.93	3.34	3.01	2.79	2.63	2.51	2.41	2.33
120	5.15	3.80	3.23	2.89	2.67	2.52	2.39	2.30	2.22
∞	5.02	3.69	3.12	2.79	2.57	2.41	2.29	2.19	2.11

Degrees of freedom for denominator

TABLE B.5   Critical Values of the *F* Distribution (α = .025) (continued)

<table>
<tr><td colspan="11" align="center">Degrees of freedom for numerator</td></tr>
<tr><td></td><td></td><td>10</td><td>12</td><td>15</td><td>20</td><td>24</td><td>30</td><td>40</td><td>60</td><td>120</td><td>∞</td></tr>
<tr><td rowspan="26">Degrees of freedom for denominator</td><td>1</td><td>969</td><td>977</td><td>985</td><td>993</td><td>997</td><td>1,001</td><td>1,006</td><td>1,010</td><td>1,014</td><td>1,018</td></tr>
<tr><td>2</td><td>39.40</td><td>39.41</td><td>39.43</td><td>39.45</td><td>39.46</td><td>39.46</td><td>39.47</td><td>39.48</td><td>39.49</td><td>39.50</td></tr>
<tr><td>3</td><td>14.42</td><td>14.34</td><td>14.25</td><td>14.17</td><td>14.12</td><td>14.08</td><td>14.04</td><td>13.99</td><td>13.95</td><td>13.90</td></tr>
<tr><td>4</td><td>8.84</td><td>8.75</td><td>8.66</td><td>8.56</td><td>8.51</td><td>8.46</td><td>8.41</td><td>8.36</td><td>8.31</td><td>8.26</td></tr>
<tr><td>5</td><td>6.62</td><td>6.52</td><td>6.43</td><td>6.33</td><td>6.28</td><td>6.23</td><td>6.18</td><td>6.12</td><td>6.07</td><td>6.02</td></tr>
<tr><td>6</td><td>5.46</td><td>5.37</td><td>5.27</td><td>5.17</td><td>5.12</td><td>5.07</td><td>5.01</td><td>4.96</td><td>4.90</td><td>4.85</td></tr>
<tr><td>7</td><td>4.76</td><td>4.67</td><td>4.57</td><td>4.47</td><td>4.42</td><td>4.36</td><td>4.31</td><td>4.25</td><td>4.20</td><td>4.14</td></tr>
<tr><td>8</td><td>4.30</td><td>4.20</td><td>4.10</td><td>4.00</td><td>3.95</td><td>3.89</td><td>3.84</td><td>3.78</td><td>3.73</td><td>3.67</td></tr>
<tr><td>9</td><td>3.96</td><td>3.87</td><td>3.77</td><td>3.67</td><td>3.61</td><td>3.56</td><td>3.51</td><td>3.45</td><td>3.39</td><td>3.33</td></tr>
<tr><td>10</td><td>3.72</td><td>3.62</td><td>3.52</td><td>3.42</td><td>3.37</td><td>3.31</td><td>3.26</td><td>3.20</td><td>3.14</td><td>3.08</td></tr>
<tr><td>11</td><td>3.53</td><td>3.43</td><td>3.33</td><td>3.23</td><td>3.17</td><td>3.12</td><td>3.06</td><td>3.00</td><td>2.94</td><td>2.88</td></tr>
<tr><td>12</td><td>3.37</td><td>3.28</td><td>3.18</td><td>3.07</td><td>3.02</td><td>2.96</td><td>2.91</td><td>2.85</td><td>2.79</td><td>2.72</td></tr>
<tr><td>13</td><td>3.25</td><td>3.15</td><td>3.05</td><td>2.95</td><td>2.89</td><td>2.84</td><td>2.78</td><td>2.72</td><td>2.66</td><td>2.60</td></tr>
<tr><td>14</td><td>3.15</td><td>3.05</td><td>2.95</td><td>2.84</td><td>2.79</td><td>2.73</td><td>2.67</td><td>2.61</td><td>2.55</td><td>2.49</td></tr>
<tr><td>15</td><td>3.06</td><td>2.96</td><td>2.86</td><td>2.76</td><td>2.70</td><td>2.64</td><td>2.59</td><td>2.52</td><td>2.46</td><td>2.40</td></tr>
<tr><td>16</td><td>2.99</td><td>2.89</td><td>2.79</td><td>2.68</td><td>2.63</td><td>2.57</td><td>2.51</td><td>2.45</td><td>2.38</td><td>2.32</td></tr>
<tr><td>17</td><td>2.92</td><td>2.82</td><td>2.72</td><td>2.62</td><td>2.56</td><td>2.50</td><td>2.44</td><td>2.38</td><td>2.32</td><td>2.25</td></tr>
<tr><td>18</td><td>2.87</td><td>2.77</td><td>2.67</td><td>2.56</td><td>2.50</td><td>2.44</td><td>2.38</td><td>2.32</td><td>2.26</td><td>2.19</td></tr>
<tr><td>19</td><td>2.82</td><td>2.72</td><td>2.62</td><td>2.51</td><td>2.45</td><td>2.39</td><td>2.33</td><td>2.27</td><td>2.20</td><td>2.13</td></tr>
<tr><td>20</td><td>2.77</td><td>2.68</td><td>2.57</td><td>2.46</td><td>2.41</td><td>2.35</td><td>2.29</td><td>2.22</td><td>2.16</td><td>2.09</td></tr>
<tr><td>21</td><td>2.73</td><td>2.64</td><td>2.53</td><td>2.42</td><td>2.37</td><td>2.31</td><td>2.25</td><td>2.18</td><td>2.11</td><td>2.04</td></tr>
<tr><td>22</td><td>2.70</td><td>2.60</td><td>2.50</td><td>2.39</td><td>2.33</td><td>2.27</td><td>2.21</td><td>2.14</td><td>2.08</td><td>2.00</td></tr>
<tr><td>23</td><td>2.67</td><td>2.57</td><td>2.47</td><td>2.36</td><td>2.30</td><td>2.24</td><td>2.18</td><td>2.11</td><td>2.04</td><td>1.97</td></tr>
<tr><td>24</td><td>2.64</td><td>2.54</td><td>2.44</td><td>2.33</td><td>2.27</td><td>2.21</td><td>2.15</td><td>2.08</td><td>2.01</td><td>1.94</td></tr>
<tr><td>25</td><td>2.61</td><td>2.51</td><td>2.41</td><td>2.30</td><td>2.24</td><td>2.18</td><td>2.12</td><td>2.05</td><td>1.98</td><td>1.91</td></tr>
<tr><td>30</td><td>2.51</td><td>2.41</td><td>2.31</td><td>2.20</td><td>2.14</td><td>2.07</td><td>2.01</td><td>1.94</td><td>1.87</td><td>1.79</td></tr>
<tr><td>40</td><td>2.39</td><td>2.29</td><td>2.18</td><td>2.07</td><td>2.01</td><td>1.94</td><td>1.88</td><td>1.80</td><td>1.72</td><td>1.64</td></tr>
<tr><td>60</td><td>2.27</td><td>2.17</td><td>2.06</td><td>1.94</td><td>1.88</td><td>1.82</td><td>1.74</td><td>1.67</td><td>1.58</td><td>1.48</td></tr>
<tr><td>120</td><td>2.16</td><td>2.05</td><td>1.94</td><td>1.82</td><td>1.76</td><td>1.69</td><td>1.61</td><td>1.53</td><td>1.43</td><td>1.31</td></tr>
<tr><td>∞</td><td>2.05</td><td>1.94</td><td>1.83</td><td>1.71</td><td>1.64</td><td>1.57</td><td>1.48</td><td>1.39</td><td>1.27</td><td>1.00</td></tr>
</table>

Table B.5   Critical Values of the F Distribution ($\alpha = .01$)

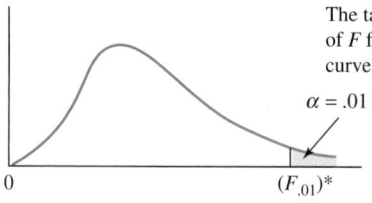

The table values are critical values of F for which the area under the curve to the right is equal to .01.

$\alpha = .01$

$(F_{.01})^*$

0

Degrees of freedom for numerator

		1	2	3	4	5	6	7	8	9
	1	4,052	5,000	5,403	5,625	5,764	5,859	5,928	5,981	6,022
	2	98.50	99.00	99.17	99.25	99.30	99.33	99.36	99.37	99.39
	3	34.12	30.82	29.46	28.71	28.24	27.91	27.67	27.49	27.35
	4	21.20	18.00	16.69	15.98	15.52	15.21	14.98	14.80	14.66
	5	16.26	13.27	12.06	11.39	10.97	10.67	10.46	10.29	10.16
	6	13.75	10.92	9.78	9.15	8.75	8.47	8.26	8.10	7.98
	7	12.25	9.55	8.45	7.85	7.46	7.19	6.99	6.84	6.72
	8	11.26	8.65	7.59	7.01	6.63	6.37	6.18	6.03	5.91
	9	10.56	8.02	6.99	6.42	6.06	5.80	5.61	5.47	5.35
	10	10.04	7.56	6.55	5.99	5.64	5.39	5.20	5.06	4.94
Degrees of freedom for denominator	11	9.65	7.21	6.22	5.67	5.32	5.07	4.89	4.74	4.63
	12	9.33	6.93	5.95	5.41	5.06	4.82	4.64	4.50	4.39
	13	9.07	6.70	5.74	5.21	4.86	4.62	4.44	4.30	4.19
	14	8.86	6.51	5.56	5.04	4.69	4.46	4.28	4.14	4.03
	15	8.68	6.36	5.42	4.89	4.56	4.32	4.14	4.00	3.89
	16	8.53	6.23	5.29	4.77	4.44	4.20	4.03	3.89	3.78
	17	8.40	6.11	5.18	4.67	4.34	4.10	3.93	3.79	3.68
	18	8.29	6.01	5.09	4.58	4.25	4.01	3.84	3.71	3.60
	19	8.18	5.93	5.01	4.50	4.17	3.94	3.77	3.63	3.52
	20	8.10	5.85	4.94	4.43	4.10	3.87	3.70	3.56	3.46
	21	8.02	5.78	4.87	4.37	4.04	3.81	3.64	3.51	3.40
	22	7.95	5.72	4.82	4.31	3.99	3.76	3.59	3.45	3.35
	23	7.88	5.66	4.76	4.26	3.94	3.71	3.54	3.41	3.30
	24	7.82	5.61	4.72	4.22	3.90	3.67	3.50	3.36	3.26
	25	7.77	5.57	4.68	4.18	3.85	3.63	3.46	3.32	3.22
	30	7.56	5.39	4.51	4.02	3.70	3.47	3.30	3.17	3.07
	40	7.31	5.18	4.31	3.83	3.51	3.29	3.12	2.99	2.89
	60	7.08	4.98	4.13	3.65	3.34	3.12	2.95	2.82	2.72
	120	6.85	4.79	3.95	3.48	3.17	2.96	2.79	2.66	2.56
	∞	6.63	4.61	3.78	3.32	3.02	2.80	2.64	2.51	2.41

TABLE B.5   Critical Values of the F Distribution ($\alpha = .01$) (continued)

		10	12	15	20	24	30	40	60	120	∞
	Degrees of freedom for numerator										
Degrees of freedom for denominator	1	6,056	6,106	6,157	6,209	6,235	6,261	6,287	6,313	6,339	6,366
	2	99.40	99.42	99.43	99.45	99.46	99.47	99.47	99.48	99.49	99.50
	3	27.23	27.05	26.87	26.69	26.60	26.50	26.41	26.32	26.22	26.13
	4	14.55	14.37	14.20	14.02	13.93	13.84	13.75	13.65	13.56	13.46
	5	10.05	9.89	9.72	9.55	9.47	9.38	9.29	9.20	9.11	9.02
	6	7.87	7.72	7.56	7.40	7.31	7.23	7.14	7.06	6.97	6.88
	7	6.62	6.47	6.31	6.16	6.07	5.99	5.91	5.82	5.74	5.65
	8	5.81	5.67	5.52	5.36	5.28	5.20	5.12	5.03	4.95	4.86
	9	5.26	5.11	4.96	4.81	4.73	4.65	4.57	4.48	4.40	4.31
	10	4.85	4.71	4.56	4.41	4.33	4.25	4.17	4.08	4.00	3.91
	11	4.54	4.40	4.25	4.10	4.02	3.94	3.86	3.78	3.69	3.60
	12	4.30	4.16	4.01	3.86	3.78	3.70	3.62	3.54	3.45	3.36
	13	4.10	3.96	3.82	3.66	3.59	3.51	3.43	3.34	3.25	3.17
	14	3.94	3.80	3.66	3.51	3.43	3.35	3.27	3.18	3.09	3.00
	15	3.80	3.67	3.52	3.37	3.29	3.21	3.13	3.05	2.96	2.87
	16	3.69	3.55	3.41	3.26	3.18	3.10	3.02	2.93	2.84	2.75
	17	3.59	3.46	3.31	3.16	3.08	3.00	2.92	2.83	2.75	2.65
	18	3.51	3.37	3.23	3.08	3.00	2.92	2.84	2.75	2.66	2.57
	19	3.43	3.30	3.15	3.00	2.92	2.84	2.76	2.67	2.58	2.49
	20	3.37	3.23	3.09	2.94	2.86	2.78	2.69	2.61	2.52	2.42
	21	3.31	3.17	3.03	2.88	2.80	2.72	2.64	2.55	2.46	2.36
	22	3.26	3.12	2.98	2.83	2.75	2.67	2.58	2.50	2.40	2.31
	23	3.21	3.07	2.93	2.78	2.70	2.62	2.54	2.45	2.35	2.26
	24	3.17	3.03	2.89	2.74	2.66	2.58	2.49	2.40	2.31	2.21
	25	3.13	2.99	2.85	2.70	2.62	2.54	2.45	2.36	2.27	2.17
	30	2.98	2.84	2.70	2.55	2.47	2.39	2.30	2.21	2.11	2.01
	40	2.80	2.66	2.52	2.37	2.29	2.20	2.11	2.02	1.92	1.80
	60	2.63	2.50	2.35	2.20	2.12	2.03	1.94	1.84	1.73	1.60
	120	2.47	2.34	2.19	2.03	1.95	1.86	1.76	1.66	1.53	1.38
	∞	2.32	2.18	2.04	1.88	1.79	1.70	1.59	1.47	1.32	1.00

TABLE B.6 Critical Values of the Studentized Range, $q(k, \nu)$, Upper 5%

ν \ k	2	3	4	5	6	7	8	9	10	11	12	13	14	15	16	17	18	19	20
1	17.97	26.98	32.82	37.08	40.41	43.12	45.40	47.36	49.07	50.59	51.96	53.20	54.33	55.36	56.32	57.22	58.04	58.83	59.56
2	6.08	8.33	9.80	10.88	11.74	12.44	13.03	13.54	13.99	14.39	14.75	15.08	15.38	15.65	15.91	16.14	16.37	16.57	16.77
3	4.50	5.91	6.82	7.50	8.04	8.48	8.85	9.18	9.46	9.72	9.95	10.15	10.35	10.52	10.69	10.84	10.98	11.11	11.24
4	3.93	5.04	5.76	6.29	6.71	7.05	7.35	7.60	7.83	8.03	8.21	8.37	8.52	8.66	8.79	8.91	9.03	9.13	9.23
5	3.64	4.60	5.22	5.67	6.03	6.33	6.58	6.80	6.99	7.17	7.32	7.47	7.60	7.72	7.83	7.93	8.03	8.12	8.21
6	3.46	4.34	4.90	5.30	5.63	5.90	6.12	6.32	6.49	6.65	6.79	6.92	7.03	7.14	7.24	7.34	7.43	7.51	7.59
7	3.34	4.16	4.68	5.06	5.36	5.61	5.82	6.00	6.16	6.30	6.43	6.55	6.66	6.76	6.85	6.94	7.02	7.10	7.17
8	3.26	4.04	4.53	4.89	5.17	5.40	5.60	5.77	5.92	6.05	6.18	6.29	6.39	6.48	6.57	6.65	6.73	6.80	6.87
9	3.20	3.95	4.41	4.76	5.02	5.24	5.43	5.59	5.74	5.87	5.98	6.09	6.19	6.28	6.36	6.44	6.51	6.58	6.64
10	3.15	3.88	4.33	4.65	4.91	5.12	5.30	5.46	5.60	5.72	5.83	5.93	6.03	6.11	6.19	6.27	6.34	6.40	6.47
11	3.11	3.82	4.26	4.57	4.82	5.03	5.20	5.35	5.49	5.61	5.71	5.81	5.90	5.98	6.06	6.13	6.20	6.27	6.33
12	3.08	3.77	4.20	4.51	4.75	4.95	5.12	5.27	5.39	5.51	5.61	5.71	5.80	5.88	5.95	6.02	6.09	6.15	6.21
13	3.06	3.73	4.15	4.45	4.69	4.88	5.05	5.19	5.32	5.43	5.53	5.63	5.71	5.79	5.86	5.93	5.99	6.05	6.11
14	3.03	3.70	4.11	4.41	4.64	4.83	4.99	5.13	5.25	5.36	5.46	5.55	5.64	5.71	5.79	5.85	5.91	5.97	6.03
15	3.01	3.67	4.08	4.37	4.59	4.78	4.94	5.08	5.20	5.31	5.40	5.49	5.57	5.65	5.72	5.78	5.85	5.90	5.96
16	3.00	3.65	4.05	4.33	4.56	4.74	4.90	5.03	5.15	5.26	5.35	5.44	5.52	5.59	5.66	5.73	5.79	5.84	5.90
17	2.98	3.63	4.02	4.30	4.52	4.70	4.86	4.99	5.11	5.21	5.31	5.39	5.47	5.54	5.61	5.67	5.73	5.79	5.84
18	2.97	3.61	4.00	4.28	4.49	4.67	4.82	4.96	5.07	5.17	5.27	5.35	5.43	5.50	5.57	5.63	5.69	5.74	5.79
19	2.96	3.59	3.98	4.25	4.47	4.65	4.79	4.92	5.04	5.14	5.23	5.31	5.39	5.46	5.53	5.59	5.65	5.70	5.75
20	2.95	3.58	3.96	4.23	4.45	4.62	4.77	4.90	5.01	5.11	5.20	5.28	5.36	5.43	5.49	5.55	5.61	5.66	5.71
24	2.92	3.53	3.90	4.17	4.37	4.54	4.68	4.81	4.92	5.01	5.10	5.18	5.25	5.32	5.38	5.44	5.49	5.55	5.59
30	2.89	3.49	3.85	4.10	4.30	4.46	4.60	4.72	4.82	4.92	5.00	5.08	5.15	5.21	5.27	5.33	5.38	5.43	5.47
40	2.86	3.44	3.79	4.04	4.23	4.39	4.52	4.63	4.73	4.82	4.90	4.98	5.04	5.11	5.16	5.22	5.27	5.31	5.36
60	2.83	3.40	3.74	3.98	4.16	4.31	4.44	4.55	4.65	4.73	4.81	4.88	4.94	5.00	5.06	5.11	5.15	5.20	5.24
120	2.80	3.36	3.68	3.92	4.10	4.24	4.36	4.47	4.56	4.64	4.71	4.78	4.84	4.90	4.95	5.00	5.04	5.09	5.13
∞	2.77	3.31	3.63	3.86	4.03	4.17	4.29	4.39	4.47	4.55	4.62	4.68	4.74	4.80	4.85	4.89	4.93	4.97	5.01

SOURCE: From E. S. Pearson and H. O. Hartley (Eds.), *Biometrika Tables for Statisticians* vol. 1, 3rd ed. (Cambridge University Press, 1970), p. 192. Reprinted by permission of the Biometrika Trustees.

TABLE B.6   Critical Values of the Studentized Range, $q(k, \nu)$, Upper 1%

ν \\ k	2	3	4	5	6	7	8	9	10	11	12	13	14	15	16	17	18	19	20
1	90.03	135.0	164.3	185.6	202.2	215.8	227.2	237.0	245.6	253.2	260.0	266.2	271.8	277.0	281.8	286.3	290.4	294.3	298.0
2	14.04	19.02	22.29	24.72	26.63	28.20	29.53	30.68	31.69	32.59	33.40	34.13	34.81	35.43	36.00	36.53	37.03	37.50	37.95
3	8.26	10.62	12.17	13.33	14.24	15.00	15.64	16.20	16.69	17.13	17.53	17.89	18.22	18.52	18.81	19.07	19.32	19.55	19.77
4	6.51	8.12	9.17	9.96	10.58	11.10	11.55	11.93	12.27	12.57	12.84	13.09	13.32	13.53	13.73	13.91	14.08	14.24	14.40
5	5.70	6.98	7.80	8.42	8.91	9.32	9.67	9.97	10.24	10.48	10.70	10.89	11.08	11.24	11.40	11.55	11.68	11.81	11.93
6	5.24	6.33	7.03	7.56	7.97	8.32	8.61	8.87	9.10	9.30	9.48	9.65	9.81	9.95	10.08	10.21	10.32	10.43	10.54
7	4.95	5.92	6.54	7.01	7.37	7.68	7.94	8.17	8.37	8.55	8.71	8.86	9.00	9.12	9.24	9.35	9.46	9.55	9.65
8	4.75	5.64	6.20	6.62	6.96	7.24	7.47	7.68	7.86	8.03	8.18	8.31	8.44	8.55	8.66	8.76	8.85	8.94	9.03
9	4.60	5.43	5.96	6.35	6.66	6.91	7.13	7.33	7.49	7.65	7.78	7.91	8.03	8.13	8.23	8.33	8.41	8.49	8.57
10	4.48	5.27	5.77	6.14	6.43	6.67	6.87	7.05	7.21	7.36	7.49	7.60	7.71	7.81	7.91	7.99	8.08	8.15	8.23
11	4.39	5.15	5.62	5.97	6.25	6.48	6.67	6.84	6.99	7.13	7.25	7.36	7.46	7.56	7.65	7.73	7.81	7.88	7.95
12	4.32	5.05	5.50	5.84	6.10	6.32	6.51	6.67	6.81	6.94	7.06	7.17	7.26	7.36	7.44	7.52	7.59	7.66	7.73
13	4.26	4.96	5.40	5.73	5.98	6.19	6.37	6.53	6.67	6.79	6.90	7.01	7.10	7.19	7.27	7.35	7.42	7.48	7.55
14	4.21	4.89	5.32	5.63	5.88	6.08	6.26	6.41	6.54	6.66	6.77	6.87	6.96	7.05	7.13	7.20	7.27	7.33	7.39
15	4.17	4.84	5.25	5.56	5.80	5.99	6.16	6.31	6.44	6.55	6.66	6.76	6.84	6.93	7.00	7.07	7.14	7.20	7.26
16	4.13	4.79	5.19	5.49	5.72	5.92	6.08	6.22	6.35	6.46	6.56	6.66	6.74	6.82	6.90	6.97	7.03	7.09	7.15
17	4.10	4.74	5.14	5.43	5.66	5.85	6.01	6.15	6.27	6.38	6.48	6.57	6.66	6.73	6.81	6.87	6.94	7.00	7.05
18	4.07	4.70	5.09	5.38	5.60	5.79	5.94	6.08	6.20	6.31	6.41	6.50	6.58	6.65	6.72	6.79	6.85	6.91	6.97
19	4.05	4.67	5.05	5.33	5.55	5.73	5.89	6.02	6.14	6.25	6.34	6.43	6.51	6.58	6.65	6.72	6.78	6.84	6.89
20	4.02	4.64	5.02	5.29	5.51	5.69	5.84	5.97	6.09	6.19	6.28	6.37	6.45	6.52	6.59	6.65	6.71	6.77	6.82
24	3.96	4.55	4.91	5.17	5.37	5.54	5.69	5.81	5.92	6.02	6.11	6.19	6.26	6.33	6.39	6.45	6.51	6.56	6.61
30	3.89	4.45	4.80	5.05	5.24	5.40	5.54	5.65	5.76	5.85	5.93	6.01	6.08	6.14	6.20	6.26	6.31	6.36	6.41
40	3.82	4.37	4.70	4.93	5.11	5.26	5.39	5.50	5.60	5.69	5.76	5.83	5.90	5.96	6.02	6.07	6.12	6.16	6.21
60	3.76	4.28	4.59	4.82	4.99	5.13	5.25	5.36	5.45	5.53	5.60	5.67	5.73	5.78	5.84	5.89	5.93	5.97	6.01
120	3.70	4.20	4.50	4.71	4.87	5.01	5.12	5.21	5.30	5.37	5.44	5.50	5.56	5.61	5.66	5.71	5.75	5.79	5.83
∞	3.64	4.12	4.40	4.60	4.76	4.88	4.99	5.08	5.16	5.23	5.29	5.35	5.40	5.45	5.49	5.54	5.57	5.61	5.65

SOURCE: From E. S. Pearson and H. O. Hartley, (Eds.), Biometrika Tables for Statisticians, vol. I, 3rd ed. (Cambridge University Press, 1970), p. 193. Reprinted by permission of the Biometrika Trustees.

Answers to Selected Odd Exercises

Your answers may not agree completely due to roundoff error. Most solutions were found with the Minitab computer package.

1.1 Experimental unit: College student;
Population: All college students at this university. The 500 students should be selected in some random manner such as selecting every 10th student from a list provided by the registrar.

1.3 Experimental unit: An accident.
The population is not clearly identified because the exercise doesn't specify what region the accidents were taken from. For example, did all accidents occur in a particular city, county, or state? Variables: Date accident occurred, How many vehicles involved, Make and model of vehicles, How many occupants, Gender of driver, Age of driver, Was alcohol involved?
The sample size is 1,000.

1.5 Population: All entering freshmen at the university;
Gender, SAT, 1st choice, 2nd choice, residence. From a list provided by the admissions office, sequentially number all freshmen from 1 to n (the last number on the list) and then with a computer randomly generate 200 numbers between 1 and n and then choose those students.

1.7 a. It would be impossible to test the drug on *all* potential users of the product.
b. If we distributed free samples to everyone then no one would be required to purchase the product.
c. It would be impossible to count every tree in the forest.
d. If we determined the life of every battery we would deplete our inventory.

1.9 a. 13.67 per 100, 7.67 per 100, 6.67 per 100, 5.11 per 100, 5.00 per 100
c. No, just because 7.6% is the average for these five stores, this doesn't mean that every store will have a 7.6% error rate. Some stores will possibly have a 0% error rate but then others will go much higher than the average.

d. No; in fact, the exercise stated that 9 stores were inspected and only 5 were fined for overcharges.

1.13 Target population: Older women. Variables of interest: Age, Weight, and Bone density.

1.17 The most general population for which this sample is representative is the population of people whose characteristics are similar to those who subscribe to *Time* magazine.

1.19 No; a large segment of rock music fans never attend concerts.

1.21 First, assign everyone in the class a number from 1 to n (the size of the class), then generate 15 random numbers from 1 to n and select those students.

1.33 It is only possible to determine the percent of the police force that is *caught* taking a bribe. It would be impossible to determine the percent that has never taken a bribe.

1.35 During this time frame it became more acceptable to report sexual abuse of children. It may not mean that there was in increase in the number of cases, but rather there was an increase in the number of *reported* cases. The size of the children in the rectangles is misleading because both the height and width are increased. In examples such as this, only one dimension should be increased.

1.37 This is a captive audience, not a random sample of disaffected Clinton supporters. They were presented only one side of the story—namely, Clinton's speech. Furthermore, it doesn't say how the 100 people were selected. This information should not be generalized to the population of all disaffected Clinton supporters.

1.41 Although the Russians were given the most tickets, they were not the worst offenders because their rate was 8.9 tickets/vehicle/month and Ukraine diplomats had a rate of 10.6 tickets/ vehicle/month.

1.43 No. The neighbor who was at home was there for a given reason; perhaps he or she was unemployed or had several small children or any number of things that would characterize them as being different from the person who was chosen for the survey. The interviewer should return at another time to obtain responses from the selected household;

otherwise, another household should be randomly selected.

1.45 A strong point is that educational information concerning nuclear reactors as a source of energy is provided. A weak point is that the results may be biased because of the pamphlet. Because the respondents choose to participate (volunteer survey) it is questionable whether the results will represent the true feeling of the population.

1.47 A variable is confounded with the treatment when its effect and the effect of the treatment cannot be distinguished from each other.

1.49 The placebo effect is a psychological effect to a treatment.

1.53 This question leads the respondents. In option **a,** the phrase "to those countries that protect the right of their citizens" should be omitted.

1.57 a. observational   **b.** randomized experiment
c. observational   **d.** randomized experiment

1.59 a. sample survey   **b.** experiment
c. sample survey

1.61 This is not a scientific poll, it is a volunteer survey. Viewers of the Weather Channel choose to participate by volunteering to respond to the survey. They certainly would not have a random sample of Weather Channel viewers.

1.65 a. Observational study
b. The treatment is employment status. The response variable is the divorce rate.
c. Extraneous variables that would affect divorce rate are years married, stress, and lifestyle.

1.67 This is convenience sampling because no scientific method is used to select the respondents.

1.69 a. telephone   **b.** mailed questionnaire
c. mailed questionnaire
d. personal interview

1.71 Systematic sampling would be the easiest and because there is no order other than being listed alphabetically, the sample would be representative of all charge accounts.

1.73 a. All oil stocks   **b.** 10 selected oil stocks
c. Price/earning ratio

1.75 Personal income

1.77 Lower-income households are omitted from the sample. Higher-income households are less likely to participate.

1.79 D   **1.81** A   **1.83** D   **1.85** treatment

1.87 confounded

1.89 a. T   **b.** F   **c.** F   **d.** F   **e.** F

1.91 a. Amount of exercise

b. Risk of heart disease
c. A bus driver or policeperson in New York City
d. Job stress and marital status
e. Observational study

1.93 The phrase "Don't you agree . . ." is leading. A better question would be "Do you agree or disagree that a farmer should be allowed to raise as much of any crop he/she chooses without any restrictions or supports from the federal government?"

1.95 a. Cluster sampling   **b.** Stratified sampling
c. Systematic sampling   **d.** Lottery sampling
e. Stratified sampling

1.97 No, because this is an observational study, we cannot conclude causation. There are too many possible confounding variables. Having a higher income and being married probably contributed to the decision to own a pet, not the other way around.

1.99 Sample survey

Chapter 2

2.1 a. All welfare recipients in the state
b. A systematic sample from a list of welfare recipients
c. Gender, race, and marital status of head-of-household
d. Annual income (continuous), number of children (discrete), and age of head-of-household (continuous)

2.3 a. Population is all third graders
b. Use a cluster sample of third grade classes.
c. Gender, race, and availability of a computer at home
d. Family income (continuous), IQ (continuous), and number of siblings (discrete)

2.5 a. Numerical—continuous   **b.** Categorical
c. Numerical—continuous
d. Numerical—continuous
e. Numerical—continuous
f. Numerical—discrete   **g.** Categorical

2.7 Gender—categorical, Major—categorical, Grade point average—numerical, Number of times using placement service—numerical, Type of employment—categorical

2.9 a. Threshold reaction time is numerical.
b. It is continuous.

2.13 a. Continuous   **b.** Continuous   **c.** Discrete
d. Continuous   **e.** Discrete   **f.** Continuous

2.15 a. Experimental unit: State
  b. Multivariate data set
  c. There are four variables.
  d. Region—categorical, Hazardous waste sites—numerical, Above average minority—numerical, Percent minority—numerical
  e. Hazardous sites—discrete, Above average minority—discrete, Percent minority—continuous

2.27 b. We know that a total of 572,032 women victims of domestic abuse and we know the rate per 1,000 women for each age group. We do not know, however, how many women are in each age group. We would need this information to find the number of women who are victims in each age group.

2.29 The percents do not add to 100%, therefore we add a category called other.

2.31 a. Larceny theft had the highest relative frequency of 54.1544%.
  b. Larceny theft and Vehicle theft account for 57.2% of all crimes.

2.35 b. Psychology, Life Sciences, Business, Physical Sci, and Engineering show a dramatic change from 1970.
  c. Comparison pie charts would not show the difference as well as the comparison bar graph.
  d. The graphical bar graph shows the differences much better than does the table.

2.37 Stem-and-leaf of miller   N = 20
Leaf Unit = 1.0

```
   2    1  67
   3    1  8
   5    2  01
  10    2  22233
  10    2  5555
   6    2  67
   4    2  9
   3    3  01
   1    3  3
```

2.41 Stem-and-leaf of score   N = 23
Leaf Unit = 1.0

```
   2    0  67
   5    0  889
   8    1  001
  (4)   1  2223
  11    1  4444555
   4    1  67
   2    1  89
```

2.43 Stem-and-leaf of loss   N = 30
Leaf Unit = 1.0

```
   2   -1  20
   4   -0  65
   8   -0  4432
  14    0  223344
 (12)   0  555556778899
   4    1  001
   1    1  5
```

2.49 Stem-and-leaf of score   N = 24
Leaf Unit = 1.0

```
   1    4  3
   2    4  8
   2    5
   5    5  589
   8    6  234
  12    6  6689
  12    7  34
  10    7  5579
   6    8  1234
   2    8  6
   1    9  1
```

2.57 placebo   treatment
  N = 20   N = 10

```
                3  |  1
                3  |  8
          4     4  |  04
         69     4  |  8
          4     5  |  0
          5     5  |
       0244     6  |  4
          6     6  |
      01112     7  |  4
         67     7  |
                8  |  1
         69     8  |
                9  |
          7     9  |
                   |  High
                   |  152
```

2.59 a. $\bar{y} = 24.4$; $\bar{y}_{T.10} = 24.25$
  b. $\bar{y} = 26.4$; $\bar{y}_{T.10} = 24.25$. The mean increased by 2 points and the trimmed mean remained the same.
  c. If the 16 is changed to 10 the mean will decrease but the trimmed mean will not be affected.

2.61 a. $\bar{y} = 33.36$; $\bar{y}_{T.10} = 34.23$

b. $\bar{y} = 23.36$; $\bar{y}_{T.10} = 24.23$. Both the mean and trimmed mean are reduced by 10 points.

c. If a constant is added to (or subtracted from) all scores, then the mean and trimmed mean will increase (or decrease) by an amount equal to the constant.

2.63 a. $\bar{y} = 24.64$; $\bar{y}_{T.10} = 25.187$

b. $\bar{y} = 2.464$; $\bar{y}_{T.10} = 2.5187$. The mean and trimmed mean are also divided by 10.

c. If each score is multiplied or divided by a constant, then the mean and trimmed mean are also multiplied or divided by the constant.

2.65

Low	Q_1	M	Q_3	High
48	93	98.5	106	149

2.67 The mean and median are both equal to 41 mpg. Because the data are very symmetrical, a trimmed mean will be very much the same.

2.69 $\bar{y} = 176,024$; $M = 163,586$. The two extreme observations on the right tail of the dotplot will increase the value of the mean but will have no effect on the median. Because they are on the right tail (positive direction), the mean will be larger than the median.

2.75 $\bar{y} = 113.45$; $M = 119$; $\bar{y}_{T.10} = 115.12$

2.79 a. Deviations from the mean are $-.7, -1.8, 1.7, .9,$ and $-.1$. Clearly, they sum to 0.

b. SS $= 7.44$ **c.** $s = 1.364$

2.81 $\bar{y} = 54.9$, $s = 9.75$. The data are close to being bell shaped so we apply the Empirical Rule. Approximately 68% of the data should be between 45.15 and 64.65. The actual percent is $18/30 = 60\%$. Approximately 95% of the data should be between 35.4 and 74.4. The actual percent is $29/30 = 96.67\%$. All the data are within three standard deviations of the mean.

2.83 $\bar{y}_1 = 78.58$, $s_1 = 6.73$; $\bar{y}_2 = 80.81$, $s_2 = 9.06$. The score of 82 is higher in class one.

2.85

Low	Q_1	M	Q_3	High
138	142	148	160.5	190

Q-spread $= 18.5$; Range $= 52$. The large difference between the Q-spread and the range is caused by the long right tail of the distribution. The Q-spread is not affected by observations on the tails of the distribution, whereas the range is completely determined by the two end observations.

2.89

Low	Q_1	M	Q_3	High
14	20	23	26	42

MidQ $= 23$; MidR $= 28$. The median and the midQ are the same because the middle 50% of the data are symmetrical. When this happens the average of Q_1 and Q_3 will be very close to the middle (median). The midRange is so much larger than the median because of the data on the right tail of the distribution. The high score of 42 increases the value of the midRange.

2.93 Because the data have a very long tail on the right, the midRange will be pulled toward that tail, causing it to be much larger than the median. For the same reason, the Range will be much larger than either the Q-spread or E-spread.

2.97 The midsummaries become progressively smaller as we scan down from the median to the midRange. This is supporting evidence that the distribution is skewed left.

2.99 The midsummaries are all about the same, suggesting that the distribution is symmetrical. We cannot deduce that the distribution has long tails based on the midsummaries. We need the boxplot to see the outliers.

2.101 These midsummaries have no distinct pattern. This is supporting evidence that the distribution is bimodal.

2.105 The test scores in class2 are more variable than those in class1. Class2 has higher scores but the median score in class1 is higher than the median of class2.

2.111 a. This distribution is bimodal.

b. Because this university has an engineering school, it would stand to reason that the two peaks are caused by the salaries of the engineering professors and the salaries of all other professors.

c. Because these data are made up of two distinct groups, it would be inadvisable to average all the scores and get a measure of a typical salary.

2.115 a. Numerical, Discrete

b. Numerical, Continuous   **c.** Categorical

d. Numerical, Continuous

e. Numerical, Continuous

2.117 a. Categorical   **b.** Numerical   **c.** Numerical

d. Categorical   **e.** Categorical   **f.** Numerical

2.123 a. Gender, Age, Marital status, Characteristics of heart received, Length of surgery

b. Categorical, Numerical, Categorical, Categorical, Numerical

2.125 parameter

2.127 a. F   **b.** F   **c.** T   **d.** T   **e.** F   **f.** F

2.129 a. $\bar{y} = 10, M = 12$ **b.** $\bar{y}_{T.10} = 10.714$
  c. $s = 4.09$
2.133 Corporate name—categorical, Market area—
  categorical, Assets—numerical, Market percent—numerical
2.135 a. Numerical **b.** Numerical **c.** Categorical
  d. Numerical **e.** Categorical
2.137 a. Continuous **b.** Continuous **c.** Discrete
  d. Continuous **e.** Continuous
2.141 a. Categorical **b.** Numerical **c.** Categorical
  d. Categorical **e.** Numerical
2.151 $\bar{y} = 1.91, M = 1.90, \bar{y}_{T.10} = 1.87$ $s = .307$

CHAPTER 3

3.1 a. $50/228 = 21.9\%$ **b.** $2/64 = 3.125\%$
  c. A convicted black defendant, when the vic-
    tim is white, is 7 times more likely to be
    given the death penalty than a convicted
    white defendant, when the victim is black.
3.5 a, b.

	upper	lower	Total
women	20	10	30
men	30	40	70
Total	50	50	100

  c. $10/30 = 33.3\%$ **d.** $40/70 = 57.1\%$
  e. $10/50 = 20\%$ **f.** men **g.** lower
3.7 a. $34/57 = 59.65\%$ **b.** $23/57 = 40.35\%$
  c. $34/61 = 55.74\%$
  d. No, in part **a** the percent is relative to the
    dealerships that replace parts unnecessarily
    and in part **c** the percent is relative to the for-
    eign dealerships.
3.9 a. $425/630 = 67.5\%$ **b.** $360/500 = 72\%$
  c. $60/630 = 9.5\%$
  d. The figures in the graph reflect frequencies
    and not relative frequencies.
3.11 a. $74/104 = 71.15\%$
  b. $74/314 = 23.57\%$
  c. The percentages in parts **a** and **b** are differ-
    ent because they are computed relative to
    different marginal totals. In part **a** the percent
    is relative to all histological type LP. In
    part **b** the percent is relative to all positive
    responses.
  d. $126/538 = 23.42\%$
3.13 a. There appears to be a straight-line relation-
    ship between X and Y.
  b. There appears to be a straight-line relation-
    ship between X and Y.

  c. There is a relationship between X and Y that
    is close to a straight line but also shows a
    little curvature.
  d. There appears to be a straight-line relation-
    ship between X and Y with a downward trend.
3.15 The relationship is a straight-line upward (positive) trend. As aptitude increases so does productivity.
3.17 The relationship between the age of the cash
  register and the maintenance cost for the first
  5 years is almost a perfect straight line. The observation at year 7 may suggest a sharp upturn in
  maintenance cost for older machines; however,
  we cannot say for sure because there is only one
  observation. If that trend continues then the relationship is nonlinear.
3.19 There is one bivariate outlier. Otherwise there is
  an upward (positive) straight-line relationship.
  Those students with high PSAT scores tend to
  have high SAT scores.
3.27 a. $r = .911$ **b.** $r = .977$ **c.** $r = .985$
  d. $r = -.991$.
  The only correlation that may seem inconsistent
  with the scatterplot is part **a**. At each x there is a
  degree of variability in the two observed y values
  that would tend to lower the correlation.
3.29 Because there are no outliers and the relationship appears linear, there is little need to calculate Spearman's correlation. As expected,
  $r_s = -.841$, which is very close to Pearson's
  correlation.
3.33 There is very little difference between Spearman's and Pearson's correlations ($r = -.884$,
  $r_s = -.885$) because there are no outliers and
  the relationship is close to a straight line.
3.35 a. Because one variable is already ranked, we
    should rank the other variable and calculate
    Spearman's correlation.
  b. Pearson's correlation is not appropriate be-
    cause one variable has already been ranked.
  c. $r_s = -.967$
3.37 Correlation of value and revenue $= .940$.
  Correlation of value and revenue without
  outlier $= .876$. The correlation is greater with
  the outlier than without because it is in the same
  straight-line plane as the rest of the data. Instead
  of the outlier having an adverse effect on the
  straight-line relationship, it actually improves
  the relationship.
3.41 Correlation of 1994 and 1995 $= .822$. Based on
  the correlation and the scatterplot, a straight-line
  will fit the data reasonably well.

3.43 a. Response variable is high blood pressure. Predictor variable is air pollution.
 b. Response variable is mental retardation. Predictor variable is lead poisoning.
 c. Response variable is death rate for automobile accidents. Predictor variable is percent of drivers wearing seatbelts.
 d. Response variable is suicide rate. Predictor variable is alcohol consumption among teenagers.
 e. Response variable is public education expenditures. Predictor variable is per capita income.

3.45 a. $Y = -1.67 + 3.00X$ **b.** $Y = .545 + .636X$
 c. $Y = -6.85 + 8.92X$ **d.** $Y = 6.00 - 1.80X$

3.47 a. $-.33333, -1.33333, -1.33333, 1.66667,$ $.66667, .66667$
 b. $-.181818, -.454545, .909091, -.363636,$ $-.636364, .727273, .454545, -.454545$
 c. $2.93082, .01258, -3.82390, 1.25786,$ $-1.82390, 1.01258, 3.33962, -2.90566$
 d. $.6, -.8, 0.0, -.2, .4$

3.51 $r^2 = 87.5\%$; that is, 87.5% of the variability in productivity is explained by the aptitude scores. The straight line is a reasonable relationship between aptitude and productivity.

3.53 a. The regression equation is $\text{cost} = -16.1 + 23.0 \text{ age}$
 b. 16.1216 and -12.8919

3.55 a. The regression equation is $\text{life} = 1,136 - 3.22 \text{ heat}$
 b. $-100.143, 69.714, 108.571, -68.571,$ $33.286, -42.857$
 c. SSE = 34,323

3.61 $\hat{y}_R = -8.67 + 1.44x$

3.63 $\hat{y}_R = 68.8 - .286x$

3.65 $\hat{y}_R = 28.8 + 3.12x$. The regression equation is $\hat{y} = 27.8 + 4.66x$.
 There are outliers in the data that will affect the regression line. Because of the way they are situated, however, their effects will tend to counteract each other (one is on one side of the regression line and another is on the other side of the regression line). Although the y-intercepts of the two lines are close (28.8 and 27.8), their slopes are somewhat different. The regression line has a steeper slope (because of the outlier at the top of the scatterplot). When income is 15 we have $\hat{y}_R = 75.6$ and $\hat{y} = 97.7$.

3.69 Only in case **c** should we use a resistant line to describe the relationship. In cases **a** and **b** there will be very little difference between the resistant line and the least squares regression line. Neither the resistant line nor the regression line will explain the relationship in part **d**.

3.71 correlated

3.73 The residual for (2, 3.8) is $e = -2.1$.

3.75 a. 70%   **b.** 45%   **c.** 75.9%   **d.** 70.9%

3.77 a.

	Science	Business	Education
Women	9	15	22
Men	17	15	5
Total	26	30	27

	Liberal Arts	Total
Women	12	58
Men	7	44
Total	19	102

 b. 25.5%   **c.** 37.9%   **d.** 81.5%

3.81 b. Correlation of Number and Cost $= -.868$
 c. There is a negative linear trend.
 d. The regression equation is $\text{Cost} = 88.6 - 4.09 \text{ Number}$.
 e. Cost per person for a family of 4 $= 88.6 - 4.09 (4) = 72.24$

3.83 a. The regression equation is $\text{number} = -0.03 + 1.64 \text{ age}$.
 c. Correlation of age and number $= .859$
 d. Strength, coordination, training

3.89 b. Correlation of students and athletes $= .920$
 c. The high correlation means that there is a strong association between the graduation rates for all students and student athletes at schools in the Big Ten Conference.
 d. No, this is just a correlation problem. We cannot say that the values of one variable depend on the values of the other variable.

UNIT 1

U1.1 a. T
 b. T
 c. F
 d. F
 e. F

U1.3 d

U1.5 b

U1.9 Stem-and-leaf of Score   N = 47
Leaf Unit = 1.0

```
 1     3   4
 1     3
 5     3   8899
13     4   00001111
19     4   222223
(9)    4   444444555
19     4   66666677
11     4   8
10     5   0000
 6     5   22
 4     5
 4     5   7
 3     5
 3     6   00
 1     6   3
```

The distribution is slightly skewed right.

U1.13 parameter

U1.15 $\bar{y} = 57.00$, $s = 17.57$

U1.19 a. Numerical   **b.** Categorical   **c.** Numerical
d. Numerical   **e.** Numerical   **f.** Categorical

U1.27 Spearman's correlation of GPA and Exam
= .893.
Spearman's is very close to Pearson's correlation.

U1.29 Spearman's correlation of X and Y = -1.000.
The outlier has no unusual effect on Spearman's
correlation.

CHAPTER 4

4.1 1/3   **4.3** 3/10   **4.5** 25/51   **4.7** 1/3
4.9 1/4   **4.11** 18/38, 2/38
4.13 $S = \{1, 2, 3, 4, 5\}$, $A = \{3\}$, $B = \{1, 2, 3\}$,
$C = \{4, 5\}$, $D = \{3, 4\}$
4.15 $S = \{(H_1, M), (H_1, F), (H_2, M), (H_2, F)\}$
4.17 No
4.19 It must be .3 so that the probabilities of all outcomes sum to 1.
4.23 $S = \{(MMM), (MMF), (MFM), (FMM), (MFF),$
$(FMF), (FFM), (FFF)\}$.
If 60% of the student body is female then the
outcomes are not equally likely. If 50% of
the student body is female then the outcomes
are equally likely, and
$P(\{(MMM)\}) = \cdots = P(\{(FFF)\}) = 1/8$.
4.27 a. 1/20   **b.** 12/20
4.29 a. $\{3, 4, 5, 6, 7\}$ **b.** $\{5\}$ **c.** $A = \{3, 5, 7\}$
d. $\{2, 8\}$ **e.** $\{4, 6\}$ **f.** $\{3, 4, 5, 6, 7\}$
g. $\{5\}$ **h.** No, see **b.**

4.31 a. .6   **b.** .1
c. No, A and B have an intersection, since
$P(A \text{ and } B) = .1$.
4.33 Each outcome has probability 1/4. $P(rr) = 1/4$.
4.35 a. Yes. A set of golf clubs cannot be made in
Denver and Phoenix and Memphis.
b. $P(M \text{ or } D) = .8$ **c.** $P(M^c) = .4$
4.37 a. .7   **b.** .6   **c.** .4   **d.** No, $P(A|B) \neq P(A)$
4.41 $P(\text{at least 2 are approved}) = .648$
4.43 $P(\text{you stop no more than one time}) = .784$
4.47 If Team A finding the hiker is independent
of Team B finding the hiker, then
$P(A \text{ and } B) = .12$, and $P(A \text{ or } B) = .58$.
4.49 a. 259.48 million   **b.** 182.15 million   **c.** .145
4.53 .106
4.55 Range of $x = \{0, 1, 2, 3, 4, 5,\}$, Discrete
4.57 Range of $w = \{t | t \text{ is a real number greater}$
than 0$\}$, Continuous
4.59 Range of $B = \{0, 1, 2, 3, ..., 15\}$, Discrete
4.61 a. continuous   **b.** discrete   **c.** discrete
d. continuous   **e.** continuous
4.63 $P(2) = 1/6$, $P(3) = 2/6$, $P(4) = 3/6$.
Yes, it is a probability function because the
total probability = 6/6 = 1.0. $P(1) = 0$.
4.65 a. 4   **b.** $P(\text{even}) = .556$; $P(\text{odd}) = .444$
4.69 $\mu \pm 2\sigma = -.5$ to 3.9 which according to
Empirical Rule contains approximately 95%
of the distribution.
4.73 $\mu = 0 + .4 + .6 + .3 + .4 = 1.7$
4.77 $\sigma^2 = 0 + .4 + 1.2 + .9 + 1.6 - (1.7)^2 = 1.21$,
$\sigma = \sqrt{1.21} = 1.1$
4.83 $\mu \pm 2\sigma = -.46$ to 4.86, which according to
Chebyshev's Rule covers at least 75% of the
distribution.
4.85 a. .201   **b.** .147   **c.** .312   **d.** .069
4.87 a. 8, 1.2649   **b.** 4.5, 1.7748   **c.** 3, 1.2247
d. 3.6, .6
4.91 a. .167   **b.** 6   **c.** 1.549
4.93 a. .304   **b.** 7.5
4.95 c and **e**
4.97 $P(\text{4 or fewer blacks selected from 12}) = .056$.
Because this probability is small, it is doubtful
that the jury was selected randomly.
4.99 $\mu = .75$, $\sigma = .75$; $\mu \pm 2\sigma = -.75$ to 2.25,
which according to Chebyshev's Rule covers at
least 75% of the distribution
4.105 a. .185   **b.** .788   **c.** 14.4
4.107 a. .9773 or 97.73%   **b.** .9953 or 99.53%
c. .9131 or 91.31%   **d.** .0228 or 2.28%
e. .8166 or 81.66%   **f.** .2786 or 27.86%
4.109 a. .5670   **b.** .0808

4.111 .0013

4.113 Its mean and standard deviation (or variance)

4.117 a. .6826 **b.** .6915 **c.** 614

4.119 a. .9050 **b.** 11.16 **c.** 18.84

4.121 a. .0668 **b.** .3830

c. 176.8 is the level at which the beach should be closed.

4.131 d **4.133** d **4.135** a

4.137 a. $S = \{$rrr, rrn, rnr, nrr, rnn, nrn, nnr, nnn$\}$

b. No, only 25% return for an advanced degree. If 50% return then the outcomes are equally likely.

c. .422 **d.** .156

4.139 1/3, 0

4.141 a. 3/5 **b.** 3/7

4.143 a. BC, BM, BP, BS, CM, CP, CS, MP, MS, PS

b. 1/10 **c.** 4/10 **d.** 3/10

4.145 .4452, .2119

4.147 a. T **b.** F **c.** F **d.** T **e.** F **f.** F

4.149 a. .0359 **b.** 3.14

4.151 a. .0004 **b.** .7967 **c.** .0475

4.153 a. .6339 **b.** .0548 **c.** 69.2

4.155 a. $L = \{(1, 1), (6, 6)\}$; $P(L) = 2/36$ **b.** 26/36

4.163 a. .376 **b.** 6 **c.** 1.55

4.165 .864 **4.167** $-\$90$

CHAPTER 5

5.1 M, the sample median

5.3 $M_1 - M_2$, the difference between two sample medians

5.5 $\bar{y}$, the sample mean

5.7 The figure $1.22 is a statistic because it was found from a sample of 6,000 service stations.

5.11 a. 347 **b.** p, the sample proportion

c. $.03/2 = .015$ **d.** very possible

e. 31% to 37%

5.13 The population of interest is all Americans. Because the figure 50% pertains to all Americans, it is a parameter.

5.15 The standard deviation of $\bar{y}$ is $\sigma/\sqrt{n}$. We generally expect that $\bar{y}$ will be within $\pm 2\sigma/\sqrt{n}$ of μ.

5.17 a. No, because 710 is greater than $690.8 + 14.14 = 704.94$

b. Yes, because 710 is less than $690.8 + 28.28 = 719.08$

c. Rarely would $\bar{y}$ exceed $\mu + 3\sigma/\sqrt{n}$ $= 690.8 + 3(100/\sqrt{200}) = 712.01$.

5.19 .1151 **5.21 a.** .2643 **b.** .0001

5.23 a. .1056

b. $P(\bar{y} > 53,000) = .0000003$, so it would be very unusual.

5.25 .0089 **5.27 a.** .8664 **b.** 1^-; almost certain.

5.29 a. 6.25 **b.** 2.5 **5.31** .0768

5.39 a. .7, .0458 **b.** .7, .0229 **c.** .7, .0145

d. .7, .0115

5.41 a. .1 **b.** .05 **c.** .0354 **d.** .0224 **e.** .0158

f. .0112

5.43 a. .2776 **b.** .1190 **c.** .1190

5.45 within $\pm .0266$ of the true value of π

5.47 .9382

5.49 a. .41 **b.** .07

c. Because the sample size is four times as large, the standard deviation will be half as much, i.e., $\sqrt{(.41)(.59)/200} = .035$

d. .0985 **e.** .0051

f. The difference is due to the smaller standard deviation that results from a larger sample size.

5.53 $P(p \le .44) = 0^+$. Because the probability that it happened by chance is so small, there is strong evidence that the aide is in error.

5.57 a. The sampling distributions of all three statistics are centered at .5.

b. When the population is uniformly distributed, all three sampling distributions will be centered at the mean of the population.

c. The median has the greatest standard deviation and the midrange has the least standard deviation.

d. The midrange provides the best estimate of the population mean.

5.59 .087. The sampling distribution of $\bar{y}_1 - \bar{y}_2$ is centered at 0 and is approximately normally distributed.

5.61 .031. The sampling distribution is centered at $-.01$ and is approximately normally distributed.

5.63 If we are interested only in automobile accidents in 1993 then it is a parameter. If the population is all automobile accidents then these data are assumed to be a sample and then 15.9 is a statistic.

5.65 .1867

5.67 a. 10, .2 **b.** 10, .4

c. 50, 1.2 **d.** 50, 2.4

5.69 The sampling distribution of $\bar{y}$ should be approximately normal, so the standard deviation of $\bar{y}$ can be approximated with Range/4, i.e., standard deviation is approximately $800/4 = 200$.

5.71 .0166   **5.75** .0354

5.77 a. .8413   **b.** .9332   **c.** .2872

5.79 a. .3707   **b.** .1825   **c.** 87.6   **d.** .0475

5.81 $n \approx 40$ **5.83** .0694   **5.85** .1562

5.87 If $p = 65/200 = .325$ then Z
$= (.325 - .29)/\sqrt{(.29)(.71)/200} = 1.09$, which
is not unusual at all. It is very conceivable that
more than 65 out of a random sample of 200
criminal victimizations would involve firearms.

5.89 $n \approx 57$

5.91 $Z = (9.2 - 8)/(3/\sqrt{100}) = 4$. The amount these
students study is a full 4 standard deviations
above what typical college students study. This
almost certainly did not happen by chance.

CHAPTER 6

6.1 Based on a midsummary analysis we should
classify this distribution as skewed right because
the midsummaries get progressively larger as we
scan down from the median to the midrange.

6.3 a. The dotplot shows two separate distributions,
one centered around 27 and the other centered around 45. This was not discovered
with either the midsummary analysis or the
boxplot.

   b. One mode is for cars and the other is for
trucks.

   c. The average mpg from the data is 36.85.
This is not representative of the data because
no data are close to this figure. The closest
observation is 41.

   d. These data should be analyzed as two separate data sets—one for cars and one for
trucks.

6.7 The normal probability plot indicates no
significant departures from normality. In Exercise 6.5 we concluded that the distribution is
symmetrical; we can now say that it is also normally distributed.

6.9 The normal probability plot pattern is that of a
skewed right distribution. This agrees with the
assessment given in Exercises 6.6 and 6.8.

6.13 The histogram shows that the waiting times are
bimodal. Normally one would expect a symmetrical unimodal distribution. The mean wait time
is 72.314 but looking at the histogram we see
that 72 is not a typical wait time. It would be
misleading to tell visitors they should expect to
wait 72 minutes. Following a short eruption, one

should expect to wait approximately 57 minutes
and following a long eruption, one should expect
to wait approximately 81 minutes before the next
eruption. These two values are very near the centers of the two groups of wait times.

6.15 σ **6.17** π **6.19** the mean

6.23 V will underestimate σ^2. For a sample size of
100, V will estimate 99% of σ^2. For a sample
size of 100 the bias is not a serious problem.

6.25 Parameter of interest is the mean time for checking reservations. Possible estimators: $\bar{y}$ or $\bar{y}_T$;
$\bar{y} = 42.35$, $\bar{y}_{T.10} = 41.87$. The stem and leaf plot
indicates no long tails so $\bar{y} = 42.35$ is the preferred statistic.

6.27 Because some animals are very slow and may
never respond, the distribution is probably
skewed right, in which case the median is the
preferred parameter.

6.31 a. The general shape is skewed right.

   b. The boxplot and normal probability plot also
indicate that the general shape is skewed
right.

   c. The median $M = 8$ is the estimate of the
center of the distribution. The Q-spread
$= 20 - 5 = 15$ is the estimate of the
variability.

6.33 The shape of the distribution is skewed left. The
median income is the preferred measure of central tendency. It is not possible to calculate the
median exactly because the raw data are not
given here. It is, however, somewhere between
$30,000 and $39,999. We could approximate its
value to be around $35,000.

6.35 a. The normal probability plot does not rule out
normality.

   b. The stem and leaf plot appears bell shaped
with one large value at 78.

   c. The median and midQ are about the same but
the midrange is greater because of the one
outlier at 78.

   d. Aside from the one observation at 78 the data
are very close to being normally distributed.
The recommended estimator of the mean
salinity level is the sample mean. The estimate is $\bar{y} = 49.54$.

6.39 a. The distribution appears to be short-tailed
and slightly skewed left.

   b. The normal probability plot has the general
appearance of a short-tailed distribution. This
agrees with the assessment given in part **a**.

c. The midrange = 255.5 is close in value to the other measures of the center of the distribution.

d. The distribution is short-tailed; any of the standard measures of the center would suffice as an estimate of the center of the distribution. For simplicity, we use $\bar{y} = 257.85$ as the estimate.

6.41 a. The distribution looks symmetrical and possibly normal.

b. The normal probability plot does not suggest any departures from normality.

c. We conclude that the distribution is close to normal and our estimate of the center is $\bar{y} = 73.1$.

6.43 An estimate of the percent of registered voters in Miami who are Puerto Rican is $p = .20$.

6.45 a. $\bar{y}$ **b.** $\bar{y}_T$ **c.** $\bar{y}$ **d.** M **e.** M

6.47 A robust estimator **6.49** μ **6.51** π

6.53 a. The distribution is skewed right.

b. The midsummaries indicate that the distribution is skewed right. Because the distribution is skewed, there is no need to check for normality because the normal distribution is symmetrical.

6.57 The normal probability plot shows no departures from normality. This agrees with our assessment in Exercises 6.55 and 6.56.

6.61 The normal probability plot shows no departures from normality. Based on Exercises 6.59 and 6.60, we should classify the distribution as normal.

6.65 If there are censored data, the population median θ is the best representation of the center of the distribution because it is not affected by the observations that are censored.

6.67 The midsummaries increase and then decrease, giving no useful information. The boxplot looks slightly skewed right. The normal probability plot shows substantial departures from normality but no distinct shape. The stem and leaf plot shows that the distribution is multimodal; this explains why the above information is inconclusive. For multimodal distributions it is recommended that we not give a single measure of the center but rather determine what is causing the different modes. If we are successful, then the data are separated into different groups and then summarized. For these data there is insufficient information to determine the cause of the different modes.

UNIT 2

U2.1 a. The parameter of interest is the Bernoulli proportion π.

b. It would be impossible to conduct a census; we should take a random sample.

c. With a sample size of 500 we should be within $\pm .045$ of the true proportion. The sample size is adequate.

U2.3 a. In the Christmas tree example π represents the proportion that will buy an artificial tree.

b. The ratio 135/500 is from the sample, therefore it is one of the possible value of p. Another sample of 500 might yield a different value of p.

c. We cannot say that 27% of all buyers will purchase artificial trees. We can say only that our estimate of the number of buyers that will purchase artificial trees is 27%.

U2.5 a. $S = \{$www, wwm, wmw, mww, wmm, mwm, mmw, mmm$\}$

b. $P(\text{www}) = (.4)^3$, $P(\text{wwm}) = (.4)^2(.6)^1$, $P(\text{wmw}) = (.4)^2(.6)^1$, $P(\text{mww}) = (.4)^2(.6)^1$, $P(\text{wmm}) = (.4)^1(.6)^2$, $P(\text{mwm}) = (.4)^1(.6)^2$, $P(\text{mmw}) = (.4)^1(.6)^2$, $P(\text{mmm}) = (.6)^3$

c. $A = \{$wwm, wmw, mww$\}$

d. $P(A) = 3(.4)^2(.6)^1 = .288$

U2.7 a. $31,892/116,564 = 27\%$

b. $(6,680 + 1,561 + 434)/31,892 = 27\%$

U2.9 $S = \{(\text{L1, L1, L1}), (\text{L1, L1, L2}), ..., (\text{L3, L3, L3})\}$;
S has 27 equally likely outcomes, so the probability of any outcome is 1/27.

U2.11 Number each president's picture from 1 to 40.
$S = \{(p_1, p_2, p_3, p_4, p_5) | p_i = 1, 2, ..., 40; i = 1, 2, 3, 4, 5\}$.
The number of outcomes in S is 40^5 and each of these is equally likely. The event A that all 5 pictures are different requires that $p_1 \neq p_2 \neq \cdots \neq p_5$. There would be 40 choices for p_1, 39 choices for p_2, and so on. Thus A has $(40)(39)(38)(37)(36)$ elements and $P(A) = (40)(39)(38)(37)(36)/40^5 = .771$.

U2.13 a. .298 **b.** .090 **c.** .945 **d.** 4.5 **e.** 6 **f.** 1.5

U2.15 Let Y be the number on welfare out of 8 families. Y is binomial with $n = 8$ and $\pi = .7$. $P(Y = 8) = .058$.

U2.17 Let X be the number of shots until he misses. Because he may never miss, the range of X is $\{1, 2, 3, 4, ...\}$. X is discrete.

U2.19 a. .2514   **b.** 20.44   **c.** .0294

U2.21 23.35   **U2.23** .1492   **U2.25** a

U2.27 d   **U2.29** d

U2.31 a. .296   **b.** .259   **c.** .037

U2.33 a. .0228   **b.** .2476   **c.** $x = 13.92$

U2.41 The probability that none of the 30 go into internal medicine is $p(0) = .015331$. The number expected is $30(.13) = 3.9$. The standard deviation is $\sqrt{30(.13)(.87)} = 1.842$.

CHAPTER 7

7.1 The length of the interval will increase.

7.3 $p = .61$. The confidence interval is .573 to .647. A 95% confidence interval would be narrower because $z^* = 1.96$ instead of 2.58.

7.5 $p = .2$; the confidence interval is .178 to .222. $n = 482$.

7.7 $p = .88$; the confidence interval is .863 to .897.

7.9 $p = .72$; the confidence interval is .697 to .743. No, either the true proportion is inside the interval or it is not. Probability is only relevant prior to sampling.

7.11 $p = .59$; the confidence interval is .567 to .613. Maximum margin of error is .024; $n = 752$.

7.13 $n = 1,448$

7.19 a. 453.15 to 478.45   **b.** 449.15 to 482.45

7.21 Interval A is the 95% interval because a 95% confidence interval is generally shorter than a 99% confidence interval.

7.23 The confidence interval is 1,316.69 to 1,333.31. We cannot say that there is a 95% chance that the mean living area is inside the interval. Either the mean is inside the interval or it is not. Probability (95% chance) is only relevant prior to sampling.

7.25 5.54 to 7.46

7.27 The confidence interval is 5.1 to 5.3. If the students had taken a random sample of 100 cups, the resulting interval would be narrower because the margin of error would have been divided by $\sqrt{100}$ instead of $\sqrt{40}$.

7.29 5.61 to 6.59; $n = 267$

7.33 65.1 to 73.3

7.35 The midsummaries are close in value, suggesting that the distribution is symmetric. With a sample size of 15 it is a good idea to check for normality. This is accomplished with a normal probability plot. The confidence interval is 51.6 to 62.7.

7.37 239.50 to 260.56

7.39 With a sample size of 48, the normality condition is not so crucial because of the Central Limit Theorem. We should, however, still be concerned with long tails and severe skewness. The confidence interval is 45.95 to 53.13.

7.41 The placebo group appears normally distributed but the treatment group appears skewed. When the one extreme outlier is removed, the data appear more normally distributed. Because the researcher stated that outliers of this type were typical, we probably should not remove the outlier.

7.43 Based on the normal probability plot, these data are severely skewed right. Because of the skewness, the median is the recommended parameter to measure a typical return on an investment. For a t-interval, the data should be more normally distributed. Clearly this is not the case.

7.47 a. 6   **b.** 5   **c.** 4   **7.49** $C_L = 20, C_H = 45$

7.51 a. The midsummaries increase and the stem and leaf plot shows severe right skewness.
   b. $C_L = 2.78, C_H = 31.75$

7.53 a. 8   **b.** 19   **c.** 41   **d.** 87

7.55 a. (27, 45)   **b.** (15, 36)   **c.** (12, 36)

7.57 (78,000, 100,000)

7.59 The normal probability plot indicates that the data are skewed. $C_L = 16, C_H = 24$

7.63 The boxplot slows a symmetric long-tailed distribution. The confidence interval is (31.12, 37.28).

7.65 The distribution appears long-tailed and therefore the confidence interval will be based on $\bar{y}_T$. $\bar{y}_{T.10} = 5.27$, $s_w = .939$, $s_T = 1.053$. The confidence interval is (4.94, 5.60).

7.67 (575, 758)   **7.69** (145.04, 160.96)

7.71 (1,436.09, 1,623.91)

7.73 a. 1.645   **b.** 2.33   **c.** 1.753   **d.** 2.602

7.75 a. (iv)   **b.** (iii)   **c.** (18, 45)

7.77 a. $n = 1,849$
   b. (.67, .73). We are very confident that the true percent of convicted felons with a history of juvenile delinquency is somewhere between 67% and 73%.

7.79 b. From the stem and leaf plot, it appears that the data have not violated the normality assumption.   **c.** (17.66, 23.54)

7.81 a. (.374, .426)   **b.** $n = 3,994$

7.83 (6.66, 6.74)

7.87 (3, 12)

7.91 $n = 385$

7.93 To construct a *t*-interval when the sample size is 16, it is important that the reading speed of fifth graders be approximately normally distributed. The confidence interval is 259.43 to 310.57.

7.95 75.683 to 89.117   **7.97** $n = 57$

7.99 .61 to .63

7.101 a. 39,698 to 44,902

   b. With a sample size of 16, it is important that the distribution of incomes for new Ph.D.s in psychology be approximately normal.

   c. By increasing the sample size

7.103 35,146.61 to 41,641.39

7.105 a. The data are moderately skewed right.

   b. With a sample size of 46 and the fact that the skewness is not severe, the normality assumption is not crucial.

   c. (469.3, 560.5)

CHAPTER 8

8.1 a. $H_0: \mu = 50$ versus $H_a: \mu > 50$

   b. $H_0: \mu = 50$ versus $H_a: \mu < 50$

   c. $H_0: \mu = 50$ versus $H_a: \mu > 50$

   d. $H_0: \mu = 50$ versus $H_a: \mu < 50$

   e. $H_0: \mu = 50$ versus $H_a: \mu \neq 50$

8.5 $H_0: \pi = .10$ versus $H_a: \pi > .10$ **8.7** c

8.9 No, there is the possibility of a Type II error.

8.11 No, Type I and II errors are not complementary events.

8.13 Both are wrong. A *p*-value is the probability that the null hypothesis is rejected *when* it is actually true.

8.15 Type I: concluding that the fire is not out and it actually is.

Type II: concluding that the fire is out and it actually is not out.

Committing a Type II error is more serious.

8.17 $H_0: \mu = .18$ versus $H_a: \mu > .18$.

A Type II error is committed if the geologist concludes that the mean porosity measurement does not exceed 18% when in fact it does.

8.21 *p*-value $= .0021$. Based on these results, there is highly significant evidence that his class is superior. Because of the sample size of 35 the normality assumption is not that crucial. We should, however, look for severe skewness or extremely long tails.

8.23 *p*-value $= .0475$. Based on a random sample of 1,600 adults, there is significant evidence that more than 10% of the adult population is illiterate.

8.25 *p*-value $= .0749$. Based on a random sample of 200 adults, there is moderately significant evidence that less than 60% of the adult population believes that there is too much violence on television.

8.27 $n = 7,203$

8.29 *p*-value $= .1492$. Based on these 600 cases, there is insignificant evidence that Cook County exceeds the national average in the number of automobile accident suits.

8.33 Based on a random sample of 200 people, there is insufficient evidence to refute the psychologist's claim. 171 out of 200.

8.35 *p*-value $= .0034$. Based on this sample of 200 prisoners, there is highly significant evidence that less than 1/3 are returning to jail within 3 years.

8.37 a. This is a left-tailed test. To conduct a *t*-test when the sample size is 25, the population distribution should be reasonably symmetrical without any outliers. To complete the test we have $t = (2.33 - 2.5)/(.67/\sqrt{25}) = -1.27$; *p*-value $= .1081$. Based on this random sample, there is insignificant evidence that the population mean is less than 2.5.

   b. This is a two-tailed test. To conduct a *t*-test when the sample size is 70, the population distribution should not be severely skewed or have extreme outliers. To complete the test we have $t = (51.6 - 50)/(5.9/\sqrt{70}) = 2.27$; *p*-value $= 2(.0132) = .0264$. Based on this random sample, there is significant evidence that the population mean is something other than 50.

   c. This is a right-tailed test. To conduct a *t*-test when the sample size is 14, the population distribution should be close to normal. To complete the test we have $t = (102.3 - 100)/(2.45/\sqrt{14}) = 3.51$; *p*-value $= .0019$. Based on this random sample, there is highly significant evidence that the population mean is greater than 100.

8.39 $H_0: \mu \leq 5.5$ versus $H_a: \mu > 5.5$.

$t_{obs} = .62$; *p*-value $= .2694$. Based on a random sample of 40 cars, there is insufficient evidence to deny the manufacturer's claim. The normality assumption is not needed here because the sample size is 40. It is important, however, that the population distribution not be severely skewed or have extreme outliers.

8.41 a. With 50 observations there is no reason to base the test on any statistic other than $\bar{y}$.
   b. $t_{obs} = -.05$; p-value $= .96$. There is no evidence whatsoever to conclude that the mean price differs from 300 pence.

8.43 $H_0: \mu \le 120$ versus $H_a: \mu > 120$.
   $t_{obs} = 1.379$; p-value $= .0903$. Based on 25 randomly selected families, there is moderately significant evidence that mean kilowatt usage exceeds 120 kilowatt hours.

8.47 $H_0: \mu \le 78$ versus $H_a: \mu > 78$.
   $t_{obs} = 2.556$; p-value $= .0097$. Based on the provided data, there is highly significant evidence that this is a superior class. Because we are using a t-test and the sample size is 20, we assume that the distribution of scores on this exam is at least symmetric without outliers.

8.49 a. The boxplot and normal probability plot indicate that the parent distribution is close to normal.
   b. $H_0: \mu = 8$ versus $H_a: \mu \ne 8$. $t_{obs} = 1.14$; p-value $= .29$. Based on the sample of 9 people, there is insufficient evidence that the mean number of trials to master a task under the influence of the drug is not 8.

8.51 a. The normal probability plot does not rule out normality.
   b. $H_0: \mu \ge 98$ versus $H_a: \mu < 98$
   c. $t_{obs} = -4.76$; p-value $= .0003$. Based on these 12 readings, there is highly significant evidence that the mean inlet oil temperature is less than 98 degrees.

8.55 $H_0: \mu = 35,000$ versus $H_a: \mu \ne 35,000$.
   $t_{obs} = -2.98$; p-value $= .0089$. Based on the information provided, there is highly significant evidence that the mean charge for a coronary bypass in North Carolina is different from $35,000.

8.59 a. p-value $= .0009$. There is highly significant evidence that the median is less than 400.
   b. p-value $= .1151$. There is insufficient evidence that the median is more than 12.5.
   c. p-value $= .0018$. There is highly significant evidence that the median is not 75.

8.63 $H_0: \theta = 100$ versus $H_a: \theta \ne 100$. p-value $= .804$. There is insufficient evidence to conclude that the median mental age is different from 100.

8.65 p-value $= .0963$. There is mildly significant evidence that the median age is not 19.

8.67 $H_0: \theta \ge 27,500$ versus $H_a: \theta < 27,500$. p-value $= .0539$. There is significant evidence that the median salary for North Carolina social workers is less than $27,500.

8.71 a. $H_0: \theta = 7$ versus $H_a: \theta \ne 7$
   b. This is a two-tailed test.
   c. The standardized test statistic is $z = (2T - n)/\sqrt{n} = 1.04$.
   d. p-value $= .2955$. The null hypothesis should not be rejected.
   e. Based on a random sample of 75 water samples, there is insignificant evidence that the median pH level is different from 7.

8.73 $z_{obs} = -1.37$; p-value $= .0853$. There is moderately significant evidence that $\mu < 20$.

8.75 $H_0: \mu = 450$ versus $H_a: \mu > 450$.
   $z_{obs} = 1.73$; p-value $= .0418$. There is sufficient evidence that the mean score is higher than 450.

8.77 $H_0: \mu = 8.50$ versus $H_a: \mu \ne 8.50$.
   $z_{obs} = 2.70$; p-value $= .007$. There is highly significant evidence that the mean cost of an evening meal is not $8.50.

8.79 $H_0: \mu = 120$ versus $H_a: \mu \ne 120$.
   $z_{obs} = 3.02$; p-value $= .0026$. There is highly significant evidence that the city official's claim is not valid.

8.81 $H_0: \mu \ge 10$ versus $H_a: \mu < 10$.
   $z_{obs} = -1.19$; p-value $= .1170$. Based on a random sample of 100 picture tubes, there is insignificant evidence that the mean life is less than 10 years. We cannot refute the manufacturer's claim.

8.85 a. The boxplot and the normal probability plot show a symmetric long-tailed distribution, which contradicts the conclusion given in Exercise 8.84.
   b. Because of the symmetric long tails $\bar{y}_T$ is the recommended estimator of the parallax of the sun.

8.87 b   **8.89** d   **8.91** c   **8.93** b

8.95 $H_0: \pi = .70$ versus $H_a: \pi > .70$

8.97 $H_0: \mu = 600$ versus $H_a: \mu \ne 600$.
   $t_{obs} = 2.20$; p-value $= .0336$. There is significant evidence that the mean is not 600.

8.99 $H_0: \mu = 2,000$ versus $H_a: \mu > 2,000$.
   $t_{obs} = 2.58$; p-value $= .0062$. There is significant evidence that the county agent is right and his county needs more than 2,000 pounds of lime per acre.

8.101 $H_0: \pi = .70$ versus $H_a: \pi \ne .70$.
   $z_{obs} = 2.16$; p-value $= .0308$. Based on a random sample of 200 convicted felons, there is significant evidence to contradict the claim that 70% have a history of juvenile delinquency.

8.105 $H_0: \mu = 40$ versus $H_a: \mu > 40$

8.107 H_0: $\mu = 800$ versus H_a: $\mu > 800$.
$z_{obs} = 1.61$; p-value $= .0537$. Based on 45 randomly selected property owners, there is moderately significant evidence to reject the government's claim.

8.109 H_0: $\mu = 36$ versus H_a: $\mu < 36$
$t_{obs} = -3$; p-value $= .0025$. Based on a random sample of 36 women, there is highly significant evidence that women on the pill have a smaller maximal oxygen uptake than women not on the pill.

8.113 H_0: $\mu = .5$ versus H_a: $\mu > .5$.
$t_{obs} = 1.32$; p-value $= .12$. There is insufficient evidence that the produced stones have a median of more than .5 carat.

8.115 H_0: $\pi \geq .7$ versus H_a: $\pi < .7$.
$z_{obs} = -1.23$; p-value $= .1093$. Based on a random sample of 200 people, there is insufficient evidence to conclude that less than 70% favor raising the drinking age.

8.117 H_0: $\theta \leq 98.5$ versus H_a: $\theta > 98.5$.
p-value $= .0195$. $T = 8$. Based on 9 trials, there is significant evidence that the median time is more than 98.5 seconds; therefore, you reject the horse. Because his lifetime median time is 100 seconds you did not make an error.

8.119 The stem and leaf plot, boxplot, and normal probability plot support the assumptions for a t-test.
H_0: $\mu = 70$ versus H_a: $\mu < 70$.
$t_{obs} = -.46$, p-value $= .3257$. Based on the results obtained from 18 volunteers, there is significant evidence that the counseling process reduces one's score on the exam.

CHAPTER 9

9.1 a. Parameters under study are the population medians.
H_0: $\theta_1 = \theta_2$ versus H_a: $\theta_1 \neq \theta_2$
  b. Parameters under study are the population means.
H_0: $\mu_1 = \mu_2$ versus H_a: $\mu_1 < \mu_2$
$\mu_1 =$ mean biodegradation rate for plastic and $\mu_2 =$ mean biodegradation rate for paper.
  c. Parameters under study are the population proportions.
H_0: $\pi_1 = \pi_2$ versus H_a: $\pi_1 < \pi_2$.
$\pi_1 =$ percent covered last year and $\pi_2 =$ percent covered this year

9.3 Controlled experiment

9.5 The null hypothesis should be rejected. Based on 100 men and women assigned to an experimental and control group there is statistical evidence that mean resting pulse of those who jog over a period of 6 months is different from those who do not jog.

9.7 To compare the relative danger of the two sports, we need the number of participants as well as the number of injuries. Parameters under study are the population proportions.
H_0: $\pi_1 = \pi_2$ versus H_a: $\pi_1 > \pi_2$.
$\pi_1 =$ proportion of injuries from in-line skating and $\pi_2 =$ proportions of injuries from skateboarding

9.9 Subjects were not randomly assigned to the female electrical workers group; this is an observational study. The study does not prove that high doses of electricity cause breast cancer.

9.13 a. $p_1 = 65/200 = .325$, $p_2 = 74/200 = .370$, and $p = 139/400 = .3475$
  b. .0492 **c.** .0476 **d.** part c
  e. $z_{obs} = -.95$, p-value $= .3422$.
  f. The fact that the data represent the percentages of men and women students with registered cars has no effect on the analysis of data. It affects the conclusion, however, because the conclusion should be written in the context of the problem. In other words, based on the provided data, there is insufficient evidence that the percentages of men and women students with registered cars are different.

9.15 a. $p_A = 19/250 = .076$ and $p_B = 27/300 = .09$
  b. For a confidence interval the sample proportions should not be pooled to estimate the standard error
  c. The confidence interval is $-.075$ to $.047$.
  d. The interval contains both positive and negative values, which means that a two-tailed test of hypothesis would not detect a significant difference between the two proportions.

9.17 p_1 estimates π_1 and p_2 estimates π_2. In hypothesis testing we evaluate the standard error of $p_1 - p_2$ under the assumption that the null hypothesis is true. Because $\pi_1 = \pi_2$ we have that p_1 and p_2 are estimating the same thing; therefore, we pool the two estimates together to get a better estimate of the common value of π. In confidence interval estimation, we do not assume that $\pi_1 = \pi_2$ and therefore we do not pool the estimates together.

9.19 $H_0: \pi_1 = \pi_2$ versus $H_a: \pi_1 < \pi_2$.
$z_{obs} = -1.69$; p-value $= .0455$. There is significant evidence that the vaccine was effective in reducing the mortality rate. This is a controlled experiment.

9.21 This should be a one-tailed test.
$H_0: \pi_2 = \pi_2$ versus $H_a: \pi_1 > \pi_2$.
$z_{obs} = 1.40$; p-value $= .0808$. There is mildly significant evidence that the course was beneficial in increasing reading scores.

9.23 $H_0: \pi_1 = \pi_2$ versus $H_a: \pi_1 \neq \pi_2$.
$\pi_1 = $ proportion of people living in cities who are worried about being a victim of crime and $\pi_2 = $ proportion of people living in suburbs who are worried about being a victim of crime.
$z_{obs} = .45$; p-value $= 2(.3264) = .6528$. There is an insignificant difference between the percents of people living in cities and suburbs who are worried about being a victim of crime.

9.29 a. 3.44 to 71.36 **b.** -4.83 to 1.83
c. No, interval b is more valid; it is a 98% confidence interval, whereas interval a is a 90% confidence interval.

9.31 a. $t_{obs} = -.94$ **b.** This is a one-tailed test. df $= 47$, p-value $> .15$
c. There is insufficient evidence to reject H_0 and conclude that $\mu_1 - \mu_2 < 0$.

9.33 $H_0: \mu_1 = \mu_2$ versus $H_a: \mu_1 > \mu_2$.
$\mu_1 = $ the mean annual income in District I and $\mu_2 = $ the mean annual income in District II.
$t_{obs} = 1.56$, $.05 < p$-value $< .10$. There is mildly significant evidence to conclude that the mean annual income in District I is greater than the mean annual income in District II.

9.35 a. Neither sample indicates that inferences should not be based on $\bar{y}_1 - \bar{y}_2$.
b. $H_0: \mu_1 = \mu_2$ versus $H_a: \mu_1 \neq \mu_2$.
$\mu_1 = $ the mean grade for a 9 A.M. class and $\mu_2 = $ the mean grade for a 2 P.M. class
c. $t_{obs} = -.73$
d. p-value $> .40$. There is insufficient evidence to conclude that the mean grade for a 9 A.M. class is different from the mean grade for a 2 P.M. class.

9.37 a. $H_0: \mu_1 = \mu_2$ versus $H_a: \mu_1 \neq \mu_2$.
$\mu_1 = $ the mean number of errors with drug and $\mu_2 = $ the mean number of errors with placebo
b. The parent distributions should not deviate substantially from normality and the samples should be independent.

c. $t_{obs} = 2.46$
d. df $= 11$, $.02 < p$-value $< .04$. There is significant evidence to conclude that the mean number of errors with the drug is different from the mean number of errors with a placebo.

9.41 Both distributions are close to symmetrical. The variability of the two groups is about the same, leading to the conclusion that the two groups are homogeneous. The middle 50% of group B is completely contained within the middle 50% of group A. It is doubtful that there is a significant difference between the two groups.

9.45 a. Both distributions appear skewed right to the point that the two-sample t-procedures may not apply.
b. The mean is much larger than the median because of the right skewness of the distribution.
c. Because of the similar skewness a comparison of population medians is more appropriate for these data.
d. The boxplots show a distinct difference between the two distributions, suggesting that the median survival time for those in group 1 exceeds the median survival time for those in group 2.

9.47 a. $H_0: \mu_1 = \mu_2$ versus $H_a: \mu_1 < \mu_2$.
$\mu_1 = $ the mean amount of coffee dispensed by vending machine A and $\mu_2 = $ the mean amount of coffee dispensed by vending machine B
b. For small samples the parent populations should be close to a normal distribution and the population variances should be homogeneous.
c. $t_{obs} = -.63$; p-value $> .10$. Insufficient evidence exists to support the conjecture that machine A dispenses significantly less than machine B.

9.49 The confidence interval is $(-3.98, .38)$. The interval contains both positive and negative values. This means that a two-tailed test of the equality of mean grades for fall and spring classes would not be rejected.

9.51 a. Based on the boxplots, there does not seem to be a difference between the mean abilities of men and women.
b. $(-4.2, 5.1)$
c. The interval contains both positive and negative values. This suggests that there is no

significant difference between the abilities of men and women.

d. Because of the long-tailed distributions a confidence interval based on trimmed means might be more appropriate.

9.55 a. $H_0: \mu_1 = \mu_2$ versus $H_a: \mu_1 \neq \mu_2$.
μ_1 = the mean number of defective gears by A and μ_2 = the mean number of defective gears by B

b. Both the normality and the homogeneous variances assumptions seem to be met.

c. $t = -2.25$; $p = .031$; df = 38. Based on a p-value = .031, there is a significant difference between the number of defective gears produced by the two manufacturers.

9.57 a. Inferences should be on the population medians.

b. $H_0: \theta_1 = \theta_2$ versus $H_a: \theta_1 > \theta_2$.
θ_1 = median survival time under Arm A and θ_2 = median survival time under Arm B

c. The pooled t-test applied to the ranks is equivalent to the Wilcoxon rank sum test. We have $t = 3.41$; $p = 0.0004$; df = 119. Based on the provided data, there is highly significant evidence that the median survival time under Arm A is greater than the median survival time under Arm B.

9.61 a. The samples are matched because the same subject is used before and after the 2-week course of instruction.

b. The normal probability plot shows no significant departures from normality.

c. $H_0: \mu_d \leq 0$ versus $H_a: \mu_d > 0$, where $\mu_d = \mu_1 - \mu_2$.
μ_1 = mean spelling score after course of instruction, μ_2 = mean spelling score before course of instruction.
$t_{obs} = 3/(3.94/\sqrt{9}) = 2.28$; p-value = .026. There is significant evidence of improvement after taking the course of instruction.

9.63 a. The samples are matched because the same subject is measured before and after viewing the movie.

b. The distribution of difference scores should be close to normal. There may be some departure from normality but the sample size is so small it is difficult to tell. There appear to be no gross departures that would preclude the paired t-test.

c. A 99% confidence interval for μ_d is $(-.369, 1.702)$.

d. The confidence interval contains both positive and negative values, thus there is no change in moral attitudes after viewing the film.

9.65 a. The samples are matched because identical twins are used in the study.

b. The dotplot shows some departure from normality but with so few observations it is difficult to tell.

c. $t = 2.40$; p-value = .043. There is significant evidence that the drug affected intelligence test scores.

9.69 a. The normality assumption may be in question; the Wilcoxon test may be more appropriate for these data.

b. $t = 3.56$; p-value = .0062. There is highly significant evidence that the mean number of errors is different before and after the course.

9.73 a. $z_{obs} = -1.57$; p-value = .0582. There is mildly significant evidence that the difference between μ_1 and μ_2 is less than 0.

b. $z_{obs} = -1.47$; p-value = $2(.0708)$ = .1416. There is insufficient evidence to conclude that the difference between μ_1 and μ_2 is not 0.

9.75 $H_0: \mu_1 = \mu_2$ versus $H_a: \mu_1 \neq \mu_2$;
$z_{obs} = 1.64$; p-value = $2(.0505)$ = .101.
There is insufficient evidence to conclude that the means are different.

9.77 $H_0: \mu_1 \leq \mu_2$ versus $H_a: \mu_1 > \mu_2$;
$z_{obs} = 4.17$; p-value = .00003.
There is highly significant evidence that the mean score for the computer group exceeds the mean score for the non-computer group.

9.79 $H_0: \mu_1 = \mu_2$ versus $H_a: \mu_1 \neq \mu_2$;
$z_{obs} = .03$; p-value = .976.
There is insufficient evidence that the mean score for the learning disabled students is different from the mean score for normal achievers.

9.81 a. The standard deviations are close enough that we can pool them together to estimate the standard error of $\bar{y}_1 - \bar{y}_2$.

b. The confidence interval is 3.87 to 22.13.

c. The interval contains only positive values. This means that a two-tailed test of the equality of the population means would be rejected at the 10% level of significance. The mean size in Region one is greater than the mean size in Region two.

9.83 $H_0: \theta_1 = \theta_2$ versus $H_a: \theta_1 \neq \theta_2$;
$t = 1.87$; $p = .069$; df = 38. There is a mildly significant difference between the two population medians.

9.85 **a.** Matched pairs t-test.
   b. The assumptions are not violated.
   c. $H_0: \mu_d \leq 0$ versus $H_a: \mu_d > 0$, where
       $\mu_d = \mu_1 - \mu_2$.
       $\mu_1 =$ mean yield for the new variety,
       $\mu_2 =$ mean yield for the standard variety.
   d. $t = 3.83$; p-value $= .0014$
   e. There is highly significant evidence that the
       yield per acre for the new variety is higher
       than the standard variety.
9.87 **a.** The assumptions are not violated.
   b. The samples are independent.
   c. The confidence interval is -1.21 to 6.65.
   d. The 98% confidence interval contains both
       positive and negative values. A null hypothe-
       sis of the equality of the population means
       cannot be rejected when the level of signifi-
       cance is 2%. Based on these results, we
       would say that there is an insignificant differ-
       ence between the number of hours spent on
       homework by students attending public and
       private high schools.
9.89 The interval is -5.39 to 18.99. The confidence
   interval based on trimmed means is recommended when the parent distributions are symmetric with long tails.
9.91 The confidence interval is $.413$ to $.623$.
9.93 $H_0: \mu_d \leq 0$ versus $H_a: \mu_d > 0$, where
   $\mu_d = \mu_1 - \mu_2$.
   $\mu_1 =$ mean number of sit-ups after the course,
   $\mu_2 =$ mean number of sit-ups before the course.
   A matched pairs t-test:
   $t = 2.75$; p-value $= .012$. There is significant
   evidence that the physical fitness course improved the number of sit-ups by the participants.
9.95 $H_0: \pi_1 = \pi_2$ versus $H_a: \pi_1 > \pi_2$.
   $\pi_1 =$ proportion interested in business and tech
   now and $\pi_2 =$ proportion interested in business
   and tech in the past.
   $z_{obs} = 1.09$; p-value $= .1379$. There is insignificant evidence that a higher proportion of students are interested in business and tech now
   than in the past.
9.99 $t = 2.30$; p-value $= .039$. Based on the p-value,
   there is a significant difference between the test
   scores before and after the vocabulary training.
   A normal probability plot does not rule out normality and therefore the matched pairs t-test is
   appropriate for analysis of these data.
9.103 A boxplot indicates no significant departures
   from assumptions for the matched pairs t-test.

$t = -.68$; p-value $= .51$. The difference between
the readings is insignificant. To check the accuracy of a machine, multiple readings would have
to be taken on the same subject.
9.105 A boxplot of the difference scores identifies one
   outlier but otherwise the distribution does not
   deviate substantially from the assumptions for
   the matched pairs t-test.
   $t = 8.51$; p-value $= .0000$. There is highly
   significant evidence that the arterial blood pressure is different after receiving oxytocin.

UNIT 3

U3.1 **a.** No, the formulas in Chapter 7 apply to simple
       random samples. This is a stratified random
       sample.
   b. The statement attempts to relate sample size
       and confidence level. However, no mention is
       made of the margin of error. In confidence in-
       terval estimation, we are attempting to esti-
       mate a certain population parameter to within
       a specified margin of error. The sample size
       is related to the confidence level, but it is also
       indirectly related to the margin of error. To
       decrease the margin of error, the sample size
       must increase. A correct statement is some-
       thing like, "Enough responses were obtained
       to be 95% confident that the estimate is
       within a specified margin of error of the true
       population parameter."
U3.3 **a.** $n \approx 49$
   b. The confidence interval is 117.62 to 138.38.
   c. The value \$135 is inside the 99% confidence
       interval and therefore the null hypothesis
       $H_0: \mu = 135$ would not be rejected at the 1%
       level of significance.
U3.5 **a.** $n \approx 1,509$
   b. $p = 94/500 = .188$. The confidence interval
       is $.147$ to $.229$.
   c. The hypothesis $H_0: \pi = .15$ would not be re-
       jected with $\alpha = .02$ because $.15$ is inside the
       98% confidence interval.
U3.7 The confidence interval is 46.8 to 55.2.
U3.9 $H_0: \mu \leq 450$ versus $H_a: \mu > 450$.
   $t_{obs} = 3.05$; $.001 < p$-value $< .005$. There is
   highly significant evidence that the applicants for
   admission are above the national norm of 450 on
   SAT math.

U3.11 $H_0: \mu = 8$ versus $H_a: \mu \neq 8$.
$t_{obs} = -2.61$; $.001 < p$-value $< .005$. There is highly significant evidence that the mean weight is not 8 ounces. Based on their established criteria, the supermarket should reject the shipment.

U3.13 $H_0: \mu_1 \leq \mu_2$ versus $H_a: \mu_1 > \mu_2$.
μ_1 = the mean life span in open environment and μ_2 = the mean life span in caged environment.
$t_{obs} = 3.23$; $.001 < p$-value $< .005$. There is highly significant evidence that the mean life span of animals is greater in an open environment than in a caged environment.

U3.15 The pooled t-test assumes that the population variances are homogeneous and that the parent populations are close to normally distributed. The larger the sample sizes the more the distributions can deviate from normality. However, we should not use a t-test if either distribution is severely skewed or is symmetric with long tails. The boxplot can identify outliers and describe the general shape of the distribution. A normal probability plot checks deviations from normality.

U3.17 a. $n \approx 849$

b. $p = 500/900 = .556$. The confidence interval is .517 to .595. We are 98% confident that the percent of smokers who would light up without asking permission is somewhere between 51.7% and 59.5%.

U3.19 $H_0: \mu \leq 20$ versus $H_a: \mu > 20$.
$t_{obs} = 2.67$; $.005 < p$-value $< .01$. There is highly significant evidence that this group of college administrators has an unusually high self-concept. To conduct a t-test with 16 observations, the parent distribution should be symmetric without outliers.

U3.21 An estimate of σ is range/4 $= 16/4 = 4$ months; $n \approx 246$

CHAPTER 10

10.1 a. 25.0 **b.** 30.6 **c.** 7.26

10.3 4.4 and 23.3

10.5 $H_0: \pi_1 = \pi_2 = \pi_3 = \pi_4 = \pi_5 = 1/5$ versus H_a: at least one is not the same.
$\chi^2_{obs} = 7.34$; p-value $> .10$. There is insufficient evidence to conclude that the classes are not equally likely.

10.7 If $H_0: \pi_1 = .40$, $\pi_2 = .45$, $\pi_3 = .07$, $\pi_4 = .05$, $\pi_5 = .03$ is true, then $e_1 = 480$, $e_2 = 540$, $e_3 = 84$, $e_4 = 60$, $e_5 = 36$, and $\chi^2_{obs} = 5.81$; p-value $> .10$. There is insufficient evidence to conclude that the percentages are different from those specified by the official.

10.9 $H_0: \pi_1 = \pi_2 = \pi_3 = \pi_4 = 1/4$ versus H_a: at least one is not the same.
$\chi^2_{obs} = 3.6$; p-value $> .10$. There is insufficient evidence to conclude that the proportions of marbles are different.

10.13 If $H_0: \pi_1 = .31$, $\pi_2 = .50$, $\pi_3 = .14$, $\pi_4 = .05$ is true, then $e_1 = 77.5$, $e_2 = 125$, $e_3 = 35$, $e_4 = 12.5$, and $\chi^2_{obs} = 206.59$; p-value $< .005$. There is highly significant evidence that the claims from the employees of this company do not follow the same distribution as given in the *Wall Street Journal*.

10.17 H_0: Variables A and B are independent.
H_a: Variables A and B are dependent.
$\chi^2_{obs} = 9.597$; df $= 2$, $p = .008$. There is highly significant evidence that the two variables are dependent.

10.19 H_0: The service rendered is independent of the type dealership.
H_a: The service rendered is dependent on the type dealership.
$\chi^2_{obs} = 30.297$, df $= 5$; $p = 0.000$. There is highly significant evidence that the service rendered is dependent on the type dealership.

10.21 H_0: The presence of toxicity and tumor regression are independent.
H_a: The presence of toxicity and tumor regression are dependent.
$\chi^2_{obs} = 16.571$; df $= 1$, $p = 0.000$. There is highly significant evidence that the presence of toxicity and tumor regression are dependent. All expected cell counts are greater than 5. This is important because the χ^2 statistic has an approximate chi-square distribution when no more than 20% of the cells have an expected count less than 5. If this condition is not met, then the approximation is not good.

10.25 H_0: A student's desire to participate in the science project program is independent of academic standing.
H_a: A student's desire to participate in the science project program is dependent on academic standing.

$\chi^2_{obs} = 5.314$; df $= 2, p = .071$. There is mildly significant evidence that participating in the science project depends on academic standing.

10.29 H_0: Injury level and wearing a seatbelt are independent.

H_a: Injury level and wearing a seatbelt are dependent.

$\chi^2_{obs} = 59.224$; df $= 3, p = 0.000$. There is highly significant evidence that the injury level is statistically related to whether or not the driver wears a seatbelt.

10.31 H_0: The proportions falling in these three categories are the same for all three populations.

H_a: The proportions falling in these three categories are different for at least one of these populations.

$\chi^2_{obs} = 4.621$; p-value $= .329$. The row marginal totals are all 80; they are fixed values. All of the expected cell counts exceed 5 (the smallest is 21.33) so the chi-square approximation should be good. Based on the p-value, there is insufficient evidence that the proportions falling in the three categories are different.

10.33 a. The classification variable is the opinion on the referendum.

b. The populations are the voters from District 1 and from District 2.

c. H_0: The proportion of people having the same opinion is the same for both districts.

H_a: The proportion of people having the same opinion is different for the two districts.

d. $\chi^2_{obs} = 4.241$; df $= 2, p = .121$. There is insufficient evidence that the proportions in the two districts are different.

e. No assumptions for the test have been violated.

10.35 a. The classification variable is the opinion of the respondent.

b. The populations are the players and members of the press.

c. H_0: Football players and members of the press have the same opinion.

H_a: Football players and members of the press have different opinions.

$\chi^2_{obs} = 25.291$; df $= 2, p = .000$. There is highly significant evidence that football players and the press have different opinions on who should receive credit for good play.

10.39 a. The classification variable is the preference of a favorite sport.

b. The populations are collegiate men and women.

c. H_0: Favorite sport is distributed the same for men and women.

H_a: Favorite sport is not distributed the same for men and women.

d. $\chi^2_{obs} = 21.553$; df $= 3, p = .000$. There is highly significant evidence that favorite sports are not distributed the same for men and women.

10.43 H_0: The row and column classifications are independent.

H_a: The row and column classifications are dependent.

$\chi^2_{obs} = 1.899$; p-value $= .594$. There is insufficient evidence to conclude that the row and column classifications are dependent. This is the chi-square test of independence.

10.45 H_0: The proportions of suicides are the same for all 4 classes.

H_a: The proportions of suicides are different for the 4 classes.

$\chi^2_{obs} = 3.6842$; df $= 3, p$-value $> .10$. There is insufficient evidence to say that the proportion in the 4 classes are different.

10.47 H_0: Moral values and opinion on the referendum are independent.

H_a: Moral values and opinion on the referendum are dependent.

$\chi^2_{obs} = 19.702$; df $= 4, p = .001$. Their moral values are related to their opinion on the referendum.

10.49 H_0: The geographical distribution of prisoners on death row is the same as the distribution of the general population.

H_a: The geographical distribution of prisoners on death row is different from the distribution of the general population.

$\chi^2_{obs} = 644.68$; df $= 3, p$-value $< .001$. There is highly significant evidence that the geographical distribution of prisoners on death row is different from the distribution of the general population.

10.51 H_0: The type of facility performing abortions is the same for 1988 and 1992.

H_a: The type of facility performing abortions is different for 1988 and 1992.

$\chi^2_{obs} = 10.514$; df $= 3, p$-value $= .015$. There is significant evidence that the type of facility performing abortions is different from 1988 and 1992.

10.53 H_0: The holdings are equally divided among the three types of stock.
H_a: The holdings are not equally divided among the three types of stock.
$\chi^2_{obs} = .93846$; p-value $> .1$. There is insufficient evidence to conclude that the stocks are not equally divided.

10.55 H_0: The sales are equally distributed among the salespeople.
H_a: The sales are not equally distributed among the salespeople.
$\chi^2_{obs} = 14.435$; $.01 < p$-value $< .025$. There is significant evidence that the sales are not equally distributed among the salespeople.

10.59 H_0: Sex of dead deer and food supply are independent.
H_a: Sex of dead deer and food supply are dependent.
$\chi^2_{obs} = .516$; df $= 2$, $p = .772$. There is insufficient evidence that an association exists between sex of the deer and the food supply.

10.63 H_0: Perceived math ability is the same for girls and boys.
H_a: Perceived math ability is different for girls and boys.
$\chi^2_{obs} = 19.869$; df $= 4$, $p = .001$.
Based on these data, there is highly significant evidence that perceived math ability is different for girls and boys.

CHAPTER 11

11.1 a. 2   **b.** -1 **c.** 0

11.3 a. 101,730
b. (97,130, 106,330). Out of 100 randomly selected homes, we expect 5 to have prices outside this interval.
c. 2,430
d. The cost 97,500 is consistent with the interval found in part **b.**

11.5 a. 3,675, 4,800.
b. (3,585, 3,765). We expect 5 to have prices outside this interval.
c. 375
d. After 3 months $3,750 is consistent with the interval found in part **b.**
e. The model is proposed for the first six months. There is no reason to believe that the same linear model will apply for a whole year.

11.7 Correlation of Rating and forty $= -.268$.
The regression equation is
Rating $= 13.7 - 1.47$ forty.
SSE $= 11.2290$, $s = .6449$. The linear relationship between rating and forty is weak.

11.11 a. A straight line could fit the data reasonably well, but a curvilinear fit seems more appropriate. There are no unusual observations other than the curvilinear pattern.
b. The regression equation is:
magnesium $= 5.77 - .000252$ distance.
The estimated standard deviation is
$s = 1.195$. A regression plot shows that the straight line fits the data reasonably well.

11.13 a. $s = 1.464$ **b.** $\hat{y} = .783 + .6848x$
c. $t_{obs} = 4.49$
d. p-value $= .004$. There is highly significant evidence that $\beta_1 \neq 0$.
e. p-value $= .695$. We cannot reject H_0: $\beta_0 = 0$. The population regression line should go through the origin.

11.15 a. There is a very definite linear trend.
b. $b_0 = 4.6748$, $b_1 = 1.2272$
c. $t_{obs} = 10.44$; p-value $= .000$. The linear relationship is highly significant. The coefficient of determination is R-sq $= 96.5\%$.
d. sales $= 4.6648 + 1.2272(6) = 12.028 \approx 12$
e. This model should not be used to predict sales for the month of January. Data for the month of January should be used to develop the model.

11.17 a. The dependent variable is sales.
b. There is a weak upward linear trend.
c. The regression equation is
sales $= 24.9 + 33.1$ nicotine.
d. $t_{obs} = 2.24$; p-value $= .067$. There is mildly significant evidence of a linear trend. The regression coefficient is positive ($+33.1$); therefore sales will increase when nicotine content increases.
e. Sales increase 33.1 ($\times \$100,000$) for each additional milligram of nicotine.

11.19 a. The regression equation is
Rating $= 13.7 - 1.47$ forty.
This agrees with the answer in Exercise 11.7.
b. $t_{obs} = 1.45$; p-value $= .159$. The null hypothesis should be accepted. This means that the proposed linear model is inadequate for explaining the relationship.

c. It seems as if the scouts should not be concerned with the time in the forty. A word of caution, however: There may be a relationship between the rating score and the time in the forty that is more complicated than a linear relationship.

11.23 a. The regression equation is SAT-M2 $= 482 + 2.48$ Profic2. $R^2 = 34.9\%$; $F_{obs} = 9.65$; p-value $= .006$. The F statistic is highly significant.

   b. The linear model found in Example 11.6 is slightly better than this model.

   c. The SAT math scores for the second group are much more variable than in the first group. There is clearly a significant difference between the two medians.

11.27 a. SSE $= (n-2)s^2 = 22(9.25)^2 = 1,882.375$

   b. For $x^* = 10$: Predicting interval is 18.1 to 31.4.
   For $x^* = 15$: Predicting interval is 23.4 to 32.9.
   For $x^* = 20$: Predicting interval is 27.5 to 35.4.
   For $x^* = 25$: Predicting interval is 30.1 to 39.6.
   For $x^* = 30$: Predicting interval is 31.5 to 44.8.

11.29 a. The t-ratio is highly significant and $R^2 = 84.4\%$, indicating that the regression line should fit the data.

   b.

	Fit	Stdev. Fit	90.0% C.I.	90.0% P.I.
For $x^* = 10$:	17.567	1.248	(15.304, 19.829)	(12.364, 22.769)
For $x^* = 20$:	22.467	.817	(20.985, 23.948)	(17.554, 27.380)
For $x^* = 25$:	24.917	.746	(23.564, 26.269)	(20.041, 29.792)
For $x^* = 30$:	27.367	.817	(25.885, 28.848)	(22.454, 32.280)
For $x^* = 40$:	32.267	1.248	(30.004, 34.529)	(27.064, 37.469)

   c. For $x^* = 25$ the confidence interval is the narrowest because $\bar{x} = 25$.

11.31 a. The t-ratio is highly significant and $R^2 = 95.5\%$, indicating that the regression line fits the data very well.

   b.

	Fit	Stdev. Fit	95.0% C.I.	95.0% P.I.
For $x^* = 2$:	1.818	.365	(.924, 2.712)	(−.010, 3.646)
For $x^* = 4$:	3.091	.286	(2.390, 3.792)	(1.350, 4.832)
For $x^* = 6$:	4.364	.237	(3.783, 4.944)	(2.667, 6.060)
For $x^* = 8$:	5.636	.237	(5.056, 6.217)	(3.940, 7.333)
For $x^* = 10$:	6.909	.286	(6.208, 7.610)	(5.168, 8.650)

11.35 a., b.

	Fit	Stdev.Fit	99.0% C.I.	99.0% P.I.
For $x^* = 1,000$:	1,686.2	35.1	(1,580.4, 1,792.0)	(1,339.8, 2,032.5)

11.39 a. $\hat{y} = -9.725$ **b.** -8.65 to 2.45

   c. $t_{obs} = -2.292$; $.01 < p$-value $< .025$. There is significant evidence that β_3 is less than 0.

11.41 $y = \beta_0 + \beta_1 x_1 + \beta_2 x_2 + \epsilon$, where $y =$ teacher's attitude, $x_1 =$ years of experience, and $x_2 =$ size of school

11.43 $y = \beta_0 + \beta_1 x_1 + \beta_2 x_2 + \epsilon$, where $y =$ family income, $x_1 =$ education level, and $x_2 =$ years residing in region

11.45 a. pulse $= 58.3 + 5.82$ dose

   b. $R^2 = 67.8\%$

   c. $t_{obs} = 5.43$; p-value $= .000$. There is highly significant evidence that dose should remain in the model.

   d. The variable dose explains 67.85% of the variability in pulse and is highly significant. This is a reasonable model to use to predict pulse.

11.49 b. There is a curvilinear appearance to the scatterplot. A straight line will not fit the data.

   c. R-sq $= 30.5\%$, $F = 7.47$, p-value $= .014$, which is significant.

   d. The answer to part **c** does not agree with the answer to part **b**.

   e. The value of R^2 increased from 30.5% to 90.9%, an enormous increase. The F-statistic $= 79.43$ increased tenfold. The second-degree polynomial model is an excellent model in comparison to the simple linear model.

11.53 It appears that the residuals are decreasing as x increases, which indicates that the variance is not constant with respect to x.

11.55 The standardized residuals increase as the fits increase. This indicates that the variance is not constant with respect to x.

11.59 a. The scatterplot indicates a fairly strong linear relationship between the number of library books checked out and the verbal test scores of the eighth graders.

   b. $R^2 = 85.2\%$ and the t-ratio is highly significant (p-value $= .000$).

   c. The standardized residual show a definite curvilinear pattern. We should consider a second-degree polynomial model.

11.61 a. The residuals are close to normally distributed.
  b. The Residuals vs. Fits plot shows a more random pattern than before.
  c. This model appears to be an improvement over the simple linear model.

11.63 The regression equation is
output = 2.98 − 6.93 recipvel.
  a. The new R^2 is 98.0%. Very much improved over the previous model.
  b. The new F-statistic is 1,128.43, again much improved over the previous model.
  c. The new residual plot shows a random pattern.

11.67 a. Independent: population density; Dependent: robbery rate
  b. Independent: attitude score; Dependent: achievement score
  c. Independent: number of fish; Dependent: growth rate
  d. Independent: expenditures per student; Dependent: teacher's salary

11.69 a. Slope = −2/3, y-intercept = 2
  b. Slope = 7.4/3.1, y-intercept = −12/3.1
  c. Slope = −5.6/4.1, y-intercept = −10/4.1

11.71 a. The relationship between x and y appears to be curvilinear.
  b. There appears to be a strong linear relationship between x and y.
  c. The relationship between x and y appears to be curvilinear.

11.73 a. $x' = 100, \hat{y} = 7.3; x' = 160, \hat{y} = 2.86$
  b. $x' = 1.5, \hat{y} = .502; x' = 5.0, \hat{y} = 1.02$
  c. $x' = 30, \hat{y} = −.96; x' = 60, \hat{y} = 4.53$

11.75 $x = 5, \hat{y} = 23.7, e = −5.7$
$x = 5, \hat{y} = 23.7, e = −3.7$

11.77 Overall, there appears to be a strong linear relationship between x and y.

11.79 a. $\mu_y = 184.25$
  b. The interval is (181.25, 187.25).
  c. −.35

11.81

Fit	Stdev.Fit	95.0% C.I.	95.0% P.I.
56.40	1.89	(52.05, 60.75)	(42.72, 70.08)

11.83 Because of the bivariate outlier in the upper left-hand corner, a resistant line would better describe the relationship between IQ and GPA.

11.87 a. The regression equation is
confid = 85.5 − 4.23 educat.
  b. SSE = 1,928.3; $s = 10.35$

c.

Predictor	Coef.	Stdev.	t-ratio	p-value
educat	−4.231	1.020	−4.15	.001

  d. R-sq = 48.9%. Thus, 48.9% of the variability in confidence is explained by education.
  e. This model adequately describes the relationship between the two variables.

11.89 The residual plot appears to have a random pattern.

11.91 a. The regression equation is
sales = 3,246 + 49.9 months.
  b. R-sq = 25.8%

c.

Predictor	Coef.	Stdev.	t-ratio	p-value
months	49.85	29.89	1.67	.134

  d. A possible violation of assumptions is that the variance does not appear to be constant.

11.97 b. The regression equation is
species = 63.8 + .0820 area.
$R^2 = 38.2\%, F = 17.29, p\text{-value} = .000$
(highly significant). Additional variables may increase R^2 and improve the model.
  c. A log transformation or square root transformation may help explain the relationship.
  d. The regression equation is
species = 45.3 + 25.7 lnarea.
The value of $R^2 = 61.6\%$ improved substantially, $F = 44.91, p\text{-value} = .000$ is highly significant. This model appears to be an improvement over the previous model.

CHAPTER 12

12.1 a. The stem and leaf plots suggest that the variances are homogeneous.
  b. Based on the stem and leaf plots, there are no serious departures from normality.
  c. The standard deviations of the three samples are very close, indicating that the variances are homogeneous.
  d. MSE = 180.4988
  e. The grand mean cannot be calculated by averaging the three means because the group sizes are different.

12.3 a. SSE = 8,662 **b.** MSE = 180
  c. SST = 1,069 **d.** MST = 534
  e. The square of the pooled standard deviation is the mean square error, MSE.

f. $H_0: \mu_1 = \mu_2 = \mu_3$ versus H_a: at least two μ_i's differ.

g. $F = 2.96$, p-value $= .061$. The null hypothesis can be rejected but the results are only mildly significant.

12.5 a. $H_0: \mu_1 = \mu_2 = \mu_3$ versus H_a: at least two μ_i's differ.

b. $F = 6.05$, p-value $= .004$. The null hypothesis should be rejected; the results are highly significant.

12.7 a. $T = 2.10$, p-value $= .047$ and $F = 4.43$, p-value $= .047$. The relationship between the two is $T^2 = F$ because there are only two groups. Because of the exact relationship between the two, their p-values are exactly the same.

b. The pooled standard deviations for the two tests are exactly the same (except for slight roundoff error). Furthermore, MSE $= 91.2$ $= s_p^2 = (9.55)^2$.

c. The independent pooled t-test and the F-test are equivalent when we have only two groups to compare.

12.9 a. Based on the boxplots, it is doubtful that there is a significant difference between the thickness of the wafers.

b. The measurements are not independent because four measurements of the thickness of the oxide layer are taken on the same wafer.

12.11 The largest standard deviation is less than twice the smallest standard deviation so we can conclude that the variances are homogeneous. Based on the p-value, the null hypothesis should be rejected. There is a highly significant difference between the number of tasks completed in a 10-minute period by subjects treated with different combinations of a drug and electroshock.

12.13 A: (12.91, 15.59); B: (17.78, 20.47); C: (9.66, 12.34); D: (15.28, 17.97)

12.19 a. 5.99 **b.** 3.70 **c.** 5.76

12.21 a. ANOVA table

Source	SS	df	MS	F	p-value
Diet	228.55	3	76.183	14.494	< .001
Error	84.1	16	5.256		

b. Based on the p-value, the difference between the treatment means is highly significant.

c. $W = 4.05 \sqrt{5.256/5} = 4.1525$

$\underline{\mu_C \quad \mu_B \quad \mu_A} \quad \mu_D$

The diet plans B, A, and D do not differ significantly. Diet plan C is significantly better than the other three diet plans.

12.23 $\mu_3 \quad \underline{\mu_2 \quad \mu_1 \quad \mu_4}$

The mean reaction effect from drug 3 is greater than all the other means.

12.25 $\underline{\mu_4 \quad \mu_2 \quad \mu_1} \quad \mu_3$

The mean from group 3 (No drug with shock) is significantly below the means for the other groups.

12.27 $\underline{\mu_3 \quad \mu_2 \quad \mu_4} \quad \mu_1$

The mean test score for those students using method III is significantly greater than those using methods I and IV.

12.31 The Kruskal–Wallis test is a nonparametric test that is an alternative to the ordinary F-test that is used to compare several population means. To execute the rank-transformation form of the test, ranks are collectively assigned to the combined samples and then the ordinary F-test is applied to the ranks.

12.33

Source	df	SS	MS	F	p-value
Treatment	2	247.65	123.825	1.67	>.10
Error	27	2,000.45	74.1		

12.35 a. All three distributions are skewed right.

b. The Kruskal–Wallis should be conducted on these data.

12.37 a. Based on the boxplots, there is no severe skewness or extreme outliers in any of the three groups. The population variances seem homogeneous. There is no serious departure from normality in any of the three distributions.

b. There are no serious departures from assumptions for the ordinary F-test. Obviously the assumptions for the Kruskal–Wallis are also satisfied. It is recommended that these data be analyzed with the ordinary F-test.

12.39 a. All three distributions are skewed right, in which case, normality is not possible. The distributions, however, are similar in shape.

b. Because the distributions are similar in shape, the Kruskal–Wallis test can be applied to the data.

12.41 If the distributions are long-tailed, the Kruskal-Wallis test is the proper procedure to use to compare the centers of the distributions. The ordinary F-test is sensitive to outliers and the K-W test is robust.

12.47 The ordinary F-test

12.49 **a.** 3.01, 3.72, 4.72   **b.** 2.53, 3.03, 3.70
   c. 2.25, 2.62, 3.12

12.51 **a.** 5.29   **b.** 3.67   **c.** 5.39

12.53

Source	SS	df	MS	F	p-value
Treatment	2,313.04	2	1,156.52	6.38	$< .01$
Error	8,154.10	45	181.2		
Total	10,467.14	47			

12.55 **a.** There are five treatment levels.
   b.

Source	SS	df	MS	F	p-value
Treatment	428	4	107	2.903	$.025 < p\text{-value} < .05$
Error	1,032	28	36.857		
Total	1,460	32			

   c. There is a significant difference between the treatment levels.

12.57

Source	DF	SS	MS	F	p-value
Factor	2	573	286	1.61	.212
Error	41	7,291	178		
Total	43	7,864			

12.61 **a.** There is probably not a significant difference between the median ages for adjacent phases but examining the medians for phase 1 and phase 4, we see a difference that is significant. There is no unusual behavior in the boxplots that would suggest that the assumptions for the ordinary F-test are not satisfied.
   b. Based on the p-value $= .000$ $(< .005)$, there is highly significant evidence that the null hypothesis should be rejected. Phases 4 and 3 are nonoverlapping, phases 3 and 1 are nonoverlapping, but phases 3 and 2 appear to overlap. It is difficult to assess whether or not phases 2 and 1 overlap.
   c. Based on the intervals produced by Tukey's multiple comparison, μ_1 and μ_2 are not statistically different and μ_2 and μ_3 are not statistically different. This suggests that phases 1 and 2 overlap and phases 2 and 3 overlap.

12.63 Recall that there was one extreme outlier in the Texas data. This one outlier distorts both the mean and standard deviation of the sample. The mean of 1,452 is not a realistic measure of the center of the distribution. The standard deviation of 6,270 is not a realistic measure of the variability of the distribution. Moreover, it causes the MSE to be unusually large, which in turn causes the F-statistic to be unusually small (MSE is in the denominator). Consequently, the mean for Texas is inflated but the F-test does not detect it because of the unusually large MSE. It is clear that we have violated two basic assumptions for the F-test: (i) the distribution is not normally distributed, and (ii) the variances are not homogeneous. It is for this reason that we recommended removing the outlier and analyzing with the Kruskal–Wallis test.

12.67 **a.** The distributions look reasonably symmetric without unusually long tails.
   b. The normal probability plots show no significant departures from normality.
   c. The assumptions for the F-test are satisfied and thus it should be used to compare the centers of the distributions.

12.69 **a.** Based on the normal probability plots and the appearance of the side-by-side boxplots, we classify the four distributions as skewed right.
   b. Because the normality assumption is in doubt and the fact that the distributions are similarly skewed, we should use the Kruskal–Wallis test to compare the centers of the distributions.

12.71 Because the difference is insignificant, there is no need to conduct a multiple comparison test.

UNIT 4

U4.1 **a.** Although there is one outlier, a straight-line should fit the data.
   b. The outlier is an influential observation.
   d. The influential observation will tend to pull the line toward it, resulting in a smaller slope.

U4.3 **a.** Chi-square test of homogeneity
   b. H_0: The reading ability of fourth grade students is the same in the two school districts.
   H_a: The reading ability of fourth grade students is different in the two school districts.

c. $\chi^2_{obs} = 4.975$, df $= 2$; $p = .084$.
There is mildly significant evidence that the reading ability of fourth grade students is different in the two school districts.

U4.5 a. 25% score below 350, 25% between 350 and 425, 25% above 560

b. Chi-square goodness of fit.

c. $H_0: \pi_1 = \pi_2 = \pi_3 = \pi_4 = 1/4$ versus H_a: at least one is not the same.
$\chi^2_{obs} = 1.8$; p-value $> .10$

d. The sample does not differ significantly from the national standard.

U4.7 $H_0: \pi_1 = \pi_2 = \pi_3 = \pi_4 = 1/4$ versus H_a: at least one is not the same.
$\chi^2_{obs} = 3.6$; p-value $> .10$. There is insufficient evidence that one wine is favored over the others.

U4.9 H_0: The mental ages are equally distributed in the intervals: less than 85, 86 to 95, 96 to 110, 111 to 125, and above 125.
H_a: The mental ages are not equally distributed in the intervals: less than 85, 86 to 95, 96 to 110, 111 to 125, and above 125.
$\chi^2_{obs} = 3.71$; p-value $> .10$. The data support the psychologist's theory; the mental ages are equally distributed in the different categories.

U4.11 H_0: The colors are distributed alike for each of the 10-year periods.
H_a: The colors are not distributed alike for each of the 10-year periods.
The test is the chi-square test of homogeneity. The p-value ($= .000$) is highly significant. Based on these results, there is highly significant evidence that color preference has changed over the years. The bay and black have gained in popularity.

U4.13 H_0: Political party and opinion on foreign policy are independent.
H_a: Political party and opinion on foreign policy are dependent.
$\chi^2_{obs} = 27.148$, df $= 2$, p-value $= 0.000$. There is highly significant evidence that political party and opinion on foreign policy are dependent.

U4.15 a. A straight line will fit the data.

b. The regression equation is verbal $= 45.3 + 2.38$ books.

c.

	Fit	Stdev.Fit	90.0% C.I.	90.0% P.I.
For $x^* = 10$:	69.17	1.94	(65.74, 72.61)	(56.83, 81.52)
For $x^* = 15$:	81.09	1.80	(77.90, 84.27)	(68.81, 93.37)
For $x^* = 20$:	93.00	2.55	(88.49, 97.52)	(80.31, 105.69)
For $x^* = 25$:	104.92	3.68	(98.40, 111.43)	(91.39, 118.45)

U4.17 b. The regression equation is
$y = -.508 + .0973\,x_1 + .0271\,x_2$

c. R-sq $= 95.2\%$, which shows that 95.2% of the variability in y is explained by x_1 and x_2.

d. $H_0: \beta_1 = \beta_2 = 0$ versus H_a: At least one $\beta_i \neq 0$.
$F_{obs} = 58.91$, p-value $= .000$. The model is significant.

e.

Predictor	Coef	Stdev.	t-ratio	p-value
Constant	$-.5079$	.6652	$-.76$	.474
x_1	.09730	.01071	9.09	.000
x_2	.02714	.02993	.91	.400

Based on the t-ratios, x_2 is insignificant. It should be removed from the model. The constant is also insignificant, indicating that the regression line should go through the origin.

U4.21 a. The distributions are slightly skewed right.

b. Because of the robustness of the F-test, it could be used in this situation. The skewness is not severe enough to cause problems with the F-test. If, on the other hand, we desired to compare the population medians, the Kruskal–Wallis test would also apply.

U4.23 $\chi^2_{obs} = 46.659$, df $= 8$, p-value $= .000$. Based on the chi-square test, the duration of time since the last visit to a physician is dependent on the income bracket.

INDEX

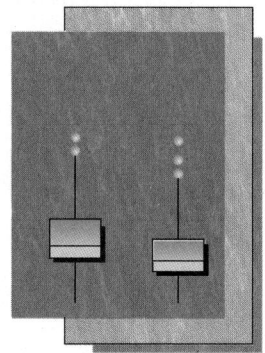

Critical Values of Student's *t* Distribution

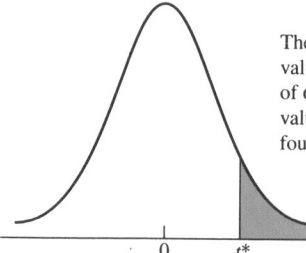

The table values are the critical values for Student's *t* for an area of α in the right-hand tail. Critical values for the left-hand tail are found by symmetry.

Amount of α in one-tail

Degrees of freedom	.2	.1	.05	.025	.01	.005	.0025	.001
1	1.376	3.078	6.314	12.706	31.821	63.657	127.3	318.3
2	1.061	1.886	2.920	4.303	6.965	9.925	14.09	22.33
3	.978	1.638	2.353	3.182	4.541	5.841	7.453	10.21
4	.941	1.533	2.132	2.776	3.747	4.604	5.598	7.173
5	.920	1.476	2.015	2.571	3.365	4.032	4.773	5.893
6	.906	1.440	1.943	2.447	3.143	3.707	4.317	5.208
7	.896	1.415	1.895	2.365	2.998	3.499	4.029	4.785
8	.889	1.397	1.860	2.306	2.896	3.355	3.833	4.501
9	.883	1.383	1.833	2.262	2.821	3.250	3.690	4.297
10	.879	1.372	1.812	2.228	2.764	3.169	3.581	4.144
11	.876	1.363	1.796	2.201	2.718	3.106	3.497	4.025
12	.873	1.356	1.782	2.179	2.681	3.055	3.428	3.930
13	.870	1.350	1.771	2.160	2.650	3.012	3.372	3.852
14	.868	1.345	1.761	2.145	2.624	2.977	3.326	3.787
15	.866	1.341	1.753	2.131	2.602	2.947	3.286	3.733
16	.865	1.337	1.746	2.120	2.583	2.921	3.252	3.686
17	.863	1.333	1.740	2.110	2.567	2.898	3.222	3.646
18	.862	1.330	1.734	2.101	2.552	2.878	3.197	3.611
19	.861	1.328	1.729	2.093	2.539	2.861	3.174	3.579
20	.860	1.325	1.725	2.086	2.528	2.845	3.153	3.552
21	.859	1.323	1.721	2.080	2.518	2.831	3.135	3.527
22	.858	1.321	1.717	2.074	2.508	2.819	3.119	3.505
23	.858	1.319	1.714	2.069	2.500	2.807	3.104	3.485
24	.857	1.318	1.711	2.064	2.492	2.797	3.091	3.467
25	.856	1.316	1.708	2.060	2.485	2.787	3.078	3.450
26	.856	1.315	1.706	2.056	2.479	2.779	3.067	3.435
27	.855	1.314	1.703	2.052	2.473	2.771	3.057	3.421
28	.855	1.313	1.701	2.048	2.467	2.763	3.047	3.408
29	.854	1.311	1.699	2.045	2.462	2.756	3.038	3.396
30	.854	1.310	1.697	2.042	2.457	2.750	3.030	3.385
40	.851	1.303	1.684	2.021	2.423	2.704	2.971	3.307
50	.849	1.299	1.676	2.009	2.403	2.678	2.937	3.261
60	.848	1.296	1.671	2.000	2.390	2.660	2.915	3.232
80	.846	1.292	1.664	1.990	2.374	2.639	2.887	3.195
120	.845	1.289	1.658	1.980	2.358	2.617	2.860	3.160
∞	.842	1.282	1.645	1.960	2.326	2.576	2.807	3.090